ADVANCES IN CHEMICAL PHYSICS

VOLUME 125

EDITORIAL BOARD

Advances in
CHEMICAL PHYSICS

Edited by

I. PRIGOGINE

Center for Studies in Statistical Mechanics and Complex Systems
The University of Texas
Austin, Texas
and
International Solvay Institutes
Université Libre de Bruxelles
Brussels, Belgium

and

STUART A. RICE

Department of Chemistry
and
The James Franck Institute
The University of Chicago
Chicago, Illinois

VOLUME 125

AN INTERSCIENCE PUBLICATION
JOHN WILEY & SONS, INC.

Published by John Wiley & Sons, Inc., Hoboken, New Jersey.
Published simultaneously in Canada.

Library of Congress Catalog Number: 58-9935

ISBN 0-471-21452-3

Printed in the United States of America

10 9 8 7 6 5 4 3 2 1

CONTRIBUTORS TO VOLUME 125

JÜRGEN GAUSS, Institut für Physikalische Chemie, Universität Mainz, Mainz, Germany

S. GUÉRIN, Laboratoire de Physique, Université de Bourgogne, Dijon, France

LAURI HALONEN, Laboratory of Physical Chemistry, University of Helsinki, Helsinki, Finland

H. R. JAUSLIN, Laboratoire de Physique, Université de Bourgogne, Dijon, France

SABRE KAIS, Department of Chemistry, Purdue University, West Lafayette, IN, U.S.A.

V. V. KRASNOHOLOVETS, Department of Theoretical Physics, Institute of Physics, National Academy of Sciences, Kyiv, Ukraine

S. P. LUKYANETS, Department of Theoretical Physics, Institute of Physics, National Academy of Sciences, Kyiv, Ukraine

JANNE PESONEN, Laboratory of Physical Chemistry, University of Helsinki, Helsinki, Finland

PABLO SERRA, Faculdad de Matemática, Astronomía y Física, Universidad Nacional de Córdoba, Ciudad Universitaria, Córdoba, Argentina

JOHN F. STANTON, Institute for Theoretical Chemistry, Departments of Chemistry and Biochemistry, University of Texas at Austin, Austin, Texas, U.S.A.

P. M. TOMCHUK, Department of Theoretical Physics, Institute of Physics, National Academy of Sciences, Kyiv, Ukraine

INTRODUCTION

Few of us can any longer keep up with the flood of scientific literature, even in specialized subfields. Any attempt to do more and be broadly educated with respect to a large domain of science has the appearance of tilting at windmills. Yet the synthesis of ideas drawn from different subjects into new, powerful, general concepts is as valuable as ever, and the desire to remain educated persists in all scientists. This series, *Advances in Chemical Physics*, is devoted to helping the reader obtain general information about a wide variety of topics in chemical physics, a field that we interpret very broadly. Our intent is to have experts present comprehensive analyses of subjects of interest and to encourage the expression of individual points of view. We hope that this approach to the presentation of an overview of a subject will both stimulate new research and serve as a personalized learning text for beginners in a field.

I. PRIGOGINE
STUART A. RICE

CONTENTS

Finite-Size Scaling for Atomic and Molecular Systems — 1
By Sabre Kais and Pablo Serra

A Discussion of some Problems Associated with the Quantum Mechanical Treatment of Open-Shell Molecules — 101
By John F. Stanton and Jürgen Gauss

Control of Quantum Dynamics by Laser Pulses: Adiabatic Floquet Theory — 147
By S. Guérin and H. R. Jauslin

Recent Advances in the Theory of Vibration-Rotation Hamiltonians — 269
By Janne Pesonen and Lauri Halonen

Proton Transfer and Coherent Phenomena in Molecular Structures with Hydrogen Bonds — 351
By V. V. Krasnoholovets, P. M. Tomchuk, and S. P. Lukyanets

Author Index — 549

Subject Index — 567

FINITE-SIZE SCALING FOR ATOMIC AND MOLECULAR SYSTEMS

SABRE KAIS

Department of Chemistry, Purdue University, West Lafayette, Indiana, U.S.A.

PABLO SERRA

Facultad de Matemática, Astronomía y Física, Universidad Nacional de Córdoba, Ciudad Universitaria, Córdoba, Argentina

CONTENTS

I. Introduction
II. Some General Results for Near-Threshold States
  A. Phase Transitions at the Large-Dimensional Limit
  B. Critical Phenomena in Spaces of Finite Dimensions
    1. One-Particle Central Potentials
    2. Few-Body Potentials
III. Finite-Size Scaling in Classical Statistical Mechanics
IV. Finite-Size Scaling in Quantum Mechanics
  A. Quantum Statistical Mechanics and Quantum Classical Analogies
  B. Finite-Size Scaling Equations in Quantum Mechanics
  C. Extrapolation and Basis Set Expansions
  D. Data Collapse for the Schrödinger Equation
V. Quantum Phase Transitions and Stability of Atomic and Molecular Systems
  A. Two-Electron Atoms
  B. Three-Electron Atoms
  C. Critical Nuclear Charges for N-Electron Atoms
  D. Critical Parameters for Simple Diatomic Molecules
  E. Phase Diagram for Three-Body Coulomb Systems

Advances in Chemical Physics, Volume 125, Edited by I. Prigogine and Stuart A. Rice.
ISBN 0-471-21452-3.

VI. Crossover Phenomena and Resonances in Quantum Systems
A. Resonances
B. Crossover Phenomena
C. Multicritical Points
VII. Spatial Finite-Size Scaling
VIII. Finite-Size Scaling and Path Integral Approach for Quantum Criticality
A. Mapping Quantum Problems to Lattice Systems
B. Quantum Criticality
IX. Finite-Size Scaling for Quantum Dots
X. Concluding Remarks
Acknowledgments
References

I. INTRODUCTION

Phase transitions and critical phenomena continue to be a subject of great interest in many fields [1]. A wide variety of physical systems exhibit phase transitions and critical phenomena, such as liquid–gas, ferromagnetic–paramagnetic, fluid–superfluid, and conductor–superconductor [2]. Over the last few decades, a large body of research has been done on this subject, mainly using classical statistical mechanics. Classical phase transitions are driven by thermal energy fluctuations, like the melting of an ice cube. Heating the ice above the freezing point causes molecules in the solid phase to break to become liquid water. If you heat them even more, they vaporize into steam. However, in the last decade, considerable attention has concentrated on a qualitatively different class of phase transitions, transitions that occur at the absolute zero of temperature. These are quantum phase transitions that are driven by quantum fluctuations as a consequence of Heisenberg's uncertainty principle [3,4]. These new transitions are tuned by parameters in the Hamiltonian. An example of this kind of transition is the melting of a Wigner crystal, orderly arrangement of electrons. As one makes the crystal more dense, the electrons become more confined, the uncertainty principle takes over, and the fluctuations in the momentum grow. Squeezing more on the crystal, the system transforms from insulator to conductor [5]. Other examples from condensed matter physics include the magnetic transitions of cuprates, superconductor–insulator transitions in alloys, metal–insulator transitions, and the Quantum–Hall transitions [4,6].

In the field of atomic and molecular physics, the analogy between symmetry breaking of electronic structure configurations and quantum phase transitions has been established at the large-dimensional limit [7]. The mapping between symmetry breaking and mean-field theory of phase transitions was shown by allowing the nuclear charge Z, the parameter that tunes the phase transition, to play a role analogous to temperature in classical statistical mechanics. For two-electron atoms, as the nuclear charge reaches a critical value $Z_c \simeq 0.911$, which is the minimum charge necessary to bind two electrons, one of the electrons jump, in a first-order phase transition, to infinity with zero kinetic energy [8]. The fact that this charge is below $Z = 1$ explains why H^- is a stable negative ion. For three-electron atoms, the transition occurs at $Z_c \simeq 2.0$ and resembles a second-order phase transition [9]; this tells us that He^- is an unstable ion. The estimated values of the critical nuclear charges for N-electron atoms show that, at most, only one electron can be added to a free atom, which means that no doubly charged atomic negative ions exist in the gas phase [10]. For simple one-electron molecular systems $Z_c \simeq 1.228$, and it follows that only the H_2^+ molecular ion is stable. The dissociation of this system occurs in a first-order phase transition [11]. The study of quantum phase transitions and critical phenomena continues to be of increasing interest in the field of atomic and molecular physics. This is motivated by the recent experimental searches for the smallest stable multiply charged anions [12,13], experimental and theoretical work on the stability of atoms and molecules in external electric and magnetic fields [14,15], design and control electronic properties of materials using quantum dots [16], the study of selectively breaking chemical bonds in polyatomic molecules [14], threshold behavior in ultracold atomic collision [17], quantum anomalies in molecular systems [18], stability of exotic atoms [19], and phase transitions of finite clusters [20,21].

Phase transitions in statistical mechanical calculations arise only in the thermodynamic limit, in which the volume of the system and the number of particles go to infinity with fixed density. Only in this limit the free energy, or any thermodynamic quantity, is a singular function of the temperature or external fields. However, real experimental systems are finite and certainly exhibit phase transitions marked by apparently singular thermodynamic quantities. Finite-size scaling (FSS), which was formulated by Fisher [22] in 1971 and further developed by a number of authors (see Refs. 23–25 and references therein), has been used in order to extrapolate the information available from a finite system to the thermodynamic limit. Finite-size scaling in classical statistical mechanics has been reviewed in a number of excellent review chapters [22–24] and is not the subject of this review chapter.

In quantum mechanics, when using variation methods, one encounters the same finite-size problem in studying the critical behavior of a quantum Hamiltonian $\mathcal{H}(\lambda_1, \ldots, \lambda_k)$ as a function of its set of parameters $\{\lambda_i\}$. In this context, critical means the values of $\{\lambda_i\}$ for which a bound-state energy is nonanalytic. In many cases, as in this study, this critical point is the point where a bound state energy becomes absorbed or degenerate with a continuum. In this case, the finite size corresponds not to the spatial dimension but to the number of elements in a complete basis set used to expand the exact wave function of a given Hamiltonian [10] and the size of a cutoff parameter [26]. The present review chapter is about finite-size scaling in quantum mechanics. Most of the work reviewed here is based on our own work on the development and application of finite-size scaling to atomic and molecular systems.

The review is arranged as follows: In the next section we present some general definitions and results for the critical behavior of quantum one-particle central potentials and few-body systems. In particular, we discuss the near-threshold behavior of quantum N-body Hamiltonians in a D-dimensional space, including the large D-limit approximation. In Section III we briefly review the main ideas of finite-size scaling in Classical Statistical Mechanics. In Section IV we present some very general features of the statistical mechanics of quantum systems and then develop the FSS equations for quantum few-body systems. To illustrate the applications of FSS method in quantum mechanics, we give an example of a short-range potential, the Yukawa potential. Finally in Section IV we examine the main assumption of FSS for quantum systems by showing data collapse for few-body problems. Applications of FSS for atomic and molecular systems is given in Section V. In Section VI we present three different phenomena: resonances, crossover, and multicritical points. We discuss in general the applicability of FSS to resonances, the existence of multicritical points, and the definition of the size of the critical region.

In previous sections, *finite size* corresponds to the number of elements of a complete basis set used in a truncated Rayleigh–Ritz expansion of an exact bound eigenfunction of a given Hamiltonian. In Section VII, we present a different FSS approach, the spatial finite-size scaling. With this method we study the scaling properties by introducing a cutoff radius in the potential. This cutoff changes the critical exponent of the energy, but, for large values of the cutoff radius, the asymptotic behavior of FSS functions is dominated by the exact critical exponent. The method gives accurate values for critical parameters and critical exponents.

To treat quantum phase transitions and critical phenomena, it seems that Feynman's path integral is a natural choice. In this approach, one can show that the quantum partition function of the system in d dimensions looks like a classical partition function of a system in $d + 1$ dimensions where the extra dimension is the time [3]. Upon doing so, and allowing the space and time variables to have discrete values, we turn the quantum problem into an effective classical lattice problem. In Section VIII we show how to carry out the mapping between the quantum problem and an effective classical space–time lattice and give an example to illustrate how the approach works. Finally we combine the finite-size scaling method with a multistage real-space renormalization group procedure to examine the Mott metal–insulator transition on a nonpartite lattice and then give discussion and conclusions.

II. SOME GENERAL RESULTS FOR NEAR-THRESHOLD STATES

Weakly bound states represent an interesting field of research in atomic and molecular physics. The behavior of systems near the threshold, which separates bound states from continuous states, is important in the study of ionization of atoms and molecules, molecule dissociation, and scattering collisions. In general, the energy is nonanalytical because a function of the Hamiltonian parameters or a bound state does not exist at the threshold energy. It has been suggested for some time that there are possible analogies between critical phenomena and singularities of the energy [27–29]. In particular, it has been noted that the energy curves of the two-electron atoms as a function of the inverse of the nuclear charge resemble the free energy curves as a function of the temperature for the van der Waals gas [27]. Using the large-dimensional limit model for electronic structure problems, we will show in this section that symmetry breaking of the electronic structure configurations for the many-electron atoms and simple molecular systems can be studied as mean-field problems in statistical mechanics. Then, we will present some general definitions and results for the critical behavior of quantum systems at finite dimensions, in particular at $D = 3$.

A. Phase Transitions at the Large-Dimensional Limit

It is possible to describe stability and symmetry breaking of electronic structure configurations of atoms and molecules as phase transitions and critical phenomena. This analogy was revealed by using the dimensional scaling method and the large-dimensional limit model of electronic structure configurations [7,30–32]. Large-dimensional models were originally developed for specific theories in the fields of nuclear physics, critical phenomena, and particle physics [33,34]. Subsequently, with the pioneering work of Herschbach et al. [29], they found wide use in the field of atomic and molecular physics [35]. In this method, one takes the dimension of space, D, as a variable, solves the problem at some dimension $D \neq 3$ where the physics becomes much simpler, and then uses perturbation theory or other techniques to obtain an approximate result for $D = 3$ [29].

To study the behavior of a given system near the critical point, one has to rely on model calculations that are simple, capture the main physics of the problem, and belong to the same universality class. For electronic structure calculations of atoms and molecules, there are three exactly solvable models: the Thomas–Fermi statistical model (the limit $N \to \infty$ for fixed N/Z, where N is the number of electrons and Z is the nuclear charge); the noninteracting electron model, the limit of infinite nuclear charge ($Z \to \infty$, for fixed N); and the large-dimensional model ($D \to \infty$ for fixed N and Z) [36]. Here we will illustrate the phase transitions and symmetry breaking using the large-dimensional model. In the application of dimensional scaling to electronic structure, the large-D limit reduces to a semiclassical electrostatic problem in which the electrons are assumed to have fixed positions relative to the nuclei and to each other in the D-scaled space [29]. This configuration corresponds to the minimum of an effective potential which includes Coulomb interactions as well as centrifugal terms arising from the generalized D-dependence kinetic energy. Typically, in the large-D regime the electronic structure configuration undergoes symmetry breaking for certain ranges of nuclear charges or molecular geometries [37].

In order to illustrate the analogy between symmetry breaking and phase transitions, we briefly review the main results for the two-electron atoms in the Hartree–Fock (HF) approximation [7]. In the HF approximation at the $D \to \infty$ limit, the dimensional-scaled effective Hamiltonian for the two-electron atom in an external weak electric field $\mathscr{E}$ can be written as [38,39]

$$\mathscr{H}_\infty = \frac{1}{2}\left(\frac{1}{r_1^2} + \frac{1}{r_2^2}\right) - Z\left(\frac{1}{r_1} + \frac{1}{r_2}\right) + \frac{1}{\left(r_1^2 + r_2^2\right)^{1/2}} - \mathscr{E}(r_1 - r_2) \tag{1}$$

where r_1 and r_2 are the electron–nucleus radii and Z is the nuclear charge. The ground-state energy at the large-D limit is then given by

$$E_\infty(Z, \mathscr{E}) = \min_{\{r_1, r_2\}} \mathscr{H}_\infty \tag{2}$$

In the absence of an external electric field, $\mathscr{E} = 0$, Goodson and Hershbach [40] have found that these equations have a symmetric solution with the two electrons equidistant from the nucleus, with $r_1 = r_2 = r$.

This symmetric solution represents a minimum in the region where all the eigenvalues of the Hessian matrix are positive, $Z \geq Z_c = \sqrt{2}$. For values of Z smaller than Z_c, the solutions become unsymmetrical with one electron much closer to the nucleus than the other ($r_1 \neq r_2$). In order to describe this symmetry breaking, it is convenient to introduce new variables (r, η) of the form

$$r_1 = r, \qquad r_2 = (1 - \eta)r \tag{3}$$

where $\eta = (r_1 - r_2)/r_1 \neq 0$ measures the deviation from the symmetric solution.

By studying the eigenvalues of the Hessian matrix, we have found that the solution is a minimum of the effective potential for the range, $1 \leq Z \leq Z_c$. We now turn to the question of how to describe the system near the critical point. To answer this question, a complete mapping between this problem and critical phenomena in statistical mechanics is readily feasible with the following analogies:

- Nuclear charge $Z \leftrightarrow$ temperature T
- External electric field $\mathscr{E} \leftrightarrow$ ordering field h
- Ground-state energy $E_\infty(Z, \mathscr{E}) \leftrightarrow$ free energy $f(T, h)$
- Asymmetry parameter $\eta \leftrightarrow$ order parameter m
- Stability limit point $(Z_c, \mathscr{E} = 0) \leftrightarrow$ critical point $(T_c, h = 0)$

Using the above scheme, we can define the critical exponents (β, $\hat{\alpha}$[1], δ, and γ) for the electronic structure of the two electron atom in the following way:

$$
\begin{aligned}
\eta(Z, \mathscr{E} = 0) &\sim (-\Delta Z)^{\beta}, & \Delta Z &\to 0^- \\
E_\infty(Z, \mathscr{E} = 0) &\sim |\Delta Z|^{2-\hat{\alpha}}, & \Delta Z &\to 0 \\
\mathscr{E}(Z_c, \eta) &\sim \eta^{\delta}\operatorname{sgn}(\eta), & \eta &\to 0 \\
\left.\frac{\partial \eta}{\partial \mathscr{E}}\right|_{\mathscr{E}=0} &\sim |\Delta Z|^{-\gamma}, & \Delta Z &\to 0
\end{aligned}
\tag{4}
$$

where $\Delta Z \equiv Z - Z_c$. These critical exponents describe the nature of the singularities in the above quantities at the critical charge Z_c. The values obtained for these critical exponents are known as classical or mean-field critical exponents with

$$
\beta = \frac{1}{2}, \qquad \hat{\alpha} = 0_{dis}, \qquad \delta = 3, \qquad \gamma = 1 \tag{5}
$$

The results of the asymmetry parameter η as a function of nuclear charge at $\mathscr{E} = 0$ is shown in Fig. 1. This curve of the asymmetry parameter shown is completely analogous to curves representing the behavior of magnetization as a function of the temperature in mean field models of ferromagnetic systems [41] as shown in Fig. 2.

The above approach, the analogy between symmetry breaking and phase transitions, was generalized to treat the large-dimensional model of the N-electron atoms [30], simple diatomic molecules [31,42], both linear and planar one-electron systems [32], and three-body Coulomb systems of the general form ABA [43].

The above simple large-D picture helps to establish a connection to phase transitions. However, the questions which remain to be addressed are: How to carry out such an analogy to the N-electron atoms at $D = 3$ and what are the physical consequences of this analogy? These questions will be examined in the following sections by developing the finite size scaling method for atomic and molecular systems.

[1] In statistical mechanics the Greek letter α is used for this exponent. We reserve α for the exponent of the energy; therefore, $\hat{\alpha} = 2 - \alpha$.

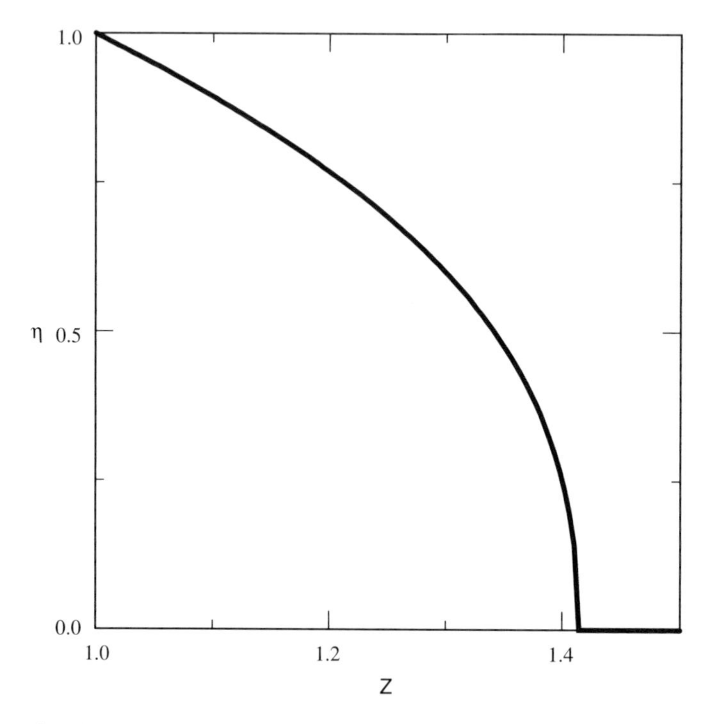

Figure 1. The asymmetry parameter η as a function of the nuclear charge Z for the Hartree–Fock two-electron atom at the large D limit.

B. Critical Phenomena in Spaces of Finite Dimensions

In this section, we will discuss the near-threshold behavior of quantum N-body Hamiltonians in a D-dimensional space. In particular we will study Hamiltonians of the form

$$\mathcal{H}(\lambda; \vec{x}_1, \ldots, \vec{x}_N) = -\frac{1}{2}\sum_{i=1}^{N} \nabla_D^2 + \lambda\, V(\vec{x}_1, \ldots, \vec{x}_N) \tag{6}$$

for different values of N and D. We will assume without loss of generality that the threshold energy is zero, and it is reached from the right $\lambda = \lambda_c > 0$:

$$E(\lambda_c) = \lim_{\lambda \to \lambda_c^+} E(\lambda) = 0 \tag{7}$$

where $E(\lambda)$ is an isolated bound-state energy for $\lambda > \lambda_c$.

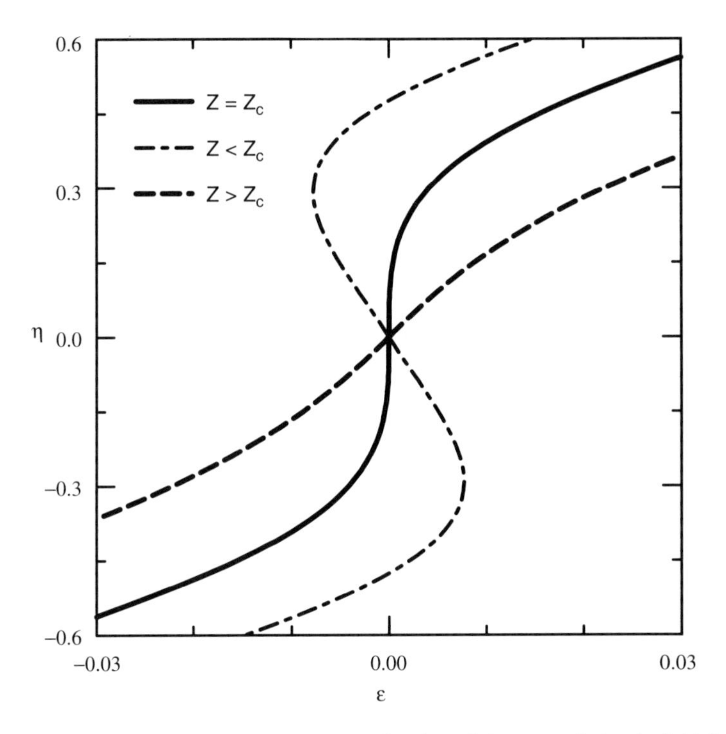

Figure 2. The asymmetry parameter η as a function of the external electric field for three values of the nuclear charge for the Hartree–Fock two-electron atom at the large D limit.

The relevant questions to be addressed at this point are related to the calculation of λ_c, the determination of the leading term in the asymptotic behavior of $E(\lambda)$ near λ_c, and the existence of a square-integrable eigenfunction at the threshold. In the forthcoming sections we will develop the finite-size scaling as a powerful method to obtain accurate numerical estimations of critical parameters. Here we will present some exact results.

Let us introduce the useful concept of a critical exponent. A function $f(x)$ has critical exponent μ at $x = x_0$ if

$$f(x) \sim (x - x_0)^{\mu} \qquad \text{for} \quad x \rightarrow x_0 \tag{8}$$

It means that the leading term obeys a power law. Sometimes the leading order is not a power law and might be logarithmic or exponential behavior. From Eq. (8) we have

$$\lim_{x \rightarrow x_0} \frac{\ln(f(x))}{\ln(x - x_0)} = \mu \tag{9}$$

For an asymptotic behavior of the form

$$f(x) \sim (x - x_0)^{\mu} \ln^{\delta}(x - x_0) \qquad \text{for} \quad x \to x_0 \tag{10}$$

the limit, as in Eq. (9), is also equal to μ. For an exponential behavior

$$f(x) \sim \exp(\pm c/(x - x_0)^{\delta}) \qquad \text{for} \quad x \to x_0 \tag{11}$$

we have

$$\lim_{x \to x_0} \frac{\ln(f(x))}{\ln(x - x_0)} = \pm\infty \tag{12}$$

We will say that a function with asymptotic behavior like Eq. (10) has a critical exponent $\mu_{\log}$ and that for a function that obeys Eq. (12), the exponent is $\mu = \mp\infty$. It is clear that these definitions of critical exponents are just a useful notation. It means that the function does not obey a power law near a given point, but has logarithmic corrections in the first case and it goes faster than any power of $(x - x_0)$ in the second case.

With these definitions we can go a step beyond Eq. (7) and define the critical exponent α for the near-threshold behavior of the bound-state energy

$$E(\lambda) \sim (\lambda - \lambda_c)^{\alpha} \qquad \text{for} \quad \lambda \to \lambda_c^{+} \tag{13}$$

There are few rigorous results for near-threshold behavior in quantum few-body systems, and most of them are one-particle central potentials.

1. One-Particle Central Potentials

The D-dimensional Hamiltonian for a one-particle system is given by

$$\mathscr{H}_D = -\frac{1}{2}\nabla_D^2 + \lambda V(r) \tag{14}$$

where $r = \| \vec{x} \|^{1/2}$, $\vec{x} = (x_1, x_2, \ldots, x_D)$, and the Laplacian operator is given by

$$\nabla_D^2 = \frac{1}{r^{D-1}} \frac{\partial}{\partial r} \left(r^{D-1} \frac{\partial}{\partial r} \right) - \frac{\mathscr{L}_{D-1}^2}{r^2} \tag{15}$$

where $\mathscr{L}_{D-1}^2$ is the total squared angular momentum operator in a D-dimensional space.

The Schrödinger equation is separable by writing the wave function as

$$\Psi(\vec{x}) = \psi(r)\, \mathscr{Y}_{D-1}(\Omega) \tag{16}$$

where $\mathcal{Y}_{D-1}(\Omega)$ is the hyperspherical harmonic, which is an eigenfunction of the generalized angular momentum operator [44]

$$\mathcal{L}_{D-1}^2 \mathcal{Y}_{D-1}(\Omega) = l\,(l + D - 2)\,\mathcal{Y}_{D-1}(\Omega) \tag{17}$$

and ψ obeys the equation

$$\left\{ -\frac{1}{2r^{D-1}} \frac{\partial}{\partial r} \left(r^{D-1} \frac{\partial}{\partial r} \right) + \frac{l(l + D - 2)}{2r^2} + \lambda V(r) \right\} \psi(r) = E(\lambda) \psi(r) \tag{18}$$

By incorporating the square root of the Jacobian into the wave function via

$$\psi(r) = r^{-(D-1)/2} \Phi(r) \tag{19}$$

Eq. (18) can be transformed to a simpler form where the centrifugal energy separates out from other kinetic terms. The resulting Hamiltonian has the following form:

$$\left\{ -\frac{1}{2} \frac{\partial^2}{\partial r^2} + \frac{(\delta - 1)(\delta - 3)}{8r^2} + \lambda V(r) \right\} \Phi = E(\lambda)\,\Phi \tag{20}$$

where $\delta = D + 2l$. Note that the boundary condition $\Phi(r = 0) = 0$ excludes the even eigenfunctions in $D = 1$, but there is nothing new in this case and the corresponding results are mentioned below. Equation (20) shows the very well known equivalence between D-dimensional s waves and three-dimensional $l = (D - 3)/2$ waves.

At this point we can establish one of the few general results for the critical exponent α in near-threshold studies due to Simon [45]; here we present it in a more restricted form with a simple proof:

Theorem. *If square-integrable solution of Eq. (20) exists at the threshold $\Phi(\lambda_c; r)$ with $\|\Phi(\lambda_c; r)\| = 1$, then $\alpha = 1$.*

Proof. Because the energy at the threshold is zero, we have

$$\begin{aligned} &E(\lambda) = \langle T \rangle + \lambda \langle V \rangle \leq 0 \quad \text{and} \quad \langle T \rangle = \frac{1}{2} \int |\hat{p}\Phi(\lambda; r)|^2 \, dr > 0 \Rightarrow \\ &\langle V \rangle(\lambda) < 0 \; \forall \; \lambda \geq \lambda_c \end{aligned} \tag{21}$$

where $\hat{p}$ is the operator momentum. Using the Hellmann–Feynman theorem and Eq. (21) we have

$$\left. \frac{dE(\lambda)}{d\lambda} \right|_{\lambda = \lambda_c} = \langle V \rangle(\lambda_c) < 0 \tag{22}$$

On the other hand, it is straightforward to show, using variational arguments, that $E(\lambda)$ is a monotonic decreasing function of $\lambda \geq \lambda_c$. Then the Tauberian theorem for differentiation of asymptotic expression [46] holds, and from Eq. (13) we obtain

$$\frac{dE(\lambda)}{d\lambda} \sim (\lambda - \lambda_c)^{\alpha - 1} \quad \text{for} \quad \lambda \to \lambda_c^+ \tag{23}$$

Equations (22) and (23) imply $\alpha - 1 = 0$. Note that the theorem is still applicable to the N-body Hamiltonian Eq. (6), and Eq. (23) is valid even when a square-integrable wave function does not exist at the threshold.

In order to obtain the critical behavior of Eq. (20), it is necessary to study short- and long-range potentials. We will define a potential as long-range if $\lim_{r\to\infty} V(r) \sim r^{-\beta}, \beta > 0$, and as short-range if $\lim_{r\to\infty} r^n V(r) \to 0 \;\forall\, n > 0$. This last case includes the important cases of exponential fall-off and compact support potentials [i.e., $V(r) = 0$ for $r > R > 0$]. In both cases we will assume regularity at $r = 0$.

Short-Range Potentials: The problem of the energy critical exponent for D-dimensional short-range central potential was studied by Klaus and Simon [47], and their main results are summarized in Table I.

It is interesting to note that for a given value of the angular momentum $l = 0$ or 1, there is a marginal dimension with logarithmic deviation from a power law, $\alpha = 1_{\log}$ and $\alpha = 1$ for all dimension D greater than the marginal dimension. The case $l \geq 2$ is equivalent to s waves in a space dimension bigger than 4, the marginal dimension for $l = 0$, and therefore the value of the critical exponent does not depend on the spatial dimension. Actually, Eq. (20) does not have two independent parameters D and l, but just δ. The exponent α depends

TABLE I
Critical Exponent α and Wave Function for Short-Range Central Potentials

Dimension D	Angular Momentum l	Critical Exponent α	Square-Integrable Wave Function at Threshold
1	0, 1	2	No
2	0	∞ (exp)	No
	1	1 (log)	No
	≥ 2	1	Yes
3	0	2	No
	≥ 1	1	Yes
4	0	1 (log)	No
	≥ 1	1	Yes
≥ 5	≥ 0	1	Yes

Source: Ref. 47.

on δ for values smaller than the marginal value $\delta_m = 4$, where it has logarithmic corrections, and it does not depend on δ for $\delta > \delta_m = 4$. We can study Eq. (20) by treating δ as a continuous parameter. Lasaut et al. [48] studied Eq. (20) for $\delta > 2$ (they considered $l > -1/2$ as continuous variable in the three-dimensional space). They have found that the critical exponent α varies continuously with l for $-1/2 < l < 1/2$. Using the symmetry of the potential at $\delta = 2$, we have

$$\alpha = \begin{cases} \infty & \delta = 2 \\ \frac{2}{|\delta - 2|} & 0 < \delta < 2 \text{ or } 2 < \delta < 4 \\ 1_{\log} & \delta = 0, 4 \\ 1 & \delta < 0 \text{ or } \delta > 4 \end{cases} \tag{24}$$

and a square-integrable wave function at the threshold exists only for values of δ corresponding to $\alpha = 1$.

For the scaling of other magnitudes, we do not have rigorous results even for one-dimensional or central potentials, but there are some exactly solvable problems that are very useful for testing the numerical methods.

A standard set of solvable potentials with critical behavior can be found in many text books on quantum mechanics [49,50], like the usual square-well potentials and other piecewise constant potentials. Also there are many potentials that are solvable only at $d = 1$ or for three-dimensional s waves like the Hulthén potential, the Eckart potential, and the Pösch–Teller potential. These potentials belong to a class of potentials, called shape-invariant potentials, that are exactly solvable using supersymmetric quantum mechanics [51,52]. There are also many approaches to make isospectral deformation of these potentials [51,53]; therefore it is possible to construct nonsymmetrical potentials with the same critical behavior as that of the original symmetric problem.

Exact solutions are useful to look for new scaling laws. For example, an interesting question is about the scaling law of the amplitude of probability. The asymptotic behavior of the normalized wave function at fixed value of the radius r_0 defines a new scaling exponent μ:

$$\Phi(\lambda; r_0) \sim (\lambda - \lambda_c)^{\mu} f(r_0) \qquad \text{for} \quad \lambda \to \lambda_c^+ \tag{25}$$

where r_0 is an arbitrary value of r except a possible node of the wave function. This exponent tells us about the localization of the particle for $\lambda \sim \lambda_c$, and therefore it is also relevant in scattering problems. Recently [54], based on the near-threshold behavior of the analytic expression for s-wave functions of the exactly solvable potentials, the Hellmann–Feynman theorem, and numerical FSS results, we assumed that the exponent μ is given by the relation

$$\mu = \frac{\alpha - 1}{2} \tag{26}$$

For short-range central potentials we will prove this relation using asymptotic expressions of the wave function. That is, assuming the short-range potential is negligible for $r > R$, we replace the potential $V(r)$ in Eq. (20) by the compact support potential

$$V_R(r) = \begin{cases} V(r) & \text{if} \quad r < R \\ 0 & \text{if} \quad r > R \end{cases} \tag{27}$$

Then we can calculate μ for a large values of $r_0 > R$.

The external wave function with the potential (27) is proportional to the exact external solution of the square-well potential. The solution, which is continuous at $r = R$, is given by

$$\Phi_R(\lambda, k; r) = \begin{cases} \mathcal{N}\, \phi_<(\lambda, k; r) & \text{if } r < R \\ \mathcal{N}\, \phi_<(\lambda, k; R) \sqrt{\dfrac{r}{R}} \dfrac{K_{\frac{\delta-2}{2}}(kr)}{K_{\frac{\delta-2}{2}}(kR)} & \text{if } r > R \end{cases} \tag{28}$$

where $K_\nu(z)$ is the modified Bessel function of the third kind [55], $k = \sqrt{-2E}$, and $\mathcal{N}$ is the normalization constant. For $E < 0$ the wave function is square-integrable and the norm of $\phi_<$ is arbitrary; in particular, we choose

$$\int_0^R |\phi_<(\lambda, k; r)|^2\, dr = 1 \tag{29}$$

Note that the specific value of R is unimportant, except that it cannot be a node of the wave function; but nodes are always localized in a finite region. Thus, we can choose R in the region where the wave function does not have nodes. The continuity of the logarithmic derivative gives the condition for the energies. Of course we need the explicit expression of the function $\phi_<$ to obtain the eigenvalues; but for our purpose, it is enough to assume that an isolated eigenvalue becomes equals to zero at $\lambda = \lambda_c$.

Equations (25) and (28) tell us that the exponent μ is defined by the asymptotic behavior of the normalization constant $\mathcal{N}$ near λ_c, and $\mathcal{N}$ is given by the usual condition $\int_0^\infty |\Phi_R|^2 dr = 1$; using Eq. (29), we obtain

$$\mathcal{N} = \frac{k\sqrt{R}\, K_{\frac{\delta-2}{2}}(kR)}{\left[k^2 R\, K^2_{\frac{\delta-2}{2}}(kR) + \phi^2_<(\lambda, k; R) \int_{kR}^\infty x\, K^2_{\frac{\delta-2}{2}}(x)\, dx\right]^{1/2}} \tag{30}$$

The function $K_\nu(z)$ [55] has no singularities for $z > 0$, and the asymptotic behavior for $z \to 0$ is

$$K_0(z) \sim -\ln(z) \quad \text{and} \quad K_\nu(z) \sim \frac{1}{2}\Gamma(\nu)\left(\frac{2}{z}\right)^\nu \qquad \text{for} \quad \nu > 0 \tag{31}$$

and for $z \to \infty$ it is

$$K_\nu(z) \sim \sqrt{\frac{\pi}{2z}} e^{-z} \tag{32}$$

As it was pointed out before, the problem is symmetrical around $\delta = 2$; thus we could analyze the asymptotic behavior of $\mathcal{N}$ for $\delta \geq 2$.

From the asymptotic behavior of the Bessel function K_ν, the leading term for $k \to 0$ of the integral in the denominator of Eq. (30) is given by Eq. (31):

$$\int_{kR}^{\infty} x K_{\frac{\delta-2}{2}}^2(x)\, dx \sim \begin{cases} A_1 & 2 \leq \delta < 4 \\ A_2 \ln(k) & \delta = 4 \\ A_3 k^{-(\delta-4)} & \delta > 4 \end{cases} \quad \text{for} \quad k \to 0 \tag{33}$$

where A_i, $i = 1, 2, 3$, are positive finite constants.

Therefore, for $\delta < 4$ the leading term of $\mathcal{N}$ is governed by the numerator in Eq. (30). Thus, for $\delta = 2$ we have $\mathcal{N} \sim k \ln(k)$, and then it goes to zero faster than any power of $\lambda - \lambda_c$. For such a behavior we say that the value of $\mu = \infty$ as predicted by Eq. (24). For $2 < \delta < 4$, the asymptotic behavior of $\mathcal{N}$ is $\mathcal{N} \sim k^{(4-\delta)/2} \Rightarrow \mu = \frac{1}{\delta-2} \frac{(4-\delta)}{2} = \frac{\alpha-1}{2}$.

For $\delta \geq 4$, the asymptotic behavior in the denominator of $\mathcal{N}$ is dominated by the divergence in the integral given by Eq. (33). For $\delta = 4$, we have $\lim_{k \to 0} k K_1(kR) \sim$ constant; thus $\mathcal{N} \sim \ln k \sim \ln(\lambda - \lambda_c) \Rightarrow \mu = 0_{\log}$, according to the value $\alpha = 1_{\log}$ for $\delta = 1$.

Finally, the proof is completed by studying the behavior of $\mathcal{N}$ for $\delta > 4$. From Eqs. (30) and (33) we obtain that $\mathcal{N}$ is a finite constant at $k = 0$; thus the critical exponent is $\mu = 0$. This last result can be obtained using the fact that for $\delta > 4$ $(\alpha = 1)$ a normalized wave function at the threshold exists [47]. Therefore, if r_0 is not equal to a node, we have $\lim_{E \to 0} \Phi_\lambda(E; r_0) = \Phi_{\lambda_c}(0; r_0) \neq 0 \Rightarrow \mu = 0$ as predicted by Eq. (26).

Long-Range Potentials: For potentials of the general form

$$V(r) = -\frac{C}{r^\beta}, \qquad C, \beta > 0 \tag{34}$$

it is known [56] that Eq. (20) always support an infinite number of bound states for $\beta < 2$. $\beta = 2$ is a marginal case with pathological behavior. In this case, there is no bound states for $C < ((\delta - 2)/2)^2$, and for $C > ((\delta - 2)/2)^2$ there are square-integrable wave functions for all negative values of the energy. Requiring orthogonality between eigenfunctions, a quantization rule emerges, but the spectrum is not bounded from below (for a renormalized solution of this problem, see Ref. 18). For $\beta > 2$ the strong singularity of the potential at $r = 0$ causes the solution to have an unclear physical meaning. If the potential V is positive, these conclusions do not necessarily hold.

Stillinger [57] presented an exact solution for a potential of the form

$$V(r) = -\frac{3}{32r^2} + \frac{b}{8r^{1/2}} - \frac{c}{8r}, \qquad b, c > 0 \tag{35}$$

By studying the ground-state energy as a function of b at a fixed value of c, he showed that there exists a critical point, the wave function is normalized at the threshold, and the critical exponent for the energy is $\alpha = 1$. A set of long-range potentials exactly solvable only at $E = 0$ ($\alpha = 1$) is discussed in Ref. 58.

The interesting case of potentials with tails falling off like $1/r^{\beta}$ was recently treated by Mortiz et al. [59]. They reported many aspects of the near-threshold properties for these potentials. However, they did not give values of the critical parameters λ_c and α. There is no general solution of the Schrödinger equation for power-law potentials, but for tails going to zero faster than $1/r^2$ and $E(\lambda_c) = 0$, the solutions are Bessel functions [60] that can be normalized for $\delta > 4$.

2. *Few-Body Potentials*

There are only few rigorous results on critical phenomena for many-body systems, and most of them are variational bounds. As an example, for N-electron atoms, experimental results [61] and numerical calculations [62] rule out the stability of doubly charged atomic negative ions in the gas phase. However, a rigorous proof exists only for the instability of doubly charged hydrogen negative ions [63].

Exact values of critical exponents are more difficult to obtain, because variational bounds do not give estimations of the exponents. Then the result presented by M. Hoffmann-Ostenhof et al. [64] for the two-electron atom in the infinite mass approximation is the only result we know for N-body problems with $N > 1$. They proved that there exists a minimum (critical) charge where the ground state degenerates with the continuum, there is a normalized wave function at the critical charge, and the critical exponent of the energy is $\alpha = 1$.

Finally, there are a series of interesting papers dealing with rigorous results, including accurate estimations for the critical lines and for the ground-state energy of three- and four-charge systems, but without mentioning the problem of the critical exponent [for more details, see a recent review article (Ref. 65) and references therein].

In Section V, we will apply FSS to the three-body Coulomb system. Therefore, it is convenient to present some rigorous results for this problem. Let us consider the stability and phase transitions of the three-body *ABA* Coulomb systems with charges (Q,q,Q) and masses (M,m,M). This system has as unique possible threshold of $AB + A$. With the scale transformation $r \to fr$, where $f = \mu|Qq|$ and $\mu = mM/(m+M)$ is the reduced mass, the scaled Hamiltonian, $\mathcal{H} \to \mu\mathcal{H}/(f)^2$, reads [11,66]

$$\mathcal{H} = -\frac{\nabla_1^2}{2} - \frac{\nabla_2^2}{2} - \frac{1}{r_1} - \frac{1}{r_2} - \kappa\, \nabla_1 \cdot \nabla_2 + \lambda \frac{1}{r_{12}} \tag{36}$$

where $0 \leq \lambda = |\frac{Q}{q}| \leq \infty$ and $0 \leq \kappa = \frac{1}{1+\frac{m}{M}} \leq 1$. Here we have formally separated the motion of the center of mass, and the reference particle is the one with mass m. With this scaling transformation, the Hamiltonian depends linearly on the parameters λ and κ.

Some theorems, valid in particular for the Hamiltonian (36), are presented by Thirring [67]. The main results of interest to our discussion can be summarized as follows:

i. If $\mathcal{H}(\beta)$ is a linear function of β, then $E_0(\beta)$ is a concave function of β.
ii. The ground-state energy of the Hamiltonian Eq. (36) is an increasing function of λ.
iii. For small values of κ (first-order perturbation result) we have $E_0(\kappa = 0) \leq E_0(\kappa)$.

We are interested in the critical line that separates the regions where the three-body system *ABA* is stable (bounded) with the region $AB + A$. Using i–iii, we can prove the following:

Lemma. *$\lambda_c(\kappa)$ is a convex function of κ.*

Proof. $0 \leq \kappa_1 < \kappa_2 \leq 1$ *and* $\lambda_1 = \lambda(\kappa_1)$; $\lambda_2 = \lambda(\kappa_2)$. *Now define a line in the κ–λ plane beginning at (κ_1, λ_1) and ending at (κ_2, λ_2). A parameterized expression of such a line is*

$$(\kappa, \lambda) = (t\kappa_2 + (1-t)\kappa_1,\ t\lambda_2 + (1-t)\lambda_1), \qquad t \in [0,1]$$

By i, for the ground-state energy over this line holds

$$E_0(t) \geq t\, E_0(t=0) + (1-t)\, E_0(t=1)$$

But $t = 0$ and $t = 1$ belong to the critical curve; thus

$$E_0(t=0) = E_0(t=1) = E_{th} = -\frac{1}{2}$$

Therefore, $E_0(t) \geq E_{th} = -\frac{1}{2}$ *together with ii yield*

$$\lambda_c(\kappa(t)) \leq t\lambda_c(\kappa_2) + (1-t)\lambda_c(\kappa_1)$$

Remarks:

- *If* $\lambda_c(\kappa)$ *is a convex function, then the critical curve has one (and only one) minimum for* $\kappa \in [0, 1]$.
- *If* $\nexists\ \kappa^* \in (0,1)/\frac{\partial\lambda}{\partial\kappa^*} = 0$, *then the minimum is located at* $\kappa = 0$ *or* $\kappa = 1$.
- *From iii we have* $\frac{\partial\lambda_c}{\partial\kappa}|_{\kappa=0} \leq 0$.
- *The exact s-state eigenfunctions of Hamiltonian Eq. (36) has the functional form*

$$\Psi_s(\vec{r_1}, \vec{r_2}) = \Psi_s(r_1, r_2, r_{12})$$

For this functional form, the correlation of fluctuations in position and momentum spaces takes the form

$$\Gamma_s(\vec{r_1}, \vec{r_2}) = \langle(\vec{r_1} - \langle\vec{r_1}\rangle_s) \cdot (\vec{r_2} - \langle\vec{r_2}\rangle_s)\rangle_s = \langle\vec{r_1} \cdot \vec{r_2}\rangle_s$$

and

$$\Gamma_s(\vec{p_1}, \vec{p_2}) = \langle(\vec{p_1} - \langle\vec{p_1}\rangle_s) \cdot (\vec{p_2} - \langle\vec{p_2}\rangle_s)\rangle_s = \langle\vec{p_1} \cdot \vec{p_2}\rangle_s$$

- *Using* $E_0(\lambda_c(\kappa), \kappa) = E_{th} = -\frac{1}{2}\ \forall\ \kappa \in [0, 1]$, *we obtain*

$$\frac{dE_0(\lambda_c(\kappa), \kappa)}{d\kappa} = 0 = \frac{\partial E_0(\lambda_c(\kappa), \kappa)}{\partial\kappa} + \frac{\partial E_0(\lambda_c(\kappa), \kappa)}{\partial\lambda}\frac{\partial\lambda_c}{\partial\kappa}$$

If there exists a minimum in the critical line, $(\kappa^*, \lambda_c(\kappa^*))$, $0 < \kappa^* < 1$, *at this minimum we have*

$$\frac{\partial\lambda_c}{\partial\kappa} = 0 \Rightarrow \partial\frac{E_0}{\partial\kappa}\Big|_* = 0$$

and using Hellman–Feynman theorem, we obtain

$$\langle\nabla_1 \cdot \nabla_2\rangle|_{(\kappa^*, \lambda_c(\kappa^*))} = 0$$

We will see in Section V that the last result plays an important role in the classification of three-body system to atom-like or to molecule-like systems.

III. FINITE-SIZE SCALING IN CLASSICAL STATISTICAL MECHANICS

In statistical mechanics, the existence of phase transitions is associated with singularities of the free energy per particle in some region of the thermodynamic space. These singularities occur only in the *thermodynamic limit* [68,69]; in this limit the volume (V) and particle number (N) go to infinity in such a way that the density ($\rho = N/V$) stays constant. This fact could be understood by examining the partition function. For a finite system, the partition function is a finite sum of analytical terms, and therefore it is itself an analytical function. It is necessary to take an infinite number of terms in order to obtain a singularity in the thermodynamic limit [68,69].

In practice, real systems have large but finite particle numbers ($N \sim 10^{23}$) and volume, and phase transitions are observed. Even more dramatic is the case of numerical simulations, where sometimes systems with only a few number (hundreds, or even tens) of particles are studied, and "critical" phenomena are still present. The question of why finite systems apparently describe phase transitions, along with the relation of this phenomenon with the true phase transitions in infinite systems, is the main subject of finite-size scaling theory [22]. However, finite-size scaling is not only a formal way to understand the asymptotic behavior of a system when the size tends to infinity. In fact, the theory gives us numerical methods capable of obtaining accurate results for infinite systems even by studying the corresponding small systems (see Refs. 23–25 and references therein).

For readers who desire more details on the development of the theory and applications, there are many excellent review articles and books on this subject in the literature [22–25]. However, in this review chapter we are going to present only the general idea of finite-size scaling in statistical mechanics, which is closely related to the application of these ideas in quantum mechanics.

In order to understand the main idea of finite-size scaling, let us consider a system defined in a D-dimensional volume V of a linear dimension L ($V = L^D$). In a finite-size system, if quantum effects are not taken into consideration, there are in principle three length scales: The finite geometry characteristic size L, the correlation length ξ, which may be defined as the length scale covering the exponential decay $e^{-r/\xi}$ with distance r of the correlation function, and the microscopic length a which governs the range of the interaction. Thus, thermodynamic quantities may depend on the dimensionless ratios ξ/a and L/a. The finite-size scaling hypothesis assumes that, close to the critical point, the microscopic length drops out.

If in the thermodynamic limit, $L \to \infty$, we consider that there is only one parameter T in the problem and the infinite system has a second-order phase transition at a critical temperature T_c, a quantity K develops a singularity as a function of the temperature T in the form

$$K(T) = \lim_{L\to\infty} K_L(T) \sim |T - T_c|^{-\rho} \tag{37}$$

whereas if it is regular in the finite system, $K_L(T)$ has no singularity.

When the size L increases, the singularity of $K(T)$ starts to develop. For example, if the correlation length diverges at T_c as

$$\xi(T) = \lim_{L\to\infty} \xi_L(T) \sim |T - T_c|^{-\nu} \tag{38}$$

then $\xi_L(T)$ has a maximum that becomes sharper and sharper, and the FSS ansatz assumes the existence of scaling function F_K such that

$$K_L(T) \sim K(T) F_K\left(\frac{L}{\xi(T)}\right) \tag{39}$$

where $F_K(y) \sim y^{\rho/\nu}$ for $y \sim 0^+$. Since the FSS ansatz, Eq. (39), should be valid for any quantity that exhibits an algebraic singularity in the bulk, we can apply it to the correlation length ξ itself. Thus the correlation length in a finite system should have the form [25]

$$\xi_L(T) \sim L\phi_\xi(L^{1/\nu}|T - T_c|) \tag{40}$$

The special significance of this result was first realized by Nightingale [70], who showed how it could be reinterpreted as a renormalization group transformation of the infinite system. The phenomenological renormalization (PR) equation for finite systems of sizes L and L' is given by

$$\frac{\xi_L(T)}{L} = \frac{\xi_{L'}(T')}{L'} \tag{41}$$

and has a fixed point at $T^{(L,L')}$. It is expected that the succession of points $\{T^{(L,L')}\}$ will converge to the true T_c in the infinite size limit.

The finite-size scaling theory combined with transfer matrix calculations had been, since the development of the phenomenological renormalization in 1976 by Nightingale [70], one of the most powerful tools to study critical phenomena in two-dimensional lattice models. For these models the partition function and all the physical quantities of the system (free energy, correlation length, response functions, etc) can be written as a function of the eigenvalues of the transfer matrix [71]. In particular, the free energy takes the form

$$f(T) = -T \ln \lambda_1 \tag{42}$$

and the correlation length is

$$\xi(T) = -\frac{1}{\ln(\lambda_2/\lambda_1)} \tag{43}$$

where λ_1 and λ_2 are the largest and the second largest eigenvalues of the transfer matrix. In this context, critical points are related with the degeneracy of these eigenvalues. For finite transfer matrix, the Perron–Frobenius theorem [72] asserts that the largest eigenvalue is isolated (nondegenerated) and phase transitions can occur only in the limit $L \to \infty$ where the size of the transfer matrix goes to infinity and the largest eigenvalues can be degenerated. It is important to note that in the Perron–Frobenius theorem all the matrix elements are positive.

For quasi-one-dimensional systems of size L, it is possible to calculate all the eigenvalues of finite transfer matrix; and therefore, using scaling ansatz like phenomenological renormalization Eq. (41), it is possible to obtain critical parameters and critical exponents for bidimensional systems. Transfer matrix with finite-size scaling theory was successfully applied to the study of a wide variety of bidimensional lattice system like magnetic models [73], modulated phases [74], percolation and self-avoiding random walks [75–81], long-range interactions [82], and many other problems of critical behavior in statistical physics of geometric nature (see Cardy [25] and the references therein).

IV. FINITE-SIZE SCALING IN QUANTUM MECHANICS

A. Quantum Statistical Mechanics and Quantum Classical Analogies

In this section, we present briefly some very general features of the statistical mechanics of quantum systems. For a given system with a Hamiltonian H, the main quantity of interest is the partition function $Z(\beta)$, where β is the inverse temperature

$$Z(\beta) = Tr\, e^{-\beta H} \tag{44}$$

From this one can calculate the expectation values of any arbitrary operator $\mathcal{O}$ by

$$\langle \mathcal{O} \rangle = \frac{1}{Z} Tr(\mathcal{O} e^{-\beta H}) \tag{45}$$

In the limit $T \to 0$, the free energy of the system, $F = -k_B T \ln Z$, becomes the ground-state energy and the various thermal averages become ground-state expectation values.

To do quantum mechanics, one starts with an orthonormal and complete set of states $|n\rangle$ that have the properties

$$I = \sum_n |n\rangle\langle n|, \qquad \langle n \mid m \rangle = \delta_{nm} \tag{46}$$

and a trace operation is given by

$$\text{Tr}\ \mathcal{O} = \sum_n \langle n|\mathcal{O}|n\rangle \tag{47}$$

Note that the density matrix operator $e^{-\beta H}$ is the same as the time evolution operator $e^{-iH\tau/\hbar}$ if we assign the imaginary value $\tau = -i\hbar\beta$ to the time interval over which the system evolves. Thus, the partition function takes the form of a sum of imaginary time transition amplitudes for the system to start and return to the same state after an imaginary time interval τ,

$$Z(\beta) = \text{Tr}\ e^{-\beta H} = \sum_n \langle n|e^{-\beta H}|n\rangle \tag{48}$$

We see that calculating the thermodynamics of a quantum system is the same as calculating transition amplitudes for its evolution in imaginary time [83].

Many problems in D-dimensional statistical mechanics with nearest-neighbor interactions can be converted into quantum mechanics problems in $(D-1)$ dimensions of space and one dimension of time [84]. The quantum theory arises here in a Feynman path integral formulation [85].

In the path integral approach, the transition amplitude between two states of the system can be calculated by summing amplitudes for all possible paths between them. By inserting a sequence of sums over sets of intermediate states into the expression for the partition function, Eq. (48) becomes

$$Z(\beta) = \sum_n \sum_{k_1,k_2,k_3,\ldots,k_N} \langle n|e^{-\delta\tau H/\hbar}|k_1\rangle\langle k_1|e^{-\delta\tau H/\hbar}|k_2\rangle \ldots \langle k_N|e^{-\delta\tau H/\hbar}|n\rangle \tag{49}$$

where $\delta\tau$ is a time interval and N is a large integer chosen such that $N\delta\tau = \hbar\beta$. This expression for the partition function has the form of a classical partition function, which is a sum over configurations expressed in terms of a transfer matrix $\langle k_i|e^{-\delta\tau H/\hbar}|k_j\rangle$, with the imaginary time as an additional dimension [3]. Thus a quantum system existed in D dimensions, and the expression for its partition function looks like a classical partition function for a system with $(D+1)$ dimensions. The extra time dimension is in units of $\hbar\beta$; when taking the limit $\beta \to \infty$, we get a truly $(D+1)$-dimensional effective classical system. Thus, one can solve many problems in one-dimensional statistical mechanics calculating partition functions; averages and correlation functions in terms of their quantum analogs as presented in Table II. This analogy opens the door to use very powerful methods developed in statistical classical mechanics to study quantum phase transitions and critical phenomena [4].

TABLE II
Quantum Classical Analogies

Quantum	Classical
D space and 1 time dimensions	$(D+1)$ space dimensions
Quantum Hamiltonian H	Transfer matrix
Generating functional	Partition function
Ground state	Equilibrium state
Expectation values	Ensemble averages
Ground state of H	Free energy
Coupling constant	Temperature
Propagators	Correlation functions
Mass gap of H $(1/\Delta E)$	Correlation length (ξ)

B. Finite-Size Scaling Equations in Quantum Mechanics

In order to apply the FSS to quantum mechanics problems, let us consider the following Hamiltonian of the form [10]

$$\mathcal{H} = \mathcal{H}_0 + V_\lambda \tag{50}$$

where $\mathcal{H}_0$ is λ-independent and V_λ is the λ-dependent term. We are interested in the study of how the different properties of the system change when the value of λ varies. A critical point λ_c will be defined as a point for which a bound state becomes absorbed or degenerate with a continuum.

Without loss of generality, we will assume that the Hamiltonian, Eq. (50), has a bound state E_λ for $\lambda > \lambda_c$ which becomes equal to zero at $\lambda = \lambda_c$. As in statistical mechanics, we can define some critical exponents related to the asymptotic behavior of different quantities near the critical point. In particular, for the energy we can define the critical exponent α as

$$E_\lambda \underset{\lambda \to \lambda_c^+}{\sim} (\lambda - \lambda_c)^\alpha \tag{51}$$

For general potentials of the form $V_\lambda = \lambda\ V$, we have shown in the previous section that the critical exponent α is equal to one if and only if $\mathcal{H}(\lambda_c)$ has a normalizable eigenfunction with eigenvalue equal to zero [45]. The existence or absence of a bound state at the critical point is related to the type of the singularity in the energy. Using statistical mechanics terminology, we can associate "first-order phase transitions" with the existence of a normalizable eigenfunction at the critical point. The absence of such a function could be related to "continuous phase transitions" [10].

In quantum calculations, the Rayleigh–Ritz variational method is widely used to approximate the solution of the Schrödinger equation [86]. To obtain exact results, one should expand the exact wave function in a complete basis set and take the number of basis functions to infinity. In practice, one truncates this expansion at some order N. In the present approach, the finite size corresponds not to the spatial dimension, as in statistical mechanics, but to the number of elements in a complete basis set used to expand the exact eigenfunction of a given Hamiltonian. For a given complete orthonormal λ-independent basis set $\{\Phi_n\}$, the ground-state eigenfunction has the following expansion:

$$\Psi_\lambda = \sum_n a_n(\lambda)\Phi_n \tag{52}$$

where n represents the set of quantum numbers. In order to approximate the different quantities, we have to truncate the series, Eq. (52), at order N. Then the Hamiltonian is replaced by $M(N) \times M(N)$ matrix $\mathscr{H}^{(N)}$, with $M(N)$ being the number of elements in the truncated basis set at order N. Using the standard linear variation method, the Nth-order approximation for the energies are given by the eigenvalues $\{\Lambda_i^{(N)}\}$ of the matrix $\mathscr{H}^{(N)}$,

$$E_\lambda^{(N)} = \min_{\{i\}}\{\Lambda_i^{(N)}\} \tag{53}$$

In order to obtain the value of λ_c from studying the eigenvalues of a finite-size Hamiltonian matrix, one has to define a sequence of pseudocritical parameters, $\{\lambda^{(N)}\}$. Although there is no unique recipe to define such a sequence, in this review we used three methods: The first-order method (FOM) can be applied if the the threshold energy is known [8,76]. In this method one defines $\lambda^{(N)}$ as the value in which the ground-state energy in the Nth-order approximation, $E_0^{(N)}(\lambda)$, is equal to the threshold energy E_T,

$$E_0^{(N)}(\lambda^{(N)}) = E_T(\lambda^{(N)}) \tag{54}$$

The second approach is the phenomenological renormalization (PR) [24,70] method, where the sequence of the pseudocritical values of λ can be calculated by knowing the first and the second lowest eigenvalues of the $\mathscr{H}$ matrix for two different orders, N and N'. The critical λ_c can be obtained by searching for the fixed point of the phenomenological renormalization equation for a finite-size system [70],

$$\frac{\xi_N(\lambda^{(N)})}{N} = \frac{\xi_{N'}(\lambda^{(N')})}{N'} \tag{55}$$

where the correlation length of the classical pseudosystem is defined as

$$\xi_N(\lambda) = -\frac{1}{\log(E_1^{(N)}(\lambda)/E_0^{(N)}(\lambda))} \tag{56}$$

and $E_0^{(N)}(\lambda)$ and $E_1^{(N)}(\lambda)$ are the ground state and the first excited eigenvalues of a sector of given symmetry of the $\mathscr{H}$ matrix [8,9,87].

The third method is a direct finite-size scaling approach to study the critical behavior of the quantum Hamiltonian without the need to make any explicit analogy to classical statistical mechanics [54,88]. The truncated wave function that approximate the eigenfunction Eq. (52) is given by

$$\Psi_\lambda^{(N)} = \sum_n^{M(N)} a_n^{(N)}(\lambda)\Phi_n \tag{57}$$

where the coefficients $a_n^{(N)}$ are the components of the ground-state eigenvector. In this representation, the expectation value of any operator $\mathcal{O}$ at order N is given by

$$\langle \mathcal{O} \rangle_\lambda^{(N)} = \sum_{n,m}^{N} a_n^{(N)}(\lambda)^* a_m^{(N)}(\lambda)\, \mathcal{O}_{n,m} \tag{58}$$

where $\mathcal{O}_{n,m}$ are the matrix elements of $\mathcal{O}$ in the basis set $\{\Phi_n\}$. In general, the mean value $\langle \mathcal{O} \rangle$ is not analytical at $\lambda = \lambda_c$, and we can define a critical exponent, $\mu_\mathcal{O}$, by the relation

$$\langle \mathcal{O} \rangle_\lambda \underset{\lambda \to \lambda_c^+}{\sim} (\lambda - \lambda_c)^{\mu_\mathcal{O}} \tag{59}$$

In statistical mechanics, the singularities in thermodynamic functions associated with a critical point occur only in the thermodynamic limit. In the variation approach, singularities in the different mean values will occur only in the limit of infinite basis functions [88].

As in the FSS ansatz in statistical mechanics [24,77], we will assume that there exists a scaling function for the truncated magnitudes such that

$$\langle \mathcal{O} \rangle_\lambda^{(N)} \sim \langle \mathcal{O} \rangle_\lambda F_\mathcal{O}(N|\lambda - \lambda_c|^\nu) \tag{60}$$

with a different scaling function $F_\mathcal{O}$ for each different operator but with a unique scaling exponent ν.

Now we are in a position to obtain the critical parameters by defining the following function [88]:

$$\Delta_\mathcal{O}(\lambda; N, N') = \frac{\ln\left(\langle \mathcal{O} \rangle_\lambda^{(N)} / \langle \mathcal{O} \rangle_\lambda^{(N')}\right)}{\ln(N'/N)} \tag{61}$$

At the critical point, the mean value depends on N as a power law, $\langle \mathcal{O} \rangle \sim N^{-\mu_{\mathcal{O}}/\nu}$; thus one obtains an equation for the ratio of the critical exponents

$$\Delta_{\mathcal{O}}(\lambda_c; N, N') = \frac{\mu_{\mathcal{O}}}{\nu} \tag{62}$$

which is independent of the values of N and N'. Thus, for three different values N, N', and N'' the curves defined by Eq. (61) intersect at the critical point

$$\Delta_{\mathcal{O}}(\lambda_c; N, N') = \Delta_{\mathcal{O}}(\lambda_c; N'', N) \tag{63}$$

In order to obtain the critical exponent α, which is associated with the energy, we can take $\mathcal{O} = \mathcal{H}$ in Eq. (62) with $\mu_{\mathcal{O}} = \alpha$:

$$\frac{\alpha}{\nu} = \Delta_{\mathcal{H}}(\lambda_c; N, N') \tag{64}$$

By using the Hellmann–Feynman theorem [86] we obtain

$$\frac{\partial E_\lambda}{\partial \lambda} = \left\langle \frac{\partial \mathcal{H}}{\partial \lambda} \right\rangle_\lambda = \left\langle \frac{\partial V_\lambda}{\partial \lambda} \right\rangle_\lambda \tag{65}$$

Taking $\mathcal{O} = \partial V_\lambda / \partial \lambda$ in Eq. (62) gives an equation for $(\alpha - 1)/\nu$ that, together with Eq. (64), gives the exponents α and ν. Now, we can define the following function:

$$\Gamma_\alpha(\lambda; N, N') = \frac{\Delta_{\mathcal{H}}(\lambda; N, N')}{\Delta_{\mathcal{H}}(\lambda; N, N') - \Delta_{\partial \mathcal{V}_\lambda / \partial \lambda}(\lambda; N, N')} \tag{66}$$

which is also independent of the values of N and N' at the critical point $\lambda = \lambda_c$ and gives the critical exponent α:

$$\alpha = \Gamma_\alpha(\lambda_c; N, N') \tag{67}$$

From Eq. (64) the critical exponent ν is readily given by

$$\nu = \frac{\alpha}{\Delta_{\mathcal{H}}(\lambda_c; N, N')} \tag{68}$$

The FSS equations are valid only as an asymptotic expression, $N \to \infty$; but with a finite basis set, unique values of λ_c, α, and ν can be obtained as a succession of values as a function of N, N', and N''. The relation between N, N', and N'' was extensively studied in FSS in statistical mechanics [24], and it is known that the fastest convergence is obtained when the difference between these numbers is as small as possible. In our work [10] we took $\Delta N = 1$, and when there are parity effects we used $\Delta N = 2$. In order to obtain the extrapolated values for $\lambda^{(N)}$, $\alpha^{(N)}$, and $\nu^{(N)}$ at $N \to \infty$, we used the algorithm of Bulirsch and Stoer [89] with $N' = N + \Delta N$ and $N'' = N - \Delta N$. This algorithm was also studied in detail and gives very accurate results for both statistical mechanics problems [90] and electronic structure critical parameters [10].

C. Extrapolation and Basis Set Expansions

To illustrate the applications of the FSS method in quantum mechanics, let us give an example of a short-range interaction, the Yukawa potential. This potential is spherically symmetric, and therefore the critical behavior can be studied for zero and nonzero angular momentum.

In atomic units the Hamiltonian of the screened Coulomb potential can be written as

$$\mathscr{H}(\lambda) = -\frac{1}{2}\nabla^2 - \lambda\frac{e^{-r}}{r} \tag{69}$$

It is well known that the perturbation expansion in $\sigma = 1/\lambda$ around the Coulombic limit, $\sigma = 0$, is asymptotic with zero radius of convergence [91]. This Hamiltonian has bound states for large values of λ and has the exact value of the critical exponent $\alpha = 2$ for states with zero angular momentum and $\alpha = 1$ for states with nonzero angular momentum [47].

A convenient orthonormal basis set that can be used is given by [88]

$$\Phi_{n,l,m}(r,\Omega) = \frac{1}{\sqrt{(n+1)(n+2)}}e^{-\frac{r}{2}}L_n^{(2)}(r)Y_{l,m}(\Omega) \tag{70}$$

where $L_n^{(2)}(r)$ is the Laguerre polynomial of degree n and order 2 and $Y_{l,m}(\Omega)$ is the spherical harmonic function of solid angle Ω [55].

In this basis set, one has to calculate the lowest eigenvalue and eigenvector of the finite Hamiltonian matrix. The matrix elements of the kinetic energy operator can be calculated analytically, and therefore the problem reduces to calculate the matrix elements of the particular potential. Now, in order to obtain the numerical values for λ_c, α, and ν we can use Eqs. (63), (64), and (68) or we can simply use $\Gamma_\alpha(\lambda; N, N')$, Eq. (66), which is independent of the values of N and N' at the critical point $\lambda = \lambda_c$. Plotting $\Gamma_\alpha(\lambda; N, N')$ as a function of λ gives a family of curves with an intersection at λ_c. At the point $\lambda = \lambda_c$, one can read the critical exponent α and the critical exponent ν is readily given by Eq. (68).

For the ground state, the curves of $\Gamma_\alpha(\lambda; N, N-2)$ as a function $1/N$ for even values of N are shown in Fig. 3 for $4 \leq N \leq 100$. From this figure, one can

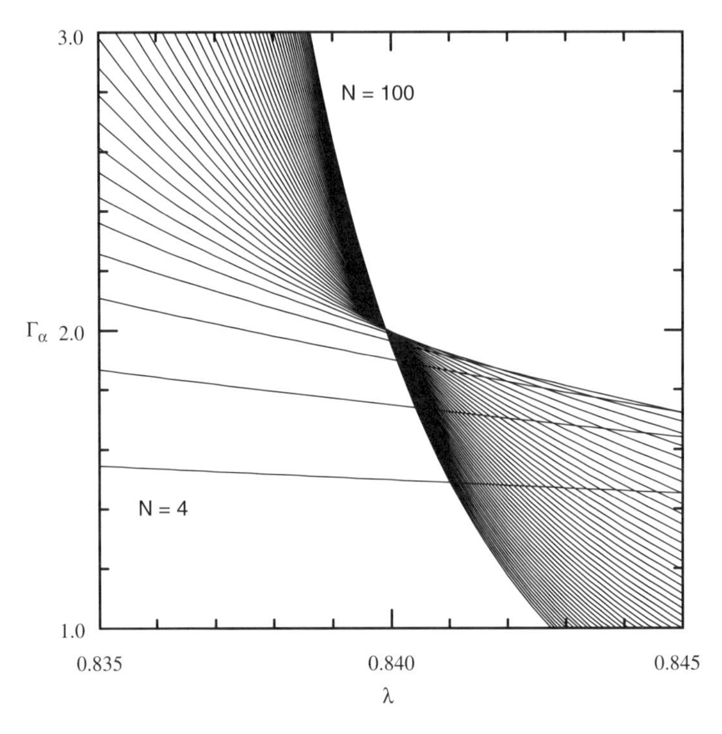

Figure 3. $\Gamma_\alpha(\lambda; N, N-2)$ as a function of λ for the ground state of the Yukawa potential for even values of $4 \leq N \leq 100$.

obtain the values of λ_c and the critical exponent α. The curves of $\lambda^{(N)}$, and $\alpha^{(N)}$ as a function of $1/N$ for $10 \leq N \leq 100$ are shown in Figs. 4 and 5. It seems that a reasonable large-N regime was obtained when $N > 20$; and finally, obtain the extrapolated values as $N \to \infty$ by using the algorithm of Bulirsch and Stoer [89]. The extrapolated values calculated with the points corresponding to $20 \leq N \leq 100$ for $l = 0$ are $\lambda_c = 0.8399039$ and $\alpha = 2$, in excellent agreement with the exact numerical value of $\lambda_c = 0.839908$ [92] and the exact value for $\alpha = 2$ [47].

The behavior of the ground-state energy, $E_0^{(N)}$, as a function of λ for different values of N is different from the state with $l = 1$. For $l = 0$ the energy curve goes smoothly to zero as a function of λ, but the second derivative function develops a discontinuity in the neighborhood of the critical point, $\lambda_c \simeq 0.8399$ and the critical exponent $\alpha = 2$. For $l = 1$, the energy curve bends sharply to zero at the critical point $\lambda_c \simeq 4.5409$ with a critical exponent $\alpha = 1$. As one should expect, there is a discontinuity in the first derivative as a function of λ [88].

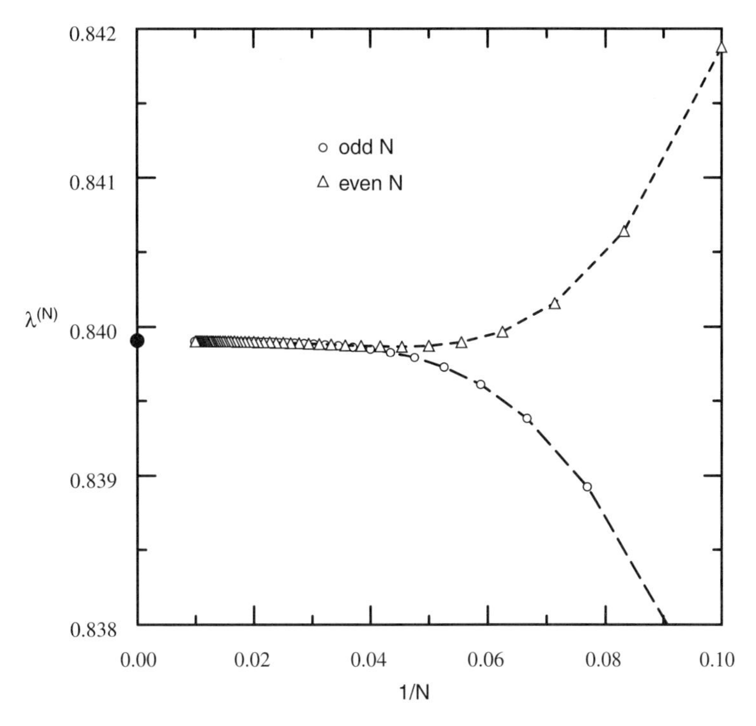

Figure 4. $\lambda^{(N)}$ as a function of $1/N$ for the ground state of the Yukawa potential. The extrapolated values are shown by dots.

D. Data Collapse for the Schrödinger Equation

In order to show the data collapse for quantum few-body problems, let us examine the main assumption we have made in Eq. (60) for the existence of a scaling function for each truncated magnitude $\langle \mathcal{O} \rangle_{\lambda}^{(N)}$ with a unique scaling exponent ν.

Since the $\langle \mathcal{O} \rangle_{\lambda}^{(N)}$ is analytical in λ, then from Eqs. (59) and (60) the asymptotic behavior of the scaling function must have the form

$$F_{\mathcal{O}}(x) \sim x^{-\mu_{\mathcal{O}}/\nu} \tag{71}$$

Eqs. (60) and (71) have the scaling form as presented in Ref. [93]. For our purposes, it is convenient to write this in a slightly different form. From Eqs. (60) and (71) we have

$$\langle \mathcal{O} \rangle^{(N)}(\lambda_c) \sim N^{-\mu_{\mathcal{O}}/\nu} \tag{72}$$

for large values of N.

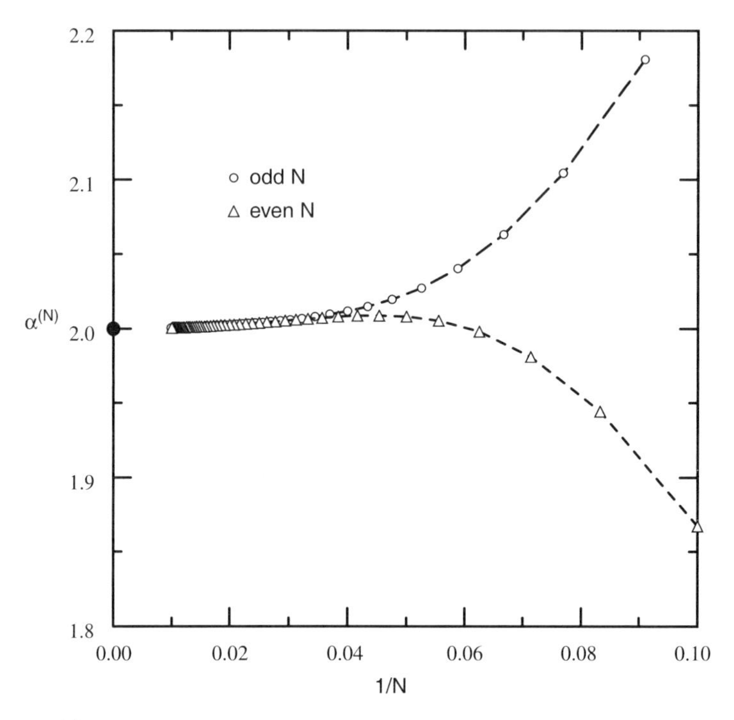

Figure 5. $\alpha^{(N)}$ as a function of $1/N$ for the ground state of the Yukawa potential. The extrapolated values are shown by dots.

Because the same argument of regularity holds for the derivatives of the truncated expectation values, we have that

$$\left.\frac{\partial^m \langle \mathcal{O} \rangle^{(N)}}{\partial \lambda^m}\right|_{\lambda=\lambda_c} \sim N^{-(\mu_{\mathcal{O}}-m)/\nu} \tag{73}$$

$\langle \mathcal{O} \rangle^{(N)}$ is analytical in λ; then using Eq. (73), the Taylor expansion could be written as [93]

$$\langle \mathcal{O} \rangle^{(N)}(\lambda) \sim N^{-\mu_{\mathcal{O}}/\nu} G_{\mathcal{O}}(N^{1/\nu}(\lambda - \lambda_c)) \tag{74}$$

where $G_{\mathcal{O}}$ is an analytical function of its argument.

This equivalent expression for the scaling of a given expectation value has a correct form to study the data collapse in order to test FSS hypothesis in quantum few-body Hamiltonians. If the scaling Eq. (60) or Eq. (74) holds, then near the critical point the physical quantities will collapse to a single universal curve when plotted in the appropriate form $\langle \mathcal{O} \rangle^{(N)} N^{\mu_{\mathcal{O}}/\nu}$ against $N^{1/\nu}(\lambda - \lambda_c)$ [93].

In order to check FSS assumptions, let us show the data collapse for the one-body Yukawa potential Eq. (69). The advantage of studying this simple model is that there exist rigorous theorems that give the exact values for energy-critical α-exponents [47] and accurate values for the critical screening length and the universal exponent ν [88] for both zero and nonzero angular momentum.

As a complete basis set, we have used the Laguerre polynomials and the spherical harmonic as given in Eq. (70). We applied the data collapse method to the ground-state energy and the lowest $l = 1$ energy. Results are shown in Fig. 6a for $l = 0$ and in Fig. 6b for $l = 1$ [93]. One can see clearly that all data for $N = 20, 40, 60, 80, 100$ collapse on one curve.

We note that in analogy with statistical mechanics, each block ($l = 0$ and $l = 1$) of the Hamiltonian matrix could be interpreted as a transfer matrix of a classical pseudosystem. Within this analogy, the lowest eigenvalue is associated with the free energy and the critical point as a first-order phase transition for

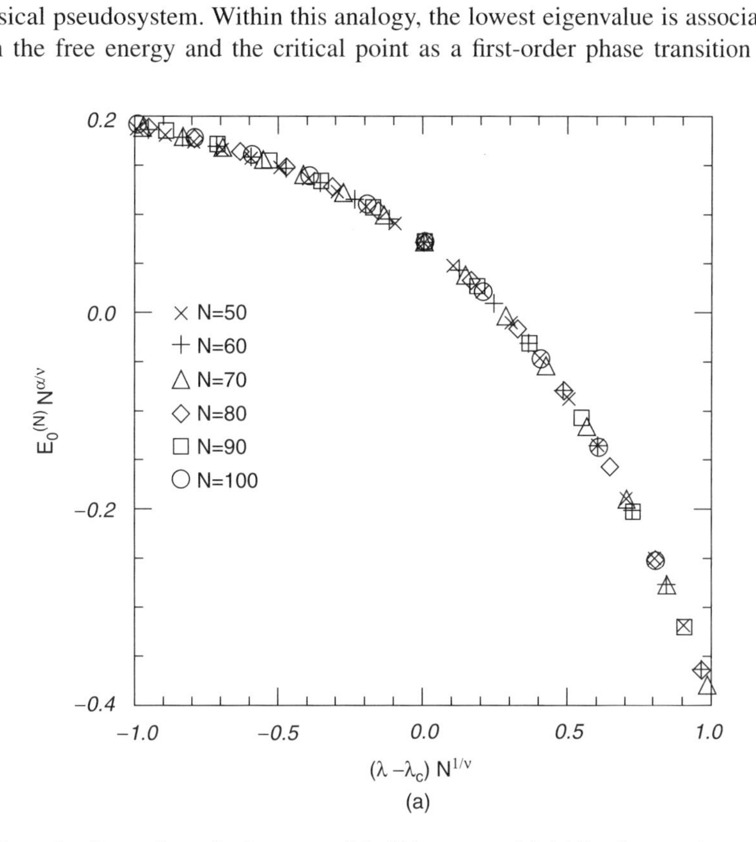

Figure 6. Data collapse for the energy of the Yukawa potential. (a) For the ground state with $\alpha = 2$ and $\nu = 1$. (b) For the lowest $l = 1$ level with $\alpha = 1$ and $\nu = 1/2$.

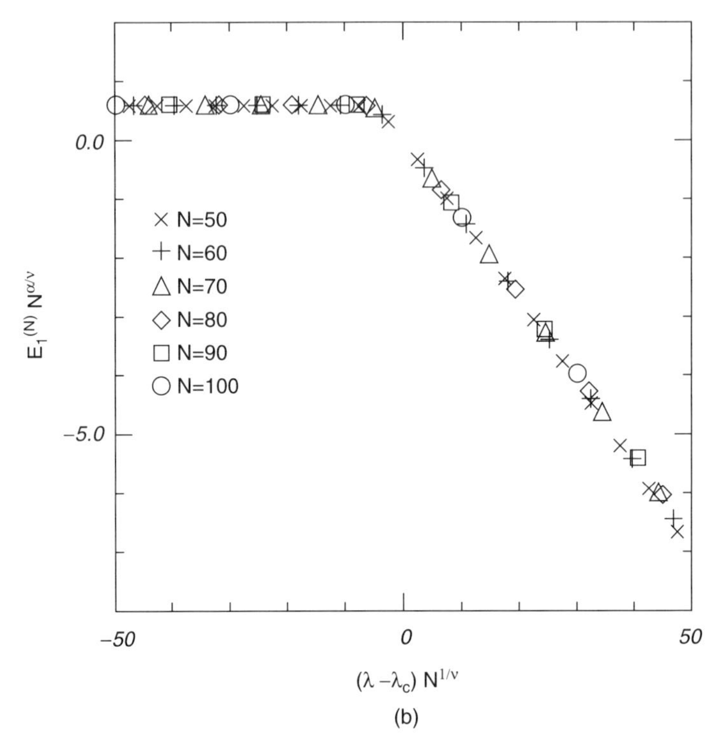

Figure 6 (*Continued*)

$\alpha = 1$ (remember that the α exponent is related to the statistical mechanics $\hat{\alpha}$ exponent for the specific heat by the relation $\hat{\alpha} = \alpha - 2$), or as a continuous phase transition for $\alpha > 1$. As a result of this analogy, we can use scaling laws from statistical mechanics to be applied to the classical pseudosystem. In particular, for continuous phase transitions we can calculate the spatial dimension d of the pseudosystem using the hyperscaling relation [94]:

$$4 - \alpha = d\nu \tag{75}$$

where d is the spatial dimension of the pseudosystem. Then for $\alpha = 2$ and $\nu = 1$ it gives a spatial dimension $d = 2$. For $l - 1$ there is a first-order phase transition; therefore the relation between the ν exponent and the spatial dimension of the pseudosystem is $d = 1/\nu$ [95], which gives again $d = 2$. The excellent collapse of the curves gives a strong support to the FSS arguments in quantum mechanics [93].

V. QUANTUM PHASE TRANSITIONS AND STABILITY OF ATOMIC AND MOLECULAR SYSTEMS

In this section we will apply the FSS method to obtain critical parameters for few-body systems with Coulombic interactions. Thus, FSS approach can be used to explain and predict stability of atomic and molecular systems.

A. Two-Electron Atoms

In this section we plan to review the analytical properties of the eigenvalues of the Hamiltonian for two-electron atoms as a function of the nuclear charge. This system, in the infinite-mass nucleus approximation, is the simplest few-body problem that does not admit an exact solution, but has well-studied ground-state properties. The Hamiltonian in the scaled variables [96] has the form

$$\mathcal{H}(\lambda) = -\frac{1}{2}\nabla_1^2 - \frac{1}{2}\nabla_2^2 - \frac{1}{r_1} - \frac{1}{r_2} + \frac{\lambda}{r_{12}} = \mathcal{H}_0 + \lambda V \tag{76}$$

where $\mathcal{H}_0$ is the unperturbed hydrogenic Hamiltonian, V is the Coulomb interelectronic repulsion, and λ is the inverse of the nuclear charge Z. For this Hamiltonian, a critical point means the value of the parameter, λ_c, for which a bound-state energy becomes absorbed or degenerate with the continuum.

An eigenvalue and an eigenvector of this Hamiltonian, Eq. (76), can be expressed as a power series in λ. Kato [97] showed that these series have a nonzero radius of convergence. This radius is determined by the distance from the origin to the nearest singularity in the complex plane λ^*. The study of the radius of convergence, λ^*, and whether or not this is the same as the critical value of λ_c, has a long history with controversial results [98–101]. Recently, Morgan and co-workers [102] have performed a 401-order perturbation calculation to resolve this controversy over the radius of convergence of the $\lambda = 1/Z$ expansion for the ground-state energy. They found numerically that $\lambda_c = \lambda^* \sim 1.09766$, which confirms Reinhardt's analysis of this problem using the theory of dilatation analyticity [101]. Ivanov [103] has applied a Neville–Richardson analysis of the data given by Morgan and co-workers and obtained $\lambda_c = 1.09766079$.

The Hamiltonian (76) commutes with the total angular momentum operator $\vec{\mathcal{L}} = \vec{\mathcal{L}}_1 + \vec{\mathcal{L}}_2$. Using a complete λ-independent basis set $\{\Phi_{\mathcal{K},\ell}\}$ with the property

$$\mathcal{L}^2\Phi_{\mathcal{K},\ell} = \ell(\ell+1)\Phi_{\mathcal{K},\ell} \tag{77}$$

where $\mathcal{K}$ represents all the quantum numbers but ℓ, we can study the spectrum of $\mathcal{H}(\lambda)$ at each block of fixed ℓ. This will allow us to study excited states of the lowest symmetry of each block.

To carry out the FSS procedure, one has to choose a convenient basis set to obtain the two lowest eigenvalues and eigenvectors of the finite Hamiltonian matrix. As basis functions for the FSS procedure, we choose the following basis set functions [104–106]:

$$\Phi_{ijk,\ell}(\vec{x}_1, \vec{x}_2) = \frac{1}{\sqrt{2}}\left(r_1^i\, r_2^j\, e^{-(\gamma r_1 + \delta r_2)} + r_1^j r_2^i\, e^{-(\delta r_1 + \gamma r_2)}\right) r_{12}^k\, F_\ell(\theta_{12}, \mathbf{\Omega}) \qquad (78)$$

where γ and δ are fixed parameters, we have found numerically that $\gamma = 2$ and $\delta = 0.15$ are a good choice for the ground state [87], r_{12} is the interelectronic distance, and $F_\ell(\theta_{12}, \mathbf{\Omega})$ is a suitable function of the angle between the positions of the two electrons θ_{12} and the Euler angles $\mathbf{\Omega} = (\Theta, \Phi, \Psi)$. This function F_ℓ is different for each orbital block of the Hamiltonian. For the ground state we have $F_0(\theta_{12}, \mathbf{\Omega}) = 1$, and for the $2p^2\ ^3P$ state we have $F_1(\theta_{12}, \mathbf{\Omega}) = \sin(\theta_{12})\cos(\Theta)$. These basis sets are complete for each ℓ-subspace [105,106]. The complete wave function is then a linear combination of these terms multiplied by variational coefficients determined by matrix diagonalization [87]. In the truncated basis set at order N, all terms are included such that $N \geq i + j + k$, so the number of trial functions $M(N)$ is

$$M(N) = \frac{1}{12}N^3 + \frac{5}{8}N^2 + \frac{17}{12}N + a_N \qquad (79)$$

where a_N is 1 $\left(\frac{7}{8}\right)$ if N is even (odd).

By diagonalizing the finite Hamiltonian matrix, one can obtain the lowest two energy eigenvalues as a function of the order of the truncated basis set, $E_0^{(N)}$ and $E_1^{(N)}$. Using the PR equation, Eq. (55), one can look for its fixed point by taking the ratio of these two eigenvalues raised to a power N as a function of λ [8]. Figure 7 shows the crossing points, which are the fixed points of Eq. (55), for $N = 6, 7, 8, \ldots, 13$. The values of the fixed points as a function of N can be extrapolated to the limit $N \to \infty$ by using the Bulirsch and Stoer algorithm [89], which is widely used for FSS extrapolation [4]. The extrapolated values of $\lambda_c = 1.0976 \pm 0.0004$ is in excellent agreement with the best estimate of $\lambda_c = 1.09766079$ [103].

The behavior of the ground-state energy, $E_0^{(N)}$, as a function of λ for different values of N is shown in Fig. 8a. When the value of N approaches the limit, $N \to \infty$, the true ground-state energy bends over sharply at λ_c to becomes degenerate with the lowest continuum at $E_0 = -\frac{1}{2}$ This behavior can be seen in the finite-order approximation: The larger the value of N, the more the energy curve bends toward a constant energy. In virtue of this behavior, we expect the first derivative of the energy with respect to λ to develop a step-like discontinuity at λ_c. The first derivative is shown in Fig. 8b for $N = 6, 7, 8, \ldots, 13$.

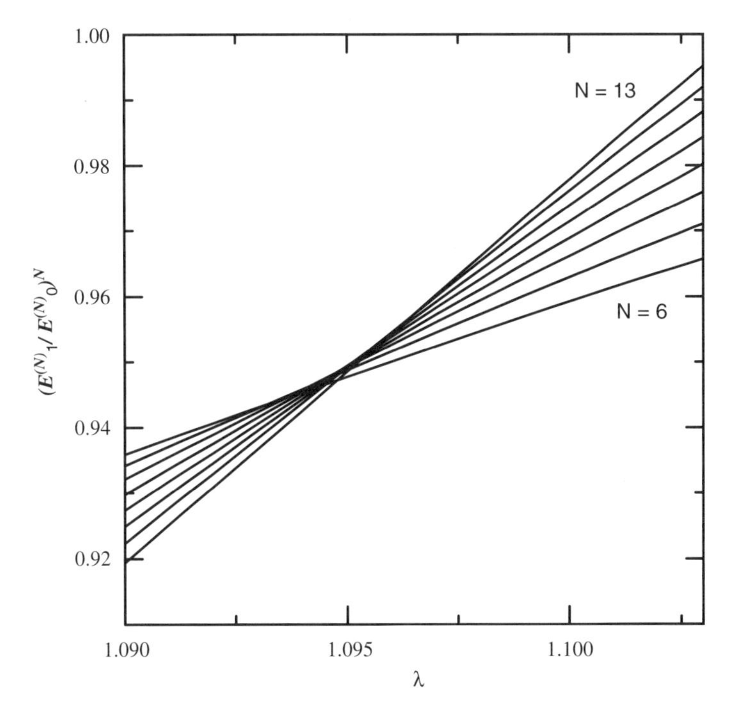

Figure 7. The ratio between the ground-state energy and the second lowest eigenvalue of the two-electron atom raised to a power N as a function of λ for $N = 6, 7, \ldots, 13$.

As expected, the second derivative will develop a delta-function-like behavior as N is getting larger, as shown in Fig. 8c.

The behavior of the ground-state energy and its first and second derivatives resembles the behavior of the free energy at a first-order phase transition. For the two-electron atoms, when $\lambda < \lambda_c$ the nuclear charge is large enough to bind two electrons; this situation remains until the system reaches a critical point λ_c, which is the maximum value of λ for which the Hamiltonian has a bound state or the minimum charge necessary to bind two electrons. For $\lambda > \lambda_c$, one of the electrons jumps to infinity with zero kinetic energy [87].

The convergence law of the results of the PR method is related to the corrections to the finite-size scaling. From Eq. (55) we expect that at the critical value of nuclear charge the correlation length is linear in N. In Fig. 9 we plot the correlation length of the finite pseudosystem (evaluated at the exact critical point λ_c) as a function of the order N. The linear behavior shows that the asymptotic equation [Eq. (60)] for the correlation length holds for very low values of N [87].

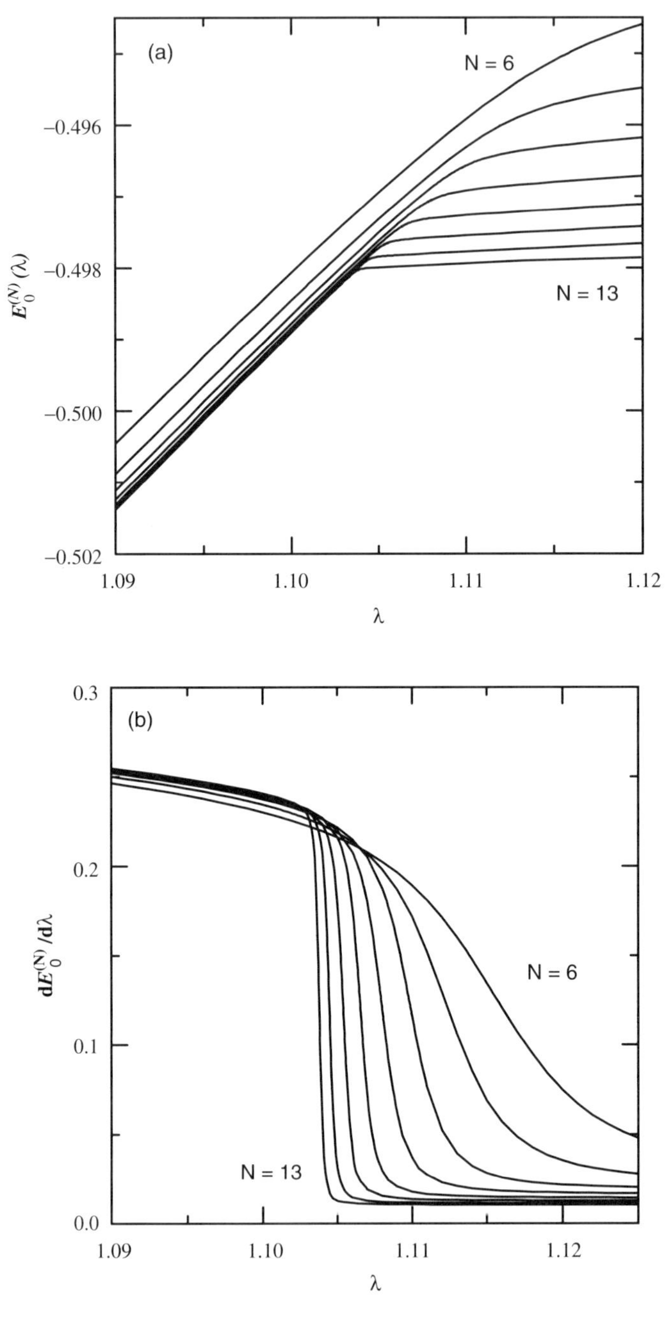

Figure 8. The two-electron atom: Variational ground-state energy (a), first derivative (b), and second derivative (c) as a function of λ for $N = 6, 7, \ldots, 13$.

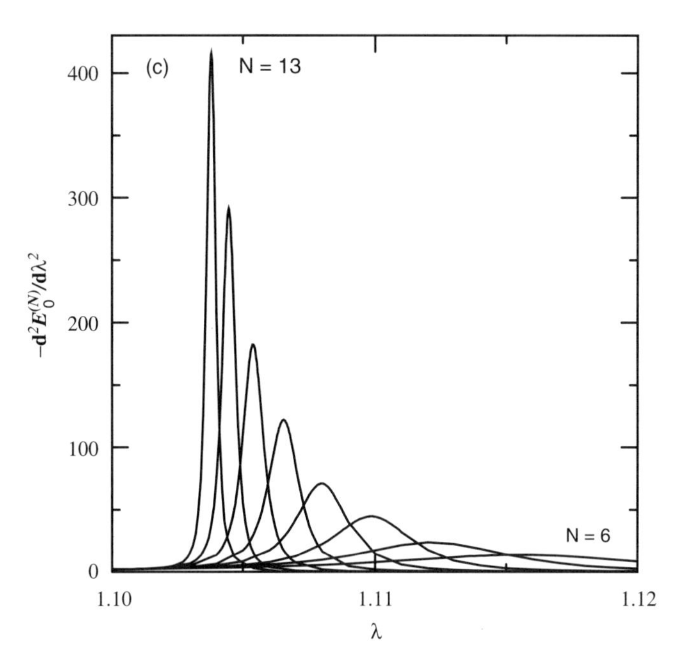

Figure 8 (*Continued*)

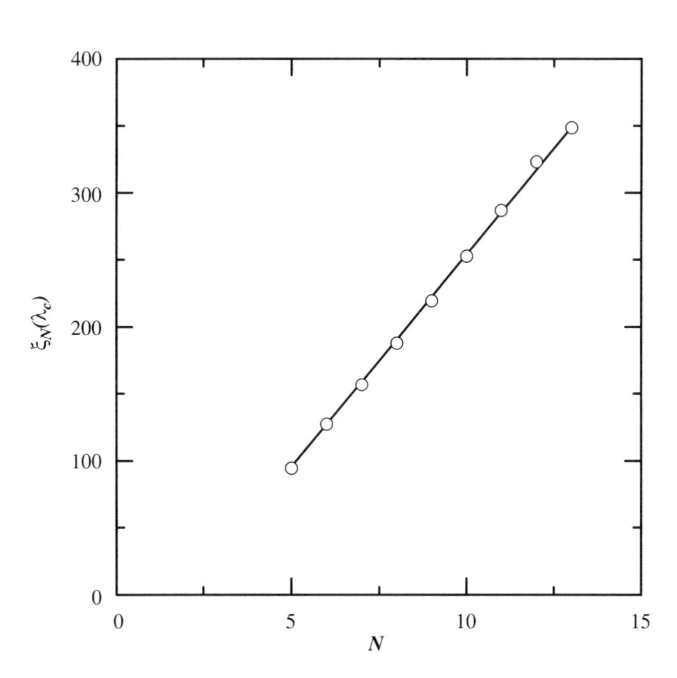

Figure 9. Correlation length for the two-electron atom evaluated at the critical point λ_c as a function of N.

The classical critical exponent ν that describes the asymptotic behavior of ξ at λ_c can be obtained for two different values of N and N' [87]:

$$\frac{1}{\nu} = \frac{\log\left[\left(\frac{d\xi_N}{d\lambda} / \frac{d\xi_{N'}}{d\lambda}\right)_{\lambda=\lambda^{(N,N')}}\right]}{\log(N/N')} - 1 \tag{80}$$

Using Eq. (80), we can estimate the critical exponent ν. The extrapolated value is $\nu = 1.00 \pm 0.02$, and the correlation length diverges as $\xi \sim |\lambda - \lambda_c|^{-\nu}$.

Hoffmann-Ostenhof et al. [64] have proved that $\mathscr{H}(\lambda_c)$ has a square-integrable eigenfunction corresponding to a threshold energy $E(\lambda_c) = -\frac{1}{2}$. They noted that the existence of a bound state at the critical coupling constant λ_c implies that for $\lambda < \lambda_c$, $E(\lambda)$ approaches $E(\lambda_c) = -\frac{1}{2}$ linearly in $(\lambda - \lambda_c)$ as $\lambda \to \lambda_c^-$. Morgan and co-workers [102] confirmed this observation by their $1/Z$ perturbation calculations. They show that $E(\lambda)$ approaches $E(\lambda_c) = -\frac{1}{2}$ as

$$E(\lambda) = E(\lambda_c) + 0.235(\lambda - \lambda_c) \tag{81}$$

From the FSS equation, Eq. (64), we obtain the ratio of the critical exponents α/ν. The extrapolated value of $\alpha = 1.04 \pm 0.07$, which is in agreement with the observation of Hoffmann-Ostenhof et al. [64].

Having reviewed the results for the critical behavior of the ground-state energy of the helium isoelectronic sequence, we may now consider other excited states. The ground state is symmetric under electronic exchange and has a natural parity, which means that its parity $\mathscr{P} = \mathscr{P}_1\mathscr{P}_2 = (-1)^{\ell_1}(-1)^{\ell_2}$ is equal to $(-1)^{\ell}$ [107], where ℓ_i is the angular momentum number of the ith particle and $|\ell_1 - \ell_2| \leq \ell \leq \ell_1 + \ell_2$. All the S states have this parity because $\ell_1 = \ell_2 = \ell = 0$. States with $\mathscr{P} = (-1)^{\ell+1}$ will be called *states of unnatural parity*. The triplet $2p^2\ ^3P$ has an unnatural parity.

For the H^- ion, Hill [108] proved that there is only one bound state with natural parity. This result, along with Kato's proof [109] that the Helium atom has an infinite number of bound states, seems to suggest that the critical point for the excited natural states is $\lambda = 1$ [87].

For the unnatural states of the H^- ion, we know from the work of Grosse and Pitner [107] that there is just one unnatural bound state $2p^2\ ^3P$ for $\lambda = 1$. An upper bound of $-0.12535\,a.u.$, which is below the threshold energy $E_T^P = -\frac{1}{8}$, was estimated for this state by Midtdal [100] and Drake [110].

From the variational calculations, the behavior of the energy for the triplet $2p^2\ ^3P$ state as a function of λ is very similar to the one found for the ground state. The curves start to bend over sharply to a constant values as N gets larger.

The true excited-state energy, in the limit $N \to \infty$, bends over sharply at λ_c^P to become degenerate with the lowest continuum at $E_T^P = -\frac{1}{8}$ [87].

Now, the PR equation, Eq. (55), can be applied for the excited state $2p^2\ ^3P$ to obtain a sequence of pseudocritical λ as a function on N, $\{\lambda_P^{(N,N')}\}$. The extrapolated value of this sequence gives $\lambda_c^P = 1.0058 \pm 0.0017$. As far as we know, the only estimate of λ_c for this triplet state is the one given by Brändas and Goscinski [111]. By applying a Darboux function ansatz [112,113] to the E_n's of Midtdal et al. [100] for n up to 27, they found $\lambda_c \sim 1.0048$, which is in good agreement with the FSS result $\lambda_c^P = 1.0058 \pm 0.0017$.

B. Three-Electron Atoms

Using the finite-size scaling method, study of the analytical behavior of the energy near the critical point shows that the open-shell system, such as the lithium-like atoms, is completely different from that of a closed-shell system, such as the helium-like atoms. The transition in the closed-shell systems from a bound state to a continuum resemble a "first-order phase transition," while for the open-shell system the transition of the valence electron to the continuum is a "continuous phase transition" [9].

To examine the behavior of open-shell systems, let us consider the scaled Hamiltonian of the lithium-like atoms:

$$\mathcal{H}(\lambda) = \sum_{i=1}^{3}\left[-\frac{1}{2}\nabla_i^2 - \frac{1}{r_i}\right] + \lambda \sum_{i<j=1}^{3} \frac{1}{r_{ij}} \tag{82}$$

where r_{ij} are the interelectron distances and λ is the inverse of the nuclear charge [9].

As a basis function for the FSS procedure, we used the Hylleraas-type functions [105] as presented by Yan and Drake [114]:

$$\Psi_{ijklmn}(\vec{x}_1, \vec{x}_2, \vec{x}_3) = \mathcal{C}\,\mathcal{A}\left(r_1^i\, r_2^j\, r_3^k r_{12}^l\, r_{23}^m\, r_{31}^n\; e^{-\alpha(r_1+r_2)} e^{-\beta r_3}\; \chi_1\right) \tag{83}$$

where the variational parameters, $\alpha = 0.9$ and $\beta = 0.1$, were chosen to obtain accurate results near the critical charge $Z \simeq 2$, χ_1 is the spin function with spin angular moment 1/2

$$\chi_1 = \alpha(1)\beta(2)\alpha(3) - \beta(1)\alpha(2)\alpha(3) \tag{84}$$

$\mathcal{C}$ is a normalization constant, and $\mathcal{A}$ is the usual three-particle antisymmetrizer operator [114].

The finite order of the basis set is allowed to be $i + j + k + l + m + n \leq N$. The maximum value of N was taken to be $N = 8$, which gives a 1589×1589 Hamiltonian matrix [9].

The general algorithm of Bulirsch and Stoer [89] was used to obtain the extrapolated value of the sequences $\lambda^{(N)}$ for lithium-like atoms. The extrapolated value from the PR method was found to be $\lambda_c = 0.48 \pm 0.03$ [9]. In the neighborhood of the critical charge, the ionization energy for lithium-like atoms, $I = E_{Li} - E_{He}$, goes smoothly to zero as a function of λ as shown in Fig. 10. In virtue of the behavior of the energy curves, the first derivative of the ionization energy with respect to λ remain continuous. This behavior is different from that of our previous results for the helium-like atoms [87] where the ionization energy bends sharply to zero at the helium critical $\lambda_c^{(He)} \simeq 1.0976$. The different behavior of the energy as a function of the Hamiltonian parameter, λ, suggests, in analogy with standard phase transitions in statistical mechanics,

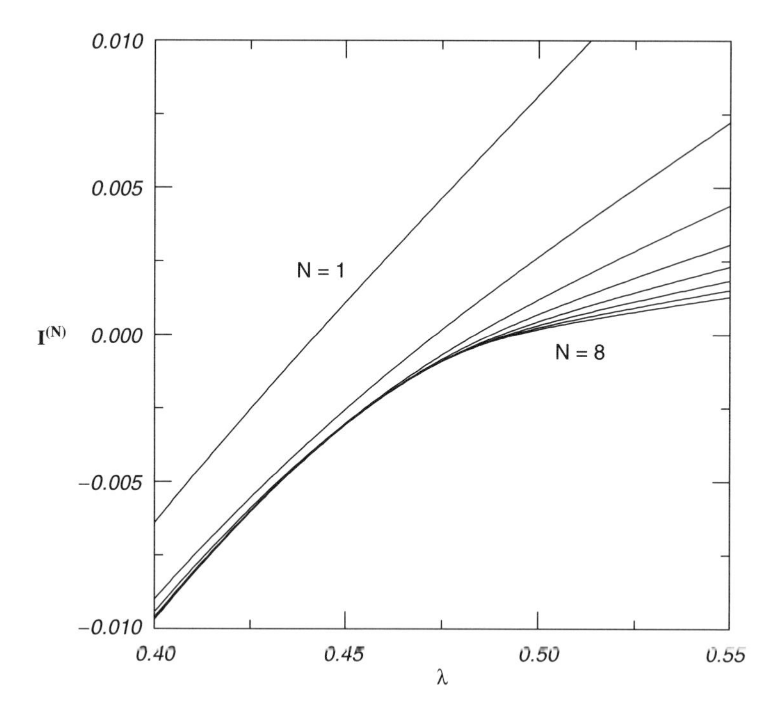

Figure 10. Ionization energy, $I^{(N)}$, for three-electron atoms as a function of λ for $N = 1, 2, \ldots, 8$. ($N = 1$ means that 5 basis functions were used in the calculations, and $N = 8$ means that 1589 basis functions were used.)

that the transition from a ground bound state to a continuum in the helium-like atoms resemble "first-order phase transitions," while for lithium-like atoms the transition is continuous.

In the previous section we showed that for the helium-like atoms the critical exponent for the energy, $E \simeq (\lambda_c - \lambda)^{\alpha}$, $\lambda \to \lambda_c^-$, is equal to one, $\alpha = 1$ [87]. For three-electron atoms, $\alpha \simeq 1.64 \pm 0.05$ and the Hamiltonian does not have a square-integrable wave function at the bottom of the continuum [9]. The behavior of the correlation length $\xi^{(N)}$ for the associated classical pseudo-system [87] as a function of λ is shown in Fig. 11. In this figure, the behavior of the correlation length is characteristic of a continuous phase transition using finite-size scaling method, which goes like an inverse power law in $(\lambda - \lambda_c)$.

Now, let us use the data collapse method to test the hypothesis of finite-size scaling used to obtain the critical parameter for this system and estimate the critical exponent ν for the lithium-like atoms. Using data collapse to the ionization energy of the three-electron atom in its ground state, $I_3(\lambda) = E_0^{Li}(\lambda) -$

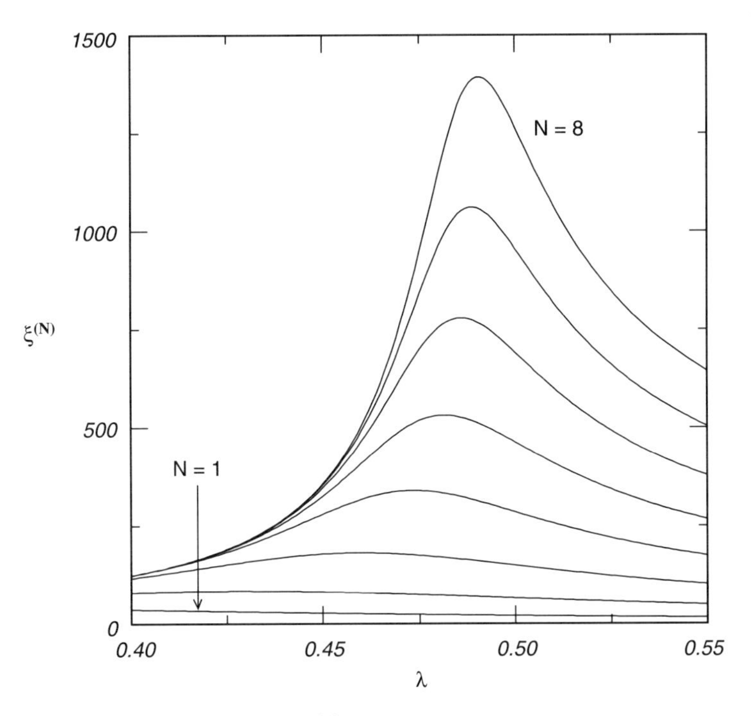

Figure 11. The correlation length, $\xi^{(N)}$, as a function of λ for the three-electron atoms for $N = 1, 2, \ldots, 8$.

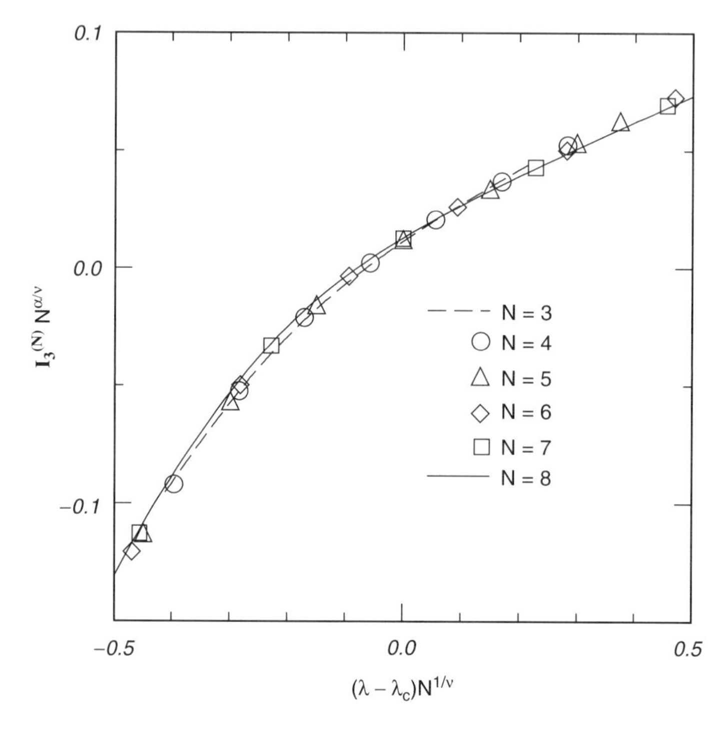

Figure 12. Data collapse for the ionization energy of the three-electron atom with $\alpha = 1.64$ and $\nu_3 = 0.8$.

$E_0^{He}(\lambda)$, one obtains $\nu = 0.8 \pm 0.1$ [93]. The calculations were done with a Hylleraas basis set with $N = 3, 4, \ldots, 8$, which represents up to 1589 Hylleraas functions. The excellent collapse of the curves for I_3 gives a strong support to the FSS hypothesis [93] as shown in Fig. 12.

C. Critical Nuclear Charges for *N*-Electron Atoms

For N-electron atoms, Lieb [115] proved that the number of electrons, N_c, that can be bound to an atom of nuclear charge, Z, satisfies $N_c < 2Z + 1$. With this rigorous mathematical result, only the instability of the dianion H^{2-} has been demonstrated [115]. For larger atoms, $Z > 1$, the corresponding bound on N_c is not sharp enough to be useful in ruling out the existence of other dianions. However, Herrick and Stillinger estimated the critical charge for a neon isoelectronic sequence, $Z_c \simeq 8.77$. Cole and Perdew [116] also confirmed this result for $N = 10$ by density functional calculations and ruled out the stability of O^{2-}. Using Davidson's tables of energies as a function of Z for atoms up to $N = 18$ [117], one can estimate the critical charges from the equality [118], $E(N, Z_c) = E(N-1, Z_c)$. However, Hogreve used large and diffuse basis sets and multireference configuration interaction to calculate the critical charges for all atoms up to $N = 19$ [119].

Let us consider an N-electron atom with a nuclear charge Z. In atomic units, the potential of interaction between the loose electron and an atomic core consisting of the nucleus and the other $N-1$ electrons tends to $-Z/r$ at small r and to $(-Z+N-1)/r$ at large r. After the scaling transformation $r \to Zr$, the potential of interaction between two electrons is λ/r_{ij} with $\lambda = 1/Z$, and the potential of interaction between an electron and the nucleus is $-1/r_i$. In these scaled units the potential of interaction between a valence electron and a core tends to $-1/r$ at small r and tends to $(-1+\gamma)/r$ with $\gamma = (N-1)\lambda$ at large r. It is easy to see that the model of the form [62]

$$V(r) = -\frac{1}{r} + \frac{\gamma}{r}\left(1 - e^{-\delta r}\right) \tag{85}$$

correctly reproduces such an effective potential both at small r and at large r.

This model has one free parameter that was fitted to meet the known binding energy of the neutral atom and its isoelectronic negative ion [62]. The critical charges are found for atoms up to Rn ($N = 86$). The critical charges are found from the following condition:

$$E_{\mathrm{I}}(\lambda_c) = 0, \qquad Z_c = 1/\lambda_c \tag{86}$$

where E_{I} is the extrapolated ionization energy. Results agree (mostly within an accuracy of 0.01 [62]) with both the ab initio multireference configuration interaction calculations of Hogreve [119] and the critical charges extracted by us from Davidson's figures of isoelectronic energies [117].

The goal here is to perform a systematic check of the stability of atomic dianions. In order to have a stable doubly negatively charged atomic ion, one should require the surcharge, $S_e(N) \equiv N - Z_c(N) \geq 2$. We have found that the surcharge never exceeds two as shown in Fig. 13. The maximal surcharge, $S_e(86) = 1.48$, is found for the closed-shell configuration of element Rn and can be related to the peak of electron affinity of the element $N = 85$. Experimental results for negative ions of lanthanides remain unreliable [120,121]. We did not calculate critical charges for lanthanides. Since the electron affinities of lanthanides are relatively small $\leq 0.5\,\mathrm{eV}$ [122,123], we expect that the surcharges will be small.

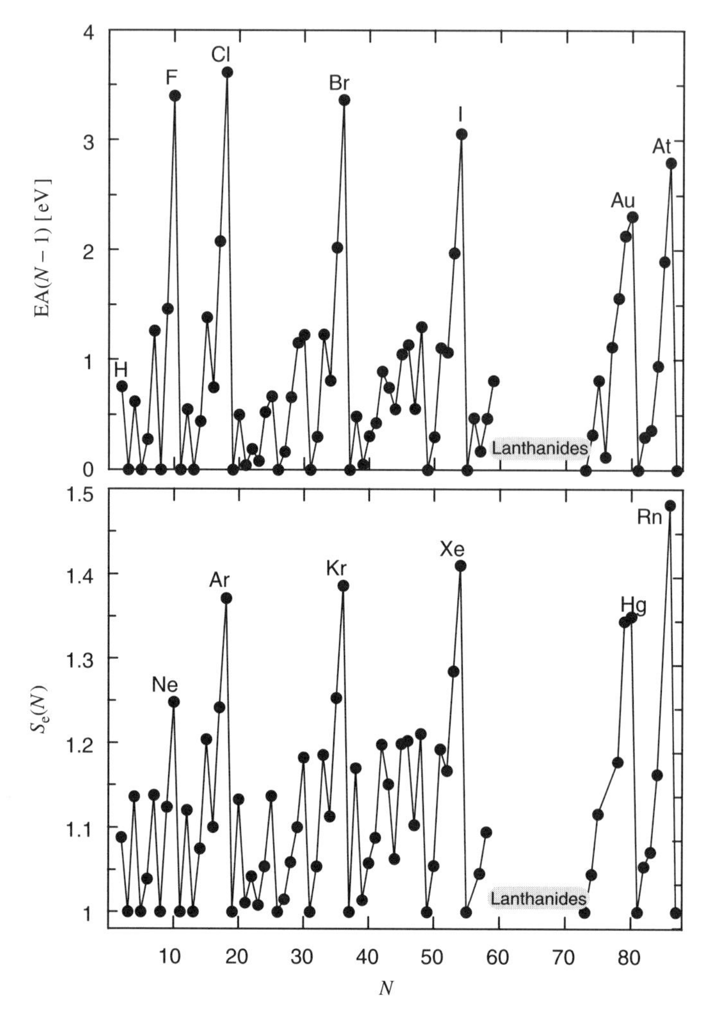

Figure 13. The calculated surcharge, $S_e = N - Z_c$, compared with the experimental electron affinity, $EA(N-1)$, as a function of the number of electrons, N.

Calculation of the critical charges for N-electron atoms is of fundamental importance in atomic physics since this will determine the minimum charge to bind N electrons. Already Kato [97] and Hunziker [124] show that an atom or ion has infinitely many discrete Rydberg states if $Z > N - 1$, and the results of Zhislin [125] show that a negative ion has only finitely many discrete states if $Z \leq N - 1$. Because experiment has yet to find a stable doubly negative atomic ion, Morgan and co-workers [102] concluded that the critical charge obeys the following inequality: $N-2 \leq Z_c \leq N-1$. The numerical results [62] confirmed this inequality and show that, at most, only one electron can be added to a free atom in the gas phase. The second extra electron is not bound by a singly charged negative ion because of the combined action of the repulsive potential surrounding the isolated negative ion and the Pauli exclusion principle [126]. However, doubly charged atomic negative ions might exist in a strong magnetic field of the order of a few atomic units, where 1 a.u. $= 2.3505 \times 10^9 G$ [62].

Recently, there has been an ongoing experimental and theoretical search for doubly charged negative molecular dianions [12,127]. In contrast to atoms, large molecular systems can hold many extra electrons because the extra electrons can stay well-separated [128]. However, such systems are challenging from both theoretical and experimental points of view. Although several authors [129–132] have studied the problem of the stability of diatomic systems as a function of the two nuclear charges, Z_1 and Z_2, there was no proof of the existence or absence of diatomic molecular dianions. The present approach might be useful in predicting the general stability of molecular dianions.

D. Critical Parameters for Simple Diatomic Molecules

Molecular systems are challenging from the critical phenomenon point of view. In this section we review the finite-size scaling calculations to obtain critical parameters for simple molecular systems. As an example, we give detailed calculations for the critical parameters for H_2^+-like molecules without making use of the Born–Oppenheimer approximation. The system exhibits a critical point and dissociates through a first-order phase transition [11].

The general Hamiltonian for A charged point particles under Coulomb interactions is represented in Cartesian coordinates by [11]

$$\mathscr{H} = \sum_{i}^{A} \frac{\mathbf{P}_i^2}{2M_i} + \sum_{i<j}^{A} \frac{Q_i Q_j}{|\mathbf{R}_i - \mathbf{R}_j|} \tag{87}$$

where $\mathbf{P}_i$, $\mathbf{R}_i = [X_i, Y_i, Z_i]$, M_i and Q_i are the momentum operator, column-vector coordinates, mass, and charge of particle i, respectively.

After separation of the translational motion [133] of the center of mass (CM) $M_0 = \sum_i^A M_i$ the Hamiltonian becomes

$$\mathscr{H} = \sum_{i=0}^{a} \frac{\mathbf{p}_i^2}{2\mu_i} + \frac{1}{2M_1} \sum_{i \neq j}^{a} \mathbf{p}_i \cdot \mathbf{p}_j + \sum_{i<j}^{a} \frac{Q_{i+1} Q_{j+1}}{r_{ij}} \tag{88}$$

where i and j go from 0 to $a = A - 1$; the μ_i are the reduced masses with $\mu_0 = M_0$ and $\mu_i = \frac{M_1 M_{i+1}}{M_1 + M_{i+1}}$; $r_{ij} = |\mathbf{r}_{ij}| = |\mathbf{R}_{i+1} - \mathbf{R}_{j+1}|$ and $r_i = |\mathbf{r}_i| = |\mathbf{R}_{i+1} - \mathbf{R}_1|$; $(\mathbf{r_0})$ is the CM vector, and $(\mathbf{r_1}, \mathbf{r_2}, \ldots, \mathbf{r_a})$ are the vectors of internal coordinates of particle $2, 3, \ldots, A$, respectively, and $(\mathbf{p_0}, \mathbf{p_1}, \mathbf{p_2}, \ldots, \mathbf{p_a})$ are their corresponding momenta.

Now, an explicit form of the Coulombic interactions V for a quantum system including B electrons and C positive-charged centers (protons) may be written as

$$V = \sum_{0=i<j}^{b} \frac{1}{r_{ij}} - \sum_{i=0}^{b} \sum_{k=b+1}^{a} \frac{Z_k}{r_{ik}} + \sum_{b+1=k<l}^{a} \frac{Z_k Z_l}{r_{kl}} \tag{89}$$

where the indices i,j are for electrons and k,l for protons with $b = B - 1$ and $a = A - 1 = B + C - 1$. The potential V includes electron–electron repulsive terms, electron–proton attractive terms and proton–proton repulsive terms.

However, for one-electron system, $B = 1, b = 0$ and $a = C$, the Z^2-scaled Hamiltonian with $\lambda = Z = Z_k$ becomes [11]

$$\mathscr{H} = \sum_{i=1}^{a} \frac{\mathbf{p}_i^2}{2\mu_i} + \frac{1}{2M_1} \sum_{0=i\neq j}^{a} \mathbf{p}_i \cdot \mathbf{p}_j - \sum_{k=1}^{a} \frac{1}{r_{0k}} + \lambda \sum_{1=k<l}^{a} \frac{1}{r_{kl}} \tag{90}$$

Therefore, the Hamiltonian of simple molecular systems can be represented as

$$\mathscr{H}(\lambda) = \mathscr{H}_0 + V_\lambda \tag{91}$$

where $\mathscr{H}_0$ is λ-independent and V_λ is the λ-dependent part. This Hamiltonian has the correct general form obtained in previous sections for the application of the FSS method to determine the critical value of the parameter λ.

Several investigators have performed calculations on the stability of H_2^+-like systems in the Born–Oppenheimer approximation. Critical charge parameters separating the regime of stable, metastable, and unstable binding were calculated using ab initio methods [134–137]. However, using the finite-size scaling approach, one can show that this critical charge is not a critical point [118]. But, without making use of the Born–Oppenheimer approximation the H_2^+-like system exhibits a critical point.

For H_2^+-like systems, Eq. (88) is used with $\lambda = Z$, $a = C = 2$, $\mu = M/(1 + M)$, and $M = 1836.152701$ a.u. The ground-state eigenfunction can be expanded in the following basis set [138] $\Phi_{(n,m,l)}$:

$$\Phi_{(n,m,l)}(r_1, r_2, r_{12}) = N_0 \phi_n(x) \phi_m(y) \phi_l(z) \tag{92}$$

where N_0 is the normalization coefficient and $\phi_n(x)$ is given in terms of Laguerre polynomials $L_n(x)$:

$$\phi_n(x) = L_n(x)e^{-x/2} \tag{93}$$

The coordinates x, y, and z are expressed in the following perimetric coordinates [139]:

$$\begin{aligned} x &= \frac{\theta}{k_x}(r_1 + r_2 - r_{12}) \\ y &= \frac{\theta}{k_y}(-r_1 + r_2 + r_{12}) \\ z &= \frac{\theta}{k_z}(r_1 - r_2 + r_{12}) \end{aligned} \tag{94}$$

where $k_x = 1 = k_y/2 = k_z/2$ and θ is an adjustable parameter that was chosen to be $\theta = 1.5$.

Calculating the matrix elements of the Hamiltonian in this basis set gives a sparse, real, and symmetric $M(N) \times M(N)$ matrix at order N. By systematically increasing the order N, one obtained the lowest two eigenvalues at different basis lengths $M(N)$. For example, $M(N) = 946$ and 20,336 at $N = 20$ and 60, respectively [11]. The symmetric matrix is represented in a sparse row-wise format [140] and then reordered [141] before triangularizations. The Lanczos method [142] of block-renormalization procedure was employed.

For the ground state of H_2^+-like molecules, the critical point λ_c of $\mathcal{H}(\lambda)$ can be obtained from finite-size scaling calculations. Using the finite-size scaling equation directly, one can obtain the fixed point from Eq. (55). Figure 14 shows the curves $E_1^{(N)}(\lambda)/E_0^{(N)}(\lambda)$ as a function of λ for $N = 31$ up to $N = 60$. By virtue of this behavior, one expect that the first derivative of the energy as a function of λ develops a step-like discontinuity at λ_c as shown in Fig. 15. The crossing points between two different sizes N and $N + 1$ give a series for $\{\lambda^{(N)}\}$. By systematically increasing the order N, one can reach a critical point $\lambda_c = 1.2286 \pm 0.0005$. Here λ_c and the error are estimated using the final minimum and maximum values and their difference over $48 < N < 60$. This value is in agreement with the calculation using the first-order method [11].

For the critical exponent α, starting from the series $\{\alpha^{(N)}(\lambda)\}$ and following the direct approach of finite-size scaling Eq. (67), one obtain the series $\alpha^{(N)}(\lambda_c)$. From these data, the estimated energy critical exponent is $\alpha = 1.000 \pm 0.005$.

Moreover, after calculating the critical exponent α, the critical exponent $\{\nu^{(N)}(\lambda)\}$ is readily given by Eq. (68). The results of the calculations for the $\nu^{(N)}(\lambda)$ series show oscillatory behavior. The data do not reach a limit at N

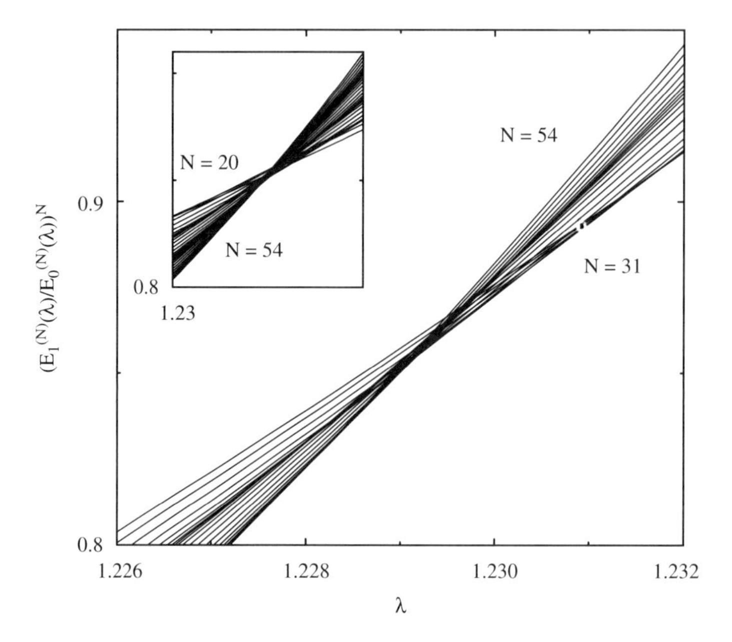

Figure 14. H_2^+-like molecule: The ratio between the ground-state energy and the second lowest eigenvalue raised to a power N as a function of λ $(= Z)$ at $\theta_t = 1.5$ for $N = 31, 32, \ldots, 60$.

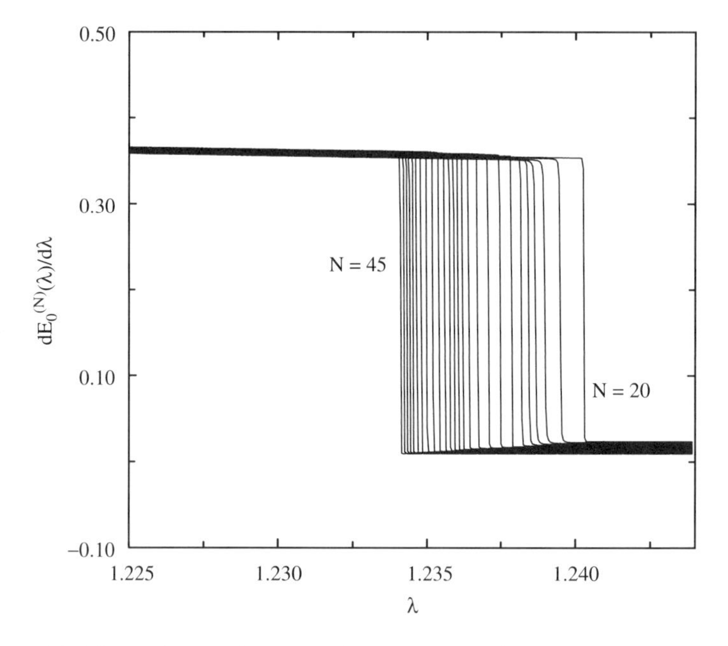

Figure 15. First derivative of the ground-state energy of the ground state of the H_2^+-like molecule as a function of λ $(= Z)$ at $\theta_t = 1.5$ for $N = 20, 21, \ldots, 45$.

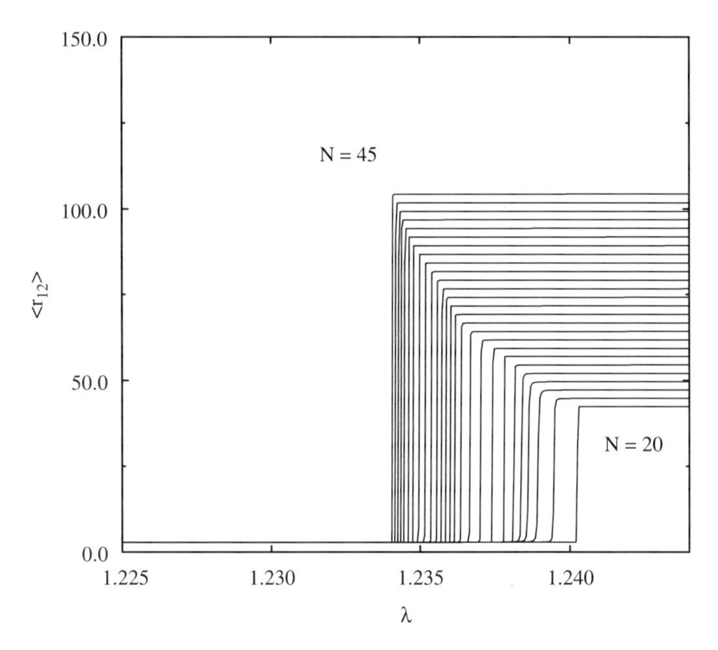

Figure 16. The expectation value of the distance between the two protons $\langle \mathbf{r}_{12} \rangle$ for the H_2^+-like molecule as a function of λ $(= Z)$ at $\theta_t = 1.5$ for $N = 20, 21, \ldots, 45$.

up to 60, but it does show that the correlation exponent is smaller than one and decreases as N increases. The estimated value using the final maximum and minimum points over $48 < N < 60$ is $\nu = 0.3 \pm 0.2$ [11].

From the present calculations, the expectation value of the operator $\mathbf{r}_{12}$ may provide a direct physical picture about the thermodynamic stability and dissociation of H_2^+-like molecules. As shown in Fig. 16, there is a vertical jump of the mean value $\mathbf{r}_{12}$ at λ_c. We note that there are similarities and differences between helium-like atoms and H_2^+-like molecules. In Section V.A of helium-like systems, based on an infinite mass assumption, we show that the electron at the critical point leaves the atom with zero kinetic energy in a first-order phase transition. This limit corresponds to the ionization of an electron as the nuclear charge varies. For the H_2^+-like molecules, the two protons move in an electronic potential with a mass-polarization term. They move apart as λ approaches its critical point and the system approaches its dissociation limit through a first-order phase transition.

Large molecular systems are challenging from the critical phenomenon point of view. In order to apply the finite-size scaling method, one needs to have a complete basis set. The present basis set used for the finite-size scaling calculations is built up on the perimetric coordinates that are only suitable for the three-body systems, thus restricting the basis set's extension to treat larger molecules. Several types of Gaussian basis sets have been tested. The first one has the following general form:

$$\Psi = (1 + \mathbf{P}_{12}) \sum_{m=1}^{M} e^{-\alpha_m r_1^2 - \beta_m r_2^2 - \gamma_m r_{12}^2} \tag{95}$$

where M is the expansion length, $\mathbf{P}_{12}$ is the exchange operator and α_m, β_m and γ_m are the variational parameters. The finite size scaling results using this basis set were strongly dependent on the initial values of the variational parameters. Another two types of Gaussian basis sets were used have the general form

$$\Psi = (1 + \mathbf{P}_{12}) \left(\sum_{m=1}^{M} e^{-\alpha_m r_1^2 - \beta_m r_2^2 - \gamma_m r_{12}^2} \right) \sum_{k=0}^{N} C_{mk} r_{12}^{2k} \tag{96}$$

where $k \leq N$, N is the order and M is expansion length for a set of given parameters $\{\alpha_m, \beta_m, \gamma_m\}$, and

$$\Psi = (1 + \mathbf{P}_{12}) \left(\sum_{m=1}^{M} e^{-\alpha_m r_1^2 - \beta_m r_2^2 - \gamma_m r_{12}^2} \right) \sum_{i,j,k}^{N} C_{mijk} r_1^{2i} r_2^{2j} r_{12}^{2k} \tag{97}$$

where $i + j + k \leq N$. These Gaussian basis sets are correlated and involve power law terms [133] that are equivalent to the power law terms used in the Laguerre polynomials in Eq. (21). This basis set can be used to recover the previous results and clearly will be suitable for larger diatomic molecules.

Qicun Shi research is still underway to combine FSS with Gaussian basis functions to treat larger molecular systems and also to investigate the effect of external fields on the molecular stability and whether or not one can use this approach to selectively break chemical bonds in polyatomic molecules.

E. Phase Diagram for Three-Body Coulomb Systems

The stability of three-body Coulomb systems is an old problem which has been treated in many particular cases [143–145] and several authors reviewed this problem [146,147]. For example, the He atom $(\alpha e^- e^-)$ and H_2^+ (ppe^-) are stable systems, H^- $(pe^- e^-)$ has only one bound state [108], and the positronium negative ion Ps^- $(e^+ e^- e^-)$ has a bound state [148], while the positron–hydrogen system $(e^- pe^+)$ is unbound and the proton–electron–negative–muon $(pe^- \mu^-)$ is an unstable system [149]. In this section, we show that all three-body ABA Coulomb systems undergo a first-order quantum phase transition from the stable phase of ABA to the unstable breakup phase of $AB + A$ as their masses and charges varies. Using the FSS method, we calculate the transition line that separates the two phases. For any combination of the three particles of the form *ABA*, one can read directly from the phase diagram if the system is stable or unstable. Moreover, the transition line has a minimum that leads to a new proposed classification of the *ABA* systems to molecule-like systems and to atom-like systems [66]. This is very important in exploring the resonance spectrum and dynamics of three particles where there is neither an obvious point of reference as the heavy nucleus in H^- nor a line of reference as the internuclear axis in H_2^+. Rost and Wintgen [150] have shown that the resonance spectrum of positronium negative ion Ps^- can be understood and classified with the molecule H_2^+ quantum numbers by treating the internuclear axis of Ps^- as an adiabatic parameter. The current approach gives a systematic classification of all *ABA* systems [66].

Let us consider the stability and quantum phase transitions of the three-body *ABA* Hamiltonian given by Eq. (36). The Hamiltonian, Eq. (36), formally separated the motion of the center of mass and depends linearly on the parameters λ and κ.

We are interested in the study of the critical behavior of the Hamiltonian as a function of both parameters λ and κ. The *ABA* system is stable if its energy is lower than the energy of the dissociation to $AB + A$. The critical behavior and stability of the ground-state energy as a function of λ for $\kappa = 0$ has been previously studied for the two-electron atoms [87] (with $Q = -1$, $M = 1$, $m = \infty$ and $\lambda = 1/q = 1/Z$, where Z is the nuclear charge) and for the hydrogen molecule-like ions [11] (with $q = -1$, $m = 1$, $M = \infty$ and $\lambda = |Q| = Z$).

In order to obtain the stability diagram for the three-body Coulomb systems in the $(\lambda - \kappa)$-plane, one has to calculate the transition line, $\lambda_c(\kappa)$, which separates the stable phase from the unstable one. To carry out the finite-size scaling calculations, the following complete basis set was used [66]:

$$\Phi_{n,m,l}(r_1, r_2, r_{12}) = \phi_n(x)\phi_m(y)\phi_l(z), \qquad \phi_n(x) = L_n(x)e^{-x/2} \tag{98}$$

where L_n is the Laguerre polynomial of degree n and order 0 and $x = \frac{\theta}{k_x}(r_1 + r_2 - r_{12})$, $y = \frac{\theta}{k_y}(-r_1 + r_2 + r_{12})$, and $z = \frac{\theta}{k_z}(r_1 - r_2 + r_{12})$ are the perimetric coordinates [151]. For faster convergence, the parameters $k_x = 1 = k_y/2 = k_z/2$, and $\theta = 1.5$ were chosen.

Solving the Schödinger equation gives a sparse, real, and symmetric $M(N) \times M(N)$ matrix of order N. The symmetric matrix is expressed in a sparse row-wise format, reordered, and LU-decomposed [152]. Then, one employs the block-renormalization Lanczos procedure [153] to obtain the eigenvalues [66]. From the leading two eigenvalues, $E_0^{(N)}(\lambda)$ and $E_1^{(N)}(\lambda)$, one can obtain the correlation length, Eq. (56), for the classical pseudosystem, $\xi_N(\lambda)$. Now we are in a position to apply the phenomenological renormalization equation, Eq. (55), to obtain a sequence of pseudocritical parameters $\lambda_c^{(N)}$ for different values of κ.

The values of the parameter $\kappa = (1 + m/M)^{-1}$ were varied in the interval [0,1] according to the different masses of the combined particles. The values, in atomic units, of the particle masses were taken from Ref. 154: for electron $m_e = 1.0$, for proton $m_p = 1836.1526675$, for deuteron $m_d = 3670.4829550$, for tritium $m_t = 5496.9216179$ [154,155], for muon $m_\mu = 206.7682657$, and for helium $m_{\text{He}} = 7296.299508$.

The numerical results for all *ABA* Coulomb systems show that the ground-state energy is a continuous function of $1.0 \leq \lambda \leq 1.25$ and $0 \leq \kappa \leq 1$, but bends over sharply at λ_c to become degenerate with the scaled lowest continuum at $E_0 = -\frac{1}{2}$. By virtue of this behavior, we expect that the first derivative of the energy with respect to λ to develop a step-like discontinuity at λ_c. $\mathscr{H}(\lambda_c)$ has a square-integrable eigenfunction as κ varies corresponding to a scaled threshold energy $E^{th}(\lambda_c) = -\frac{1}{2}$ [64]. $E(\lambda)$ approaches $E^{th}(\lambda_c)$ linearly in $(\lambda - \lambda_c)$ as $\lambda \to \lambda_c^-$ [64,102].

For the *ABA* Coulomb systems when $1 \leq \lambda \leq \lambda_c$, the ratio of charges is sufficiently small enough to keep the three particles bound and the system at least has one bound state [108]. This situation remains until the system reaches a critical point λ_c, which is the maximum value of λ for which the Hamiltonian has a bound state. For $\lambda \geq \lambda_c$, one of the particles jumps to infinity with zero kinetic energy. In Fig. 17 the transition line, λ_c, is shown as a function of κ. The parameter κ changes between $\kappa = 1$, which corresponds to the H_2^+-like systems in the Born–Oppenheimer approximation, and $\kappa = 0$, which corresponds to the He-like atoms in the infinite mass approximation. Between the two limits $\kappa = 0$ and $\kappa = 1$, there are many stable three-particle systems as shown in the figure. The transition line separates the three-particle systems into stable systems (with at least one bound state) and unstable systems. These numerical results confirm the general properties of the stability domain discussed by Martin [149], and the instability region in the λ–κ plane should be convex. Particularly, the transition curve has a minimum, which occurs at $\kappa_m = 0.35$, and hence all possible bound three-body systems are divided into two branches in the λ–κ plot, one with $0 \leq \kappa < \kappa_m$ and the other with $\kappa_m < \kappa \leq 1$. The two systems closest to κ_m are Ps^- with $\kappa = 0.5$ and $\bar{p}\bar{p}$d with $\kappa = 0.33$. The window in Fig. 17 shows the two different branches in their ground-state energy as a function of the location of the pseudocritical points $\lambda_c^{(N)}(\kappa)$.

The observation of two different branches leads us to investigate the similarity between the molecule-like systems of the right branch, $\kappa > \kappa_m$, such as the Ps^-, and the atom-like systems of the left branch, $\kappa < \kappa_m$, such as $\bar{p}\bar{p}$d. The parameter κ measures the strength of the mass polarization term, which is due to the motion of the two identical particles with respect to the third particle. The mass polarization term is then a measure of the momentum correlation of the two identical particles with respect to the third particle. If $\kappa \gg \kappa_m$, as in the case of a molecule such as H_2^+, then the light particle with mass m and charge q

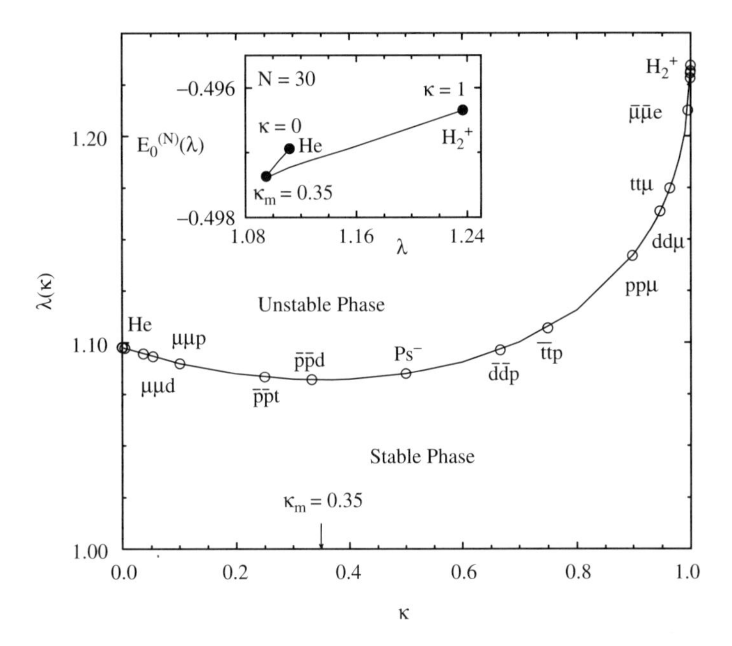

Figure 17. The critical parameter λ_c as a function of κ in the range $0 \leq \kappa \leq 1$. The different three-body systems are shown along the transition line that separates the stable phase from the unstable one. Note the minimum value of $\kappa_m = 0.35$; for $\kappa > \kappa_m$ we have molecule-like systems, while for $\kappa < \kappa_m$ the systems behave like atoms. The inset in this figure shows the two branches in the $(E_0^{(N)} - \lambda)$-plane as κ varies in the interval [0, 1].

tends to stay in the middle of the two heavy particles to achieve bonding, while for $\kappa \ll \kappa_m$, as in the case of the He atom, each light particle with mass M and charge Q is less localized and thus the momentum correlation is smaller. The fact that the resonance spectrum and dynamics of Ps^- $(\kappa = 0.5)$ was understood and classified with the H_2^+ quantum numbers [150] is very encouraging and shows that the above proposed classification might shed some light on a systematic and concise picture of the dynamics of all *ABA* Coulomb systems.

VI. CROSSOVER PHENOMENA AND RESONANCES IN QUANTUM SYSTEMS

In this section we present three different phenomena, resonances, crossover and multicritical points. We will discuss in general the applicability of FSS to resonances, the existence of multicritical points and the definition of the size of the critical region. Application of FSS to a specific Hamiltonian that presents the three phenomena shows the relation between them [156].

A. Resonances

We will study, in particular, resonances that may occur for potentials that have a barrier over the threshold value separating an inner region from an outer region where the potential goes asymptotically to the threshold value. These resonances are known in the literature as *shape-type* [157] or *potential* [158] resonances.

Resonances are not truly bound states, but they are interpreted as metastable states. Because of the boundary conditions of resonances, the problem is not Hermitian even if the Hamiltonian is (that is, for square integrable eigenfunctions). Resonances are characterized by complex eigenvalues (complex poles of the scattering amplitude)

$$E = E_n + \frac{i}{2}\Gamma_n \tag{99}$$

where the real part E_n of an eigenvalue is the energy of the system and the imaginary part Γ_n is the inverse of the lifetime of the corresponding *quasi-bound* state. The corresponding eigenfunctions are nonnormalizable. The classical book by Newton [159] presents an excellent discussion on this subject.

The nature of the resonance states, narrow or broad, crucially depends on the behavior of the corresponding bound eigenvalue in the neighborhood of the threshold. There is no rigorous definition of *narrow* and *broad* resonances, but the former has a long lifetime and is accessible for observation. For a broad resonance, the practical definition of its energy and width becomes a difficult problem [158].

For Hamiltonian (6), as we showed in Section II, if there is a true bound state at the threshold, then the eigenvalue hits the continuum linearly in $(\lambda - \lambda_c)$; that is, in a "first-order phase transition," a bound state and a virtual state coexists at the threshold and a sharp resonance (i.e., $\Gamma_n \ll E_n$) will develop for $\lambda < \lambda_c$. However, if there is no bound state at the threshold, then the eigenvalue merges into the continuum quadratically in $(\lambda - \lambda_c)$—that is, in a "second-order phase transition." The eigenvalue is analytical at $\lambda = \lambda_c$, and a virtual state emerges for $\lambda < \lambda_c$. Therefore, if a resonance appears, it will be a virtual resonance ($E_n < 0$) and will turn to a truly resonance for $\lambda < \lambda_r$, where λ_r is defined by the condition $E_n(\lambda_r) = 0$. Now, $\Gamma(\lambda_r) > 0$ and therefore this kind of resonance comes out broad. That is, for one-dimensional systems and s waves of three-dimensional radial potentials, all resonances, including the most narrow one, are not obtained by a bound–resonance transition mechanism. They are obtained, however, due to the sequence bound–virtual–virtual resonance–resonance transition, where virtual states are associated with (real) negative energies of non-normalizable eigenfunctions and virtual resonances are complex eigenvalues with real parts embedded below the threshold energy; that is, the real parts are negative when the threshold energy is equal to zero.

If a virtual–virtual resonance transition exists, it occurs at λ_v, where $\lambda_r < \lambda_v < \lambda_c$. Such virtual–virtual resonance transition is associated with a branch point with exponent one-half:

$$E_n(\lambda) \sim (\lambda - \lambda_v)^{1/2}, \qquad \lambda \to \lambda_v^+ \tag{100}$$

Note that Eq. (100) is valid only for the real virtual energy.

To illustrate the appearance of resonances when a potential parameter is varied we consider the one-dimensional Hamiltonian [156],

$$\mathcal{H}(a,J) = \frac{p^2}{2} + \left(\frac{x^2}{2} - J\right) e^{-ax^2} \tag{101}$$

where a and J are free parameters (instead λ, we use the usual notation (a, J) for this Hamiltonian [157]). The potential in Eq. (101) exhibits predissociation resonances analogous to those found in diatomic molecules [160,161] and was used as a model potential to check the accuracy of different methods for the calculations of resonances [162,163].

We studied the critical behavior of the eigenfunctions and resonances of the Hamiltonian equation, Eq. (101), using the FSS method described in Section IV. As a basis function for the finite-size scaling procedure, we used the orthonormalized eigenfunctions of the harmonic oscillator with mass equal to 1 and frequency equal to a:

$$\Psi_n(a;x) = \left(\frac{a}{\pi}\right)^{1/4} \frac{1}{\sqrt{2^n n!}} e^{-ax^2/2} H_n(\sqrt{a}x) \tag{102}$$

where $H_n(x)$ are the Hermite polynomials of order n [55].

In order to obtain the matrix elements of the Hamiltonian equation, Eq. (101), we need to calculate the kinetic energy terms

$$\begin{aligned} T_{m,n} &\equiv \langle m | \frac{p^2}{2} | n \rangle \\ &= \frac{a}{2}\left[\left(n + \frac{1}{2}\right)\delta_{m,n} - \frac{\sqrt{(n+1)(n+2)}}{2}\delta_{m-2,n} - \frac{\sqrt{(n-1)n}}{2}\delta_{m+2,n}\right] \end{aligned} \tag{103}$$

and the potential energy terms [60]

$$e_{m,n} \equiv \langle m | e^{-ax^2} | n \rangle = \begin{cases} (-1)^{n+\frac{n+m}{2}} \dfrac{\Gamma(\frac{m+n+1}{2})}{\sqrt{2\pi m! n!}}, & m+n \text{ even} \\ 0, & m+n \text{ odd} \end{cases} \tag{104}$$

The other matrix elements, $\langle m|x^k e^{-ax^2}|n\rangle$ for any value of k, can be evaluated using the recurrence relations between Hermite polynomials [55]. By diagonalizing the Hamiltonian in the above basis set, we obtained the eigenstates as a function of both parameters a and J.

One of the most powerful tools to study resonances is *complex scaling* techniques (see Ref. 157 and references therein). In complex scaling the coordinate $\vec{x}$ of the Hamiltonian was rotated into the complex plane; that is, $H(\vec{x}) \rightarrow H(\vec{x}\, e^{i\phi/2})$. For resonances that have $\theta_{res} = \tan^{-1}[\mathrm{Im}(E^{(res)})/\mathrm{Re}(E^{(res)})] < \phi$ the wave functions of both the bound and resonance states are represented by square-integrable functions and can be expanded in standard L^2 basis functions.

As an example, the Hamiltonian equation, Eq. (101), with $a = 0.2$ and $J = 0.2$, presents a sharp resonance $E_0^{(res)} \simeq 0.23676931 - i\,0.98613158\,10^{-03}$, and its second resonance is $E_1^{(res)} \simeq 1.2516098 - i\,0.48039784$. Therefore we have $\theta_0 \simeq 0.004 < \pi/10$ and $\pi/10 < \theta_1 \simeq 0.366484 < \pi/5$. In Fig. 18 we

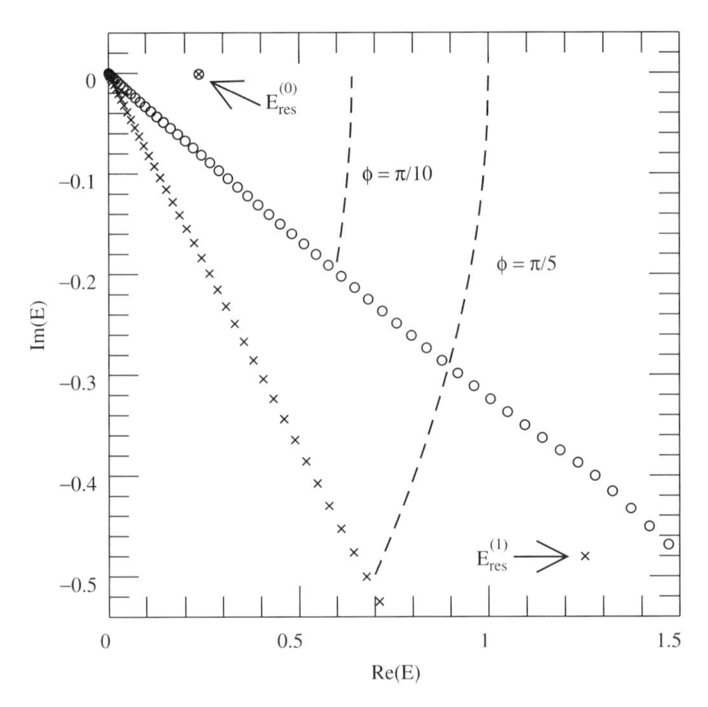

Figure 18. Im(E) versus Re(E) for complex-rotated $N = 100$ diagonalization of Hamiltonian equation (101), with $a = 0.2$, $J = 0.2$, $\phi = \pi/10$ (circles), and $\phi = \pi/5$ (crosses). The continuous line occurs for an angle ϕ with the real axis, the first (sharp) resonance appears for both calculations, and the second resonance appears only for $\phi = \pi/5 > \theta_1$.

show all the eigenvalues for $\phi = \pi/10$ and $\phi = \pi/5$ obtained with a complex-rotated 500-function expansion. The rotated continuum forms an angle ϕ with the $\mathrm{Re}(E)$ axis, and two resonances appear as isolated eigenvalues. Note that the sharp resonance appears for both $\phi = \pi/10$ and $\phi = \pi/5$ curves, but the broad resonance cannot be obtained with the $\phi = \pi/10 < \theta_1$ expansion.

It is well known that complex-rotated basis-set expansions give very accurate values for complex energies [157]. But in order to apply FFS to obtain critical exponents, we observe that the scaling function $F_{\mathcal{O}}(x)$ in the scaling relation (60) has to be replaced by a complex function of a ϕ-dependent complex argument for both resonances and bound states. Then it is necessary to introduce new scaling functions and critical exponents. The convergence process with the number N of basis functions is not uniform, and therefore it is very difficult to make extrapolations from the numerical data.

To visualize the phenomenon, the (real) ground-state energy of Hamiltonian (101) with $a = 0.2$ and $J = 1$ is $E_0 \simeq -0.4770651355$. We show in Fig. 19

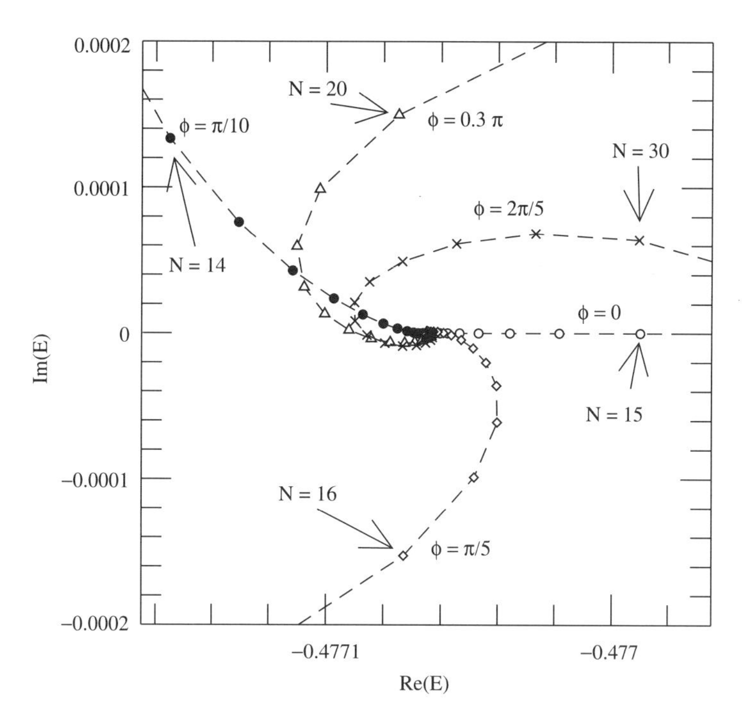

Figure 19. $\mathrm{Im}(E)$ versus $\mathrm{Re}(E)$ for the ground-state energy of Hamiltonian equation (101), with $a = 0.2$, $J = 1$ from complex-rotated diagonalization with $\phi = n\pi/10$; $n = 0(\circ)$, $1(\bullet)$, $2(\diamond)$, $3(\triangle)$, and $4(\times)$ for increasing values of N. The minimum value of N for each angle is indicated by an arrow.

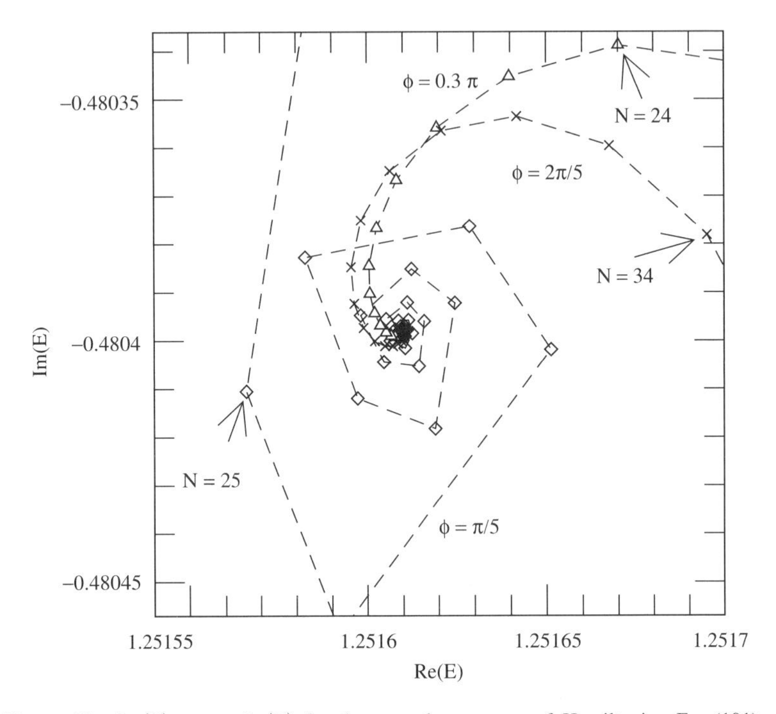

Figure 20. $\mathrm{Im}(E)$ versus $\mathrm{Re}(E)$ for the second resonance of Hamiltonian Eq. (101) with $a = 0.2$, $J = 0.2$ from complex-rotated diagonalization with $\phi = n\pi/10$; $n = 2(\Diamond)$, $3(\triangle)$, and $4(\times)$ for increasing values of N. The minimum value of N for each angle is indicated by an arrow.

$\mathrm{Re}(E_0^{(N)})$ versus $\mathrm{Im}(E_0^{(N)})$ for large values of N and five different values of ϕ. It is clear that the process converges, but convergence is not uniform (except for $\phi = 0$) even for $|E_0^{(N)}|^2$. The same picture occurs for resonances, as shown in Fig. 20 for the second resonance $E_2^{(res)} \simeq 1.25161 - i\,0.4803978$ of Hamiltonian (101) with $a = 0.2$ and $J = 0.2$.

To summarize, for our model Hamiltonian, resonances appear after a bound–virtual and a virtual–virtual resonance transition. There is no method to obtain virtual energies using a square-integrable basis set, even in the complex-rotated formalism. Then, at this point we can ask if FSS is a useful method to study this kind of resonance. As we will show in the next subsection, the answer is yes; FSS is a method to obtain near-threshold properties, and with FSS we can characterize the near-threshold resonances by solving the Hermitian (not complex-rotated) Hamiltonian using a real square-integrable basis-set expansion. Moreover, the critical point of the virtual resonance–resonance transition, λ_r, could also be obtained using FSS.

B. Crossover Phenomena

In general, FSS uses scaling laws to describe asymptotic behavior of many-body systems near the threshold energy. An important question is, What is the meaning of the word *near* in the preceding sentence. We will call a crossover phenomena to a phenomena related with the failure of the system to attain its asymptotic scaling regime [94]. Even when crossover phenomena is defined in relation with FSS expansions, as we will see, the *size* of the critical region, where asymptotic regime holds, has possible experimental consequences.

Therefore we are going to examine the critical behavior of the system defined by Hamiltonian (101) using the FSS method described in Section IV. That is, we are going to calculate the values of a and J for which a bound-state energy becomes absorbed or degenerate with a continuum. We define $J_c^{(n)}(a)$ as the value of J for which the n-bound-state energy becomes equal to zero (the threshold energy is set at zero)

$$J_c^{(n)}(a) \equiv \inf_{\{J\}}\{E_n(a,J) < 0\} \tag{105}$$

and the related critical exponent for the energy α_n is given by

$$E_n(a,J) \underset{J \to J_c^{(n)+}}{\sim} (J - J_c^{(n)})^{\alpha_n} \tag{106}$$

The FSS calculation of critical parameters was done using the functions Eq. (102) and the matrix elements Eqs.(103) and (104). The critical line $J_c^{(n)}(a)$ and the critical exponent α_n were calculated using Eqs. (66) and (67).

Figures 21 and 22 show the "phase diagrams" $J^{(n)}$ as a function of a for the ground state $n = 0$; and for several states, $n = 0, 1, 2, 3$ and $n = 4$. From a theorem proved by Klaus and Simon [47] valid for Hamiltonian (101), we know that the critical exponent for the energy is $\alpha_n = 2\ \forall n, a > 0$, where n denotes the isolated bound states (see Section II).

In both figures the critical line separates the resonance region from the bound region. As will be shown later, the transition from bound to resonance states go through virtual states. There is no bound states with positive energy for $a > 0$, but there is at least one bound state for $J \geq J^*(a) \equiv 1/(4a)$ [164]. This result shows that the solid line in Fig. 21 does not cross the curve $J^*(a)$ (the Simon line). Numerical results show that $J_c(a)$ goes asymptotically to $J^*(a)$ when $a \to \infty$. For fixed a, the ground state $E_0(J,a)$ is concave, nonincreasing, and continuous as a function of J, and it is decreasing for $J \geq 1/(4a)$ [165].

In the limit $a = 0$, Hamiltonian (101) reduces to the Hamiltonian of a harmonic oscillator with frequency equals one plus a constant J. Then the

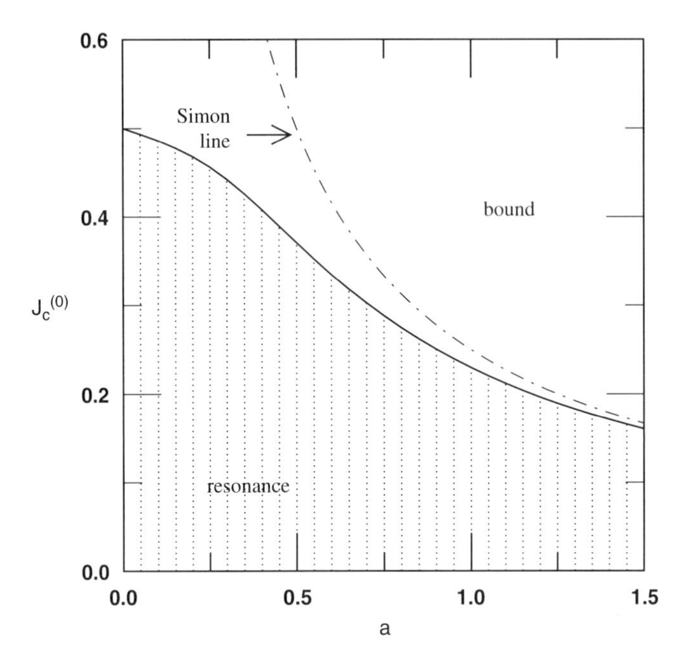

Figure 21. Phase diagram for the ground-state energy of Hamiltonian equation (101), with the critical J as a function of a for $N = 300$. The solid line separates the bound-state energies from the resonance energies. The simon line is shown by the dashed line.

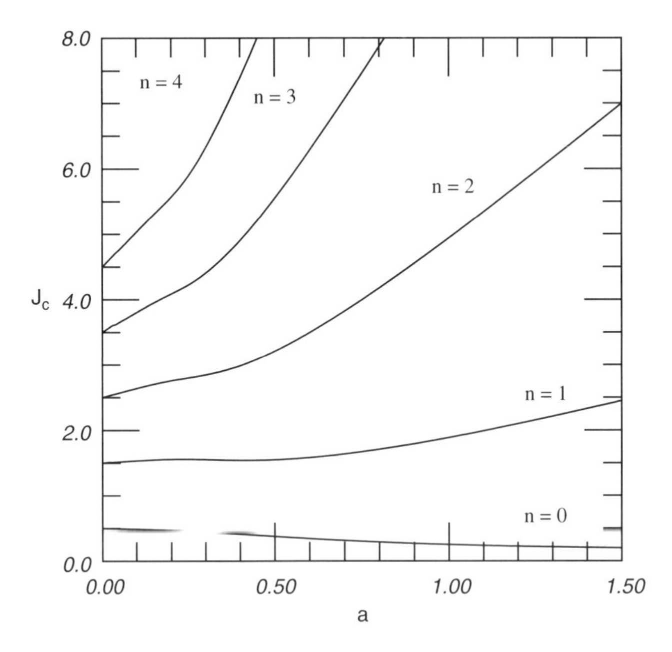

Figure 22. Phase diagram for several energy states of Hamiltonian equation (101), with the critical J as a function of a for $N = 300$.

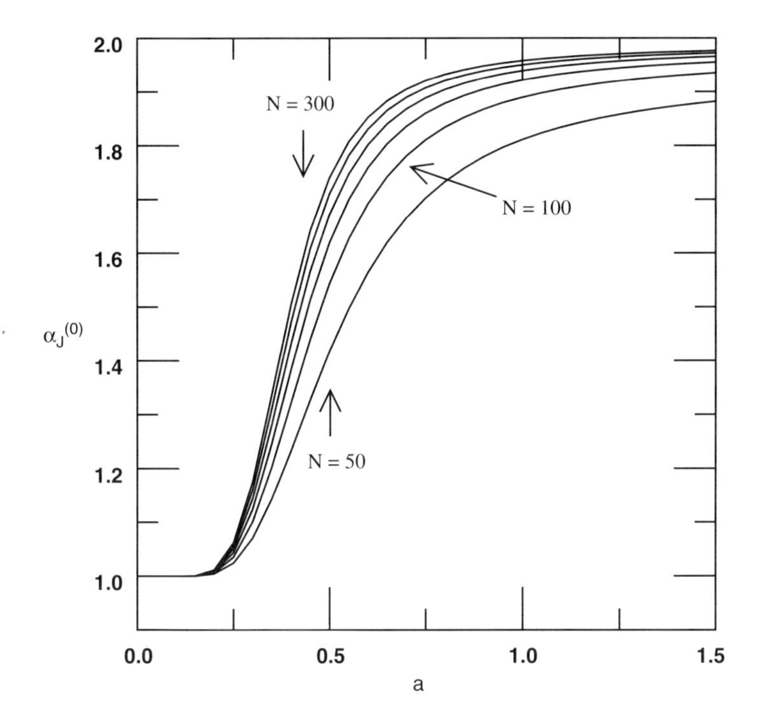

Figure 23. The critical exponent for the ground-state energy of Hamiltonian equation (101) as a function of a for different values of $N = 50, 100, \ldots, 300$.

ground state is $E_0 = 1/2 - J$, and therefore the "critical" value of the parameter is $J_c(a = 0) = 1/2$. Of course, for $a = 0$ there is no "true" critical parameters, because the system has an infinite number of bound states for all values of J. But $J = 1/2$ corresponds to the value of J where the ground-state energy is equal to zero, in agreement with the definition, Eq. (105). For all values of $a > 0$, there are no bound states with positive energy and the phase curve goes continuously to the point $(a = 0, J = 1/2)$ when $a \to 0$. Simple variational bounds show that the slope of the critical line $J_c(a)$ at $a = 0$ is smaller than $-1/8$ (numerical results give a slope value near -0.14).

In Figs. 23 and 24 the critical exponent α is shown as a function of a for the ground state and first even excited state, respectively. A crossover phenomenon appears for this exponent. For large values of a, we obtain $\alpha_n \simeq 2$:

$$E_n(a, J) \underset{J \to J_c^{(n)+}}{\sim} (J - J_c^{(n)})^2, \qquad a \text{ fixed}, \ n = 0, 2 \tag{107}$$

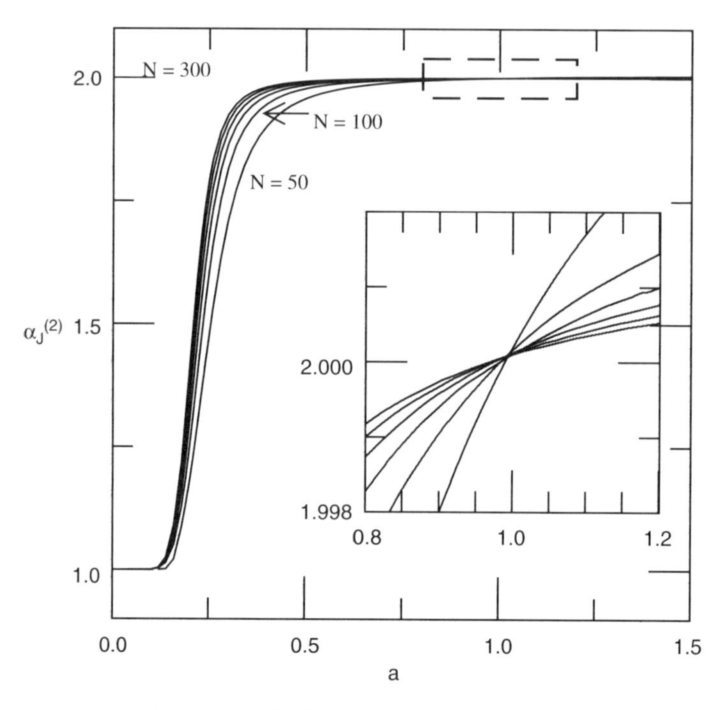

Figure 24. The critical exponent for the first even excited state of Hamiltonian equation (101) as a function of a for different values of $N = 50, 100, \ldots, 300$. Note the appearance of a special point at about $a = 1$.

But for small values of a we have $\alpha_n \simeq 1$

$$E_n(a, J) \underset{J \to J_c^{(n)+}}{\sim} (J - J_c^{(n)}), \qquad a \text{ fixed}, \; n = 0, 2 \tag{108}$$

As a matter of fact, the sharp transition from $\alpha_0 = 2$ to $\alpha_0 = 1$ at $a \sim 0.2$–0.5 in Fig. 23 and at $a \sim 0.1$–0.3 in Fig. 24 is a result of the truncation of the Hamiltonian matrix. As the size of the basis set that is used to represent the Hamiltonian is increased, this transition occurs at $a \to 0$. Namely, the crossover phenomena disappear as the size of the basis set is increased. The exact result for this short-range potential is $\alpha_n = 2$ for $a > 0$, and bound states become virtual states as J is reduced.

For large values of a the exponent is $\alpha = 2$, but for small values of a the exponent is one. The crossover region $\alpha_n\colon 2 \to 1$ is related to the characteristic length of the finite basis set. The larger the number N of functions, the larger the region with $\alpha_n^{(N)} \sim 2$. This fact is characteristic of this kind of scaling phenomena. For small values of a the behavior of the system is linear except in a small neighborhood of the critical point, not accessible to a "small" N expansion. This phenomenon has practical consequences: Even the exact transition is continuous with $\alpha = 2$, and for small values of a this asymptotic behavior could not be seen in an experiment if the critical region is smaller than the working appreciation. Therefore an effective exponent $\alpha = 1$ and a sharp resonance with $J_{res} = J_c$ will be observed.

Note that it is not a *true* change in the phase transition (second order to first order). If such a change occurs, a new scaling relation appears and the curves with different N should cross at approximately the same point. This point is a particular case of critical point, called multicritical point in theory of phase transition [25]. Multicritical points in few-body systems is the subject of the next subsection.

C. Multicritical Points

We studied in the previous section several types of phase transition, namely, bound–virtual, bound–resonance, and so on. A characteristic of a phase transition is that two different solutions merge ($\alpha \neq 1$), or coexist at the critical point ($\alpha = 1$). Many-body and multiparameter Hamiltonians could present more complicated transitions, and we will call them multicritical points.

One kind of a multicritical point is a point over a critical line where more than two different states coalesce. The common multicritical points in statistical mechanics theory of phase transition are tricritical points (the point that separates a first order and a continuous line) or bicritical points (two continuous lines merge in a first order line) (see, for example, Ref. 166). These multicritical points were observed in quantum few-body systems only in the large dimension limit approximation for small molecules [10,32]. For three-dimensional systems, this kind of multicritical points was not reported yet.

On the other hand, the two-parameter Hamiltonian, Eq. (101), presents a multicritical point that has, to the best of our knowledge, no classical statistical mechanic analogy [156]. A crossing point appears for α_2 as shown in the window of Fig. 24. This special point is a multicritical point, but with no change in the value of the critical exponent $\alpha_2 = 2$. Even when we are using real square-integrable functions, it is necessary to study virtual and resonances states to explain this crossing point. Because complex rotating methods cannot give the virtual states, we use a numerical integration procedure to solve the Scrhödinger equation.

In order to obtain virtual and resonances states, we have to find eigenfunctions of Hamiltonian (101) which grows up exponentially when $|x| \to \infty$. Using the fact that the potential goes to zero very fast, we can obtain accurate results for Siegert states [167]. $V(x)$ is assumed to be zero for $|x| > x_0$:

$$V(x) = \begin{cases} \left(\frac{x^2}{2} - J\right)e^{-ax^2}, & |x| < x_0 \\ 0, & |x| > x_0 \end{cases} \tag{109}$$

Following Meyer and Walter [168] instead of the Hamiltonian defined in Eq. (101), we solved the non-Hermitian eigenvalue problem

$$\left[-\frac{1}{2}\frac{d^2}{dx^2} + V(x)\right]\Psi_s(x) = p^2\Psi_s(x) \tag{110}$$

With the energy-dependent boundary conditions at $x = 0, x_0$, we obtain

$$\left.\frac{d\Psi_s}{dx}\right|_{x=0} = 0, \qquad \left.\frac{d\Psi_s}{dx}\right|_{x=x_0} = ik\Psi_s(x_0) \tag{111}$$

where $k = \sqrt{2E}$ and the energy is determinate by the condition

$$p = k \tag{112}$$

Eigenfunction expansions as used in Ref. 168 are not accurate near the critical point. Instead, we developed a shooting point method in order to make a direct numerical integration of Eq. (110) with the condition Eq. (112). Real energies (bound and virtual) were found by bisection methods, and for complex energies it was necessary to combine the Newton–Raphson and grid methods.

As we show in Fig. 25, the multicritical point is related to the crossing of the bound state $n = 2$ line with the resonance $n = 4$ at the critical $\text{Re}(E) = 0$ energy. In this figure we show also the results for $n = 3$. The dashed lines in Fig. 25 describe virtual states. The cusp behavior is a reflection of a transition through a branch point with exponent of one-half from a virtual state associated with a real eigenvalue to a virtual state that is associated with a *complex* eigenvalue.

It is important to emphasize that the $n = 4$ virtual–virtual resonance and resonance curves in Fig. 25 cannot be obtained by using basis-set expansion. Note that the resonance is broad; therefore it cannot be calculated even by using stabilization methods [169]. However, the FSS method gives the localization of the virtual resonance–resonance transition.

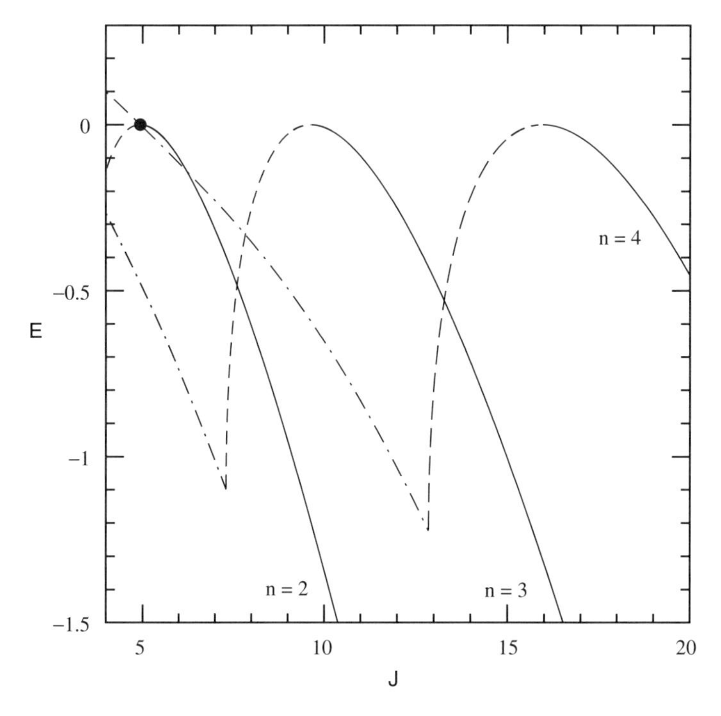

Figure 25. Energies for the states $n = 2$, $n = 3$, and $n = 4$ (continuous line), and virtual state energies (dashed lines) and the real part of the complex energy for $n = 3$ and $n = 4$ (dot–dashed line) of Hamiltonian equation (101), calculated using Eqs. (110) and (111) with $a = a_c = 1.027$, $x_0 = 6$. The multicritical point (•) is located at $(a, J) \sim (1.027, 4.932)$.

VII. SPATIAL FINITE-SIZE SCALING

In previous sections, *finite size* corresponds to the number of elements of a complete basis set used in a truncated Rayleigh–Ritz expansion of an exact bounded eigenfunction of a given Hamiltonian. In this section we present a different FSS approach. In this case, we will confine the system inside a box of size R. The box could be penetrable ($V(r) = 0$ for $r > R$) or impenetrable ($V(r) = \infty$ for $r > R$). Different approaches to solve the Schrödinger equation, for which the system is confined, have been developed from the 1930s to nowadays (Refs. 167 and 170 and references therein). These methods were used as approximations to the corresponding "free system" [170] or to calculate bound states of "true confined systems" [171]. But confining the potential could change drastically the behavior for large value of r; and then, as we see in Section II, the critical properties of the bound states. Therefore these approximate solutions, capable of giving accurate results for "deep" bound states, could give an erroneous near-threshold behavior even for very large values of R.

We will study the scaling properties by introducing a cutoff radius R in the potential. This cutoff changes the critical exponent of the energy, but, for large values of the cutoff radius, the asymptotic behavior of FSS functions is dominated by the exact critical exponent [26]. The method gives accurate values for critical parameters and critical exponents.

We will develop the theory for one-body central potentials described by Eq. (20), confined to a penetrable box. The main difference between a penetrable and an impenetrable box is the boundary condition at $r = R$. The former has the advantage that it is useful for calculation of bound states, but also for resonance and virtual states.

We will use the fact that the potential goes to zero for $r \to \infty$ to introduce the scaling length R defined for a compact support potential $V_R(\lambda; r)$ as

$$V_R(\lambda; r) = \begin{cases} \lambda V(r) + \dfrac{(\delta - 1)(\delta - 3)}{8r^2} & \text{if } r \leq R \\ 0 & \text{if } r > R \end{cases} \tag{113}$$

Replacing the potential plus the centrifugal term in Eq. (20) by potential (113) gives

$$\mathscr{H}_R(\lambda; r)\Phi_R(\lambda; r) = \left[-\frac{1}{2}\frac{d^2}{dr^2} + V_R(\lambda; r)\right]\Phi_R(\lambda, r) = E_R(\lambda)\Phi_R(\lambda; r) \tag{114}$$

with the boundary conditions $\Phi_R(\lambda; r = 0) = 0$, and $d\ln(\Phi_R(\lambda; r))/dr$ is continuous at $r = R$.

Hamiltonian (114) could be solved with a prefixed precision using the Siegert method [167]. Consequently, the scaling ansatz also gives a powerful numerical tool useful for calculating critical parameters related to the bound states, resonances, and virtual states [167,172].

As we pointed out in Section II, for $\delta = 3$, a Hamiltonian with a compact support potential has a critical exponent $\alpha = 2$ and the energy is analytical at the critical point defined by the condition $E_R(\lambda_R) = 0$. Therefore near the critical point $\lambda = \lambda_R$ of Hamiltonian (114) the asymptotic form of the lowest eigenvalue is

$$E_R(\lambda) \sim a_R\,(\lambda - \lambda_R)^2 \qquad \text{for} \quad \lambda \to \lambda_R,\ a_R \neq 0\ \forall\ R < \infty \tag{115}$$

where $E_R(\lambda)$ corresponds to a bound (virtual) state if $\lambda > (<)\lambda_R$.

We will assume that the ground-state energy of the Hamiltonian (20) has a critical exponent $\alpha \neq 2$ (for example, a short-range potential $V(r)$ and $2 < \delta \neq 3$). The main hypothesis of the spatial finite-size scaling (SFSS) ansatz, which makes the different values of α compatible, is that the coefficient a_R, analytical for finite values of R, has to develop a singularity at the exact critical value λ_c when $R \to \infty$ as

$$a_R \sim -\frac{a}{|\lambda_c - \lambda_R|^{\eta}} \qquad \text{for} \quad R \to \infty \tag{116}$$

where a is a positive constant. Then, the asymptotic behavior of the energy near λ_R for large values of R is

$$E_R(\lambda) \sim -a \frac{(\lambda - \lambda_R)^2}{|\lambda_c - \lambda_R|^{\eta}} \tag{117}$$

Therefore, we can evaluate the asymptotic expression of the energy (117) at $\lambda = \lambda_c$:

$$E_R(\lambda_c) \sim -a\,|\lambda_c - \lambda_R|^{2-\eta} \tag{118}$$

This relation gives us the correction to the finite-size critical exponent obtaining the exact α exponent for Hamiltonian (20) as

$$\alpha = 2 - \eta \tag{119}$$

In a very different context, in statistical mechanics theory of critical phenomena, corrections to classical exponents are calculated using a systematic series of mean field approximations. In this case, the deviation η from the mean-field value of a critical exponent is called *coherent anomaly* [173].

Remember that $E_R(\lambda)$ in Eqs. (115)–(118) corresponds to a bound state if $\lambda_R < \lambda_c$ and corresponds to a virtual state if $\lambda_R > \lambda_c$. Note that there is no other formal difference between bound and virtual states other than the sign in the logarithmic derivate of the wave function at $r = R$. Therefore there are no technical problems related with this fact. A relation between λ_R and λ_c can be established for compact support potentials. In this case, using variational arguments, we obtain

$$\lambda_R(\delta) \begin{cases} > \lambda_c & \text{if } \delta < 3 \\ = \lambda_c & \text{if } \delta = 3 \\ < \lambda_c & \text{if } \delta > 3 \end{cases} \tag{120}$$

Generally, η, λ_c, λ_R, and a in Eqs. (118) and (119) are unknown parameters. In order to calculate them, we will study asymptotic expressions of diverse magnitudes for large values of the cutoff radius R.

We assume that the critical parameter for Hamiltonian (114) goes to the exact critical parameter for Hamiltonian (20) for large values of R in the form

$$\lambda_R \sim \lambda_c - \frac{\Delta}{R^{\mu}} \qquad \text{for} \quad R \to \infty \tag{121}$$

where Δ is a positive constant and μ is an unknown scaling exponent. The energy at $\lambda = \lambda_c$ Eq. (118) takes the form

$$E_R(\lambda_c) \sim -\frac{\text{constant}}{R^{(2-\eta)\mu}} \qquad \text{for} \quad R \to \infty \tag{122}$$

We can calculate the energy for two different values of R, and the logarithm of the quotient gives

$$\ln\left(\frac{E_{R_1}(\lambda_c)}{E_{R_2}(\lambda_c)}\right) \sim (2-\eta)\mu \ln\left(\frac{R_2}{R_1}\right) \qquad \text{for} \quad R_1, R_2 \to \infty \tag{123}$$

In order to obtain a second relation between the exponents α and μ using the Hellmann–Feynman theorem, Eqs. (23) and (117), we have

$$\frac{\partial E_R(\lambda)}{\partial \lambda} = \left\langle \frac{\partial \mathscr{H}_R}{\partial \lambda} \right\rangle \sim -2a \frac{(\lambda - \lambda_R)}{|\lambda_c - \lambda_R|^{\eta}} \tag{124}$$

In a similar way, we calculate this quantity for different values of R, and the quotient gives

$$\ln\left(\frac{\frac{\partial}{\partial\lambda} E_{R_1}(\lambda)}{\frac{\partial}{\partial\lambda} E_{R_2}(\lambda)}\right)\Bigg|_{\lambda=\lambda_c} \sim (1-\eta)\mu \ln\left(\frac{R_2}{R_1}\right) \qquad \text{for } R_1, R_2 \to \infty \tag{125}$$

Now, we can eliminate the exponent μ by defining a function $\Gamma(R_1, R_2; \lambda)$ as

$$\Gamma(R_1, R_2; \lambda) = \ln\left(\frac{E_{R_1}(\lambda)\left(\frac{\partial}{\partial\lambda} E_{R_2}(\lambda)\right)^2}{E_{R_2}(\lambda)\left(\frac{\partial}{\partial\lambda} E_{R_1}(\lambda)\right)^2}\right) \Bigg/ \ln\left(\frac{E_{R_1}(\lambda)\, \frac{\partial}{\partial\lambda} E_{R_2}(\lambda)}{E_{R_2}(\lambda)\, \frac{\partial}{\partial\lambda} E_{R_1}(\lambda)}\right) \tag{126}$$

In particular, for $\lambda = \lambda_c$ we have

$$\Gamma(R_1, R_2; \lambda_c) = \eta \qquad \text{for} \quad R_1, R_2 \to \infty \tag{127}$$

independent of the values of R_1 and R_2. Therefore, two curves calculated with different values of R will cross at the same point $\lambda = \lambda_c$. The value of the function Γ at this point is the critical exponent η. Actually, this is an asymptotic result, and we will have to calculate a set of values $\{\lambda_c^{(i)}, \eta^{(i)}\}_{i=1,N}$ for different values of R. Final estimation of (λ_c, η) has to be obtained by performing an extrapolation of the data for $1/R \to 0$ as we did in previous sections.

The numerical approach is as follows: Take $\{R_i\}$ $i = 1, \ldots, N$ an arbitrary set of (large) values of R, and let $\Delta \ll R_i$ be a fixed parameter. We can obtain a set of values $\{\lambda_c^{(i)}, \eta^{(i)}\}_{i=1,N}$ by using Eq. (127) as

$$\Gamma(R_i - \Delta, R_i; \lambda_c^{(i)}) = \Gamma(R_i, R_i + \Delta; \lambda_c^{(i)}) = \eta^{(i)} \tag{128}$$

In previous sections the FSS parameter was a discrete variable, the number of functions in a basis-set expansion. In this case the most accurate results were obtained by searching for crossing points of curves with minimum difference between the FSS parameters (in general, 1; or 2 for problems with parity effects). In the present case, the parameter R is a real variable. Therefore, the minimum difference between parameters is given by the limit $\Delta \to 0$. This limit introduces derivatives of the functions $E_R(\lambda)$ and $\partial E_R(\lambda)/\partial\lambda$ with respect to R. In practice the derivatives have to be calculated numerically, and then it is convenient to use a finite value of Δ, which is fixed by numerical stability studies.

To illustrate the method, we apply SFSS to the usual spherical square well of width $r_0 = 1$ and depth equal to -1, and then the potential Eq. (113) takes the form

$$V_R(\lambda; r) = \begin{cases} -\lambda + \dfrac{(\delta-1)(\delta-3)}{8r^2} & \text{if } r \leq 1 \\ \dfrac{(\delta-1)(\delta-3)}{8r^2} & \text{if } 1 < r \leq R \\ 0 & \text{if } r > R \end{cases} \tag{129}$$

In this case, the exponent is $\alpha = 2$ for finite values of R and for all values of δ. The exact critical exponent α for $R = \infty$ obeys Eq. (24).

According to Eq. (120), we have to calculate bound-state energies for $\delta > 3$ and virtual-state energies for $2 < \delta < 3$. The Siegert method [167] assumes the exact boundary conditions at $r = 0$ and $r = R$ and gives exact results for the Hamiltonian equation (114). A Siegert state is a solution of the Schrödinger equation (114) defined for $0 \leq r \leq R$ with the boundary conditions $\Phi_R(\lambda; r = 0) = 0$ and $\Phi_R(\lambda; r = R) = \mp k\Phi'_R(\lambda; r = R)$, where $k = \sqrt{-2E_R}$ and the upper (lower) sign is for bound (virtual) states.

The usual technique is to perform a numerical integration of Eq. (114) using boundary conditions at $r = 0$ and searching for the eigenvalue $E_R(\lambda)$ by the shooting method applied iteratively until the boundary condition at $r = R$ is obtained. For the potential equation (129), we will write down the exact transcendental equation for the energy for all values of R, δ, and λ, and no numerical integration is needed.

The (unnormalized) lowest-energy wave function of Hamiltonian (114) with the potential (129) is

$$\Phi_R(\lambda; r) = \begin{cases} \sqrt{r}\, J_{\frac{\delta-2}{2}}(\kappa r) & \text{if } r \leq 1 \\ \sqrt{r}\left(A\, I_{\frac{\delta-2}{2}}(kr) + B\, K_{\frac{\delta-2}{2}}(kr)\right) & \text{if } 1 < r \leq R \end{cases} \tag{130}$$

where $\kappa = \sqrt{2(\lambda + E_R)}$ and $J_\nu(z)$, $I_\nu(z)$, and $K_\nu(z)$ are Bessel functions of the first kind and modified Bessel functions of the first and third kind, respectively [55]. Continuity of the logarithmic derivative at $r = 1$ plus the Siegert boundary condition at $r = R$ give the transcendental equation for the eigenvalues

$$\begin{aligned} &\left[I_{\frac{\delta-2}{2}}(kR) + 2kR\left(\pm I_{\frac{\delta-2}{2}}(kR) + I'_{\frac{\delta-2}{2}}(kR)\right)\right]\left[kK'_{\frac{\delta-2}{2}}(k)\, J_{\frac{\delta-2}{2}}(\kappa) - \kappa\, J'_{\frac{\delta-2}{2}}(\kappa) K_{\frac{\delta-2}{2}}(k)\right] \\ &\quad = \left[K_{\frac{\delta-2}{2}}(kR) + 2kR\left(\pm K_{\frac{\delta-2}{2}}(kR) + K'_{\frac{\delta-2}{2}}(kR)\right)\right] \\ &\quad\quad \times \left[kI'_{\frac{\delta-2}{2}}(k)\, J_{\frac{\delta-2}{2}}(\kappa) - \kappa\, J'_{\frac{\delta-2}{2}}(\kappa) I_{\frac{\delta-2}{2}}(k)\right] \end{aligned} \tag{131}$$

where the prime ($'$) means derivatives with respect to the argument, the upper signs give bound-state energies, and the lower signs give virtual state energies. The critical parameter λ_R is obtained taking the limit $E_R \to 0$ in Eq. (131). Using standard relations between Bessel functions and their derivatives [55], we get

$$(\delta - 1)\, R^{\delta-2} J_{\frac{\delta-4}{2}}\left(\sqrt{2\lambda_R}\right) - (\delta - 3)\, J_{\frac{\delta}{2}}\left(\sqrt{2\lambda_R}\right) = 0 \tag{132}$$

We look for numerical solutions of Eq. (128) with different values of Δ. Results are numerically stable in an acceptable range, and the plots we show here were done with $\Delta = 1$.

SFSS gives corrections to critical exponents and also very accurate estimations of the critical parameter λ_c. In Fig. 26 we plot λ_c, $\lambda_{R=10}$, and $\lambda_c^{R=10}$ versus δ for small values of δ. Note the discrepancy between the exact $R = \infty$ and $R = 10$ lines; the curve obtained from SFSS with $R = 10$ is indistinguishable from the exact $R = \infty$ line at the graphic resolution.

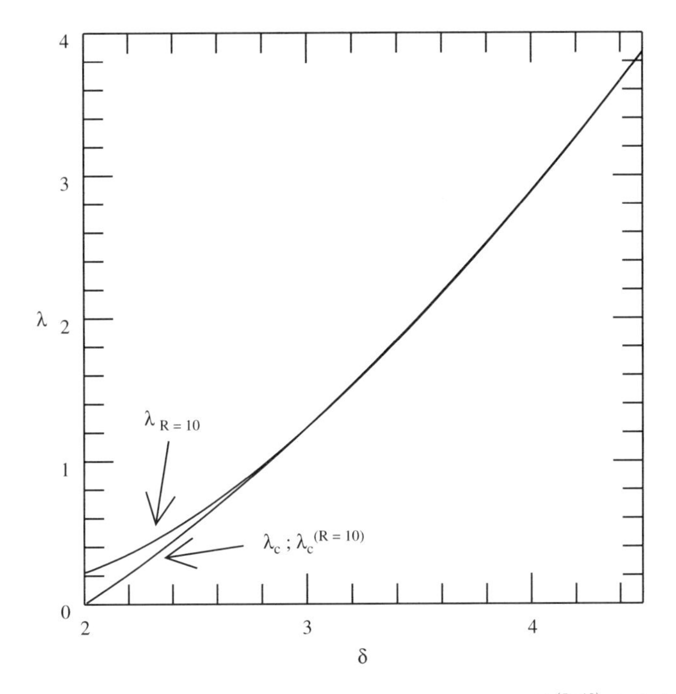

Figure 26. Exact values of λ_c for $R = \infty$ and SFSS calculations of $\lambda_c^{(R=10)}$ (indistinguishable lines) and $\lambda_{R=10}$ as a function of δ for the minimum energy state of the square-well potential.

In Fig. 27 we compare the exact values of the exponent $\alpha(\delta)$ given in Eq. (24) and $\alpha^{(R=20)}$. As expected, we obtained good numerical results except near $\delta = 4$, where the exact exponent has logarithmic deviations and the curve is nondifferentiable. This zone around $\delta = 4$ is magnified in Fig. 28, comparing the exact exponent with the approximations obtained with $R = 10, 20, 30, 40, 50$. Note from Fig. 26 that the values obtained for $\lambda_c^{(R)}$ are accurate even near $\delta = 4$.

For the important case of odd values of δ, corresponding to three-dimensional l waves $l = 0, 1, 2, \ldots$ for $\delta = 3, 5, 7, \ldots$, the Bessel functions reduce to the usual trigonometric and hyperbolic functions times powers of r [55]. In this case, everything but the solution of the transcendental equation for the energy can be done analytically. The advantage is that we can go further to very large values of R, and the convergence for $R \to \infty$ could be studied without any extrapolation process.

In particular, we study the case $\delta = 5$ corresponding to the three-dimensional $(l = 1)$ waves. The exact values of the critical parameters are $\lambda_c = \pi^2/2 \simeq 4.9348022$ and $\alpha = 1$.

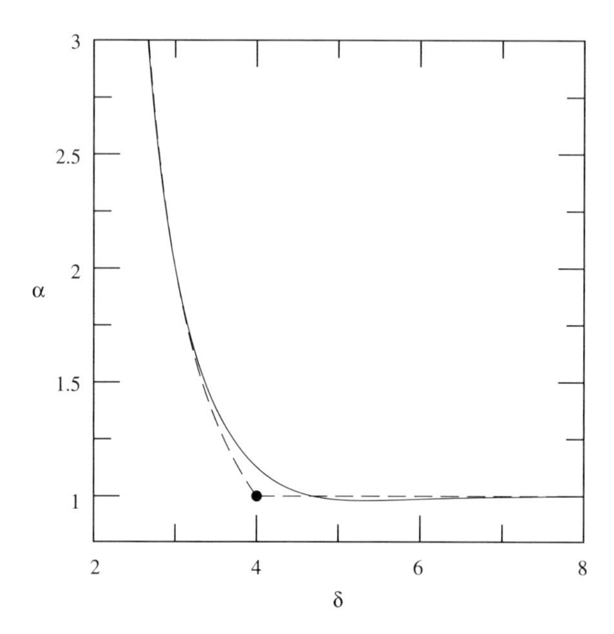

Figure 27. Exact values of α for $R = \infty$ (dashed line) and SFSS calculations of $\lambda^{(R=20)}$ (continuous line) as a function of δ for the minimum energy state of the square-well potential. The point where the exponent has a logarithmic deviation $\alpha(\delta = 4) = 1_{\log}$ is shown by a dot.

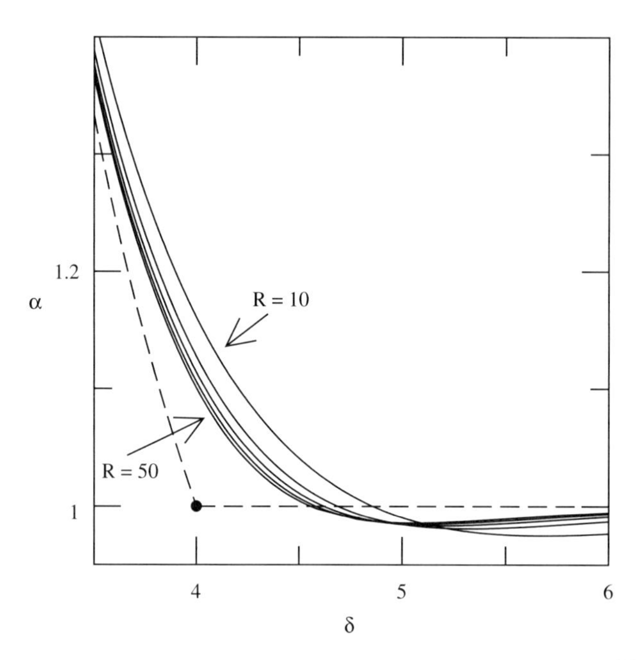

Figure 28. Exact values of α for $R = \infty$ (dashed line) and SFSS calculations of $\lambda^{(R)}$ for $R = 10$, $20, 30, 40, 50$ (continuous lines) as a function of δ for the minimum energy state of the square-well potential. The point where the exponent has a logarithmic deviation $\alpha(\delta = 4) = 1_{\log}$ is shown by a dot.

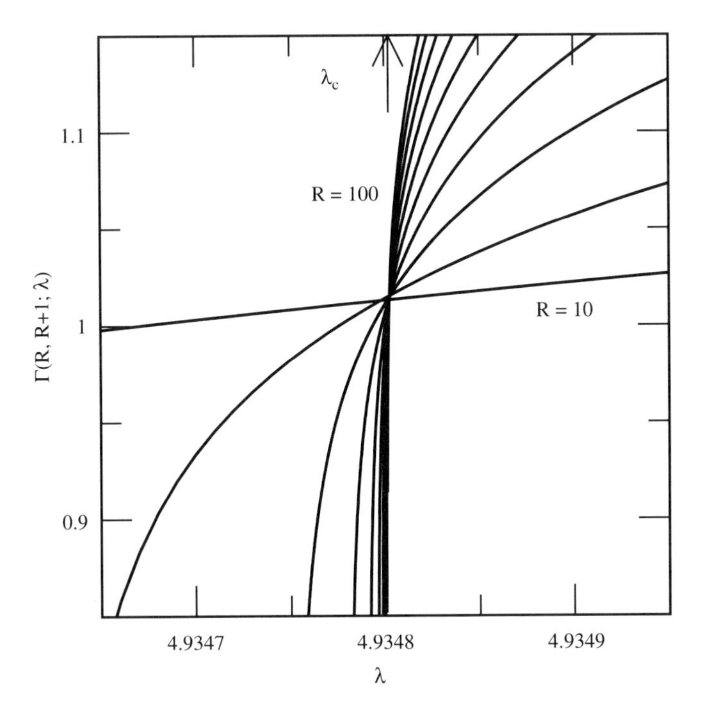

Figure 29. $\Gamma(R, R+1; \lambda)$ as a function of λ for the minimum $\delta = 5$ $(l = 1)$ bound state of the square-well potential for values of $R = 10, 20, \ldots, 100$.

In Fig. 29, we show $\Gamma(R, R+1; \lambda)$ against λ for $R = 10, 20, \ldots, 100$ for the lowest-energy $\delta = 5$ $(l = 1)$ bound state. As the asymptotic result, Eq. (127) predicts the curves cross at (approximately) the same point.

In Fig. 30 the critical point $\lambda_c^{(R)}$ for finite values of R is plotted against $1/R$, obtaining an excellent agreement with the exact value showed in the plot by a dot. Finally, $\alpha^{(R)}$ against $1/R$ is shown in Fig. 31. Note the minimum occur at $R \simeq 25$. Figures 30 and 31 also show that, as usual in FSS [10,24], the convergence process is faster for the critical parameter calculation than for the critical exponent calculation.

In summary, in this section we presented a SFSS approach to study the critical behavior of bound and virtual states of the radial Schrödinger equation. The scaling is done by introducing a cutoff radius in the potential. This cutoff changes the critical exponent of the energy, but, for large values of the cutoff radius, the asymptotic behavior of FSS functions is dominated by the exact critical exponent. The method gives accurate values for critical parameters and critical exponents, even when the eigenfunctions are not square-integrable.

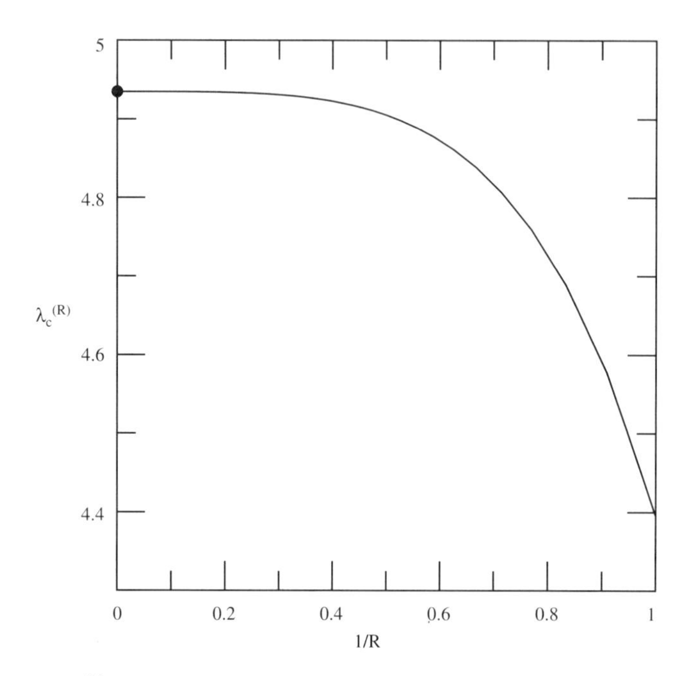

Figure 30. $\lambda_c^{(R)}$ versus $1/R$ for the minimum $\delta = 5$ ($l = 1$) bound state of the square-well potential. The exact value of $\lambda_c = \pi^2/2$ is also shown by a dot.

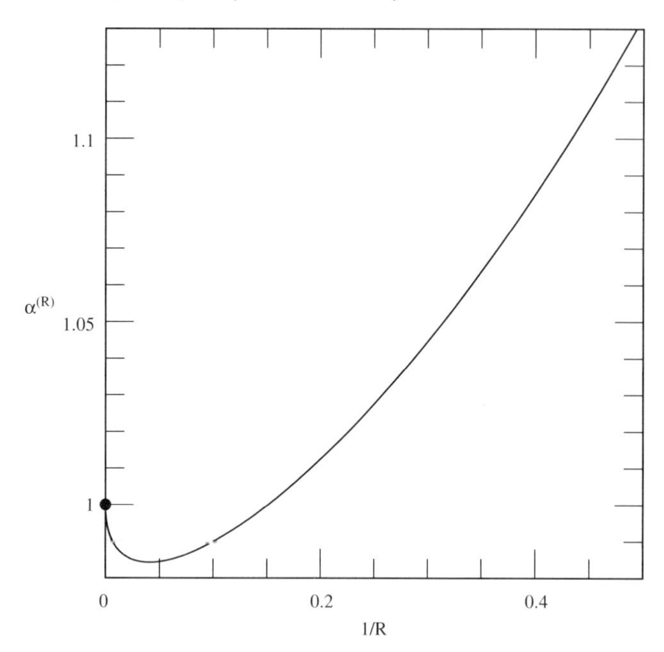

Figure 31. $\alpha^{(R)}$ versus $1/R$ for the minimum $\delta = 5$ ($l = 1$) bound state of the square-well potential. The exact value of $\alpha = 1$ is also shown by a dot.

The main idea of SFSS is similar to the other FSS approaches presented in this review, but the SFSS ansatz is not supported by variational principles and does not use basis-set expansions. It is more difficult to apply it to many-body systems, but it still might be a powerful tool to study bound–virtual–resonance transitions [156] and other analytical properties of the complex energy plane.

VIII. FINITE-SIZE SCALING AND PATH INTEGRAL APPROACH FOR QUANTUM CRITICALITY

In Section IV.A, we have shown that the quantum partition function in D dimensions looks like a classical partition function of a system in $(D+1)$ dimensions, with the extra dimension being the time. With this mapping and allowing the space and time variables to have discrete values, we turn the quantum problem into an effective classical lattice problem.

Having a classical pseudosystem connected to the original quantum problem allows us to go further in the phase transition analogies by realizing that the divergences in thermodynamic quantities are consequences of a more fundamental phenomena [174]. The divergence of the correlation length when the system is critical is due to the fact that the classical lattice shows fractal patterns; in other words, the classical lattice becomes self-similar in all length scales. Thus it is not necessary to limit the phase transition analogies to the search for points where the correlation length diverges. Any quantity that changes its scaling behavior in a phase transition can be used.

A. Mapping Quantum Problems to Lattice Systems

In the path integral approach, the analytical continuation of the probability amplitude to imaginary time $t = -i\tau$ of closed trajectories, $x(t) = x(t')$, is formally equivalent to the quantum partition function $Z(\beta)$, with the inverse temperature $\beta = -i(t'-t)/\hbar$. In path integral discrete time approach, the quantum partition function reads [175–177]

$$Z(\beta) = \lim_{\Delta\tau\to 0}\left(\frac{m}{2\pi\Delta\tau\hbar}\right)^{N_\tau/2}\int\prod_{\ell=0}^{N_\tau} dx_\ell \\ \times \exp\left[-\frac{\Delta\tau}{\hbar}\left(\sum_{\ell=0}^{N_\tau-1} m\frac{(x_{\ell+1}-x_\ell)^2}{2(\Delta\tau)^2} + \lambda\sum_{\ell=0}^{N_\tau-1}\frac{V(x_{\ell+1})+V(x_\ell)}{2}\right)\right] \tag{133}$$

where λ is the strength of the one-dimensional potential $V(x)$, $\Delta\tau = \beta/N_\tau$ is the regular grid spacing between N_τ points along the imaginary time axis indexed by $\ell = 0, 1, \ldots, N_\tau$. The closed path is made by a periodic boundary condition in the time direction such that $x_{N_\tau} = x_0$.

In order to obtain a connection with statistical mechanics lattice systems, we discretize the position space. The position in time slice ℓ is given by

$$x_{i_\ell} = x_0 + i_\ell \Delta L, \qquad \text{with} \quad i_\ell = 1, 2, \ldots, N_q \tag{134}$$

where ΔL is the regular grid spacing of the position axis which has a total of N_q points and q_0 is a constant used to adjust the origin of the coordinate system. Moreover, the size of the space is limited by $L = N_q \, \Delta L$.

Now we to concentrate on the properties of the two-dimensional lattice, the space–time lattice. The partition function, Eq. (133), shows that there is coupling only in the time direction and only between nearest-neighbor time slices. This allows us to use the statistical mechanics technique of writing the partition function Z of the finite system as the trace of a matrix T to the power N_τ.

$$Z(\Delta L, \Delta\tau)_{N_q, N_\tau} = \mathrm{Tr}(T^{N_\tau}) \tag{135}$$

The matrix T is called the transfer matrix. Its elements are given by

$$T(i_\ell, i_{\ell+1}) = \Delta L \left(\frac{m}{2\pi\hbar\Delta\tau} \right)^{1/2} \tag{136}$$

$$\times \exp\left\{ -\frac{\Delta\tau}{\hbar} \left[\frac{m}{2} \left(\frac{\Delta L}{\Delta\tau} \right)^2 (i_{\ell+1} - i_\ell)^2 + \lambda \frac{V_{i_{\ell+1}} + V_{i_\ell}}{2} \right] \right\} \tag{137}$$

where $V_{i_\ell} = V(q_\ell)$ is the potential energy of time slice ℓ evaluated at the space point i_ℓ.

The above transfer matrix can be seen as a transfer matrix of a classical pseudosystem. One can draw the analogy between this classical pseudosystem and a polymer that is constrained to lie in a two-dimensional lattice, each time slice being a polymer bead as shown in Fig. 32. The Hamiltonian of the pseudosystem reads

$$\mathscr{H} = \sum_{\ell=1}^{N_\tau} \mathscr{H}_{\ell,\ell+1} \tag{138}$$

where

$$\mathscr{H}_{\ell,\ell+1} = \frac{m}{2} \left(\frac{\Delta L}{\Delta\tau} \right)^2 (i_{\ell+1} - i_\ell)^2 + \lambda \frac{V_{i_{\ell+1}} + V_{i_\ell}}{2} + C \tag{139}$$

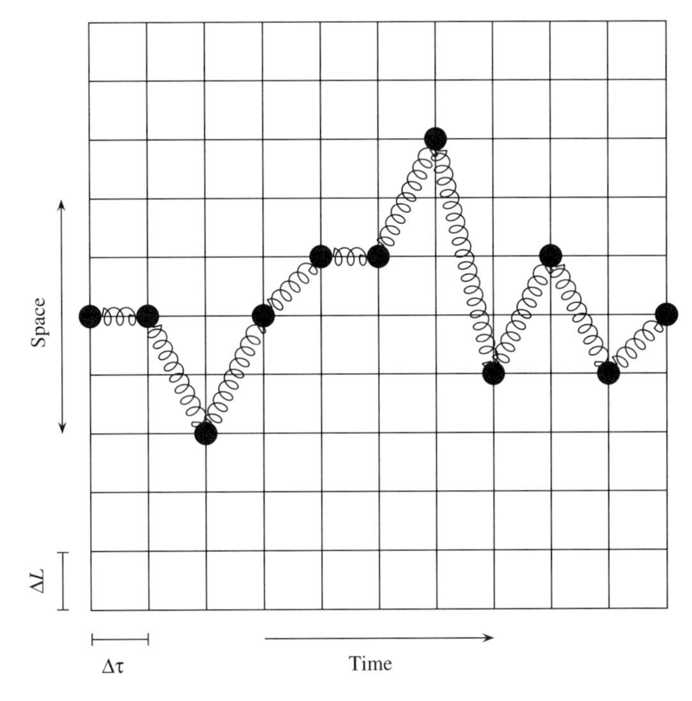

Figure 32. Mapping the quantum problem to a space–time lattice. The analogy to a polymer that is constrained to lie in a two-dimensional lattice is shown. Thus each time slice represents a polymer bead while the coupling between neighbor beads is connected by springs. For each time slice there is only one possible bead.

with $C = -\hbar \ln\{[m/(2\pi\hbar\Delta\tau)]^{1/2}\Delta L\}/\Delta\tau$ being a constant independent of the state of the lattice. With the above polymer analogy, the first term represents the harmonic coupling between neighboring beads connected by a spring and the second term is the interaction of each bead with some "local external field."

The partition function of the classical pseudosystem becomes

$$T(i_\ell, i_{\ell+1}) = \exp[-(\Delta\tau/\hbar)\mathscr{H}_{\ell,\ell+1}] \tag{140}$$

If the classical pseudosystem Hamiltonian were independent of $\Delta\tau$, the classical pseudosystem would behave as a statistical mechanics lattice system with inverse temperature $\Delta\tau/\hbar$.

In order to complete the mapping between the quantum problem and the classical pseudosystem, one must address the problems of both the continuum and the infinite limits. The ground-state properties of the original system are obtained by taking both the continuum limit, ($\Delta\tau$, $\Delta L \to 0$), and the thermodynamic limit, (β, $L \to \infty$) [174].

B. Quantum Criticality

In the path integral lattice definition, the time t is defined only in the sites of a regular time lattice with lattice constant $\Delta T = T/N_T$. The initial lattice point is $t_0 = t'$, the last one is $t_{N_T} = t''$, and any intermediate instant is given by $t_j = t' + j\Delta T,\ j = 0, 1, \ldots, N_T$. The position in each instant is given by $x_j = x(t_j)$. In the absence of the potential, the path integral summation is a summation over Brownian paths. Given the position x_{j-1} and the time interval ΔT, the increment $y_j = x_j - x_{j-1}$ is a random variable with probability proportional to $e^{-my^2/(2\hbar\Delta T)}$. The Brownian paths have a fractal nature and are self-similar as long as one scales the space and time direction with [178]

$$x \to bx, \qquad t \to b^2 t \tag{141}$$

where the scale factor b is any real positive number.

The classical system, whose states are given by the Brownian paths (the system can be also interpreted as a Gaussian polymer [179]) can be rescaled by the relations given in Eq. (141) to preserve the same structure. This fact makes the time lattice critical in the sense of renormalization group theory [25,180]. When the strength of the attractive potential is $\lambda \to \infty$, the particle must be bound and the contribution to the path integral summation of the Brownian paths are weighted by the factor $\exp(-\frac{1}{\hbar}\int_{t'}^{t''} V(x(t); \lambda)\, dt)$. Thus, the paths in the neighborhood of the origin contribute much more than paths filling uniformly the whole space. The system is not scaling-invariant anymore because we can devise two regions in the space, one where the particle is likely to be localized and the other where it is not. If the system is scale-invariant, the system is in a critical phase. If the scaling invariance is broken by the potential, we have a noncritical phase. The transition between these two phases at a finite value of lambda ($\lambda = \lambda_c$) will be properly called a *phase transition*.

The numerical study of the phase transition is made by fixing the grid spacing ΔL and ΔT and the discretization number N_L, which sets the rank of the transfer matrix to be diagonalized. The numerical calculation in a discrete and finite system gives an estimate of the actual values of all observable. The ground-state energy $E_L^{(0)}$ is given by

$$e^{-N_T \Delta T E_L^{(0)}/\hbar} = Z = \mathrm{Tr}[\mathrm{T}^{N_T}] \tag{142}$$

where the transfer matrix T is defined as

$$T(x_j, x_{j-1}) = \left(\frac{m\Delta L^2}{2\pi\hbar\Delta T}\right)^{1/2} \exp\left[-\frac{1}{\hbar}\left(\frac{m}{2\Delta T}(x_j - x_{j-1})^2 + \Delta T V(x_j; \lambda)\right)\right] \tag{143}$$

For a large N_T, the trace is dominated by the leading eigenvalue of the transfer matrix $Z \approx (a_L^{(0)})^{N_T}$ and the ground-state energy is given by

$$E_L^{(0)} = -\frac{\hbar}{\Delta T}\ln(a_L^{(0)}) \tag{144}$$

where $a_L^{(0)}$ is the leading eigenvalue of the transfer matrix. Having the leading eigenvector of the transfer matrix, one can evaluate any other ground-state expectation value. Because all geometric properties of the fractal Brownian paths are preserved in the critical region, the root mean square displacement $R_L = \langle x^2 \rangle^{1/2}$ must scale with the macroscopic dimension L. If the particle is bound, R_L must achieve a finite value independent of L. Hence R_L has a different scaling behavior if the particle is free or bound. This in principle can be used to determine the phase transition point.

The correlation length ξ along the imaginary time direction is the other quantity we can use to determine the critical region. The correlation length is defined as the asymptotic behavior of the correlation function

$$C(j\Delta T) = \langle x_0 x_j \rangle - \langle x_0 \rangle^2 \approx \exp\left(-\frac{j\Delta T}{\xi}\right), \qquad j \to \infty \tag{145}$$

The correlation length can be written in terms of the two leading eigenvalues $a_L^{(0)}$ and $a_L^{(1)}$ of the transfer matrix

$$\xi_L = -\Delta T \frac{1}{\ln(a_L^{(1)}/a_L^{(0)})} \tag{146}$$

When the system is critical, the quantum states must be correlated in all length scales along the time direction and thus the correlation length must scale with $\xi \sim T$. Hence in the true free particle case with $T \to \infty$ and $L \to \infty$ the correlation length diverges. Since L is finite, one cannot have a true divergence, but the scaling relations presented in Eq. (141) should still apply if L is finite and sufficiently large. Thus, the correlation length must scale as $\xi \sim L^2$ in the critical region.

For a given value of the critical parameter λ, we perform calculations with different system sizes. If $\xi_L(\lambda)$ scales with L^2 and $R_L(\lambda)$ scales with L, we call the system critical because the particle behaves like a free particle. When the strength of the potential breaks down this scaling behavior, the system is not critical and the particle is bound. The value of $\lambda = \lambda_c$ is the transition point.

As an example to illustrate this method, we study the case of a single particle in the presence of the Pöschl–Teller potential [181]

$$V(x;\lambda) = -\lambda(\lambda-1)/\cosh^2(x), \qquad \lambda \geq 1 \tag{147}$$

This problem has an exact solution [50]

$$E_n = -(\lambda - 1 - n)^2, \qquad n \leq \lambda - 1, \quad n = 0, 1, 2, \ldots \tag{148}$$

In the one-dimensional case, there is always a bound solution unless $\lambda = 1$ when the potential vanishes, and the particle is free. In three dimensions the behavior is much more interesting. Regardless of the presence of an attractive potential in the interval $1 < \lambda < 2$, there is no bound solution until $\lambda \geq 2$. Hence there is a finite value of the potential strength parameter, $\lambda_c = 2$, that defines the stability limit of the bound solution. In the present approach, this point can be obtained by investigating the scaling properties of the correlation length ξ_L and the mean radial distance R_L.

The one-dimensional case is a straightforward application of Eq. (143). The lattice in the position space is defined by picking evenly spaced points in the interval $x \in [-L/2, L/2]$. The results are obtained by exact diagonalization of the transfer matrix for every system size defined by $L = N_L \Delta L$. To investigate the trivial transition at $D = 1$, it is enough to consider only one grid spacing $\Delta L = 0.03$. For the remainder of this section we return the atomic units particle mass equal to one. Thus the values of ξ_L and R_L are calculated with 10 different system sizes, $L = 100, 200, \ldots, 1000$, where L is measure in units of ΔL. The results are shown in Fig. 33. In Fig. 33a we plot ξ_L/L^2, and in Fig. 33b we plot R_L/L as a function of λ. The curves correspond to different system sizes $L = 100, 200, \ldots, 1000$. For a given value of λ, it is clear that the only point where $R_L \sim L$ and $\xi_L \sim L^2$ is $\lambda = 1$. Thus the one-dimensional system is critical only when the potential vanishes, and the particle is free. So $\lambda = 1$ is regarded as a trivial critical point. In Figs. 33c and 33d, we show ξ_L and R_L as a function of L^2 and L for fixed value of λ. The curve with circles corresponds to $\lambda = 1$, and the other five correspond to small deviations from the free particle case with $\lambda = 1.02, 1.04, 1.06, 1.08$, and 1.10. The only case that can be adjusted to a straight line is $\lambda = 1$. In the presence of a weak potential, the scaling of ξ_L and R_L deviates from the critical scaling represented by the straight lines.

In summary, this approach is based on the breakdown of the free-particle scaling properties as the strength of the attractive external potential is made strong enough. This general idea can certainly be applied to systems with more than one particle as long as the unbound solution can be well represented by noninteracting free particles. This approach is general and might be used with other simulation techniques, such as Monte Carlo methods [182], to obtain critical parameters for few-electron atoms and simple molecular systems.

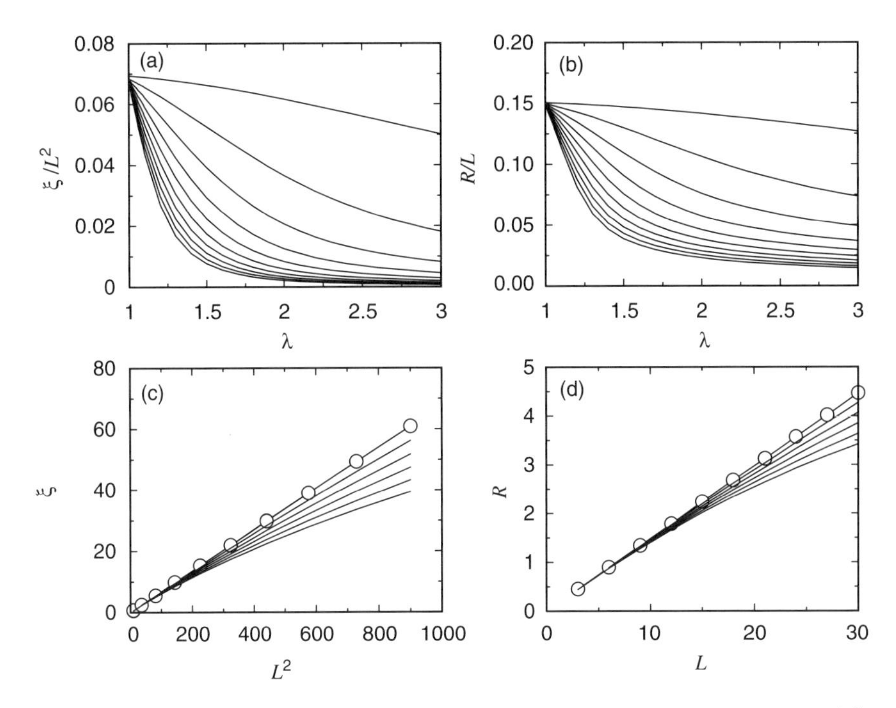

Figure 33. Pöschl–Teller potential: Panels (a) and (b) show the scaled correlation length ξ/L^2 and the scaled radial mean distance R/L as a function of the potential strength λ for different system sizes with $L = 3, 6, 9, \ldots, 30$. The grid spacing is kept fixed $\Delta L = 0.03$, so the smallest system has $N_L = 100$ and the largest has $N_L = 1000$ points. Panels (c) and (d) illustrate the scaling of $a\xi$ and R with the system size L for different values of $\lambda = 1(\bigcirc), 1.02, 1.04, 1.06, 1.08, 1.10$. All numerical values are in atomic units.

IX. FINITE-SIZE SCALING FOR QUANTUM DOTS

Quantum dots are a cluster (hundreds or few thousands) of atoms or molecules (Cds, CdSe, ... GaAs, InAs, ...) and are small enough that their electronic states are discrete. They present the opportunity to synthesize atomlike building blocks so that we can measure electronic properties, and they have generated much current experimental and theoretical interest [183–193]. Quantum dots of nearly identical sizes self-assemble into a planar array. (The dots become passivated against collapse by coating them with organic ligands.) For Ag nanodots (for example) the packing is hexagonal. The lower-lying electronic states of an isolated dot are discrete, being determined by the confining potential (and therefore the size) of the dot. Because of their larger size, it takes only a relatively low energy to add another electron to a dot, as revealed by scanning tunneling microscopy [195,196]. This energy is much lower than the corresponding energy for ordinary atoms and most molecules. It follows that when dots are close enough to be exchange coupled, which is the case in an array, the charging energy can be quite low. Here we review a computational method that allows the contributions of such ionic configurations even for extended arrays. The technical problem is that the Coulombic repulsion between two electrons (of opposite spins) that occupy the same dot cannot be described in a one-electron approximation. It requires allowing for correlation of electrons. Most methods that explicitly include correlation effects scale as some high power of the number of atoms (here, dots) and are computationally intractable. For example, a hexagonal array of only 19 dots, 3 dots per side, has already 2,891,056,160 low electronic configurations. Earlier exact computations including charging energy were limited to a hexagonal array of only 7 dots, 2 dots per side [197]. Yet current measurements of both static [198] and transport [199] properties use arrays of at least 100 dots per side. The simplest Hamiltonian that includes both the Coulombic (or charging energy) effects and the exchange coupling is the Hubbard model [200]. This model can be solved exactly for a one-dimensional chain, but for a two-dimensional array it is, so far, analytically intractable. In the absence of a closed solution, various methods have been developed [201]. Renormalization group (RG) methods are receiving increasing attention because of their nonperturbative nature, which allows application to the intermediate-to-strong coupling regime. In the following, we will show the applications of a real-space block renormalization group (BRG) method [202–204] on a two-dimensional triangular lattice with hexagonal blocks as shown in Fig. 34. Specifically, the size-dependence of the Mott metal–insulator transition (MIT) is studied.

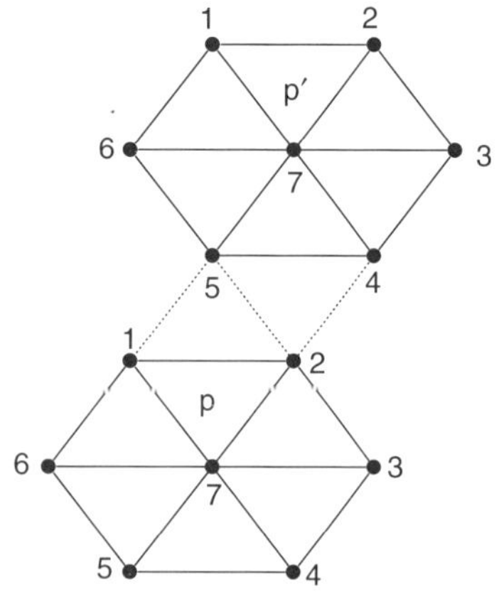

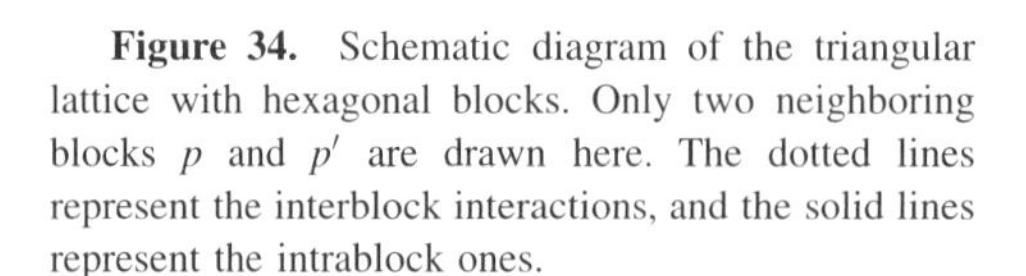

Figure 34. Schematic diagram of the triangular lattice with hexagonal blocks. Only two neighboring blocks p and p' are drawn here. The dotted lines represent the interblock interactions, and the solid lines represent the intrablock ones.

The used model is the Hubbard model [203], and the Hamiltonian is written as

$$\begin{aligned}\mathcal{H} = -t \sum_{\langle i,j \rangle,\sigma} [c_{i\sigma}^{+} c_{j\sigma} + \text{H.c.}] + U \sum_{i} n_{i\uparrow} n_{i\downarrow} \\ - \mu \sum_{i} (n_{i\uparrow} + n_{i\downarrow})\end{aligned} \tag{149}$$

where t is the nearest-neighbor hopping term, U is the local repulsive interaction, and μ is the chemical potential. $c_{i\sigma}^{+}(c_{i\sigma})$ creates (annihilates) an electron with spin σ in a Wannier orbital located at site i; the corresponding number operator is $n_{i\sigma} = c_{i\sigma}^{+} c_{i\sigma}$ and $\langle\ \rangle$ denotes the nearest-neighbor pairs. H.c. denotes the Hermitian conjugate.

Only the half-filled system is considered since the corresponding electron interactions are most prominent in this case, which leads to $\mu = U/2$. Hence, Eq. (149) can be rewritten as

$$\begin{aligned}\mathcal{H} = -t \sum_{\langle i,j \rangle,\sigma} [c_{i\sigma}^{+} c_{j\sigma} + \text{H.c.}] \\ + U \sum_{i} \left(\frac{1}{2} - n_{i\uparrow}\right)\left(\frac{1}{2} - n_{i\downarrow}\right) + K \sum_{i} I_i\end{aligned} \tag{150}$$

where $K = -U/4$ and I_i is the unit operator. As to the lattice structure, we use the nonpartite triangular lattice, because MIT emerges at finite $U = U_c$. It is well known that MIT on a square lattice can only take place at $U = 0$ due to the perfect nesting of the Fermi surface. If we don't study the exotic case with $U < 0$, which is possible in a strongly polarizable medium, the square lattice is not an optimal option for our purpose. The physical quantity we are concerned will be the charge gap, $\triangle_g$, which is defined as

$$\triangle_g = E(N_e - 1) + E(N_e + 1) - 2E(N_e) \tag{151}$$

where $E(N_e)$ denotes the lowest energy for an N_e–electron system. In our case, N_e is equal to the site number N_s of the lattice. This quantity is the discretized second derivative of the ground-state energy with respect to the number of particles—that is, the inverse compressibility.

The essence of the BRG method is to map the above many-particle Hamiltonian on a lattice to a new one with fewer degrees of freedom and with the same low-lying energy levels [205]. Then the mapping is repeated, leading to a final Hamiltonian for which an exact solution can be obtained. The procedure can be divided into three steps: First divide the N–site lattice into appropriate n_s–site blocks labeled by $p(p = 1, 2, \ldots, N/n_s)$ and separate the Hamiltonian $\mathscr{H}$ into intrablock part $\mathscr{H}_B$ and interblock $\mathscr{H}_{IB}$:

$$\mathscr{H} = \mathscr{H}_B + \mathscr{H}_{IB} = \sum_p H_p + \sum_{\langle p,p' \rangle} V_{p,p'} \tag{152}$$

where

$$\mathscr{H}_p = -t \sum_{\langle i^{(p)}, j^{(p)} \rangle} [c^+_{i^{(p)}\sigma} c_{j^{(p)}\sigma} + \text{H.c.}] + U \sum_{i^{(p)}} n_{i^{(p)}\uparrow} n_{i^{(p)}\downarrow} - \mu \sum_{i^{(p)}} (n_{i^{(p)}\uparrow} + n_{i^{(p)}\downarrow}) \tag{153}$$

and

$$V_{p,p'} = -t \sum_{\langle i^{(p,b)}, j^{(p',b)} \rangle} [c^+_{i^{(p,b)}\sigma} c_{j^{(p',b)}\sigma} + \text{H.c.}] \tag{154}$$

in which $i^{(p)}$ denotes the *ith* site on the *pth* block and $i^{(p,b)}$ denotes the border site of the block p.

The second step is to solve $\mathscr{H}_p$ exactly to get the eigenvalues E_{p_i} and eigenfunctions $\Phi_{p_i}(i = 1, 2, \ldots, 4^{n_s})$. Then we can build the eigenfunctions of $\mathscr{H}_B$ by direct multiplication of Φ_{p_i}, which can be written as $|\Psi_B(i_1, i_2, \ldots, i_{N/n_s})\rangle = |\Phi_{1i_1}\rangle|\Phi_{2i_2}\rangle \ldots |\Phi_{N/n_s} i_{N/n_s}\rangle (i_1, i_2, \ldots \in \{1, 2, \ldots, 4^{n_s}\})$.

The last step is to treat each block as one site on a new lattice and treat the correlations between blocks as hopping interactions. The original Hilbert space has four states per site. By following the above procedure, one obtains an equivalent Hamiltonian with $(4^{n_s})^{N/n_s} = 4^N$, degrees of freedom, which is the same as the original Hamiltonian. But in the realistic case, if we only care about the properties related to some special energy levels of the system, it is not necessary to keep all the states for a block to obtain the new Hamiltonian. For example, when studying the metal–insulator–transition [206], we may only need to consider the ground-state and the first excited-state energies.

The above scheme is a general procedure for applying the BRG method. In order to make the new Hamiltonian more tractable, it is desirable to make it have the same structure as the original one; that is, the reduction in size should not be accompanied by a proliferation of new couplings. Then we can use the iteration procedures to solve the model. To achieve this goal, it is necessary to keep only four states in step 2, which can be understood from the following renormalized intrasite Hamiltonian. The four selected states are taken to be

$$|\Phi_{p1}\rangle \equiv |0\rangle'_p \tag{155}$$
$$|\Phi_{p2}\rangle \equiv c'^{+}_{p\uparrow\downarrow}|0\rangle'_p = |\uparrow\downarrow\rangle'_p \tag{156}$$
$$|\Phi_{p3}\rangle \equiv c'^{+}_{p\uparrow}|0\rangle'_p = |\uparrow\rangle'_p \tag{157}$$
$$|\Phi_{p4}\rangle \equiv c'^{+}_{p\downarrow}|0\rangle'_p = |\downarrow\rangle'_p \tag{158}$$

where $c'^{+}_{p\sigma}(c'_{p\sigma})$ is the creation (annihilation) operator of the block state $|\sigma\rangle'_p$ and their corresponding energies are E_i $(i = 1, 2, 3, 4)$.

Our next task is to rewrite the old Hamiltonian $\mathscr{H} = \mathscr{H}_B + \mathscr{H}_{IB}$ in the space spanned by the truncated basis

$$\mathscr{H}' = \sum_{\Psi_B^{\text{Truncated}}\bar{\Psi}_B^{\text{Truncated}}} |\Psi_B^{\text{Truncated}}\rangle\langle\Psi_B^{\text{Truncated}}|H|\bar{\Psi}_B^{\text{Truncated}}\rangle\langle\bar{\Psi}_B^{\text{Truncated}}| \tag{159}$$

where the truncated basis is given by

$$|\Psi_B^{\text{Truncated}}(i_1, i_2, \ldots, i_{N/n_s})\rangle = |\Phi_{1i_1}\rangle|\Phi_{2i_2}\rangle \ldots |\Phi_{N/n_s}i_{N/n_s}\rangle \\ \times (i_1, i_2, \ldots \in \{1, 2, 3, 4\}) \tag{160}$$

In order to avoid proliferation of additional couplings in $\mathscr{H}'$, the four states kept from the block cannot be arbitrarily chosen. Some definite conditions must be satisfied in order to make $\mathscr{H}'$ have the same structure as $\mathscr{H}$. Substituting $\mathscr{H}$ into $\mathscr{H}'$ and using the product of different operators (see Table III), we can get the expression for $\mathscr{H}_p$:

$$\begin{aligned}\mathscr{H}_p &= |0\rangle'_p E_1 \langle 0|'_p + |\uparrow\downarrow\rangle'_p E_2 \langle\downarrow\uparrow|'_p + |\downarrow\rangle'_p E_4 \langle\downarrow|'_p + |\uparrow\rangle'_p E_3 \langle\uparrow|'_p \\ &= E_1 + (E_3 - E_1)n'_{p,\uparrow} + (E_4 - E_1)n'_{p,\downarrow} + (E_1 + E_2 - E_3 - E_{4)}n'_{p,\uparrow}n'_{p,\downarrow}\end{aligned} \tag{161}$$

TABLE III
The Internal Product of Different Operator Transformations[a]

	$\langle 0\|'$	$\langle\uparrow\|'$	$\langle\downarrow\|'$	$\langle\uparrow\downarrow\|'$
$\|0\rangle'$	$1 - n'_\uparrow - n'_\downarrow + n'_\uparrow n'_\downarrow$	$c'_\uparrow - n'_\downarrow c'_\uparrow$	$c'_\downarrow - n'_\uparrow c'_\downarrow$	$c'_\downarrow c'_\uparrow$
$\|\uparrow\rangle'$	$c'^{+}_\uparrow - c'^{+}_\uparrow n'_\downarrow$	$n'_\uparrow - n'_\uparrow n'_\downarrow$	$c'^{+}_\uparrow c'_\downarrow$	$-n'_\uparrow c'_\downarrow$
$\|\downarrow\rangle'$	$c'^{+}_\downarrow - c'^{+}_\downarrow n'_\uparrow$	$c'^{+}_\downarrow c'_\uparrow$	$n'_\downarrow - n'_\downarrow n'_\uparrow$	$n'_\downarrow c'_\uparrow$
$\|\uparrow\downarrow\rangle'$	$c'^{+}_\uparrow c'^{+}_\downarrow$	$-n'_\uparrow c'^{+}_\downarrow$	$c'^{+}_\uparrow n'_\downarrow$	$n'_\uparrow n'_\downarrow$

[a] In this table the product reads $|0\rangle'\langle 0|' = 1 - n'_\uparrow - n'_\downarrow + n'_\uparrow n'_\downarrow$, etc.

Note that by keeping only four states from the block states in the beginning gives no other extra couplings in the new Hamiltonian.

Comparing the above intrasite Hamiltonian with Eq. (149), we get the next conditions in order to copy the intrasite structure of the old Hamiltonian; that is, $E_3 = E_4$. Because of the additional vacuum energy E_1 in the new Hamiltonian, we rewrite the intrasite part of Eq. (149) as

$$\mathscr{H}_B = U \sum_i n_{i\uparrow} n_{i\downarrow} - \mu \sum_i (n_{i\uparrow} + n_{i\downarrow}) + K \sum_i I_i \tag{162}$$

where we introduce another parameter K to the original system and I_i is a unit operator. The new intrasite Hamiltonian is given by

$$\mathscr{H}'_B = (E_1 + E_2 - 2E_3) \sum_p n'_{p\uparrow} n'_{p\downarrow} - (E_1 - E_3) \sum_p (n'_{p\uparrow} + n'_{p\downarrow}) + E_1 \sum_p I_p \tag{163}$$

Then the renormalized parameters U, μ, and K can be obtained from the following relations:

$$U' = E_1 + E_2 - 2E_3 \tag{164}$$

$$\mu' = E_1 - E_3 \tag{165}$$

$$K' = E_1 \tag{166}$$

in which E_1, E_2, and E_3 are functions of the old parameters t, U, μ, K.

For the half-filled case, μ is fixed to be $U/2$. Moreover, by using the particle–hole symmetry, $E_1 = E_2$, the renormalization group equations for U and K take the form

$$U' = 2(E_1 - E_3) \tag{167}$$

$$K' = (E_1 + E_3)/2 \tag{168}$$

To illustrate this procedure, let us consider the triangular lattice with hexagonal blocks as shown in Fig. 34. For this nonbipartite lattice the interaction between blocks can be written as

$$\begin{aligned} V_{pp'} = (-t) \sum_{\sigma, i_1, i'_1 i_2, i'_2} \{ & [|\Phi_{pi_1}\rangle \langle \Phi_{pi_1}| c^{+}_{1(p)\sigma} |\Phi_{pi'_1}\rangle \langle \Phi_{pi'_1}|] \\ & \times [|\Phi_{p'i_2}\rangle \langle \Phi_{p'i_2}| c_{5(p')\sigma} |\Phi_{p'i'_2}\rangle \langle \Phi_{p'i'_2}|] \\ & + [|\Phi_{pi_1}\rangle \langle \Phi_{pi_1}| c^{+}_{2(p)\sigma} |\Phi_{pi'_1}\rangle \langle \Phi_{pi'_1}|] \\ & \times [|\Phi_{p'i_2}\rangle \langle \Phi_{p'i_2}| c_{4(p')\sigma} |\Phi_{p'i'_2}\rangle \langle \Phi_{p'i'_2}|] \\ & + [|\Phi_{pi_1}\rangle \langle \Phi_{pi_1}| c^{+}_{2(p)\sigma} |\Phi_{pi'_1}\rangle \langle \Phi_{pi'_1}|] \\ & \times [|\Phi_{p'i_2}\rangle \langle \Phi_{p'i_2}| c_{5(p')\sigma} |\Phi_{p'i'_2}\rangle \langle \Phi_{p'i'_2}|] + \text{H.c.} \} \end{aligned} \tag{169}$$

Since we would like to keep $V_{pp'}$ of the form

$$V_{pp'} = (-t')\sum_{\sigma}[c'^{+}_{p\sigma}c'_{p'\sigma} + \text{H.c.}] \tag{170}$$

we use the product transformation in Table III to simplify Eq. (169):

$$\begin{aligned} V_{pp'} = \sum_{\sigma,\langle i,j\rangle} &\{\langle\sigma|'_p c^{+}_{i(p)\sigma}|0\rangle'_p + [\langle-\sigma,\sigma|'_p c^{+}_{i(p)\sigma}| - \sigma\rangle'_p - \langle\sigma|'_p c^{+}_{i(p)\sigma}|0\rangle'_p]n'_{p-\sigma}\}c'^{+}_{p\sigma} \\ &\times \{\langle 0|'_{p'} c_{j(p')\sigma}|\sigma\rangle'_{p'} + [\langle-\sigma|'_{p'} c_{j(p')\sigma}|\sigma, -\sigma\downarrow\uparrow\rangle'_{p'} - \langle\sigma|'_{p'} c_{j(p')\sigma}|0\rangle'_{p'}]n'_{p'-\sigma}\}c'_{p'\sigma} + \text{H.c.} \\ &(\langle ij\rangle = \langle 1,5\rangle, \langle 2,4\rangle, \langle 2,5\rangle) \end{aligned} \tag{171}$$

It can be easily seen now that in order to make all the extra couplings vanish, it is necessary to make further restrictions upon the selected states,

$$\langle-\sigma,\sigma|'_p c^{+}_{i(p)\sigma}|_p - \sigma\rangle'_p = \langle\sigma|'_p c^{+}_{i(p)\sigma}|0\rangle'_p \tag{172}$$

$$\langle-\sigma|'_{p'} c_{j(p')\sigma}|\sigma, -\sigma\rangle'_{p'} = \langle 0|'_{p'} c_{j(p')\sigma}|\sigma\rangle'_{p'} \tag{173}$$

Using calculations similar to those of the other neighboring interactions of the block, we can finally obtain the following conditions:

$$\langle-\sigma,\sigma|'_p c^{+}_{i(p)\sigma}|_p - \sigma\rangle'_p = \langle\sigma|'_p c^{+}_{i(p)\sigma}|0\rangle'_p = \lambda \tag{174}$$

for all the border sites on the block. Then the new hopping term becomes

$$V_{pp'} = \nu\lambda^2\sum_{\sigma} c'^{+}_{p\sigma}c'_{p'\sigma} \tag{175}$$

where ν represents the number of couplings between neighboring blocks. In Fig. 34, $\nu = 3$. The last renormalization group equation is readily obtained:

$$t' = \nu\lambda^2 t \tag{176}$$

Up to now, we have given a general discussion of the conditions under which no proliferation of couplings results from the application of the BRG method to nonpartite lattice. Because on the border of a nonpartite lattice block there is only one type of site, the above procedures can be extended to other lattices with different dimensions or blocks without much difficulty.

After deriving the conditions for the renormalization group equations, the next task is to select states that satisfy these conditions. At this stage the symmetry properties of the lattice play an important role. From Eqs. (170) and (171), it can be easily seen that if we assume the particle number in the state $|0\rangle'$ to be $N_e - 1$, then in $|\uparrow\rangle', |\downarrow\rangle'$ and $|\uparrow\downarrow\rangle'$ there should be $N_e, N_e, N_e + 1$ particles, respectively. Moreover, if the spin in $|0\rangle'$ is S_z, the spins for $|\uparrow\rangle', |\downarrow\rangle'$, and $|\uparrow\downarrow\rangle'$ should be $S_z + 1/2, S_z - 1/2$, and S_z. The total electron number N_e and the spin S_z for each block are good quantum numbers since their corresponding operators commute with the Hubbard Hamiltonian. So when we diagonalize the Hubbard Hamiltonian of the selected block, we keep N_e and S_z fixed to be $(N_e - 1,\ S_z)$, $(N_e,\ S_z + 1/2)$, $(N_e,\ S_z - 1/2)$, and $(N_e + 1,\ S_z)$, respectively. Thus we obtain four groups of eigenenergies and eigenstates corresponding to the above quantum numbers. From each group, we select the lowest-energy state to form the final required four states. It should be mentioned that the lowest-energy state has to be selected according to definite special symmetry considerations, which shall be discussed in the next paragraph. In order to obtain the insulating to conducting gap, which is defined to be the energy difference between extracting one electron from the system and adding one electron to it, N_e is selected to be equal to N_s. For S_z, we choose it to be zero so as to make the block have the same spin property as the one-site. So now the renormalized lattice will be composed of N/n_s renormalized "sites" with N/n_s "particles."

Instead of forcing the above conserved quantities upon the selected states in analogy to the one site properties in a consistent way, here we get them directly from the no-coupling-proliferation conditions. λ does not depend on σ in Eq. (174); this can be guaranteed by the particle–hole symmetry, which means that only in half-filled lattice can the renormalized Hamiltonian have exactly the same form as the original one [207]. Moreover, the irrelevance of λ with respect to the border site $i^{(p,b)}$ can be shown by requiring the selected states to belong to the same irreducible representation of the spatial group of the lattice. For the triangular lattice with hexagonal blocks, the Hamiltonian is invariant under C_{6v} [208]. So if we choose the same one-dimensional irreducible representation of the group C_{6v} for $|0\rangle'$, $|\uparrow\rangle'$, $|\downarrow\rangle'$, and $|\uparrow\downarrow\rangle'$, the conditions can be satisfied.

Eqs. (167), (168), and Eq. (176) are the so-called RG flow equations. Usually, they are iterated until we get the fixed point. The charge gap for an infinite lattice can then be written as

$$\Delta_g = \lim_{n\to\infty} U^{(n)} \tag{177}$$

Because of the implicit functional in RG flow equations, it is very difficult to obtain any other useful information except the critical transition point U_c from the above procedures. In the following, we will handle these procedures in another way, namely, instead of letting RG flow to infinity for a fixed initial parameters (U, k, t), we can stop the RG flow at some stage. Thus the energy gap obtained from Eq. (151) will correspond to a system of fixed size. For example, if we stop the RG flow at the first iteration, then obtained t' and U' will be for hexagonal block mapped from a system of 7^2 sites. Because we can solve a hexagonal block Hamiltonian exactly, the energy gap for a system of 49 sites can be obtained easily.

In this way, we can study the variations of Δ_g against the system size of $7^2, 7^3, 7^4, 7^5, 7^6 \ldots$. We call this procedure a multistage real-space RG method, which is well-adapted to start the finite-size scaling analysis.

In Fig. 35a, the size dependence of relationship between $\triangle_g$ and U in the unit of t is presented. But from this figure, it is somewhat difficult to decide the transition point for $\triangle_g$ to reach zero as U is decreased from big values. To explicitly display the critical phenomenon, we construct Fig. 35b by scaling $\triangle_g$ with respect to N at first. Here $N = N_e = N_s$. Now it is very easy to get the critical value of $(U/t)_c$ at the crossing point of all the curves corresponding to different system size, which is found to be 12.5. The same value is obtained by letting RG equations flow to infinity, which should be expected. But this kind of scaling gives us more. In Fig. 35c, all the data collapse to one curve once we carry out a second step of scaling with $U/t - (U/t)_c$ by N. It is an obvious evidence for the occurrence of a quantum phase transition with U to be the

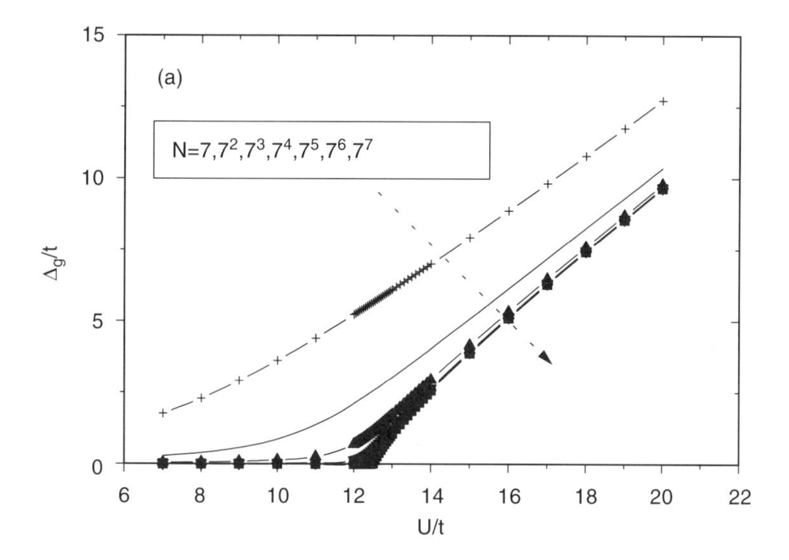

Figure 35. Variations of the charge gap Δ_g against the on-site electron interaction U for different system sizes, that is, the number of sites: $7(+)$, $7^2(\times)$, 7^3(▲), 7^4(▼), 7^5(◆), 7^6(●), 7^7(■). More points are calculated around the transition point. In (a), no scaling is utilized. In (b), the charge gap is scaled by $1/N^{0.405}$ to display clearly the phase transition. In (c), all the data are collapsed onto one curve by scaling both axis with respect to N.

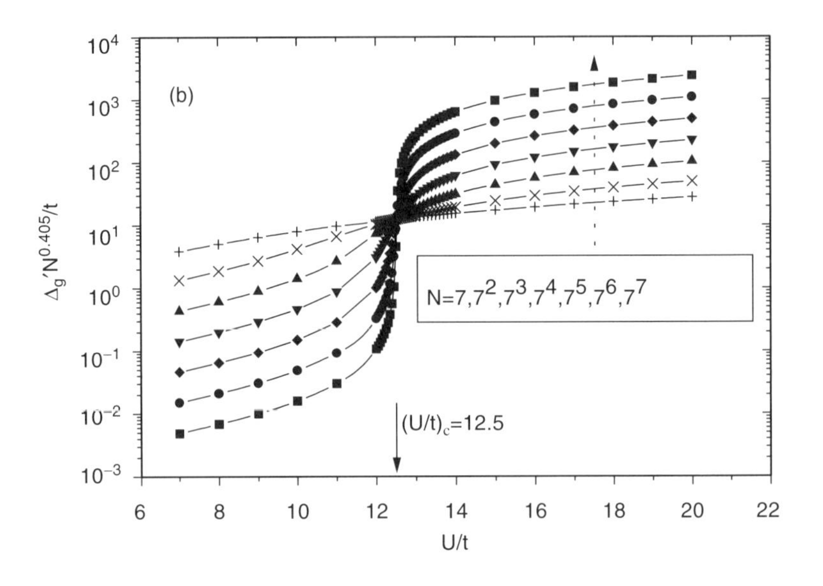

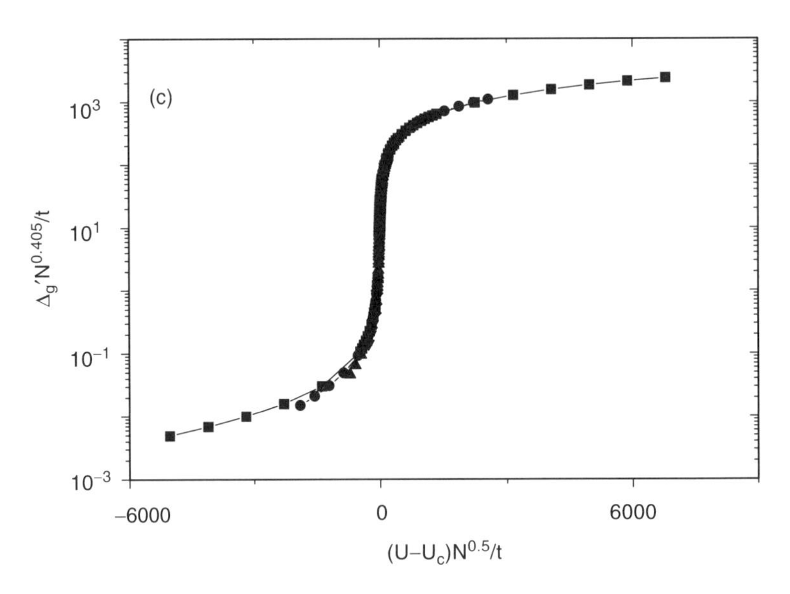

Figure 35 (*Continued*)

tuning parameter. Hence we can write down the following equation:

$$\triangle_g N^{0.405} = f\left[qN^{0.5}\right] \tag{178}$$

where $f(x)$ is a universal functions independent of the size and $q = U/t - (U/t)_c$ measures the distance of the electron correlations from its critical value for MIT. By using $N = L^2$ for 2D systems, the above equation can be rewritten as

$$\triangle_g = L^{-0.91} f[qL] \tag{179}$$

from which we can get two scaling relationship for the charge gap. One is the finite-size scaling at the transition point–that is, when $q = 0$, $\wedge_g \sim 1/L^{0.91}$. As shown in Ref. 209, in Anderson MIT, when the electron correlation energy dominate the Fermi energy, the average inverse compressibility ($= \Delta_g$) exhibits a scaling as $1/L$ with respect to the system size. Here it shows a slower decay as L increases. The other one is the bulk scaling around the transition point for infinite systems, $\triangle_g \sim q^{0.91}$. According to the scaling analysis of the Gutzwiller solution for the Mott MIT in the Hubbard model [210], $\triangle_g \sim q^{0.5}$. Since the Gutzwiller solution is a mean field approximation and the upper critical dimension for it to give a correct description of this critical phenomenon is $d_c = 3$, it is understandable that our above 2D results cannot be merged into the mean-field theory.

By introducing a critical exponent $y_\triangle$ for $\triangle_g$, from the one-parameter scaling theory, we can have

$$\triangle_g = q^{y_\triangle} f\left(\frac{L}{\xi}\right) \tag{180}$$

in which $\xi = q^{-\upsilon}$ is the correlation length with υ to be the corresponding critical exponent and L denotes the system size. By using $N = L^2$ for 2D systems, the above equation can be rewritten as

$$\triangle_g = N^{-(y_\triangle/2\upsilon)} f\left(qN^{1/2\upsilon}\right) \tag{181}$$

Comparing Eq. (180) and Eq. (181), we can easily get

$$y_\triangle = 0.91, \qquad \upsilon = 1 \tag{182}$$

By relating $y_\triangle$ to the dynamic exponent z with $y_\triangle = z\nu$, we can have $z \approx 0.91$. It should be meaningful to compare the obtained results here with those from other kinds of MIT in 2D systems.

For filling-control or density-driven MIT, there are two types of universality classes. One is characterized by $z = 1/\nu = 2$, which is the case for all 1D systems as well as for several transitions at higher dimensions, such as transitions between insulator with diagonal order of components and metal with diagonal order components and small Fermi volume. Another one is characterized by $z = 1/\nu = 4$. Numerical calculations have shown that the Hubbard model on a square lattice is an example of this class. The large dynamic exponent is associated with suppression of coherence, which is associated with strong incoherent scattering of charge by a large degeneracy of component excitations.

For Anderson MIT, most of the analytical, numerical, and experimental research have produced $\upsilon > 1$, such as $\upsilon = 1.35$ [211], 1.54 [212], and 1.62 [213]. Our work leads to $z = 0.91$ and $\nu = 1$, which might imply that Mott MIT does belong to any universality class mentioned above. But because of the approximations involved in real-space RG method and the sensitivity of the critical exponents upon the used approaches, more work, especially analytical work, is much needed to reveal the implications and the underlying mechanism for $z = 0.91$ and $\upsilon = 1$. Since we lack in the results from other approaches for comparisons, z and ν are far from decided. Our work tends to show $z \approx \nu \approx 1$. The research aiming to check this guess is still in progress.

In summary, by using a multistage real-space renormalization group method, we show that the finite-size scaling can be applied in Mott MIT. And the dynamic and correlation length critical exponents are found to be $z = 0.91$ and $\upsilon = 1$, respectively. At the transition point, the charge gap scales with size as $\triangle_g \sim 1/L^{0.91}$.

X. CONCLUDING REMARKS

In this review chapter, we show how the finite-size scaling ansatz can be combined with the variational method to extract information about critical behavior of quantum Hamiltonians. This approach is based on taking the number of elements in a complete basis set as the size of the system. As in statistical mechanics, the finite-size scaling can then be applied directly to the Schrödinger equation. This approach is general and gives very accurate results for the critical parameters, for which the bound-state energy becomes absorbed or degenerate with a continuum. To illustrate the applications in quantum calculations, we present detailed calculations for both short- and long-range potentials, atomic and simple molecular systems, resonances, and quantum dots.

The field of quantum critical phenomena in atomic and molecular physics is still in its infancy; and there are many open questions about the interpretations of the results, including whether or not these quantum phase transitions really do exist. The possibility of exploring these phenomena experimentally in the field of quantum dots offers an exciting challenge for future research. This finite-size scaling approach is general and might provide a powerful way in determining critical parameters for the stability of atomic and molecular systems in external fields, for selectively breaking chemical bonds, and for design and control electronic properties of materials using artificial atoms.

Acknowledgments

We would like to thank our collaborators Qicun Shi for his major contributions in applying FSS to molecular systems, Juan Pablo Neirotti for his contributions to the studies of few-electron atoms, Alexie Sergeev for his work on generalizing the calculations to N-electron atoms, Ricardo Sauerwein for developing the path integral for quantum criticality, and Jiaxiang Wang for his work on quantum dots.

We would like also to acknowledge the financial support of the Office of Naval Research (N00014-97-1-0192) and The National Science Foundation (NSF-501-1393-0650). One of us (PS.) thanks CONICET and SECYTUNC for partial financial support.

References

1. B. Benderson, *More Things in Heaven and Earth, A Celebration of Physics at the Millennium*, Springer, APS, New York, 1999, pp. 501–665.
2. J. J. Binney, N. J. Dowrick, A. J. Fisher, and M. E. J. Newman, *The Theory of Critical Phenomena: An Introduction to the Renormalization Group*, Oxford University Press, New York, 1992.
3. S. L. Sondhi, S. M. Girvin, J. P. Carini, and D. Shahar, *Rev. Mod. Phys.* **69**, 315 (1997).
4. S. Sachdev, *Quantum Phase Transitions*, Cambridge University Press, New York, 1999.
5. D. Voss, *Science* **282**, 221 (1998).
6. H. L. Lee, J. P. Carini, D. V. Baxter, W. Henderson, and G. Gruner, *Science* **287**, 633 (2000).
7. P. Serra and S. Kais, *Phys. Rev. Lett.* **77**, 466 (1996).
8. J. P. Neirotti, P. Serra, and S. Kais, *Phys. Rev. Lett.* **79**, 3142 (1997).
9. P. Serra, J. P. Neirotti, and S. Kais, *Phys. Rev. Lett.* **80**, 5293 (1998).
10. S. Kais and P. Serra, *Int. Rev. Phys. Chem.* **19**, 97 (2000).
11. Q. Shi and S. Kais, *Mol. Phys.* **98**, 1485 (2000).
12. M. K. Scheller, R. N. Compton, and L. S. Cederbaum, *Science* **270**, 1160 (1995).
13. X. Wang and L. Wang, *Phys. Rev. Lett.* **83**, 3402 (1999).
14. R. Rost, J. Nygard, A. Pasinski, and A. Delon, Phys. Rev. Lett. **78**, 3093 (1997).
15. Y. P. Kravchenko and M. A. Lieberman, *Phys. Rev. A* **56**, R2510 (1997).
16. F. Remacle and R. D. Levine, *Proc. Natl. Acad. Sci. USA* **97**, 553 (2000).
17. H. R. Sadeghpour, J. L. Bohn, M. J. Cavagnero, B. D. Esry, I. I. Fabrikant, J. H. Macek, and A. R. P. Rau, *J. Phys. B* **33**, R93 (2000).
18. H. E. Camblong, L. N. Epele, H. Fanchiotti, and C. A. García Canal, *Phys. Rev. Lett.* **87**, 220402 (2001).
19. J. Zs. Mezei, J. Mitroy, R. G. Lovas, and K. Varga, *Phys. Rev. A* **64**, 032501 (2001).
20. R. S. Berry, *Nature (London)* **393**, 212 (1998).
21. P. Nigra, M. Carignano, and S. Kais, *J. Chem. Phys.* **115**, 2621 (2001).

22. M. E. Fisher, in *Critical Phenomena*, Proceedings of the 51st Enrico Fermi Summer School, Varenna, Italy, M. S. Green, ed., Academic Press, New York, 1971.
23. M. N. Barber, Finite-Size Scaling, in *Phase Transitions and Critical Phenomena*, C. Domb and J. L. Lebowitz, eds., Academic Press, New York, 1983.
24. V. Privman, ed., *Finite Size Scaling and Numerical Simulations of Statistical Systems*, World Scientific, Singapore, 1990.
25. J. L. Cardy, *Finite-Size Scaling*, Elsevier Science Publishers, New York, 1988.
26. P. Serra, Preprint, 2002.
27. F. H. Stillinger and D. K. Stillinger, *Phys. Rev. A* **10**, 1109 (1974).
28. J. Katriel and E. Domany, *Int. J. Quantum Chem.* **8**, 559 (1974).
29. D. R. Hershbach, J. Avery, and O. Goscinsky, *Dimensional Scaling in Chemical Physics*, Kluwer, Dordrecht, 1993.
30. P. Serra and S. Kais, *Phys. Rev. A* **55**, 238 (1997).
31. P. Serra and S. Kais, *Chem. Phys. Lett.* **260**, 302 (1996).
32. P. Serra and S. Kais, *J. Phys. A* **30**, 1483 (1997).
33. For reviews see A. Chartterjee, *Phys. Rep.* **186**, 249 (1990).
34. E. Witten, *Phys. Today* **33** (7), 38 (1980).
35. C. A. Tsipis, V. S. Popov, D. R. Hershbach, and J. S. Avery, *New Methods in Quantum Theory*, Kluwer, Dordrecht, 1996.
36. S. Kais, S. M. Sung, and D. R. Hershbach, *Int. J. Quantum Chem.*, **49**, 657–674 (1994).
37. D. D. Frantz and D. R. Hershbach, *Chem. Phys.* **126**, 59 (1988); *J. Chem. Phys.* **92**, 6688 (1990).
38. J. G. Loeser, *J. Chem. Phys.* **86**, 5635 (1987).
39. M. Cabrera, A. L. Tan, and J. G. Loeser, *J. Phys. Chem.* **97**, 2467 (1993).
40. D. Z. Goodson and D. R. Hershbach, *J. Chem. Phys.* **86**, 4997 (1987); D. J. Doren and D. R. Hershbach, *J. Phys. Chem.* **92**, 1816 (1988).
41. H. E. Stanley, *Introduction to Phase Transitions and Critical Phenomena*, Oxford University Press, New York, 1971.
42. Q. Shi, S. Kais, F. Remacle, and R. D. Levine, *CHEMPHYSCHEM* **2**, 434–442 (2001).
43. Q. Shi and S. Kais, *Int. J. Quantum Chem.* **85**, 307 (2001).
44. J. Avery, *Hyperspherical Harmonics; Applications in Quantum Theory*, Kluwer Academic Publishers, Dordrecht, Netherlands, 1989.
45. B. Simon, *J. Funct. Analysis* **1977**, 25, 338.
46. N. G. Bruijn, *Asymptotic Methods in Analysis*, Dover, New York, 1981.
47. M. Klaus and B. Simon, *Ann. Phys. (NY)* **130**, 251 (1980).
48. M. Lassaut, I. Bulboaca, and R. J. Lombard, *J. Phys. A* **29**, 2175 (1996).
49. See, for example, A. Galindo and P. Pascual, *Quantum Mechanics*, Vol. I, Springer-Verlag, Berlin (1990).
50. S. Flügge, *Practical Quantum Mechanics*, Springer-Verlag, Berlin (1999).
51. F. Cooper, A. Khare, and U. Sukhatme, *Phys. Rep.* **251**, 267 (1995).
52. S. Kais, *Exact Models*, in *The Encyclopedia of Computational Chemistry*, P. v. R. Schleyer, N. L. Allinger, T. Clark, J. Gasteiger, P. A. Kollman, H. F Schaefer III, and P. R. Schreiner, eds., John Wiley & Sons, Chichester, 1998, pp. 959–963.
53. V. M. Eleonsky and V. G. Korolev, *Phys. Rev. A* **55**, 2580 (1997).
54. P. Serra, J. P. Neirotti, and S. Kais, *J. Phys. Chem. A* **102**, 9518 (1998).

55. M. Abramowitz and I. A. Stegun, eds., *Handbook of Mathematical Functions*, Dover, New York, 1972.
56. P. M. Morse and H. Feschbach, *Methods of Theoretical Physics*, Vol. II, McGraw-Hill, New York, 1953.
57. F. H. Stillinger, *J. Math. Phys.* **20**, 1891 (1979).
58. W. M. Frank and D. J. Land, *Rev. Mod. Phys.* **43**, 36 (1971).
59. M. J. Mortiz, C. Eltschka, and H. Friedrich, *Phys. Rev. A* **63**, 042102 (2001); *ibid.* **64**, 022101 (2001).
60. I. S. Gradshteyn and I. M. Ryzhik, *Table of Integrals, Series and Products*, 5th ed., Academic Press, San Diego, 1994.
61. M. K. Scheller, N. Compton, and L. S. Cederbaum, *Science* **270**, 1160 (1995).
62. A. V. Sergeev and S. Kais, *Int. J. Quantum Chem.* **75**, 533 (1999).
63. E. H. Lieb, *Phys. Rev. Lett.* **52**, 315 (1982).
64. M. Hoffmann-Ostenhof, T. Hoffmann-Ostenhof, and B. Simon, *J. Phys. A* **16**, 1125 (1983).
65. J.-M. Richard, *Few-Body Systems*, preprint: physics/0111111, in press.
66. S. Kais and Q. Shi, *Phys. Rev. A* **62**, 060502(R) (2000).
67. Walter Thirring, *A Course in Mathematical Physics*, Vol. III, Springer-Verlag, New York, 1986.
68. C. N. yang and T. D. Lee, *Phys. Rev.* **87**, 404 (1952).
69. T. D. Lee and C. N. Yang, *Phys. Rev.* **87**, 410 (1952).
70. M. P. Nightingale, *Physica* **83A**, 561 (1976).
71. See, for example, C. J. Thompson, *Classical Statistical Mechanics*, Clarendon Press, London, 1988.
72. F. R. Gantmacher, *The Theory of Matrices*, Chelsea, New York, 1959.
73. M. P. Nightingale, *Proceedings of the Koninklijke Nederlandse Akademie van Wetenschappen*, series B, Vol. 82, 1979.
74. P. D. Beale, P. M. Duxbury, and J. Yeomans, *Phys. Rev. B* **31**, 7166 (1985).
75. B. Derrida, *J. Phys. A* **14**, L5 (1981); B. Derrida and H. Saleur, *J. Phys. A* **18**, L1075 (1985).
76. P. Serra and J. F. Stilck, *Europhys Lett.* **17**, 423 (1992); *Phys. Rev. E* **49**, 1336 (1994).
77. B. Derrida and L. De Seze, *J. Phys.* **43**, 475 (1982).
78. P. J. Reynolds, H. E. Stanly, and W. Klein, *Phys. Rev. B* **21**, 1223 (1980).
79. H. Nakanishi and P. J. Reynolds, *Phys. Lett. A* **71**, 252 (1979).
80. H. Nakanishi and H. E. Stanley, *Phys. Rev. B* **22**, 2466 (1980).
81. S. B. Lee and H. Nakanishi, *J. Phys. A* **33**, 2943 (2000).
82. Z. Glumac and K. Uzelac, *J. Phys. A* **22**, 4439 (1989).
83. L. P. Kadanoff and G. Baym, *Quantum Statistical Mechanics*, Addison-Wesley, Reading, MA, 1989.
84. L. P. Kadanoff, *Statistical Physics: Statics, Dynamics and Renormalization*, World Scientific, Singapore, 2000.
85. B. Hatfield, *Quantum Field Theory of Point Particles and Strings*, Addison-Wesley, Reading, MA, 1992.
86. See, for example, in E. Merzbacher, *Quantum Mechanics*, 2nd ed., John Wiley & Sons, New York, 1998.
87. J. P. Neirotti, P. Serra, and S. Kais, *J. Chem. Phys.* **108**, 2765 (1998).

88. P. Serra, J. P. Neirotti, and S. Kais, *Phys. Rev. A* **57**, R1481 (1998).
89. R. Bulirsch and J. Stoer, *Numer. Math.* **6**, 413 (1964).
90. M. Henkel in Ref. 24, Chapter VIII.
91. V. M. Vainberg, V. L. Eletskii, and V. S. Popov, *Sov. Phys. JETP* **1981**, 54, 833.
92. F. J. Rogers, H. C. Graboske, and D. J. Harwood, *Phys. Rev. A* **1**, 1577 (1970).
93. P. Serra, and S. Kais, *Chem. Phys. Lett.* **319**, 273 (2000).
94. See, for example, N. Goldenfeld, *Lectures on Phase Transitions and the Renormalization Group*, Addison-Wesley, Reading, MA, 1992.
95. B. Nienhuis and M. Nauenberg, *Phys. Rev. Lett.* **35**, 477 (1975).
96. E. A. Hylleraas, *Z. Phys.* **65**, 209 (1930).
97. T. Kato, in *Perturbation Theory for Linear Operators*, 2nd ed., Springer, New York, 1976.
98. C. W. Scherr and R. E. Knight, *Rev. Mod. Phys.* **35**, 436 (1963).
99. F. M. Stillinger Jr., *J. Chem. Phys.* **45**, 3623 (1966).
100. J. Midtdal, *Phys. Rev. A* **138**, 1010 (1965).
101. W. P. Reinhardt, *Phys. Rev. A* **15**, 802 (1977).
102. J. D. Baker, D. E. Freund, R. N. Hill, and J. D. Morgan III, *Phys. Rev. A* **41**, 1247 (1990).
103. I. A. Ivanov, *Phys. Rev. A* **51**, 1080 (1995).
104. G. W. F. Drake and Zong-Chao Yan, *Chem. Phys. Lett.* **229**, 489 (1994).
105. E. A. Hylleraas, *Z. Physik* **48**, 469 (1928); *ibid.* **54**, 347 (1929).
106. A. K. Bhatia and A. Temkin, *Rev. Mod. Phys.* **36**, 1050 (1964).
107. H. Grosse and L. Pittner, *J. Math. Phys.* **24**, 1142 (1983).
108. R. N. Hill, *Phys. Rev. Lett.* **38**, 643 (1977).
109. T. Kato, *Trans. Am. Math. Soc.* **70**, 212 (1951).
110. G. W. F. Drake, *Phys. Rev. Lett.* **24**, 126 (1970).
111. E. Brändas and O. Goscinski, *Int. J. Quantum Chem.* **4**, 571 (1970).
112. E. Brändas and O. Goscinski, *Int. J. Quantum Chem. Symp.* **6**, 59 (1972).
113. G. A. Arteca, F. M. Fernández, and E. A. Castro, *J. Chem. Phys.* **84**, 1624 (1986).
114. Z. Yan and G. W. F. Drake, *Phys. Rev. A* **52**, 3711 (1995).
115. E. H. Lieb, *Phys. Rev. A* **29**, 3018 (1984).
116. L. A. Cole and J. P. Perdew, *Phys. Rev. A* **25**, 1265 (1982).
117. S. J. Chakravorty and E. R. Davidson, *J. Phys. Chem.* **100**, 6167 (1996).
118. S. Kais, J. P. Neirotti, and P. Serra, *Int. J. Mass Spectrom.* **182/183**, 23–29 (1999).
119. H. Hogreve, *J. Phys. B: At. Mol. Opt. Phys.* **31**, L439 (1998).
120. D. R. Bates, *Adv. At. Mol. Opt. Phys.* **27**, 1 (1991).
121. M.-J. Nadeau, M. A. Garwan, X.-L. Zhao, and A. E. Litherland, *Nucl. Instrum. Methods Phys. Res. B* **123**, 521 (1997).
122. H. Hotop and W. C. Lineberger, *J. Phys. Chem. Ref. Data* **14**, 731 (1985).
123. D. Berkovits and E. Boaretto et al., *Phys. Rev. Lett.* **75**, 414 (1995).
124. W. Hunziker, *Helv. Phys. Acta* **48**, 145 (1975).
125. G. M. Zhislin, *Theor. Math. Phys.* **7**, 571 (1971).
126. R. Benguria and E. H. Lieb, *Phys. Rev. Lett.* **50**, 1771 (1983).
127. Q. Shi and S. Kais, *Mol. Phys.* **100**, 475 (2002).

128. X. B. Wang, C. F. Ding, and L. S. Wang, *Phys. Rev. Lett.* **81**, 3351 (1998).
129. J. Ackermann and H. Hogreve, *J. Phys. B* **25**, 4069 (1992).
130. B. J. Laurenzi and M. F. Mall, *Int. J. Quant. Chem.* **xxx**, 51 (1986).
131. T. K. Rebane, *Sov. Phys. JETP* **71**, 1055 (1990).
132. Z. Chen and L. Spruch, *Phys. Rev. A* **42**, 133 (1990).
133. D. B. Kinghorn and L. Adamowicz, *J. Chem. Phys.* **110**, 7166 (1999).
134. B. J. Laurenzi, *J. Chem. Phys.* **65**, 217 (1976).
135. T. K. Rebane, *Sov. Phys. JETP* **71**, 1055 (1990).
136. H. Hogreve, *J. Chem. Phys.* **98**, 5579 (1993).
137. A. Martin, J. Richard, and T. T. Wu, *Phys. Rev. A* **52**, 2557 (1995).
138. B. Grémaud, D. Delande, and N. Billy, *J. Phys. B* **31**, 383 (1998).
139. C. L. Pekeris, *Phys. Rev.* **112**, 1649 (1958).
140. For example, A. R. Curtis and J. K. Reid, *J. Inst. Math. Appl.* **8**, 344 (1971).
141. P. R. Amestony, T. A. Davis, and I. S. Duff, *SIAM J. Matrix Anal. Appl.* **17**, 886 (1996).
142. C. Lanczos, *J. Res. Natl. Bur. Stand. B* **45**, 255 (1950).
143. Y. K. Ho, *Phys. Rev. A* **48**, 4780 (1993).
144. A. M. Frolov, *Phys. Rev. A* **60**, 2834 (1999).
145. R. Krivec, V. B. Mandelzweig, and K. Varga, *Phys. Rev. A* **61**, 062503 (2000).
146. C. D. Lin, *Phys. Rep.* **257**, 1 (1995).
147. E. A. G. Armour and W. B. Brown, *Acc. Chem. Res.* **26**, 168 (1993).
148. A. P. Mills, Jr. *Phys. Rev. Lett.* **46**, 717 (1981).
149. A. Martin, *Heavy Ion Phys.* **8**, 285 (1998).
150. J. M. Rost and D. Wintgen, *Phys. Rev. Lett.* **69**, 2499 (1992).
151. C. L. Pekeris, *Phys. Rev.* **112**, 1649 (1958).
152. J. R. Rice, *Matrix Computations and Mathematical Software*, McGraw-Hill, New York, 1981.
153. B. N. Parlett, *The Symmetric Eigenvalue Problem*, Prentice-Hall, Englewood Cliffs, NJ, 1980.
154. E. R. Cohen, and B. N. Taylor, *Phys. Today* **8**, BG 7 (1998).
155. R. S. Dyck, D. L. Farnham, Jr., and P. B. Schwinberg, *Phys. Rev. Lett.* **70**, 2888 (1993).
156. P. Serra, S. Kais, and N. Moiseyev, *Phys. Rev. A* **64**, 062502 (2001).
157. N. Moiseyev, *Phys. Rep.* **302**, 211 (1998).
158. H. Friedrich, *Theoretical Atomic Physics*, 2nd ed., Springer, Berlin, 1998.
159. R. G. Newton, *Scattering Theory of Waves and Particles*, McGraw-Hill, New York, 1966.
160. N. Moiseyev, P. R. Certain, and F. Weinhold, *Mol. Phys.* **36**, 1613 (1978).
161. N. Moiseyev, P. Froelich, and E. Watkins, *J. Chem. Phys.* **80**, 3623 (1984).
162. E. Narevicius, D. Neuhauser, H. J. Korsc, and N. Moiseyev, *Chem. Phys. Lett.* **276**, 250 (1997).
163. M. Rittby, N. Elander, and E. Brandas, *Phys. Rev. A* **24**, 1636 (1981).
164. B. Simon, *Ann. Phys.* **97**, 279 (1976).
165. G. Raggio, private communication.
166. C. Domb and J. L. Lebowitz, *Phase Transition and Critical Phenomena*, Vol. 9, Academic Press, London, 1992.
167. A. F. J. Siegert, *Phys. Rev.* **56**, 750 (1939)
168. H.-D. Meyer and O. Walter, *J. Phys. B* **15**, 3647 (1982).

169. H. S. Taylor and L. D. Thomas, *Phys. Rev. Lett.* **28**, 1091 (1972).
170. C. Laughlin, B. L. Burrows, and M. Cohen, *J. Phys. B* **35**, 701 (2002).
171. B. Szafran, J. Adamowsky and S. Bednarek, *Physica E* **4**, 1 (1999).
172. M. Čížek and H. Horáček, *J. Phys. A* **29**, 6325 (1996).
173. M. Suzuki, X. Hu, M. Katori, A. Lipowski, N. Hatano, K. Minami, and Y. Nomnomura, *Coherent Anomaly Method; Mean Field, Fluctuations and Systematics*, World Scientific, Singapore, 1995.
174. R. A. Sauerwein and S. Kais, *Chem. Phys. Lett.* **333**, 451 (2001).
175. H. Kleinert, *Path Integrals*, World Scientific, Singapore, 1995.
176. N. Makri, *Annu. Rev. Phys. Chem.* **50**, 167 (1999).
177. M. L. Bellac, *Quantum and Statistical Field Theory*, Oxford University Press, New York, 1998.
178. A. Lasota and M. C. Mackey, *Chaos, Fractals and Noise*, Springer, Berlin, 1991.
179. M. Doi and S. F. Edwards, *The Theory of Polymer Dynamics*, Oxford Press, New York, 1989.
180. G. Jongeward and P. G. Wolynes, *J. Chem. Phys.* **79**, 3517 (1983).
181. G. Pöschl and E. Teller, *Z. Phys.* **83**, 143 (1933).
182. D. Bressanini and P. J. Reynolds, *in advances in Chemical Physics*, Vol. 105, Wiley, New York, 1999, p. 37.
183. F. Remacle and R. D. Levine, *CHEMPHYSCHEM* **2**, 20 (2001).
184. L. Jacak, P. Hawrylak, and A. Wols, *Quantum Dots*, Springer, Berlin, 1998.
185. T. Chakraborty, *Quantum Dots: A Survey of the Properties of Artificial Atoms*, Elsevier, New York, 1999.
186. A. P. Alivisatos, *Science* **271**, 933 (1996).
187. C. B. Murray, C. R. Kagan, and M. G. Bawendi, *Science* **270**, 1335 (1995).
188. R. L. Whetten, J. T. Khoury, M. M. Alvarez, et al., *Adv. Mater.* **8**, 428 (1996).
189. G. Markovich, C. P. Collier, S. E. Henrichs, et al., *Acc. Chem. Res.* **32**, 415 (1999).
190. C. A. Stafford and S. DasSarma, *Phys. Rev. Lett.* **72**, 3590 (1994).
191. C. R. Kagan, C. B. Murray, M. Nirmal, et al., *Phys. Rev. Lett.* **76**, 1517 (1996).
192. G. C. Schatz, *J. Mol. Strut. (Theochem.)* **573**, 73 (2001).
193. K. C. Beverly, J. L. Sample, J. F. Sampaio, et al., *Proc. Natl. Acad. Sci. USA, in press (2002).*
194. M. A. Kastner, *Phys. Today* **46**, 24 (1993).
195. R. C. Ashoori, *Nature* **379**, 413 (1996).
196. U. Banin, Y. W. Cao, D. Katz, and O. Millo, *Nature* **400**, 542 (1999).
197. F. Remacle and R. D. Levine, *J. Phys. Chem. A* **104**, 10435 (2000).
198. J. L. Sample, K. C. Berverly, P. R. Chaudhari, et al., *Adv. Mater.* **14**, 124 (2002).
199. K. C. Beverly, J. F. Sampaio, and J. R. Heath, *J. Phys. Chem. B* **106**, 2131 (2002).
200. J. Hubbard, *Proc. R. Soc. London A* **276**, 238 (1963).
201. P. Fulden, *Electron Correlation in Molecules and Solids*, Springer-Verlag, Berlin, 1995.
202. J. E. Hirsch, *Phys. Rev. B* **22**, 5259 (1980).
203. J. Perez-Conde and P. Pfeuty, *Phys. Rev. B* **47**, 856 (1993).
204. B. Bhattacharyya and S. Sil, *J. Phys. Condens. Matter* **11**, 3513 (1999).
205. J. X. Wang, S. Kais, and R. D. Levine, *Int. J. Mol. Sci.* **3**, 4 (2002).
206. T. W. Burkhardt and J. M. J. van Leuven, in *Topics in Current Physics*, Springer-Verlag, Berlin, 1982.

207. F. Remacle, C. P. Collier, J. R. Heath, et al., *Chem. Phys. Lett.* **291**, 453 (1998).

208. N. F. Mott, *Metal–Insulator Transitions*, Taylor & Francis, London, 1990.

209. G. Benenti, X. Waintal, J.-L. Pichard and D. L. Shepelyansky, arXiv:cond/0003208.

210. M. A. Continentino, *Quantum Scaling in Many-Body Systems*, World Scientific Lecture Notes in Physics, Vol. 67, World Scientific, Singapore,

211. M. Schreiber, B. Kramer, and A. MacKinnon, *Physica Scripta* **T25**, 67 (1989); E. Hofstetter and M. Schreiber, *Europhys. Lett.* **21**, 933 (1993).

212. A. MacKinnon, *J. Phys. Condens. Matter* **6**, 2511 (1994).

213. F. Milde, R. A. Romer, M. Schreiber, and V. Uski, *J. Eur. Phys. B* **15**, 685 (2000).

A DISCUSSION OF SOME PROBLEMS ASSOCIATED WITH THE QUANTUM MECHANICAL TREATMENT OF OPEN-SHELL MOLECULES

JOHN F. STANTON

Institute for Theoretical Chemistry, Departments of Chemistry and Biochemistry, University of Texas at Austin, Austin, Texas, U.S.A.

JÜRGEN GAUSS

Institut für Physikalische Chemie, Universität Mainz, Mainz, Germany

CONTENTS

I. Introduction
II. Overview of Quantum Chemical Approaches
III. The Problem of Spin Contamination
IV. Response of Molecular Orbitals and the Issue of "Symmetry Breaking"
V. The Pseudo-Jahn–Teller Effect in Quantum Chemistry
VI. Skirting the Reference Function Issue
VII. Summary
Acknowledgments
References

I. INTRODUCTION

To some extent, the field of theoretical chemistry can be partitioned into three principal domains: statistical mechanics, dynamics, and the field of "electronic structure theory," or quantum chemistry. Among these, quantum chemistry is arguably the most fundamental field, since most applications of statistical

Advances in Chemical Physics, Volume 125, Edited by I. Prigogine and Stuart A. Rice.
ISBN 0-471-21452-3.

mechanics as well as classical and quantum dynamics ultimately require the potential energy surfaces that are obtained in quantum chemical calculations. Tremendous progress was made on several frontiers in quantum chemistry during the last three decades. During this time, the vital importance of electron correlation for obtaining high accuracy in the calculation of molecular properties was recognized, and several approaches for treating correlation were explored. These included straightforward diagonalization of the electronic Hamiltonian over a (necessarily) restricted basis (configuration interaction, or CI) [1,2], as well as methods that are based on many-body, or Møller–Plesset, perturbation theory (MBPT and MPPT, respectively; these methods are precisely the same when a Hartree–Fock self-consistent field (SCF) zeroth-order wave function is used) [3,4] and the well-known coupled-cluster (CC) approximation [5–7].

Since it was recognized long ago that the calculation of molecular properties is much more useful than that of the energy itself, efficient methods for calculating derivatives of the energy were developed during this "golden age," beginning with Pulay's seminal work on HF-SCF derivatives [9]. In subsequent work, analytic first and second derivatives have been formulated and implemented efficiently for essentially all CI [10,11], MBPT [12–27] and CC [28–37] methods that are used in chemical applications, culminating recently with analytic first [36] and second [37] derivatives for the full coupled-cluster singles, doubles, and triples (CCSDT) treatment [38–40] of correlation. In addition to the obvious application of using analytic energy derivatives to locate stationary points on potential energy surfaces and to calculate force constants, the same basic technology has been exploited for properties such as electric polarizabilities [41] and hyperpolarizabilities [42], chemical shift tensors [43], dipole moment derivatives, indirect nuclear spin–spin coupling constants [44], and nonadiabatic coupling matrix elements [45].

In recent years, high-level treatments of electron correlation in conjunction with various basis-set extrapolation methods [46–55] have been used to calculate energies and properties of small molecules at an extremely high level of accuracy. The goal of "chemical accuracy" (within 1 kcal/mol) has been achieved, but "spectroscopic accuracy" (within 1 cm^{-1}) is (and will continue to be) elusive. The development of a standard hierarchy of basis sets by Dunning and co-workers [56–59] that has been embraced by the quantum chemistry community provides a systematic approach to the basis set limit for molecular calculations. HF-SCF energies converge much more rapidly with respect to basis-set expansion than do correlation energies, which is due to the difficulty of describing cusps in the wave function at points in configuration space where the interelectron separation (r_{12}) vanishes [60]. Another approach that has gained some favor in the last decade is the explicit inclusion of r_{12} terms in the wavefunction, using either the strategy of gaussian geminals [61–65] or the so-called R12 method [66–68]. Calculations that treat electron correlation within this framework have been reported using both MBPT and CC approaches [69], and they provide values that are among the most accurate in the literature. Other, more traditional approaches for obtaining high accuracy include HF-SCF-based calculations together with correlation treatments such as CCSD(T) [70,71], CCSDT [38–40], and approximations to CCSDTQ [72] together with basis-set extrapolation; multireference configuration interaction and related methods [2]; schemes involving empirical corrections such as G1 [73], G2 [74], G3 [75], W1 [76],W2 [76] and CBS [77–79] "theories"; and quantum Monte Carlo calculations [80–82].

For relatively small molecules (those having up to four nonhydrogen atoms, particularly when these are found in the second row of the periodic table), it is now fairly straightforward to use most of the approaches mentioned above to calculate static properties such as equilibrium geometries, electric moments, thermodynamic stabilities, and force constants at a level of accuracy rivaling that which can be obtained experimentally. This is particularly true for systems that are well-represented by a single-determinant (molecular orbital) picture. For such molecules, the barriers to high accuracy are clear: The basis set must be expanded until the *correlated* results converge, and the treatment of electron correlation must be adequate to account for the so-called dynamical correlation effects associated with problems of this type. For the latter, it is generally believed that the CCSD(T) method is sufficient; in the last several years, calculations of properties and thermodynamic stabilities for well-behaved small molecules comprising light atoms using CCSD(T) and various basis-set extrapolation methods have been shown to be remarkably accurate [83].

In order to obtain equally reliable results for molecules that contain more light ($Z < 10$) atoms that but are still satisfactorily represented by single-determinant descriptions, the obstacles are technological rather than intellectual. The methods cited in the previous two paragraphs will be sufficient, and molecules with 10 or so "heavy" (nonhydrogen) atoms will become feasible when and if the rather startling advances in computational power today continue throughout the next two decades. Moreover, recent work in algorithm development has led to the successful implementation of CC methods (including the vitally important CCSD(T) model) based on localized orbitals [84–86], in which the cost of the calculations is substantially reduced for large molecules. This advance, which builds upon pioneering work in which the same idea was used in conjunction with MBPT [87], has been efficiently implemented and represents an important achievement in computational quantum chemistry. Results have already been reported for systems described by more than a thousand basis functions [88]; in the next decade, we believe that this approach will become increasingly useful for highly accurate calculations.

Hence, for the accurate theoretical treatment of "well-behaved" molecules containing light atoms, the situation is apparently well in hand. Methods capable of nearly quantitative recovery of electron correlation effects are available, and analytic first and second derivative methods have been implemented for essentially all of them. It is precisely these techniques that will be most valuable for calculating properties of these systems, and advances in computational power will allow the scope of investigations to be extended to larger molecules. While there is little doubt that even more sophisticated treatments of correlation will be implemented in the next decade (a general-purpose implementation of CCSDTQP has already been reported [89], and a number of groups now have the capabilities to do even higher-level CC calculations [90–92]), it is unlikely that these will have a major impact on the field of chemistry as a whole. Analytic gradient methods for methods such as CCSDT(Q) and, in particular, the explicitly correlated geminal and R12 approaches will probably prove more useful.

In our view, the most important area for future method development and application in quantum chemistry is in the treatment of open-shell molecules. For these systems, the single-determinant starting point is often inadequate; several types of problems arise that are infrequently or never encountered in treating closed-shell molecules. These include spin contamination and the phenomenon usually called "symmetry breaking" in the reference function [93,94], instabilites and near-singularities of the HF-SCF solution [95–97], strong (nondynamical) electron correlation effects, and adiabatic potential energy surfaces that exhibit many complicated features such as loci of conical intersections and avoided crossings. Some of the more sophisticated methods from the standard toolkit of quantum chemistry have not yet been implemented for open-shell systems (no coupled-cluster calculations with an explicit treatment of quadruple excitations have been reported for an open-shell molecule, and there is but a single example where the R12 approach has been used for an open-shell system [98]), while the performance of others is either satisfactory, surprisingly poor, or disastrous. While it is now possible to "push a button" and obtain highly accurate results for a closed-shell system, the same is not true and perhaps never will be true for radicals. Each system presents its own unique set of problems and requires careful study. The selection of a technique can involve a nontrivial analysis, and experience is invaluable for making a judicious choice.

In addition to the intrinsic challenge to theory posed by open-shell systems, a considerable amount of motivation for studying them also stems from pragmatic considerations. First, the reactive nature of radicals makes them extraordinarily difficult to study in the laboratory, and the presence of several low-lying excited states tends to make their electronic spectroscopy complicated. In fact, it has even been stated that the assistance of quantum chemistry is absolutely *necessary* to properly interpret many experimental studies of these systems [99]. The importance of open-shell molecules is obvious in areas that include atmospheric chemistry, interstellar chemistry, and theories of the origin of life [100]. In recent years, there has been an explosion of interest in the chemistry of radical cations and anions since the importance of these species in a wide variety of organic and biological reactions, radiation chemistry, and single electron transfer processes has begun to be appreciated [101,102]. Due to their ephemeral nature and other problems associated with experimental studies, theory can and will play a vital role in unraveling their properties and therefore facilitate an understanding of these important issues.

This review focuses on the quantum chemical treatment of open-shell molecules, with particular emphasis on difficulties that may be encountered in studying them. The scope of this work is limited to radicals that have "high-spin" electronic configurations ($M_s = S$). The subject of "biradicals," which are systems having energetically proximate singlet and triplet states corresponding to a particular electronic configuration, poses an altogether different set of problems, and their computational study has been adequately reviewed elsewhere [103,104]. In addition, excited states of closed-shell systems (properly known as open-shell singlets) are also excluded. Instead, we focus mostly on the treatment of doublet and triplet radicals in their electronic ground state, extending the discussion to excited states only insofar as these perturb or otherwise play a role on the ground-state properties. Perhaps the principal aim of this work is to give the reader some familiarity with the strengths and limitations of various "black box" quantum chemical methods for treating open-shell systems, provide some guidance for how to identify and diagnose problems, and outline potential strategies for overcoming them.

II. OVERVIEW OF QUANTUM CHEMICAL APPROACHES

The techniques used to study the electronic structure of molecules can be grouped into five separate categories. The first, and most commonly used for applications, involves the construction of a single Slater determinant to describe the electronic wavefunction with or without a treatment of electron correlation. The orbitals that make up the Slater determinant are usually obtained in Hartree–Fock self-consistent field calculations, but need not be. Since the exact (full configuration interaction, or FCI) energy is necessarily independent of the choice of orbitals, it can be expected that highly accurate calculations exhibit a pronounced insensitivity to the orbitals, and this property can be exploited in cases where there are problems associated with the HF-SCF solution. A specific realization of this idea is the use of Brueckner orbitals [105,106], which satisfy a "generalized Brillouin condition" and are determined self-consistently along with a treatment of electron correlation (usually CC theory [107–111]). Indeed, later in this review, it will become apparent that such calculations can be advantageous in certain situations involving open-shell systems, underscoring the comment made in the previous section about the difficulties associated with choosing a particular theoretical approach to study a problem involving radicals.

Treatments of electron correlation built upon a single determinant involve CI, CC, and MBPT methods. Due to its variational property and relative simplicity, CI was the first method to be widely exploited in chemistry [1]. In such methods, the electronic Hamiltonian is projected onto the space of n electron Slater determinants obtained by distributing the n electrons amongst the N orbitals that are either occupied or unoccupied (virtual) in the zeroth-order description[1] and then diagonalized. If all possible determinants are included in the basis, this approach (FCI) provides the exact answer in the space spanned by the atomic orbital basis set used for the calculation. However, even for a small molecule such as water, the dimension of the FCI basis when a moderate to large atomic orbital basis set is used is on the order of 10^{15} or greater, and the method is clearly impractical. However, to the extent that the problem under consideration is reasonably well treated by a single Slater determinant, it can be expected that determinants having a large number of electrons assigned to virtual orbitals make negligible contributions to the wave function. This can be exploited by partitioning the determinantal basis into the HF-SCF solution and those that differ from it by specific "excitation" levels, where the level of excitation is equal to the number of electrons in virtual orbitals. Straightforward application of perturbation theory shows that doubly excited determinants are expected to make the dominant corrections to the zeroth-order HF-SCF wave function, while response theory suggests that singly excited determinants are important for the description of molecular properties. Diagonalization of the Hamiltonian in the basis comprising the HF-SCF determinant as well as single and double excitations is known as CISD, which was widely used to study a number of chemical problems in the 1970s and early 1980s and is the subject of a comprehensive early review [1].

In the last 20 years, CI calculations based on a single reference function have lost favor among practitioners. The principal shortcoming of these approaches is that they do not satisfy the property of "size-consistency," which means that the CI energy does not scale properly with the size of the system [112]. It is fairly easy to see why this is so. Consider two beryllium atoms, separated by a distance sufficiently large that the true physical interaction between the atoms vanishes. In a CISD description of this system, contributions to the wave function are excluded in which two electrons on each beryllium atom are in virtual orbitals, since these correspond to quadruply excited determinants and would require a method such as CISDQ or CISDTQ for their inclusion. However, the CISD wave function for a single beryllium atom contains all determinants with two electrons in virtual orbitals. Since CI methods involve

[1] N is precisely equal to the number of basis functions used in the calculation.

diagonalization of the Hamiltonian and therefore are variational, the energy of the separated beryllium atoms is above twice that of an individual atom. This formal shortcoming causes the quality of the correlation treatment to be systematically degraded as the size of the system increases. However, the lack of size consistency plagues applications to even small molecules when reaction energies or properties such as excitation energies and force constants are determined.

Size-consistent methods based on many-body techniques were developed in the field of nuclear physics during the 1950s [105,113–115] and were first imported into quantum chemistry a decade later [116,117]. However, widespread application of these methods came only with their implementation in general-purpose programs toward the end of the 1970s and had to overcome a reluctance of the theoretical chemistry community to part with the variational nature of SCF and CI theories. Conceptually simplest of these is the many-body perturbation theory of Brueckner and Goldstone [105,115], especially when carried out to second order. Defining the Fock operator as zeroth-order (Møller–Plesset partitioning [118]), one then obtains a very simple formula for the correlation energy:

$$E = \frac{1}{4}\sum_{abij} \frac{|\langle ab \| ij \rangle|^2}{\varepsilon_i + \varepsilon_j - \varepsilon_a - \varepsilon_b} \tag{1}$$

where indices $i, j, k, \ldots$ represent orbitals that are occupied in the HF-SCF reference and $a, b, c, \ldots$ are virtual orbitals, and the ε_q represent the HF-SCF orbital energies.[2] This theory (known as both MP2 or MBPT(2) in the literature) is quite successful in accounting for the dominant contributions to the dynamic correlation energy, and it has probably been used more than any other method for the treatment of electron correlation [3]. MBPT has been implemented through sixth order[3] [119–121], but fifth- and sixth-order treatments are exorbitantly expensive, and third order does not represent a systematic improvement over MBPT(2) despite a considerably higher cost. However, at fourth order, where the effects of singly, triply, and quadruply excited determinants first come into play,[4] significant improvement relative to MBPT(2) is sometimes achieved [3]. Calculations using fourth order with [MBPT(4)] and without [SDQ-MBPT(4)] the relatively expensive triple excitation contribution have also been used in a

[2]Indices $p, q, r, \ldots$ are reserved for orbitals that may be either occupied or unoccupied.

[3]This statement refers to general-purpose implementations. For benchmark studies, codes developed for FCI calculations have been used to calculate terms in the perturbation expansion to very high orders [122–124].

[4]This statement is strictly true only when an RHF or ROHF reference function is used; single excitations enter at second order for ROHF.

significant number of chemical applications. However, the popularity of MBPT has faded substantially in the last several years for a number of reasons. These include (a) an explicit demonstration that the perturbation expansion is actually divergent for systems as simple as the neon atom when certain basis sets are used [124,125], a rather damning criticism of MBPT for its treatment of open-shell problems [126] (which will be discussed later in this review), and (b) the burgeoning popularity of the related CC technique.

The coupled-cluster approximation was brought to chemistry in the mid-1960s [116]. However, the diagrammatic many-body framework of the theory was sufficiently difficult and unfamiliar that it did not achieve significant popularity until the early 1980s, when the CC singles and doubles (CCSD) method was implemented in a general-purpose program [127]. The general idea of CC theory is built upon the so-called exponential *ansatz*, which requires that the wave function be represented by

$$|\psi\rangle = \exp(T)|0\rangle \tag{2}$$

where $|0\rangle$ is a Slater determinant (usually, but not necessarily, that corresponding to the HF-SCF solution) and T is an operator that generates excited determinants with corresponding weights, namely,

$$T = T_1 + T_2 + \cdots + T_k \tag{3}$$

$$T_1 = \sum_{ai} t_i^a a^\dagger i \tag{4}$$

$$T_2 = \frac{1}{4}\sum_{abij} t_{ij}^{ab} a^\dagger b^\dagger ji \tag{5}$$

$$\vdots$$

The creation and annihilation operators that follow the scalar coefficients (amplitudes) in Eqs. (4) and (5) have the effect of promoting an electron from orbital i to orbital a [Eq. (4)] as well as j to b [Eq. (5)] when acting on the reference function $|0\rangle$. Inserting Eq. (2) into the Schrödinger equation followed by projection against $\langle 0|$ and Hermitian conjugates of the excited determinants $|\Phi\rangle$ gives a hierarchy of coupled nonlinear equations for the energy and the cluster amplitudes t, which is conveniently decoupled when the left- and right-hand sides of the Schrödinger equation are premultiplied by the inverse of the cluster operator $\exp(-T)$ prior to projection:

$$\langle 0|\exp(-T)H\exp(T)|0\rangle = E \tag{6}$$

$$\langle \Phi|\exp(-T)H\exp(T)|0\rangle = 0 \tag{7}$$

Note that the composite operator $\exp(-T)H\exp(T)$ is a similarity transformation of the electronic Hamiltonian (usually denoted as $\bar{H}$), which turns out to play the central role in the EOM-CC method discussed later in this section. If the cluster operator T is not truncated [k in Eq. (3) is equal to the number of electrons in the molecule of interest], the method is equivalent to FCI and equally impractical. The benefits of CC theory are realized only when k is restricted to a certain excitation level, the most computationally tractable of which is $k = 2$. This defines the CCSD model [127], which led to the first significant interest in using CC methods for chemical problems [128]. The cost of CCSD calculations increases roughly with the sixth power of the basis set size, which is the same scaling associated with fourth-order MBPT without triples [SDQ-MBPT(4)] and CISD. Relative to CISD, the advantages [129] of CCSD are that the correlated wave function is generated by application of the *exponential* operator $\exp(T)$. This means that determinants of all excitation levels are present in the wave function. Even though the coefficients of determinants corresponding to higher levels of excitation are not optimized (for example, quadruples in CCSD), arguments of perturbation theory show that the approximations are reasonably good [130]. In particular, the treatment of quadruple excitations in CCSD is correct through fourth order in perturbation theory, and therefore recovers most of the effects included in a full treatment (i.e., that from CCSDQ or CISDQ) without requiring steps with a *tenth*-power dependence that are needed to treat all quadruple excitation effects. In addition, CCSD methods are—like MBPT—rigorously size-consistent.[5]

While CCSD is undoubtedly an important quantum-chemical method, it is widely appreciated that some explicit treatment of triple excitation effects is necessary for quantitative accuracy. The most straightforward means to achieve this is the full CCSDT method, in which k is equal to 3 in Eq. (3) and the nonlinear system of equations for the amplitudes is solved without further approximation [38–40]. However, the cost of CCSDT scales with the eighth power of the basis-set size (in contrast to the N^6 scaling of CCSD), and the method necessitates storage of T_3 amplitudes; the latter restriction is one that can easily exhaust the storage capacity of most computers. Hence, there has been a significant amount of work aimed at approximating the contribution of triple excitation effects, where one of two strategies is adopted. In the first, the Eq. (8) is simplified by excluding the most expensive and other specific terms; some of the resulting iterative approaches are known as CCSDT-n ($n = 1a$ [132], $1b$ [132], $1c$ [133], 2 [132], 3 [132]) and CC3 [134] and involve N^7

[5]The two properties mentioned here are not entirely independent, since it is precisely the incorporation of higher levels of excitation via nonlinear terms in the exponential expansion that act to preserve the property of size consistency. For an authoritative discussion of this subject, see Ref. 131.

steps in each iteration of the equations. The second class treats the effects of triple excitations *ex post facto* and relies on arguments that are loosely based on perturbation theory to estimate the contribution to the energy. These "noniterative treatments" include CCSD + T(4) [135], CCSD + T(CCSD) [135], CCSD(T) [70,71], and a-CCSD(T) [136] (or Λ-CCSD(T) [137]), but CCSD(T) is the most popular of these. In fact, all evidence is that CCSD(T) is superior to any of the considerably more expensive iterative methods for calculating the correlation energy and molecular properties [138,139]. Although this can be rationalized to a degree by recognizing that the iterative methods are not perfectly balanced and that CCSD(T) is in fact a proper lowest-order correction to CCSD [140], the success of CCSD(T) is truly miraculous. This method has enjoyed tremendous popularity since its formulation at the close of the 1980s. While CCSD(T) has been called "the gold standard of quantum chemistry" [141] and is the method usually chosen for high accuracy calculations, CCSDT is used to some degree in applications. More sophisticated methods such as CCSDT(Q) [72], CCSDTQ [142,143] and most recently CCSDTQP [89] have also been implemented, albeit only for closed-shell systems.

The second most popular class of methods are those based on density functional theory (DFT) [144]. Although the basic idea of DFT goes back to the dawn of quantum mechanics, the area was largely ignored until the (first) Hohenberg–Kohn theorem [145] was formulated in the 1960s. This states that the exact ground-state energy of an n-particle system can be expressed as a functional of the one-electron density. Although this is simply an existence theorem no different in principle from that which applies to equations of state for single-component systems due to the Gibbs phase rule, it has generated a enormous amount of interest in chemistry during the last 10 years. Another important step was that Kohn and Sham [146] showed that the n-electron problem can be treated in an exact manner by solving n. Sham [146] states that the exact energy can be obtained by solving n one-electron equations, the so-called Kohn–Sham self-consistent-field (KS or KS-SCF) equations. The development of functionals that relate the energy to the density is an extremely fertile and active area of research at this time, and the accuracy obtained with the best functionals is comparable to (and often better than) second-order MBPT [147]. DFT has been applied to a wide variety of chemical systems, including "simple" closed-shell molecules, radicals, biradicals, transition states, and even low-spin systems that contain one or more metal atoms; its success in these areas has been remarkable. For ballpark estimates of reaction energies, thermodynamic stabilities, geometries, and a number of other properties, the combination of accuracy and favorable computational cost associated with DFT is such that it will likely remain the method of choice for the foreseeable future. It should be emphasized that DFT methods are somewhat semiempirical in nature (there is no well-defined way to develop the functionals) and occasionally fail to produce even qualitatively correct results. However, the same can be said of any quantum chemical method. Nevertheless, unlike other approaches, there is no systematic means for improving the accuracy of a given DFT calculation. Hence, for highly accurate calculations of molecular properties, prospects for DFT are not particularly bright.

Next are methods based on "zeroth-order" wave functions that are linear combinations of Slater determinants that again may or may not include a treatment of residual electron correlation effects. In these multiconfigurational (MC) approaches, the starting wave function is presupposed to be given by a determinantal expansion

$$|\psi\rangle = \sum_i C_i |\Phi_i\rangle \tag{8}$$

where the coefficients C_i and the orbitals in Φ_i are simultaneously optimized [148]. If the sum is carried over the full space of n electron determinants, the method is again equivalent to FCI,[6] but truncation is required in practical calculations. In general, the "most important" configurations must be included in the expansion, the identification of which is usually not obvious *a priori*. It is common to partition the orbitals into active and inactive spaces, the former including either the valence region in its entirety or some appropriately chosen subset and the corresponding virtual levels. Determinants representing all possible occupation schemes within this active space are then included in Eq. (8). Such calculations go by various names, including fully optimized reaction space and complete active space self-consistent field (FORS-SCF [149] and CASSCF [150]). These are special cases of the multiconfigurational SCF (MCSCF) label that can be applied to any calculation that determines a fully optimized wave function of the form given by Eq. (8). The great virtue of MCSCF calculations is that they are well-suited to handle cases in which more than one configuration makes a substantial contribution to the wave function. A balanced treatment of these determinants is guaranteed by the nature of the wave-function parameterization, unlike single-determinant CI, PT, and CC methods in which bias toward a specific determinant is inherent to the method.

Despite an effective treatment of nondynamical electron correlation, MCSCF calculations carried out in feasible active spaces[7] do not produce quantitatively reliable results. The reason for this is the neglect of dynamical electron correlation involving the inactive occupied and virtual orbitals. Qualitatively,

[6] In this limit, the equations are singular, since the same energy can be obtained for any set of orbitals.

[7] The size of the valence space grows linearly with the size of the molecule, which translates to an exponential growth in the cost of the calculation. This severely compromises the extent to which MCSCF-based methods can be used for modestly sized and larger molecules.

the effects of antibonding orbitals are typically exaggerated in MCSCF calculations, with the result that force constants are usually too small and bond lengths too long. To treat residual correlation effects, a number of methods have been used. Most straightforward of these is CI, in which these effects are treated by matrix diagonalization. If used in conjunction with large active spaces, these "MRCI" calculations are extremely accurate [2] and are even used as benchmark values for calibrating the performance of other methods. Unfortunately, however, if one wishes to maintain a uniform level of accuracy, the cost of MRCI calculations (like MCSCF itself) increases exponentially with the size of the system. Applications are accordingly limited to small molecules. While all MRCI calculations suffer from a size-consistency error,[8] this can become appreciable for molecules containing more than three nonhydrogen atoms because of the necessarily small size of the MCSCF expansion in these cases. This can be treated by *ad hoc* corrections with rather dubious theoretical foundations or, preferably, in a self-consistent fashion by methods such as "averaged coupled pair functional" (MR-ACPF) [151] or "approximate quadratic coupled cluster" (MR-AQCC)[9] [152]. While the domain of applicability is limited due to their expensive nature, highly accurate results can be achieved in these calculations.

Another way of treating residual correlation effects beyond the MCSCF level is by means of perturbation theory. Indeed, if the size of the active space is large, it follows that the amount of correlation not already treated in zeroth order is small, and it is sensible to appeal to perturbation theory. While MRCI predated MRPT by many years, the latter has become increasingly fashionable in the last 10 years. Several variants of second-order perturbation theory have been reported [153], the most popular of which is the CASPT2 method [154,155]. Many of the applications of CASPT2 have focused on excited states of closed-shell molecules and biradical systems that are of considerable interest in organic chemistry [104]. For ground-state treatments of closed- and high-spin open-shell systems, the method has seen considerably less application. CASPT2 is not a size-consistent method [156], but the practical import of this shortcoming is not clear at this time. However, in applications to large molecules that necessarily involve small active spaces, the suitability of using second-order perturbation theory to treat residual correlation effects is called into question. Hence, the benefits of CASPT2 relative to a more elaborate treatment based on MRCI are vastly reduced in these cases.

The fourth method used for quantum chemical calculations is the quantum Monte Carlo (QMC) method, in which the Schrödinger equation is solved numerically. There are three general variants of QMC: variational MC (VMC), diffusion QMC (DQMC), and Green's function QMC (GFQMC), all of which

[8]Unless, of course, the calculation is equivalent to FCI.

[9]Despite its name, MR-AQCC is *not* a coupled-cluster method.

adopt a different strategy. In the first, a trial form is imposed on the wave function and the requisite integrations are done by Monte Carlo techniques; this allows the use of complicated trial functions that explicitly include the interelectronic coordinate. In DMC, a formal relationship between the electronic Schrödinger equation and the diffusion equation in imaginary time is exploited. The GFQMC method is related to DMC but is somewhat more simple to use. The latter two schemes offer solutions to the electronic problem that are exact in principle, but there are issues related to discontinuities in the gradient of the potential energy as well as nodes for multielectron systems that have restricted their potential. For very small systems (H_3 being perhaps the most well-known example), QMC methods are the most accurate technique available, but they have not yet been successfully applied to larger molecules at comparable levels of accuracy.

The last category of quantum chemical methods, and probably the least familiar to most readers, are those in which the reference wave function is calculated for an electronic state that is *different* from that which is ultimately of interest. These include propagator methods [157–179], which include electron propagator theory (EPT) where the state under investigation has one more or one fewer electrons than the reference state; the polarization propagator approximation (PP), which focuses on excited states relative to a reference ground state; the group of increasingly popular equation of motion or linear response CC [180–196] (EOM-CC and LRCC, respectively[10]) methods; and other less well-characterized but promising approaches such as "similarity transformed EOM-CC" (STEOM-CC) [197] and the spin-flip (SF) model [198–200]. Common to all of these computational methods is a potentially balanced treatment of strongly mixed configurations, precisely the same feature that makes MC methods attractive for selected applications. More will be said about this class of methods in Section VI.

III. THE PROBLEM OF SPIN CONTAMINATION

There is generally little ambiguity associated with calculations of molecular orbitals for closed-shell molecules. The Hartree–Fock method (as appoximated by self-consistent field calculations in necessarily finite atomic orbital basis sets[11]) provides a solution that obeys some of the symmetry properties that must

[10]These two categories of methods are essentially identical, differing only in the way some transition properties are defined. Energies given by the two approaches are the same and are obtained by diagonalization of a matrix that can be "derived" from the EOM point of view, or time-dependent and time-independent response theory.

[11]The term Hartree–Fock, without qualification, should be reserved for calculations that are numerically exact, rather than those based on a finite basis set. For the latter, we prefer the designations Hartree–Fock self-consistent field (HF-SCF).

be satisfied by the exact wave function. Specifically, the density calculated from the Slater determinant comprising the molecular orbitals transforms according to the totally symmetric representation of the molecular point group, and the wave function is an eigenfunction of the spin-squared S^2 operator with eigenvalue zero. More often than not, the treatment of electron correlation that is necessary to achieve high accuracy for molecular properties is relatively straightforward since correlation effects are most often found to be of the "dynamical" type. Hence, the mathematical structure of the exact wave function (when written in the FCI expansion) is dominated by the reference Slater determinant, with small but important contributions from doubly and (to a lesser extent) singly excited determinants.

To be sure, there are exceptions to the simple picture presented above for closed-shell molecules. Those categorized as biradicals usually have wave functions that involve strong mixing of two determinants which are related by a double excitation, while molecules with multiple bonds can also have substantial contributions from single, double, and higher excitations. In addition, HF-SCF-based MBPT, CI, and CC calculations can be inadequate when trying to treat closed-shell molecules that are far from the equilibrium geometry, where several Slater determinants make substantial contributions to the wave function. However, for the calculation of most molecular properties and parameters relevant for the treatment of spectroscopic problems, these regions of the potential energy surface are less important. Nevertheless, a great deal of work has gone into this so-called "bond-breaking problem" in recent years. Coupled-cluster methods that are intended to approximate CASSCF calculations have been developed [200–202], and the treatment of very high levels of excitation [203–206] within CC theory has also been advocated for problems of this type.[12]

HF-SCF calculations followed by a thorough treatment of dynamical correlation such as CCSD(T) are almost always a satisfactory means of achieving high accuracy for problems involving closed-shell molecules in their ground electronic state at or near equilibrium geometries. The generally amenable nature of closed-shell molecules to quantum chemical calculation has led to a considerable success of "theoretical model chemistries" for studying these species. For example, the G1, G2, G3, W1, W2, and CBS treatments are all based on HF-SCF treatments followed by MBPT or CC corrections for

[12]Due to the inherent multireference nature of "bond-breaking" problems, it seems rather unlikely that single-determinant CC methods of any sort will ever provide a satisfactory solution to these problems. It seems much more sensible to use MR approaches to study problems of this sort when they are relevant for a particular chemical application. For small molecules, MRCI offers a suitable option, while CASSCF calculations corrected by perturbation theory are more generally useful for larger systems.

correlation, with additional empirical corrections to account for high-level correlation and basis set effects. The main goal of this sort of calculation is to obtain highly accurate thermochemical parameters; these methods have been enormously successful in this area. Hence, accurate results for closed-shell molecules can now almost always be obtained by well-defined and straightforward procedures; the requisite calculations can be performed by someone with little to no knowledge in the field of quantum chemistry.

Open-shell molecules are another issue altogether, and most of the problems associated with their theoretical description is associated with the choice of a starting point for higher-level calculations. In the HF-SCF method, the Fock operator that becomes diagonal at convergence is itself a function of its solution. Hence, unlike eigenfunctions of the Hamiltonian, which must obey certain spin and symmetry properties, the Slater determinant made up of molecular orbitals obtained in an SCF calculation need not [93]. By relaxing constraints on the molecular orbital solutions, lower HF-SCF energies often can be obtained at the expense of violating fundamental symmetry properties. While such treatments have been advocated and said to include some treatment of correlation, their supposed merits must be regarded with equivocation.

The most well-known problem associated with HF-SCF solutions for radicals is that known as "spin contamination" [207,208] and occurs when the unrestricted Hartree–Fock (UHF) appoximation [209] is made. This idea was introduced long ago in quantum chemistry and allows distinct sets of molecular orbitals for electrons of different spin. Unlike the paradigm whereby a radical is described by a set of doubly occupied orbitals and then one half-filled orbital for each unpaired electron, all of the orbitals in a UHF calculation are singly occupied. The result is that the wave function is no longer an eigenfunction of the S^2 operator with eigenvalue $S(S+1)$.[13] If resolved into eigenfunctions of particular spin states, the UHF wave function contains components of the appropriate spin multiplicity $(2S+1)$ plus "contamination" from those with higher levels of spin multiplicity. For example, the UHF description of a doublet contains doublet, quartet, sextet, and so on, components, while that of a triplet is contaminated by pentet, septet, and so on, contributions.[14]

[13] S is the total electronic spin, so that S^2 eigenvalues for pure spin states are $\frac{3}{4}$ for doublets, 2 for triplets, $\frac{15}{4}$ for quartets, and so on.

[14] That doublets and triplets are not contaminated by triplets and quartets, respectively, is simply due to the fact that the contaminating multiplicities must be associated with determinants containing the same number of electrons. Hence, doublet states (which must have an odd number of electrons) can only be contaminated by states with multiplicities four, six, and so on. Similarly, triplet states (which must have an even number of electrons) can only be contaminated by states with higher and odd multiplicities.

Spin-contaminated solutions to the HF-SCF equations *always* occur for open-shell systems, but the extent of contamination (the difference between the expectation value of S^2 and the nominal value corresponding to the state of interest) varies widely. To gain some idea of the complexity that can occur with UHF calculations, it is sufficient to consider a very simple diatomic radical: CN. Using a standard basis set of double-zeta plus polarization quality, at least three different UHF solutions are found at internuclear separations greater than 1.27 Å. At 1.28 Å, they have $\langle S^2 \rangle$ values of 1.620, 1.125, and 0.859; the relative energies corresponding to these solutions (which will hereafter be referred to as A, B, and C) are 0, 22.02, and 22.52 kcal mol^{-1}, respectively. Despite their large qualitative and energetic differences, all three solutions correspond to the same molecular state; any doubt concerning this issue is removed when highly correlated calculations are carried out. At the CCSD level, the relative energies are a mere 0, 0.96, and 0.15 kcal mol^{-1}; these are reduced still further to 0, 0.15, and -0.01 kcal mol^{-1} when the full CCSDT model is used. Thus, there is no question that a full CI calculation based on any of the three solutions would give the same energy, but the results of any approximate calculation depend on the choice of the reference function. Which should be used? If one is concerned solely with the spin contamination issue, solution C would seem to be the logical choice. However, a geometry optimization starting at 1.30 Å using solution C quickly fails, because the SCF equations fail to converge. The underlying problem is that solutions B and C cease to exist at bond lengths less than $\sim$1.27 Å. More correctly, this pair of solutions goes through a degeneracy near this internuclear separation and corresponds to a complex conjugate pair of solutions at shorter bond lengths! Since the HF-SCF and CC energies are monotonically increasing functions of the bond length over the domain where B and C are real solutions to the HF-SCF equations, the only choice to use in a geometry optimization based on real orbitals is solution A, which is severely spin contaminated. Since this solution is qualitatively far from the exact wave function (which of course is an eigenfunction of S^2), it presents a challenge to the treatment of electron correlation. Using the simple MBPT(2) approximation, the equilibrium bond length *shortens* from 1.167 to 1.139 Å. This is strikingly different from the normal case, in which bond lengths are increased when electron correlation effects are introduced[15] [210], and satisfactory results are obtained only when CC methods are used. CCSD and

[15]This can be rationalized in one of two ways. First, the effect of electron correlation is to keep electrons apart, which therefore tends to increase internuclear distances to accommodate the "expansion" of the electron distribution. The other equally satisfactory argument is that antibonding orbitals are mixed into the wave function by the correlation treatment.

CCSD(T) give 1.187 and 1.193 Å, respectively, while a very high-level calculation gives 1.196 Å.[16]

To avoid spin contamination, another type of reference function can be used, specifically that of the restricted open-shell Hartree–Fock (ROHF) type [211,212]. In ROHF, maximum double occupancy of spatial orbitals is enforced, and the resulting Slater determinant solution is an eigenfunction of the S^2 operator. Hence, for CN, the ROHF solution has an S^2 value of 0.75 and in this respect is far superior to the UHF[A] solution. At the SCF level, the equilibrium bond length of CN is 1.140 Å; this changes to 1.215, 1.188, and 1.197 Å at the MBPT(2), CCSD, and CCSD(T) levels, respectively. While much has been made of the "wrong" direction of the UHF[A] bond length change for CN in going from SCF to MBPT(2), the corresponding ROHF-based correction has received less attention in the literature, even though it overshoots the CCSD value by roughly 40%. However, at the CCSD and CCSD(T) levels, ROHF and UHF[A] structures are nearly identical; this attests to the high-quality treatment of correlation provided by CC methods. One subtlety that is not widely appreciated is that most implementations of open-shell MP and CC theories based on an ROHF determinant do not give results that are free from spin contamination [213]. The most common realizations of ROHF-based MP and CC theories are carried out in a spin orbital basis, and the correlation treatment introduces a small amount of spin contamination. For the CN example at 1.28 Å, S^2 calculated as expectation values from the CC wave functions are 0.7516 (ROHF), 0.7584 (UHF[A]), 0.7501 (UHF[B]), and 0.7497 (UHF[C]). Thus, the extent of spin contamination that persists at the CCSD level is relatively small for all three UHF reference functions, even the badly contaminated UHF[C]. In general, reference functions with $\langle S^2 \rangle$ values below 0.9 generally give $\langle S^2 \rangle_{CCSD}$ somewhere in the range 0.748–0.752,[17] which is roughly the same range of values found for ROHF reference states. Hence, in cases where small to modest spin contamination is found in the UHF calculation, CC calculations based on ROHF functions actually offer no advantage with regard to a proper treatment of spin.

Nevertheless, some effort has been extended toward developing coupled-cluster treatments that are either exactly or partially spin-adapted. Rigorously spin-adapted methods have been presented in the literature [214–216], but most applications have used either the partially spin-adapted variants [217,218] and the spin-restricted approach [219] in which the expectation value of $\langle S^2 \rangle$ is

[16]The calculation was performed at the EOMIP-CCSDT level, which provides a highly correlated, spin pure, and balanced description of the final state. See Section VI of this chapter for a brief discussion.

[17]See footnote 19.

constrained to the correct value[18] [219]. For ground states of radicals, these methods appear to offer negligible benefits relative to the more standard UHF and ROHF-based spin-orbital methods, but they may prove advantageous when extended to excited state treatments [220].

While the discussion above is intended to communicate that spin contamination does not represent a major problem in CC calculations [221], it is not always feasible to carry out large-scale electron correlation treatments for larger molecules. In these cases, the question of reference function might become relevant if the treatment of correlation is omitted or simply restricted to the MBPT(2) model. Superficially, one would think that ROHF would be the preferred choice, since it is "closer" (in the sense of spin properties) to the exact wave function. Indeed, in the early years of the last decade, much work was expended to develop low-order treatments of correlation (specifically MBPT(2)) based on ROHF reference determinants [223–227] and the corresponding energy derivatives [225,228]; these calculations were widely advocated for radicals at that time. In our opinion, however, the issue is not clear-cut. For the vast majority of open-shell molecules, spin contamination is not extensive and UHF-MBPT(2) and ROHF-MBPT(2) results generally do not differ significantly. Since MBPT(2) does not offer even a nearly quantitative treatment of electron correlation, there is no reason to prefer properties calculated by either of the two methods. While ROHF-MBPT(2) gives better results than UHF-MBPT(2) for the CN radical [223,224], there are other examples where UHF-MBPT(2) significantly outperforms ROHF-MBPT(2) [104]. The problem of spin contamination is certainly not the cause of the problem with the ROHF-MBPT(2) results. Rather, ROHF methods are susceptible to another problem (usually termed *symmetry breaking* but in fact a more general and pernicious phenomenon than the name indicates) that involves the way SCF

[18]A seeming contradiction occurs here. For a doublet state, $\exp(T)|0\rangle$ contains contributions for doublet, quartet, and so on, spin multiplicities, so it might seem that an expectation value of the S^2 operator could yield values only equal to or greater than 0.75, with the exact value obeyed only when the CC wave function is an exact eigenfunction. However, the definition of an expectation value in CC theory is somewhat ambiguous. The theory is inherently non-Hermitian, and the ground state is noninteracting and diagonal only in a biorthogonal framework [190]. Hence, the energy and other properties are given by a nonsymmetric expectation value of the form $\langle\tilde{\Psi}_{CC}|\mathcal{O}|\Psi_{CC}\rangle$ where $\langle\tilde{\Psi}_{CC}|$ is given by $\langle 0|L\,\exp(-T)$, and L is the left eigenvector of the similarity transformed Hamiltonian $\bar{H}$ (see Section II) corresponding to the state. Thus the theory admits to a paradoxical situation whereby the expectation of an operator might give the exact value even though the right- and left-hand CC wave functions are not eigenfunctions. Others have defined the coupled-cluster expectation value in the usual sense with $\langle\tilde{\Psi}_{CC}| \equiv [|\Psi_{CC}\rangle]^{\dagger}$, but this has the distinct disadvantage that the "expectation value" of the Hamiltonian in this method does not return the corresponding CC energy. We prefer the nonsymmetric formulation, which is perhaps most comprehensively described in Ref. 222. The asymmetric nature of the expectation value therefore also admits to cases where the expectation value is *below* the exact value, as found for some of the examples discussed in this work.

orbitals respond to various perturbations. This issue, which we believe to be at least an order of magnitude more important than spin contamination in computational studies of open-shell molecules, is dealt with in some detail in the next section.

Before closing, two points should be addressed. First, MC methods are generally formulated in a basis of spin-adapted configurations and are free from spin contamination. Both MCSCF and MRCI wave functions are rigorous eigenfunctions of S^2, and the problem of multiple solutions of different spin multiplicity that nonetheless describe the same molecular state occurs less frequently. Nevertheless, MC calculations on open-shell molecules require a careful choice of the configurational basis. This is especially true in MRCI, where the lack of size-extensivity can cause problems until extremely large reference spaces (that often include configurations one would not choose on the basis of chemical intuition) are used. In addition, something should be said about reference functions for closed-shell systems. Although a restricted (RHF) calculation in which each molecular orbital is doubly occupied yields a pure singlet wave function at both the HF-SCF and correlated levels, it is not uncommon for singlet states to admit to lower-energy UHF solutions. These occur in the presence of so-called "triplet instabilities" [96,229], the meaning of which will become clear in the next section. In fact, for ethylene and all higher polyenes, UHF solutions with energies below the RHF solution exist. In the literature, it is sometimes argued that any system with a triplet instability requires a multireference treatment, an assertion that we regard as extreme. There is little evidence that RHF-based CC calculations for ethylene and other polyenes are in any way inadequate, and the importance of triplet instabilities for the vast majority of molecular properties is essentially nil.[19]

IV. RESPONSE OF MOLECULAR ORBITALS AND THE ISSUE OF "SYMMETRY BREAKING"

The first-order response of the exact adiabatic wave function for a many-electron system subjected to the perturbation χ obeys the following equation:

$$\left|\frac{\partial\Psi}{\partial\chi}\right\rangle = \sum_k{}' \frac{|\Psi_k\rangle\langle\Psi_k|\frac{\partial H}{\partial\chi}|\Psi\rangle}{E - E_k} \tag{9}$$

[19]An exception is the so-called triplet-type properties such as the Fermi-contact and spin–dipole contributions to indirect spin–spin coupling constants (for a discussion, see Ref. 44).

where the sum extends over all other states.[20] Hence, in regions where two or more adiabatic potential energy surfaces are in close proximity, the wave function can change rapidly. Second-order properties, defined as those that correspond to second derivatives of the energy, depend upon the response of the wave function and are given by

$$\frac{\partial^2 E}{\partial\chi\partial\lambda} = \langle\Psi|\frac{\partial^2 H}{\partial\chi\partial\lambda}|\Psi\rangle + 2{\sum_k}' \frac{\langle\Psi|\frac{\partial H}{\partial\chi}|\Psi_k\rangle\langle\Psi_k|\frac{\partial H}{\partial\lambda}|\Psi\rangle}{E - E_k} \tag{10}$$

The first contribution depends upon the second-order behavior of the Hamiltonian operator and the unperturbed reference state wave function, while the second term (which will be subsequently be referred to as the "relaxation contribution" or "relaxation term") depends on the derivative of the wave function. This is perhaps most easily appreciated by inserting the equation for the wave-function derivative into that for the second derivative of the energy, giving

$$\frac{\partial^2 E}{\partial\chi\partial\lambda} = \langle\Psi|\frac{\partial^2 H}{\partial\chi\partial\lambda}|\Psi\rangle + \left(\left\langle\Psi\left|\frac{\partial H}{\partial\chi}\right|\frac{\partial\Psi}{\partial\lambda}\right\rangle + \left\langle\frac{\partial\Psi}{\partial\chi}\left|\frac{\partial H}{\partial\lambda}\right|\Psi\right\rangle\right) \tag{11}$$

Considering the above, it should be clear that second-order molecular properties including force constants, electric polarizabilties, NMR chemical shifts, magnetic susceptibilities, and a host of others can exhibit a quite dramatic dependence with respect to nuclear geometry when the relevant adiabatic potential energy surface lies near that of another state. In the immediate vicinity of degeneracies, a plot of a second-order property versus reciprocal excitation energy will exhibit a first-order pole.[21] Properties calculated in the adiabatic (Born–Oppenheimer) approximation in such regions of the potential energy surface generally have little physical relevance per se, but they can be used to calculate nonadiabatic coupling terms for treatments of spectroscopy or dynamics that go beyond the simple Born–Oppenheimer treatment [230]. Hence, force constants, polarizabilities, or other properties can take on extremely large and "unphysical" magnitudes only in the vicinity of potential surface crossings

[20] If the wave function describes the ground state, k represents all the excited states of the system; if the system in question is an excited state, k includes the ground state but excludes the state of interest, which is the meaning usually denoted by the primed summation symbol.

[21] The same is approximately true for a plot of the property versus an appropriate geometrical coordinate, in the approximation that the two states have approximately equal force constants and behave harmonically along the coordinate of interest. It should be emphasized that the coordinate discussed in this context is not the coupling coordinate, but rather a totally symmetric coordinate over which the gap between the coupled states varies. Such a coordinate is often called a "tuning mode" [230].

where the adiabatic treatment is not particularly useful, anyway. However, anyone with experience in calculations on open-shell systems or who has read the corresponding literature knows that strange results are sometimes found for molecules when there is not a low-lying state of interest. Moreover, the behavior can be wholly inconsistent with that predicted by Eq. (10), in the sense that the sign of the anomalous property is not consistent with the relative position of the "perturbing" state in question. For example, vibrational frequencies for the asymmetric C–C stretch of the ground-state ($^2B_2$) cyclic C_3H radical vary from roughly 10,000 to 220,000 cm^{-1} at the usually reliable CCSD(T) level [231], the variation in numbers being that associated with the choice of basis set. All other states of C_3H calculated with correlated methods at the same geometry lay above the $^2B_2$ state [232], so any effect that they might have through the second term of Eq. (10) would be to make the force constant smaller than normal (and perhaps negative). Hence, an anomalously small or perhaps imaginary vibrational frequency would make sense, but not large and real values.

Largely, these anomalous results have historically been attributed to nebulous "symmetry breaking" effects. The common wisdom that has arisen in quantum chemistry is that such strange results are somehow due to so-called Hartree–Fock instabilties, where the lowest-energy ROHF or UHF solution is one that provides a density that does not transform as the totally symmetric irreducible representation of the point group. It is widely believed that results are suspect when these instabilities exist and all manner of catastrophe might befall the unsuspecting user of quantum chemistry. In fact, a statement to this effect is made in the manual that accompanies one of the most quantum chemistry popular program packages. Unfortunately, this contention is at best misleading and at worst wrong. In order to explain what is going on, however, it is first necessary to show how the exact quantum mechanical relationships given by Eqs. (9) and (10) are manifested in the necessarily inexact science of quantum chemistry.

At the HF-SCF level of theory, the wave function is determined completely by the molecular orbitals. In the vast majority of cases, these are given by linear combinations of atom-centered basis functions, and these "MO coefficients" are obtained by the self-consistent field procedure. The first-order change to the wave function is therefore governed by the first-order change in the MO coefficients. It is not difficult to work out expressions for the derivatives of the MO coefficients,[22] and one obtains

$$\frac{\partial c_{\mu p}}{\partial \chi} = \sum_q U_{pq}^{\chi} c_{\mu q} \tag{12}$$

[22]This is the basis of the so-called coupled perturbed Hartree–Fock (CPHF) theory, which, although generally traced to Ref. 233, actually is presented in essentially its complete form in Ref. 235. In Eq. (12), the Greek symbol μ refers to an atomic basis function.

where the CPHF coefficients U^{χ}_{ai},[23] U^{χ}_{ij} and U^{χ}_{ab} can be written as[24]

$$U^{\chi}_{ai} = \sum_{bj} (\mathbf{A}^{-1})_{ai;bj} b^{\chi}_{bj} \tag{13}$$

$$U^{\chi}_{ij} = -\frac{1}{2} S^{\chi}_{ij} \tag{14}$$

$$U^{\chi}_{ab} = -\frac{1}{2} S^{\chi}_{ab} \tag{15}$$

where the perturbation-dependent $\mathbf{b}^{\chi}$ describes changes in the Hamiltonian and orthogonality constraint with the perturbation, $\mathbf{S}^{\chi}$ is the derivative of the atomic orbital overlap matrix transformed into the molecular orbital representation, and the $\mathbf{A}$ matrix is discussed below.

Strong changes in the wave function akin to those that occur in an exact theory near state crossings occur at the SCF level when the $\mathbf{A}$ matrix is nearly singular. Elements of this matrix are given by

$$\mathbf{A}_{ai;bj} = (\varepsilon_i - \varepsilon_a)\delta_{ij}\delta_{ab} + \langle ai \| bj \rangle \pm \langle ab \| ij \rangle \tag{16}$$

where the HF-SCF eigenvalues are denoted by ε_q and δ_{pq} is the usual Kronecker delta. Of course, the difference between occupied and virtual HF-SCF eigenvalues offers only a crude approximation to the excitation energies found in the denominator of Eq. (10), and these never become degenerate in practical calculations. However, since $\mathbf{A}$ is not a diagonal matrix, the conditions under which it becomes singular are not associated with degeneracies in HF-SCF eigenvalues. Rather, the quantities that play the role of excitation energies in determining the response of the molecular wave function are the eigenvalues of $\mathbf{A}$. There is a rather annoying ambiguity here, which will be discussed in slightly more detail later. For now, simply note that the third term is presented without a definite sign. If the perturbation under consideration is of the usual kind that does not cause the orbital coefficients to become complex numbers under its influence, the appropriate sign is plus. If, however, the perturbation is a formally imaginary quantity (such as a magnetic field), the sign changes.

At this point, it is appropriate to return to the issue of Hartree–Fock instabilities, which are perhaps most easily discussed within this framework. Let us consider the somewhat esoteric perturbation χ which is simply a rotation that

[23]The coefficients U^{χ}_{ia} are equal to $S^{\chi}_{ai} - U^{\chi}_{ai}$.

[24]Since the HF-SCF energy and density are invariant with respect to unitary transformations amongst occupied or amongst virtual orbitals, it is only the mixing of occupied and virtual spaces with the perturbation that is of importance. The occupied–occupied and virtual–virtual CPHF coefficients are not uniquely defined, and they only need to preserve orthogonality restrictions. The form given here is the simplest such representation.

mixes occupied and virtual orbitals, and investigate how the energy is affected by this perturbation. If the molecular orbitals satisfy the HF-SCF equations, the first-order change in energy vanishes. In second order, it is fairly simple to show that the change is

$$\frac{\partial^2 E}{\partial \chi^2} = 2 \sum_k{}' \frac{\langle 0| \frac{\partial H}{\partial \chi} |k\rangle \langle k| \frac{\partial H}{\partial \chi} |0\rangle}{d_k} \tag{17}$$

where the sum is taken for convenience over the diagonal representation of orbital rotations described by **A** and the denominators are the corresponding eigenvalues. It should be clear that the presence of a single negative eigenvalue means that a lower-energy solution can be obtained by an appropriate rotation of the molecular orbitals. However, the corresponding numerators must not vanish, a necessary (but insufficient) condition being that the direct product of the irreducible representations of the states $|0\rangle$, $|k\rangle$ and the sense of orbital rotation χ must contain the totally symmetric representation. It is precisely when these negative eigenvalues occur that one has a so-called Hartree–Fock instability.[25] If the associated orbital rotation is not totally symmetric, there is a lower-lying solution to the HF-SCF equations in which the symmetry properties of the overall wave function are corrupted. Less common but relevant in subsequent discussion are cases where the orbital rotation is totally symmetric; when present, there is a lower-energy solution that maintains the same spatial symmetry properties as $|0\rangle$.

Force constants and other second-order properties calculated for molecules at the HF-SCF level contain terms that roughly correspond to the two contributions in Eq. (11), except that it is the eigenvalues of **A** rather than true excitation energies[26] which determine the magnitude of the relaxation term. If none of the eigenvalues of **A** are too small in magnitude, then the presence of an HF-SCF instability does not have foreboding consequences for the results. For example, the UHF[A] **A** matrix for our CN example at 1.28 Å has a pair of negative eigenvalues corresponding to a rotation of π symmetry, and their

[25]For singlets described at the RHF level, the **A** matrix has eigenvectors that correspond to singlet and triplet states (the sense of rotation of orbitals of different spin is the same for the singlet solutions and apposite for the triplets) just as a singlet state has singlet and triplet excited states. In general, these obey a Hund's rule of sorts, and the triplets lay below the singlets. When there is one or more negative eigenvalues corresponding to triplets, then one has the so-called "triplet instability" mentioned in the previous section and it is possible to find a lower-energy UHF solution to the HF-SCF equations.

[26]The set of excitation energies consistent with the SCF parameterization of the wave function are those given by the random-phase approximation (RPA), in the sense that these are the locations of poles in the dynamic electric polarizability. The eigenvalues of **A** are *not* the RPA energy differences, and they often deviate significantly from them.

magnitude is relatively small (slightly greater than -1 eV). This shows up to some degree in the perpendicular polarizability, where the HF-SCF value is calculated to be 4 atomic units, while the corresponding CCSD result is about three times higher. However, the ROHF polarizability is anomalously high (30 atomic units) even though the solution is stable. What is the problem with ROHF? Certainly it is not spin contamination, and there are none of the dreaded "Hartree–Fock instabilities" either. The issue is simply that the ROHF **A** matrix has an eigenvalue of 0.005 atomic units (roughly 0.1 eV), and the orbital relaxation contribution to the property is anomalously large and positive. While this point has been made in the literature before [126,236], it warrants repetition and emphasis:

It is not the presence of wave-function instabilities that poses problems for the calculation of molecular properties, but rather the presence of near-instabilities. It is entirely possible to obtain quite satisfactory results with a highly unstable HF-SCF solution, and it is also entirely possible to obtain completely absurd results with a stable solution. The important issue is the smallest eigenvalue of the A matrix in magnitude, irrespective of sign.[27]

Truly strange things can occur when electron correlation is included [236]. Short of an FCI treatment, the total energy is dependent upon the choice of orbitals. Therefore, properties calculated as second derivatives also involve contributions from the relaxation of the molecular orbitals. For methods like CC theory that show relatively strong invariance of the total energy to orbital changes (see the results for CN presented in the previous section), the effects are rather minimal and tend to be conspicuous if and only if the orbital relaxation is very strong. Specifically, problems occur in regions of the potential energy surface where there are very small eigenvalues of **A**. Nevertheless, even at the CC level, one is sometimes unlucky. A case in point is the ground $^2A_1$ state of NO_2, where a seemingly adequate calculation using the fashionable cc-pVTZ basis set [56] and the CCSD(T) treatment of correlation yields an asymmetric stretching frequency in excess of $17{,}000\,\text{cm}^{-1}$! There is negligible spin contamination, and the first excited ($^2B_2$) state is a few electron volts *higher* in energy.[28] Moreover, there is no Hartree–Fock instability [231]. Aficionados of coupled-cluster theory could look at the unperturbed wave function and see no

[27]Some qualification is in order here, because although good results can be obtained with highly unstable wave functions (particularly at the correlated level), it can be difficult to obtain them in practice. The reason is that a numerical calculation of the property of interest necessitates that the wave function be evaluated in the presence of the perturbation. It can then be difficult to obtain the "correct" solution—the one that correlates with the unperturbed and unstable solution—because it is impossible to impose the symmetry constraints of the unperturbed problem. Nevertheless, this solution does exist; it is simply a matter of obtaining the correct solution to the HF-SCF equations. However, problems of this sort are avoided when derivatives are calculated analytically.

[28]Note the apparent contradiction with the direction of the anomaly.

sign of "multireference" behavior. So what is wrong? The problem is again a near-singularity of the UHF wave function; considerably more sensible results can be obtained with an ROHF reference. Having said that, for every case where one faces a UHF near-instability and can obtain improved results with another UHF solution or an ROHF reference, there is another case where ROHF fails and UHF offers a practical advantage. *A priori*, it is generally not possible to know which sort of reference function is best to study a particular open-shell molecule, to say nothing of the subsequent question of how the electron correlation effects should be treated. If one does not know how to best start the calculation, how can the study of open-shell molecules ever be relegated to a black box procedure?

Another oddity is that the sign of anomalous property values calculated at correlated levels does not give away the sign of the corresponding small **A** matrix eigenvalue. Because the total energy is not stationary with respect to orbital rotations, it can be shown [236] that the properties blow up quadratically with reciprocal eigenvalue rather than linearly. As a result, if the magnitude of a second-order property is plotted as a function of a geometrical coordinate, it will blow up to plus or minus infinity on both sides of the point where **A** is precisely singular. The sign can be demonstrated to be that of the second derivative of the total energy with respect to an orbital rotation, and the magnitude of the singularity (essentially the region of the potential surface over which the property is strongly affected by the instability) correlates with the magnitude of the second derivative. For methods that are highly sensitive to orbital choice (perturbation theory, especially the popular MBPT(2) variant), the presence of the singularity can cause serious problems for the calculation of second-order properties even far from where they actually diverge. In general, our experience has shown that the total energy that corresponds to a symmetry-broken reference function is generally higher than that calculated from the symmetry-constrained function (exactly the opposite of the relative HF-SCF energy ordering), so properties typically diverge to large positive values on both sides of the singularity.

Finally, something should be said about "symmetry-breaking" effects in molecules where there is no symmetry, when the same fundamental problem that causes the instability does not lead to a wave function having lower symmetry. Essentially, the reason that HF-SCF calculations often give symmetry-broken solutions at lower energies than constrained solutions is that in many symmetric open-shell molecules (NO_2, LiO_2, and others) the broken symmetry solution allows the unpaired electron spin to be almost entirely located on a single atomic center [94]. This increases the exchange contribution to the energy and therefore decreases the total energy. However, the same unphysical localization phenomenon can also occur in molecules where the localization does not break the framework molecular symmetry. This minefield for computational chemists is quite simply exemplified by comparing the isoelectronic CN and N_2^+ molecules. If one blindly runs UHF calculations for both species, it appears that they behave quite differently. At and near the equilibrium geometry of CN, the UHF[A] solution discussed in the previous section is obtained, and the spin contamination if quite significant. For N_2^+ at about the same internuclear distances, the UHF equations converge to a solution that is only weakly contaminated.[29] MBPT(2) behaves normally when this reference function is used (the bond length increases relative to the uncorrelated calculation) and N_2^+ is apparently quite well-behaved. This of course seems rather odd, but is easily explained. When the **A** matrix is constructed and diagonalized, one finds a large negative eigenvalue (just above $-3\,\text{eV}$) of σ_u symmetry. When the symmetry constraint of the wave function is lowered, the SCF procedure converges to a solution that is badly spin-contaminated ($\langle S^2 \rangle = 1.66$). The corresponding spin density on one of the nitrogen nuclei is 1.65 a.u. (the other is -0.35 a.u.), two orders of magnitude greater than the value of 0.016 a.u. computed with the symmetry-constrained solution. CCSD(T) correlation corrections change these values to 0.29 and 0.20 for the symmetry-broken solution and 0.30 for the pure symmetry solution. The asymmetry in the spin distribution as well as the residual dipole moment at the CCSD(T) level (0.10 Debye versus 1.58 Debye at the UHF[A]-SCF level) gives some measure of the error in the CCSD(T) treatment of correlation. In any event, the symmetry-broken solution is definitely one that exaggerates unpaired spin localization and is precisely the solution that corresponds to the UHF[A] solution for CN! In that case, the spin density on the carbon atom at the HF-SCF level is 1.239 a.u., a value that is reduced to 0.50 a.u. at the CCSD(T) level.

The fundamental difference between CN and N_2^+ is simply that one molecule is centrosymmetric while the other is not. The lowest-energy UHF wave function in both cases suffers from unphysical spin localization, and it is illusory to believe that N_2^+ is the "easier" of the two molecules to calculate. The low spin-contamination solution to the UHF equations exists simply because of the molecular symmetry, while in CN the lower symmetry of the molecule allows the equations to converge to the lowest energy solution. This is a somewhat unappreciated difficulty in calculations on open-shell molecules. If one has appropriate elements of symmetry, then unphysical solutions can be avoided by enforcing the constraints on the wave function. Even if the constraints are not enforced, problems with the reference function are easily identifiable: nonzero dipole moments along directions where the exact value must vanish by symmetry, unsymmetric spin densities, and so on. However, the issue is more diabolical in lower-symmetry species where localization does not break the framework molecular symmetry. In these cases, UHF and ROHF

[29]Using the DZP basis set and a bond distance of 1.20 Å, $\langle S^2 \rangle_{SCF}$ is 0.771.

solutions can be quite unphysical. However, it is considerably more difficult to recognize the problem (especially with ROHF since it is necessarily an eigenfunction of the S^2 operator) and might be impossible to prevent. Therefore, the term "symmetry-breaking" is, in our view, not entirely appropriate. Rather, it applies to a larger class of problems than just those to which the name literally applies.

V. THE PSEUDO-JAHN–TELLER EFFECT IN QUANTUM CHEMISTRY

In the previous section, strongly varying wave functions and anomalously large relaxation contributions to molecular second-order properties were discussed. In the context of near-singularities of the reference function, these are entirely artifactual in nature and are best avoided. However, it is important that one not lose sight of the fact that these effects *should* occur in other circumstances, namely where the spectrum of the exact electronic Hamiltonian has a near-degeneracy. An excellent early review of the symmetry-breaking issue [94] distinguished loosely between "artifactual" and "real" effects. In the language of the present work, the former are those arising from problems associated with the reference function, while "real" effects are those due to true nearby states. When there is a strong coupling with a nearby state, force constants can be profoundly affected and otherwise symmetric molecules assume nonrigid behavior (such as NO_3 [237–239]) and might even adopt equilibrium geometries with lower symmetry. Systems so affected are said to experience a second-order or pseudo-Jahn–Teller (PJT) effect [240]; this is an important topic in the study of open-shell molecules, electron transfer, and a number of other issues in physical chemistry. The question of how various theories used in quantum chemistry account for PJT effects is surprisingly complicated and will be discussed briefly in this section.

It should be clear that SCF calculations cannot account for these effects in anything resembling an appropriate manner unless by sheer chance the **A** matrix eigenvalues are close to the exact excitation energies and that they vanish where there is a true degeneracy of states. As an example, the $^2A_1$ and $^2B_2$ states of NO_2 cross in C_{2v} symmetry at an ONO bond angle near 110°, but there is no singularity in the **A** matrix in this region. Hence, the relatively small force constant associated with the b_2 stretching mode in NO_2 obtained at very high levels of theory [231,241] is significantly overestimated in HF-SCF calculations owing to the poor treatment of the relaxation contribution. Again, that HF-SCF should fail to describe a situation of this sort is not particularly surprising. Significantly more interesting is the behavior of perturbation theory. MBPT(2) and higher-order calculations have been, and continue to be, used to study radicals that are subject to PJT effects, but the fact that perturbation theory is intrinsically unable to describe this behavior *irrespective of order* [126,242] is not widely appreciated. The reasons are relatively simple. In addition to the molecular orbital coefficients, the only additional parameters that come into play in the MBPT wave-function expansion are those associated with denominators formed from differences of orbital energies. In practice, these never vanish, and the only poles that occur in property calculations are those that are already found at the HF-SCF level of theory. The exact poles that reflect the true spectrum of the Hamiltonian will *never* be observed in calculations based on perturbation theory, and a rather persuasive argument can be formulated that SCF may be superior to MBPT for studying PJT effects. At the very least, poles that occur at the HF-SCF level are of the proper first-order behavior, even though the energy differences that the properties depend on are clearly only crude approximations to the excitation energies. However, the high sensitivity of the MBPT(2) energy to orbital rotations and the second-order nature of instability poles at correlated methods generally mean that the domain of nuclear configuration space affected by the singularity is much larger than at the HF-SCF level. Accordingly, the chance of obtaining an anomalous result is greater.[30]

Coupled cluster calculations represent a qualitative improvement over perturbation theory in treating PJT effects. The principal difference between CC and MBPT in this context is that the CC wavefunction involves an additional nontrivial set of parameters, specifically the T amplitudes [see Eq. (2)]. Furthermore, it can be shown [126] that force constants in CC theory can be written as

$$\begin{aligned}\frac{\partial^2 E}{\partial \chi^2} &= \langle \tilde{\Psi}_{CC} | \frac{\partial^2 H}{\partial \chi^2} | \Psi_{CC} \rangle - 2 \sum_k{}' \frac{\langle \tilde{\Psi}_{CC} | \frac{\partial H}{\partial \chi} | \Psi^k_{EOM} \rangle \langle \tilde{\Psi}^k_{EOM} | \frac{\partial H}{\partial \chi} | \Psi_{CC} \rangle}{E_k - E} \\ &\quad + \langle 0 | L\bar{H} | \mathbf{q} \rangle \langle \mathbf{q} | \frac{\partial T}{\partial \chi} \frac{\partial T}{\partial \chi} | 0 \rangle \end{aligned} \tag{18}$$

In the equation above, the symbol $\mathbf{q}$ designates determinants that lie outside the principal projection space used in the CC equations [Eq. (7)]—that is, triples, quadruples, and so on, for CCSD. Note that the first two terms strongly resemble those in the exact expression [Eq. (10)]. In fact, to the extent that the right and left CC wave functions approximate the exact wave function and its Hermitian conjugate and that the excited states of the system are represented by the

[30] In addition, the affected force constant of second-order property will likely have an artificially large *positive* value, irrespective of whether the state in question is "interacting" (through the small **A** matrix eigenvalue with a higher or lower state, in contrast to the HF-SCF case where the sign of the energy difference determines the direction of the relaxation contribution to the property).

EOM appoximation, these terms correspond precisely to the exact quantum-mechanical result. However, there is an additional term that spoils the fundamental description of PJT effects within coupled-cluster theory and warrants discussion. This term (which vanishes in the FCI limit) depends quadratically on the derivative of the T amplitudes, which in turn diverge when another state in the EOM-CC spectrum becomes degenerate with the state of interest. Hence the force constant[31] blows up quadratically in the immediate vicinity of the state crossing, which is clearly not the correct sort of behavior. Moreover, the direction of the second-order contribution is given by $\langle 0|L\bar{H}|\mathbf{q}\rangle$, which resists a detailed mathematical analysis; indeed, we know of examples in which the second-order contribution goes in each direction. Fortunately, the residue of the second-order contribution can be shown to vanish in the limit of a pure single excitation process [126]. Since most excited states of interest are those that have largely single excitation character, the region of the potential surface strongly affected by the second term is small. The relaxation term that depends on the EOM excited states, however, has an appropriate magnitude *and sign* so standard CC methods[32] represent a qualitative improvement over MBPT. Therefore, at distances relatively far from the crossing, the sign and magnitude of the relaxation effect on the force constant (or other second-order property) is generally fairly accurate. In this context, it should be emphasized that the equilibrium geometry of most open-shell molecules is generally not close to adiabatic surface crossings, even though they might experience a PJT effect that is not negligible. In these cases, CC methods can provide reliable results, as seen in a number of examples in the literature [231,241].

In the last decade, a few approaches have been developed and applied in which the orbitals used in the CC calculation are optimized self-consistently with the energy. These include (a) the Brueckner CC (B-CC) methods in which the orbitals are chosen so that the T_1 (single excitation) amplitudes vanish [107,108,110] and (b) the optimized-orbital CC (OO-CC) doubles approach in which the CCD energy is minimized [201,244]. Since the parameters used in these methods (orbital rotation coefficients and cluster amplitudes) are

[31]The discussion here applies to any second-order property, indcidentally. Force constants are simply used as an example, but the troublesome second term affects polarizabilities, dipole moment derivatives, and so on, as well.

[32]Standard CC methods, which have been termed "plain old" CC (POCC) in the literature [231], are those in which the orbital optimization and correlation steps of the calculation are performed separately. POCC calculations therefore suffer from instability poles in addition to the appropriately located EOM poles, but the width of the former are quite small because of the approximate orbital invariance of CC methods that include single excitations [243]. These methods offer some advantages in treating PJT effects relative to CC approaches in which orbitals and cluster amplitudes are determined simultaneously, as discussed briefly in the next section.

determined in a coupled fashion, there is just one set of linear equations that must be satisfied by the perturbed wave function. Hence, the problem of artifactual instability poles is alleviated, and these methods can be used to study open-shell problems in cases where there is a nearly singular **A** matrix based on HF-SCF orbitals. Additionally, they offer advantages in numerical calculations of properties or scans of potential energy surfaces where traditional HF-SCF-based approaches are impossible to use because of problems associated with converging to the appropriate solution when lower solutions arising from instabilities are present. For these purposes, B-CC and OO-CC can be very useful indeed, since it has been demonstrated that artifactual symmetry breaking and spin contamination effects are greatly ameliorated by these approaches [109,245–247].

However, there is some danger in using these methods to study radicals, which has been noted in the literature only quite recently [241]. When there is a potential for a true PJT effect, these methods appear to be somewhat less adequate than one might hope. Although the technical reasons for the problem are beyond the scope of this presentation, they have been discussed elsewhere.[33] Briefly, the orbital relaxation in B-CC and OO-CC is governed by linear equations that involve something akin to the **A** matrix, again with a sign that is dependent on the type of perturbation (real or imaginary). Hence, the effective denominators in the response equations are *not* the excitation energies of the system, and in fact are again dependent on the type of perturbation [248]. So one can obtain results for an electric polarizability that are relatively unaffected by a singularity and magnetizabilities that are strongly affected (or vice versa), an unphysical and aesthetically unappealing feature of these theories. Of course, if the CC treatment is taken very far (CCSDT and higher levels), the difference between magnetic and electric properties[34] becomes smaller and ultimately vanishes. At the computationally convenient doubles level, however, a recent paper shows that B-CC fails qualitatively in describing true PJT effects in a number of small radicals [241]. In each case, the magnitude of the effect is grossly exaggerated by B-CC. Moreover, a formal analysis [249] as well as direct computation [241] shows that the poles in plots of second-order properties versus a geometric coordinate are strongly second order in nature and unreasonably large in extent. Hence, we conclude that these methods should be used only with caution for radicals and are by no means a solution to the "symmetry breaking problem."

[33]For B-CC. The problems with OO-CC have not been mentioned elsewhere, but fundamentally are due to the same causes as B-CC.

[34]Which arise from the so-called **B** matrix of response theory. It is easily shown that **B** vanishes in the full B-CC or full OO-CC limits.

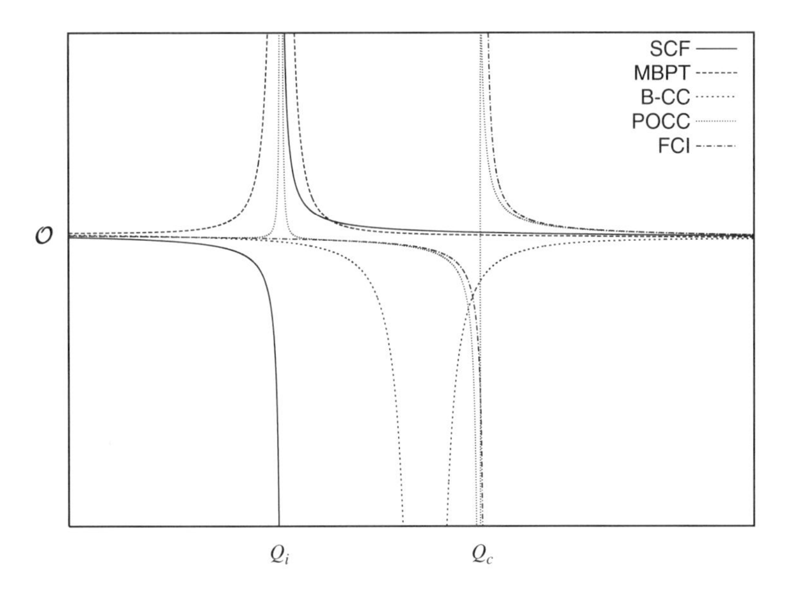

Figure 1. Schematic representation of the second-order property $\mathcal{O}$ as a function of a totally symmetric coordinate. The points Q_i and Q_c represent values of the totally symmetric coordinate Q where the SCF **A** matrix is singular and where the state of interest becomes degenerate with another state in the exact spectrum. See text for details.

Figure 1 shows a schematic representation of the situation discussed in this section. The ordinate in the plot is a generic "tuning mode" [230] along which the gap between the state of interest and an excited state varies. At the point Q_c, the states cross and the exact value of the second-order property $\mathcal{O}$ diverges. To the left of Q_c, the perturbing state lays above the state of interest, while it falls below on the other side of the intersection. The pole exhibits first-order behavior. The HF-SCF solution does not exhibit any irregular behavior in this vicinity, but rather has a first-order pole of its own at the point Q_i which is where the **A** matrix is exactly singular. The eigenvalue that is zero at Q_i is positive to the left of Q_i and negative to the left Q_i. Hence at $Q > Q_i$, the HF-SCF solution is unstable. The perturbation theory result has a singularity only at Q_i, but the pole exhibits second-order behavior with one or the other sign, but which is usually positive. The relative width shown in the figure is typical of MBPT(2). For CC calculations based on an HF-SCF reference function, there are two features; one of these occurs at Q_i and has pronounced second-order character but is much narrower than the corresponding feature in the perturbation theory calculation. The second feature occurs at a value of Q where the excited state given by the corresponding EOM-CC calculation is degenerate with the reference state. If EOM-CC is offering a good treatment of the excited-state spectrum,[35] then this should be close to the location of the exact intersection. However, the behavior of the pole is not consistent with the exact theory, because it exhibits a second-order contribution. As it is plotted here, the second-order feature goes off toward positive values, but here the direction of the pole is very difficult to predict *a priori*. Finally, the B-CCD or OO-CCD calculation gives a very wide second-order feature at a point that lies intermediate between Q_i and Q_c. Note that B-CCD or OO-CCD calculations of properties at Q_i should be relatively accurate, but that these methods are apt to fail for any geometry close to Q_i.

It is clear that none of the methods described here are entirely satisfactory in all respects for PJT problems. HF-SCF and perturbation theory are wholly inadequate, while CC methods represent a substantial but still incomplete improvement. Single-reference CI calculations have not been discussed in this context, but suffice it to say that the interacting state is that described as a higher root of the CI matrix. Since the CISD method that gives a satisfactory treatment of ground-state correlation effects performs much less well for excited states,[36] the associated poles do not occur where they should, although they do have the correct first-order behavior (which is due to the variational nature of the method). Furthermore, CISD is very sensitive to orbital rotations, and the instability poles (which are second order in CI) will be extremely wide. In conclusion, none of the Type I methods (see Section II) gives a qualitatively satisfactory description of PJT effects. The remainder of this section gives a very brief overview of the way these effects are treated by other types of calculation, with the caveat that our expertise lays in the areas of MBPT, CC, and related methods and that our knowledge of how PJT effects are treated in other areas of quantum chemistry is not comprehensive.

Density functional calculations are in many respects similar to SCF in implementation, and the PJT effect is again carried by the **A** matrix, although the terms before and after the indeterminate $\pm$ sign in Eq. (16) and the eigenvalues are of course different in magnitude. Although the theoretical basis is unclear, experience indicates that the eigenvalues of the **A** matrix in DFT are actually relatively close to the excitation energies of the system in many cases [250,251], and that the term following the $\pm$ sign is relatively small in

[35]The EOM-CCSD method offers an accurate treatment of states that are largely described by single excitations, and it gives a considerably poorer description of doubly excited states. For a discussion, see Ref. 190.

[36]Only one root of the matrix is correct through third order, in contrast to EOM-CCSD where "single excitations" are correct through thrid order, as is the underlying CCSD energy. The result of this is that EOM-CCSD excitation energies are vastly better than those given by higher roots of the CISD matrix [184].

magnitude. Hence, instability poles for real and imaginary perturbations are spatially proximate, and these occur closer to the "right place" than do those of HF-SCF theory. Moreover, they are first order in character. Hence, DFT is probably a better way to treat PJT effects than is HF-SCF, even when correlation is treated at any order of perturbation theory beyond the SCF approximation, as attested to by a few examples in the literature [252,253]. However, it is necessary to be cautious because DFT is of course a semiempirical theory in construction, and this rather weak recommendation is based on the tacit assumption that the spectrum of the **A** matrix is a fairly faithful representation of the exact energy level structure. Counterexamples are known where DFT fails qualitatively to describe symmetry-breaking effects [254], so it is best to simply point out that this class of method might give good results for problems where other methods fail.

QMC methods (type III) involve a direct numerical solution of the Schrödinger equation, subject to restrictions associated with the placement of nodes in nontrivial multielectron systems. Hence, they potentially provide an exact treatment of PJT effects, just as they provide a potentially exact treatment of all other molecular properties. However, there seems to have been very little work done in using QMC to study problems involving potential energy surfaces of radicals, possibly because of the numerical uncertainty issues associated with these calculations. Nevertheless, the potential for such applications is vast, and we encourage the QMC community to explore this challenging and important area of application.

Methods based on MC starting points (type III) are similar to type I methods in many respects. The response of the MCSCF wave function is also governed by a generalized form of the **A** matrix [255] and differs for both real and imaginary perturbations. Just like UHF- and ROHF-SCF calculations, MCSCF calculations are subject to, and often afflicted by, instabilities and near instabilities for radicals[37] and therefore the entire set of issues associated with "artifactual symmetry breaking." The only difference is that the MCSCF instability poles should be closer to the correct locations than the corresponding features obtained in single-reference calculations, insofar as the MCSCF solution approaches the exact (FCI) result as the size of the configurational basis is increased. Nevertheless, the treatment of true PJT effects in MCSCF calculations is a very difficult business, and even the qualitative nature of a particular stationary point can be strongly dependent upon the choice of active space [257]. The treatment of these problems within an MCSCF framework is far from straightforward and should probably only be tackled by experts in the field.

[37]Excellent discussions of this issue can be found in Refs. 255 and 256.

When residual correlation effects are built upon the MCSCF starting point with CI, then the MCSCF instability poles will become second order in character.[38] Additional poles will then be located where the state of interest becomes degenerate with another eigenvalue of the CI matrix which is diagonalized. This will be first order in character and, in contrast to single-reference CI, should be in approximately the right region of space. Hence, large-scale MRCI calculations should be an excellent means of treating PJT effects. The only qualifications that we wish to express are those associated with the cost of performing such calculations for polyatomic molecules (the effects discussed here are relatively uninteresting for atoms and diatomic molecules, and then only for properties other than force constants), and there might be "interesting" interference effects associated with the MCSCF and MRCI poles if they are located close together. As far as we know, nothing at all has been done to investigate these issues to date, but such a study would be most interesting.

VI. SKIRTING THE REFERENCE FUNCTION ISSUE

Since so many problems are encountered in the construction of a zeroth-order representation for many open-shell systems, it is logical to wonder if the entire issue can somehow be avoided. Indeed, it is possible to circumvent problems of instabilities, near-instabilites, and spin contamination by using reference determinants comprising orbitals that are not variationally optimized for the state of interest. Such a reference is generally somewhat "far" from the exact wave function, and a high-level treatment of correlation is needed for such approaches to be useful. Accordingly, little work has been done in this area within the context of perturbation theory, but there is an approach that has been advocated for CC calculations. In the "quasi-restricted" QRHF model [258], orbitals are first obtained in an HF-SCF calculation for a nearby closed-shell state (one generally having more or fewer electrons). After this, the CC equations are solved for the radical, using the orbitals of the closed-shell system. If the latter behaves normally, there should be no reference function (near-) instabilities in the region of interest. Hence any strong changes in the wave function or large values of second-order properties must be due to a "real" curve crossing.[39] Hence, artifactual symmetry breaking is completely avoided in QRHF-CC calculations. Regrettably, however, a very limited amount of computational evidence suggests that the QRHF-CCSD method

[38]They are strictly first order at the simple MCSCF level, just as the UHF or ROHF-SCF poles are first order when no correlation corrections are included.

[39]Here the interaction state is one given by the EOM-CC calculations based on the QRHF-CC wave function [259].

overestimates PJT effects [238,260] and that some treatment of triples is generally necessary to achieve qualitative accuracy. Almost nothing has been done to investigate the source of these problems, although we conjecture that they arise from a large (second order) contribution to the property through the last term in Eq. (18), presumably with a negative sign.[40] Finally, like ROHF-CC calculations that also start with a spin-adapted reference function, the nonlinear parameterization of the wave function leads to a small amount of spin contamination in the correlated wave function.

In a related class of approaches, the HF-SCF wave function is calculated for the "nearby" closed-shell state (usually that with one additional electron in studies of doublet radicals). The electronic Hamiltonian is then projected onto determinants that are appropriate for the state of interest. The most common realization of this strategy is the so-called 2*hp*–TDA (Tamm–Dancoff approximation) [157,261–263], in which the determinants used in the CI calculation are those in which one electron has been removed from the closed-shell reference (*h* type) and those obtained from these by promotion of one electron (2*hp*). Final states so obtained are rigorously spin-adapted, and all determinants in the *h* set are treated in a balanced fashion. The balanced nature of the treatment is, of course, the desirable feature associated with MR approaches,[41] and 2*hp*–TDA is capable of treating cases involving strong configuration mixing. Furthermore, since the reference is a closed-shell state, there should only be "real" singularities in the property surfaces, and these should be properly first order in nature. However, the method offers a minimal treatment of dynamical electron correlation effects and moreover is not size-consistent. The latter two deficiences have tended to reduce the scope of its chemical applications.

Conceptually related and numerically superior to 2*hp*–TDA are the EOMIP-CC [264] and related EOMEA-CC [265] methods.[42] In these approaches, the HF-SCF *and* coupled-cluster wave functions are obtained for the closed-shell state,[43] and the similarity transformed Hamiltonian $\bar{H}$ (see Section II) parameterized by the closed-shell T amplitudes rather than the bare Hamiltonian

[40]Examples where QRHF-CCSD seems to fail qualitatively include ground states of the NO_2 and NO_3 radicals.

[41]It effectively accounts for nondynamical correlation effects.

[42]Results given by the EOMEA and EOMIP methods are equivalent to those of certain variants of the Fock-space coupled-cluster (FSCC) method. For a discussion of this correspondence, as well as an overview of and references to the general FSCC approach, see Ref. 266.

[43]It is not necessary to base the calculation on a singlet closed-shell reference state, but the vast majority of chemical applications choose this option. Open-shell reference EOM-CC and EOM-CC methods are, however, relevant to the discussion of "real" singularities, since it is the (ROHF)UHF-EOM-CC excitation energies relative to the open-shell state of interest that give the energy differences in the denominators of Eq. (18).

is diagonalized in the $h, 2hp$ basis. Like $2hp$–TDA, EOMIP and EOMEA offer nicely balanced and spin-adapted descriptions of final states, but the latter methods are also rigorously size-consistent and effectively treat dynamical as well as nondynamical correlation effects. EOMEA-CC differs from EOMIP-CC only in that the closed-shell reference state has one fewer (rather than one more) electron than the state that is ultimately of interest. The pole structure in EOMIP-CC and EOMEA-CC methods is interesting and potentially complicated [126]. The principal feature is that associated with degeneracies in the spectrum of the n electron $\bar{H}$ operator, which correspond to "real" PJT effects. They have the proper first-order behavior and are located in approproximately the correct positions. These methods offer a treatment of the PJT phenomenon that is formally superior to all of the methods discussed previously, with the possible exception of large-scale MRCI calculations. However, the other types of poles that can occur in EOMIP and EOMEA—those corresponding to **A** matrix singularities and degeneracies in the $n + 1$ (EOMIP) or $n - 1$ (EOMEA) $\bar{H}$ spectrum (which both have second-order contributions)—are unlikely to appear unless the chosen closed-shell reference state is pathological. Thus, the interference effects between poles arising from different sources that may appear in MRCI are a most unlikely event in EOMIP and EOMEA.

Although the vast majority of applications using EOMIP-CC methods in the literature have been carried out at the CCSD level, it can be argued that these methods treat dynamical electron correlation somewhat less effectively than the standard open-shell UHF and ROHF-based CCSD methods. Thus, for well-behaved problems, using the standard approaches certainly offers advantages, especially when the noniterative (T) correction is applied. However, some work has recently been devoted to extending the EOMIP and EOMEA methods to CC treatments that include some treatment of triple excitation effects [231,267,268]. Although there does not appear to be a truly reliable noniterative approach, these methods have been implemented with CC3, CCSDT-3, and (recently) CCSDT [269] correlation treatments. Calculations using these methods, while expensive, offer a promising means to routinely treat even the most difficult problems involving doublet radicals at very high levels of theory. However, experience suggests that even the simple CCSD-based models are quite reliable in determining qualitative features of the PJT effect and to describe the topology of strongly interacting potential energy surfaces.[44]

Other methods that share the same fundamental properties as EOMIP-CC and EOMEA-CC (balanced treatment, spin-adapted, and size-consistent)

[44]The "interaction" referred to here as well as other places in the text is between the electronic states at a reference nuclear geometry—that is, those that appear in the Herzberg–Teller expansion of the potential and which form the basis used in Eqs. (9) and (10).

include methods based on electron propagator theory [174,178,179] and Green's functions [270–272]. Indeed, such approaches have been advocated for many years to treat problems involving symmetry breaking in radicals by groups that with a long-standing interest in this area. This insightful choice of methods and the subsequent development that went on was, however, not widely adopted in the chemical community. Nevertheless, several variants of EPT have recently been included in popular quantum chemistry packages, and we suspect that the use of these methods as well as EOMIP-CC and EOMEA-CC will increase for applications to radicals in the future.

All of the methods discussed thus far in this section apply only to doublet radicals. While it is fair to say that doublets pose more problems than radicals of higher spin multiplicity, the question arises as to how one can best treat triplets, quartets, and so on. Type I methods can be applied to these states just as easily as doublets (and in fact, spin contamination tends not to be so bad for higher multiplicities), but they are plagued by precisely the same set of problems that is encountered for doublets. Similarly, DFT and MR methods can also be applied to these cases, presumably with equal degrees of success. At present, MRCI calculations are probably the best option for very high accuracy treatments in which the issue of interacting states and PJT effects arise, while CCSD(T) and higher-level single-reference methods are competitive for less difficult examples.

The development of type V methods for higher spin multiplicities would be highly useful. To some degree, it is already achievable for triplets through the EOMEE-CC and related approaches designed for excitation of closed-shell systems. Triplet excitation energies can be calculated with these methods, and then the final-state energies can be used to describe potential energy surfaces. Such a method is spin-adapted, balanced, and comparable in theoretical rigor to the EOMIP-CC and EOMEA-CC approaches. It has, however, seen very little application in the literature despite the recent formulation of a computationally efficient method for triplets that fully exploits the spin symmetry of $\bar{H}$ [273].

VII. SUMMARY

Molecules with unpaired electron spins play a role in several important areas of chemistry, including biological and atmospheric reactions, the process of radiation damage, and as products of combustion reactions. Because of their highly reactive nature, radicals are often very difficult to study in the laboratory, and theory can play a pivotal role in gaining an understanding of open-shell molecules. In this review, we have tried to make several points. First, the quantum-chemical calculation of open-shell molecules is far from straightforward, and efforts to study any such system must be carefully thought out and examined. Unlike closed-shell molecules where various black-box quantum chemical procedures almost always give quite predictable levels of accuracy, the same is not true for open-shell molecules. Second, the vast majority of problems associated with the theoretical description of radicals involves the selection of an appropriate reference function for the calculation. The problems that occur might or might not be obvious to the casual user of quantum chemistry programs, and they might even be so subtle that they are not noticed by even experienced practitioners in the field. Finally, we have attempted to describe the problems alluded to above in a fair amount of detail, and we have given an overview of strategies that can be adopted in difficult cases.

Acknowledgments

This work was supported by grants from the National Science and Robert A. Welch Foundations (JFS) as well as the Fonds der Chemischen Industrie (JG). We are also indebted to T. Daniel Crawford (Virginia Tech) and P. G. Szalay (ELTE, Budapest) for helping to shape our perspective on open-shell problems through both discussions and research collaborations.

References

1. I. Shavitt, in *The Electronic Structure of Atoms and Molecules*, H. F. Schaefer, ed., Addison-Wesley, Reading, MA, 1972.
2. For an up-to-date overview of developments in the field of MRCI methods, see H. Lischka, R. Shepard, R. M. Pitzer, I. Shavitt, M. Dallos, T. Müller, P. G. Szalay, M. Seth, G. S. Kedziora, S. Yabjshita, and Z. Zhang, *Phys. Chem. Chem. Phys.* **3**, 664 (2001).
3. R. J. Bartlett, *Annu. Rev. Phys. Chem.* **32**, 359 (1981).
4. D. Cremer, in *Encyclopedia of Computational Chemistry*, P. v. R. Schleyer, N. L. Allinger, T. Clark, J. Gasteiger, P. Kollman, H. F. Schaefer, and P. R. Schreiner, eds., John Wiley & Sons, New York, 1998.
5. R. J. Bartlett, in *Modern Electronic Structure Theory, Part II*, D. R. Yarkony, ed., World Scientific, Singapore, 1995.
6. T. J. Lee and G. E. Scuseria, in *Quantum Mechanical Electronic Structure Calculations with Chemical Accuracy*, S. R. Langhoff, ed., Kluwer, Dordrecht, 1995.
7. J. Gauss, in *Encyclopedia of Computational Chemistry*, P. v. R. Schleyer, N. L. Allinger, T. Clark, J. Gasteiger, P. Kollman, H. F. Schaefer, and P. R. Schreiner, eds., John Wiley & Sons, New York, 1998.
8. T. D. Crawford and H. F. Schaefer, *Rev. Comp. Chem.* **14**, 33 (2000).
9. P. Pulay, *Mol. Phys.* **17**, 197 (1969).
10. B. R. Brooks, W. D. Laidig, P. Saxe, J. D. Goddard, Y. Yamaguchi, and H. F. Schaefer, *J. Chem. Phys.* **72**, 4652 (1980).
11. R. Krishnan, H. B. Schlegel, and J. A. Pople, *J. Chem. Phys.* **72**, 4654 (1980).
12. J. A. Pople, R. Krishnan, H. B. Schlegel, and J. S. Binkley, *Int. J. Quantum Chem.* **S13**, 255 (1979).
13. G. Fitzgerald, R. Harrison, W. D. Laidig, and R. J. Bartlett, *J. Chem. Phys.* **82**, 4379 (1985).
14. J. Gauss and D. Cremer, *Chem. Phys. Lett.* **138**, 131 (1987).
15. I. L. Alberts and N. C. Handy, *J. Chem. Phys.* **89**, 2107 (1987).

16. E. A. Salter, G. W. Trucks, G. Fitzgerald, and R. J. Bartlett, *Chem. Phys. Lett.* **141**, 61 (1987).
17. J. Gauss and D. Cremer, *Chem. Phys. Lett.* **153**, 303 (1988).
18. J. D. Watts, G. W. Trucks, and R. J. Bartlett, *Chem. Phys. Lett.* **164**, 502 (1989).
19. N. C. Handy, R. D. Amos, J. F. Gaw, J. E. Rice, E. D. Simandiras, T. J. Lee, R. J. Harrison, W. D. Laidig, G. B. Fitzgerald, and R. J. Bartlett, in *Geometrical Derivatives of Energy Surfaces and Molecular Properties*, P. Jørgensen and J. Simons, eds., Reidel, Dordrecht, 1986.
20. N. C. Handy, R. D. Amos, J. F. Gaw, J. E. Rice, and E. D. Simandiras, *Chem. Phys. Lett.* **120**, 151 (1985).
21. R. J. Harrison, G. B. Fitzgerald, W. D. Laidig, and R. J. Bartlett, *Chem. Phys. Lett.* **124**, 291 (1986).
22. J. Gauss, *Chem. Phys. Lett.* **191**, 614 (1992).
23. J. Gauss, *J. Chem. Phys.* **99**, 3629 (1993).
24. J. Gauss, *Chem. Phys. Lett.* **229**, 198 (1994).
25. P. Jørgensen and T. U. Helgaker, *J. Chem. Phys.* **89**, 1560 (1988).
26. P. Jørgensen and J. Simons, *J. Chem. Phys.* **79**, 334 (1983).
27. R. J. Bartlett, in *Geometrical Derivatives of Energy Surfaces and Molecular Properties*, P. Jørgensen and J. Simons, eds., Reidel, Dordrecht, 1986.
28. A. C. Scheiner, G. E. Scuseria, J. E. Rice, T. J. Lee, and H. F. Schaefer, *J. Chem. Phys.* **87**, 5361 (1987).
29. E. A. Salter, G. W Trucks, and R. J. Bartlett, *J. Chem. Phys.* **90**, 1752 (1989).
30. J. Gauss and J. F. Stanton, *J. Chem. Phys.* **103**, 3561 (1995).
31. J. Gauss and J. F. Stanton, *J. Chem. Phys.* **102**, 251 (1995).
32. J. Gauss and J. F. Stanton, *J. Chem. Phys.* **104**, 2574 (1996).
33. J. F. Stanton and J. Gauss, in *Recent Advances in Coupled-Cluster Methods*, R. J. Bartlett, ed., World Scientific Publishing, Singapore, 1997.
34. J. Gauss and J. F. Stanton, *Chem. Phys. Lett.* **276**, 70 (1997).
35. P. G. Szalay, J. Gauss, and J. F. Stanton, *Theor. Chim. Acc.* **100**, 5 (1998).
36. J. Gauss and J. F. Stanton, *J. Chem. Phys.* **116**, 1773 (2002).
37. J. Gauss, *J. Chem. Phys.* **116**, 4773 (2002).
38. J. Noga and R. J. Bartlett, *J. Chem. Phys.* **86**, 7041 (1987).
39. G. E. Scuseria and H. F. Schaefer, *Chem. Phys. Lett.* **152**, 382 (1988).
40. J. D. Watts and R. J. Bartlett, *J. Chem. Phys.* **93**, 6104 (1990).
41. For an overview, see: R. J. Bartlett, in *Nonlinear Optical Materials*, S. Karna, ed., American Chemical Society, Washington, D. C., 1996, pp. 23–57.
42. J. Gauss, O. Christiansen, and J. F. Stanton, *Chem. Phys. Lett.* **296**, 117 (1998).
43. J. Gauss and J. F. Stanton, *Adv. Chem. Phys.*, **123**, 355 (2002).
44. A. A. Auer and J. Gauss, *J. Chem. Phys.* **115**, 1619 (2001) and references therein.
45. See, for example, B. H. Lengsfield and D. R. Yarkony, Electronic structure aspects of nonadiabatic processes in polyatomic systems, in *Modern Electronic Structure Theory*, D. R. Yarkony, ed., World Scientific, Singapore, 1995, p. 642.
46. D. Feller, *J. Chem. Phys.* **96**, 6104 (1992).
47. S. S. Xantheas and T. H. Dunning, Jr., *J. Phys. Chem.* **97**, 18 (1993).
48. J. M. L. Martin, *Chem. Phys. Lett.* **259**, 669 (1996).

49. A. K. Wilson and T. H. Dunning, Jr., *J. Chem. Phys.* **106**, 8718 (1997).
50. A. Halkier, T. Helgaker, P. Jorgensen, W. Klopper, H. Koch, J. Olson and A. K. Wilson, *Chem. Phys. Lett.* **286**, 243 (1998).
51. K. A. Peterson, D. E. Woon, and T. H. Dunning, Jr., *J. Chem. Phys.* **100**, 7410 (1994).
52. D. G. Truhlar, *Chem. Phys. Lett.* **294**, 45 (1998).
53. Y.-Y. Chuang and D. G. Truhlar, *J. Phys. Chem. A* **103**, 651 (1999).
54. C. Schwartz, *Phys. Rev.* **126**, 1015 (1962).
55. S. L. Mielke, B. C. Garrett, and K. A. Peterson, *J. Chem. Phys.* **111**, 3806 (1999).
56. T. H. Dunning, Jr., *J. Chem. Phys.* **90**, 1007 (1989).
57. R. A. Kendall, T. H. Dunning, Jr., and R. J. Harrison, *J. Chem. Phys.* **96**, 6796 (1992).
58. D. E. Woon and T. H. Dunning, Jr., *J. Chem. Phys.* **103**, 4572 (1995).
59. K. A. Peterson, preprint.
60. T. Helgaker, P. Jørgensen, and J. Olsen, *Molecular Electronic-Structure Theory*, Wiley, New York, 2000.
61. K. Wenzel, K. Szalewicz, J. G. Zabolitzky, B. Jeziorski, and H. J. Monkhorst, *J. Chem. Phys.* **85**, 3964 (1986). P. M. Kozlowski, and L. Adamowicz, *J. Chem. Phys.* **95**, 6681 (1991).
62. P. M. Kozlowski and L. Adamowicz, *J. Comput. Chem.* **13**, 602 (1992).
63. E. Schwegler, P. M. Kozlowski, and L. Adamowicz, *J. Comput. Chem.* **14**, 566 (1993).
64. B. J. Persson and P. R. Taylor, *J. Chem. Phys.* **105**, 5915 (1996).
65. B. J. Persson and P. R. Taylor, *Theor. Chem. Acc.* **97**, 240 (1997).
66. W. Klopper, in *Modern Methods and Algorithms of Quantum Chemistry*, J. Grotendorst, ed., John von Neumann Institute for Computing, Jülich, 2000, pp. 153–201.
67. W. Klopper and W. Kutzelnigg, *J. Mol. Struct. (Theochem.)* **135**, 339 (1986).
68. W. Klopper and W. Kutzelnigg, *Chem. Phys. Lett.* **134**, 17 (1987).
69. J. Noga, W. Kutzelnigg, and W. Klopper, *Chem. Phys. Lett.* **199**, 497 (1992).
70. K. Raghavachari, G. W. Trucks, J. A. Pople, and M. Head-Gordon, *Chem. Phys. Lett.* **157**, 479 (1989).
71. R. J. Bartlett, J. D. Watts, S. A. Kucharski, and J. Noga, *Chem. Phys. Lett.* **165**, 513 (1990).
72. S. A. Kucharski and R. J. Bartlett, *J. Chem. Phys.* **108**, 9221 (1998).
73. J. A. Pople, M. Head-Gordon, D. J. Fox, K. Raghavachari, and L. A. Curtiss, *J. Chem. Phys.* **90**, 5622 (1989).
74. L. A. Curtiss, K. Raghavachari, G. W. Trucks, and J. A. Pople, *J. Chem. Phys.* **94**, 7221 (1991).
75. L. A. Curtiss, K. Raghavachari, P. C. Redfern, V. Rassolov, and J. A. Pople, *J. Chem. Phys.* **109**, 7764 (1998).
76. J. M. L. Martin, *J. Chem. Phys.* **111**, 1843 (1999).
77. J. A. Montgomery, Jr., J. W. Ochterski, and G. A. Petersson, *J. Chem. Phys.* **101**, 5900 (1994).
78. J. W. Ochterski, G. A. Petersson, and J. A. Montgomery, Jr., *J. Chem. Phys.* **104**, 2598 (1996).
79. J. A. Montgomery, Jr., M. J. Frisch, J. W. Ochterski, and G. A. Petersson, *J. Chem. Phys.* **110**, 2822 (1999).
80. J. B. Anderson, *Rev. Comp. Chem.* **13**, 133 (1999).
81. W. A. Lester, Jr., and R. N. Barnett, in *Encyclopedia of Computational Chemistry*, Vol. 3, P. v. R. Schleyer, N. L. Allinger, T. Clark, J. Gasteiger, P. A. Kollman, H. F. Schaefer III, and P. R. Schreiner, eds., John Wiley & Sons, Chichester, 1998, p. 1735.

82. W. A. Lester, Jr., ed., *Recent Advances in Quantum Monte Carlo Methods*, World Scientific, Singapore, 1997.
83. See, for example, K. A. Peterson and T. H. Dunning, Jr., *J. Chem. Phys.* **106**, 4119 (1997).
84. C. Hampel and H.-J. Werner, *J. Chem. Phys.* **104**, 6286 (1996).
85. M. Schütz and H.-J. Werner, *J. Chem. Phys.* **114**, 661 (2001).
86. M. Schütz, *J. Chem. Phys.* **116**, 8772 (2002).
87. S. Saebø and P. Pulay, *Annu. Rev. Phys. Chem.* **44**, 213 (1993).
88. M. Schütz, *J. Chem. Phys.* **113**, 9986 (2000).
89. M. Musial, S. A. Kucharski, and R. J. Bartlett, *J. Chem. Phys.* **116**, 4382 (2002).
90. J. Olsen, *J. Chem. Phys.* **113**, 7140 (2000).
91. S. Hirata, M. Nooijen, and R. J. Bartlett, *Chem. Phys. Lett.* **326**, 255 (2000).
92. M. Kallay and P. R. Surjan, *J. Chem. Phys.* **113**, 1359 (2000).
93. P. O. Löwdin, *Rev. Mod. Phys.* **35**, 496 (1963).
94. E. R. Davidson and W. D. Borden, *J. Phys. Chem.* **87**, 4783 (1983).
95. D. J. Thouless, *The Quantum Mechanics of Many-Body Systems*, Academic Press, New York, 1961, p. 126.
96. J. Paldus and J. Cizek, *J. Chem. Phys.* **47**, 3976 (1967).
97. H. Fukutome, *Int. J. Quantum Chem.* **20**, 955 (1981).
98. J. Noga and P. Valiron, *Chem. Phys. Lett.* **324**, 166 (2000).
99. H. Reisler, personal communication.
100. C. M. Dobson, G. B. Ellison, A. F. Tuck, and V. Vaida, *Proc. Natl. Acad. Sci. USA* **97**, 11864 (2000).
101. See, for example, A. F. Parsons, *An Introduction to Free Radical Chemistry*, Blackwell, Oxford, 2000.
102. T. Bally, in *Ionic Molecular Systems*, A. Lund and M. Shiotani, eds., Kluwer, Dordrecht, 1991.
103. W. T. Bordon, in *Encyclopedia of Computational Chemistry*, P. v. R. Schleyer, ed., John Wiley & Sons, New York, 1998, pp. 708–722.
104. T. Bally and W. T. Borden, *Rev. Comp. Chem.* **13**, 1 (1999).
105. K. A. Brueckner, *Phys. Rev.* **96**, 508 (1954).
106. R. K. Nesbet, *Phys. Rev.* **109**, 1632 (1958).
107. R. A. Chiles and C. E. Dykstra, *J. Chem. Phys.* **74**, 4544 (1981).
108. N. C. Handy, J. A. Pople, M. Head-Gordon, K. Raghavachari, and G. W. Trucks, *Chem. Phys. Lett.* **164**, 185 (1989).
109. J. F. Stanton, J. Gauss, and R. J. Bartlett, *J. Chem. Phys.* **97**, 5554 (1992).
110. C. Hampel, K. A. Peterson, and H.-J. Werner, *Chem. Phys. Lett.* **190**, 1 (1992).
111. T. D. Crawford, T. J. Lee, and H. F. Schaefer, *J. Chem. Phys.* **107**, 9980 (1997).
112. E. R. Davidson, in *The World of Quantum Chemistry*, R. Daudel and B. Pullman, eds., Reidel, Dordrecht, 1974, p. 17.
113. F. Coester, *Nucl. Phys.* **1**, 421 (1958).
114. F. Coester and H. Kümmel, *Nucl. Phys.* **17**, 477 (1960).
115. J. Goldstone, *Proc. R. Soc. (London)* **A239**, 267 (1957).
116. J. Cizek, *J. Chem. Phys.* **42**, 4256 (1966).
117. J. Cizek, *Adv. Chem. Phys.* **14**, 35 (1969).

118. C. Møller and M. S. Plesset, *Phys. Rev.* **46**, 618 (1934).
119. K. Raghavachari, J. A. Pople, E. S. Replogle, and M. Head-Gordon, *J. Phys. Chem.* **94**, 5579 (1990).
120. S. A. Kucharski and R. J. Bartlett, *Chem. Phys. Lett.* **237**, 264 (1993).
121. D. Cremer and Z. He, *J. Phys. Chem.* **100**, 6173 (1996).
122. N. C. Handy, P. J. Knowles, and K. Somasundram, *Theor. Chim. Acta* **68**, 87 (1985).
123. J. Olsen, O. Christiansen, H. Koch, and P. Jørgensen, *J. Chem. Phys.* **105**, 5082 (1996).
124. M. L. Leininger, W. D. Allen, H. F. Schaefer, and C. D. Sherrill, *J. Chem. Phys.* **112**, 9213 (2000).
125. J. Olsen, O. Christiansen, H. Koch, and P. Jørgensen, *J. Chem. Phys.* **105**, 5082 (1996).
126. J. F. Stanton, *J. Chem. Phys.* **115**, 10382 (2001).
127. G. D. Purvis and R. J. Bartlett, *J. Chem. Phys.* **76**, 1910 (1982).
128. R. J. Bartlett, *J. Phys. Chem.* **93**, 1697 (1989).
129. G. E. Scuseria, A. C. Scheiner, T. J. Lee, J. E. Rice, and H. F. Schaefer, *J. Chem. Phys.* **86**, 2881 (1987).
130. O. Sinanoglu, *Adv. Chem. Phys.* **6**, 315 (1964).
131. R. J. Bartlett and D. M. Silver, *Int. J. Quantum Chem.* **9**, 183 (1975).
132. J. Noga, R. J. Bartlett and M. Urban, *Chem. Phys. Lett.* **134**, 126 (1987).
133. J. D. Watts and R. J. Bartlett, *Spectrochim. Acta* **A55**, 495 (1999).
134. O. Christiansen, H. Koch, and P. Jørgensen, *J. Chem. Phys.* **103**, 7429 (1995).
135. M. Urban, J. Noga, S. J. Cole, and R. J. Bartlett, *J. Chem. Phys.* **83**, 4041 (1985).
136. T. D. Crawford and J. F. Stanton, *Int. J. Quantum Chem.* **70**, 601 (1998).
137. S. A. Kucharski and R. J. Bartlett, *J. Chem. Phys.* **108**, 5255 (1998).
138. G. E. Scuseria and T. J. Lee, *J. Chem. Phys.* **93**, 489 (1990).
139. J. Gauss and J. F. Stanton, *Phys. Chem. Chem. Phys.* **2**, 2047 (2000).
140. J. F. Stanton, *Chem. Phys. Lett.* **281**, 130 (1997).
141. J. M. L. Martin, personal communication.
142. N. Oliphant and L. Adamowicz, *J. Chem. Phys.* **95**, 6645 (1991).
143. S. A. Kucharski and R. J. Bartlett, *J. Chem. Phys.* **97**, 4282 (1992).
144. R. G. Parr and W. Yang, *Density-Functional Theory of Atoms and Molecules*, Oxford University Press, New York, 1989.
145. P. Hohenberg and W. Kohn, *Phys. Rev. B* **136**, 864 (1964).
146. W. Kohn and L. S. Sham, *Phys. Rev. A* **140**, 1133 (1965).
147. W. Koch and M. C. Holthausen, *A Chemist's Guide to Density Functional Theory*, John Wiley & Sons, Weinheim, 2001.
148. For reviews of MCSCF theories, see H.-J. Werner (p. 1), R. Shepard (p. 62), and B. O. Roos (p. 399), in *Ab Initio Methods in Quantum Chemistry, Part II* K. P. Lawley, ed., John Wiley & Sons, New York, 1987.
149. K. Ruedenberg, M. W. Schmidt, M. M. Gilbert, and S. T. Elbert, *Chem. Phys.* **71**, 41 (1982).
150. B. Ross, P. R. Taylor, and P. E. M. Siegbahn, *Chem. Phys.* **48**, 157 (1980).
151. R. Gdanitz and R. Ahlrichs, *Chem. Phys. Lett.* **143**, 413 (1988).
152. P. G. Szalay and R. J. Bartlett, *J. Chem. Phys.* **103**, 3600 (1995).

153. A good overview is *Recent Advances in Multireference Methods*, K. Hirao, ed., World Scientific, Singapore, 1999.
154. K. Anderson, P.-A. Malmqvist, B. O. Roos, A. Sadlej, and K. Wolinski, *J. Phys. Chem.* **94**, 5483 (1990).
155. K. Anderson, P.-A. Malmqvist, and B. O. Roos, *J. Phys. Chem.* **96**, 1218 (1992).
156. See Ref. 60, pp. 798–800.
157. J. Linderberg and N. Y. Öhrn, *Propagators in Quantum Chemistry*, Academic Press, London, 1973.
158. B. T. Pickup and O. Goscinski, *Mol. Phys.* **26**, 1013 (1973).
159. P. Jørgensen and J. Simons, *J. Chem. Phys.* **63**, 5302 (1975).
160. P. Jørgensen, *Annu. Rev. Phys. Chem.* **26**, 359 (1975).
161. J. Simons, *J. Chem. Phys.* **64**, 4541 (1976).
162. J. Simons, *Int. J. Quantum Chem.* **12**, 227 (1977).
163. L. S. Cederbaum and W. Domcke, *Adv. Chem. Phys.* **36**, 205 (1977).
164. J. Oddershede and P. Jørgensen, *J. Chem. Phys.* **66**, 1541 (1977).
165. J. Simons, *Theor. Chem. Adv. Perspect.* **3**, 1 (1978).
166. V. McKoy, *Physica Scripta* **21**, 238 (1979).
167. K. F. Freed, M. F. Herman, and D. L. Yeager, *Physica Scripta* **21**, 243 (1979).
168. J. V. Ortiz and Y. Öhrn, *J. Chem. Phys.* **72**, 5744 (1980).
169. J. V. Ortiz and Y. Öhrn, *Chem. Phys. Lett.* **72**, 548 (1981).
170. W. Von Niessen, J. Schirmer, and L. S. Cederbaum, *Comput. Phys. Rep.* **1**, 57 (1984).
171. L. S. Cederbaum, W. Domcke, J. Schirmir, and W. von Niessen, *Adv. Chem. Phys.* **65**, 115 (1986).
172. J. T. Golab and D. L. Yeager, *J. Chem. Phys.* **87**, 2925 (1987).
173. J. Geertsen, J. Oddershede, and G. E. Scuseria, *Int. J. Quantum Chem.* **S21**, 475 (1987).
174. J. V. Ortiz, *Int. J. Quantum Chem.* **S25**, 35 (1991).
175. J. Cioslowski and J. V. Ortiz, *J. Chem. Phys.* **96**, 8379 (1992).
176. D. L. Yeager, in *Applied Many-Body Methods in Spectroscopy and Electronic Structure*, D. Mukherjee, ed., Plenum, New York, 1992.
177. B. Datta, D. Mukhopadhyay, and D. Mukherjee, *Phys. Rev.* **A47**, 3632 (1993).
178. J. V. Ortiz, *J. Chem. Phys.* **108**, 1008 (1998).
179. J. V. Ortiz, *Adv. Quantum Chem.* **35**, 33 (1999).
180. H. J. Monkhorst, *Int. J. Quantum Chem. (Symp.)* **11**, 421 (1977).
181. D. Mukherjee and P. K. Mukherjee, *Chem. Phys.* **39**, 325 (1979).
182. K. Emrich, *Nucl. Phys. A* **351**, 379 (1981).
183. E. Dalgaard and H. J. Monkhorst, *Phys. Rev.* **A28**, 1217 (1983).
184. S. Ghosh and D. Mukherjee, *Proc. Ind. Acad. Sci.* **93**, 947 (1984).
185. H. Sekino and R. J. Bartlett, *Int. J. Quantum Chem. (Symp.)* **18**, 255 (1984).
186. M. Takahashi and J. Paldus, *J. Chem. Phys.* **85**, 1486 (1986).
187. J. Geertsen, M. Rittby, and R. J. Bartlett, *Chem. Phys. Lett.* **164**, 57 (1989).
188. H. Koch and P. Jørgensen, *J. Chem. Phys.* **93**, 3333 (1990).
189. H. Koch, H. J. A. Jensen, T. Helgaker, and P. Jørgensen, *J. Chem. Phys.* **93**, 3345 (1990).
190. J. F. Stanton and R. J. Bartlett, *J. Chem. Phys.* **98**, 7029 (1993).

191. D. C. Comeau and R. J. Bartlett, *Chem. Phys. Lett.* **207**, 414 (1993).
192. R. J. Rico and M. Head-Gordon, *Chem. Phys. Lett.* **213**, 224 (1993).
193. R. J. Rico, T. J. Lee, and M. Head-Gordon, *Chem. Phys. Lett.* **218**, 139 (1994).
194. H. Nakatsuji, *Chem. Phys. Lett.* **39**, 562 (1978).
195. H. Nakatsuji, *Acta Chim. Hungarica* **129**, 719 (1992).
196. M. Nooijen and R. J. Bartlett, *J. Chem. Phys.* **102**, 3629 (1995).
197. M. Nooijen and R. J. Bartlett, *J. Chem. Phys.* **106**, 6441 (1997).
198. A. I. Krylov, *Chem. Phys. Lett.* **338**, 375 (2001).
199. A. I. Krylov, *Chem. Phys. Lett.* **350**, 522 (2001).
200. A. I. Krylov and C. D. Sherrill, *J. Chem. Phys.* **116**, 3194 (2002).
201. C. D. Sherrill, A. I. Krylov, E. F. C. Byrd, and M. Head-Gordon, *J. Chem. Phys.* **109**, 4171 (1998).
202. A. I. Krylov, C. D. Sherrill, E. F. C. Byrd, and M. Head-Gordon, *J. Chem. Phys.* **109**, 10669 (1998).
203. T. Van Voorhis and M. Head-Gordon, *Chem. Phys. Lett.* **330**, 585 (2000).
204. T. Van Voorhis and M. Head-Gordon, *J. Chem. Phys.* **113**, 8873 (2000).
205. K. Kowalski and P. Piecuch, *Chem. Phys. Lett.* **344**, 165 (2001).
206. P. Piecuch, S. A. Kucharski, V. Spirko, and K. Kowalski, *J. Chem. Phys.* **115**, 5796 (2001).
207. W. J. Hehre, L. Radom, P. v. R. Schleyer, and J. A. Pople, *Ab Initio Molecular Orbital Theory*, John Wiley & Sons, New York, 1986.
208. J. S. Andrews, D. Jayatilaka, R. G. A. Bone, N. C. Handy, and R. D. Amos, *Chem. Phys. Lett.* **183**, 423 (1991).
209. J. A. Pople and R. K. Nesbet, *J. Chem. Phys.* **22**, 571 (1954).
210. R. J. Bartlett and J. F. Stanton, *Rev. Comp. Chem.* **14**, 33 (2000).
211. R. McWeeny and G. Diercksen, *J. Chem. Phys.* **49**, 4852 (1968).
212. E. R. Davidson, *Chem. Phys. Lett.* **21**, 565 (1973).
213. J. F. Stanton, *J. Chem. Phys.* **101**, 371 (1994).
214. C. J. Janssen and H. F. Schaefer, *Theor. Chim. Acta* **79**, 1 (1991).
215. X. Li and J. Paldus, *J. Chem. Phys.* **101**, 8812 (1994).
216. J. Gauss, M. Heckert, O. Heun, and P. G. Szalay, to be published.
217. P. J. Knowles, C. Hempel, and H.-J. Werner, *J. Chem. Phys.* **99**, 5219 (1993).
218. P. Neogrady, M. Urban, and I. Hubac, *J. Chem. Phys.* **100**, 3706 (1994).
219. P. G. Szalay and J. Gauss, *J. Chem. Phys.* **107**, 9028 (1997).
220. P. G. Szalay and J. Gauss, *J. Chem. Phys.* **111**, 4027 (1999).
221. H. B. Schlegel, *J. Phys. Chem.* **92**, 3075 (1988).
222. M. Nooijen, Ph. D. thesis, Vrije Universiteit, Amsterdam, 1992.
223. W. J. Lauderdale, J. F. Stanton, J. Gauss, J. D. Watts, and R. J. Bartlett, *Chem. Phys. Lett.* **187**, 21 (1991).
224. W. J. Lauderdale, J. F. Stanton, J. Gauss, J. D. Watts, and R. J. Bartlett, *J. Chem. Phys.* **97**, 6606 (1992).
225. P. J. Knowles, J. S. Andrews, R. D. Amos, N. C. Handy, and J. A. Pople, *Chem. Phys. Lett.* **186**, 130 (1991).
226. R. D. Amos, J. S. Andrews, N. C. Handy, and P. J. Knowles, *Chem. Phys. Lett.* **185**, 256 (1991).

227. C. Murray and E. R. Davidson, *Chem. Phys. Lett.* **187**, 451 (1991).
228. J. Gauss, J. F. Stanton, and R. J. Bartlett, *J. Chem. Phys.* **97**, 7825 (1992).
229. J. Paldus and J. Cizek, *Phys. Rev. A* **2**, 2268 (1970).
230. H. Köppel, W. Domcke, and L. S. Cederbaum, *Adv. Chem. Phys.* **57**, 59 (1984).
231. J. C. Saeh and J. F. Stanton, *J. Chem. Phys.* **111**, 8275 (1999).
232. J. F. Stanton, *Chem. Phys. Lett.* **237**, 20 (1995).
233. J. Gerratt and I. M. Mills, *J. Chem. Phys.* **49**, 1719 (1967).
234. A. A. Auer, Ph.D. thesis, Universität Mainz, 2002.
235. R. M. Stevens, R. M. Pitzer, and W. N. Lipscomb, *J. Chem. Phys.* **38**, 550 (1963).
236. T. D. Crawford, J. F. Stanton, W. D. Allen, and H. F. Schaefer, *J. Chem. Phys.* **107**, 10626 (1997).
237. A. Weaver, D. W. Arnold, S. E. Bradforth, and D. M. Neumark, *J. Chem. Phys.* **94**, 1740 (1991).
238. J. F. Stanton, J. Gauss, and R. J. Bartlett, *J. Chem. Phys.* **94**, 4084 (1991).
239. M. Mayer, L. S. Cederbaum, and H. Köppel, *J. Chem. Phys.* **100**, 899 (1994).
240. R. G. Pearson, *Symmetry Rules for Chemical Reactions* (Wiley, New York, 1976).
241. T. D. Crawford and J. F. Stanton, *J. Chem. Phys.* **112**, 7873 (2000).
242. T. D. Crawford, E. Kraka, J. F. Stanton, and D. Cremer, *J. Chem. Phys.* **114**, 10638 (2001).
243. E. A. Salter, H. Sekino, and R. J. Bartlett, *J. Chem. Phys.* **87**, 502 (1987).
244. G. E. Scuseria and H. F. Schaefer, *Chem. Phys. Lett.* **142**, 354 (1987).
245. L. A. Barnes and R. Lindh, *Chem. Phys. Lett.* **223**, 207 (1994).
246. Y. Xie, Y. Yamaguchi, and H. F. Schaefer, *J. Chem. Phys.* **104**, 7615 (1996).
247. J. Hrusak and S. Iwata, *J. Chem. Phys.* **106**, 4877 (1997).
248. H. Koch, R. Kobayashi, and P. Jørgensen, *Int. J. Quantum Chem.* **49**, 835 (1994).
249. J. F. Stanton, unpublished data.
250. S. Hirata and M. Head-Gordon, *Chem. Phys. Lett.* **302**, 375 (1999).
251. S. Hirata, T. J. Lee, and M. Head-Gordon, *J. Chem. Phys.* **111**, 8904 (1999).
252. C. D. Sherrill, M. S. Lee, and M. Head-Gordon, *Chem. Phys. Lett.* **302**, 425 (1999).
253. R. D. Cohen and C. D. Sherrill, *J. Chem. Phys.* **114**, 8257 (2001).
254. See, for example, K. R. Asmis, T. R. Taylor, and D. M. Neumark, *J. Chem. Phys.* **111**, 8838 (1999).
255. A. Rizzo and D. L. Yeager, *J. Chem. Phys.* **93**, 8011 (1990).
256. D. L. Yeager and P. Jørgensen, *Chem. Phys. Lett.* **65**, 77 (1979).
257. See, for example, A. D. McLean, B. H. Lengsfeld, J. Pacansky, and Y. Ellinger, *J. Chem. Phys.* **83**, 3567 (1983) and Ref. 104.
258. M. Rittby and R. J. Bartlett, *J. Phys. Chem.* **92**, 3033 (1989).
259. S. Williams, L. B. Harding, J. F. Stanton, and J. C. Weisshaar, *J. Phys. Chem. A* **104**, 9906 (2000).
260. J. Gauss, J. F. Stanton, and R. J. Bartlett, *J. Chem. Phys.* **95**, 2629 (1991).
261. M. Deleuze, J. P. Denis, J. Delhalle, and B. T. Pickup, *J. Phys. Chem.* **97**, 5115 (1993).
262. M. Deleuze and B. T. Pickup, *Int. J. Quantum Chem.* **63**, 483 (1997).
263. R. L. Martin and E. R. Davidson, *Chem. Phys. Lett.* **51**, 237 (1977).
264. J. F. Stanton and J. Gauss, *J. Chem. Phys.* **101**, 8938 (1994).

265. M. Nooijen and R. J. Bartlett, *J. Chem. Phys.* **102**, 3629 (1995).
266. M. Nooijen, *J. Chem. Phys.* **104**, 2638 (1996).
267. J. F. Stanton and J. Gauss, *Theor. Chim. Acta* **93**, 303 (1996).
268. J. F. Stanton and J. Gauss, *J. Chem. Phys.* **111**, 8275 (1999).
269. Y. Bomble and J. F. Stanton, unpublished data.
270. J. Schirmer, L. S. Cederbaum, and O. Walter, *Phys. Rev. A* **28**, 1237 (1983).
271. W. von Niessen, J. Schirmer, and L. S. Cederbaum, *Comput. Phys. Rep.* **1**, 57 (1984).
272. H.-G. Weikert, H.-D. Meyer, L. S. Cederbaum and F. Tarantelli, *J. Chem. Phys.* **104**, 7122 (1994).
273. K. Hald, C. Hättig, and P. Jørgensen, *J. Chem. Phys.* **113**, 7765 (2000).

CONTROL OF QUANTUM DYNAMICS BY LASER PULSES: ADIABATIC FLOQUET THEORY

S. GUÉRIN and H. R. JAUSLIN

Laboratoire de Physique, Université de Bourgogne, Dijon, France

CONTENTS

I. Introduction
II. Floquet Theory
  A. Floquet Formalism from the Semiclassical Point of View
  B. Floquet Formalism from Quantized Cavity Dressed States
  C. Connection with the Semiclassical Formulation: Interaction Representation and Coherent States
    1. Interaction Representation
    2. Coherent States
    3. Expectation Values for General Initial States of the Photon Field
    4. Expectation Values on Coherent States; Relation with the Semiclassical Model
  D. Emission and Absorption of Photons in Floquet Theory
    1. Exchanges of Photons in Floquet Theory
    2. Invariance with Respect to the Choice of the Origin of the Relative Photon Number
    3. Number of Exchanged Photons in Adiabatic Passage with Photons in Coherent States
  E. Floquet Representation with Two or More Lasers
III. Effective Hamiltonian—Dynamical Resonances
  A. Perturbation Theory Formulated as an Iteration of Unitary Transformations—Nonresonant Case: KAM Techniques
    1. Iterative Perturbation Algorithm
    2. Construction of the Contact Transformations
    3. Interpretation as an Averaging Procedure
  B. High-Frequency Perturbation Theory
  C. Nonperturbative Treatment of Resonances—Resonant Transformations
    1. Zero-Field Resonances
    2. Dynamical Resonances

Advances in Chemical Physics, Volume 125, Edited by I. Prigogine and Stuart A. Rice.
ISBN 0-471-21452-3.

D. Effective Dressed Hamiltonians by Partitioning of Floquet Hamiltonians
1. Partitioning: General Formulation
2. Relation with Adiabatic Elimination
3. High-Frequency Partitioning
4. Effective Dressed Hamiltonians: Partitioning in the Enlarged Space
E. Effective Hamiltonian for Two-Photon Quasi-Resonant Processes in Atoms: The Two-Photon RWA
F. Effective Hamiltonian for Rotational Excitations in Diatomic Molecules
1. The Born–Oppenheimer Floquet Hamiltonian
2. Raman Processes in the Ground Vibronic State by a Single Laser: Rotational Excitations
IV. Adiabatic Floquet Theory
A. The Dressed Schrödinger Equation for Chirped Laser Pulses
B. Adiabatic and Diabatic Evolution of Floquet Dressed States
1. Adiabatic Evolution
2. Nonresonant Deviations from Adiabaticity: Perturbation Theory, Superadiabatic Schemes, and Dykhne–Davis–Pechukas Formula
3. Resonant Laser Fields—Lifting of Degeneracy
4. Diabatic Versus Adiabatic Dynamics Around Eigenenergy Crossings and Avoided Crossings
V. Topology of the Dressed Eigenenergy Surfaces
A. Topology of Adiabatic Passage by a Chirped Pulse and SCRAP
B. Robustness of Adiabatic Passage as a Consequence of the Topological Properties of the Eigensurfaces
C. Optimization of Adiabatic Passage
D. Resonant Processes—Creation of Coherent Superposition of States—Half-Scrap
E. Topology of Stimulated Raman Adiabatic Passage (STIRAP) and STIRAP-like Processes
1. Transfer to a Unique State
2. Transfer to a Coherent Superposition of States
VI. State-Selectivity by Bichromatic Pulses
A. Two-Level Systems—Topological Quantization of Atomic Beam Deflection
1. The Effective Hamiltonian
2. Eigenenergy Surface Topology
3. Analytical Construction of the Dressed Eigenenergies
4. Dynamics and Topological Quantization of the Number of Exchanged Photons
B. Three-Level Systems
1. The Effective Hamiltonian
2. Eigenenergy Surface Topology
3. Dynamics
Appendix A: Mathematical Complements
1. Relation Between the Semiclassical and the Floquet Evolution
2. The Structure of Eigenvectors and Eigenvalues of Floquet Hamiltonians—The Concept of Dressed Hamiltonian
3. Relation Between Eigenvectors and Diagonalization Transformations
Appendix B: Coherent States in the Floquet Representation
Appendix C: The Adiabatic Theorem for Floquet Hamiltonians
Acknowledgments
References

I. INTRODUCTION

The development of laser sources yielding pulses of high intensity and short duration has opened up possibilities to manipulate atoms and molecules with high precision. The aim of this chapter is to present in a self-contained way a set of theoretical tools that allow us to analyze the quantum dynamics of atoms or molecules driven by laser pulses and to determine pulse designs that lead to specific effects. In particular we will discuss processes to control the internal excitation of states, like electronic states or vibrational and rotational states of molecules. Many of the tools apply indifferently to atoms or molecules; often we will use one or the other term interchangeably. The idea is to design the characteristics of the laser pulse in such a way that if initially the molecule is in a given state, at the end of the pulse the population will be completely transferred to a selected target state. The parameters of a single pulse that can be designed are the peak intensity, the shape of the pulse envelope, the carrier frequency, and the chirp, which is a slow variation of the carrier frequency during the pulse. Furthermore, one can use sequences of two or several pulses of different characteristics, acting simultaneously or with a well-defined delay. An important condition for the successful implementation of a control process is that it should be robust with respect to variations or imprecisions in the values of the parameters. For instance, the usual resonant π-pulse technique is not robust for a complete transfer, since it is achieved only for precise values of the total pulse area that interacts with the molecule, which is very difficult to fix in an experimental setup. A different type of technique that yields much more robust results is based on adiabatic passage. The analysis that we present here yields an explanation of the principles on which this robustness is based.

The models for the control processes start with the Schrödinger equation for the molecule in interaction with a laser field that is treated either as a classical or as a quantized electromagnetic field. In Section II we describe the Floquet formalism, and we show how it can be used to establish the relation between the semiclassical model and a quantized representation that allows us to describe explicitly the exchange of photons. The molecule in interaction with the photon field is described by a time-independent Floquet Hamiltonian, which is essentially equivalent to the time-dependent semiclassical Hamiltonian. The analysis of the effect of the coupling with the field can thus be done by methods of stationary perturbation theory, instead of the time-dependent one used in the semiclassical description. In Section III we describe an approach to perturbation theory that is based on applying unitary transformations that simplify the problem. The method is an iterative construction of unitary transformations that reduce the size of the coupling terms. This procedure allows us to detect in a simple way dynamical or field induced resonances—that is, resonances that appear only beyond a threshold of the field amplitudes. If resonances are present, the simple perturbative approach is not enough to capture the relevant effects. It can, however, be improved by performing a different kind of unitary transformation that is nonperturbative and is specifically adapted to the relevant resonances. This can be interpreted as a generalization of rotating wave transformations to strong coupling regimes. The method of iterative unitary transformations can furthermore be adapted to the construction of effective models by partitioning of degrees of freedom. The idea is to simplify the problem by determining the most relevant subspace in Hilbert space and to construct a simplified Hamiltonian in which the coupling with the complement subspace is reduced by suitably chosen unitary transformations.

In Section IV we describe the main ideas of adiabatic dynamics and combine it with the Floquet approach. The essential idea is that if some parameters like the pulse envelope or the frequency vary sufficiently slowly compared with the other characteristic times of the system, the time evolution will follow instantaneous eigenstates of the Floquet Hamiltonian. The analysis of adiabatic dynamics is thus reduced to the determination of eigenvalues and eigenvectors of Floquet Hamiltonians, as a function of the parameters of the pulse. In Section V we describe how the possible transfers of population by adiabatic passage are determined by the topology of the eigenenergy surfaces defined by varying the parameters of the pulses. The topology is, in turn, determined by resonances and quasi-resonances. This topological aspect is the basis of the robustness of the adiabatic transfer. An important conclusion of this analysis is that in the adiabatic regime the final result of the process is determined almost exclusively by the resonances of the Floquet Hamiltonian. The perturbative corrections lead only to small deformations of the path in Hilbert space that is followed while the pulse is in interaction, but they do not change the target state that is reached at the end. The different aspects of this approach are illustrated with some simple examples in the corresponding sections. In Section VI we describe a more elaborate application of the method to state selectivity by bichromatic pulses in two- and three-level systems. The appendixes contain some mathematical complements.

II. FLOQUET THEORY

We will discuss the Floquet approach from two different points of view. In the first one, discussed in Section II.A, the Floquet formalism is just a mathematically convenient tool that allows us to transform the Schrödinger equation with a time-dependent Hamiltonian into an equivalent equation with a time-independent Hamiltonian. This new equation is defined on an enlarged Hilbert space. The time dependence has been substituted by the introduction of one auxiliary dynamical variable for each laser frequency. The second point of view, described in Section II.B, consists of constructing the Floquet representation starting from a model in which both the molecule and the field are quantized.

The initial photon state can be a number state (with a not well-defined phase) or a linear combination of number states, for instance a coherent state. We formulate the construction of coherent states in the Floquet theory and show that choosing one as the initial photon state allows us to recover the usual semiclassical time dependent Schrödinger equation, with a classical field of a well-defined phase (see Section II.C).

This Floquet approach provides a physical interpretation of the dynamics in terms of photons in interaction with the molecule, which is in close analogy to the theory of dressed states in a cavity (see Section II.D).

The formalism is developed for the case of an interacting field of a single frequency. It can be easily extended to the multifrequency case as shown in Section II.E.

A. Floquet Formalism from the Semiclassical Point of View

In the semiclassical model the molecule is treated quantum mechanically whereas the field is represented classically. The field is an externally given function of time F that is not affected by any feedback from the interaction with the molecule. We consider the simplest case of a dipole coupling. The formalism is easily extended to other types of couplings. The time dependence of the periodic Hamiltonian is introduced through the time evolution of the initial phase: $F = F(\theta + \omega t) = \mathcal{E} \cos(\theta + \omega t)$ [1–6], where $\mathcal{E}$ is the amplitude of the electric field and ω is its frequency. The semiclassical Hamiltonian can be, for example, written as

$$H = H(x, \theta + \omega t) = H_0(x) - \mu(x)\mathcal{E} \cos(\theta + \omega t) \tag{1}$$

where x symbolizes the degrees of freedom of the atom or molecule, $\mu(x)$ is its dipole moment, and $H_0(x)$ is the Hamiltonian of the free molecule. The semiclassical Schrödinger equation

$$i\hbar \frac{\partial}{\partial t} \phi = H(x, \theta + \omega t)\phi, \qquad \phi \in \mathcal{H} \tag{2}$$

is defined on a Hilbert space $\mathcal{H}$, which can be of infinite dimension (e.g., the space of square-integrable functions $\mathcal{H} = L_2(\mathbb{R}^n, d^n x)$, where n is the number of the degrees of freedom of the molecule) or of finite dimension (e.g., in N-level models $\mathcal{H} = \mathbb{C}^N$). The initial phase θ appears as a parameter. One can think of Eq. (2) as a family of equations parameterized by the angle θ. We denote the corresponding family of propagators by $U(t, t_0; \theta)$, which describe the time evolution of arbitrary initial conditions $\phi(t_0)$:

$$\phi(t) = U(t, t_0; \theta)\phi(t_0) \tag{3}$$

and satisfy

$$i\hbar \frac{\partial}{\partial t} U(t, t_0; \theta) = H(\theta + \omega t) U(t, t_0; \theta), \qquad U(t, t; \theta) = \mathbb{1}_{\mathscr{H}} \tag{4}$$

The Floquet Hamiltonian K, also called the quasienergy operator, is constructed as follows: We define an enlarged Hilbert space

$$\mathscr{K} := \mathscr{H} \otimes \mathscr{L} \tag{5}$$

where $\mathscr{L} := L_2(\mathbb{S}^1, \ d\theta/2\pi)$ denotes the space of square-integrable functions on the circle $\mathbb{S}^1$ of length 2π, with a scalar product

$$\langle \xi_1 | \xi_2 \rangle_{\mathscr{L}} := \int_{\mathbb{S}^1} \frac{d\theta}{2\pi} \xi_1^*(\theta) \xi_2(\theta) \tag{6}$$

This space is generated by the orthonormal basis $\{e^{ik\theta}\}, \quad k \in \mathbb{Z}$ (i.e., all integers). On the enlarged Hilbert space $\mathscr{K}$ the Floquet Hamiltonian is defined as

$$K = -i\hbar\omega \frac{\partial}{\partial \theta} + H(\theta) \tag{7}$$

In this expression, $H(\theta)$ is just the semiclassical Hamiltonian (1) but with the phase $\Theta(t)$ taken at the (fixed) initial value θ corresponding to $t = 0$. The usefulness of the Floquet Hamiltonian comes from the fact that it is time-independent and that the dynamics it defines on $\mathscr{K}$ is essentially equivalent with the one of Eq. (2). This can be formulated as follows. The Floquet Hamiltonian K defines a time evolution in $\mathscr{K}$ through the equation

$$i\hbar \frac{\partial}{\partial t} \psi = K\psi, \qquad \psi \in \mathscr{K} = \mathscr{H} \otimes \mathscr{L} \tag{8}$$

This time evolution can be expressed in terms of a propagator $U_K(t, t_0)$ characterized by

$$i\hbar \frac{\partial}{\partial t} U_K(t, t_0) = K U_K(t, t_0), \qquad U_K(t, t) = \mathbb{1}_{\mathscr{K}} \tag{9}$$

that is, $\psi(t) = U_K(t, t_0)\psi(t_0)$ (where $\mathbb{1}_{\mathscr{K}}$ is the identity operator in $\mathscr{K}$). Since K is time-independent, the propagator can be written as

$$U_K(t, t_0, \theta) = U_K(t - t_0, \theta) = e^{-iK(t-t_0)/\hbar} \tag{10}$$

In order to establish a relation between U and U_K, we define the phase translation operator $\mathscr{T}_{\omega t}$, which acts on $\xi \in \mathscr{L}$:

$$\mathscr{T}_{\omega t}\,\xi(\theta) = \xi(\theta + \omega t) \tag{11}$$

which can be expressed as

$$\mathscr{T}_{\omega t} = e^{\omega t \partial/\partial\theta} \tag{12}$$

We first lift the family of operators $U(t, t_0; \theta)$ (defined on $\mathscr{H}$) into an operator acting on the enlarged space $\mathscr{K}$ by treating the dependence on θ as a multiplication operator. This operator is unitary in $\mathscr{K}$. The relation between U and U_K can then be expressed as

$$\mathscr{T}_{-\omega t}\, U(t, t_0; \theta)\, \mathscr{T}_{\omega t_0} = U_K(t - t_0, \theta) \equiv e^{-iK(t-t_0)/\hbar} \tag{13}$$

The proof of this relation is given in Appendix A. It implies that if $\psi(t, x, \theta)$ is a solution of (8) then we can obtain a solution of Eq. (2) by $\phi(t, x) = \mathscr{T}_{\omega t}\psi(t, x, \theta) = \psi(t, x, \theta + \omega t)$.

The fact that K is time independent opens the possibility to work with eigenfunction expansions. We consider the case in which K has pure point spectrum, i.e. no continuum. This is always the case for N-level models with periodic time dependent fields. Further remarks on other cases are given in Appendix A.

First we lift the initial condition $\phi(t_0)$ for Eq. (2) to the enlarged space $\mathscr{K}$ by taking $\phi(t_0) \otimes 1_{\mathscr{L}}$. This form reflects the fact that the initial condition is the same for the whole family of equations (2); that is, it does not depend on the phase θ. With the eigenvalues and eigenvectors of

$$K\psi_\nu = \lambda_\nu \psi_\nu \tag{14}$$

using the inverse of (13), the time evolution can be expressed by the eigenfunction expansion

$$\begin{aligned}
\phi(t) &= U(t, t_0; \theta)\phi(t_0) \\
&= \mathscr{T}_{\omega t} e^{-iK(t-t_0)/\hbar} \mathscr{T}_{-\omega t_0} \phi(t_0) \otimes 1_{\mathscr{L}} \\
&= \sum_\nu c_\nu e^{-i\lambda_\nu(t-t_0)/\hbar} \psi_\nu(x, \theta + \omega t)
\end{aligned} \tag{15}$$

where the coefficients c_ν are determined by the scalar product

$$c_\nu = \langle \psi_\nu, \phi(t_0) \otimes 1 \rangle_{\mathscr{K}} = \langle \bar{\psi}_\nu, \phi(t_0) \rangle_{\mathscr{H}} \tag{16}$$

where the subindices $\mathscr{K}$ and $\mathscr{H}$ specify to which space the scalar product corresponds, and $\bar{\psi}_\nu := \int_{\mathbb{S}^1} d\theta/2\pi \; \psi_\nu(\theta)$ is the average of $\psi_\nu(\theta)$ over the phase, or, equivalently, its constant Fourier component. Thus, the determination of the Floquet eigenvectors and eigenvalues allows us to solve the dynamics of the semiclassical model.

The Floquet eigenelements have a periodic structure: $\psi_\nu \equiv \psi_{n,k} = \psi_{n,0} e^{ik\theta}$, and $\lambda_\nu \equiv \lambda_{n,k} = \lambda_{n,0} + k\hbar\omega$, where the index n refers to the molecule's Hilbert space $\mathscr{H}$ (i.e. $n = 1, \ldots, N$ if $\mathscr{H} = \mathbf{C}^N$), and k are all positive or negative integers. This allows to classify the Floquet eigenstates in families labeled by n. The individual members within one family are distinguished by the index k.

$$\phi(t) = \sum_n \tilde{c}_n(\theta + \omega t) e^{-i\lambda_{n,0}(t-t_0)/\hbar} \psi_{n,0}(x, \theta + \omega t) \tag{17}$$

with

$$\tilde{c}_n(\theta) := \langle \psi_{n,0}(\theta), \phi(t_0) \rangle_{\mathscr{H}} \tag{18}$$

The coefficients $\tilde{c}_n(\theta + \omega t)$ are functions of θ, and become thus time dependent.

B. Floquet Formalism from Quantized Cavity Dressed States

Although in the semiclassical model the only dynamical variables are those of the molecule, and the extended Hilbert space $\mathscr{K} = \mathscr{H} \otimes \mathscr{L}$ and the Floquet Hamiltonian K can be thought as only mathematically convenient techniques to analyze the dynamics, it was clear from the first work of Shirley [1] that the enlarged Hilbert space should be related to photons. This relation was made explicit by Bialynicki-Birula and co-workers [7,8] and completed in [9]. The construction starts with a quantized photon field in a cavity of finite volume in interaction with the molecule. The limit of infinite volume with constant photon density leads to the Floquet Hamiltonian, which describes the interaction of the molecule with a quantized laser field propagating in free space. The construction presented below is taken from Ref. 9, where further details and mathematical precisions can be found.

We consider a quantized photon field in a cavity of volume V, of single frequency ω and polarized in the $\vec{e}$ direction, described by the Hamiltonian H_{L}, in interaction with a molecule characterized by the Hamiltonian H_{M}. For simplicity we consider the simplest situation of a dipole interaction described by the Hamiltonian [10,11]

$$H_{\mathrm{ML}} = H_{\mathrm{M}} + H_{\mathrm{L}} + H_{\mathrm{int}} \tag{19}$$

with

$$H_{\rm L} = \hbar\omega a^{\dagger} a \tag{20a}$$

$$H_{\rm int} = -\mu \otimes \mathscr{E}_V (a + a^{\dagger}) \tag{20b}$$

and $\mu = \vec{\mu} \cdot \vec{e}$, where $\vec{\mu}$ is the dipole moment of the molecule. The mode of the laser with frequency ω is described by the number operator of a harmonic oscillator, which can be expressed in terms of the annihilation and creation operators a, $a^{\dagger}$. They act on the Fock space $\mathscr{F}$ generated by the stationary states $|n\rangle$, $n = 0, 1, 2, \ldots$ of the harmonic oscillator. The coupling constant is given by

$$\mathscr{E}_V = \sqrt{\frac{\hbar\omega}{2\varepsilon_0 V}} \tag{21}$$

where ε_0 is the permeability of the vacuum. The states of the coupled system evolve in the Hilbert space

$$\mathscr{H}_{\rm ML} = \mathscr{H} \otimes \mathscr{F} \tag{22}$$

where we call $\mathscr{H}$ the Hilbert space of the molecule and $\mathscr{F}$ the Hilbert space of the photons.

We will establish a precise relation between dressed states in a cavity and the Floquet formalism. We show that the Floquet Hamiltonian K can be obtained exactly from the dressed Hamiltonian in a cavity in the limit of infinite cavity volume and large number of photons: K represents *the Hamiltonian of the molecule interacting in free space with a field containing a large number of photons.* We establish the physical interpretation of the operator

$$N_r = -i\frac{\partial}{\partial\theta} \tag{23}$$

in the limit of large number of photons as the *relative photon number operator.* It characterizes the relative photon number of the field with respect to the average $\bar{n}$. The variation of the average of N_r in the Floquet formalism gives the number of photons gained or lost (depending on the sign) by the field.

We remark that, with the cavity dressed state model (19), the field intensity does not appear explicitly. It depends on the average number of photons contained in the initial state of the field. The connection between this model and the Floquet formulation is given by the following property: Since the radiation is not confined in a cavity, but propagates and interacts with the molecule in free space, we have to take the limit

$$\begin{aligned} V &\to \infty \quad &&\text{(infinite cavity volume)} \\ \bar{n} &\to \infty \quad &&\text{(large photon number average)} \\ \rho = \bar{n}/V &= \text{const} \quad &&\text{(constant photon density)} \end{aligned}$$

In this limit, the Hamiltonian $H_{\rm ML}$ is identical, up to an additive constant, to the Floquet Hamiltonian K

$$H_{\rm ML} - \hbar\omega\bar{n} \to -i\hbar\omega\frac{\partial}{\partial\theta} + H_0 - \mu\mathscr{E}\cos\theta \equiv K \tag{24}$$

where

$$\mathscr{E} = \sqrt{\frac{2\rho\hbar\omega}{\varepsilon_0}} \tag{25}$$

To show this relation, we use the phase representation of $H_{\rm LM}$, as formulated by Bialynicki-Birula [7,8,12,13]. We construct an isomorphism between the Fock space and the space $\mathscr{L}_{\bar{n},\theta}$ defined as a subspace of $\mathscr{L} = L_2\left(\mathbb{S}^1, \frac{d\theta}{2\pi}\right)$, generated by the basis functions $\{|e^{ik\theta}\rangle; -\bar{n} \le k < +\infty\}$:

$$|n\rangle \in \mathscr{F} \leftrightarrow |e^{ik\theta}\rangle \in \mathscr{L}_{\bar{n},\theta} \quad \text{with } \bar{n} + k = n, \qquad \text{i.e.,} \quad k \in [-\bar{n}, \infty) \tag{26}$$

In the limit $\bar{n} \to \infty$ we obtain the whole space

$$\mathscr{L}_{\bar{n},\theta} \xrightarrow{\bar{n}\to\infty} \mathscr{L} \quad \text{and} \quad \mathscr{H}_{\rm LM} \xrightarrow{\bar{n}\to\infty} \mathscr{K} = \mathscr{H} \otimes \mathscr{L} \tag{27}$$

By this isomorphism, the creation, annihilation, and photon number operators ($a^\dagger$, a, and N) have a corresponding representation acting on $\mathscr{L}_{\bar{n},\theta}$, which we denote, respectively, $a^\dagger_{\bar{n},\theta}$, $a_{\bar{n},\theta}$ and $N_{\bar{n},\theta}$:

$$a^\dagger|n\rangle = \sqrt{n+1}|n+1\rangle \longleftrightarrow a^\dagger_{\bar{n},\theta} = \sqrt{\bar{n} - i\frac{\partial}{\partial\theta}}e^{i\theta}P_{\bar{n}} \tag{28a}$$

$$a|n\rangle = \sqrt{n}|n-1\rangle \longleftrightarrow a_{\bar{n},\theta} = e^{-i\theta}\sqrt{\bar{n} - i\frac{\partial}{\partial\theta}}P_{\bar{n}} \tag{28b}$$

$$N|n\rangle = a^\dagger a|n\rangle = n|n\rangle \longleftrightarrow N_{\bar{n},\theta} = \left(\bar{n} - i\frac{\partial}{\partial\theta}\right)P_{\bar{n}} \tag{28c}$$

where $P_{\bar{n}} = \sum_{k=-\bar{n}}^{\infty} |e^{ik\theta}\rangle\langle e^{ik\theta}|$ is the projector on $\mathscr{L}_{\bar{n},\theta}$. The operator in the coupling term becomes

$$a_{\bar{n},\theta} + a_{\bar{n},\theta}^{\dagger} = P_{\bar{n}} \left(e^{-i\theta} \sqrt{\bar{n} - i\frac{\partial}{\partial\theta}} + \sqrt{\bar{n} - i\frac{\partial}{\partial\theta}}\, e^{i\theta} \right) P_{\bar{n}} \tag{29}$$

and the Hamiltonian reads

$$H_{\mathrm{LM}}^{(\bar{n})} = H_0(x) \otimes P_{\bar{n}} + \mathbb{1}_{\mathscr{H}} \otimes \hbar\omega N_{\bar{n},\theta} - \mu(x) \otimes \mathscr{E}_V \left(a_{\bar{n},\theta} + a_{\bar{n},\theta}^{\dagger} \right) \tag{30}$$

We remark that this is an exact correspondence, which is just a precise expression of Dirac's transformation formalism of quantum mechanics [14,15]. The explicit writing of the projector $P_{\bar{n}}$ in Eqs. (28a)–(28c) is motivated by the fact that in this way the operators $H_0(x) \otimes P_{\bar{n}}$, $N_{\bar{n},\theta}$, $a_{\bar{n},\theta}$, $a_{\bar{n},\theta}^{\dagger}$, and $H_{\mathrm{LM}}^{(\bar{n})}$ are also well-defined in the total space $\mathscr{L} = L_2(\mathbb{S}^1, d\theta/2\pi)$, and the discussion of the limit $\bar{n} \to \infty$ becomes conceptually clearer.

In Refs. 7 and 8 the formal hypothesis

$$-i\frac{\partial}{\partial\theta} \ll \bar{n} \tag{31}$$

is invoked to approximate

$$\sqrt{\bar{n} - i\frac{\partial}{\partial\theta}} = \sqrt{\bar{n}}\sqrt{1 - \frac{i}{\bar{n}}\frac{\partial}{\partial\theta}} = \sqrt{\bar{n}} + \mathscr{O}\left(\frac{1}{\sqrt{\bar{n}}}\right)$$

which leads to

$$\left(a_{\bar{n},\theta} + a_{\bar{n},\theta}^{\dagger} \right) \Big/ \sqrt{\bar{n}} \;\xrightarrow{\bar{n}\to\infty}\; (e^{-i\theta} + e^{i\theta}) = 2\cos\theta \tag{32}$$

In the limit $V \to \infty$, $\bar{n} \to \infty$, keeping the photon density $\rho = \bar{n}/V$ constant, we obtain the interaction term

$$\mathscr{E}_V \left(a_{\bar{n},\theta} + a_{\bar{n},\theta}^{\dagger} \right) \to \sqrt{\frac{2\rho\hbar\omega}{\varepsilon_0}} \cos\theta \tag{33}$$

Introducing the laser intensity per unit surface I

$$I = \frac{1}{2}\varepsilon_0 c \mathscr{E}^2 = \hbar\omega\Phi_{\mathrm{ph}} \tag{34}$$

with the photon velocity c, the field amplitude $\mathscr{E}$, and the photon flow $\Phi_{\rm ph} = \bar{n}c/V$, allows us to identify the interaction constant of Eq. (33) with $\mathscr{E}$ of Eq. (24) as $\mathscr{E} = \sqrt{2\rho\hbar\omega/\varepsilon_0}$. We obtain thus the Floquet Hamiltonian K of Eq. (24).

The formal hypothesis (31) must be interpreted in relation to the functions on which $-i\partial/\partial\theta$ acts. The statement is that if all the states $\{|e^{ik\theta}\rangle\}$ that are relevant in the dynamics are such that $|k| \ll \bar{n}$—that is if only few photons are exchanged between light and matter compared to the average photon number $\bar{n}$ contained in the laser field — then *the coupled Hamiltonian* $H_{LM}^{(\bar{n})}$ *can be identified with the Floquet Hamiltonian* K.

One can give a more precise formulation of this construction, based on the dynamics of the coupled system. Since $H_{\rm LM}^{(\bar{n})}$ and K are both well-defined on $\mathscr{H} \otimes \mathscr{L}$, we can compare the time evolutions generated by the two Hamiltonians of any initial state $\psi_0 \in \mathscr{H} \otimes \mathscr{L}$: *For* N*-level models* $(\mathscr{H} = \mathbb{C}^N)$*, given any initial state* $\psi_0 \in \mathscr{H} \otimes \mathscr{L}$*, the limit of the cavity dressed state dynamics is identical to the Floquet dynamics*:

$$\lim_{\substack{V,\bar{n}\to\infty \\ V/n=\rho}} e^{-i(H_{\rm LM}^{(\bar{n})}/\hbar - \bar{n}\omega)t}\psi_0 = e^{-iKt/\hbar}\psi_0 \tag{35}$$

The proof of this statement is given in Ref. 9.

C. Connection with the Semiclassical Formulation: Interaction Representation and Coherent States

From the formulation of the Floquet formalism given above, we can establish the precise connection between the dynamics in the enlarged space $\mathscr{K}$ defined by the Floquet Hamiltonian K, and the one defined by the semiclassical Hamiltonian in $\mathscr{H}$ with a classical description of the electric field:

> The Schrödinger equation of the Floquet Hamiltonian in $\mathscr{K}$, where θ is a dynamical variable, is equivalent, in an interaction representation, to the semiclassical Schrödinger equation in $\mathscr{H}$, where θ is considered as a parameter corresponding to the fixed initial phase. The dynamics of the two models are identical if the initial photon state in the Floquet model is a coherent state.

1. *Interaction Representation*

The Schrödinger equation of the Floquet Hamiltonian in $\mathscr{K}$

$$i\hbar\frac{\partial}{\partial t}\psi(t) = K\psi(t) \tag{36}$$

can be expressed equivalently in an interaction representation defined by the unitary transformation

$$\phi(t) = U_{0\mathrm{r}}^{\dagger}(t)\psi(t) \tag{37}$$

where

$$U_{0\mathrm{r}}(t) = e^{-\omega t\partial/\partial\theta} \equiv \mathscr{T}_{-\omega t} \tag{38}$$

is the free photon field propagator, which is just the translation operator (12) used in the Floquet construction of Section II.A. Using Eq. (13), we obtain

$$\begin{aligned}\phi(t) &= \mathscr{T}_{\omega t}\psi(t) = \mathscr{T}_{\omega t}\, U_K(t - t_0, \theta)\, \mathscr{T}_{-\omega t_0}\phi(t_0)\\ &= U(t, t_0; \theta)\phi(t_0)\end{aligned}$$

and the evolution equation in this representation becomes

$$i\hbar\frac{\partial}{\partial t}\phi(t) = H(\theta + \omega t)\phi(t) \tag{39}$$

where we still have $\phi(t) \in \mathscr{K}$; that is, $H(\theta + \omega t)$ is still interpreted as an operator acting on the enlarged Hilbert space $\mathscr{K}$, which with respect to the variable θ is a multiplication operator.

Although this equation looks formally like the semiclassical Schrödinger equation (2), we emphasize that it is still different because it is defined in the enlarged Hilbert space $\mathscr{K}$ and the phase θ does not have a definite value, since it is a dynamical variable on the same footing as x. In order to recover the semiclassical equation from Eq. (39), we have to reduce it to an equation defined in the Hilbert space $\mathscr{H}$. From a mathematical point of view, this can be done by fixing a particular value of θ, as we did in Section II.A. Physically, this can be achieved, as we show in the following, by choosing the initial condition of the photon field as a coherent state.

2. *Coherent States*

The coherent states of the photon field can be defined as the eigenvectors of the annihilation operator

$$a|\alpha\rangle = \alpha|\alpha\rangle, \qquad \alpha = |\alpha|e^{-i\theta_0} \tag{40}$$

In the usual Fock number state representation they are given, up to a phase factor, by

$$|\alpha\rangle = e^{-|\alpha|^2/2}\sum_{n=0}^{\infty}\frac{\alpha^n}{\sqrt{n!}}|n\rangle \tag{41}$$

In the phase representation they can be written as

$$\Phi_{\theta_0}^{(\bar{n})}(\theta) = e^{i\zeta} e^{-|\alpha|^2/2} \sum_{n=0}^{\infty} \frac{\alpha^n}{\sqrt{n!}} e^{i(n-\bar{n})\theta}$$
$$= e^{-|\alpha|^2/2} \sum_{n=0}^{\infty} \frac{|\alpha|^n}{\sqrt{n!}} e^{i(n-\bar{n})(\theta-\theta_0)} \tag{42}$$

(where ζ is an arbitrary constant phase that we have chosen as $\zeta = \bar{n}\theta_0$). In order to obtain the representation of coherent states in Floquet theory, we have to take $|\alpha| = \sqrt{\bar{n}}$, since the average photon number in a coherent state is given by $|\alpha|^2$, and then apply the limit $\bar{n} \to \infty$. In Appendix B we show that in this limit the coherent states are represented by a generalized function $\Phi_{\theta_0}(\theta)$, which is real and depends on $\theta - \theta_0$, where $\theta_0 \in \mathbb{S}^1$ is a fixed angle, and

$$(\Phi_{\theta_0}(\theta))^2 = 2\pi\delta(\theta - \theta_0) \tag{43}$$

3. *Expectation Values for General Initial States of the Photon Field*

For a general initial condition of the photon field $\xi(\theta) \in \mathscr{L}$, we first remark that the evolution of the initial condition (that we take here at $t = t_0 = 0$) $\phi(x) \otimes \xi(\theta)$ can be obtained from the one of the initial condition $\phi(x) \otimes 1$ (where the constant function $1 \equiv e^{i(k=0)\theta}$ is the relative number state of zero photons):

$$U_K(t,\theta)\,(\phi(x) \otimes \xi(\theta)) = \mathscr{T}_{-\omega t}\, U(t,0;\theta)\,(\phi(x) \otimes \xi(\theta))$$
$$= \xi(\theta - \omega t)\, U(t,0;\theta - \omega t)(\phi(x) \otimes 1)$$
$$= \xi(\theta - \omega t)\, U_K(t,\theta)(\phi(x) \otimes 1) \tag{44}$$

(since $U(t,0;\theta)$ is a multiplication operator with respect to θ).

As a consequence, for any observable $M(\theta)\colon \mathscr{K} \to \mathscr{K}$ that with respect to θ is a multiplication operator, using Eq. (13), we can write the expectation value as

$$\langle M\rangle(t) := \langle \phi \otimes \xi|\, U_K^{\dagger}(t,\theta)\, M(\theta)\, U_K(t,\theta)\, |\phi \otimes \xi\rangle_{\mathscr{K}}$$
$$= \int_0^{2\pi} \frac{d\theta}{2\pi} |\xi(\theta)|^2 \langle \phi|\, U^{\dagger}(t,0;\theta) M(\theta + \omega t)\, U(t,0;\theta) |\phi\rangle_{\mathscr{H}}$$
$$= \int_0^{2\pi} \frac{d\theta}{2\pi} |\xi(\theta)|^2 \langle \phi(t;\theta)|M(\theta + \omega t)|\phi(t;\theta)\rangle_{\mathscr{H}} \tag{45}$$

where we denote by $\phi(t;\theta) \equiv U(t,0;\theta)\phi$ the semiclassical evolution with initial phase θ of the initial condition $\phi \in \mathscr{H}$.

In particular, for an observable A of the molecule (i.e., $A \otimes \mathbb{1}_{\mathscr{L}}$) we have

$$\langle A\rangle(t) = \int_0^{2\pi} \frac{d\theta}{2\pi} |\xi(\theta)|^2 \langle\phi(t;\theta)|A|\phi(t;\theta)\rangle_{\mathscr{H}} \tag{46}$$

4. *Expectation Values on Coherent States; Relation with the Semiclassical Model*

We have stated that we can recover the evolution of the semiclassical model from the Floquet evolution in the interaction representation by taking initial states in which the photon field is in a coherent state. This can be formulated more precisely by the following statements.

If we take an initial condition of the form $\psi(t=0) = \phi(x) \otimes \Phi_{\theta_0}(\theta)$, then:

i. If $A : \mathscr{H} \to \mathscr{H}$ is an observable of the molecule, then according to Eqs. (46) and (43) we have

$$\langle A\rangle(t) = \langle\phi(t;\theta_0)|A|\phi(t;\theta_0)\rangle_{\mathscr{H}} \tag{47}$$

The last expression is the expectation value calculated with the semiclassical model with initial phase θ_0. We thus conclude that, *if one considers only observables of the molecule, the Floquet evolution with a coherent state in the initial condition is equivalent to the semiclassical model*. We remark that a somewhat related construction, linking the evolution from cavity dressed states directly to the semiclassical model (i.e., without the intermediate level of Floquet states as we do here), was established in Ref. 16.

ii. More generally, if $M(\theta)\colon \mathscr{K} \to \mathscr{K}$ is an observable that with respect to θ is a multiplication operator, continuous in θ, then taking for θ a particular value θ_0 defines a family of operators $M(\theta_0)\colon \mathscr{H} \to \mathscr{H}$, parameterized by θ_0. Then, according to Eqs. (45) and (43) we have

$$\langle M\rangle(t) = \langle\phi(t;\theta_0)|M(\theta_0+\omega t)|\phi(t;\theta_0)\rangle_{\mathscr{H}} \tag{48}$$

It was remarked in Refs. 1 and 7 that in the semiclassical model, if the initial phase θ_0 is not known, one can take a statistical average over the initial phases, with uniform distribution:

$$\overline{A_{\text{sc}}}(t) := \int_0^{2\pi} \frac{d\theta_0}{2\pi} \langle\phi(t;\theta_0)|A|\phi(t;\theta_0)\rangle_{\mathscr{H}} \tag{49}$$

From the discussion above, this coincides with the expectation value $\langle A\rangle(t)$ calculated with the evolution in the Floquet picture of an initial condition of the photon field that is a photon number eigenstate $e^{ik\theta}$ (with arbitrary k); that is, $\langle A\rangle(t) = \int_0^{2\pi} \frac{d\theta_0}{2\pi} \langle\phi(t;\theta_0)|A|\phi(t;\theta_0)\rangle_{\mathscr{H}}$ according to Eq. (46). We have seen on the other hand that the semiclassical evolution with an initial phase θ_0 corresponds, in the Floquet picture, to a coherent state initial condition for the photon field.

This property is quite remarkable: *In the large photon number regime the coherent quantum average on a number state gives the same result as the incoherent statistical average over coherent states.*

D. Emission and Absorption of Photons in Floquet Theory

1. Exchanges of Photons in Floquet Theory

In Floquet theory the exchange of photons can be analyzed from the temporal variation of the relative photon number. In experiments, one measures for instance the difference in intensity of the laser pulse before and after the interaction with the molecules. Denoting the initial condition (at $t = t_0 = 0$) by $\phi(x) \otimes \xi(\theta)$, we describe the exchange of photons by

$$\delta\langle N\rangle(t) := \langle\phi\otimes\xi|U_K^{\dagger}(t)\left(-i\frac{\partial}{\partial\theta}\right)U_K(t)|\phi\otimes\xi\rangle_{\mathscr{K}} - \langle\phi\otimes\xi| - i\frac{\partial}{\partial\theta}|\phi\otimes\xi\rangle_{\mathscr{K}} \tag{50}$$

and we show below that

$$\delta\langle N\rangle(t) = \int_0^{2\pi} \frac{d\theta}{2\pi\hbar\omega}|\xi(\theta)|^2[\langle\phi|H(\theta)|\phi\rangle_{\mathscr{H}} - \langle\phi(t;\theta)|H(\theta+\omega t)|\phi(t;\theta)\rangle_{\mathscr{H}}] \tag{51}$$

In particular, if the photon field is initially in a photon number eigenstate $|e^{ik\theta}\rangle$, we obtain

$$\delta\langle N\rangle(t) = \int_0^{2\pi} \frac{d\theta}{2\pi\hbar\omega}[\langle\phi|H(\theta)|\phi\rangle_{\mathscr{H}} - \langle\phi(t;\theta)|H(\theta+\omega t)|\phi(t;\theta)\rangle_{\mathscr{H}}] \tag{52}$$

We remark that $\delta\langle N\rangle(t)$ is independent of the particular k we take, in accordance with the interpretation as relative photon number.

If the photon field is initially in a coherent state $\Phi_{\theta_0}(\theta) - (2\pi)^{1/2}$ $\delta_{1/2}(\theta - \theta_0)$, then

$$\delta\langle N\rangle_{\text{cs}}(t) = \frac{1}{\hbar\omega}[\langle\phi|H(\theta_0)|\phi\rangle_{\mathscr{H}} - \langle\phi(t;\theta_0)|H(\theta_0+\omega t)|\phi(t;\theta_0)\rangle_{\mathscr{H}}] \tag{53}$$

Again, if the precise initial phase θ_0 of the coherent state is not known, one can take the (incoherent) statistical average over all phases θ_0:

$$\overline{\delta\langle N\rangle_{\rm cs}}(t) = \int_0^{2\pi} \frac{d\theta_0}{2\pi}\,\delta\langle N\rangle_{\rm cs}(t)$$
$$= \int_0^{2\pi} \frac{d\theta_0}{2\pi\hbar\omega}\left[\langle\phi|H(\theta_0)|\phi\rangle_{\mathscr{H}} - \langle\phi(t;\theta_0)|H(\theta_0+\omega t)|\phi(t;\theta_0)\rangle_{\mathscr{H}}\right] \quad (54)$$

This incoherent statistical average over the phases also gives exactly the same result as the coherent average (52) in a photon number state.

We can obtain these relations as follows: Using the definition of the quasi-energy operator (7), we can express $\delta\langle N\rangle(t)$ in terms of quantities that do not involve the derivative $-i\partial/\partial\theta$:

$$\delta\langle N\rangle(t) = \langle\phi\otimes\xi|U_K^\dagger(t)\frac{K}{\hbar\omega}U_K(t)|\phi\otimes\xi\rangle_{\mathscr{K}} - \langle\phi\otimes\xi|U_K^\dagger(t)\frac{H(\theta)}{\hbar\omega}U_K(t)|\phi\otimes\xi\rangle_{\mathscr{K}}$$
$$- \langle\phi\otimes\xi| - i\frac{\partial}{\partial\theta}|\phi\otimes\xi\rangle_{\mathscr{K}} \quad (55)$$

Using the fact that $[K, U_K] = 0$, and $U_K^\dagger U_K = \mathbb{1}$ and using Eq. (7), we can write

$$\delta\langle N\rangle(t) = \langle\phi\otimes\xi|\frac{H(\theta)}{\hbar\omega}|\phi\otimes\xi\rangle_{\mathscr{K}} - \langle\phi\otimes\xi|U_K^\dagger(t)\frac{H(\theta)}{\hbar\omega}U_K(t)|\phi\otimes\xi\rangle_{\mathscr{K}} \quad (56)$$

and since

$$U_K^\dagger(t,\theta)\,H(\theta)\,U_K(t,\theta) = U^\dagger(t,0;\theta)\mathscr{T}_{\omega t}\,H(\theta)\,\mathscr{T}_{-\omega t}U(t,0;\theta)$$
$$= U^\dagger(t,0;\theta)\,H(\theta+\omega t)\,U(t,0;\theta) \quad (57)$$

we obtain Eq. (51).

We can also get more precise information on the probability $P(L,t)$ that L photons are exchanged: If at time $t=0$ the photon field is in a photon number eigenstate $e^{ik\theta}$ and $\psi(t=0)=\psi_0=\phi\otimes e^{ik\theta}$, then the probability that a measurement performed at time t yields that L photons have been exchanged is given by

$$P(L,t) = \langle U_K(t)\psi_0|[\mathbb{1}_{\mathscr{H}}\otimes|e^{i(L+k)\theta}\rangle\langle e^{i(L+k)\theta}|]|U_K(t)\psi_0\rangle_{\mathscr{K}}$$
$$= \sum_n |\langle\phi_n\otimes e^{i(k+L)\theta}|U_K(t)(\phi\otimes e^{ik\theta})\rangle_{\mathscr{K}}|^2 \quad (58)$$

where $\{\phi_n\}$ is an arbitrary basis of $\mathscr{H}$.

2. *Invariance with Respect to the Choice of the Origin of the Relative Photon Number*

Due to the relative character of the number operator $-i\partial/\partial\theta$, all the physical predictions of the Floquent model must be invariant with respect to a global translation of the relative photon numbers. We show that this is indeed the case for the properties discussed above. The propability $P(L, t)$ is independent of the particular initial photon number state chosen; that is, it is independent of k since

$$U_K(t)(\phi \otimes e^{ik\theta}) = U(t, 0; \theta - \omega t)(\phi \otimes e^{ik(\theta-\omega t)}) \tag{59}$$

and thus

$$P(L, t) = \sum_n |\langle \phi_n \otimes e^{iL\theta} | U_K(t)(\phi \otimes 1)\rangle_{\mathscr{K}}|^2 \tag{60}$$

For the average number of exchanged photons $\delta\langle N\rangle(t)$, it is straightforward to verify that one obtains the same result for the choice of any initial condition of the photon field of the form

$$\xi = \sum_k c_k e^{i(k+m)\theta}, \qquad \text{with arbitrary translation } m \tag{61}$$

3. *Number of Exchanged Photons in Adiabatic Passage with Photons in Coherent States*

In adiabatic passage processes with pulsed lasers, as we will discuss in the forthcoming sections, one often encounters the following particular situation: If the initial condition of the photon field were a number state, that is,

$$\psi_i = \phi_i(x) \otimes e^{ik\theta} \tag{62}$$

then at the end of the pulse, the final state would be

$$\psi_f = \phi_f(x) \otimes e^{i(k-m)\theta} \tag{63}$$

that is, the photon field would be again in a well-defined number state, and one can state that m photons had been adsorbed, since according to Eq. (60) we have $P(L, t_f) = \delta_{L,-m}$. Since k is the relative number of photons, if these relations are satisfied for one choice of the initial k, then they are also satisfied for all other choices of $k \in \mathbb{Z}$. However, in the actual experimental realizations the initial states of the photon field are coherent states instead of number states. The coherent states can be considered as a coherent superposition of number states of the form

$$\Phi = \sum_k c_k e^{ik\theta}, \qquad \text{with} \quad \sum_k |c_k|^2 = 1 \tag{64}$$

In the preceding sections we have taken the limit $\bar{n} \to \infty$ in which the coherent states become a $\delta_{1/2}(\theta - \theta_0)$ function. For the discussion of the exchanged photons we consider a large but finite $\bar{n}$, such that the coherent state is represented by a sharply peaked function that can be written as a superposition (64). Under this condition, the relations (62) and (63) imply that the initial condition

$$\psi^{(i)} = \phi_i(x) \otimes \xi^{(i)}, \qquad \text{with} \quad \xi^{(i)} = \sum_k c_k e^{ik\theta} \tag{65}$$

evolves at the end of the pulse to

$$\psi^{(f)} = \phi_f(x) \otimes \sum_k c_k e^{i(k-m)\theta} = \phi_f(x) \otimes e^{-im\theta} \xi^{(i)}(\theta) \tag{66}$$

Our aim here is to give a precise meaning to the statement that, also in this process involving coherent states, m photons have been absorbed: The probability to observe $\bar{n} + k$ photons at the initial time t_i is

$$P^{(i)}(\bar{n} + k) = |c_k|^2 \tag{67}$$

and at the end of the pulse

$$P^{(f)}(\bar{n} + k) = |c_{k+m}|^2 = P^{(i)}(\bar{n} + k + m) \tag{68}$$

which implies $P^{(f)}(\bar{n} + k - m) = P^{(i)}(\bar{n} + k)$; that is, the probability to measure $\bar{n} + k - m$ photons at the end is equal to the probability to measure $\bar{n} + k$ at the beginning of the pulse. In terms of averages and moments of photon numbers, one can make the following statement: Equations (65) and (66) imply that the photon expectation number changes by $-m$

$$\delta\langle N \rangle = \langle \psi^{(f)} | - i\frac{\partial}{\partial\theta} | \psi^{(f)} \rangle - \langle \psi^{(i)} | - i\frac{\partial}{\partial\theta} | \psi^{(i)} \rangle = -m \tag{69}$$

and the second moment of the relative number of photons at the end of the process is equal to the one at the beginning:

$$\langle\psi^{(f)}|\left(-i\frac{\partial}{\partial\theta}\right)^2|\psi^{(f)}\rangle = \langle\psi^{(i)}|\left(-i\frac{\partial}{\partial\theta}\right)^2|\psi^{(i)}\rangle \tag{70}$$

E. Floquet Representation with Two or More Lasers

The treatment described in the preceding sections can be easily generalized to the case in which two (or several) lasers of frequencies ω_j, $j = 1, \ldots, d$, act on the molecule. We introduce the notation $\boldsymbol{\omega} = (\omega_1, \ldots, \omega_d)$, and $\boldsymbol{\theta} = (\theta_1, \ldots, \theta_d)$, which represents the phases at time $t = 0$ of the d lasers. The semiclassical Schrödinger equation reads

$$i\hbar\frac{\partial\phi}{\partial t} = H(x, \boldsymbol{\theta} + \boldsymbol{\omega}t)\phi \tag{71}$$

with, for example in a dipole coupling model with two lasers,

$$H(x, \boldsymbol{\theta} + \boldsymbol{\omega}t) = H_0(x) - \mu(x)\mathscr{E}_1\cos(\theta_1 + \omega_1 t) - \mu(x)\mathscr{E}_2\cos(\theta_2 + \omega_2 t) \tag{72}$$

where x symbolizes the degrees of freedom of the molecule, $\mu(x)$ is its dipole moment, $\mathscr{E}_1$ and $\mathscr{E}_2$ are the respective amplitudes of the two lasers, and $H_0(x)$ the Hamiltonian of the free molecule. The corresponding Floquet Hamiltonian is defined as [5,6,17–20]

$$K = -i\hbar\boldsymbol{\omega}\cdot\frac{\partial}{\partial\boldsymbol{\theta}} + H(x, \underline{\theta}), \tag{73a}$$

$$= -i\hbar\sum_{j=1}^{d}\omega_j\frac{\partial}{\partial\theta_j} - \mu(x)\mathscr{E}_1\cos\theta_1 - \mu(x)\mathscr{E}_2\cos\theta_2 \tag{73b}$$

and acts on the enlarged Hilbert space

$$\mathscr{K} = \mathscr{H} \otimes \underbrace{\mathscr{L} \otimes \cdots \otimes \mathscr{L}}_{d\ \text{products}} \tag{74}$$

where $\mathscr{L} :- L_2(\mathbb{S}^1, d\theta_j/2\pi)$ denotes the space of square-integrable functions on the circle $\mathbb{S}^1$ of length 2π. The tensor product $\mathscr{L} \otimes \cdots \otimes \mathscr{L}$ is equivalent to $\mathscr{L}_2(\mathbf{T}^d, d\boldsymbol{\theta}/2\pi)$—that is, the square-integrable functions on the unit torus $\mathbf{T}^d$. The relations we have described for the single laser case extend in most practical cases to the d-laser case just by adapting the notation.

III. EFFECTIVE HAMILTONIAN—DYNAMICAL RESONANCES

Since in the Floquet representation the Hamiltonian K defined on the enlarged Hilbert space $\mathscr{K}$ is time-independent, the analysis of the effect of perturbations (like, e.g., transition probabilities) can be done by stationary perturbation theory, instead of the usual time-dependent one. Here we will present a formulation of stationary perturbation theory based on the iteration of unitary transformations (called contact transformations or KAM transformations) constructed such that the form of the Hamiltonian gets simplified. It is referred to as the KAM technique. The results are not very different from the ones of Rayleigh–Schrödinger perturbation theory, but conceptually and in terms of speed of convergence they have some advantages.

We first formulate this KAM technique in a general setting (see Section III.A). Next, in Section III.B, we apply it to the case of an interaction of high frequency with respect to the energy differences of the free system and show its connection with the standard Born–Oppenheimer approximation.

Resonant effects that prevent convergence of the perturbation theory and that appear as small denominators will be treated specifically by rotating wave transformations (RWT) in Section III.C.

Combining the KAM techniques with the RWT will allow us to construct effective Hamiltonians in a systematic way and to estimate the order of the neglected terms (see Section III.D). We show that the KAM technique allows us to partition at a desired order operators in orthogonal Hilbert subspaces. We adapt this partitioning technique to treat Floquet Hamiltonians. Its connection with the standard adiabatic elimination is shown at a second-order approximation.

In Section III.E this partitioning technique is illustrated for two-photon processes in atoms. It is next applied in Section III.F to construct an effective Hamiltonian relevant for the rotational excitation in diatomic molecules.

A. Perturbation Theory Formulated as an Iteration of Unitary Transformations—Nonresonant Case: KAM Techniques

1. Iterative Perturbation Algorithm

We start with an unperturbed Hamiltonian

$$K_0 = -i\hbar\omega\frac{\partial}{\partial\theta} + H_0 \tag{75}$$

where H_0 is a θ-independent Hamiltonian that represents the free molecule. The coupling with the field is represented by a perturbation $\varepsilon V_1(\theta)$:

$$K = K_0 + \varepsilon V_1(\theta) \tag{76}$$

The purpose of ε is only to keep track of the different orders, and at the end we can set $\varepsilon = 1$. The method is defined quite generally, independently of the particular form of K_0 and V_1. We assume that $V_1(\theta)$ is a bounded operator. The idea is to construct a unitary transformation $e^{\varepsilon W_1(\theta)}$, with $W_1^\dagger = -W_1$ such that

$$e^{-\varepsilon W_1} K e^{\varepsilon W_1} = K_0 + \varepsilon D_1 + \varepsilon^2 V_2 =: K_2 \tag{77}$$

where D_1 is a θ-independent operator satisfying $[K_0, D_1] = 0$. Thus the perturbation will be reduced from order ε to order ε^2. Once this is achieved, the approximation of order ε of the eigenvalues and the eigenvectors is obtained from $K_0 + \varepsilon D_1$ — that is, neglecting $\varepsilon^2 V_2$. The eigenvectors of $K_0 + \varepsilon D_1$ are the same as those of K_0, since the two operators commute. If the eigenvalues $\lambda_{m,k}^{(0)}$ of K_0 are nondegenerate, and we denote $D_1 = \mathrm{diag}\{d_m\}$, then the perturbed eigenvalues of first order (i.e., neglecting corrections of second order) are

$$\lambda_{m,k}^{(1)} = \lambda_{m,k}^{(0)} + d_m \tag{78}$$

and the corresponding eigenvectors are

$$|\psi_{m,k}^{(1)}\rangle = e^{\varepsilon W_1} |\psi_{m,k}^{(0)}\rangle \tag{79}$$

with $|\psi_{m,k}^{(0)}\rangle = |e_m \otimes e^{ik\theta}\rangle$, where $\{|e_m\rangle\}$ is the eigenbasis of H_0 in $\mathscr{H}$.

If some eigenvalues of K_0 are degenerate, the addition of D_1 can lift the degeneracy.

Since the transformed Hamiltonian K_2 is of the same general form as the one we started with [Eq. (76)], this procedure can be iterated. The order of the perturbation can thus be reduced successively from ε to ε^2, to $\varepsilon^4, \ldots$. After N iterations the remaining perturbation is of order $\varepsilon^{2^N} = \varepsilon^{e^{N \ln 2}}$; that is, we have a superexponential decrease. This type of iterative algorithm is therefore called *superconvergent*. We call this procedure a quantum KAM algorithm, since it is the quantum analogue of the Kolmogorov–Arnold–Moser (KAM) transformations developed in classical mechanics [21–27]. This procedure is also known as the van Vleck perturbation theory [28]. The transformations $e^{\varepsilon W_1}$ are called *contact transformations* or *KAM transformations*. One step of the algorithm is roughly equivalent to first-order perturbation theory. The idea is that instead of performing a perturbation calculation of high order, one can perform several times a calculation of first order. The acceleration of convergence can be explained by the fact that at each step of the iteration, one develops around a different effective nonperturbed Hamiltonian that already contains the corrections found in the previous iterations.

We remark that D_N, the diagonal part obtained after N iterations, is a function of ε that is not a polynomial, since (as we will see below) the construction involves rational functions and exponentials. However, if one expands D^N as a power series in ε up to a certain order (smaller than ε^{2^N}), the result must coincide with the Rayleigh–Schrödinger power series of that order, since the coefficients of this expansion are unique (even in the case when the series is only asymptotic, i.e., nonconvergent). In practice, performing one or two iterations gives already the main information of the processes we will study.

2. *Construction of the Contact Transformations*

We discuss now how one can construct the transformation for one step in the iterative algorithm. An equation to determine D_1 and W_1 can be obtained by expanding Eq. (77) in powers of ε and requiring that the θ-dependent terms of order ε cancel out. This leads to the two equations

$$[K_0, W_1] + V_1 - D_1 = 0 \tag{80a}$$

$$[K_0, D_1] = 0 \tag{80b}$$

The solution of these equations can be given using the eigenvalues and eigenvectors of K_0, which we will denote by λ_ν^0 and $|\nu, j\rangle$ [we represent the indices m, k of Eqs. (78) and (79) by a single index ν that labels the different eigenvalues, and j distinguishes different basis vectors corresponding to a degenerate eigenvalue]. We define a projection operator Π_{K_0} that extracts from the perturbation V_1 the diagonal component with respect to the eigenbasis of K_0:

$$\Pi_{K_0} V_1 = \sum_{\nu, j, j'} |\nu, j\rangle\langle \nu, j|V_1|\nu, j'\rangle\langle \nu, j'| \tag{81}$$

With this notation a solution of Eq. (80) can be written as

$$D_1 = \Pi_{K_0} V_1 = \text{diagonal part of } V_1 \tag{82a}$$

$$W_1 = -\sum_{\nu, j, j', \nu' \neq \nu} \frac{|\nu, j\rangle\langle \nu, j|V_1|\nu', j'\rangle\langle \nu', j'|}{\lambda_\nu^0 - \lambda_{\nu'}^0} \tag{82b}$$

The solution W_1 is not unique, since if A is any operator such that $[K_0, A] = 0$, then $W_1 + A$ is also a solution. The solution (82a) is singled out as the unique solution with zero diagonal blocks, $\Pi_{K_0} W_1 = 0$ [29].

There are two ways to proceed, depending on what we do with the diagonal part $\Pi_{K_0} V_1$ of the perturbation. It can be added to the unperturbed Hamiltonian either after or before the transformation.

(i) In the first case we take K_0 as the unperturbed Hamiltonian and the perturbation V_1 has a nonzero projection $\Pi_{K_0} V_1$, which leads to a term

$D_1 = \Pi_{K_0} V_1$ in the solution (82a). In this case the second-order perturbation that remains at the end of the transformation takes the form

$$\varepsilon^2 V_2 = \frac{\varepsilon^2}{2}[V_1, W_1] + \frac{\varepsilon^3}{3}[[V_1, W_1], W_1] + \cdots + \varepsilon^M \frac{(M-1)}{M!} \underbrace{[\ldots[[V_1, W_1], W_1], \ldots]}_{M-1 \text{ commutators}} + \cdots$$
$$+ \frac{\varepsilon^2}{2!}[D_1, W_1] + \frac{\varepsilon^3}{3!}[[D_1, W_1], W_1] + \cdots + \frac{\varepsilon^M}{M!} \underbrace{[\ldots[D_1, W_1], W_1], \ldots]}_{M-1 \text{ commutators}} + \cdots \tag{83}$$

We remark that defining an operator $L_{W_1} : V \mapsto L_{W_1}(V) := [V, W_1]$, the transformation (77) can be expressed as

$$e^{-\varepsilon W_1} K e^{\varepsilon W_1} = e^{L_{\varepsilon W_1}}(K) = \sum_{M=0}^{\infty} \varepsilon^M \frac{1}{M!} (L_{W_1})^M (K) \tag{84}$$

and the above expression (83) can be written as

$$\varepsilon^2 V_2 = \sum_{M=2}^{\infty} \varepsilon^M \frac{(M-1)}{M!} (L_{W_1})^{(M-1)}(V_1) + \sum_{M=2}^{\infty} \varepsilon^M \frac{1}{M!} (L_{W_1})^{(M-1)}(D_1) \tag{85a}$$

$$= \sum_{M=2}^{\infty} \varepsilon^M \frac{1}{M!} (L_{W_1})^{(M-1)}((M-1)V_1 + D_1) \tag{85b}$$

(ii) The second possibility is to define a new unperturbed Hamiltonian $\tilde{K}_0^D$ in which the projection $\Pi_{K_0} V_1$ of the perturbation is already absorbed:

$$\tilde{K}_0^D = K_0 + \Pi_{K_0} V_1 \tag{86}$$

The remaining perturbation $V_1 - \Pi_{K_0} V_1$ has zero projection and thus there is no supplementary $\tilde{D}_1$ to be added. In this case the second-order perturbation that remains at the end of the transformation takes the somewhat simpler form

$$\varepsilon^2 V_2 = \frac{\varepsilon^2}{2}[V_1, W_1] + \frac{\varepsilon^3}{3}[[V_1, W_1], W_1] + \cdots + \varepsilon^M \frac{(M-1)}{M!} \underbrace{[\ldots[[V_1, W_1], W_1], \ldots]}_{M-1 \text{ commutators}} + \cdots \tag{87}$$

Both alternatives can be useful; depending on the particular problem, one of them can be more convenient than the other one. As we will see in Section III.D, the second version is particularly adapted to the construction of effective Hamiltonians by the partitioning technique.

3. *Interpretation as an Averaging Procedure*

The perturbation theory outlined above can be interpreted as an averaging procedure [21–25,29]: The projector can be expressed as

$$D_1 = \Pi_{K_0} V_1 = \lim_{\tau \to \infty} \frac{1}{\tau} \int_0^{\tau} ds\, e^{-iK_0 s} V_1 e^{iK_0 s} \tag{88}$$

and

$$W_1 = \lim_{\tau \to \infty} \frac{-i}{\tau} \int_0^{\tau} ds' \int_0^{s'} ds\, e^{-iK_0 s} (V_1 - \Pi_{K_0} V_1) e^{iK_0 s} \tag{89}$$

The term $e^{-iK_0 s} V_1 e^{iK_0 s}$ in Eq. (88) is equal to the inverse time evolution that the operator V_1 would have in the Heisenberg picture for the dynamics generated by K_0. Thus D_1, which is the term that is added to constitute the approximate effective Hamiltonian $K_{01} := K_0 + D_1$, can be interpreted as the average of the perturbation with respect to the dynamics generated by K_0.

Another equivalent expression is [30,31]

$$D_1 = \Pi_{K_0} V_1 = \lim_{\beta \to 0^+} \beta \int_0^{\infty} ds\, e^{-\beta s} e^{-iK_0 s} V_1 e^{iK_0 s} \tag{90}$$

and

$$W_1 = -i \lim_{\beta \to 0^+} \beta \int_0^{\infty} ds'\, e^{-\beta s'} \int_0^{s'} ds\, e^{-iK_0 s} (V_1 - \Pi_{K_0} V_1) e^{iK_0 s} \tag{91}$$

The relation between these two expression can be thought of as two equivalent realizations of the time average

$$\text{Average}(f) = \lim_{\beta \to 0^+} \beta \int_0^{\infty} ds' e^{-\beta s'} f(s') = \lim_{\tau \to \infty} \frac{1}{\tau} \int_0^{\tau} ds' f(s') \tag{92}$$

A third alternative expression for W_1 is [30,31]

$$W_1 = \lim_{\beta \to 0^+} -i \int_0^{\infty} ds\, e^{-\beta s} e^{-iK_0 s} (V_1 - \Pi_{K_0} V_1) e^{iK_0 s} \tag{93}$$

B. High-Frequency Perturbation Theory

A variation of the procedures described above can be applied to situations in which the frequency ω of the perturbation is high with respect to the internal frequencies of the considered system [32]. We start with a Floquet Hamiltonian of the form

$$K = -i\hbar\omega\frac{\partial}{\partial\theta} + H_0(x) + V_1(x, \theta) \tag{94}$$

where x symbolizes the degrees of freedom of the molecule. Since we are interested in the limit $\hbar\omega \to \infty$, we define a small parameter $\epsilon := 1/(\hbar\omega)$ and we rewrite

$$K = \hbar\omega\hat{K} \tag{95}$$

with

$$\hat{K} = -i\frac{\partial}{\partial\theta} + \epsilon(H_0 + V_1) \tag{96}$$

The eigenvectors of K are the same ones as those for $\hat{K}$, and the eigenvalues just have to be multiplied by $\hbar\omega$. The difference with the preceding discussion is that here H_0 and V are both of order ϵ. Thus we take as the unperturbed Floquet Hamiltonian just

$$\hat{K}_0 := -i\frac{\partial}{\partial\theta} \tag{97}$$

If the frequency is large compared with the frequencies of the system, there will not be any resonances. We can thus proceed with the iterative perturbative KAM algorithm by first determining a unitary transformation $e^{\epsilon W_1(x,\theta)}$, with $W_1^\dagger = -W_1$ such that

$$e^{-\epsilon W_1}\hat{K}e^{\epsilon W_1} = \hat{K}_0 + \epsilon D_1 + \epsilon^2 V_2 =: K_2 \tag{98}$$

where D_1 is a θ-independent operator such that $[\hat{K}_0, D_1] = 0$. Thus the perturbation will be reduced from order ϵ to order ϵ^2. The generator W_1 of the contact transformation is determined by the equations

$$[K_0, W_1] + H_0 + V_1 - D_1 = 0 \tag{99a}$$

$$[K_0, D_1] = 0 \tag{99b}$$

The Eq. (99a) can be written as

$$-i\frac{\partial W_1}{\partial\theta} + H_0 + V_1 - D_1 = 0 \tag{100}$$

whose general solution is given by

$$W_1 = -i\int^{\theta} d\theta\ (H_0 + V_1 - D_1) + C \tag{101}$$

where C is an arbitrary θ-independent operator acting on $\mathscr{H}$, which one can choose as $C = 0$. Since $W_1(x, \theta)$ is a multiplication operator acting on functions of the angle θ, it must be necessarily 2π-periodic. This condition determines D_1 uniquely in terms of the average

$$\overline{V_1} := \int_0^{2\pi} \frac{d\theta}{2\pi} V_1(\theta) \tag{102}$$

as

$$D_1 = H_0 + \overline{V_1} \tag{103}$$

Thus we obtain

$$W_1(x, \theta) = -i\int^{\theta} d\theta\ (V_1(x, \theta) - \overline{V_1}(x)) \tag{104}$$

We remark that the solution (103), (104) of Eqs. (99) can also be obtained from the general equations (88), (89). This contact transformation $e^{\epsilon W_1}$ can be interpreted, in the case where $[V, (x, \theta), V_1(x, \theta')] = 0$ for all θ, θ', as the unitary transformation that allows us to diagonalize exactly the Hamiltonian $\hbar\omega\hat{K}_0 + V_1(x, \theta)$ with respect to θ, taking x as a parameter. We obtain $\lambda_k^{(0)} = \overline{V_1} + k\hbar\omega$ (k positive or negative integer) for the eigenvalues associated to the eigenvectors $\chi(\theta, x) = \exp(\epsilon W_1(x, \theta) + ik\theta)$. We remark that if $\overline{V_1} = 0$, the eigenvalues do not depend on the variable x.

Adapting Eq. (85b), the remaining perturbation of order ϵ^2 can be written as

$$\begin{aligned} \epsilon^2 V_2 &= \sum_{M=2}^{\infty} \varepsilon^M \frac{1}{M!} (L_{W_1})^{(M-1)}((M-1)(H_0 + V_1) + D_1) \\ &= \frac{\epsilon^2}{2!}[V_1 + \overline{V_1} + 2H_0, W_1] + \epsilon^3 \ldots \end{aligned} \tag{105}$$

In the particular case where $[V, (x, \theta), V_1(x, \theta')] = 0$ Eq. (105) reduces to $\epsilon^2 V_2 = \epsilon^2 [H_0, W_1] + \epsilon^3 \ldots$ we can apply a second contact transformation $e^{\epsilon^2 W_2}$ (with respect to $\hat{K}_0$), $W_2 = -i\int^{\theta} d\theta (V_2 - \overline{V_2})$, which averages Eq. (105) with respect to θ and leads to correction of order ϵ^3. In the basis of the eigenvectors $\chi(\theta, x)$ of $\hbar\omega\hat{K}_0 + V_1(x, \theta)$, the nondiagonal terms $V_2 - \overline{V_2}$ of this remaining term (105) can be seen as the couplings between the states corresponding to the eigenvalues $\lambda_k^{(0)}$ and the diagonal terms $\overline{V_2}$ will lead to geometrical phases for the dynamics (see Section IV.B.1). We thus obtain the effective high-frequency Hamiltonian H^{HF} (independent of θ) of order ϵ^2

$$H^{\mathrm{HF}}(x) = H_0(x) + \overline{V_1}(x) + \overline{V_2}(x) \tag{106}$$

We remark that the standard Born–Oppenheimer approximation, allowing us to separate the fast electronic motion with respect to the slow vibrational motion of the nuclei of molecules, can be thought of as a high-frequency perturbation theory. For the Born–Oppenheimer approximation, we formally identify K_0 with the kinetic energy of the electrons and identify ϵ with the mass of the electrons. As we stated above, the approximation consists in first applying a contact transformation that diagonalizes exactly with respect to the electronic coordinates, keeping the nuclei coordinates as parameters, and next neglecting the nondiagonal coupling between the eigenvalues (which are associated to the electronic states).

C. Nonperturbative Treatment of Resonances—Resonant Transformations

The properties that we have stated in the last section allow us to analyze the situation in which there are resonances. The analysis of resonances involves two aspects: The first one is the determination of degenerate eigenvalues of an unperturbed Hamiltonian K_0, and the second one is the detection of terms in the perturbation V_1 that couple these degenerate modes. We show that the projectors of type Π_{K_0} can be used to detect resonant terms in the coupling operators V_1. This is an alternative to another formulation that consists in writing down formally a Fourier series of the generator W_1 of the KAM transformation and detecting diverging terms—that is, terms with zeros in the denominator and a finite numerator.

This leads to distinguish two types of resonances: the resonances induced by the field that occur beyond a threshold of the field, and resonances that occur for an arbitrary small value of the field. These are called, respectively, (i) the *dynamical resonances* (or equivalently *field induced resonances* or *nonlinear resonances*) and (ii) the *zero-field resonances*.

The zero-field resonances can be identified with respect to the system energy levels and the field frequency when the field is off. They are usually one- or two-photon resonances. The one-photon resonance is of first order with respect to the field amplitude in the sense that the degeneracy of the eigenvalues is lifted linearly with the field amplitude. The two-photon resonance is of second order since the degeneracy of the eigenvalues is lifted quadratically with the field amplitude. Multiphoton resonances (more than two-photon) are more complicated since they are generally accompanied by dynamical shifts of second order before the actual occurrence of the resonance at a higher order. They are, in general, dynamical.

For very small field amplitudes, the multiphoton resonances can be treated by time-dependent perturbation theory combined with the rotating wave approximation (RWA) [10]. In a strong field, all types of resonances can be treated by the concept of the rotating wave transformation, combined with an additional stationary perturbation theory (such as the KAM techniques explained above). It will allow us to construct an effective Hamiltonian in a subspace spanned by the resonant dressed states, degenerate at zero field.

To illustrate the effects of these two types of resonances, we consider the simplest concrete example of a two-level system driven by a strong field:

$$K = K_0 + V, \quad K_0 = -i\hbar\omega\frac{\partial}{\partial\theta} + \frac{\beta}{2}\begin{pmatrix} 1 & 0 \\ 0 & -1 \end{pmatrix}, \quad V = \varepsilon\cos(\theta)\begin{pmatrix} 0 & 1 \\ 1 & 0 \end{pmatrix} \tag{107}$$

with ε and β real and positive.

1. Zero-Field Resonances

We consider the case $\beta = \omega$. There is a one-photon resonance in K_0, since its eigenvalues are $\lambda^0_{m,k} = m\frac{\beta}{2} + k\omega$, $m \in \{-1, +1\}$, $k \in \mathbb{Z}$, and therefore $\lambda^0_{-1,k+1} = \lambda^0_{+1,k}$; that is, all the eigenvalues are degenerate of order two. The degeneracy eigenspaces are spanned by the vectors

$$\psi^0_{-1,k+1} = e^{i(k+1)\theta} \otimes \begin{pmatrix} 0 \\ 1 \end{pmatrix} \quad \text{and} \quad \psi^0_{+1,k} = e^{ik\theta} \otimes \begin{pmatrix} 1 \\ 0 \end{pmatrix} \tag{108}$$

The projector Π_{K_0} applied on the coupling term yields

$$\Pi_{K_0} V \equiv V_{resonant} = \frac{\varepsilon}{2}\begin{pmatrix} 0 & e^{-i\theta} \\ e^{i\theta} & 0 \end{pmatrix} \tag{109}$$

This resonant term cannot be eliminated by the KAM transformation. Instead we can treat it with a different type of transformation. We define a unitary transformation

$$R_1 = \begin{pmatrix} 1 & 0 \\ 0 & e^{i\theta} \end{pmatrix} \tag{110}$$

As opposed to the KAM-type transformation $e^{\varepsilon W}$, the transformation R_1 is not close to the identity. It is named *rotating wave transformation* (RWT) (or equivalently *resonant transformation*) in contrast with the usual RWA for which only the resonant terms are kept, and the counterrotating terms are neglected. It is defined in such a way that

$$R_1^\dagger V_{resonant} R_1 = \frac{\varepsilon}{2}\begin{pmatrix} 0 & 1 \\ 1 & 0 \end{pmatrix} \tag{111}$$

that is, the resonant term becomes θ-independent. Thus

$$R_1^\dagger K R_1 = -i\omega\frac{\partial}{\partial\theta} + \frac{\varepsilon}{2}\begin{pmatrix} 0 & 1 \\ 1 & 0 \end{pmatrix} + \frac{\varepsilon}{2}\begin{pmatrix} 0 & e^{+i2\theta} \\ e^{-i2\theta} & 0 \end{pmatrix} + \frac{\omega}{2}\mathbb{1}_{\mathscr{K}} \tag{112}$$

This Floquet Hamiltonian can be further simplified by diagonalizing the constant matrix with $T_1 = \frac{1}{\sqrt{2}}\begin{pmatrix} 1 & -1 \\ 1 & 1 \end{pmatrix}$:

$$\begin{aligned} K' = T_1^\dagger R_1^\dagger \; K \; R_1 T_1 = -i\omega\frac{\partial}{\partial\theta} + \frac{\varepsilon}{2}\begin{pmatrix} 1 & 0 \\ 0 & -1 \end{pmatrix} \\ + \frac{\varepsilon}{2}\begin{pmatrix} \cos(2\theta) & i\sin(2\theta) \\ -i\sin(2\theta) & -\cos(2\theta) \end{pmatrix} + \frac{\omega}{2}\mathbb{1}_{\mathscr{K}} \end{aligned} \tag{113}$$

This transformed Floquet Hamiltonian can now be decomposed into a *renormalized* unperturbed part

$$K_0'(\varepsilon) = -i\omega\frac{\partial}{\partial\theta} + \frac{\varepsilon}{2}\begin{pmatrix} 1 & 0 \\ 0 & -1 \end{pmatrix} + \frac{\omega}{2}\mathbb{1}_{\mathscr{K}} \tag{114}$$

(which is explicitly ε-dependent) and a perturbation

$$V' = \frac{\varepsilon}{2}\begin{pmatrix} \cos(2\theta) & i\sin(2\theta) \\ -i\sin(2\theta) & -\cos(2\theta) \end{pmatrix} \tag{115}$$

In this form, the part of the perturbation that is left is not resonant anymore (for small ε), and we can apply KAM-type transformations to eliminate it iteratively. This procedure is an adaptation of the technique developed by H. Eliasson to study the problem of localization in quasi-periodic potentials [33] and extended further to problems that are close to the one discussed here [34–38].

The RWA consists in neglecting V'. If we consider additionally a detuning $\Delta = \beta - \omega$, Eq. (112) becomes (before diagonalization)

$$R_1^\dagger K R_1 = -i\omega\frac{\partial}{\partial\theta} + \frac{1}{2}\begin{pmatrix} \Delta & \varepsilon \\ \varepsilon & -\Delta \end{pmatrix} + \frac{\varepsilon}{2}\begin{pmatrix} 0 & e^{+i2\theta} \\ e^{-i2\theta} & 0 \end{pmatrix} + \frac{\omega}{2}\mathbb{1}_{\mathscr{K}} \tag{116}$$

and the effective RWA Hamiltonian reads (see, e.g., Refs. 39 and 40)

$$H_{\mathrm{RWA}} = \frac{1}{2}\begin{pmatrix} \Delta & \varepsilon \\ \varepsilon & -\Delta \end{pmatrix} \tag{117}$$

2. *Dynamical Resonances*

As we have stated, the Floquet Hamiltonian (113) has no terms that are resonant if we take small enough ε, and the iteration of the KAM procedure converges. However, if we take ε large enough, we encounter new resonances that are not present at zero or small fields; that is, they are not related to degeneracies of the unperturbed eigenvalues of K_0 that lead to the zero-field resonances we have discussed in the previous subsection. These new resonances are related to degeneracies of the new effective unperturbed operator $K_0'(\varepsilon)$, which appear at some specific finite values of ε. These are the dynamical resonances.

In the present case, the first nontrivial dynamical resonance (for $\omega > 0$) appears at $\varepsilon = 2\omega = 2\beta$, since the eigenvalues of $K_0'(\varepsilon)$ are of the form $\lambda^{0'}_{m,k}(\varepsilon) = m\varepsilon/2 + k\omega$, $m \in \{-1, +1\}$, $k \in \mathbb{Z}$, and thus $\lambda^{0'}_{-1,k+2}(\varepsilon = 2\omega) = \lambda^{0'}_{+1,k}(\varepsilon = 2\omega)$; that is, all the eigenvalues are degenerate of order two at the resonant amplitude $\varepsilon = 2\omega$. This dynamical resonance can be interpreted as a two-photon resonance with respect to the effective Hamiltonian K_0', and a three-photon resonance with respect to the original Hamiltonian K_0. The degeneracy eigenspaces are spanned by the vectors

$$\psi^0_{-1,k+2} = e^{i(k+2)\theta} \otimes \begin{pmatrix} 0 \\ 1 \end{pmatrix} \quad \text{and} \quad \psi^0_{+1,k} = e^{ik\theta} \otimes \begin{pmatrix} 1 \\ 0 \end{pmatrix} \tag{118}$$

Again, we can detect the resonant terms of the perturbation V' by applying the projector $\Pi_{K_0'(\varepsilon=2\omega)}$:

$$\Pi_{K_0'(\varepsilon=2\omega)} V' \equiv V'_{resonant} = -\frac{\varepsilon}{4}\begin{pmatrix} 0 & e^{-i2\theta} \\ e^{i2\theta} & 0 \end{pmatrix} \tag{119}$$

We notice that the eigenvalues of $K_0'(\varepsilon)$ are also degenerate at $\varepsilon = \omega$. But since there are no terms in V' with modes $e^{\pm i\theta}$, we have $\Pi_{K_0'(\varepsilon=\omega)} V' = 0$. This degeneracy does not give rise to an actual resonance. One can call it an inactive resonance. An equivalent way of stating this is that in the calculation of W for the perturbation analysis, the degeneracy of the eigenvalues leads to a zero in the denominator, but the corresponding numerator is identically zero.

As before, if we want to eliminate the perturbation by a KAM iteration for values of $\varepsilon \gtrsim 2\omega$, we first have to deal with the resonant term (119). This can be done as above by using a transformation (that is not close to the identity) of the form

$$R_2 = \begin{pmatrix} 1 & 0 \\ 0 & e^{i2\theta} \end{pmatrix} \tag{120}$$

The transformed Floquet Hamiltonian becomes

$$R_2^\dagger K' R_2 = -i\omega\frac{\partial}{\partial\theta} + H_0'' + \frac{\varepsilon}{4}\begin{pmatrix} 2\cos(2\theta) & e^{i4\theta} \\ e^{-i4\theta} & -2\cos(2\theta) \end{pmatrix} + \frac{3}{2}\omega \mathbb{1}_{\mathscr{K}} \tag{121}$$

with

$$H_0'' = \frac{\varepsilon}{2}\begin{pmatrix} (1-2\omega/\varepsilon) & -1/2 \\ -1/2 & -(1-2\omega/\varepsilon) \end{pmatrix} \tag{122}$$

As before, the constant part H_0'' can be diagonalized by a transformation (which depends on ε, but not on θ):

$$T_2 = \frac{1}{d}\begin{pmatrix} \alpha_+ + \varepsilon/2 - \omega & \varepsilon/4 \\ -\varepsilon/4 & \alpha_+ + \varepsilon/2 - \omega \end{pmatrix} \tag{123}$$

where $\alpha_\pm = \pm(\omega^2 + \varepsilon^2 5/16 - \varepsilon\omega)^{1/2}$ are the eigenvalues of H_0'' and $d = [(\alpha_+ + \varepsilon/2 - \omega)^2 + \varepsilon^2/16]^{1/2}$. The transformed Floquet Hamiltonian can thus be written as

$$\begin{aligned} K'' &= T_2^\dagger R_2^\dagger T_1^\dagger R_1^\dagger \; K \; R_1 T_1 R_2 T_2 \\ &= -i\omega\frac{\partial}{\partial\theta} + (\omega^2 + 5\varepsilon^2/16 - \varepsilon\omega)^{1/2}\begin{pmatrix} 1 & 0 \\ 0 & -1 \end{pmatrix} + \frac{\varepsilon}{4}V'' + \frac{3\omega}{2}\mathbb{1}_{\mathscr{K}} \end{aligned} \tag{124}$$

where

$$V'' = T_2^\dagger \begin{pmatrix} 2\cos(2\theta) & e^{i4\theta} \\ e^{-i4\theta} & -2\cos(2\theta) \end{pmatrix} T_2 \tag{125}$$

This transformed Floquet Hamiltonian is nonresonant for values of ε up to a certain amplitude $\varepsilon > 2\omega$, and the KAM iteration based on Eq. (124) can be expected to converge.

D. Effective Dressed Hamiltonians by Partitioning of Floquet Hamiltonians

In this subsection we will combine the general ideas of the iterative perturbation algorithms by unitary transformations and the rotating wave transformation, to construct effective models. We first show that the preceding KAM iterative perturbation algorithms allow us to partition at a desired order operators in orthogonal Hilbert subspaces. Its relation with the standard adiabatic elimination is proved for the second order. We next apply this partitioning technique combined with RWT to construct effective dressed Hamiltonians from the Floquet Hamiltonian. This is illustrated in the next two Sections III.E and III.F for two-photon resonant processes in atoms and molecules.

1. *Partitioning: General Formulation*

We develop the partitioning technique with the use of the iterative KAM perturbation algorithms. We derive an effective Hamiltonian of second order. The scheme we show can be easily extended to higher orders.

We consider the dynamics of a system defined on a Hilbert space $\mathscr{H}$ of dimension N, by a time-independent Hamiltonian H. We consider situations in which the Hilbert space can be split into two orthogonal subspaces $\mathscr{H} = \mathscr{H}^0 \oplus \mathscr{H}^1$, that are only weakly coupled by H. Introducing the projectors P^j into these subspaces, $\mathscr{H}^0 = P^0\mathscr{H}$ and $\mathscr{H}^1 = P^1\mathscr{H}$, the Hamiltonian can be separated into four parts:

$$H = H^{00} + H^{11} + H^{01} + H^{10} \quad \text{with} \quad H^{ij} := P^i H P^j, \quad H^{01} = (H^{10})^\dagger \tag{126}$$

We can represent this partition symbolically in matrix form as

$$H = \begin{pmatrix} H^{00} & H^{01} \\ H^{10} & H^{11} \end{pmatrix} \tag{127}$$

The idea is that the coupling $H^{01} + H^{10}$ is small with respect to the other relevant energies of $|H^{11} - H^{00}|$. We think of it as being of order ε. We introduce the notation

$$\varepsilon V_1 := H^{01} + H^{10} = \varepsilon(\tilde{H}^{01} + \tilde{H}^{10}) \tag{128}$$

We will show that the *effective Hamiltonian of second order* (connected with an initial condition in $\mathscr{H}^0$) reads as

$$H^{00}_{\text{eff}} = H^{00} + \frac{\varepsilon^2}{2} P^0[\tilde{H}^{01} W_1^{10} + (\tilde{H}^{01} W_1^{10})^\dagger] P^0 \tag{129}$$

with W_1^{10} defined in Eq. (140c). In terms of the eigenvalues and eigenvectors of H^{00} restricted to the subspace $P^0\mathscr{H}$, which we denote by λ_n^{00} and $|n^{00}\rangle$, and those of H^{11} restricted to the subspace $P^1\mathscr{H}$, which we denote by λ_m^{11} and $|m^{11}\rangle$, it becomes

$$H_{\text{eff}}^{00} = H^{00} - \frac{1}{2}\sum_{n,\tilde{n}} |\tilde{n}^{00}\rangle \times \left[\sum_m \langle\tilde{n}^{00}|H^{01}|m^{11}\rangle\langle m^{11}|H^{10}|n^{00}\rangle\left(\frac{1}{\lambda_m^{11}-\lambda_n^{00}} + \frac{1}{\lambda_m^{11}-\lambda_{\tilde{n}}^{00}}\right)\right]\langle n^{00}| \tag{130}$$

The goal is to find a unitary transformation S that transforms H into block-diagonal form, at least to some order of approximation:

$$S^\dagger HS = \begin{pmatrix} H^{00} + D_1^{00} & 0 \\ 0 & H^{11} + D_1^{11} \end{pmatrix} \tag{131}$$

The idea is in general that instead of diagonalizing by perturbation methods the complete Hamiltonian, one first reduces it approximately to block-diagonal form, singling out a block $H^{00} + D_1^{00}$, of small dimension, that is the most relevant part for the dynamics of a particularly chosen initial condition. The Hamiltonian $H^{00} + D_1^{00}$ is called the *effective Hamiltonian* for the considered process. Since it is of small dimension, it can often be analyzed in detail with nonperturbative methods (for example, by exact diagonalization). The sub-block should contain all the states with which the initial state is mainly coupled by the dynamics. These states are called *essential states*. In other words, within each of the initial diagonal blocks the couplings can be strong, but the couplings between the blocks should be small. We assume in particular that the spectrum of H^{00} is well separated from the one of H^{11} by a minimal distance between eigenvalues denoted $\Delta\lambda_{min}$. We require that the norm of the coupling $H^{01} + H^{10}$ between the two blocks is small with respect to $\Delta\lambda_{min}$.

We thus construct a unitary transformation of the form $e^{\varepsilon W_1}$, with $W_1^\dagger = -W_1$ such that

$$e^{-\varepsilon W_1}(H^{00} + H^{11} + \varepsilon V_1)e^{\varepsilon W_1} = H^{00} + H^{11} + D_1 + \varepsilon^2 V_2 \tag{132}$$

where $D_1 = D_1^{00} + D_1^{11}$, with $D_1^{ii} = P^i D_1 P^i$, $i = 0, 1$, and V_2 is a remaining coupling term that is of order ε^2. We will use the notation $D_1 =: \varepsilon\tilde{D}_1$. Expansion of the exponentials and extractions of the non-block-diagonal terms leads to the equations that D_1 and W_1 are required to fulfill:

$$[H^{00} + H^{11}, W_1] + V_1 - \tilde{D}_1 = 0 \tag{133a}$$

$$[P^0, D_1^{00}] = 0 \tag{133b}$$

$$[P^1, D_1^{11}] = 0 \tag{133c}$$

Defining

$$W_1^{ij} := P^i W_1 P^j \tag{134}$$

and acting with the projectors P^0 and P^1 from the left and from the right in the four possible combinations, Eq. (133a) can be decomposed in four independent equations:

$$H^{00} W_1^{01} - W_1^{01} H^{11} + \tilde{H}^{01} = 0 \tag{135a}$$

$$H^{11} W_1^{10} - W_1^{10} H^{00} + \tilde{H}^{10} = 0 \tag{135b}$$

$$D_1^{00} = 0 \tag{135c}$$

$$D_1^{11} = 0 \tag{135d}$$

Equations (135c) and (135d) are a direct consequence of the fact that $V_1 \equiv H^{01} + H^{10}$ have zero diagonal blocks and mean that $D_1 = 0$. We remark that Eq. (135a)–(135d) do not impose any condition on the components W_1^{00} and W_1^{11}, which can therefore be chosen arbitrarily. The choice that leads to the simplest expressions is $W_1^{00} = 0$ and $W_1^{11} = 0$. The two other components, W_1^{10} and W_1^{01}, are uniquely determined. In close analogy with the construction of Section III.A.2, we define the projector Π by

$$\Pi V := P^0 V P^0 + P^1 V P^1 \tag{136}$$

A solution of Eqs. (133a) and (133b) is given by

$$\tilde{D}_1 = \Pi V_1 = 0 \tag{137}$$

and

$$W_1^{01} = \lim_{\tau \to \infty} \frac{-i}{\tau} \int_0^\tau ds' \int_0^{s'} ds\, e^{-iH^{00}s} \tilde{H}^{01} e^{iH^{11}s} \tag{138a}$$

$$= \lim_{\beta \to 0^+} -i \int_0^\infty ds\, e^{-\beta s}\, e^{-iH^{00}s} \tilde{H}^{01} e^{iH^{11}s} \tag{138b}$$

with

$$W_1^{10} = -(W_1^{01})^\dagger \tag{139}$$

This solution can be expressed in terms of the eigenvalues and eigenvectors of H^{00} and of H^{11}:

$$W_1 = W_1^{01} + W_1^{10} \tag{140a}$$

$$W_1^{01} = -\sum_{n,m} \frac{|n^{00}\rangle\langle n^{00}|\tilde{H}^{01}|m^{11}\rangle\langle m^{11}|}{\lambda_n^{00} - \lambda_m^{11}} \tag{140b}$$

$$W_1^{10} = -\sum_{n,m} \frac{|m^{11}\rangle\langle m^{11}|\tilde{H}^{10}|n^{00}\rangle\langle n^{00}|}{\lambda_m^{11} - \lambda_n^{00}} \tag{140c}$$

Since $D_1 = 0$, we conclude that there is no contribution to the diagonal blocks in the first iteration (i.e., of first order in ε); and in analogy with Eq. (87), the remaining coupling term can be written as

$$\varepsilon^2 V_2 = \frac{\varepsilon^2}{2}[V_1, W_1] + \frac{\varepsilon^3}{3}[[V_1, W_1], W_1] + \cdots + \varepsilon^M \frac{(M-1)}{M!} \underbrace{[\ldots[[V_1, W_1], W_1], \ldots]}_{M-1 \text{ commutators}} + \cdots \tag{141}$$

We can obtain an effective Hamiltonian of second order with little supplementary effort: We first extract from the block-diagonal part of $\varepsilon^2 V_2$ the term

$$B_2 := \frac{\varepsilon^2}{2}[V_1, W_1] \tag{142}$$

which is the only one that carries the lowest power ε^2. Since $P^1[V_1, W_1]P^0 = 0$ and $P^0[V_1, W_1]P^1 = 0$, the term of power ε^2 has no off-block-diagonal part. Thus we can write

$$e^{-\varepsilon W_1}(H^{00} + H^{11} + \varepsilon V_1)e^{\varepsilon W_1} = H^{00} + H^{11} + \frac{\varepsilon^2}{2}[V_1, W_1] + \varepsilon^3 V_3 \tag{143}$$

with

$$\varepsilon^3 V_3 = \frac{\varepsilon^3}{3}[[V_1, W_1], W_1] + \cdots + \varepsilon^M \frac{(M-1)}{M!} \underbrace{[\ldots[[V_1, W_1], W_1], \ldots]}_{M-1 \text{ commutators}} + \cdots \tag{144}$$

or, symbolically, in matrix notation

$$
\begin{aligned}
H_2 &= e^{-\varepsilon W_1} \begin{pmatrix} H^{00} & H^{01} \\ H^{10} & H^{11} \end{pmatrix} e^{\varepsilon W_1} \\
&= \begin{pmatrix} H^{00} + \frac{\varepsilon^2}{2}(\tilde{H}^{01} W_1^{10} - W_1^{01} \tilde{H}^{10}) & 0 \\ 0 & H^{11} + \frac{\varepsilon^2}{2}(\tilde{H}^{10} W_1^{01} - W_1^{10} \tilde{H}^{01}) \end{pmatrix} + \varepsilon^3 V_3 \\
&= \begin{pmatrix} H^{00} + \frac{\varepsilon^2}{2}[\tilde{H}^{01} W_1^{10} + (\tilde{H}^{01} W_1^{10})^\dagger] & 0 \\ 0 & H^{11} - \frac{\varepsilon^2}{2}[(W_1^{10} \tilde{H}^{01})^\dagger + W_1^{10} \tilde{H}^{01}] \end{pmatrix} + \varepsilon^3 V_3
\end{aligned}
\tag{145}
$$

Thus we find the effective Hamiltonian (129) that gives the eigenvectors and eigenvalues of order ε^2 (and corrections of order ε^3).

Remark. In fact we can show that while the next order correction for the eigenvectors is indeed of order ε^3, the one for the eigenvalues is of order ε^4. The term of order ε^3 in (144) is $[[V_1, W_1], W_1]$ which has zero block-diagonal projection, since the product of two off-block-diagonal operators is block-diagonal, and the product of an off-block-diagonal operator with a block-diagonal operator is off-block-diagonal. Symbolically, we can represent this by

$$
\begin{pmatrix} & \blacksquare \\ \blacksquare & \end{pmatrix}\begin{pmatrix} & \blacksquare \\ \blacksquare & \end{pmatrix} = \begin{pmatrix} \blacksquare & \\ & \blacksquare \end{pmatrix}, \qquad \begin{pmatrix} \blacksquare & \\ & \blacksquare \end{pmatrix}\begin{pmatrix} & \blacksquare \\ \blacksquare & \end{pmatrix} = \begin{pmatrix} & \blacksquare \\ \blacksquare & \end{pmatrix}
\tag{146}
$$

and for arbitrary operators A and B we can show this by

$$
(P^0 A P^1 + P^1 A P^0)(P^0 B P^1 + P^1 B P^0) = P^0 A P^1 B P^0 + P^1 A P^0 B P^1 \tag{147}
$$

and

$$
(P^0 A P^0 + P^1 A P^1)(P^0 B P^1 + P^1 B P^0) = P^0 A P^0 B P^1 + P^1 A P^1 B P^0 \tag{148}
$$

The fact that the term of order ε^3 in Eq. (144) is off-block-diagonal implies that if we perform a second unitary transformation $e^{\varepsilon^3 W_3}$, there will be no term of order ε^3 in the diagonal block projection D_3, and thus the next order correction for the diagonal block, and therefore for eigenvalues, will be of order ε^4 (given by $\varepsilon^4[[[V_1, W_1], W_1], W_1]/8$).

2. *Relation with Adiabatic Elimination*

In the literature a different technique has been widely used to construct effective Hamiltonians, based on the partitioning technique combined with an approximation procedure known as adiabatic elimination for the time-dependent Schrödinger equation (see Ref. 39, p. 1165). In this section we show that the effective Hamiltonian constructed by adiabatic elimination can be recovered from the above construction by choosing the reference of the energy appropriately. Moreover, our stationary formulation allows us to estimate the order of the neglected terms and to improve the approximation to higher orders in a systematic way.

The idea in the method of adiabatic elimination is that the time evolution of the components in $\mathscr{H}^1$ oscillates very rapidly with respect to the evolution of the components in $\mathscr{H}^0$. This justifies the substitution of the time-dependent components in $\mathscr{H}^1$ by some average values. This leads then to an effective Hamiltonian in $\mathscr{H}^0$ that takes the form [see Ref. 39, p. 1166, Eq. (18.7-7), where there is a sign misprint]

$$H^{00}_{\text{eff,ae}} = H^{00} - H^{01}(H^{11})^{-1}H^{10} \tag{149}$$

This equation can be obtained from Eq. (129) as follows. Denoting by λ^{00}_{max} the largest eigenvalue of H^{00}, we can write the denominator of W^{10}_1 [Eq. (140c)] as

$$\lambda^{11}_m - \lambda^{00}_n = (\lambda^{11}_m - \lambda^{00}_{max}) + (\lambda^{00}_{max} - \lambda^{00}_n) \tag{150}$$

The condition that the time evolution of the components in $\mathscr{H}^1$ oscillate very rapidly with respect to the evolution of the components in $\mathscr{H}^0$ can be formulated by an inequality between the eigenvalues:

$$\lambda^{00}_{max} - \lambda^{00}_n \ll \lambda^{11}_m - \lambda^{00}_{max} \tag{151}$$

Thus the expression (140c) can be approximated by

$$\begin{aligned} W^{10}_1 &\approx -\sum_{n,m} \frac{|m^{11}\rangle\langle m^{11}|\tilde{H}^{10}|n^{00}\rangle\langle n^{00}|}{\lambda^{11}_m - \lambda^{00}_{max}} \\ &= -\sum_{n,m} |m^{11}\rangle\langle m^{11}|(H^{11} - \lambda^{00}_{max}\mathbb{1}_{11})^{-1}\tilde{H}^{10}|n^{00}\rangle\langle n^{00}| \\ &= -(H^{11} - \lambda^{00}_{max}\mathbb{1}_{11})^{-1}\tilde{H}^{10} \end{aligned} \tag{152}$$

which gives for the effective Hamiltonian (129)

$$H^{00}_{\text{eff}} \approx H^{00} - H^{01}(H^{11} - \lambda^{00}_{max}\mathbb{1}_{11})^{-1}H^{10} \tag{153}$$

Choosing the reference of energy $\lambda^{00}_{max} = 0$ allows us to recover Eq. (149).

We remark that in the approach by adiabatic elimination a further approximation is implicitly made, since the eigenvectors (or the initial conditions) when it is applied to dynamics are not transformed with $e^{\varepsilon W_1}$. This amounts to the approximation $e^{\varepsilon W_1} = \mathbb{1} + \varepsilon W_1 + \cdots \approx \mathbb{1}$. This does not produce a big difference when adiabatic elimination is applied to adiabatic processes with laser pulses, since, as we will see in Section V, the initial and final eigenvectors of the perturbed Hamiltonian coincide with those of the unperturbed one.

3. *High-Frequency Partitioning*

We can adapt the method described for high-frequency perturbation theory to the partitioning setup, which is another way to obtain the result of adiabatic elimination. We consider a partition represented symbolically in matrix form as

$$H = \begin{pmatrix} H^{00} - \lambda^{00}_{max}\mathbb{1}_{00} & H^{01} \\ H^{10} & f(H^{11} - \lambda^{00}_{max}\mathbb{1}_{11}) \end{pmatrix} \tag{154}$$

in a regime where $f \to \infty$. Defining $\epsilon := 1/f$, we decompose accordingly as

$$\hat{H} := H/f = \begin{pmatrix} 0 & 0 \\ 0 & \hat{H}^{11} \end{pmatrix} + \epsilon \begin{pmatrix} \hat{H}^{00} & H^{01} \\ H^{10} & 0 \end{pmatrix} =: \hat{H}_0 + \epsilon \hat{V}_1 \tag{155}$$

where we have simplified the notation by defining $\hat{H}^{00} := H^{00} - \lambda^{00}_{max}\mathbb{1}_{00}$ and $\hat{H}^{11} := H^{11} - \lambda^{00}_{max}\mathbb{1}_{11}$. We construct a unitary transformation $\exp(\epsilon W_1)$ such that

$$e^{-\epsilon W_1}(\hat{H}_0 + \epsilon \hat{V}_1)e^{\epsilon W_1} = \hat{H}_0 + \epsilon D_1 + \epsilon^2 V_2 \tag{156}$$

with the condition

$$[P^1, D_1] = 0 \tag{157}$$

where P^1 is the projection into the subspace corresponding to the 11-block. We remark that defining the projector into the orthogonal complement, $P^0 := \mathbb{1} - P^1$, the condition (157) implies that also $[P^0, D_1] = 0$, which leads to

$$D_1 = P^1 D_1 P^1 + P^0 D_1 P^0 \tag{158}$$

As before, the generator W_1 can be chosen such that $P^1 W_1 P^1 = 0$, $P^0 W_1 P^0 = 0$, and

$$D_1 = P^1 \hat{V}_1 P^1 + P^0 \hat{V}_1 P^0 = P^0 \hat{V}_1 P^0 = \begin{pmatrix} \hat{H}^{00} & 0 \\ 0 & 0 \end{pmatrix} \tag{159}$$

The generator W_1 is determined by the equation

$$[\hat{H}_0, W_1] + \hat{V}_1 - D_1 = 0 \tag{160}$$

From the general procedure described in the preceding sections, $W_1 = W_1^{01} + W_1^{10}$ can be written, for example, as

$$W_1^{01} = \lim_{\beta \to 0^+} -i \int_0^{\infty} ds\, e^{-\beta s} H^{01} e^{i\hat{H}^{11} s}, \qquad W_1^{10} = -(W_1^{01})^{\dagger} \tag{161}$$

Alternatively, we can write explicitly the equation (160), which in the present case becomes

$$\begin{aligned} 0 &= \left[\begin{pmatrix} 0 & 0 \\ 0 & \hat{H}^{11} \end{pmatrix}, \begin{pmatrix} 0 & W_1^{01} \\ W_1^{10} & 0 \end{pmatrix} \right] + \begin{pmatrix} 0 & H^{01} \\ H^{10} & 0 \end{pmatrix} \\ &= \begin{pmatrix} 0 & -W_1^{01} \hat{H}^{11} \\ \hat{H}^{11} W_1^{10} & 0 \end{pmatrix} + \begin{pmatrix} 0 & H^{01} \\ H^{10} & 0 \end{pmatrix} \end{aligned} \tag{162}$$

which leads to the solution

$$W_1^{01} = H^{01} (\hat{H}^{11})^{-1} \tag{163a}$$

$$W_1^{10} = -(W_1^{01})^{\dagger} = -(\hat{H}^{11})^{-1} H^{10} \tag{163b}$$

and thus to the remaining corrections of order ϵ^2 of the form

$$\begin{aligned} H_2 &= e^{-\epsilon W_1} \begin{pmatrix} \hat{H}^{00} & H^{01} \\ H^{10} & f\hat{H}^{11} \end{pmatrix} e^{\epsilon W_1} \\ &= \begin{pmatrix} H^{00} + \frac{\epsilon^2}{2} (H^{01} W_1^{10} - W_1^{01} H^{10}) & \frac{\epsilon^2}{2} \hat{H}^{00} W_1^{01} \\ -\frac{\epsilon^2}{2} W_1^{10} \hat{H}^{00} & H^{11} + \frac{\epsilon^2}{2} (H^{10} W_1^{01} - W_1^{10} H^{01}) \end{pmatrix} + \epsilon^3 V_3 \end{aligned} \tag{164}$$

After a second transformation of the form $\exp(\epsilon^2 W_2)$ designed to eliminate the non-block-diagonal terms of order ϵ^2, one obtains an effective Hamiltonian of second order for the 00-block of the form

$$H_{\text{eff}}^{00} = H^{00} - \lambda_{max}^{00} \mathbb{1}_{11} - H^{01} (H^{11} - \lambda_{max}^{00} \mathbb{1}_{11})^{-1} H^{10} \tag{165}$$

We remark that, as opposed to Eqs. (145)–(129), in this construction two unitary transformations are needed to obtain the effective eigenvectors to second order (after the first transformation, where keeping only the diagonal blocks in Eq. (164) yields the eigenvalues to second order, but not the eigenvectors).

4. *Effective Dressed Hamiltonians: Partitioning in the Enlarged Space*

When considering the stationary Floquet Hamiltonian $K = -i\hbar\omega\frac{\partial}{\partial\theta} + H(\theta)$, $H(\theta) = H_0 + \varepsilon V(\theta)$, describing the dynamics of a quantum (atomic or molecular) system H_0, illuminated by a strong photon field (of one frequency), we have to extend the preceding partitioning to the enlarged space $\mathscr{K} = \mathscr{H} \otimes \mathscr{L}$. This is in practice done in two steps:

1. First we identify a set of atomic (or molecular) essential states, connected with the initial condition, whose population will be appreciable during the dynamics. This means that these states are in multiphoton resonance (or quasi-resonance). This allows us to split the Hilbert space into two orthogonal subspaces $\mathscr{H} = \mathscr{H}^0 \oplus \mathscr{H}^1$, and thus the enlarged Hilbert space also splits into two orthogonal subspaces:

$$\mathscr{K} = (\mathscr{H}^0 \otimes \mathscr{L}) \oplus (\mathscr{H}^1 \otimes \mathscr{L}) = \mathscr{K}^0 \oplus \mathscr{K}^1 \tag{166}$$

   We next partition the Floquet Hamiltonian with respect to these atomic blocks. We obtain effective Floquet Hamiltonians inside each block.
2. The second step is the construction of an effective dressed Hamiltonian, independent of the θ-variable, inside the block connected to the initial condition. This can be done by the KAM iterations combined by the RWT techniques to treat the resonances. The second step depends on the specific problem that is treated.

Two-photon process examples will be illustrated below.

a. Partitioning of the Floquet Hamiltonian. We formulate in a general way the first step for partitioning the Floquet Hamiltonian up to the second order. As before, the free atomic Hamiltonian is defined on a Hilbert space $\mathscr{H}$ of dimension N. The Hamiltonian $H(\theta)$ can be separated into four parts such that the Floquet Hamiltonian can be represented symbolically in matrix form as

$$K = \begin{pmatrix} K_0^{00} + \varepsilon V^{00}(\theta) & \varepsilon V^{01}(\theta) \\ \varepsilon V^{10}(\theta) & K_0^{11} + \varepsilon V^{11}(\theta) \end{pmatrix} \tag{167}$$

with

$$K_0^{ii} = -i\hbar\omega \frac{\partial}{\partial\theta} \otimes \mathbb{1}_{\mathscr{H}^i} + H_0^{ii} \tag{168a}$$

$$V = V^{00} + V^{11} + V^{01} + V^{10} \tag{168b}$$

$$H_0 = H_0^{00} + H_0^{11} \tag{168c}$$

In the enlarged space, the splitting can be interpreted as two weakly coupled subspaces $\mathscr{K}^0$ and $\mathscr{K}^1$.

As before, we construct a unitary transformation of the form $e^{\varepsilon W_1}$, with $W_1^\dagger = -W_1$ such that

$$e^{-\varepsilon W_1}(K_0^{00} + K_0^{11} + \varepsilon V)e^{\varepsilon W_1} = K_0^{00} + K_0^{11} + \varepsilon D_1(\theta) + \varepsilon^2 V_2(\theta) \tag{169}$$

where $D_1 = D_1^{00} + D_1^{11}$, with $D_1^{ii} = P^i D_1 P^i$, $i = 0, 1$, and giving the remaining coupling

$$\begin{aligned}\varepsilon^2 V_2 = {} & \frac{\varepsilon^2}{2}[V, W_1] + \frac{\varepsilon^3}{3}[[V, W_1], W_1] + \cdots + \varepsilon^M \frac{(M-1)}{M!} \underbrace{[\ldots[[V, W_1], W_1], \ldots]}_{M-1 \text{ commutators}} + \cdots \\ & + \frac{\varepsilon^2}{2!}[D_1, W_1] + \frac{\varepsilon^3}{3!}[[D_1, W_1], W_1] + \cdots + \frac{\varepsilon^M}{M!} \underbrace{[\ldots[D_1, W_1], W_1], \ldots]}_{M-1 \text{ commutators}} + \cdots\end{aligned} \tag{170}$$

Since we have here a block-diagonal perturbation, we choose

$$D_1 = V^{00} + V^{11} \tag{171}$$

which in this case is not necessarily zero, as opposed to Eqs. (135a)–(135d). This leads to $W^{00} = 0 = W^{11}$. We obtain a Floquet Hamiltonian of second order, in matrix notation:

$$\begin{aligned}K_2 &= e^{-\varepsilon W_1}(K_0^{00} + K_0^{11} + \varepsilon V)e^{\varepsilon W_1} \\ &= \begin{pmatrix} K^{00} + \varepsilon V^{00} + \frac{\varepsilon^2}{2}[V^{01} W_1^{10} + (V^{01} W_1^{10})^\dagger] & \epsilon^2(V^{00} W_1^{01} - W_1^{01} V^{11}) \\ \epsilon^2(V^{11} W_1^{10} - W_1^{10} V^{00}) & K^{11} + \varepsilon V^{11} - \frac{\varepsilon^2}{2}[(W_1^{10} V^{01})^\dagger + W_1^{10} V^{01}] \end{pmatrix} \\ &\quad + \mathscr{O}(\varepsilon^3)\end{aligned} \tag{172}$$

We obtain the effective Floquet Hamiltonian (assuming that the initial condition is connected to the block $\mathscr{K}_0$) of second order:

$$K_{\text{eff}}^{00} = -i\hbar\omega \frac{\partial}{\partial\theta} \otimes \mathbb{1}_{\mathscr{H}^0} + H_0^{00} + \varepsilon V^{00} + \frac{\varepsilon^2}{2}[V^{01} W_1^{10} + (V^{01} W_1^{10})^\dagger] \tag{173}$$

The next corrections of order ε^3 are given by $\frac{\varepsilon^3}{3}[[V, W_1], W_1] + \frac{\varepsilon^3}{3!}[[D_1, W_1], W_1]$. The particular case

$$V^{00} + V^{11} = 0 \tag{174}$$

leads to $D_1 = 0$, and to corrections for the eigenvalues of order ε^4, given by $\varepsilon^4[[[V, W_1], W_1], W_1]/8$.

b. Calculation of W_1. It is convenient to calculate the term of order ε^2 of the effective Floquet Hamiltonian by expansion of V^{01} and W_1^{10} in Fourier series. The perturbation can be indeed written as

$$V(\theta) = \sum_{\ell} \tilde{V}_{\ell}\, e^{i\ell\theta} \tag{175}$$

Writing W_1 in the basis of eigenvectors $|n, k\rangle$ of K_0 and denoting the corresponding eigenvalues $\lambda_{n,k}^0 = E_n + k\hbar\omega$ with E_n the eigenvalues of H_0 in Eq. (176)

$$W_1 = \sum_{n,m,k,k',(m,k')\neq(n,k)} \frac{|n, k\rangle\langle n, k|V(\theta)|m, k'\rangle\langle m, k'|}{\lambda_{m,k'}^0 - \lambda_{n,k}^0} \tag{176}$$

gives

$$\begin{aligned} W_1 &= \sum_{n,m,k,k',\ell} \frac{|n\rangle\langle n|\tilde{V}_{\ell}|m\rangle\langle m| \otimes [|k\rangle\langle k|e^{i\ell\theta}|k'\rangle\langle k'|]}{E_m + k'\hbar\omega - (E_n + k\hbar\omega)} \\ &= \sum_{n,m,\ell} \frac{|n\rangle\langle n|\tilde{V}_{\ell}|m\rangle\langle m|}{E_m - E_n - \ell\hbar\omega} \otimes e^{i\ell\theta} \end{aligned} \tag{177}$$

since

$$\sum_{k} |k\rangle\langle k - \ell| = e^{+i\ell\theta} \tag{178}$$

Thus we can expand W_1 in the Fourier modes

$$W_1 = \sum_{\ell} \tilde{W}_{1,\ell}\, e^{i\ell\theta}, \qquad \tilde{W}_{1,\ell} = \sum_{n,m} \frac{|n\rangle\langle n|\tilde{V}_{\ell}|m\rangle\langle m|}{E_m - E_n - \ell\hbar\omega} \tag{179}$$

Defining $W_1^{ij} := P^i W_1 P^j$ and choosing $W_1^{00} = 0$ and $W_1^{11} = 0$, we obtain as before, in terms of the eigenvalues and eigenvectors (in the enlarged space) of K^{00} and of K^{11},

$$W_1 = W_1^{01} + W_1^{10} \tag{180a}$$

$$W_1^{01} = \sum_{\ell} \tilde{W}_{1,\ell}^{01}\, e^{i\ell\theta}, \qquad \tilde{W}_{1,\ell}^{01} = \sum_{n,m} \frac{|n^{00}\rangle\langle n^{00}|\tilde{V}_\ell^{01}|m^{11}\rangle\langle m^{11}|}{E_m^{11} - E_n^{00} - \ell\hbar\omega} \tag{180b}$$

$$W_1^{10} = \sum_{\ell} \tilde{W}_{1,\ell}^{10}\, e^{i\ell\theta}, \qquad \tilde{W}_{1,\ell}^{10} = \sum_{n,m} \frac{|m^{11}\rangle\langle m^{11}|\tilde{V}_\ell^{10}|n^{00}\rangle\langle n^{00}|}{E_n^{00} - E_m^{11} - \ell\hbar\omega} \tag{180c}$$

with

$$\tilde{V}_\ell^{10} = (\tilde{V}_{-\ell}^{01})^\dagger, \qquad \tilde{W}_{1,\ell}^{10} = -(\tilde{W}_{1,-\ell}^{01})^\dagger \tag{181}$$

We remark that the denominators of W_1 allow us to detect resonant states of the subspace $\mathscr{H}^1$ that should be thus included in the subspace $\mathscr{H}^0$ in the partitioning.

Using a perturbation satisfying $\tilde{V}_{-\ell} = \tilde{V}_\ell$ (such as perturbation proportional to $\cos\theta$), we obtain for the second-order effective Floquet Hamiltonian

$$\begin{aligned} K_{\text{eff}}^{00} &= -i\hbar\omega\frac{\partial}{\partial\theta} \otimes \mathbb{1}_{\mathscr{H}^0} + H_0^{00} + \varepsilon V^{00} \\ &\quad - \frac{\varepsilon^2}{2} \sum_{n,\tilde{n},\ell,\ell',m} |\tilde{n}^{00}\rangle\langle\tilde{n}^{00}|\tilde{V}_{\ell'}^{01}|m^{11}\rangle\langle m^{11}|\tilde{V}_\ell^{10}|n^{00}\rangle \\ &\quad \times \left(\frac{e^{i(\ell+\ell')\theta}}{E_m^{11} - E_n^{00} + \ell\hbar\omega} + \frac{e^{-i(\ell+\ell')\theta}}{E_m^{11} - E_{\tilde{n}}^{00} + \ell'\hbar\omega} \right) \langle n^{00}| \end{aligned} \tag{182}$$

We can easily extend this formula to the multifrequency case by adapting the notations $\omega \to \boldsymbol{\omega}$, $\theta \to \boldsymbol{\theta}$, $\ell \to \boldsymbol{\ell}$, and $\ell' \to \boldsymbol{\ell}'$ (see also Section II.E).

E. Effective Hamiltonian for Two-Photon Quasi-Resonant Processes in Atoms: The Two-Photon RWA

We illustrate the preceding formulation to construct an effective Hamiltonian for the example of a two-photon transition in atoms by a laser pulse, with no single-photon resonances.

Since this effective Hamiltonian will be parameterized by the laser amplitude and its frequency, it will be relevant for processes with chirped laser pulses.

We consider an atom, of Hamiltonian H_0, illuminated by a laser of polarization $\vec{e}$, amplitude $\mathscr{E}$, and frequency ω, such that two atomic states $|a\rangle$

and $|b\rangle$ (of respective energies E_a and E_b giving $H_0 = \text{diag}[E_a, E_b]$ in the basis $\{|a\rangle, |b\rangle\}$) are in two-photon quasi-resonance:

$$E_a + 2\hbar\omega + \hbar\Delta = E_b \tag{183}$$

One considers the electric dipole moment $\vec{\mu} = e\sum_i \vec{r}_i$, with $\vec{r}_i$ the position of each electron i and e the elementary electric charge. The Floquet Hamiltonian reads

$$K = -i\hbar\omega\frac{\partial}{\partial\theta} + H_0 - \vec{\mu}\cdot\vec{e}\mathscr{E}\cos\theta \tag{184}$$

We partition the Floquet Hamiltonian such that the states $|a\rangle$ and $|b\rangle$ span the Hilbert subspace $\mathscr{H}^0$ and the other atomic states $\{|1\rangle, \ldots, |N\rangle\}$ span the Hilbert subspace $\mathscr{H}^1$.

Denoting $\mu_{nn'} = \langle n|\vec{\mu}\cdot\vec{e}|n'\rangle_{\mathbf{r}}$, with $\mathbf{r}$ standing for the electron coordinates and applying Eq. (182), we obtain for the second-order effective Floquet Hamiltonian

$$K_{\text{eff}} = -i\hbar\omega\frac{\partial}{\partial\theta} + H_0 + \mathscr{E}V(\theta) \tag{185}$$

with

$$\mathscr{E}V(\theta) = -\mathscr{E}\begin{pmatrix} \mu_{aa} & \mu_{ab} \\ \mu_{ba} & \mu_{bb} \end{pmatrix}\cos\theta - \frac{\mathscr{E}^2}{2}\begin{pmatrix} \mathscr{V}_{aa}(\theta) & \mathscr{V}_{ab}(\theta) \\ \mathscr{V}_{ba}(\theta) & \mathscr{V}_{bb}(\theta) \end{pmatrix} \tag{186}$$

and

$$\mathscr{V}_{nn} = \cos^2\theta\sum_m\left(\frac{|\mu_{nm}|^2}{E_m - E_n - \hbar\omega} + \frac{|\mu_{nm}|^2}{E_m - E_n + \hbar\omega}\right) \tag{187a}$$

$$\mathscr{V}_{ab} = \frac{1}{4}\sum_m \mu_{am}\mu_{mb}\sum_{\ell,\ell'\in\{-1,1\}}\left(\frac{e^{i(\ell+\ell')\theta}}{E_m - E_b + \ell\hbar\omega} + \frac{e^{-i(\ell+\ell')\theta}}{E_m - E_a + \ell'\hbar\omega}\right) \tag{187b}$$

To extract from the effective Floquet Hamiltonian (185) an effective dressed Hamiltonian independent of θ, we can apply a contact transformation consisting in averaging the Hamiltonian (185) with respect to $K_0 := -i\hbar\omega\,\partial/\partial\theta + H_0$—that is, in diagonalizing it with respect to 0 and to the basis $\{|a\rangle, |b\rangle\}$. Since $V(\theta) = \sum_{\ell\in\{-2,\ldots,2\}}\tilde{V}_\ell e^{i\ell\theta}$, this could be done using $W(\theta) = \sum_\ell \tilde{W}_\ell e^{i\ell\theta}$ satisfying Eq. (179) with, for example, the a,b component of the mode $\ell = 2$

$$\varepsilon\langle a|\tilde{W}_2|b\rangle = \frac{\Omega_{ab}}{E_b - E_a - 2\hbar\omega} \tag{188}$$

where we denote the effective two-photon Rabi frequency

$$\Omega_{ab} = -\frac{\mathscr{E}^2}{8}\sum_m \mu_{am}\mu_{mb}\left(\frac{1}{E_m - E_a - \hbar\omega} + \frac{1}{E_m - E_b + \hbar\omega}\right) \tag{189}$$

This leads to a *quasi-divergence* of the W which occurs *when the numerator is equal to or larger than the denominator in absolute value*. It is due to the quasi-resonance $E_b - E_a - 2\hbar\omega = \hbar\Delta$ and thus occurs when $|\Omega_{ab}| \gtrsim \hbar|\Delta|$. Thus we apply instead the two-photon RWT $R = \mathrm{diag}[1, \exp(-2i\theta)]$, giving the Hamiltonian $R^\dagger K_{\mathrm{eff}} R$. We are now able to average this Floquet Hamiltonian with respect to K_0 since the small denominators of the associated W have been removed. Instead of diagonalizing fully this Hamiltonian by the contact transformation, we construct an effective dressed Hamiltonian by just averaging with respect to θ [this corresponds to applying the contact transformation $\exp(\mathscr{E}W)$ with $W = -i\int^\theta (R^\dagger V(\theta) R - \overline{R^\dagger V(\theta) R})$ and $\overline{f(\theta)} = (\frac{1}{2\pi})\int_0^{2\pi} d\theta\, f(\theta)$] and we retain terms up to $\mathscr{E}^2$. The effective dressed Hamiltonian (independent of θ) obtained reads (using the energy reference $E_a = 0$)

$$H_{\mathrm{eff}} = \begin{pmatrix} S_a & \Omega_{ab} \\ \Omega_{ab}^* & S_b + \hbar\Delta \end{pmatrix} \tag{190}$$

with the effective two-photon Rabi frequency (189) and the dynamical Stark shift of the state n:

$$S_n = -\frac{\mathscr{E}^2}{4}\sum_m \left(\frac{|\mu_{nm}|^2}{E_m - E_n - \hbar\omega} + \frac{|\mu_{nm}|^2}{E_m - E_n + \hbar\omega}\right) \tag{191}$$

The effective two-photon Rabi frequency is well approximated by

$$\Omega_{ab} \approx -\frac{\mathscr{E}^2}{4}\sum_m \frac{\mu_{am}\mu_{mb}}{E_m - E_a - \hbar\omega} \tag{192}$$

which is an expression equivalent to Eq. (189) with a correction of the order $\hbar\Delta/(E_m - E_a - \hbar\omega)$.

We remark that this effective Hamiltonian (190) constructed by the combination of a partitioning of the Floquet Hamiltonian, a two-photon RWT, and a final θ-averaging can be seen as a *two-photon RWA*, which extends the usual (one-photon) RWA [39,40]. We have thus rederived a well-known result, using stationary techniques that allow us to estimate easily the order of the neglected terms. This method allows us also to calculate higher order corrections. We apply it in the next subsection to calculate effective Hamiltonians for molecules illuminated by strong laser fields.

F. Effective Hamiltonian for Rotational Excitations in Diatomic Molecules

To determine an effective dressed Hamiltonian characterizing a molecule excited by strong laser fields, we have to apply the standard construction of the free effective Hamiltonian (such as the Born–Oppenheimer approximation), taking into account the interaction with the field nonperturbatively (if resonances occur). This leads to four different time scales in general: (i) for the motion of the electrons, (ii) for the vibrations of the nuclei, (iii) for the rotation of the nuclei, and (iv) for the frequency of the interacting field. It is well known that it is a good strategy to take into account the time scales from the fastest to the slowest one.

We consider below an example in diatomic molecules where the field frequency is low with respect to the electronic excitation, but high with respect to the vibrations and consequently also with respect to the rotations. This will lead to Raman rotational excitations in the ground vibrational state of the ground electronic state (i.e. the ground vibronic state). The laser frequency is such that the ground electronic state is not one- or two-photon resonant with other electronic states.

The resulting effective Hamiltonian [see Eq. (216)] has been used to show the adiabatic alignment of molecules by a laser field (see, for example, Refs. 32 and 41). The same technique can be used to construct an effective Hamiltonian, but with a field frequency resonant with vibration. This Hamiltonian has been used to show the adiabatic orientation of polar molecules by a laser field and its second harmonic in quasi-resonance with the first excited vibrational state [42].

1. The Born–Oppenheimer Floquet Hamiltonian

The Born–Oppenheimer approximation allows us to decouple the electronic and nuclear motions of the free molecule of the Hamiltonian H_0. Solving the Schrödinger equation $H_0\Psi = E\Psi$ with respect to the electron coordinates $\mathbf{r} = \{\vec{r}_1, \vec{r}_2, \dots\}$ gives rise to the electronic states $\Psi_n(\mathbf{r}, \mathbf{R}) = \langle \mathbf{r} | n(\mathbf{R}) \rangle$, $n = 0, \dots,$ N_e, of respective energies $\{E_n^{(e)}(\mathbf{R})\}$ as functions of the nuclear coordinates $\mathbf{R}$, with the electronic scalar product defined as $\langle n(\mathbf{R}) | n'(\mathbf{R}) \rangle_{\mathbf{r}} = \int d\mathbf{r}\ \Psi_n^*(\mathbf{r}, \mathbf{R}) \Psi_{n'}$ $(\mathbf{r}, \mathbf{R})$. We assume N_e bound electronic states. The Floquet Hamiltonian of the molecule perturbed by a field (of frequency ω, of amplitude $\mathscr{E}$, and of linear polarization $\vec{e}$), in the dipole coupling approximation, and in a coordinate system of origin at the center of mass of the molecule can be written as

$$K = -i\hbar\omega \frac{\partial}{\partial\theta} + H_0 - \vec{\mu} \cdot \vec{e}\mathscr{E} \cos\theta, \qquad \vec{\boldsymbol{\mu}} = e \sum_i \vec{r}_i \tag{193}$$

with $\vec{r}_i$ the position of each electron i and e the elementary electric charge. The Born–Oppenheimer approximation allows us to write H_0 as decoupled electronic states:

$$H_0 = T^{(\mathrm{n})} + \mathrm{diag}[E_0^{(\mathrm{e})}(\mathbf{R}), E_1^{(\mathrm{e})}(\mathbf{R}), \ldots, E_{N_e}^{(\mathrm{e})}(\mathbf{R})] \tag{194}$$

where $T^{(\mathrm{n})}$ is the kinetic energy of the nuclei. We can write the dipole moment in the basis of these electronic states:

$$\vec{\mu} = \sum_{n,n'} |n(\mathbf{R})\rangle \vec{\mu}_{nn'}(\mathbf{R}) \langle n'(\mathbf{R})|, \qquad \vec{\mu}_{nn'}(\mathbf{R}) = \langle n(\mathbf{R})|\vec{\mu}|n'(\mathbf{R})\rangle_{\mathbf{r}} \tag{195}$$

$\vec{\mu}_{00}(\mathbf{R})$ is the permanent dipole moment of the ground electronic state, also denoted as $\vec{\mu}_{00}(\mathbf{R}) \equiv \vec{\mu}_0(\mathbf{R})$.

For a process with two lasers, the Floquet Hamiltonian is (see Section II.E)

$$K = -i\hbar\boldsymbol{\omega} \cdot \frac{\partial}{\partial \boldsymbol{\theta}} + H_0 - \vec{\mu} \cdot [\vec{e}_1 \mathscr{E}_1 \cos\theta_1 + \vec{e}_2 \mathscr{E}_2 \cos\theta_2] \tag{196}$$

2. *Raman Processes in the Ground Vibronic State by a Single Laser: Rotational Excitations*

We assume that the frequency of the laser is such that no excited electronic state is coupled by a one- or two-photon resonance with the ground electronic state.

a. Effective Hamiltonian in the ground electronic surface. The partitioning is in this case as follows: The Electronic State $|n = 0\rangle$ spans the Hilbert subspace $\mathscr{H}_e^0$ and the other electronic states $\{|1\rangle, \ldots, |N_e\rangle\}$ the Hilbert subspace $\mathscr{H}_e^1$. The dipole moment can be decomposed into diagonal and nondiagonal parts: respectively $\vec{\mu}_{nn}$ and $\vec{\mu}_{nn'}, n \neq n'$. Denoting $\vec{\mu}_{nn'} = [\mu_{1,nn'}, \mu_{2,nn'}, \mu_{3,nn'}]$ and $\vec{e} = [e_1, e_2, e_3]$ and applying Eq. (182), we obtain for the second-order effective Floquet Hamiltonian connected to the ground electronic state

$$K_{\mathrm{eff}}^{(\mathrm{e})} = -i\hbar\omega \frac{\partial}{\partial\theta} + T^{(\mathrm{n})} + E_0^{(\mathrm{e})} - \vec{\mu}_0 \cdot \vec{e}\mathscr{E}\cos\theta - \frac{1}{2}\mathscr{E}^2\cos^2\theta \sum_{i,j,m\neq 0}{}' \left(\frac{e_i \mu_{i,0m} e_j \mu_{j,m0}}{E_m^{(\mathrm{e})} - E_0^{(\mathrm{e})} + \hbar\omega} + \frac{e_i \mu_{i,0m} e_j \mu_{j,m0}}{E_m^{(\mathrm{e})} - E_0^{(\mathrm{e})} - \hbar\omega} \right) \tag{197a}$$

$$= -i\hbar\omega \frac{\partial}{\partial\theta} + T^{(\mathrm{n})} + E_0^{(\mathrm{e})} - \vec{\mu}_0 \cdot \vec{e}\mathscr{E}\cos\theta - \frac{1}{2}\vec{e} \cdot (\vec{\vec{\alpha}}\vec{e})\mathscr{E}^2\cos^2\theta \tag{197b}$$

with the *dynamical electronic polarizability* tensor $\vec{\vec{\alpha}}$ of components

$$\alpha_{ij} = \sum_{m \neq 0} \left(\frac{\mu_{i,0m}\mu_{j,m0}}{E_m^{(e)} - E_0^{(e)} + \hbar\omega} + \frac{\mu_{i,0m}\mu_{j,m0}}{E_m^{(e)} - E_0^{(e)} - \hbar\omega} \right) \tag{198}$$

This allows us to define an effective dipole moment $\vec{\mu}_{\text{eff}}$ as the sum of the permanent and induced dipole moments

$$\vec{\mu}_{\text{eff}} = \vec{\mu}_0 + \frac{1}{2}\vec{\vec{\alpha}}\,\vec{e}\mathcal{E}\cos\theta \tag{199}$$

such that the effective Floquet Hamiltonian can be written as

$$K_{\text{eff}}^{(e)} = -i\hbar\omega\frac{\partial}{\partial\theta} + T^{(n)} + E_0^{(e)} - \vec{\mu}_{\text{eff}} \cdot \vec{e}\mathcal{E}\cos\theta \tag{200}$$

For a linear molecule, when the system of coordinates is chosen such that the third axis is along the molecular axis, the electronic polarizability tensor (in the molecular frame) is diagonal:

$$\vec{\vec{\alpha}} = \text{diag}[\alpha_\perp, \alpha_\perp, \alpha_\parallel] \tag{201}$$

Transforming the polarizability into the laboratory frame and using Θ to denote the angle between the molecular axis and the polarization axis of the laser, we obtain the effective Floquet Hamiltonian in the ground electronic state of the linear molecule:

$$\begin{aligned} K_{\text{eff}}^{(e)} &= -i\hbar\omega\frac{\partial}{\partial\theta} + T^{(n)} + E_0^{(e)}(R) \\ &\quad - \mu_0(R)\mathcal{E}\cos\Theta\cos\theta - \frac{1}{2}(\alpha_\parallel(R) + \Delta\alpha(R)\sin^2\Theta)\mathcal{E}^2\cos^2\theta \end{aligned} \tag{202}$$

with $\Delta\alpha(R) = \alpha_\perp(R) - \alpha_\parallel(R)$, where $\Theta \in [0, \pi]$, $\varphi \in [0, 2\pi]$ are the angles of the usual spherical coordinates with origin at the center of mass of the molecule, and R is the internuclear distance.

It is usually a good approximation to consider the *static electronic polarizability* [i.e., Eq. (198) with $\omega = 0$] instead of the dynamical one when we are in the limit of a low frequency with respect to the electronic states:

$$E_n^{(e)} - E_0^{(e)} \gg \hbar\omega \tag{203}$$

b. Effective Hamiltonian in the ground vibrational state. Using the Born–Oppenheimer approximation (or equivalently the high-frequency approximation; see Section III.B) to effectively eliminate the fast vibrational motion with respect to the slow rotational one, we can obtain the Hamiltonian for the rotation of the free molecule in the ground electronic state by diagonalizing with respect to the vibrations of the nuclei (described here by the internuclear distance R), taking the angles Θ and φ as parameters. To this end, we split the kinetic energy of the nuclei into vibrational and rotational parts $T^{(\mathrm{n})} = T^{(\mathrm{n})}_{\mathrm{vib}} + T^{(\mathrm{n})}_{\mathrm{rot}}$, where $T^{(\mathrm{n})}_{\mathrm{rot}}$ depends also on the vibrational coordinate R. We diagonalize $T^{(\mathrm{n})}_{\mathrm{vib}} + E^{(\mathrm{e})}_0(R)$, denote the eigenvalues by $E^{(v)}_0, \ldots, E^{(v)}_{N_v}$, and denote the basis of eigenvectors by $\{|v=0\rangle, |v=1\rangle, \ldots, |v=N_v\rangle\}$, where we assume N_v bound vibrational states. The Born–Oppenheimer approximation consists in neglecting the terms of $T^{(\mathrm{n})}_{\mathrm{rot}}$ that are nondiagonal with respect to this vibrational basis of eigenvectors.

We assume that the laser frequency is far from any resonance between the ground vibrational state and the excited ones, such that the partitioning is very similar to the one made with the electronic states: the vibrational state $|v=0\rangle$ spans the Hilbert subspace $\mathscr{H}^0_v$ and the other vibrational states $\{|1\rangle, \ldots, |N_v\rangle\}$ the Hilbert subspace $\mathscr{H}^1_v$. We obtain for the second-order effective Floquet Hamiltonian connected to the ground vibrational state of the ground electronic state

$$K^{(v)}_{\mathrm{eff}} = -i\hbar\omega\frac{\partial}{\partial\theta} + T^{(\mathrm{n})}_{\mathrm{rot},00} + E^{(v)}_0 - \mu_{0,00}\mathscr{E}\cos\Theta\cos\theta$$
$$-\frac{1}{2}(\alpha_{\|,00} + \Delta\alpha_{00}\sin^2\Theta + \alpha^{(v)}_0\cos^2\Theta)\mathscr{E}^2\cos^2\theta \tag{204}$$

where $\alpha^{(v)}_0$ is the effective polarizability in the ground vibrational state induced by the other vibrational states, given by

$$\alpha^{(v)}_0 = \sum_{v\neq 0}\left(\frac{|\mu_{0,0v}|^2}{E^{(v)}_v - E^{(v)}_0 + \hbar\omega} + \frac{|\mu_{0,0v}|^2}{E^{(v)}_v - E^{(v)}_0 - \hbar\omega}\right) \tag{205}$$

where we have denoted $T^{(\mathrm{n})}_{\mathrm{rot},vv'} := \langle v|T^{(\mathrm{n})}_{\mathrm{rot}}|v'\rangle$, $\mu_{0,vv'} = \langle v|\mu_0(R)|v'\rangle_R$, $\alpha_{\|,vv'} = \langle v|\alpha_\|(R)|v'\rangle_R$, $\alpha_{\perp,vv'} = \langle v|\alpha_\perp(R)|v'\rangle_R$ and $\Delta\alpha_{vv'} = \alpha_{\perp,vv'} - \alpha_{\|,vv'}$. It is usually a good approximation to take only the contribution of the first excited vibrational state $E^{(v)}_1$ in the summation of $\alpha^{(v)}_0$ since the higher couplings are much smaller: $\mu_{0,01} \gg \mu_{0,02} \gg \mu_{0,03}\cdots$.

c. Effective Hamiltonian for Rotational Excitations. We now finally consider the rotational coordinates for the free Hamiltonian in the ground vibrational state of the ground electronic state and will consider that the field is able to populate a priori many rotational states, which we treat thus as essential states. Neglecting $T^{(\mathrm{n})}_{\mathrm{rot},vv'}$, for $v \neq v'$, according to the Born–Oppenheimer approximation, we approximate the ground vibronic state as a rigid rotor with the vibrational energies independent of the rotational coordinates (higher corrections corresponding for example to centrifugal distortion can be obtained from the non-diagonal terms $T^{(\mathrm{n})}_{\mathrm{rot},vv'}$, $v \neq v'$):

$$T^{(\mathrm{n})}_{\mathrm{rot},00} := B_0\,\hat{J}^2, \qquad B_0 = \langle v=0|B|v=0\rangle \tag{206}$$

with $B(R)$ the rotational constant, and the angular momentum operator

$$\hat{J}^2 = -\frac{1}{\sin\Theta}\frac{\partial}{\partial\Theta}\left(\sin\Theta\frac{\partial}{\partial\Theta}\right) - \frac{1}{\sin^2\Theta}\frac{\partial^2}{\partial\varphi^2} \tag{207}$$

Choosing $E_0^{(v)} = 0$, we obtain the effective Floquet Hamiltonian

$$K^{(v)}_{\mathrm{eff}} = -i\hbar\omega\frac{\partial}{\partial\theta} + B_0\,\hat{J}^2 + V^{(v)}_{\mathrm{eff}}(\theta,\Theta) \tag{208}$$

with

$$V^{(v)}_{\mathrm{eff}}(\theta,\Theta) = -\mu_{0,00}\mathscr{E}\cos\Theta\cos\theta - \frac{1}{2}(\alpha_{\|,00} + \Delta\alpha_{00}\sin^2\Theta + \alpha_0^{(v)}\cos^2\Theta)\mathscr{E}^2\cos^2\theta \tag{209}$$

We consider the *high-frequency limit* with respect to the rotation

$$\hbar\omega \gg B_0 \tag{210}$$

and apply the results of Section III.B with the Hamiltonian written as

$$K^{(v)}_{\mathrm{eff}}/\hbar\omega = -i\frac{\partial}{\partial\theta} + \epsilon[B_0\hat{J}^2 + V^{(v)}_{\mathrm{eff}}(\theta,\Theta)] \tag{211}$$

with the small parameter $\epsilon := 1/\hbar\omega$ and identifying the quantities of Section III.B: $x \equiv \Theta$, $H_0 \equiv B_0\hat{J}^2$, $V_1 \equiv V^{(v)}_{\mathrm{eff}}$. We first apply the contact transformation $S_1 = \exp(\epsilon W_1)$ with $W_1(\Theta,\theta) = -i\int^{\theta}(V^{(v)}_{\mathrm{eff}} - \overline{V}^{(v)}_{\mathrm{eff}})d\theta$ and the average with respect to $\theta : \overline{V}^{(v)}_{\mathrm{eff}}(\Theta) = \frac{1}{2\pi}\int_0^{2\pi} d\theta\, V^{(v)}_{\mathrm{eff}}(\theta,\Theta)$. Splitting $\hat{J}^2 := T_\Theta + T_\varphi$, we obtain the exact result (i.e., the terms of order ϵ^n, $n > 3$, are exactly zero):

$$S_1^\dagger\frac{K^{(v)}_{\mathrm{eff}}}{\hbar\omega}S_1 = -i\frac{\partial}{\partial\theta} + \epsilon(B_0\hat{J}^2 + \overline{V}^{(v)}_{\mathrm{eff}}) + \epsilon^2[B_0T_\Theta, W_1] + \frac{\epsilon^3}{2}[[B_0T_\Theta, W_1], W_1] \tag{212}$$

We apply again a contact transformation $S_2 = \exp(\epsilon^2 W_2)$ with $W_2(\Theta,\theta) = -i\int^{\theta}(V_2 - \overline{V}_2)d\theta$, and

$$V_2 = [B_0T_\Theta, W_1] + \frac{\epsilon}{2}[[B_0T_\Theta, W_1], W_1] \tag{213}$$

which averages with respect to θ and gives, to second order in $1/\hbar\omega$,

$$K_{\text{eff}}^{(\text{ro})} = -i\hbar\omega\frac{\partial}{\partial\theta} + B_0\hat{J}^2 + \overline{V}_{\text{eff}}^{(v)} + \epsilon\overline{[B_0T_\Theta, W_1]} + \frac{\epsilon^2}{2}\overline{[[B_0T_\Theta, W_1], W_1]} \quad (214)$$

We calculate

$$\overline{V}_{\text{eff}}^{(v)} = -\frac{\mathscr{E}^2}{4}(\alpha_{\|,00} + \Delta\alpha_{00}\sin^2\Theta + \alpha_0^{(v)}\cos^2\Theta), \qquad \overline{[T_\Theta, W_1]} = 0 \quad (215a)$$

$$W_1 = \frac{i\mathscr{E}^2}{8}(\alpha_{\|,00} + \Delta\alpha_{00}\sin^2\Theta + \alpha_0^{(v)}\cos^2\Theta)\sin 2\theta + i\mathscr{E}\mu_0\cos\Theta\cos\theta \quad (215b)$$

$$\overline{[[T_\Theta, W_1], W_1]} = -2\overline{\left(\frac{\partial W_1}{\partial\Theta}\right)^2} \quad (215c)$$

$$= \frac{\mathscr{E}^4}{64}(\Delta\alpha_{00} - \alpha_0^{(v)})^2\sin^2 2\Theta + \mu_0^2\mathscr{E}^2\sin^2\Theta \quad (215d)$$

and obtain the effective Floquet Hamiltonian of second order in $1/\hbar\omega$ and of second order in field amplitude $\mathscr{E}$ [43]:

$$K_{\text{eff}}^{(\text{ro})} = -i\hbar\omega\frac{\partial}{\partial\theta} + B_0\hat{J}^2 - \frac{\mathscr{E}^2}{2}\left[\frac{\alpha_{\|,00} + \alpha_0^{(v)}}{2} + \left(\frac{\Delta\alpha_{00} - \alpha_0^{(v)}}{2} - \frac{B_0\mu_0^2}{(\hbar\omega)^2}\right)\sin^2\Theta\right] \quad (216)$$

Since we consider a linear polarization, the coupling does not depend on the angle φ and we can thus consider this Hamiltonian (216) acting on a state of the form $\psi(\theta, \varphi; t) = \phi(\theta; t)e^{iM\varphi}/\sqrt{2\pi}$, with M the projection of the angular momentum $\hat{J}$ on the field polarization axis, which allows us to identify $\hat{J}^2$ to $B_0(T_\Theta + M^2/\sin^2\Theta)$.

IV. ADIABATIC FLOQUET THEORY

The models we have discussed so far correspond to continuous (CW) lasers with a fixed sharp frequency and constant intensity. They can be easily adapted to the case of pulsed lasers that have slowly varying envelopes. They can furthermore have a chirped frequency—that is, a frequency that changes slowly with time. For periodic (or quasi-periodic) semiclassical Hamiltonians, the Floquet states are the stationary states of the problem. Processes controlled by chirped laser pulses include additional time-dependent parameters (the pulse envelopes and swept frequencies), whose time scales are slow with respect to the optical frequencies. A first step is to relate the usual semiclassical time-dependent Schrödinger equation to the time-dependent dressed Schrödinger equation, defined by the Floquet Hamiltonian. Since this equation does not have the fast optical time dependence, we can treat it with adiabatic principles, by studying the properties of the spectrum of the Floquet Hamiltonian as a function of the slow parameters.

We first derive the time-dependent dressed Schrödinger equation generated by the Floquet Hamiltonian, relevant for processes induced by chirped laser pulses (see Section IV.A). The adiabatic principles to solve this equation are next described in Section IV.B.

A. The Dressed Schrödinger Equation for Chirped Laser Pulses

We consider here for simplicity one chirped laser mode. Extension to multimode process is direct. The slow parameters of characteristic time τ are the laser pulse envelope $\Lambda(t)$ and frequency $\omega(t)$. The time-dependent phase can be written as

$$\theta + g(t), \qquad \text{with} \quad g(t) = \omega(t)t \tag{217}$$

We consider a semiclassical Hamiltonian that depends on these slow parameters:

$$\hat{H}^{[\Lambda(t),\omega(t)]}(t) = H^{\Lambda(t)}(\theta + g(t)) \tag{218}$$

For instance, for the dipole coupling of Eq. (1) with a pulse envelope

$$\mathscr{E}\left(\frac{t}{\tau}\right) = \Lambda\left(\frac{t}{\tau}\right)\mathscr{E}_{max} \tag{219}$$

and chirped frequency $\omega(t/\tau)$, we take the semiclassical Hamiltonian

$$\hat{H}^{[\Lambda(\frac{t}{\tau}),\omega(\frac{t}{\tau})]}(t) = H_0 - \mathscr{E}\left(\frac{t}{\tau}\right)\mu\cos\left(\theta + \omega\left(\frac{t}{\tau}\right)t\right) \tag{220}$$

The parameter τ is a measure of the total duration of the pulse. For instance, we can take Gaussian pulses

$$\Lambda_{\text{Gaussian}}\left(\frac{t}{\tau}\right) = e^{-(t/\tau)^2} \tag{221}$$

If we take a pulse of the form

$$\Lambda_{\text{trig}}\left(\frac{t}{\tau}\right) = \begin{cases} \sin^2\left(\pi\frac{t-t_i}{\tau}\right) & \text{if } t \in [t_i, t_i + \tau] \\ 0 & \text{elsewhere} \end{cases} \tag{222}$$

usually called *trig pulse*, the parameter τ is the total length of the pulse. We use the square to ensure the continuity of the first derivative, which avoids additional nonadiabatic losses as we will discuss.

Remark. *For some arguments involving adiabatic evolution pulses that have a well-defined beginning and end present conceptual advantages, since they allow clear-cut statements. For other considerations (like the Dykhne–Davis–Pechukas analysis of Section V.C), one needs real analytic pulse shapes, which excludes shapes that are identically zero on the complement of a finite interval.*

We derive the Floquet Hamiltonian K associated to this semiclassical Hamiltonian by starting with the following definition of the corresponding propagator, which is the natural generalization of (13)

$$U^K(t,t_0;\theta) := \mathscr{T}_{-g(t)} U(t,t_0;\theta) \mathscr{T}_{g(t_0)} \tag{223}$$

where $\mathscr{T}_{g(t)}$ is the translation operator that acts on $\mathscr{L}_2(S^1, d\theta)$ as $\mathscr{T}_{g(t)}\xi(\theta) = \xi(\theta + g(t))$.

The operator U is the propagator of the Schrödinger equation

$$i\hbar \frac{\partial}{\partial t} U(t,t_0;\theta) = H^{\Lambda(t)}(\theta + g(t)) U(t,t_0;\theta) \tag{224}$$

if and only if U^K satisfies the dressed Schrödinger equation

$$i\hbar \frac{\partial}{\partial t} U^K(t,t_0;\theta) = K^{[\Lambda(t),\omega_{\text{eff}}(t)]} U^K(t,t_0;\theta) \tag{225}$$

where

$$\omega_{\text{eff}}(t) = \frac{d\Theta(t)}{dt} = \frac{dg(t)}{dt} = \omega(t) + \dot{\omega} t \tag{226}$$

and

$$K^{[\Lambda(t),\omega_{\text{eff}}(t)]}(\theta) = H^{\Lambda(t)}(\theta) - i\hbar\omega_{\text{eff}}(t) \frac{\partial}{\partial\theta} \tag{227}$$

In terms of states, Eq. (223) gives the correspondence between $\phi(t;\theta)$, the solution of Eq. (224), and $\psi(t,\theta)$, the solution of Eq. (225):

$$\phi(t;\theta) = \mathscr{T}_{g(t)}\psi(t,\theta) = \psi(t,\theta + g(t)) \tag{228}$$

The above result is proved in Appendix A. We point out the appearance of an *effective* instantaneous frequency in the Floquet Hamiltonian (227).

B. Adiabatic and Diabatic Evolution of Floquet Dressed States

The preceding analysis is well adapted when one considers slowly varying laser parameters. One can study the dressed Schrödinger equation invoking adiabatic principles by analyzing the Floquet Hamiltonian as a function of the slow parameters.

We distinguish the adiabatic evolution for nonresonant processes, for resonant processes at zero field, and for processes with a dynamical resonance. For nonresonant processes the adiabatic transport of the dressed states is simple: The dynamics follows, up to a phase, the instantaneous dressed state whose eigenenergy is continuously connected to the one associated to the initial dressed state. This adiabatic transport will be generalized if more than one dressed state is involved in the dynamics.

A zero-field resonance can be explored adiabatically by the dynamics in different ways. One considers first a laser pulse of resonant (or quasi-resonant) carrying frequency leading to two degenerate (or quasi-degenerate) dressed states at early times, such that the dynamics has to be described in the subspace spanned by these two states. The process can be described as follows: When the field rises, this degeneracy is *dynamically lifted*, which induces a sharing of the population between the two instantaneous dressed states whose eigenenergies are continuously connected to the initial degenerate ones. This process will be the origin of the creation of coherent superposition of states by adiabatic passage (see Section V.D). We will see that this lifting of degeneracy is instantaneous for one- and two-photon exact resonance processes. These two branches are next *followed adiabatically* by the dynamics if the pulse envelopes are slow enough. When the pulse later falls, the dynamics goes through the inverse process of *creation of degeneracy* which induces the interference of the two branches at the very end of the process.

One can see in this example the necessity to consider an adiabatic transport along more than one dressed state.

An alternative way to explore the zero-field resonance is to chirp a laser pulse that is switched on and off adiabatically, sufficiently far from the resonance. The chirp is such that the frequency is swept through the resonance when the field is on. The resonance appears in a dressed eigenenergy diagram (as a function of time or as a function of the field parameters) as an *avoided crossing*.

An avoided crossing will mainly limit the application of the adiabatic theorem: If the dynamics is not slow enough, dynamical transitions, so-called nonadiabatic transitions, will be induced between the dressed states forming the avoided crossing. A local Landau–Zener analysis can be invoked to determine this local nonadiabatic transition. Adiabatic passage along the avoided crossing will induce in general a transition in the bare states.

The dynamical resonances, which occur beyond a threshold of field amplitude, also usually appear as avoided crossings (see Section VI).

All these different types of resonances will be characterized geometrically in the next section.

1. Adiabatic Evolution

It is convenient to consider explicitly the time scale in the slow parameters: $\Lambda(s)$ and $\omega_{\text{eff}}(s)$, where $s = t/\tau$ is a reduced time, τ a characteristic time for the slow parameters, and t is the physical time. The slow parameters are gathered in a formal vector $\mathbf{r}(s) = [\Lambda(s), \omega_{\text{eff}}(s)]$. The dressed Schrödinger equation reads

$$i\hbar \frac{\partial \psi(\theta, t)}{\partial t} = K^{\mathbf{r}(s)} \psi(\theta, t) \tag{229}$$

We denote $P(t) = \sum_{m \in \mathcal{S}} |\psi_m^{\mathbf{r}(s)}\rangle\langle\psi_m^{\mathbf{r}(s)}|$ the projector at time t in $\mathcal{S}$, a subspace in $\mathcal{K}$, in which we want to apply the adiabatic evolution. The adiabatic theorem can be formulated as

$$\lim_{\tau \to \infty} U^K(t, t_0; \theta) P(t_0) = \lim_{\tau \to \infty} P(t) U^K(t, t_0; \theta) \tag{230}$$

if the instantaneous eigenenergies $\lambda_m^{\mathbf{r}(s)}, m \in \mathcal{S}$, are far enough from the other eigenenergies for all time $t \geq t_0$. In terms of eigenvectors, if one assumes a unique eigenvector $|\psi_m^{\mathbf{r}(s)}\rangle$ associated to the eigenenergy $\lambda_m^{\mathbf{r}(s)}$, one has $P(t) = |\psi_m^{\mathbf{r}(s)}\rangle\langle\psi_m^{\mathbf{r}(s)}|$ and the preceding formulation becomes the following:

If the system is at time $t_0 = \tau s_0$ in the Floquet instantaneous eigenstate $\psi(\theta, t_0) = \psi_m^{\mathbf{r}(s_0)}(\theta)$, then in the adiabatic limit $\tau \to \infty$ the state solution $\psi(\theta, t)$ of Eq. (229) is up to a phase given by the instantaneous Floquet state whose eigenenergy is continuously connected to the initial one at t_0:

$$\psi(\theta, t) \simeq \exp[i\delta_m^{\mathbf{r}(s)}(t)] \psi_m^{\mathbf{r}(s)}(\theta) \tag{231}$$

where

$$\delta_m^{\mathbf{r}(s)}(t) = -\frac{1}{\hbar} \int_{t_0}^{t} du\, \lambda_m^{\mathbf{r}(u/\tau)} + i \int_{\mathbf{r}(s_0)}^{\mathbf{r}(s)} d\mathbf{r} \cdot \langle \psi_m^{\mathbf{r}} | \nabla_{\mathbf{r}} \psi_m^{\mathbf{r}} \rangle_{\mathcal{K}} \tag{232}$$

is the sum of, respectively, the dynamical phase and the Berry geometrical phase.

The dynamical phase depends on the trajectory followed in the space parameter and on its speed. The geometrical phase does not depend on its speed.

Since the phase of the instantaneous Floquet states is not uniquely specified at each $\mathbf{r}$, one can always choose the geometrical phase as zero by requiring that at each time $\dot{\mathbf{r}} \cdot \langle \psi_m^{\mathbf{r}} | \nabla_{\mathbf{r}} \psi_m^{\mathbf{r}} \rangle_{\mathscr{K}} = 0$. However, if one follows a closed loop in parameter space the eigenvector at the end of the loop will differ from the initial one by a phase. This phase, which depends only on the geometry of the loop but not on the speed, is the geometrical Berry phase [44]. If only one parameter is varied, the closed-loop geometrical phase is 0. If two parameters are varied, it can be 0 or π. If more than two parameters are varied, it can take any value.

The adiabatic theorem is valid in two quite different situations, which can be illustrated with a two-level example:

1. Well-separated instantaneous eigenvalues. In this case, if the initial condition is an instantaneous eigenstate, the evolution in the adiabatic limit follows the corresponding branch.
2. Exact crossing of eigenvalues. In this case the adiabatic evolution follows the initial branch across the intersection.

Thus, in both cases in the adiabatic limit the population is carried at all times by a single branch of instantaneous eigenstates. A quite general formulation of the adiabatic theorem, which imposes only smoothness conditions on the instantaneous eigenprojections, has been presented recently by Avron and Elgart [45,46].

In the applications to the control of molecular processes the property that is most often required is that the population stays on a single branch. The case that is most detrimental is when two or several branches do not cross but come close to each other—for example, in the form of narrowly avoided crossings. In this case, for a finite speed of the parameters there are *nonadiabatic transitions* between the branches; that is, the population is spread among them. This behavior will be discussed in more detail below.

More than one Floquet state can be involved in the dynamics—for example, if the initial condition is a linear combination of the instantaneous eigenvectors. These Floquet states span a subspace $\mathscr{S}$, and the adiabatic transport can be formulated in terms of eigenvectors:

$$\psi(\theta, t) \simeq \sum_{m \in \mathscr{S}} c_m \exp[i\delta_m^{\mathbf{r}(s)}(t)]\psi_m^{\mathbf{r}(s)}(\theta) \tag{233}$$

where the c_m are complex numbers determined by the initial condition

$$c_m = \langle \exp(i\delta_m^{\mathbf{r}(s_0)}(t_0))\psi_m^{\mathbf{r}(s_0)}(\theta) | \psi(\theta, t_0) \rangle \tag{234}$$

A sketch of an argument that leads to the adiabatic theorem for the Floquet Hamiltonian of an N-level system is given in Appendix C.

2. *Nonresonant Deviations from Adiabaticity: Perturbation Theory, Superadiabatic Schemes, and Dykhne–Davis–Pechukas Formula*

Deviations from strict adiabatic evolution given by the adiabatic theorem are of the order of $1/\tau$ (in amplitude) and can thus be estimated for short time by time-dependent perturbation theory in the adiabatic basis, which does not diverge for nonresonant processes—that is, if there is no degeneracy nor quasi-degeneracy (appearing as avoided crossings) of the quasi-energies. These deviations can be due to (i) the fact that τ is finite and (ii) the possible nonsmoothness of the parameters. These cases have been considered in Ref. 47 for nonsmooth pulse ends in a two-level model driven by a nonresonant field. This study shows that adiabatic evolution is a good approximation even for a pulse of a few optical cycles. It also shows that the first-order correction, which involves the coupling of the Floquet zone considered with the other zones, captures well the small deviations that appear as oscillations (see Section 3.1 of Ref. 47).

It is known, however, that in fact the adiabatic passage is in general much more efficient when considered at the end of the process. This is well understood in two-level systems for which one has the following result: *For smooth parameters and nondegenerate eigenvalues, the nonadiabatic corrections in the asymptotic adiabatic limit* $\tau \to \infty$ *are of order* $e^{-|\text{const.}|\tau}$, *exponential in* τ*—that is, beyond all orders in* $1/\tau$ *at the end of the process.*

This result is due to Dykhne [48] and Davis and Pechukas [49] and has been extended to N-level systems [50]. The conditions of validity of the so-called Dykhne–Davis–Pechukas (DDP) formula has been established in [51,52]. This formula allows us to calculate in the adiabatic asymptotic limit the probability of the nonadiabatic transitions. This formula captures, for example, the result of the Landau–Zener formula, which we study below.

An alternative interpretation of this exponential nonadiabatic corrections has been given through *superadiabatic schemes*. The superadiabatic schemes allow us to transform the problem to more adapted new basis (the so-called superadiabatic basis), where transition amplitudes proportional to powers of $1/\tau$ are removed. The scheme can be either iterative or by expansion in power series of $1/\tau$. This series expansion has been introduced by Berry for parametrically time-dependent quantum systems [53]. The iterative scheme [54–57] consists in constructing iteratively Schrödinger equations by successive appropriate unitary transformations of the effective dressed Hamiltonians. The first step corresponds to the instantaneous diagonalization (C4) of the Hamiltonian giving the new exact Schrödinger equation (C6), containing nonadiabatic couplings of first order in $1/\tau$. The next steps are diagonalizations of the new Hamiltonians which reduce the nonadiabatic couplings to higher orders. Neither the series nor the iterations converge in general. However, Berry showed that for this asymptotic series in two-level systems, an optimal order, corresponding to the minimization of the nonadiabatic couplings, gives an optimal superadiabatic basis with respect to which the transition amplitude acquires a universal error-function-like form. It is universal in the sense that it does not depend on the details of the Hamiltonian. This optimal superadiabatic basis coincides with the free basis (and also the adiabatic basis) at the beginning and at the end of the process when the fields are off. This means that if there are no degeneracies or quasi-degeneracies at the beginning and the end of the process, *the adiabatic passage is in fact supported by a superadiabatic transport between the beginning and the end of the process*. The nonadiabatic corrections are then given at the end of the process by the Berry's universal error function times an exponential in τ, in agreement with the DDP analysis [53,57]. This approach has been successfully applied for a two-level atom strongly perturbed by a nonresonant field, in the full Floquet representation [47]. The resonant stimulated Raman adiabatic passage (STIRAP) process in a three-level system (that is studied below) is an example that has degeneracies of the eigenvalues at the beginning and the end of the process. An effective two-level model has shown [57] that in this case the nonadiabatic correction at the end of the process is given, in addition to the DDP exponential term, by a *perturbative* term, whose dominant contribution is of first order in $1/\tau$.

Degeneracy and avoided crossing of the eigenvalues can be treated in a specific way, as shown below. Optimization of adiabatic evolution will be studied in Section V.C with the use of geometric arguments.

Nonsmoothness of the parameters usually leads to nonadiabatic corrections in transition amplitude of order p if the nonsmoothness is characterized by a discontinuous pth derivative. Nonsmooth pulse ends can be investigated [58] in the simplified model

$$H(t) = \frac{\hbar}{2}\begin{bmatrix} 0 & \Omega(t) \\ \Omega(t) & 2\Delta(t) \end{bmatrix} \tag{235}$$

where we assume a real and positive Ω, which represents a two-level atom (with states $|1\rangle$ and $|2\rangle$) interacting with a one-photon quasi-resonant pulse in the rotating wave approximation [39,40]. We consider a coupling characterized by a Rabi frequency $\Omega(t) = \Omega_0 \sin^2(\pi t/\tau)$, having discontinuous second derivatives at the beginning and at the end of the pulse. The frequency of the field is chirped in such a way that the distance between the two eigenenergies is kept constant: $\Delta(t) = \frac{|t|}{t}\sqrt{\Omega_0^2 - \Omega^2(t)}$ (this choice will appear clearer in Section V.C). With this choice the nonadiabatic corrections are due uniquely to the nonsmoothness at the beginning and at the end. In the adiabatic limit the population mostly transferred from $|1\rangle$ to $|2\rangle$. For the nonadiabatic corrections in probability at the end of the pulse (i.e., the probability of the population to return to state $|1\rangle$), the first-order nonadiabatic corrections in the adiabatic asymptotic limit $\tau \to \infty$, are given, after integrating twice by parts, by

$$P_1 \approx \frac{1}{4}\left(\frac{\pi}{\tau\Omega_0}\right)^4 \sin^2(\Omega_0\tau) \tag{236}$$

This gives, as expected, asymptotic nonadiabatic corrections in probability that scale as $(1/\tau)^4$, since the discontinuity is in the second derivatives.

3. *Resonant Laser Fields—Lifting of Degeneracy*

Processes that are resonant at zero field (i.e., with a atomic Bohr frequency that is an integer multiple of the laser frequency) can be investigated through an effective Hamiltonian of the model constructed from a multilevel atom driven by a quasi-resonant pulsed and chirped radiation field (referred to as a pump field). If one considers an n-photon process between the considered atomic states $|1\rangle$ and $|2\rangle$ (of respective energy E_1 and E_2), one can construct an effective Hamiltonian with the two dressed states $|1;0\rangle$ (dressed with 0 photon) and $|2;-n\rangle$ (dressed with $-n$ photons) coupled by the n-photon Rabi frequency $\Omega(t)$ (of order n with respect to the field amplitude and that we assume real and positive) and a dynamical Stark shift of the energies. It reads in the two-photon RWA [see Section III.E and the Hamiltonian (190)], where we assume Ω real and positive for simplicity,

$$H(t) = \frac{\hbar}{2}\begin{bmatrix} 0 & \Omega(t) \\ \Omega(t) & 2\Delta(t) \end{bmatrix} \tag{237}$$

with the effective detuning

$$\Delta(t) = \Delta_0(t) + S(t) \tag{238}$$

where $S(t)$ is the relative dynamical Stark shift (of second order) due to the contribution of the other states and $\Delta_0(t)$ the detuning associated to the multiphoton near-resonant process, which is time-dependent if a chirp is applied. The population resides initially in the atomic state $|1\rangle$.

Note that the Hamiltonian (237) is a good approximation for the one- and two-photon processes (as shown in Section III.E for the two-photon case), but that it is only a rough approximation for higher multiphoton processes, since the Stark shifts should contain additional terms of higher order to be consistent with the order of the effective Rabi frequency.

There is a quasi-resonance in this system if there exist times t for which $\Omega(t)$ is of the same order as $\Delta(t)$, as described more precisely in Section III.E.

a. Adiabatic Evolution. It is important to note that for this model (237) we can write the conditions for adiabatic behavior in detail using the procedure described in Appendix C. One obtains the transformed Schrödinger equation (C6):

$$i\hbar \frac{\partial}{\partial t}\tilde{\psi}(t) = \begin{bmatrix} \lambda(t) & i\gamma(t) \\ -i\gamma(t) & -\lambda(t) \end{bmatrix} \tilde{\psi}(t) \tag{239}$$

with

$$\lambda(t) = \frac{\hbar}{2}\sqrt{\Delta^2(t) + \Omega^2(t)} \tag{240}$$

and the nonadiabatic coupling

$$\gamma(t) = \frac{\hbar}{2} \frac{\Omega(t)\dot{\Delta}(t) - \dot{\Omega}(t)\Delta(t)}{\Delta^2(t) + \Omega^2(t)} \tag{241}$$

The conditions for adiabatic evolution are satisfied if the nonadiabatic coupling is much smaller than the separation of the eigenvalues

$$|\gamma(t)| \ll \hbar\sqrt{\Omega^2(t) + \Delta^2(t)} \tag{242}$$

If the detuning is constant, estimating $\dot{\Omega}(t) \sim \Omega_0/\tau$ with $\Omega_0 = \max_t \Omega(t)$ and taking $\Omega_0 \sim \Delta$, we obtain that *the dynamics is adiabatic if one assumes a large detuning with respect to* $1/\tau$, where τ characterizes the length of the pulse:

$$\tau\Delta \gg 1 \tag{243}$$

Thus in this case the dynamics is at all times adiabatic in the sense that it mainly follows the dressed eigenstate whose eigenvalue is continuously connected to the one associated to the initial dressed state. This adiabatic transport results at the end of the pulse in an (almost) complete return in the initially populated state. It is important to point out that the dynamics is affected by the resonance in the sense that the excited bare state $|2\rangle$ is highly populated during the pulse if Ω is of the same order as Δ or larger at the peak laser amplitude. For two-level systems, the nonadiabatic small corrections lost to the other eigenstate have been extensively studied (see, for example, Ref. 59 and references therein).

b. Instantaneous Lifting and Creation of Degeneracy. In the opposite case

$$\tau\Delta \ll 1 \tag{244}$$

a dynamical sharing of the dynamics between the resonant dressed states occurs. One can reinterpret the well-known π-pulse formula for a one-photon process and extend it for multiphoton processes [60,61]. In the case of an exact n-photon resonance, defined by (244), the two relevant dressed states $|1;0\rangle$ and $|2;-n\rangle$ can be considered, before the rising of the pulse, as exactly degenerate with respect to the dynamics, associated to the dressed energy $E_1 = E_2 - n\hbar\omega$. The pulse rising induces a *dynamical splitting of the population* along two eigenstate branches. The splitting is *instantaneous* only in the case of exact one-photon resonance $n = 1$ and in the exact two-photon case $n = 2$. These cases are simple since the nonadiabatic coupling is exactly zero. Thus we can calculate exactly the solution of the Schrödinger equation for the two-level effective Hamiltonian.

The one-photon resonance case induces an equal sharing of the dynamics along the two eigenstate branches, which allows to recover the π-pulse formula

$$P_2 \equiv |\langle 2;-1|\psi(t_f)\rangle|^2 = \sin^2 \frac{1}{2}\int_{t_i}^{t_f} |\Omega(t)|\, dt \tag{245}$$

One can generalize it for the exact resonant case $n = 2$: one has (for α real)

$$\Delta(t) = S(t) = \alpha\mathcal{E}^2(t), \quad \Omega(t) = \beta\mathcal{E}^2(t), \quad \beta = |\beta|e^{-i\varphi} \tag{246}$$

and the effective Hamiltonian (237) written as

$$H(t) = \frac{\hbar}{2}\mathcal{E}^2(t)\begin{bmatrix} 0 & \beta \\ \beta^* & 2\alpha \end{bmatrix} \tag{247}$$

At each time, the time-independent unitary transformation (having on its column the dressed states $|\psi_+\rangle$ and $|\psi_-\rangle$)

$$T = \begin{bmatrix} \cos(\zeta/2) & -\sin(\zeta/2) \\ e^{i\varphi}\sin(\zeta/2) & e^{i\varphi}\cos(\zeta/2) \end{bmatrix} \tag{248}$$

with

$$\tan\zeta = -\frac{|\beta|}{\alpha}, \qquad 0 \le \zeta < \pi \tag{249}$$

diagonalizes the Hamiltonian $H(s)$.

$$T^\dagger H(t) T = \begin{bmatrix} \lambda_+^{\mathcal{E}} & 0 \\ 0 & \lambda_-^{\mathcal{E}} \end{bmatrix} \equiv D(t) \tag{250}$$

with

$$\lambda_{\pm}^{\mathscr{E}} = \frac{\hbar}{2}\mathscr{E}^2(\alpha \pm \sqrt{\alpha^2 + |\beta|^2}) \tag{251}$$

This can be interpreted as the lifting of the degeneracy

$$|1;0\rangle = \cos(\zeta/2)|\psi_+\rangle - \sin(\zeta/2)|\psi_-\rangle \tag{252a}$$

$$|2;-2\rangle = e^{-i\varphi}[\sin(\zeta/2)|\psi_+\rangle + \cos(\zeta/2)|\psi_-\rangle] \tag{252b}$$

Since the population resides initially in the atomic state $|1\rangle$, one obtains a sharing of the dynamics along the two dressed states (with nonequal weight in general):

$$|\psi(t)\rangle = \cos(\zeta/2)\exp\left[-\frac{\mathrm{i}}{\hbar}\int_{t_i}^{t} du\, \lambda_{+}^{\mathscr{E}(u)}\right]|\psi_+\rangle - \sin(\zeta/2)\exp\left[-\frac{\mathrm{i}}{\hbar}\int_{t_i}^{t} du\, \lambda_{-}^{\mathscr{E}(u)}\right]|\psi_-\rangle \tag{253}$$

Note that Eq. (253) is exact, not an adiabatic approximation, since for the Hamiltonian (247) the nonadiabatic coupling is exactly zero. At the end of the pulse, the inverse mechanism of *instantaneous creation of degeneracy* occurs

$$|\psi(t_f)\rangle = \left[\cos\gamma(t_f) + \mathrm{i}\frac{\alpha}{\sqrt{\alpha^2 + |\beta|^2}}\sin\gamma(t_f)\right]|1;0\rangle - \mathrm{i}\frac{e^{i\varphi}|\beta|}{\sqrt{\alpha^2 + |\beta|^2}}\sin\gamma(t_f)|2;-2\rangle \tag{254}$$

with the phase

$$\gamma(t_f) = \frac{1}{2\hbar}\int_{t_i}^{t_f} du\,[\lambda_{+}^{\mathscr{E}(u)} - \lambda_{-}^{\mathscr{E}(u)}] = \frac{1}{2}\sqrt{\alpha^2 + |\beta|^2}\int_{t_i}^{t_f} dt\, \mathscr{E}^2(t) \tag{255}$$

It can be interpreted as an interference of the two branches, with relative weight determined by $\gamma(t_f)$, which is equal to half the area between the two eigenvalues. One obtains the generalized two-photon π-pulse formula:

$$P_2 = \frac{|\beta|^2}{\alpha^2 + |\beta|^2}\sin^2\left[\frac{1}{2}\sqrt{\alpha^2 + |\beta|^2}\int_{t_i}^{t_f} \mathscr{E}^2(t)\, dt\right] \tag{256}$$

which means that if

$$\sqrt{\alpha^2 + |\beta|^2}\int_{t_i}^{t_f} \mathscr{E}^2(t)\, dt = \pi \tag{257}$$

one obtains the maximal population transfer to the state $|2;-2\rangle$. There is no complete transfer except in the limiting case $\alpha/|\beta| \to 0$. This process is not robust with respect to the pulse amplitude, since any deviation will in general change the area.

If one takes an area different from π, it leads to a final coherent superposition of states. For example, in the case of a one-photon process, one has [using formula (254) with $\alpha = 0$ and $\Omega = \beta\mathscr{E}$]

$$|\psi(t_f)\rangle = \cos\left(\int_{t_i}^{t_f} \frac{|\Omega(t)|}{2} dt\right)|1;0\rangle - ie^{i\varphi} \sin\left(\int_{t_i}^{t_f} \frac{|\Omega(t)|}{2} dt\right)|2;-1\rangle \quad (258)$$

and the area $\int_{t_i}^{t_f} |\Omega(t)|dt = \pi/2$ leads to a superposition of states with equal sharing in probability

$$|\psi(t_f)\rangle = \frac{1}{\sqrt{2}}(|1;0\rangle - ie^{i\varphi}|2;-1\rangle) \quad (259)$$

Again, this creation of superposition of states is not robust with respect to the pulse amplitude. We will study in Section V.D a way to create a superposition of states whose coefficients are robust in probability (i.e., squared absolute value of the coefficients).

We have described the picture with the simplified effective two-level model (237). However, it is still valid in the more general case of a strong field resonant n-multiphoton ($n > 2$) process, although the effective two-level model is rather inaccurate as already mentioned. In this case, one has to consider the full eigenenergies of the Floquet Hamiltonian that are relevant for the process (for example calculated numerically), denoted as $\lambda_+^{\mathscr{E}(t)}$ and $\lambda_-^{\mathscr{E}(t)}$, continuously connected to the degenerate energies $\lambda_+^{\mathscr{E}(t_i)} = \lambda_-^{\mathscr{E}(t_i)} = E_1 = E_2 - n\hbar\omega$. The associated Floquet states, which in general depend on the pulse amplitude, are denoted $|\psi_+^{\mathscr{E}(t)}\rangle$ and $|\psi_-^{\mathscr{E}(t)}\rangle$. The splitting of the dynamics along the two branches in this case is not instantaneous but in general dynamical. This splitting does not necessarily coincide with the lifting of degeneracy. We assume here, however, that it occurs approximately *instantaneously* at a time $t_s > t_i$ (associated with the complex coefficients denoted a_+ and a_- such that $|a_+|^2 + |a_-|^2 = 1$), and that before this time, the solution is first shifted adiabatically into the Floquet state that is continuously connected to the initial one [60]:

$$|\psi(t_s)\rangle \simeq \exp\left(-\frac{i}{\hbar}\int_{t_i}^{t_s} du\, \lambda_-^{\mathscr{E}(u)}\right)|\psi_1^{\mathscr{E}(t_s)}\rangle \quad (260a)$$

$$\simeq \exp\left(-\frac{i}{\hbar}\int_{t_i}^{t_s} du\, \lambda_-^{\mathscr{E}(u)}\right)(a_+|\psi_+^{\mathscr{E}(t_s)}\rangle + a_-|\psi_-^{\mathscr{E}(t_s)}\rangle) \quad (260b)$$

where we have denoted before the sharing of the dynamics, for $t < t_s$, the Floquet states $|\psi_1^{\mathcal{E}(t)}\rangle$ and $|\psi_2^{\mathcal{E}(t)}\rangle$, continuously connected to $|1;0\rangle$ and $|2;-n\rangle$, respectively, and assumed associated to the eigenenergies $\lambda_-^{\mathcal{E}(t)}$ and $\lambda_+^{\mathcal{E}(t)}$, respectively. These two branches are next *followed adiabatically* if the pulse envelopes are slow enough, which gives (up to an irrelevant global phase)

$$|\psi(t)\rangle \simeq a_+ \exp\left(-\frac{\mathrm{i}}{\hbar}\int_{t_s}^{t} du\,\lambda_+^{\mathcal{E}(u)}\right)|\psi_+^{\mathcal{E}(t)}\rangle + a_- \exp\left(-\frac{\mathrm{i}}{\hbar}\int_{t_s}^{t} du\,\lambda_-^{\mathcal{E}(u)}\right)|\psi_-^{\mathcal{E}(t)}\rangle \tag{261}$$

When the pulse later falls, the dynamics goes through the inverse process of creation of degeneracy which occurs symmetrically at time $t_s' = t_f - t_s$ (if we assume a symmetric envelope) such that

$$|\psi_+^{\mathcal{E}(t_s')}\rangle = a_+^*|\psi_1^{\mathcal{E}(t_s')}\rangle - a_-|\psi_2^{\mathcal{E}(t_s')}\rangle \tag{262a}$$

$$|\psi_-^{\mathcal{E}(t_s')}\rangle = a_-^*|\psi_1^{\mathcal{E}(t_s')}\rangle + a_+|\psi_2^{\mathcal{E}(t_s')}\rangle \tag{262b}$$

The resulting final transfer can be written as [60,62]

$$P_2 \simeq 4|a_+a_-|^2\sin^2\left[\frac{1}{2}\int_{t_s}^{t_f-t_s} du\,(\lambda_+^{\mathcal{E}(u)} - \lambda_-^{\mathcal{E}(u)})\right] \tag{263}$$

Since in general the relative distance $\lambda_+^{\mathcal{E}} - \lambda_-^{\mathcal{E}}$ between the eigenenergies is small at the beginning and at the end of the pulse (before the population and after the recombination), the formula is well approximated by

$$P_2 \simeq 4|a_+a_-|^2\sin^2\left[\frac{1}{2}\int_{t_i}^{t_f} du\,(\lambda_+^{\mathcal{E}(u)} - \lambda_-^{\mathcal{E}(u)})\right] \tag{264}$$

Thus the transfer probability depends on (i) the way in which the population is split and recombined and (ii) the difference of the dynamical phases. In the context of complete transfer, the process has been named generalized or multiphoton π-pulse and has been tested numerically for a five-photon resonance in a Morse potential [60]. This formula (264) also displays generalized Rabi oscillations.

We remark that this multiphoton process, like the one-photon process described by Eq. (245) and the two-photon process described by (256), are not robust with respect to the pulse area.

c. *Intermediate Quasi-Resonant Regimes—Dynamical Splitting of Population.* The intermediate quasi-resonant regimes, defined as

$$\tau\Delta \sim 1 \tag{265}$$

leading to a lifting (or creation) of a quasi-degeneracy, is still an open question. In this case, the unitary transformation T (248) is time-dependent and as a consequence the splitting of population is not instantaneous and leads to a nontrivial dynamics. It has only been studied numerically in, for example, Refs. 62 and 63.

4. *Diabatic Versus Adiabatic Dynamics Around Eigenenergy Crossings and Avoided Crossings*

As mentioned above, an avoided crossing can result from a chirping process or from a dynamical resonance induced by a field. An avoided crossing appears *locally* in the spectrum between two dressed states. One considers in general that the dynamics is *globally* adiabatic with respect to the other states in the subspace spanned by the dressed states forming the avoided crossing. The adiabatic approximation might fail *inside this subspace* when the dynamics encounters this avoided crossing.

If the coupling between the two dressed eigenvectors is zero, the eigenvalue crossing appears as a true crossing. This means that the branches ignore each other and the adiabatic approximation still holds through the true crossing.

If the coupling is different from zero, the dynamics can either *follow* the avoided crossing (*adiabatic* evolution), *cross* it (*diabatic* evolution), or partially cross it, depending on the speed of the dynamics with respect to the shape of the avoided crossing.

We assume that the model depends on one slow time-dependent parameter, denoted $r(t)$, and that the shape of the avoided crossing as a function of r is well described around the avoided crossing $r = r_c$ (occurring at time $t = t_c$) by its width h and its curvature C (see Fig. 1). We choose the parameterization such that $r_c = 0$. The eigenenergies read

$$\lambda_{\pm}^{r} = \pm h\sqrt{1 + \left(\frac{r}{\delta r}\right)^2} \tag{266}$$

The curvature is defined by

$$C = \left|\frac{\partial^2 \lambda_{\pm}}{\partial r^2}(r = 0)\right| = \frac{h}{(\delta r)^2} \tag{267}$$

We label these two continuous branches by the instantaneous Floquet states $\psi_{-}^{r(t)}$ and $\psi_{+}^{r(t)}$. The two eigenvalues $\lambda_{\pm}^{r}$ can be deduced from an effective local dressed Hamiltonian

$$H_{\text{eff}} = h\begin{bmatrix} r/\delta r & 1 \\ 1 & -r/\delta r \end{bmatrix} \tag{268}$$

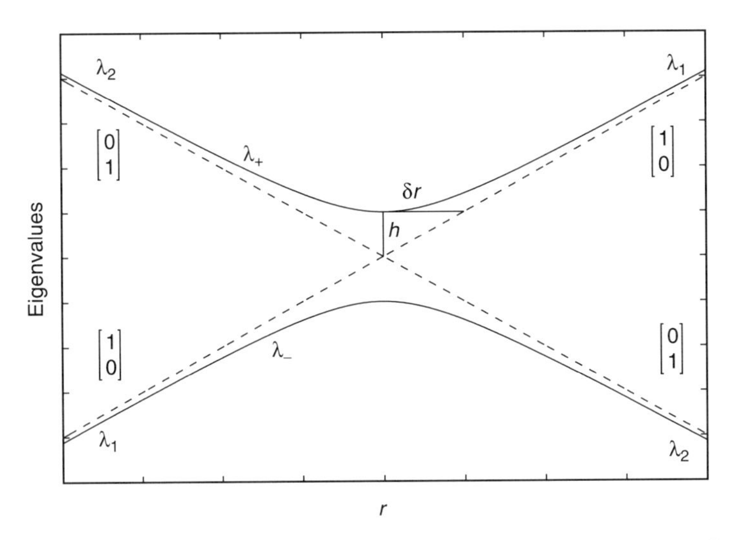

Figure 1. Diagram of an avoided crossing of width $2h$ and of curvature $C = h/(\delta r)^2$.

in the basis

$$\psi_1 \equiv \psi_-^{r(t\to-\infty)} \equiv \begin{bmatrix} 1 \\ 0 \end{bmatrix}, \qquad \psi_2 \equiv \psi_+^{r(t\to-\infty)} \equiv \begin{bmatrix} 0 \\ 1 \end{bmatrix} \tag{269}$$

We can approximately characterize the dynamics by linearizing it in time around the avoided crossing $r(t) = \dot{r}_c(t - t_c)$ and apply the Landau–Zener formula [64,65] to calculate the probability to jump from the branch ψ_- to ψ_+ [66]:

$$P_{-\to+} = \exp\left(-\pi \frac{h \times \delta r}{\dot{r}_c}\right) \tag{270}$$

The asymptotic transition probability $P_{-\to+}$ is higher for (i) a thinner avoided crossing, (ii) a steeper curvature, and (iii) a faster passage.

An avoided crossing gives rise to three qualitatively different regimes that we can analyze considering the system starting in the initial state $\psi(t \to -\infty) = \psi_1$:

(a) If the speed is slow enough ($\dot{r}_c \ll h \times \delta r$), the dynamics is *adiabatic*, that is, $P_{-\to+} \approx 0$ (meaning that the system goes into the state ψ_2 far after the avoided crossing);

(b) If the speed is fast enough ($\dot{r}_c \gg h \times \delta r$), the dynamics is *diabatic*, i.e. $P_{-\to+} \approx 1$ (meaning that the system stays in the state ψ_1 far after the avoided crossing);

(c) Any intermediate speed leads to a sharing of the dynamics into the two branches, which gives rise afterwards to two dynamical states that have their own adiabatic evolution.

Formula (270) thus defines the efficiency of the diabatic passage. We remark that the Landau–Zener formula gives the information for the whole range of gap distances, from the limit of exact crossings to widely separated ones.

If we apply this analysis locally in the Floquet spectrum, it provides the matching between the adiabatic evolution far from any avoided crossings and a local adiabatic or diabatic behavior around them.

We conclude that the only unfavorable situation for adiabatic following is the intermediate regime in which the population is split among the branches. In systems encountered in applications, there are often many levels that display exact and avoided crossings of different sizes. In order to have an adiabatic transfer concentrated on a single branch, it is required to choose the adiabatic speed in such a way that the population either goes completely across (in narrow avoided crossings) or completely stays in the same branch by going slowly enough at wide avoided crossings. *This gives a strategy for the design of adapted laser pulses.* We remark that the preceding argument is based on the hypothesis that successive avoided crossings, involving the same or different branches, can be treated sequentially, independently of each other. There are cases where several avoided crossings interfere with each other, and the simple sequential Landau–Zener analysis does not apply. (See for example Ref. [66a]).

V. TOPOLOGY OF THE DRESSED EIGENENERGY SURFACES

In order to achieve a given population transfer between the initial and target states by adiabatic passage, one has additionally to develop a global picture showing the possible paths that link these states, to design the appropriate field parameters as a function of time which will allow the desired adiabatic passage. We will describe how these connectivity properties of the dressed states are determined by the topology of the dressed state energy surfaces as a function of the time-dependent external field parameters [67,68].

Adiabatic passage can result in a robust population transfer if one uses adiabatic variations of at least two *effective* parameters of the total laser fields. They can be the amplitude and the detuning of a single laser (chirping) or the amplitudes of two delayed pulses [stimulated Raman adiabatic passage (STIRAP); see Ref. 69 for a review]. The different eigenenergy surfaces are connected to each other by conical intersections, which are associated with resonances (which can be either zero field resonances or dynamical resonances appearing beyond a threshold of the the field intensities). The positions of these intersections determine the possible sets of paths that link an initial state and the different target states. The paths can be classified into topological equivalence classes. Two paths are topologically equivalent if one can be deformed into the other without cutting it or leaving the surfaces. All paths linking the initial and target states that are in the same topology class are equivalent in the adiabatic limit. The topological aspect is the key of the robustness of the process in the sense that the final transfer does not depend on the precise shapes or areas of the laser pulses nor on precise tuning of laser frequencies.

The topology of the surfaces is essentially determined by the resonances, which produce avoided crossings of surfaces and conical intersections. When the surfaces do not interact (zero coupling), one can also observe one-dimensional intersections. The main ingredients of adiabatic transport are a *global adiabatic passage* along *one* eigenstate combined with *local diabatic evolution* near conical intersections (or with local adiabatic evolution through the exact conical intersections). We will illustrate these properties using several simple examples.

A. Topology of Adiabatic Passage by a Chirped Pulse and SCRAP

The concept of the topology of adiabatic passage can be illustrated in the simple model of an effective two-level atom interacting with a one-photon or two-photon near-resonant pulse in the two-photon RWA [see Section III.E and the Hamiltonian (190)]

$$H(t) = \frac{\hbar}{2}\begin{bmatrix} 0 & \Omega(t) \\ \Omega(t) & 2\Delta(t) \end{bmatrix} \tag{271}$$

where $\Omega(t)$ (assumed here a positive real) stands for a one- or two-photon Rabi frequency of a pump laser and $\Delta(t) = \Delta_0(t) + S(t)$ is the sum of $\Delta_0(t)$, the detuning from the one- or two-photon resonance, and $S(t)$ is the dynamical Stark shift produced by the other states. One can consider two types of processes occurring in this effective two-level system:

(a) *Direct Chirping*. The detuning from the resonance $\Delta_0(t)$ is time-dependent due to an active sweeping of the laser frequency (see, e.g., Ref. 70 for an effective two-photon chirping). Moreover, if one considers a one-photon chirp, the dynamical Stark shift $S(t)$ can be neglected.

(b) *Stark Chirped Rapid Adiabatic Passage (SCRAP)*. The quasi-resonant laser frequency (pump laser) is not chirped (the detuning Δ_0 is time independent). The effective chirping results from a total dynamical Stark shift $S(t) = S_S(t) + S_P(t)$, with $S_S(t)$ due to an auxiliary laser field (nonresonant with any levels of the system), referred to as a Stark laser [71,72] and $S_P(t)$ due to the pump laser itself. If one considers a one-photon quasi-resonance for the pump laser, the dynamical Stark shift $S_P(t)$ can be neglected.

The processes associated to this Hamiltonian (271) can be completely described in the diagram of the two eigensurfaces

$$\lambda_{\pm}(\Omega, \Delta) = \frac{\hbar}{2}\left(\Delta \pm \sqrt{\Omega^2 + \Delta^2}\right) \tag{272}$$

which represent the eigenenergies of (271) as functions of the instantaneous effective Rabi frequency Ω and the detuning Δ (see Fig. 2). All the quantities are normalized with respect to a characteristic detuning denoted $|\Delta_{\text{in}}|$. They display a conical intersection at $\Omega = 0, \Delta = 0$ induced by the crossing of the lines corresponding to the states $|1;0\rangle$ and $|2;-1\rangle$ for $\Omega = 0$ and varying Δ. In the plane $\Omega = 0$, the states $|1;0\rangle$ and $|2;-1\rangle$ do not interact. The crossing of these states in this plane $\Omega = 0$ can be seen consequently as a *mute resonance*. Thus *diabatic passage through the intersection leaves the system in the same state.* The way of passing around or through this conical intersection is the key of the successful transfer. Three generic curves representing all the possible passages with a negative initial detuning $-|\Delta_{\text{in}}|$ are shown in Fig. 2. Note that the three other equivalent curves with a positive initial detuning have not been drawn. The path (a) corresponds to a direct chirping of the laser frequency from the initial detuning $-|\Delta_{\text{in}}|$ to the final one $+|\Delta_{\text{in}}|$. The paths (b) and (c) correspond to

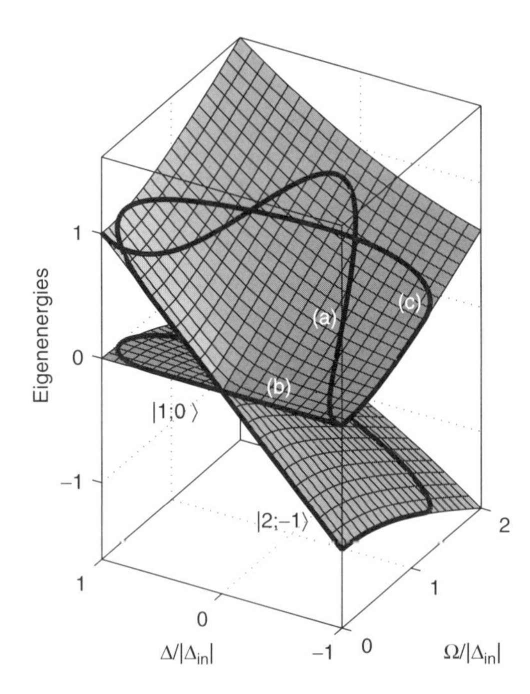

Figure 2. Surafaces of eigenenergies (in units of $|\Delta_{\text{in}}|$) as functions of $\Omega/|\Delta_{\text{in}}|$ and $\Delta/|\Delta_{\text{in}}|$ (dimensionless). Three different paths, denoted (a), (b), and (c), are depicted: path (a) corresponds to a direct chirping and paths (b) and (c) correspond to SCRAP.

SCRAP with $\Delta_0 = -|\Delta_{\text{in}}|$ for the case of a one-photon resonant pump. For the path (b), while the quasi-resonant pump pulse is off, another laser pulse (the Stark pulse, which is far from any resonance in the system) is switched on and induces positive Stark shifts $S(t) > 0$ (the Stark pulse frequency is chosen with this aim). Thus the Stark pulse makes the eigenstates get closer and induces a resonance with the pump frequency. This resonance is mute since the pump pulse is still off, which results in the true crossing in the diagram. The pump pulse is switched on after the passage through the crossing while the Stark pulse decreases. This induces the passage through the nonmute resonance, generally characterized by an avoided crossing (which is located behind the true crossing in Fig. 2; see also Fig. 3b). Finally the pump pulse is switched off. As shown in the diagram, the adiabatic following of the path (b), combining the passage through the true crossing and through an avoided crossing, induces the complete population transfer from state $|1\rangle$ to state $|2\rangle$. The path (c) is similar to the path (b) but with the pulse sequence reversed: It leads exactly to the same effect. In this case, the pump pulse is indeed switched on first (making the eigenstates repel each other as shown in the diagram) before the Stark pulse $S(t) > 0$, which is then switched off after the pump pulse.

In summary, the three paths (a), (b), and (c) represent fully adiabatic passage from state $|1;0\rangle$ to state $|2;-1\rangle$; (a) passes around the conical intersection, and (b) and (c) pass both once around the conical intersection and once through it.

We remark that the topology gives information on the dynamics for purely adiabatic passage. For real pulses of finite duration, one has to complement this information with the analysis of the effects of nonadiabatic corrections.

B. Robustness of Adiabatic Passage as a Consequence of the Topological Properties of the Eigensurfaces

The process (a) is robust with respect to fluctuations of the two parameters since it is based on the passage of the dynamics around the conical intersection. Thus neither a precise path nor any phase condition is required. Adiabaticity conditions have additionally to be fulfilled for the success of a path. Here they are given by (242). In the next subsection, we discuss optimized paths that minimize this nonadiabatic loss.

The processes (b) and (c) require an additional analysis around the crossing when the dynamics slightly misses it. One has to consider the neighborhood of the conical intersection as a thin avoided crossing. In this case, the dynamics meets locally a thin avoided crossing instead of an exact crossing. This avoided crossing has to be passed *diabatically* for the success of the process. The Landau–Zener analysis of Section IV.B.4 gives an estimation of the efficiency of the diabatic passage through Eq. (270) approximating the local dynamics around an avoided crossing with the linear time-dependent detuning $\Delta(t) = \dot{\Delta}(t_c)(t - t_c) \equiv \dot{\Delta}_c(t - t_c)$ and the coupling $\Omega(t) = \Omega(t_c) \equiv \Omega_c$, with t_c the time

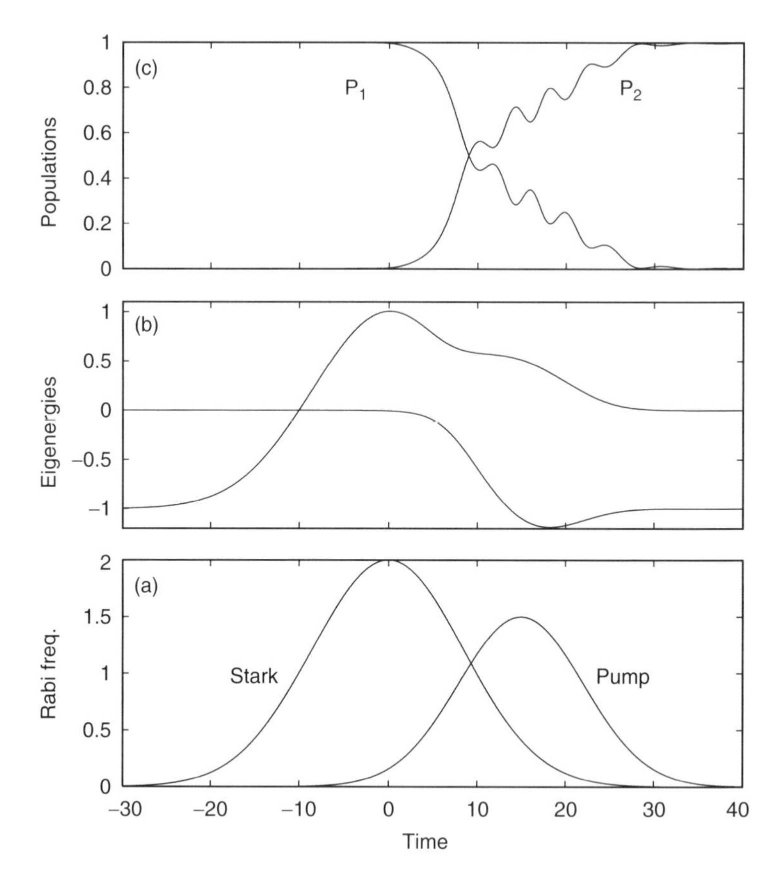

Figure 3. Numerical calculation illustrating the scrap process with Gaussian pulses for the pulse peaks $\Omega_0/|\Delta_{\text{in}}| = 1.5, S_0/|\Delta_{\text{in}}| = 2$, the delay $t_d|\Delta_{\text{in}}| = 15$, and the pulse lengths $\tau_p|\Delta_{\text{in}}| = 10, \tau_s|\Delta_{\text{in}}| = 12$ (corresponding approximately to path (b) of Fig. 2): (a) Population histories $P_n(t) = |\langle n|\psi\rangle|^2$ for $n = 1, 2$ as a function of time in units of $1/|\Delta_{\text{in}}|$. Population transfer $P_2(\infty)$ to the bare state $|2\rangle$ is nearly complete. (b) The associated eigenenergies $\lambda_\pm/\hbar$ (272) in units of $|\Delta_{\text{in}}|$. (c) The Rabi frequencies $\Omega(t)$ and $S(t)$ (in units of $|\Delta_{\text{in}}|$), respectively, associated to pump and Stark pulses.

when the avoided crossing is passed. The condition to achieve the diabatic passage locally can thus be formulated as

$$\dot{\Delta}_c \gg \pi\Omega_c^2/2 \tag{273}$$

Thus adiabatic passage in multilevel systems can be considered in general as *global adiabatic passage combined with local diabatic evolutions near conical intersections*. In multilevel systems, near a conical intersection, where one considers a local ideal diabatic evolution, it is essential that the evolution also be adiabatic with respect to the other states.

The peak amplitudes, the delay between the two fields, and the pulse shapes are chosen such that the conditions (242) and (273) are met in the concerned regions. Detailed conditions to achieve diabatic and adiabatic passage can be found in [72,73] for the example of delayed Gaussian pulses.

We remark that if condition (273) is not satisfied, which is the case if one misses the conical intersection in an intermediate regime ($\Omega_c^2 \approx \dot{\Delta}_c$), the Landau–Zener formula shows that the dynamics splits the population into the two surfaces near the intersection. This gives rise afterwards to two states that will have their own adiabatic evolution.

In Fig. 3, we have performed a numerical calculation of the dynamics corresponding to the path (b) and which confirms the preceding analysis. We have solved the time-dependent Schrödinger equation $i\partial\psi/\partial t = (H/\hbar)\psi$ in units of a characteristic detuning $|\Delta_{\text{in}}|$, with Gaussian pulses for the pump and Stark lasers of respective characteristic length τ and τ_s: $\Omega(t) = \Omega_0 \exp[-(t-t_d)^2/\tau^2]$ $S(t) = S_0 \exp[-t^2/\tau_s^2]$. The pump is time-shifted by t_d. Figure 3b clearly shows a crossing followed by the avoided crossing. Using Gaussian pulses (which are never zero) imply that we never have true crossings. However, the delay is chosen such that one has a diabatic passage through the thin avoided crossing which thus appears as a true crossing in Fig. 3b for the scale of the dynamics. We have used $\tau_p|\Delta_{\text{in}}| = 10 \gg 1$ and $\tau_s|\Delta_{\text{in}}| = 12 \gg 1$ to ensure the adiabatic passage condition (243) far after the crossing.

C. Optimization of Adiabatic Passage

Since in real experiments, pulses of finite area are used, it is useful to analyze the conditions that will optimize the adiabatic passage—that is, the conditions that will allow us to minimize the nonadiabatic losses with a minimal pulse area. Adiabatic passage can be optimized by inspection of the eigenenergy surfaces as functions of the time-dependent parameters of the coupling. A contour plot of the difference of the eigenenergy surfaces exhibits *level lines*. In Fig. 4, we have displayed level lines (as contours) corresponding to the eigenenergy surfaces of Fig. 2. For this example, they are half-circles given by

$$\Omega^2 + \Delta^2 = \Delta_0^2 \tag{274}$$

with radius Δ_0 and center $\Omega = 0$, $\Delta = 0$. The radius Δ_0 corresponds to the chirp width for a one-photon chirping process in a two-level atom.

In Ref. 58, it was shown, for a class of two-level models, that *the passage along these level lines in the adiabatic regime minimizes the nonadiabatic correction*. The analysis is based on the Dykhne–Davis–Pechukas (DDP)

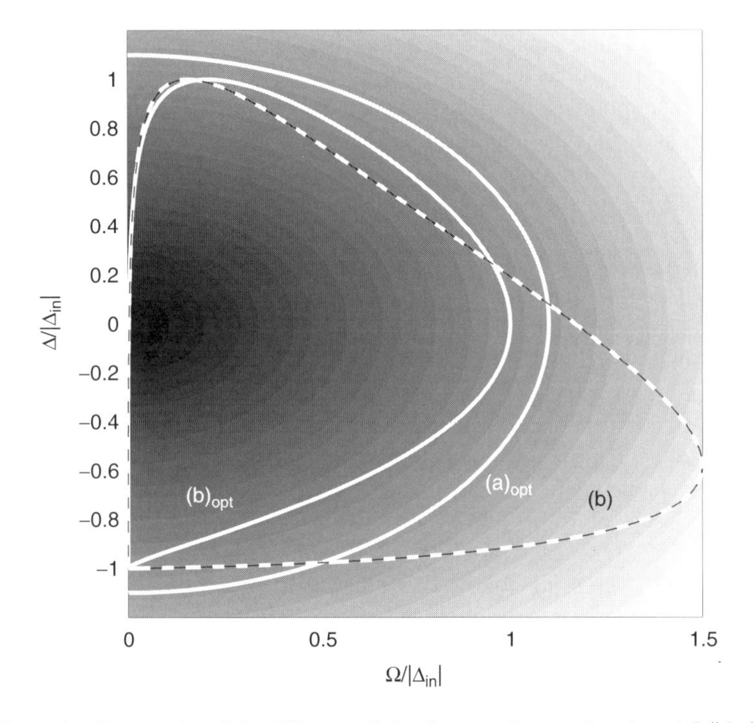

Figure 4. Contour plot of the difference of the eigenenergies as a function of $\Omega/|\Delta_{\rm in}|$ and $\Delta/|\Delta_{\rm in}|$. Three different paths have been drawn: an optimal path $(a)_{\rm opt}$ (i.e., on a level line) corresponding to the topologically equivalent path (a) of Fig. 2 (with the final detuning $\Delta_0 = 1.1|\Delta_{\rm in}|$), path (b) corresponding to the numerical calculation of Fig. 3, and path $(b)_{\rm opt}$ close to an optimal one corresponding to the numerical calculation of Fig. 5.

formula [48,49,51,52] (requiring analytic pulses). The nonadiabatic correction that is minimized is the dominant contribution given by the DDP formula [58].

This means that for a one-photon chirping process, if we choose the pulse shape $\Lambda(t)$, giving the Rabi frequency

$$\Omega(t) = \Omega_0 \Lambda(t) \tag{275}$$

the optimized detuning is then given by

$$\Delta(t) = \Delta_0 \frac{|\tau|}{\tau} \sqrt{1 - \Lambda^2(\tau)}, \qquad \Delta_0 = \Omega_0 \tag{276}$$

such that the distance between the quasi-energies is kept constant and equal to $\hbar\Delta_0$ during the process.

The adiabatic criterion is thus reduced to the choice of a level line, which has to be far enough from the origin. For a one-photon chirping process, this corresponds to the choice of the chirp width Δ_0 such that

$$\Delta_0 \tau \gg 1 \tag{277}$$

with τ the length of the pulse, according to (243). In practice, one observes that adiabaticity can be achieved for quite small pulse areas. For instance, for Gaussian pulses, the precise adiabaticity condition is $\Delta_0\tau \gg 1/(2\sqrt{2}) \approx 0.35$. For the choice $\Delta_0\tau = 1.75 (= 5 \times 0.35)$, one observes a nonadiabatic loss in probability of only $P_1(+\infty) \approx 0.0015$ [58].

This result gives the strategy to choose appropriate time-dependent parameters to achieve the adiabatic passage with a minimum pulse area.

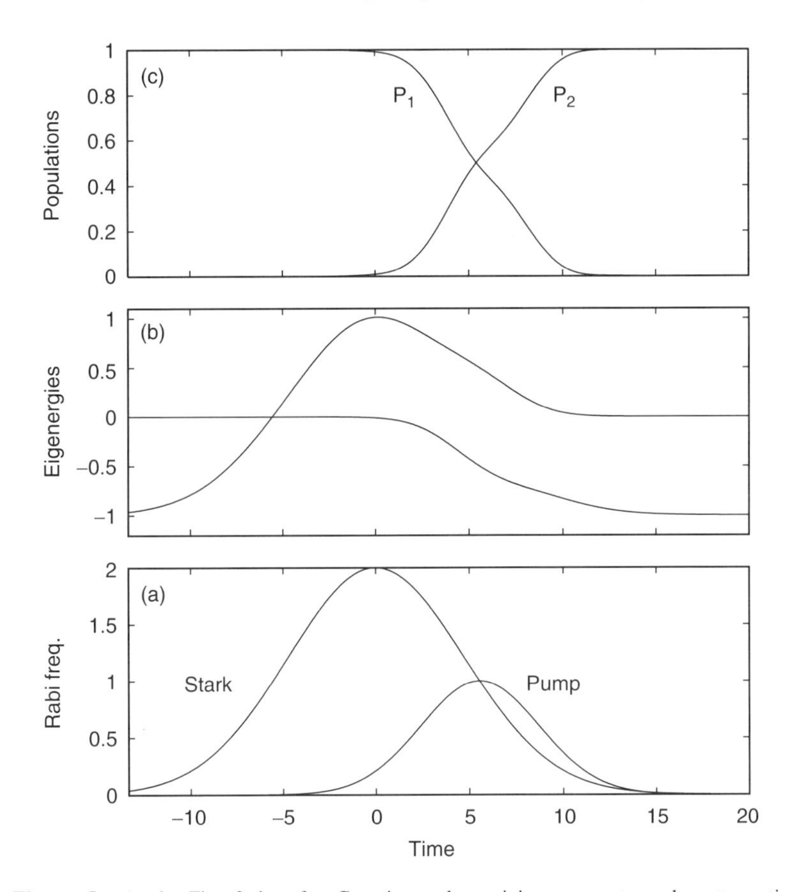

Figure 5. As in Fig. 3 but for Gaussian pulses giving parameters close to optimal, corresponding to the path $(b)_{\text{opt}}$ in Fig. 4: pulse peaks $\Omega_0/|\Delta_{\text{in}}| = 1, S_0/|\Delta_{\text{in}}| = 2$, the delay $t_d|\Delta_{\text{in}}| = 8.3$, and the (smaller) pulse lengths $\tau_p|\Delta_{\text{in}}| = 6.7$ $\tau_s|\Delta_{\text{in}}| = 10$. One can note the almost parallel eignenergies after the crossing.

This can be applied for the scrap process described above. We choose again Gaussian pulses, but with parameters such that the path is now close to the optimal one—that is, a level line. The numerical calculation is shown in Fig. 5, and the associated path $(b)_{opt}$ is shown in Fig. 4. We can see that for smaller pulse areas compared to the ones used in Fig. 3, we obtain a better population transfer, which, moreover, is monotonic.

D. Resonant Processes—Creation of Coherent Superposition of States—Half-Scrap

When the processes involve a zero-field resonance, one has to add the ingredient of lifting of degeneracy. This means that we have to consider the dynamics starting (or ending) near the conical intersection in a *direction not parallel to the* $\Omega = 0$*-plane*. This can be seen in Fig. 6 where the surfaces of Fig. 2 have been redrawn for positive detunings (case of a one-photon resonance). When the dynamics starts this way, it is characterized by two adiabatic paths, one on each surface. They will lead in general to coherent superpositions of states.

An essential point is that the result of the lifting (or creation) of degeneracy depends on the direction of the dynamics as already discussed in Section IV.B.3 for the case of one laser.

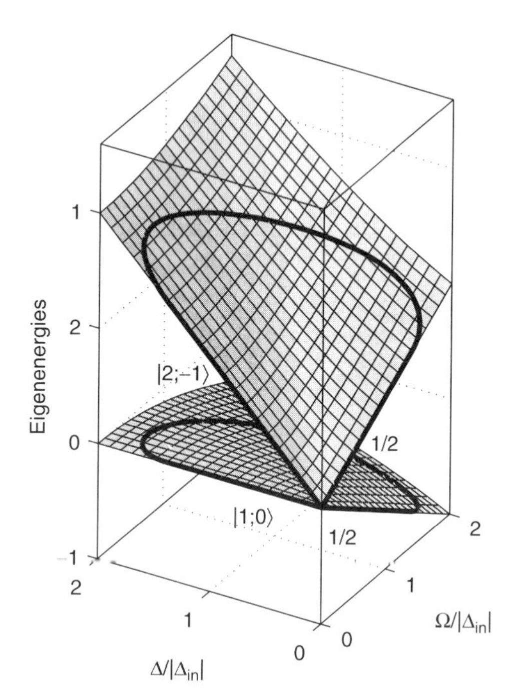

Figure 6. Surfaces of eigenenergies (in units of $|\Delta_{in}|$) as functions of $\Omega/|\Delta_{in}|$ and $\Delta/|\Delta_{in}|$ for positive Δ. The coefficients 1/2 refer to the equal sharing of the lifting of degeneracy in the $\Delta = 0$-plane.

Thus we study an exact resonant process ($\Delta_0 = 0$) starting (or ending) exactly at the conical intersection. We consider the cases that can be calculated analytically, with the Hamiltonian

$$H(t) = \frac{\hbar}{2}\begin{bmatrix} 0 & \Omega(t) \\ \Omega^*(t) & 2\Delta(t) \end{bmatrix} \tag{278}$$

(i) a one-photon resonant pump (in this case, the Stark shift $S_P(t)$ can be neglected) giving the detuning $\Delta(t) = S_S(t)$ and the Rabi frequency $\Omega(t) = -\mu\mathscr{E}(t)/\hbar,\ \mu = |\mu|e^{-i\varphi}$;

(ii) a two-photon resonant pump giving $\Delta(t) = S_P(t) + S_S(t)$ with the effective two-photon Rabi frequency $\Omega(t) = \beta\mathscr{E}^2(t),\ \beta = |\beta|e^{-i\varphi}$, and its associated relative Stark shift $S_P(t) = \alpha\mathscr{E}^2(t)$.

We denote the eigenvalues

$$\lambda_\pm = \frac{\hbar}{2}\left(\Delta \pm \sqrt{\Delta^2 + |\Omega|^2}\right) \tag{279}$$

and the corresponding dressed eigenvectors $|\psi_\pm\rangle$. In the Δ (assumed positive) direction (i.e., for $\Omega = 0$), the lifting or creation of degeneracy occurs trivially along a unique surface:

$$|1;0\rangle = |\psi_-\rangle \tag{280a}$$

$$|2;-n\rangle = |\psi_+\rangle \tag{280b}$$

All the other directions lead to a lifting of degeneracy with a sharing of population into the two surfaces. The one-photon process gives in the Ω direction, or equivalently in the field-amplitude $\mathscr{E}$ direction, the following:

$$|1;0\rangle = \frac{1}{\sqrt{2}}(|\psi_+\rangle - |\psi_-\rangle) \tag{281a}$$

$$|2;-1\rangle = \frac{e^{-i\varphi}}{\sqrt{2}}(|\psi_+\rangle + |\psi_-\rangle) \tag{281b}$$

which implies a lifting of the degeneracy occurring along the two surfaces with an equal weight. The lifting of degeneracy of the two-photon process in the field-amplitude $\mathscr{E}$ direction does not occur in the Ω direction due to the Stark shifts, as seen in Section IV.B.3:

$$|1;0\rangle = \cos(\zeta/2)|\psi_+\rangle - \sin(\zeta/2)|\psi_-\rangle \tag{282a}$$
$$|2;-2\rangle = e^{-i\varphi}[\sin(\zeta/2)|\psi_+\rangle + \cos(\zeta/2)|\psi_-\rangle] \tag{282b}$$

with

$$\tan\zeta = -\frac{|\beta|}{\alpha}, \qquad 0 \leq \zeta < \pi \tag{283}$$

The creation of degeneracy of the one-photon process is conversely given by

$$|\psi_+\rangle = \frac{1}{\sqrt{2}}(|1;0\rangle + e^{i\varphi}|2;-1\rangle) \tag{284a}$$
$$|\psi_-\rangle = \frac{1}{\sqrt{2}}(-|1;0\rangle + e^{i\varphi}|2;-1\rangle) \tag{284b}$$

while that of the two-photon process is given by

$$|\psi_+\rangle = \cos(\zeta/2)|1;0\rangle + e^{i\varphi}\sin(\zeta/2)|2;-2\rangle \tag{285a}$$
$$|\psi_-\rangle = -\sin(\zeta/2)|1;0\rangle + e^{i\varphi}\cos(\zeta/2)|2;-2\rangle \tag{285b}$$

to which dynamical phases have to be added.

We analyze two kinds of paths which, starting in state $|1;0\rangle$, will lead to a coherent superposition of states:

(a) First, lifting of degeneracy in the Δ direction [according to (280a)] giving one dressed state involved in the dynamics, next adiabatic following on this dressed state along the lower surface with $\Omega \neq 0$, and finally creation of degeneracy for decreasing Ω [according to (284a) for the one-photon process and to (285a) for the two-photon process]. In Fig. 6, a particular path of the one-photon process has been drawn (i.e., creation of degeneracy for $\Delta = 0$), yielding a coherent superposition of states with equal weights in absolute value.

(b) First, lifting of degeneracy in a direction not parallel to the $\Omega = 0$ plane, which gives two dressed states involved in the dynamics [according to (281a) for the one-photon process and to (282a) for the two-photon process], next *independent* adiabatic following on these dressed states (along both the lower and upper surface), and finally creation of degeneracy in the Δ direction [according to Eqs. (280)]. In Fig. 6, one can see the two paths associated with this (one-photon) case, yielding also a coherent superposition of states with equal weights in absolute value.

These two cases are produced by two different sequences: respectively (a) first the Stark pulse and next the pump pulse (referred to as Stark-pump sequence) and (b) first the pump pulse and next the Stark pulse (referred to as pump-Stark sequence). The phases associated to the superposition of states resulting from these two different sequences are not identical. For the Stark-pump sequence, we start (at time t_i) with the lifting of degeneracy $|\psi(t_i)\rangle = |\psi_+\rangle$, which leads at the final time t_f to (up to an irrelevant global phase)

$$|\psi(t_f)\rangle = \cos(\zeta/2)|1;0\rangle + e^{i\varphi}\sin(\zeta/2)|2;-n\rangle \tag{286}$$

For the pump-Stark sequence, we start with the lifting of degeneracy $|\psi(t_i)\rangle = \cos(\zeta/2)|\psi_+\rangle - \sin(\zeta/2)|\psi_-\rangle$. Using the adiabatic transport for each branch, the state solution reads at the end

$$|\psi(t_f)\rangle = \cos(\zeta/2)|2;-n\rangle - e^{i\int_{t_i}^{t_f} ds(\lambda_+(s)-\lambda_-(s))}\sin(\zeta/2)|1;0\rangle \tag{287}$$

Thus the two sequences lead to the same superposition in probabilities but with different phases. The pump-Stark sequence leads to an additional *nonrobust* phase difference $\int_{t_i}^{t_f} ds[\lambda_+(s) - \lambda_-(s)]$ coinciding with the dynamical phase difference.

If one considers an initial coherent state for the photon field instead of a photon-number state, the superpositions of states have the additional optical phase, giving for (286)

$$|\phi(t_f)\rangle = \cos(\zeta/2)|1\rangle + e^{i\varphi}\sin(\zeta/2)e^{-in\omega t}|2\rangle \tag{288}$$

and for (287)

$$|\phi(t_f)\rangle = \cos(\zeta/2)e^{-in\omega t}|2\rangle - e^{i\int_{t_i}^{t_f} ds(\lambda_+(s)-\lambda_-(s))}\sin(\zeta/2)|1\rangle \tag{289}$$

with $n = 2$ for the two-photon process.

This process leading to a coherent superposition of states has been suggested in Ref. 63 and named *half-scrap*, since it is very similar to the scrap process except it starts (or ends) in resonance.

The question of robustness with respect to the detuning for resonant processes is not yet completely clarified (see numerical studies in Ref. 63).

This half-scrap could be generalized for a n-multiphoton process ($n > 2$) in a multilevel system, with the use of the full quasi-energies and Floquet states (calculated numerically).

E. Topology of Stimulated Raman Adiabatic Passage (STIRAP) and STIRAP-like Processes

The adiabatic passage induced by two delayed laser pulses, the well-known process of STIRAP [69], produces a population transfer in Λ systems (see Fig. 7a). The pump field couples the transition 1–2, and the Stokes field couples the transition 2–3. It is known that, with the initial population in state $|1\rangle$, a complete population transfer is achieved with delayed pulses, either (i) with a so-called counterintuitive temporal sequence (Stokes pulse before pump) for various detunings as identified in Refs. 73 and 74 or (ii) with two-photon resonant (or quasi-resonant) pulses but far from the one-photon resonance with the intermediate state $|2\rangle$, for any pulse sequence (demonstrated in the approximation of adiabatic elimination of the intermediate state [75]). Here we analyze the STIRAP process through the topology of the associated surfaces of eigenenergies as functions of the two field amplitudes. Our results are also valid for ladder and V systems.

We also obtain the following results related to STIRAP: (i) We can explain the transfer of population to state $|3\rangle$ with intuitive (as well as with counterintuitive) specific quasi-resonant pulses *without invoking the approximation of adiabatic elimination.* (ii) With specific quasi-resonant pulses, we can *selectively* transfer the population to state $|2\rangle$ for an *intuitive* sequence or

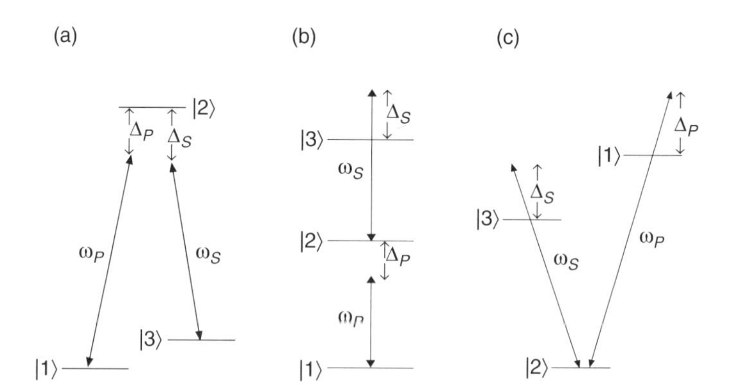

Figure 7. Diagram of linkage patterns between three atomic states showing pump (P) and Stokes (S) transitions and the various detunings for (a) Λ, (b) ladder, and (c) V systems.

to state $|3\rangle$ for a *counterintuitive* sequence, and (iii) with an intuitive or counterintuitive sequence, we can *selectively* transfer the population to state $|2\rangle$ or to state $|3\rangle$ playing on the *detunings* and on the *ratio of the peak pulse amplitudes*.

We also analyze the counterpart of the preceding processes in V systems (see Fig. 7c): The initial population being in state $|2\rangle$, we show that with specific nonresonant pulses, (i) we can *selectively* transfer the population to state $|1\rangle$ for an intuitive sequence or to state $|3\rangle$ for a counterintuitive sequence and (ii) we can *selectively* transfer the population to state $|1\rangle$ or to state $|3\rangle$ playing on the ratio of the peak pulse amplitudes.

The topology will allow us to classify all the possibilities of complete population transfer by adiabatic passage for a three-level system interacting with two delayed pulses, as it was done for the two-level system interacting with a chirped laser pulse.

The most general dressed Hamiltonian in the rotating wave approximation for these processes reads [69]

$$H(t) = \frac{\hbar}{2}\begin{bmatrix} 0 & \Omega_P(t) & 0 \\ \Omega_P(t) & 2\Delta_P & \Omega_S(t) \\ 0 & \Omega_S(t) & 2(\Delta_P - \Delta_S) \end{bmatrix} \tag{290}$$

where $\Omega_j(t)$, $j = P, S$ (assumed real and positive), are the one-photon Rabi frequencies associated respectively to the pump pulse (of carrier frequency ω_P) and the Stokes pulse (of carrier frequency ω_S). We have assumed that the states $|1\rangle$ and $|3\rangle$ have no dipole coupling and that spontaneous emission is negligibly small on the time scale of the pulse duration. The rotating wave approximation is valid if $\hbar\Omega_P(t) \ll |E_2 - E_1|$ and $\hbar\Omega_S(t) \ll |E_3 - E_2|$, where E_j, $j = 1, 2, 3$, are the energies associated to the bare states $|j\rangle$.

The detunings Δ_P and Δ_S are one-photon detunings with respect to the pump and Stokes frequencies respectively and

$$\delta = \Delta_P - \Delta_S \tag{291}$$

is the two-photon detuning.

For Λ, ladder and V systems (see, respectively, Figs. 7a, 7b, and 7c), the one-photon detunings Δ_P, Δ_S are, respectively, defined as

$$\Lambda: \quad \hbar\Delta_P = E_2 - E_1 - \hbar\omega_P, \quad \hbar\Delta_S = E_2 - E_3 - \hbar\omega_S \tag{292a}$$

$$\text{Ladder:} \quad \hbar\Delta_P = E_2 - E_1 - \hbar\omega_P, \quad \hbar\Delta_S = E_2 - E_3 + \hbar\omega_S \tag{292b}$$

$$V: \quad \hbar\Delta_P = E_2 - E_1 + \hbar\omega_P, \quad \hbar\Delta_S = E_2 - E_3 + \hbar\omega_S \tag{292c}$$

which determines the dressed basis in which the dressed Hamiltonian (290) has been written: respectively, $\{|1;0,0\rangle, |2;-1,0\rangle, |3;-1,1\rangle\}$, $\{|1;0,0\rangle, |2;-1,0\rangle, |3;-1,-1\rangle\}$, and $\{|1;0,0\rangle, |2;1,0\rangle, |3;1,-1\rangle\}$. In the following, we consider the population of the atomic states.

In what follows we study the topology of the eigenenergy surfaces for various generic sets of the parameters. The topology depends on the detunings that determine the relative position of the energies at the origin. We study various *quasi-resonant* pulses in the sense that the detunings are small then or of the order of the associated peak Rabi frequencies, that is,

$$\Delta_P \lesssim \max_t(\Omega_P), \qquad \Delta_S \lesssim \max_t(\Omega_S), \tag{293a}$$

$$\delta \lesssim \max_t(\Omega_P), \qquad \delta \lesssim \max_t(\Omega_S) \tag{293b}$$

1. *Transfer to a Unique State*

Allowing large enough amplitudes leads to three generic cases for $\delta_2 > 0$ and three other cases for $\delta_2 < 0$, which are equivalent by symmetry.

Two typical examples are displayed in Figs. 8 and 9. In both cases, the surface continuously connected to the state $|2\rangle$ is isolated from the two other surfaces that present a conical intersection for $\Omega_S = 0$ and for $\Omega_P = 0$, respectively. This crossing corresponds to a mute resonance as described above for chirping. (A resonance is called mute if the frequencies are in resonance but the corresponding coupling term is zero or small in the Hamiltonian.)

The topologies shown in the respective Figs. 8 and 9 are generic for the condition

$$\Delta_P \Delta_S > 0 \tag{294}$$

with, respectively,

$$|\Delta_P| < |\Delta_S| \quad \text{and} \quad |\Delta_P| > |\Delta_S| \tag{295}$$

In the following, we describe in detail the case of Fig. 8. For the process in Λ or ladder systems, where the initial population resides in state $|1\rangle$, two different adiabatic paths lead to the complete population transfer, depending on the pulse sequence. The path denoted (a) corresponds to an intuitive sequence for the rise of the pulses. The pump pulse is switched on first, making the levels connected to the states $|1\rangle$ and $|2\rangle$ repel each other (dynamical Stark shift) until the level connected to $|1\rangle$ crosses the level connected to $|3\rangle$. The Stokes pulse is switched on after the crossing. Next the two pulses can decrease in any order. Path (b) is associated to a counterintuitive sequence for the decrease of the pulses. The two

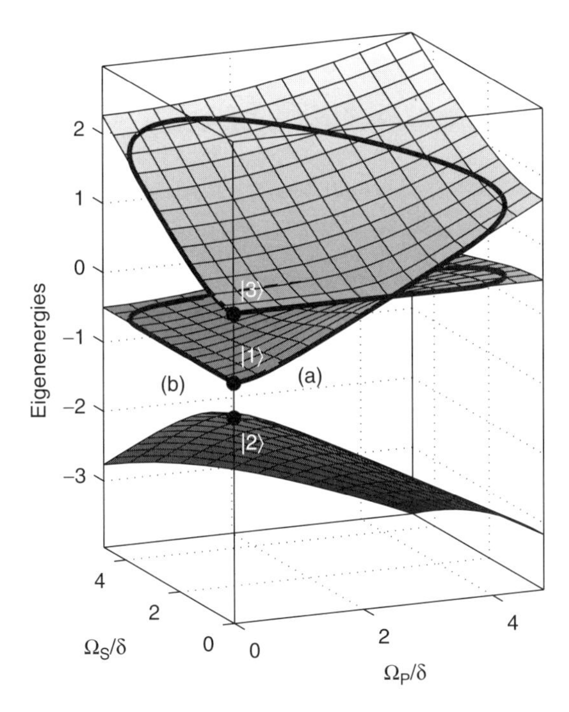

Figure 8. Surfaces of eigenenergies (in units of δ) as functions of Ω_P/δ and Ω_S/δ when the dressed states (denoted λ_1, λ_2, and λ_3, respectively connected to E_1, E_2, and E_3 for fields off) are such that $\lambda_2 < \lambda_1 < \lambda_3$, with $\Delta_p = -\delta/2$ and $\Delta_S = -3\delta/2$. Paths (a) and (b) (constructed with delayed pulses of the same length and peak amplitude) correspond respectively to the intuitive and counterintuitive pulse sequences in Λ or ladder systems (for which the initial population resides in state $|1\rangle$).

pulses can be switched on in any order. The pump pulse has to decrease through the crossing when the Stokes pulse is already off. These two results are valid even without application of adiabatic elimination. The conditions of global adiabaticity are similar to the ones of the chirped frequency case (242). As discussed in Section IV.B.4, an analysis of the diabatic evolution near the conical intersections can be made locally with the Landau–Zener approximation (270).

The V systems are uninteresting in these cases since the final population comes back to the state $|2\rangle$ for any pulse sequence.

Another typical example is displayed in Fig. 10. The topology shown on this figure is generic for the condition

$$\Delta_P \Delta_S < 0 \tag{296}$$

In this configuration, two conical intersections involve the intermediate surface, one with the lower surface and the other one with the upper surface. This topology gives here more possibilities for transfer: *The combined choice of the pulse sequence and the ratio of the peak amplitudes allows the selective transfer into the two other states.*

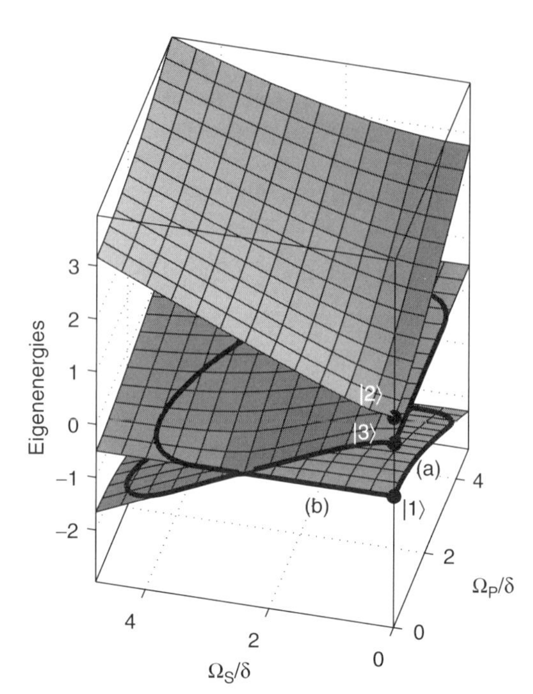

Figure 9. Surfaces of eigenenergies (in units of δ) as functions of Ω_P/δ and Ω_S/δ for the case $\lambda_1 < \lambda_3 < \lambda_2$, with $\Delta_P = 3\delta/2$ and $\Delta_S = \delta/2$. Paths (a) and (b) (with pulses of the same length and peak amplitude) correspond respectively to the intuitive and counterintuitive pulse sequences in Λ or ladder systems.

Figure 10 shows that, for the process in Λ (or ladder) systems, two different adiabatic paths lead to different complete population transfers, depending on the pulse sequence. Path (a) corresponds to an intuitive pulse sequence (for the decrease of the pulses) and allows pulses to populate at the end the state $|2\rangle$. The Stokes and pump pulses can be switched on in any sequence and the pump pulse is switched off before the Stokes one. The path (b) corresponds to a counterintuitive pulse sequence (for the rise of the pulses) and allows pulses to populate at the end the state $|3\rangle$. The Stokes pulse is switched on before the pump, and the stokes pulse has to be switched off before the pump. We can thus selectively populate the states $|2\rangle$ or $|3\rangle$, provided that the peak amplitudes are sufficiently strong to induce the adiabatic path to cross the intersection involved.

For the process in V systems, paths (a) and (c) of Fig. 10 show the respective selective transfer into the states $|1\rangle$ or $|3\rangle$.

Figure 11 corresponds to the same topology of Fig. 10 but with a different path (a). Figure 11 shows that, for Λ (or ladder) systems, we can selectively populate the states $|2\rangle$ or $|3\rangle$ if the pulse sequences are designed differently in their order and their peak amplitude. The path (b) corresponds to path (b) of Fig. 10 and allows pulses to populate at the end the state $|3\rangle$. Path (a)

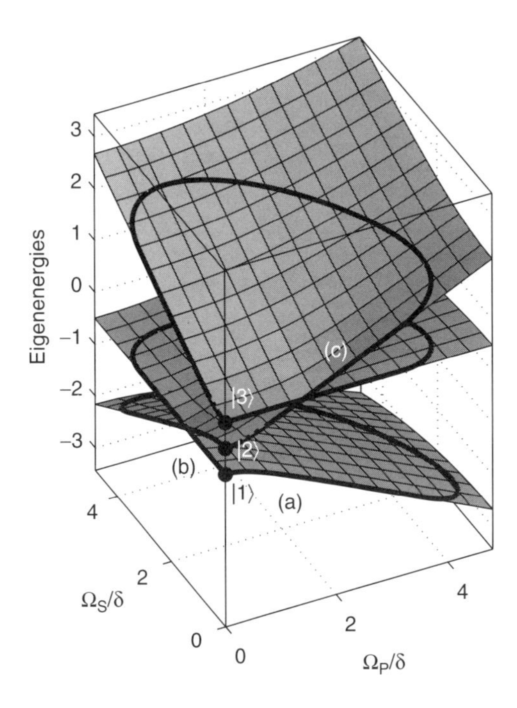

Figure 10. Surfaces of eigenenergies (in units of δ) as functions of Ω_P/δ and Ω_S/δ for the case $\lambda_1 < \lambda_2 < \lambda_3$, with $\Delta_P = \delta/2$ and $\Delta_S = -\delta/2$. Paths (a) and (b) (with pulses of the same length and peak amplitude) correspond respectively to the intuitive (transfer to $|2\rangle$) and counterintuitive (transfer to $|3\rangle$) pulse sequences in Λ or ladder systems leading to the selective transfer. Paths (a) and (c) correspond to the selective transfer in V systems (for which the initial population resides in $|2\rangle$), respectively to $|1\rangle$ and $|3\rangle$.

corresponds to a longer pump pulse (still switched on after the Stokes pulse) of smaller peak amplitude which allows pulses to populate at the end the state $|2\rangle$. Note that we can obtain a similar path (a) with a counterintuitive pulse sequence and equal peak amplitudes if the detuning Δ_P is taken smaller so that the crossing for $\Omega_S = 0$ is pushed to a higher pump pulse amplitude Ω_P.

For V systems, Fig. 11 shows that this selectivity [paths (a) and (c)] also occurs (for any sequence of the pulses).

Figure 12 shows numerical calculations that illustrate some of the predictions of the above analysis. It displays the populations of the states $|2\rangle$ and $|3\rangle$ at the end of the pulses for intuitive and counterintuitive sequences with a large pulse area. The boundaries of the areas of efficient transfer (black areas) are predicted quite accurately by the topology analysis: They are determined by (i) the straight lines (thick full lines) $\Delta_P = 0$ and $\Delta_S = 0$ coming from the inequalities (294) and (296) and (ii) the branches of the hyperbolas (dashed lines)

$$\Delta_S = \Delta_P - \frac{(\Omega_{\max})^2}{4\Delta_P} \tag{297a}$$

$$\Delta_P = \Delta_S - \frac{(\Omega_{\max})^2}{4\Delta_S} \tag{297b}$$

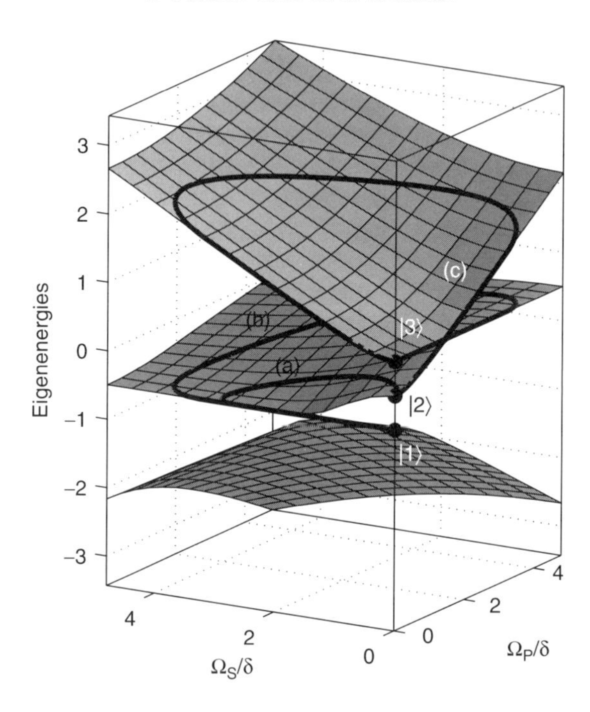

Figure 11. Surfaces of eigenenergies (in units of δ) with the same parameters as Fig. 10 showing the selective transfer with pulses of different peak amplitudes and length. For counterintuitive sequences in Λ or ladder systems, path (b) [corresponding to path (b) of Fig. 10] shows the transfer to $|3\rangle$, and path (a) (with pulses of different length and peak amplitude) characterizes the transfer to $|2\rangle$. Paths (a) and (c) correspond to the selective transfer in V systems.

which are determined from the positions of the conical intersections. Figure 12 shows that the efficiency of the robust population transfer to the states $|2\rangle$ or $|3\rangle$ is identical for the intuitive and counterintuitive sequences except in two regions: (i) areas bounded by $\Delta_P\Delta_S < 0$ and the branches of the hyperbolas, where the population is transferred in a robust way to state $|2\rangle$ for the intuitive sequence or to state $|3\rangle$ for the counterintuitive sequence, and (ii) an area (smaller for longer pulse areas) near the origin where *nonadiabatic effects* are strong for the intuitive sequence and where the population transfer depends precisely on the pulse areas of this intuitive sequence (see the comments below). *Nonadiabatic effects*, which are smaller for larger pulse areas, also occur near the straight-line boundary regions. *Nondiabatic effects* arise as well near the hyperbolic boundaries.

For the concrete realization with finite pulses of moderate areas, we have to analyze the precise influence of nonadiabatic and nondiabatic effects. We discuss here these nonadiabatic effects referring to Fig. 8, supposing that the

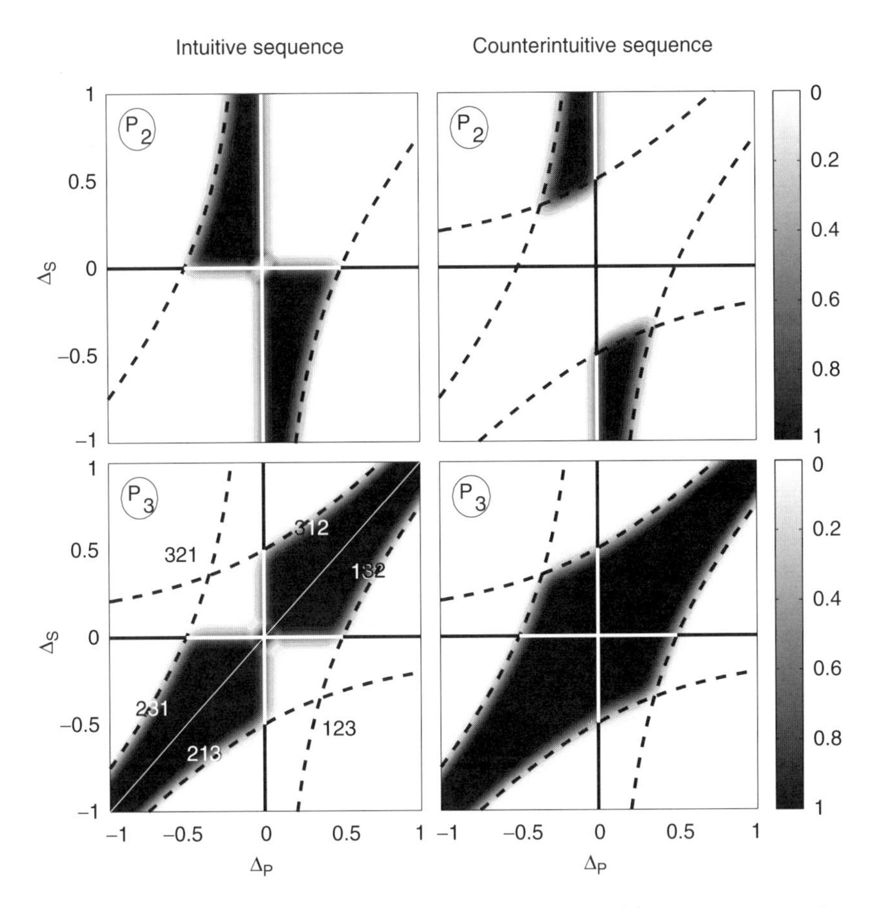

Figure 12. Transfer efficiencies P_2 to $|2\rangle$ (upper row) and P_3 to $|3\rangle$ (lower row) as functions of the detunings Δ_P and Δ_S (in units of $\Omega_{\max}$) at the end of the pulses for the intuitive (left column) and counterintuitive (right column) sequences of delayed sine-squared pulses eith the same peak amplitude $\Omega_{\max}$ and a large temporal area $\Omega_{\max}\tau = 500$ (τ is the pulse length and the delay is $\tau/2$). The efficient pupulation transfer are bounded by $\Delta_P = 0$ and $\Delta_S = 0$ (thick full lines) and the branches of hyperbolas (dashed lines). The areas bounded by the full lines are labeled by the cases 213, 132, 123, ... (These number sets are associated with the eigenenergies for zero field amplitudes from the smallest to the biggest; for example, 213 means $\lambda_2 < \lambda_1 < \lambda_3$.) The three first ones correspond respectively to Figs. 8, 9, and 10.

detunings are small enough with respect to the speed of the process to yield nonadiabatic transitions.

In the intuitive case, at the beginning of the process, the states $|1\rangle$ and $|2\rangle$ are coupled by the pump pulse, and thus nonadiabatic transitions can occur near the origin between the surfaces connected to $|1\rangle$ and $|2\rangle$. In the counterintuitive case, at the beginning of the process, state $|1\rangle$ is not coupled to the other levels and there are no nonadiabatic transitions near the origin. At the end of the process, the adiabatic path ending in $|3\rangle$ is not coupled to the other levels, implying again absence of nonadiabatic transitions near the origin. We thus recover the well-known fact that resonant STIRAP (with very small detunings) is more favorable with a counterintuitive pulse sequence and leads to Rabi oscillations in the intuitive case.

2. *Transfer to a Coherent Superposition of States*

We study the various superpositions of states that can be created by adiabatic passage in a robust way with respect to variations of the field amplitude, using the topological analysis with resonances of Section V.D (see also Section IV.B.3 for the case of one laser). We assume that one starts (at time $t = t_i$) with a coherent state for the photon field and in the atomic state $|1\rangle$. We study here the Λ-system. Our results are easily extended to the other system (ladder and V), using the appropriate signs accompanying the field frequencies. We study the creation of a superposition of states at the final time $t = t_f$.

This can be analyzed with the help of Fig. 8, where we have taken $\Delta_P = 0$. The intuitive pump-Stokes sequence induces first a lifting of degeneracy with equal sharing between the dressed states $|\psi_+\rangle$ (the upper one, associated to the eigenenergy λ_+) and $|\psi_-\rangle$ (the lower one, associated to the eigenenergy λ_-) initially connected to $|1\rangle$ and $|2\rangle$. If we assume that the peak pump field amplitude is *beyond* the conical intersection, then the branches $|\psi_-\rangle$ and $|\psi_+\rangle$, respectively, connect $-|2\rangle$ and $|3\rangle$ at the end. When $\Delta_S < 0$ (as in Fig. 8), this leads at the end of the process to the coherent superposition with a dynamical phase (up to an irrelevant global phase)

$$|\psi(t_f)\rangle = \frac{1}{\sqrt{2}}\left[|3\rangle - e^{i\int_{t_i}^{t_f} ds(\lambda_+(s)-\lambda_-(s))} e^{-i\omega_S(t_i-t_f)}|2\rangle\right] \tag{298}$$

When $\Delta_S > 0$, we obtain

$$|\psi(t_f)\rangle = \frac{1}{\sqrt{2}}\left[|3\rangle + e^{-i\int_{t_i}^{t_f} ds(\lambda_+(s)-\lambda_-(s))} e^{-i\omega_S(t_i-t_f)}|2\rangle\right] \tag{299}$$

The counterintuitive Stokes-pump sequence leads to the transfer to the unique state $|3\rangle$.

If we assume that the peak pump field amplitude is *below* the conical intersection, then the branches $|\psi_-\rangle$, and $|\psi_+\rangle$, respectively, connect $-|2\rangle$ and $|1\rangle$ at the end. This leads to coherent superpositions between the states $|1\rangle$ and $|2\rangle$.

If we have additionally $\Delta_S = 0$, the counterintuitive sequence gives the standard STIRAP (transfer to state $|3\rangle$) and the intuitive sequence induces interferences of the branches at the end of the processes, which do not lead to robust superposition of states.

We now analyze Fig. 9, where we assume $\Delta_S = 0$. When $\Delta_P > 0$ (as in Fig. 9), this leads at the end of the process to the coherent superposition without dynamical phase (but still with the optical phase):

$$|\psi(t_f)\rangle = \frac{1}{\sqrt{2}}[|3\rangle - e^{-i\omega_S(t_i - t_f)}|2\rangle] \tag{300}$$

When $\Delta_S > 0$, we obtain

$$|\psi(t_f)\rangle = \frac{1}{\sqrt{2}}[|3\rangle + e^{-i\omega_S(t_i - t_f)}|2\rangle] \tag{301}$$

The topological analysis thus shows that *with two quasi-resonant delayed lasers it is not possible to end in a superposition of states between the lowest states* $|1\rangle$ *and* $|3\rangle$ *in a robust way.* We can remark that in Ref. 76, it has been shown that one can create such a superposition—however, in a nonrobust way but still by adiabatic passage, by modifying the end of the STIRAP process (with the counterintuitive sequence), maintaining a fixed ratio of Stokes and pump pulse amplitudes.

The numerical calculations of Fig. 12 show the predicted superpositions of states at $\Delta_P = 0$ and $\Delta_S = 0$. They also show that a final superposition between $|1\rangle$ and $|3\rangle$ is possible on some pieces of the hyperbolas (dashed lines). However, they are not robust since the equation of these hyperbolas (297) depend on the peak field amplitudes.

VI. STATE-SELECTIVITY BY BICHROMATIC PULSES

The tools described in the preceding sections have been applied to investigate different kinds of control processes, such as the control of tunneling [62,77,78], the vibrational state selectivity in molecules (see, e.g., Refs. 47 and 61) up to its dissociation in the ground electronic state [79], the alignment [32,41] and orientation [42,43] of molecules, the deflection of atomic beams [67,80], and the creation of entangled states in two coupled spins 1/2 [81,82]. These processes are based on the possibility to select a given dressed state in the system during or after the laser pulses.

In this section we will describe the application of these methods to some examples of population transfer by delayed bichromatic pulses in two- and three-level systems. Bichromatic effects with CW lasers in population trapping have been also investigated in Ref. 83.

We study processes with two fields of different carrier frequencies ω_1 and ω_2 which act in resonance (or in quasi-resonance) on the same atomic transition, which are referred to as *bichromatic processes*. They induce dynamical resonances in the system due to the beat frequency

$$\delta = \omega_1 - \omega_2 \tag{302}$$

We consider Hamiltonians of the form

$$H(\boldsymbol{\omega} t + \boldsymbol{\theta}) = H_0 - \mathbf{d} \cdot \left[\sum_{j=1}^{2} \mathbf{e}_j \mathscr{E}_j(t) \cos(\omega_j t + \theta_j) \right] \tag{303}$$

where H_0 is the Hamiltonian of the free atomic system of energies E_ℓ, $\ell = 1, \ldots, N$ associated to the states $\{|\ell\rangle\}$ spanning the Hilbert space $\mathscr{H} = \mathbb{C}^N$ on which H_0 and the dipole moment operator $\mathbf{d}$ act. The total electric field, containing two carrier frequencies $\boldsymbol{\omega} = (\omega_1, \omega_2)$, is characterized by unit polarization vectors $\mathbf{e}_j$, smooth pulse-shaped envelope functions of time $\boldsymbol{\mathscr{E}}(t) = [\mathscr{E}_1(t), \mathscr{E}_2(t)]$, and the initial phases $\boldsymbol{\theta} = (\theta_1, \theta_2)$. The interaction is thus characterized by the time-dependent Rabi frequencies $\Omega_j^{(m\ell)}(t) = -\langle m | \mathbf{d} \cdot \mathbf{e}_j | \ell \rangle \mathscr{E}_j(t)/\hbar$, $j = 1, 2$ when the frequency ω_j is quasi-resonant between the states $|m\rangle$ and $|\ell\rangle$. One additionally assumes that the fields are weak enough such that $|\hbar\Omega_j^{(m\ell)}(t)| \ll |E_\ell - E_m|$ for all times, meaning that the nonresonant terms can be neglected. The fields are, however, sufficiently strong such that, for some $|m\rangle$ and $|\ell\rangle$, the peak Rabi frequency is comparable to the beat frequency: $\max_t |\Omega_j^{(m\ell)}(t)| \sim |\delta|$. The resonant terms with respect to the frequency difference δ will be kept, since they will produce dynamical resonances.

In the Floquet representation

$$K = -i\hbar\boldsymbol{\omega} \cdot \frac{\partial}{\partial \boldsymbol{\theta}} + H(\boldsymbol{\theta}) \tag{304}$$

the effective Hamiltonian will be derived with the following change of variables:

$$\begin{aligned} \theta &= \theta_1 - \theta_2 \\ \theta_a &= \theta_2 \end{aligned} \tag{305}$$

giving

$$\partial/\partial\theta_1 = \partial/\partial\theta$$
$$\partial/\partial\theta_2 = \partial/\partial\theta_a - \partial/\partial\theta$$

and

$$K = -i\hbar\omega_2 \frac{\partial}{\partial\theta_a} - i\hbar\delta\frac{\partial}{\partial\theta} + H(\boldsymbol{\theta}) \tag{306}$$

We will apply specific rotating wave transformations R that will allow us to identify resonant terms and to eliminate the nonresonant ones. We obtain an effective one-mode Floquet Hamiltonian of the form

$$K_{\text{eff}} = -i\hbar\delta\frac{\partial}{\partial\theta} + H_{\text{eff}}(\theta) \simeq R^{\dagger}KR \tag{307}$$

that will take into account nontrivial resonant bichromatic effects with respect to the frequency δ. Although the field intensities are moderate, the system exhibits dynamical resonances that in the case of a single laser are usually encountered only in a strong-field regime.

The dynamics of the dressed atom is determined by the effective one-mode time-dependent Schrödinger equation

$$i\hbar\frac{\partial}{\partial t}\psi(\theta, t) = K_{\text{eff}}(t)\psi(\theta, t) \tag{308}$$

where $\psi(\theta, t)$ is an N-element column vector. The exact solution will thus be approximated by $R\psi(\theta, t)$.

These bichromatic resonances and their consequences will be studied in two- and three-level systems.

A. Two-Level Systems—Topological Quantization of Atomic Beam Deflection

1. *The Effective Hamiltonian*

We consider a two-level system ($E_2 > E_1$) [67] driven by two quasi-resonant fields of detunings $\Delta_j \equiv (E_2 - E_1)/\hbar - \omega_j$, $j = 1, 2$, as depicted in Fig. 13. We assume that the dipole moment couples only the two levels: $\langle 1|\mathbf{d}\cdot\mathbf{e}_j|1\rangle = \langle 2|\mathbf{d}\cdot\mathbf{e}_j|2\rangle = 0$. The two characteristic Rabi frequencies denoted $\Omega_j(t) =$

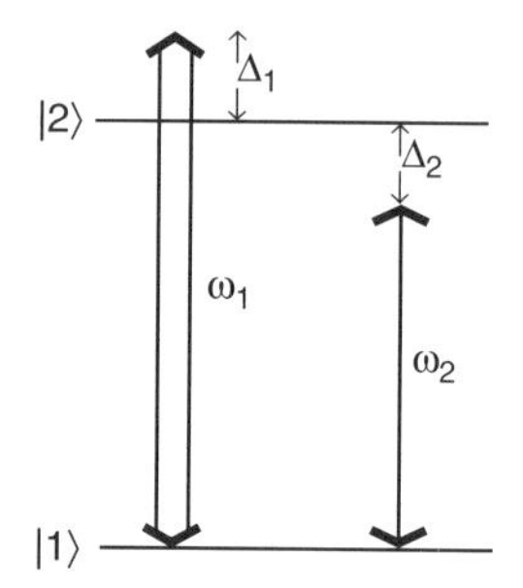

Figure 13. Diagram of linkage patterns between two atomic states.

$-\langle 1|\mathbf{d}\cdot\mathbf{e}_j|2\rangle\mathscr{E}_j(t)/\hbar$, $j=1,2$, involve the same transition 1–2. In the basis $\{|1\rangle,|2\rangle\}$, the Hamiltonian reads

$$H(\boldsymbol{\theta})=\begin{bmatrix}E_1 & 0\\ 0 & E_2\end{bmatrix}+\hbar(\Omega_1(t)\cos\theta_1+\Omega_2(t)\cos\theta_2)\begin{bmatrix}0 & 1\\ 1 & 0\end{bmatrix} \tag{309}$$

Here the beat frequency is $\delta\equiv\omega_1-\omega_2=\Delta_2-\Delta_1$. We study the nonperturbative regime $|\delta|\lesssim\max_t|\Omega_j(t)|\ll(E_2-E_1)/\hbar$, $j=1,2$.

We use the rotating wave transformation dressing the state $|2\rangle$ with minus one ω_1 photon:

$$R=\begin{bmatrix}1 & 0\\ 0 & e^{-i\theta_1}\end{bmatrix} \tag{310}$$

Applying this transformation to the Floquet Hamiltonian (304) gives

$$\begin{aligned}R^\dagger KR=&-i\hbar\boldsymbol{\omega}\cdot\frac{\partial}{\partial\boldsymbol{\theta}}+\frac{\hbar}{2}\begin{bmatrix}0 & \Omega_1\\ \Omega_1 & 2\Delta_1\end{bmatrix}+\frac{\hbar\Omega_2}{2}\begin{bmatrix}0 & e^{-i(\theta_1-\theta_2)}\\ e^{i(\theta_1-\theta_2)} & 0\end{bmatrix}\\&+\frac{\hbar\Omega_1}{2}\begin{bmatrix}0 & e^{-2i\theta_1}\\ e^{2i\theta_1} & 0\end{bmatrix}+\frac{\hbar\Omega_2}{2}\begin{bmatrix}0 & e^{-i(\theta_1+\theta_2)}\\ e^{i(\theta_1+\theta_2)} & 0\end{bmatrix}\end{aligned} \tag{311}$$

which can be approximated, after the transformation (305), by the effective Floquet Hamiltonian [80]

$$K_{\text{eff}}=-i\hbar\delta\frac{\partial}{\partial\theta}+\frac{\hbar}{2}\begin{bmatrix}0 & \Omega_1\\ \Omega_1 & 2\Delta_1\end{bmatrix}+\frac{\hbar\Omega_2}{2}\begin{bmatrix}0 & e^{-i\theta}\\ e^{i\theta} & 0\end{bmatrix} \tag{312}$$

Since $-i\hbar\omega_2\partial/\partial\theta_a$ is decoupled from the rest of the Floquet Hamiltonian, it acts trivially and can be omitted. This effective model is valid only if two different frequencies are assumed. The derivative term represents the relative number of photon pairs, one ω_1 photon minus one ω_2 photon. Thus the absorption of one "effective photon" of frequency δ in the effective model (312) corresponds in the complete model (304) to the absorption of one photon of frequency ω_1 and the emission of one photon of frequency ω_2. If the two laser fields are counterpropagating, perpendicularly to the atomic beam, this double photon exchange results in a net transfer of momentum to the atom of $\hbar(\omega_1 + \omega_2)/c$ which manifests as a deflection of the beam.

The second term of the effective Hamiltonian (312) is the usual RWA Hamiltonian (associated to the ω_1 field), with eigenvalues $2\lambda_{\pm}^0 = \hbar\Delta_1 \pm \hbar\sqrt{(\Delta_1)^2 + (\Omega_1)^2}$. The third term can be viewed as a perturbation of this RWA Hamiltonian.

The analysis of the dynamics consists of (i) the calculation of the dressed eigenenergy surfaces of the effective quasienergy operator as a function of the two Rabi frequencies Ω_1 and Ω_2, (ii) the analysis of their topology, and (iii) the application of adiabatic principles to determine the dynamics of processes in view of the topology of the surfaces.

2. *Eigenenergy Surface Topology*

In the following, we will consider for simplicity the case $\Delta_1 = -\Delta_2$ so that $\delta = -2\Delta_1$. For frozen values of the two fields Ω_1 and Ω_2, we calculate dressed states and dressed energies by diagonalizing K_{eff}. The eigenelements can be labeled with two indices: One, denoted n, refers to the levels of the atom; and another one, denoted k, refers to the relative photon numbers. The index k stands for the number of the ω_1 photons absorbed and the number of ω_2 photons emitted. The eigenvalues and eigenvectors have the following property of periodicity:

$$\lambda_{n;k,-k} = \lambda_{n;0,0} + k\hbar\delta, \qquad |n; k, -k\rangle_{\text{eff}} = |n; 0, 0\rangle_{\text{eff}} \exp(ik\theta)$$

The eigenelements appear as two families, each of which consists of an infinite set of eigenvalues with equal spacing $\hbar\delta$. The eigenstates of K_{eff}, $|1; k, -k\rangle_{\text{eff}}$ and $|2; k, -k\rangle_{\text{eff}}$, can thus be labeled by $|1; k, -k\rangle$ and $|2; -1 + k, -k\rangle$, $k \in \mathbb{Z}$, in the original basis of (304). If one starts with the initial state $|n_i; 0, 0\rangle$, the state $|n_f; k_1, k_2\rangle$ at the end of the process will characterize the atom in the state $|n_f\rangle$ with emission of k_i photons of frequency ω_i if $k_i > 0$ or absorption of k_i photons if $k_i < 0$, $i = 1, 2$.

In Fig. 14, we display eigenenergy surfaces, calculated numerically, as functions of the scaled Rabi frequencies Ω_1/δ and Ω_2/δ, assumed positive

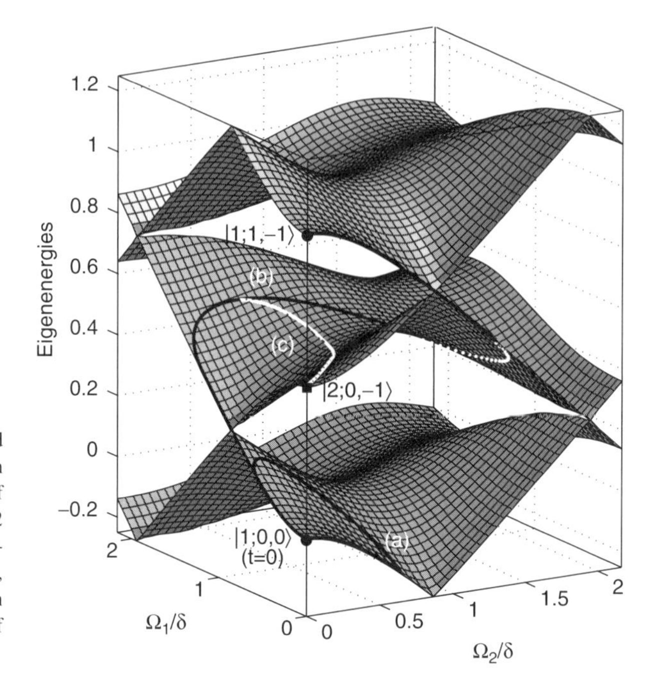

Figure 14. Dressed eigenenergy surfaces (in units of δ) as functions of Ω_1 and Ω_2 for $\delta = -2\Delta_1 = 2\Delta_2$. Three different paths [denoted (a), (b), and (c)] depending on the temporal evolution of the pulses are depicted.

without loss of generality. Together with the adiabatic analysis, the topology of these surfaces gives insight into the various atomic population and photon transfers that can be produced by choosing appropriately the temporal evolution of the pulses. The process starts in the dressed state $|1;0,0\rangle$—that is, the lowest atomic state with zero ω_1 and ω_2 photons. Its energy is shown as the starting point of various paths. We define the *transfer state* as the Floquet eigenvector that is adiabatically followed—that is, on which the dressed population resides during the dynamics. There are two infinite families of quasi-energy surfaces, which are constructed by the translations by $\hbar\delta k$, $k \in \mathbb{Z}$, of two surfaces. Any two neighboring surfaces have points of contact that are conical intersections. In the present model all the points of intersection are located either at the line $\Omega_1 = 0$ or at the line $\Omega_2 = 0$, corresponding to the situations where only one of the laser fields is interacting with the atom. Besides these true crossings, the quasi-energy surfaces display avoided crossings. These true crossing and avoided crossings are associated with dynamical resonances as we will show below.

Three kinds of adiabatic paths can occur in this topology, as shown in Fig. 14, which displays three examples of adiabatic paths leading to three different final atomic population and photon transfers. They are labeled (a), (b), and (c).

For curve (a), the shifts of the eigenvalues are smaller than the energy of the first intersections. As a consequence, the path stays on a single surface, and at the end the system it returns to the initial state, without any final transfer of photons nor of the atomic population.

Curve (b) corresponds to shifts that are larger than the first intersections. The crossing of the first intersection as Ω_1 increases with $\Omega_2 = 0$ brings the dressed system into the first upper quasi-energy surface. Turning on and increasing the amplitude Ω_2 (while Ω_1 decreases) moves the path across this surface. When the second field Ω_2 decreases, the curve crosses an intersection (with $\Omega_1 = 0$) that brings the system to the third-level surface, on which the curve stays until the end of the pulse Ω_2. The transfer state is finally connected to state $|1; 1, -1\rangle$: There is no transfer of atomic population, but one ω_2 photon has been absorbed and one ω_1 photon has been emitted at the end of the process. This path is produced with two delayed pulses of approximately the same peak amplitudes. Two dynamical resonances occur in this system. Each is crossed twice, appearing as one true crossing and one avoided crossing. This appears clearer in the temporal representation of the quasi-energies shown in Fig. 16b (see below for details of the dynamics). They can be described as follows: Field 1 dynamically shifts the eigenvalues that become resonant with field 2. This resonance is mute when field 2 is off (left true crossing) and becomes effective when field 2 is on (left avoided crossing). The second dynamical resonance occurs symmetrically from dynamical Stark shift due to field 2, which makes the eigenvalues resonant with the field 1. The topology shows the connection of the adiabatic paths related to the preceding dynamical resonances.

If the field amplitudes are taken even larger, such that two dynamical resonances are crossed (corresponding to the true crossings when Ω_1 rises with $\Omega_2 = 0$ and when Ω_2 decreases with $\Omega_1 = 0$), the final state is $|1; 2, -2\rangle$; that is, there is no atomic population transfer but an absorption of two ω_2 photons and an emission of two ω_1 photons. This path is shown in Fig. 15. This kind of process can be generalized to paths yielding the connectivity of the transfer state to $|1; k, -k\rangle$—that is, the emission of k ω_1 photons and the absorption of k ω_2 photons (k positive for pulse 1 before pulse 2 and negative for pulse 2 before pulse 1), with no atomic population transfer.

Path (c) in Fig. 14 involves a crossing of one conical intersection of the two described above. The first resonance is crossed by the rising pulse 1 (with $\Omega_2 = 0$). The second pulse is chosen with a smaller peak amplitude in order to avoid the passage through the resonance that would lead the system to the third level surface. This leads to an atomic population transfer, accompanied by absorption of one ω_2 photon, since the path ends at $|2; 0, -1\rangle$. This can be generalized for upper and lower paths: The connectivity leads to $|2; -1 + k, -k\rangle$, with k positive (pulse 1 before smaller pulse 2 amplitude) or negative or zero integer (pulse 2 before smaller pulse 1 amplitude).

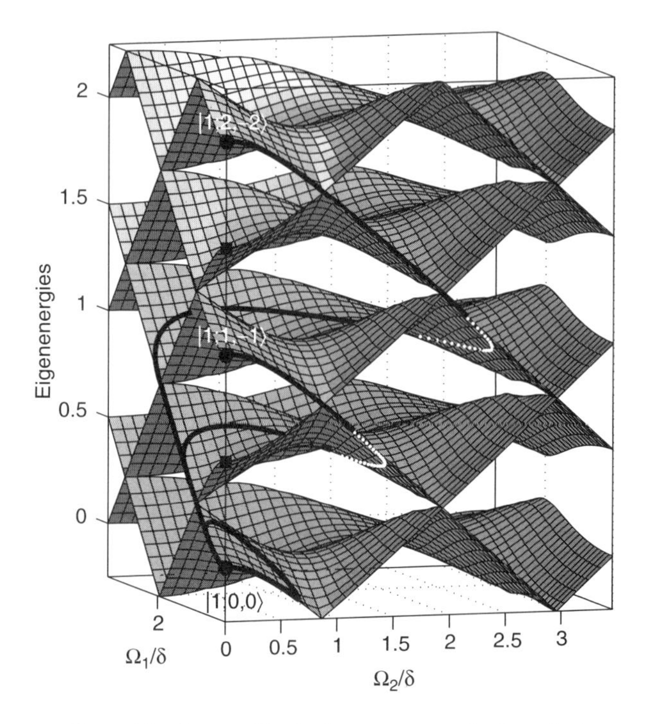

Figure 15. Same as Fig. 14, but for stronger field amplitudes.

The topology of the quasi-energy surfaces thus shows which appropriate delays and peak amplitudes induce desired atomic population and photon transfers. In the adiabatic regime, these loops can be classified into topologically inequivalent classes. If the evolution is adiabatic, all paths of a given class lead to the same end effect. This property underlies the robustness of the process.

3. *Analytical Construction of the Dressed Eigenenergies*

With the technique combining the rotating wave transformations and contact transformations developed in Section III.C, one can treat accurately the dynamical resonances and construct approximately the quasi-energies. If we take into account the first two dynamical resonances by appropriate RWTs [associated with path (b)], one obtains the following explicit expression for the dressed energy surfaces:

$$\frac{\lambda_{\pm,k}}{\hbar} = \frac{\Delta_1}{2} + k\delta \mp \left[\frac{1}{4}(\sqrt{A} - \delta)^2 + \frac{(\varepsilon^2\Omega_1\lambda_-^0)^2}{\hbar^2 A}\right]^{1/2} \tag{313}$$

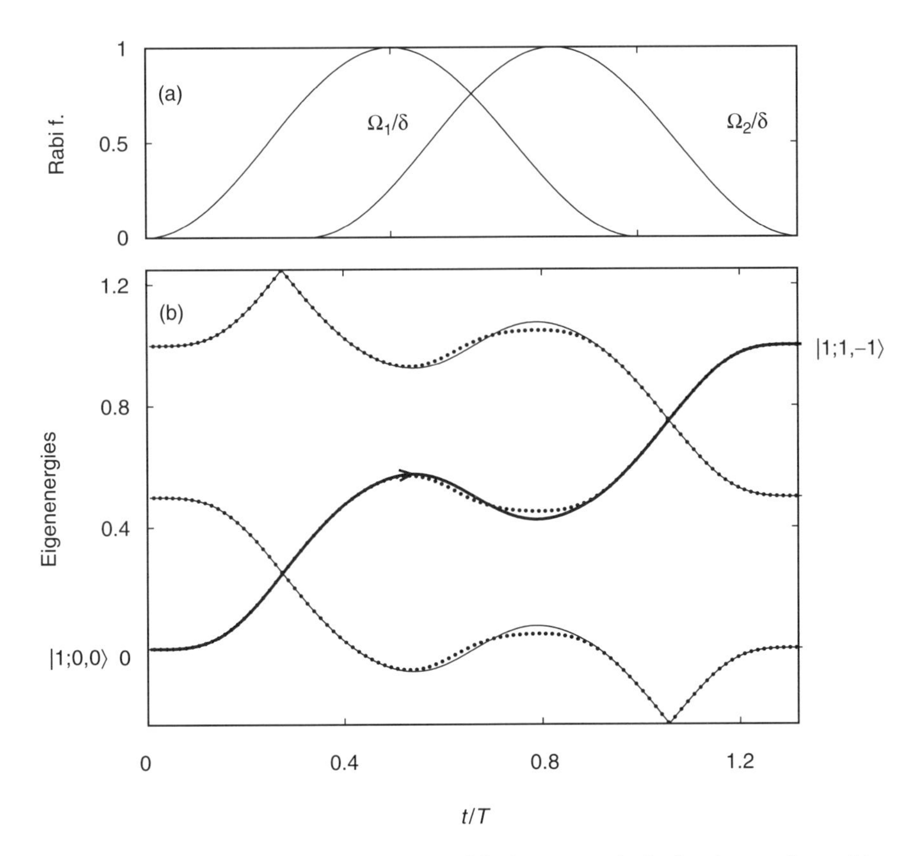

Figure 16. (a) Rabi frequencies (in units of δ) from squared trig function envelopes. (b) Dressed eigenenergy curves (in units of δ), corresponding to path (b) of Fig. 14 ($\Omega_{\max} = 1.5\delta$) from formula (313) (dotted lines) and exact numerical result (full line). The arrow indicates the adiabatic path (big line).

with $A = \{[(\Delta_1)^2 + (\Omega_1)^2]^{1/2} - \delta\}^2 + 4(\varepsilon\lambda_-^0/\hbar)^2$ and $2\varepsilon = -\Omega_2/\sqrt{(\Delta_1)^2 + (\Omega_1)^2}$. Figure 16b displays these eigenvalues as functions of time for the dynamics described below. They are in close agreement with the exact eigenvalues calculated numerically from the Hamiltonian (312). The explicit consideration of the small perturbative corrections from the full model (304) by contact transformations does not change the topology of the surfaces in the sense that the conical intersections are not removed but only slightly shifted.

This systematic method can also be applied to treat the next dynamical resonances occurring for higher field amplitudes.

4. *Dynamics and Topological Quantization of the Number of Exchanged Photons*

The path described above can be constructed by two smooth pulses, associated with the Rabi frequencies $\Omega_1(t)$ and $\Omega_2(t)$, with a time delay τ. To a sequence of such pulses corresponds a closed loop in the parameter plane Ω_1 and Ω_2. Each of the two black curves [labeled (a) and (b)] correspond to a sequence of two smooth pulses of equal length T and equal peak Rabi frequencies $\Omega_{\max} \equiv \max_t[\Omega_1(t)] = \max_t[\Omega_2(t)]$, separated by a delay such that the pulse 1 is switched on before the pulse 2. This path has been redrawn as a function of time on Fig. 16b, using $\sin^2$ envelopes of length $T = 100/\delta$ and a delay of $\tau = T/3$, shown in Fig. 16a. Details of this dynamics of bichromatic processes, in particular in relation with the initial condition for the photon field, are given and discussed in the next subsection. Path (c) needs two pulses with different peak amplitudes.

For equal peak amplitudes, we display in Fig. 17 the final average effective number k of exchanged photons as a function of the peak Rabi frequencies, calculated numerically by solving the dressed time-dependent Schrödinger equation. This shows the consequence of the topology described above. Since the connectivity of the transfer state to $|1; k, -k\rangle$ is based on the crossings, we can determine analytically the final number of effective photons k as a function of the peak Rabi frequencies $\Omega_{\max}/\delta$ (taken equal) in the purely adiabatic regime:

$$k = \text{Integer part of} \sqrt{(\Omega_{\max}/\delta)^2 + (\Delta_1/\delta)^2} \tag{314}$$

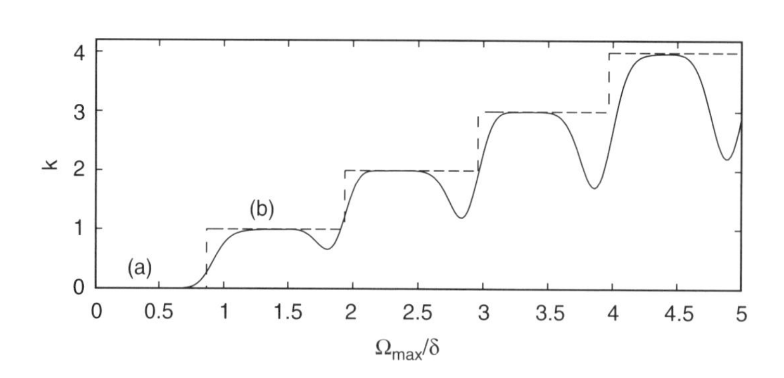

Figure 17. Comparison of the number k of effective photons emitted at the end of the process [Eq. (314)] (dashed line) with the average number of effective photons from the exact numerical result (full line). The plateaus labeled (a) and (b) refer to the two paths of Fig. 14 for pulse length $T = 100/\delta$ and delay $\tau = T/3$.

It predicts the adiabatic plateaus of Fig. 17, which can be interpreted as a topological quantization of the number of exchanged photons. The dips are due to nonadiabatic Landau–Zener transitions when the pulse overlap is in the neighborhood of the intersections. With a configuration of counterpropagating laser fields, perpendicular to an atomic beam, this translates into the possibility of deflection of the beam by the quantized transfer of a momentum $k\hbar(\omega_1 + \omega_2)/c$.

B. Three-Level Systems

The full semiclassical Hamiltonian (303) contains

$$H_0 = \begin{bmatrix} E_1 & 0 & 0 \\ 0 & E_2 & 0 \\ 0 & 0 & E_3 \end{bmatrix}, \qquad \mathbf{d} = \begin{bmatrix} 0 & \mathbf{d}_{12} & 0 \\ \mathbf{d}_{21} & 0 & \mathbf{d}_{23} \\ 0 & \mathbf{d}_{32} & 0 \end{bmatrix} \tag{315}$$

which are, respectively, the Hamiltonian of the free three-level system, acting on the Hilbert space $\mathscr{H} = \mathbb{C}^3$ spanned by the vector set $\{|1\rangle, |2\rangle, |3\rangle\}$, and the dipole moment operator (coupling transitions 1–2 and 2–3, but not 1–3). For simplicity, we take equal coupling for the transitions 1–2 and 2–3. The system is characterized by the time-dependent Rabi frequencies $\hbar\Omega_1(t) = -\langle 1|\mathbf{d}\cdot\mathbf{e}_1|2\rangle$, $\mathscr{E}_1(t) = -\langle 2|\mathbf{d}\cdot\mathbf{e}_1|3\rangle\mathscr{E}_1(t)$, and $\hbar\Omega_2(t) = -\langle 1|\mathbf{d}\cdot\mathbf{e}_2|2\rangle\mathscr{E}_2(t) = -\langle 2|\mathbf{d}\cdot\mathbf{e}_2|3\rangle\mathscr{E}_2(t)$. We consider the situation where the frequency ω_1 is one-photon quasi-resonant with the 1–2 transition, and the frequency ω_2 is one-photon quasi-resonant with the 2–3 transition. We study the intermediate field intensities regime

$$|\delta| \lesssim \max_t[|\Omega_1(t)|, |\Omega_2(t)|] \ll (E_2 - E_1)/\hbar, (E_3 - E_2)/\hbar \tag{316}$$

In particular we consider Λ-systems, depicted in Fig. 18, where the lasers 1 and 2 are, respectively, called *pump* and *Stokes* lasers. We use here resonant

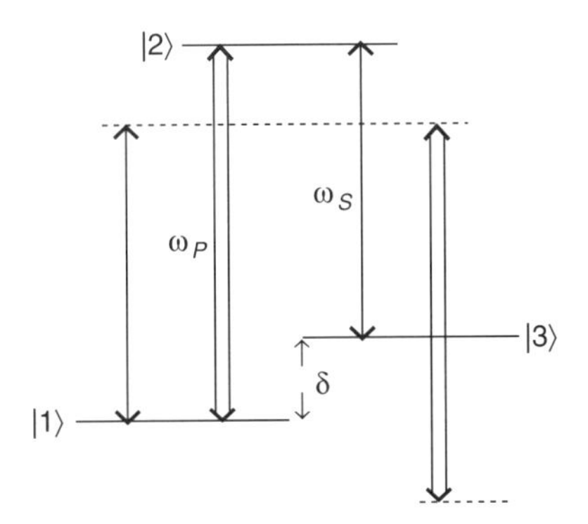

Figure 18. Diagram of linkage patterns between three atomic states (full horizontal lines), showing pump (P or 1) and Stokes (S or 2) laser frequencies.

frequencies $\hbar\omega_1 = E_2 - E_1$, $\hbar\omega_2 = E_2 - E_3$ so that the two-field combination maintains the two-photon resonance between the states $|1\rangle$ and $|3\rangle$, as for the usual STIRAP process [69]. Thus we have here $\delta \equiv \omega_1 - \omega_2 = (E_3 - E_1)/\hbar$. The main results of this part can be found in Refs. 38, 84, and 85.

1. The Effective Hamiltonian

To obtain the effective Floquet Hamiltonian, we apply the rotating wave transformation (RWT)

$$R(\boldsymbol{\theta}) = \begin{bmatrix} 1 & 0 & 0 \\ 0 & e^{-i\theta_1} & 0 \\ 0 & 0 & e^{i(\theta_2-\theta_1)} \end{bmatrix} \tag{317}$$

to obtain (setting $E_1 = 0$ as the reference of the energies)

$$R^\dagger K R = -i\hbar\boldsymbol{\omega} \cdot \frac{\partial}{\partial\boldsymbol{\theta}} + \frac{\hbar}{2}\begin{bmatrix} 0 & \Omega_1 & 0 \\ \Omega_1 & 0 & \Omega_2 \\ 0 & \Omega_2 & 0 \end{bmatrix} + V_1(\boldsymbol{\theta}) \tag{318}$$

with

$$V_1(\boldsymbol{\theta}) = \frac{\hbar}{2}\begin{bmatrix} 0 & \Omega_1 e^{-2i\theta_1} & 0 \\ \Omega_1 e^{2i\theta_1} & 0 & \Omega_2 e^{2i\theta_2} \\ 0 & \Omega_2 e^{-2i\theta_2} & 0 \end{bmatrix} + \frac{\hbar}{2}\begin{bmatrix} 0 & \Omega_2 e^{-i(\theta_1+\theta_2)} & 0 \\ \Omega_2 e^{i(\theta_1+\theta_2)} & 0 & \Omega_1 e^{i(\theta_1+\theta_2)} \\ 0 & \Omega_1 e^{-i(\theta_1+\theta_2)} & 0 \end{bmatrix}$$
$$+ \frac{\hbar}{2}\begin{bmatrix} 0 & \Omega_2 e^{-i(\theta_1-\theta_2)} & 0 \\ \Omega_2 e^{i(\theta_1-\theta_2)} & 0 & \Omega_1 e^{-i(\theta_1-\theta_2)} \\ 0 & \Omega_1 e^{i(\theta_1-\theta_2)} & 0 \end{bmatrix} \tag{319}$$

The usual RWA consists in neglecting the $\boldsymbol{\theta}$-dependent operator V_1. The first term of V_1 (319) contains the counterrotating terms of the pump laser on the 1–2 transition and of the Stokes laser on the 2–3 transition. The next two terms correspond to the interactions of the pump laser on the 2–3 transition and of the Stokes laser on the 1–2 transition. Following the hypothesis (316), we neglect the first two terms and keep the last term, which becomes large (see Ref. 38 for details) when $\max_t[|\Omega_1(t)|, |\Omega_2(t)|]$ approaches or overcomes $|\delta|$. The (approximate) effective one-mode Floquet Hamiltonian is thus

$$K_{\text{eff}} = -i\hbar\delta\frac{\partial}{\partial\theta} + \frac{\hbar}{2}\begin{bmatrix} 0 & \Omega_1 & 0 \\ \Omega_1 & 0 & \Omega_2 \\ 0 & \Omega_2 & 0 \end{bmatrix} + \frac{\hbar}{2}\begin{bmatrix} 0 & \Omega_2 e^{-i\theta} & 0 \\ \Omega_2 e^{i\theta} & 0 & \Omega_1 e^{-i\theta} \\ 0 & \Omega_1 e^{i\theta} & 0 \end{bmatrix} \tag{320}$$

The derivation term is the relative number operator for pairs of photons, one pump-field photon minus one Stokes-field photon. The second term is the well-known RWA Hamiltonian (dressed Hamiltonian used in the usual STIRAP), and the third one can be viewed as a perturbation of this RWA Hamiltonian.

We choose to have zero pump and Stokes photons at the beginning of the process. The initial condition is thus $|1;0,0\rangle$, which corresponds here to $|\psi(t=t_i)\rangle = |1;0\rangle$ for the effective one-mode Schrödinger equation (308). At each value of Ω_1 and Ω_2, the eigenvalues of K_{eff} can be decomposed as $\lambda_{n;-k,k} = \lambda_{n;0,0} - k\delta = \lambda_{n;0,0} - k\omega_1 + k\omega_2$ and their respective eigenvectors can be decomposed as $|n;-k,k\rangle_{\text{eff}} = |n;0,0\rangle_{\text{eff}} \exp[-ik\theta]$. The eigenstates of K_{eff} $|1;-k,k\rangle_{\text{eff}}$, $|2;-k,k\rangle_{\text{eff}}$, and $|3;-k,k\rangle_{\text{eff}}$ can thus be respectively labeled by $|1;-k,k\rangle$, $|2;-1-k,k\rangle$, and $|3;-1-k,k+1\rangle$, $k \in \mathbb{Z}$, in the original basis of (304). If one starts with the initial state $|n_i;0,0\rangle$, the state $|n_f;k_1,k_2\rangle$ at the end of the process will characterize the atom in the state $|n_f\rangle$ with emission of k_i photons of frequency ω_i if $k_i > 0$ or absorption of k_i photons if $k_i < 0$, $i = 1, 2$. The eigenvalues appear as three families with periodic replicas (with period $2\pi/\delta$) and yield one-mode Floquet zones that can interact each other.

2. *Eigenenergy Surface Topology*

In Fig. 19, we display quasi-energy surfaces, calculated numerically, as functions of the scaled Rabi frequencies Ω_1/δ and Ω_2/δ (assumed positive without loss of generality). The process starts in the dressed state $|1;0,0\rangle$—that is, the lowest atomic state with zero ω_1 and ω_2 photons. Its energy (which is zero in Fig. 19) is shown as the starting point of various paths. There are three infinite families of quasi-energy surfaces, constructed by the translations by $\hbar\delta k$, $k \in \mathbb{Z}$, of three surfaces. The surfaces exhibit conical intersections between two neighbors. In the present model all the points of intersection are located either at the line $\Omega_1 = 0$ or at the line $\Omega_2 = 0$, corresponding to the situations where only one of the laser fields is interacting with the atom. Besides these true crossings, the quasi-energy surfaces display avoided crossings. These crossings and avoided crossings are also associated with dynamical resonances of the same type as the ones shown above in the two-level system.

The lifting of the degeneracy in the Ω_1 or Ω_2 directions is the same as in the case of the usual STIRAP, since this lifting of degeneracy occurs for very small field intensities: The lifting of degeneracy is such that the state solution is adiabatically connected to $|1;0,0\rangle$ in the Ω_2 direction. We study below the conditions yielding complete population transfer from $|1\rangle$ to $|3\rangle$, for pulses in counterintuitive orders, that is, for delayed Ω_1 (pump pulse) switching on *after* Ω_2 (Stokes pulse).

Figure 19 shows that for $\max_t \Omega_1(t) \sim \max_t \Omega_2(t) < \delta$, we recover a STIRAP-type path (denoted as path a)—that is, connecting $|1;0,0\rangle$ to $|3;-1,1\rangle$. The creation of degeneracy is indeed such that the middle state (connected to the

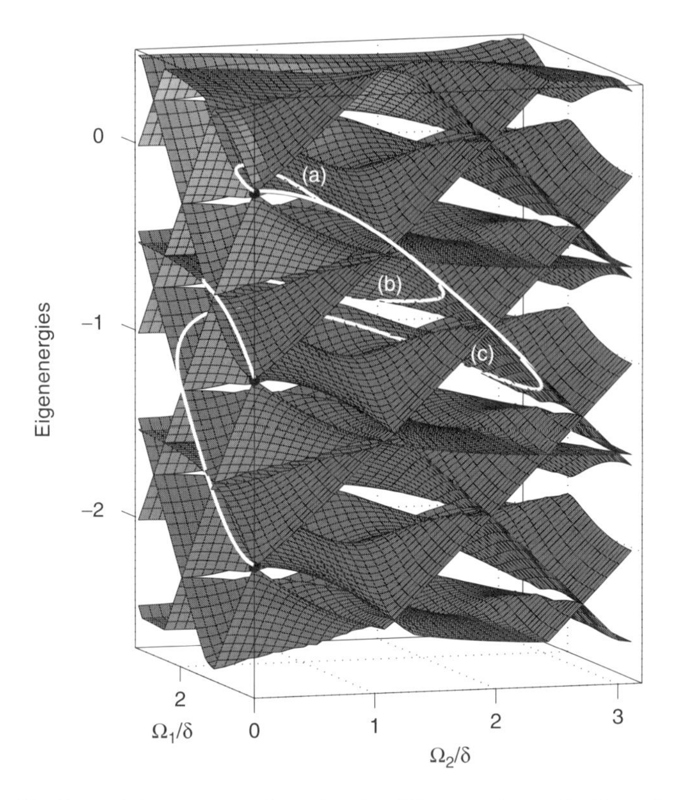

Figure 19. Dressed eigenenergy surfaces (in units of δ) as functions of Ω_1/δ and Ω_2/δ. Three characteristic paths are shown, all starting at the same state of energy zero and with a sequence of pulses Ω_1–Ω_2 of different peak amplitudes: Path (a), corresponding to a STIRAP-like process, whose dynamics is shown in Figs. 20 and 21, is of relatively small amplitude and makes a small loop in a single eigenenergy surface; path (b) gives a superposition of states $(|1;-1,1\rangle + |2;-2,1\rangle)/\sqrt{2}$ (of energy $-\delta$); and path (c) gives a STIRAP-like process accompanied by three ω_1 photons absorbed and three ω_2 photons emitted, whose dynamics is shown in Figs. 22 and 23.

energy zero) is adiabatically connected to $|3;-1,1\rangle$ in the Ω_1 direction. The upper and lower states (connected to the energy zero) are, respectively, connected to the superpositions of states: $(|1;0,0\rangle + |2;-1,\ 0\rangle)/\sqrt{2}$ and $(|1;0,0\rangle - |2;-1,0\rangle)/\sqrt{2}$, in the Ω_1 direction. In the case of path (a), the last term of the Hamiltonian (320) can be seen as a small perturbation of the RWA Hamiltonian of the standard resonant STIRAP. This has been studied in detail in Ref. 38. The effect of this perturbation is a distortion of the path (see Fig. 21b for a time evolution of this path). The dynamics associated to this path is studied below.

Increasing the intensity of Ω_2, we obtain a path [path (b) in Fig. 19] connecting $|1;0,0\rangle$ to the superposition of states $(|1;-1,1\rangle + |2;-2,1\rangle)/\sqrt{2}$ (of energy $-\delta$). Increasing again the intensity of Ω_2, we obtain a path [path (c)] connecting $|1;0,0\rangle$ to $|3;-3,3\rangle$ (of energy -2δ). This is similar to the usual STIRAP in the sense that this path allows transfer of the atomic population from $|1\rangle$ to $|3\rangle$—however, with the nontrivial effect of an absorption of three ω_1 photons and an emission of three ω_2 photons. For higher intensities of Ω_2 we can generalize the preceding connections. In summary, the topology shows two kinds of adiabatic connections: (i) from $|1;0,0\rangle$ to $(|1;-(2k+1),2k+1\rangle + |2;-(2k+2),2k+1\rangle)/\sqrt{2}$ and (ii) from $|1;0,0\rangle$ to $|3;-(2k+1),2k+1\rangle$, $k \geq 0$.

3. *Dynamics*

We study the dynamics for the complete transfer to state $|3\rangle$. The dynamics is considered either with the semiclassical Schrödinger equation

$$i\hbar\frac{\partial}{\partial t}\phi(t) = H_{\text{eff}}(t)\phi(t) \tag{321}$$

with the effective time-dependent Hamiltonian, constructed with K_{eff} (320)

$$H_{\text{eff}} = \frac{\hbar}{2}\begin{bmatrix} 0 & \Omega_1 + \Omega_2 e^{-i\delta t} & 0 \\ \Omega_1 + \Omega_2 e^{i\delta t} & 0 & \Omega_2 + \Omega_1 e^{-i\delta t} \\ 0 & \Omega_2 + \Omega_1 e^{i\delta t} & 0 \end{bmatrix} \tag{322}$$

or with the dressed Schrödinger equation (308) with the effective time-dependent quasienergy Hamiltonian (320). We recall that the semiclassical Schrödinger equation (321) is equivalent to the dressed Schrödinger equation (308) with a coherent state as the initial condition for the photon field. Studying the dressed Schrödinger equation (308) with a number state as the initial condition for the photon field allows to characterize the dynamics by each path considered above. It is important to note that in these examples *the information on the number of photons exchanged with the system, obtained from the calculation with the number state as the initial condition, is still valid at the end of the pulse for a coherent state as an initial condition.*

We consider two specific conditions, one for which the semiclassical and the dressed approaches are equivalent (which is the case for the STIRAP configuration) with respect to the number of photons exchanged at the end of the process, and another one for which the dressed theory brings the additional information of multiphoton processes.

To ensure that the interactions have a finite duration, we consider truncated $\sin^2$ envelopes. Time and frequency are scaled with respect to δ. The scaled pulse length is set to $T = 100/\Omega_0$ and the delay $\tau = 0.33\,T$. The pulses have to be applied in the so-called counterintuitive order: The ω_2 Stokes pulse precedes the ω_1 pump pulse with the delay τ. To fulfill the standard adiabatic condition, the relevant Rabi frequencies Ω have to be sufficiently large: $\Omega T \gg 1$.

For the parameters $\delta = 2\Omega_0$ and $\Omega_{\max} = \Omega_0$, corresponding to the path (a) on the surfaces in Fig. 19, we show in Fig. 20 the solution of the semiclassical Schrödinger equation (321). It features a STIRAP-like process inducing a complete population transfer for this choice of the delays. Two zones of the quasi-energy spectrum associated with the surfaces of Fig. 19 are pictured as a function of time in Fig. 21b. We notice that the state $|1;0,0\rangle$ is

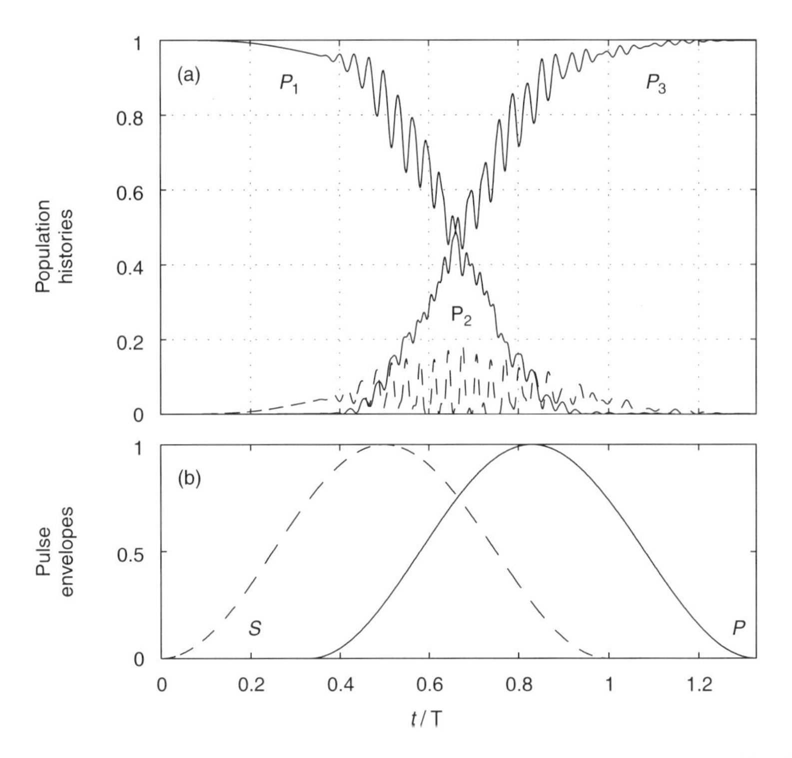

Figure 20. From the semiclassical Schrödinger equation. (a) Population histories $P_n(t)$ for $n = 1, 2, 3$ with $\delta = 2\Omega_0$ and $\Omega_{\max} = \Omega_0$ (top frame) and excitation by trig function pulse envelopes (of length $T = 100/\Omega_0$ and delay $0.33T$) with pump (full line) before Stokes (dashed line) shown in the bottom frame (b). Population transfer $P_3(\infty)$ to bare state $|3\rangle$ is nearly complete.

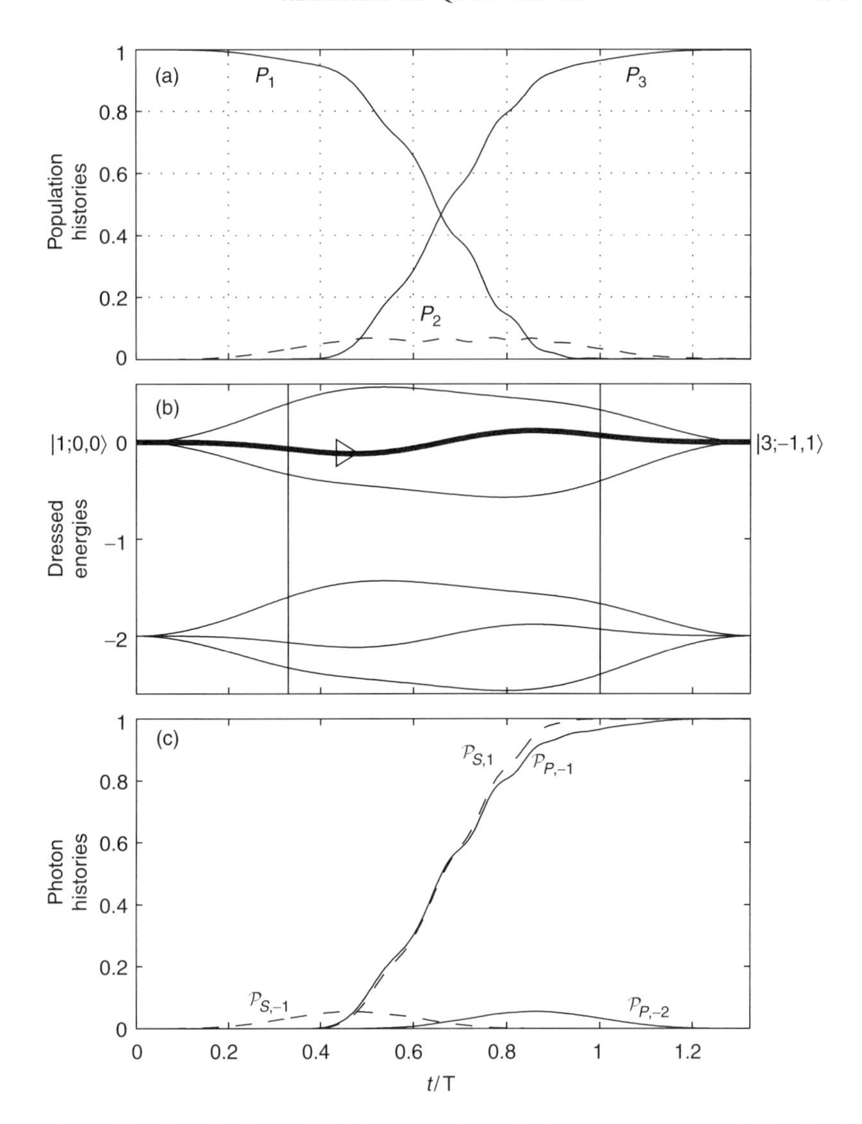

Figure 21. From the dressed Schrödinger equation, with the same parameters as in Fig. 20. (a) Population histories $P_n(t)$ for $n = 1, 2, 3$ (top frame), (c) photon histories (bottom frame), associated with the dressed spectrum in middle frame (b). The arrow characterizes the transfer eigenvector. Vertical lines indicate where the pump pulse starts and the Stokes pulse ends.

adiabatically connected to the final target state $|3; -1, 1\rangle$. This implies a complete population transfer from the bare state $|1\rangle$ to the bare state $|3\rangle$ with absorption of one pump photon and emission of one Stokes photon at the end of the process.

This is confirmed by the numerical solution of the dressed Schrödinger equation (308) with a number state as the initial condition for the photon field $|1;0,0\rangle$: It shows that the solution dressed state vector $\psi(t)$ (the transfer state, which in the bare basis is given by $R\psi(\theta,t)$) mainly projects on the transfer eigenvector during the process. Additional data of the dressed solution during time are shown in Fig. 21a and 21c. Figure 21a displays the probabilities of being in the bare states 1, 2, and 3:

$$P_n = \sum_{k_P,k_S} |\langle n;k_P,k_S|R|\psi(t)\rangle_{\mathscr{K}}|^2, \qquad n=1,2,3 \tag{323a}$$

$$= \frac{1}{(2\pi)^2}\int_0^{2\pi} d\theta_1 \int_0^{2\pi} d\theta_2\ |\langle n|R\psi(t)\rangle_{\mathscr{H}}|^2 \tag{323b}$$

Figure 21c shows the respective probabilities of one and two ω_1 pump photon absorption $\mathscr{P}_{P,-1}$, $\mathscr{P}_{P,-2}$, and of one ω_2 Stokes photon emission and absorption $\mathscr{P}_{S,1}$, $\mathscr{P}_{S,-1}$, defined with the respective formulas of the probabilities of ℓ ω_1 photons emissions and of ℓ ω_2 photons emissions

$$\mathscr{P}_{P,\ell} = \sum_{n,k_2} |\langle n;\ell,k_2|R\psi(t)\rangle_{\mathscr{K}}|^2 \tag{324a}$$

$$\mathscr{P}_{S,\ell} = \sum_{n,k_1} |\langle n;k_1,\ell|R\psi(t)\rangle_{\mathscr{K}}|^2 \tag{324b}$$

The other probabilities of photon emissions or absorptions are negligible. During the process, we remark that small transient ω_1 and ω_2 photon absorption probabilities arise. An early ω_2 photon absorption is observed, coinciding exactly with the (negative) shift of the transfer eigenvector. The first effects of the ω_2 pulse are indeed to (i) split the unpopulated dressed states connected to $|2\rangle$ and $|3\rangle$ and (ii) produce a Stark shift of the dressed state connected to $|1\rangle$ (the early part of the transfer state), which is equivalent to a partial absorption of a ω_2 photon. Symmetrically, a late ω_1 photon absorption occurs. It is due to a (positive) Stark shift of the dressed state connected to $|3\rangle$ (the late part of the transfer state). Arising near the end of the process, for which one ω_1 photon has already been absorbed, it leads to a partial absorption of a second ω_1 photon. At the end of the process the complete population transfer from state $|1\rangle$ to state $|3\rangle$ is accompanied by the loss of a ω_1 photon and the gain of a ω_2 photon. Thus the final result is not different from the semiclassical result.

Comparing Figs. 20a and 21a, we notice that, as expected, the solution of the dressed Schrödinger equation, with a number state as initial condition for the

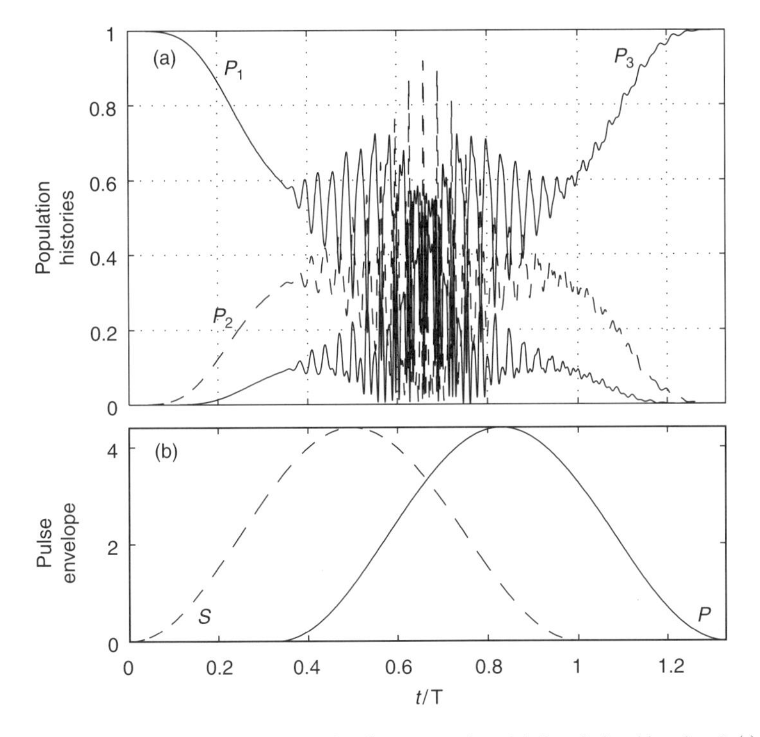

Figure 22. From the semiclassical Schrödinger equation. (a) Population histories $P_n(t)$ for $n = 1, 2, 3$ with $\delta = 2\Omega_0$ and $\Omega_{\max} = 4.4\Omega_0$ and (b) pulse excitation. Population transfer $P_3(\infty)$ to bare state $|3\rangle$ is nearly complete.

photon field, *averages* the solution of the semiclassical Schrödinger equation, with respect to the formula (323b).

We now study the situation when the detuning from the transition frequencies satisfies $\delta < \Omega_{\max}$ so that different Floquet zones cross.

To that effect, we choose the parameters $\delta = 2\Omega_0$ and $\Omega_{\max} = 4.4\,\Omega_0$, corresponding to the path (c) on the surfaces in Fig. 19. As shown in Fig. 22, the solution of the semiclassical Schrödinger equation (321) leads to nearly complete population transfer from state $|1\rangle$ to state $|3\rangle$. The analysis of the surfaces shows that the state $|1; 0, 0\rangle$ connects $|3; -3, 3\rangle$. Thus the complete population transfer from the bare state $|1\rangle$ to the bare state $|3\rangle$ must be accompanied with absorption of three pump photons and emission of three Stokes photons at the end of the process. This is confirmed by the numerical solution of the dressed Schrödinger equation (308) with the initial state as a number state for the photon field $|1; 0, 0\rangle$, shown in Fig. 23a: the dressed state vector $\psi(t)$ approximately projects on the transfer eigenvectors during the process. It shows the probabilities of being in the bare states 1, 2, and 3. Figure 23c shows the respective probabilities of one, two, three, and four ω_1 photon absorptions, of one ω_2 photon absorption and of one, two, and three ω_2 photon emissions, calculated with the formulas (324). The other probabilities of photon emissions

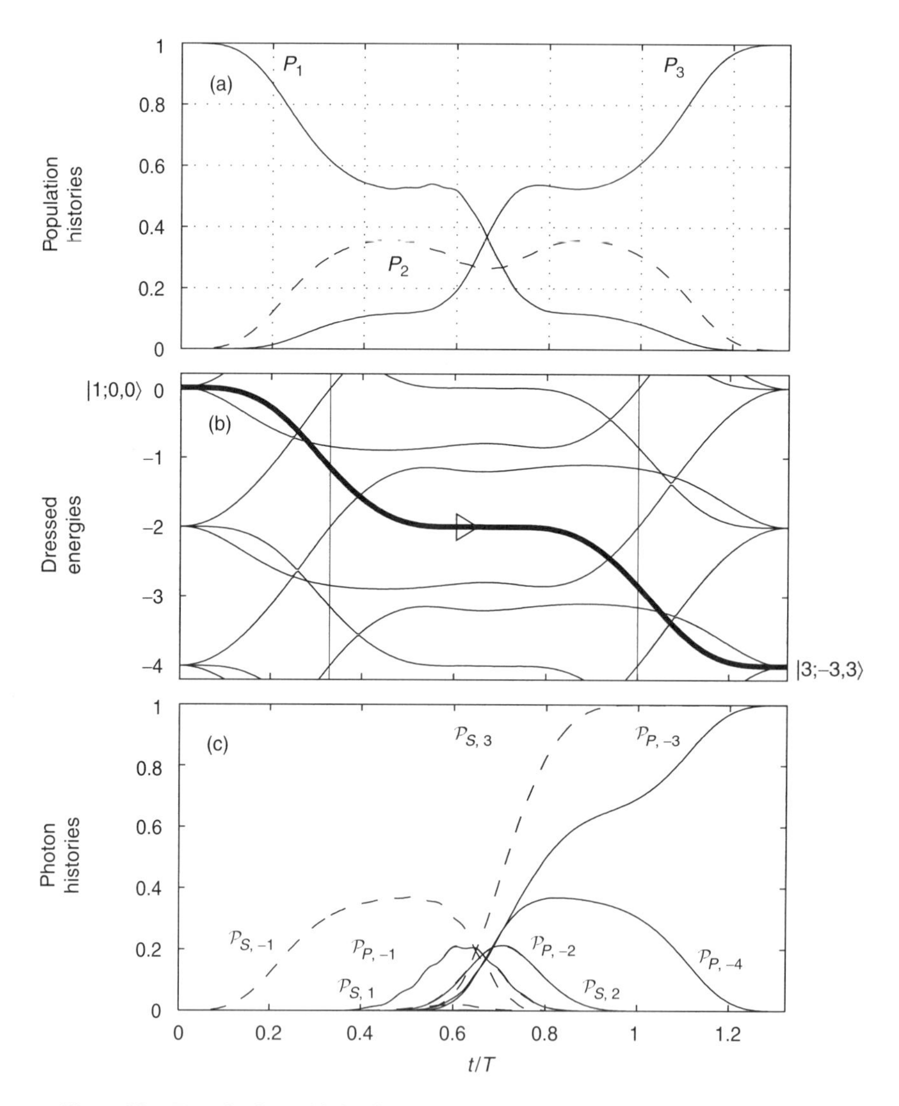

Figure 23. From the dressed Schrödinger equation, with the same parameters as in Fig. 22. (a) Population histories $P_n(t)$ for $n = 1, 2, 3$ (top frame) and (c) photon histories (with $\mathscr{P}_{P,-1} \approx \mathscr{P}_{S,1}$ and $\mathscr{P}_{P,-2} \approx \mathscr{P}_{S,2}$), associated with the dressed spectrum in middle frame (b).

or absorptions are negligible. As in the preceding case, we observe an early Stokes photon absorption and a late pump photon absorption characterizing Stark shifts of the dressed state connected to $|1\rangle$ and the one connected to $|3\rangle$, respectively. Moreover, in this case the field is so strong that it induces absorption (respectively emission) of one, two, and then three pump (respectively Stokes) photons. The complete population transfer from state $|1\rangle$ to state $|3\rangle$ is now accompanied by the loss of three ω_1 photons and the gain of three ω_2 photons at the end of the process.

APPENDIX A: MATHEMATICAL COMPLEMENTS

1. Relation Between the Semiclassical and the Floquet Evolution

We want to show the relation (13) of Section II.A in a more general setting in which the semiclassical Hamiltonian can have other time-dependent parameters, and in which the frequency can be chirped. This is the general setting needed for the treatment of adiabatic evolution with chirped pulses in Section IV.A, Eq. (223). We consider a semiclassical Hamiltonian of the form

$$H^{\mathbf{r}(\mathbf{t})}(\boldsymbol{\Theta}(t)) \tag{A1}$$

where $\mathbf{r}(\mathbf{t})$ represents a set of parameters that can have an arbitrary time dependence, and $\boldsymbol{\Theta}(t) = (\Theta_1(t) \ldots \Theta_d(t))$ represents d phases corresponding to d lasers acting on the molecule.

Furthermore, we consider a more general form of the time dependence of the phase

$$\boldsymbol{\Theta}(t) = \boldsymbol{\theta} + \mathbf{g}(t) \tag{A2}$$

In the case of a chirped frequency we have e.g. $\boldsymbol{\Theta}(t) = \boldsymbol{\theta} + \boldsymbol{v}(t)\, t$. As mentioned in Section IV.A the effective instantaneous frequency is defined as $\boldsymbol{\omega}_{\text{eff}}(t) = d\boldsymbol{\Theta}(t)/dt = d\mathbf{g}(t)/dt = \boldsymbol{v}(t) + \dot{\boldsymbol{v}}(t)\, t$. If we define the generalized translation operator

$$(\mathscr{T}_{\mathbf{g}(t)}\Psi)(\boldsymbol{\theta}) = (e^{\mathbf{g}(t)\frac{\partial}{\partial\boldsymbol{\theta}}}\Psi)(\boldsymbol{\theta}) = \Psi(\boldsymbol{\theta} + \mathbf{g}(t)) \tag{A3}$$

the semiclassical Hamiltonian can be written as

$$H^{\mathbf{r}(\mathbf{t})}(\boldsymbol{\Theta}(t)) = \mathscr{T}_{\mathbf{g}(t)} H^{\mathbf{r}(t)}(\boldsymbol{\theta}) \mathscr{T}_{-\mathbf{g}(t)} \tag{A4}$$

Proposition. *The operator U is the propagator of the semiclassical Schrödinger equation (lifted to the enlarged space $\mathscr{K}$)*

$$i\hbar\frac{\partial}{\partial t}U(t,t_0;\boldsymbol{\theta}) = H^{\mathbf{r}(\mathbf{t})}(\boldsymbol{\Theta}(t))U(t,t_0;\boldsymbol{\theta}), \qquad U(t,t;\boldsymbol{\theta}) = \mathbb{1}_{\mathscr{H}} \tag{A5}$$

if and only if the operator U_K, defined by

$$U_K(t,t_0;\boldsymbol{\theta}) = \mathscr{T}_{-\mathbf{g}(t)}U(t,t_0;\boldsymbol{\theta})\mathscr{T}_{\mathbf{g}(t_0)} \tag{A6}$$

satisfies

$$i\hbar\frac{\partial}{\partial t}U_K(t,t_0;\boldsymbol{\theta}) = K^{\mathbf{r}(\mathbf{t})}U_K(t,t_0;\boldsymbol{\theta}), \qquad U_K(t,t;\boldsymbol{\theta}) = \mathbb{1}_{\mathscr{K}} \tag{A7}$$

with

$$K^{\mathbf{r}(\mathbf{t})} = -i\hbar\boldsymbol{\omega}_{\text{eff}}(t)\cdot\frac{\partial}{\partial\boldsymbol{\theta}} + H^{\mathbf{r}(\mathbf{t})}(\boldsymbol{\theta}) \tag{A8}$$

where $\boldsymbol{\omega}_{\text{eff}}(t)$ are the effective instantaneous frequencies

$$\boldsymbol{\omega}_{\text{eff}}(t) = d\boldsymbol{\Theta}(t)/dt = d\mathbf{g}(t)/dt \tag{A9}$$

The Floquet Hamiltonian $K^{\mathbf{r}(\mathbf{t})}$ acts on the enlarged Hilbert space $\mathscr{K} = \mathscr{H}\otimes \mathscr{L}_2(\mathbf{T}^d, d\boldsymbol{\theta}/2\pi)$, where $\mathbf{T}^d$ is the d-dimensional unit torus.

Proof: *We first remark that*

$$\frac{\partial}{\partial t}\mathscr{T}_{\mathbf{g}(t)} \equiv \frac{\partial}{\partial t}e^{\mathbf{g}(t)\cdot\frac{\partial}{\partial\boldsymbol{\theta}}} = \frac{d\mathbf{g}(t)}{dt}\cdot\frac{\partial}{\partial\boldsymbol{\theta}}\mathscr{T}_{\mathbf{g}(t)} \equiv \mathscr{T}_{\mathbf{g}(t)}\boldsymbol{\omega}_{\text{eff}}\cdot\frac{\partial}{\partial\boldsymbol{\theta}} \tag{A10}$$

We start with Eq. (A5) and invert Eq. (A6):

$$U(t,t_0;\boldsymbol{\theta}) = \mathscr{T}_{\mathbf{g}(t)}U_K(t,t_0;\boldsymbol{\theta})\mathscr{T}_{-\mathbf{g}(t_0)} \tag{A11}$$

The time derivative of U can be expressed as

$$\frac{\partial}{\partial t}U = \frac{\partial}{\partial t}\mathscr{T}_{\mathbf{g}(t)}U_K\mathscr{T}_{-\mathbf{g}(t_0)} = \mathscr{T}_{\mathbf{g}(t)}\boldsymbol{\omega}_{\text{eff}}\cdot\frac{\partial}{\partial\boldsymbol{\theta}}U_K\mathscr{T}_{-\mathbf{g}(t_0)} + \mathscr{T}_{\mathbf{g}(t)}\frac{\partial U_K}{\partial t}\mathscr{T}_{-\mathbf{g}(t_0)} \tag{A12}$$

which, after insertion into Eq. (A5) and using Eq. (A4), yields

$$i\hbar\mathscr{T}_{\mathbf{g}(t)}\left(\boldsymbol{\omega}_{\text{eff}}\cdot\frac{\partial}{\partial\boldsymbol{\theta}}U_K + \frac{\partial U_K}{\partial t}\right)\mathscr{T}_{-\mathbf{g}(t_0)} = \mathscr{T}_{\mathbf{g}(t)}H^{\mathbf{r}(\mathbf{t})}(\boldsymbol{\theta})\mathscr{T}_{-\mathbf{g}(t)}\mathscr{T}_{\mathbf{g}(t)}U_K\mathscr{T}_{-\mathbf{g}(t_0)} \tag{A13}$$

which by multiplication from the left by $\mathscr{T}_{-\mathbf{g}(t)}$ and from the right by $\mathscr{T}_{\mathbf{g}(t_0)}$ yields Eq. (A7). The inverse implication follows from the same argument run backwards.

2. The Structure of Eigenvectors and Eigenvalues of Floquet Hamiltonians—The Concept of Dressed Hamiltonian

In this section we show that the Floquet eigenvectors have the following general structure:

$$\psi_{m,\mathbf{k}}(x, \boldsymbol{\theta}) = C(x, \boldsymbol{\theta})[\phi_m^B(x) \otimes e^{i\mathbf{k}\cdot\boldsymbol{\theta}}] \tag{A14}$$

where $C(x, \boldsymbol{\theta})$ is a unitary operator in $\mathscr{K}$, and $\phi_m^B(x) \in \mathscr{H}$ are the eigenvectors of a time- and $\boldsymbol{\theta}$-independent operator B acting on $\mathscr{H}$. The eigenvalues can be written in the form

$$\lambda_{m,\mathbf{k}} = \lambda_m^B + \hbar\mathbf{k}\cdot\boldsymbol{\omega} \tag{A15}$$

where λ_m^B are the eigenvalues of B. The eigenelements can thus be classified by two labels: m, related to the molecule, and $\mathbf{k}$, related to the photon field.

We remark that if λ is an eigenvalue of K with eigenvector ψ, then for any $\mathbf{k} \in \mathbb{Z}^d$, $\lambda + \hbar\mathbf{k}\cdot\boldsymbol{\omega}$ is also an eigenvalue with corresponding eigenvector $e^{i\mathbf{k}\cdot\boldsymbol{\theta}}\psi$. This is an immediate consequence of the form of $K = -i\hbar\boldsymbol{\omega}\cdot\frac{\partial}{\partial\boldsymbol{\theta}} + H(x, \boldsymbol{\theta})$. In the periodic case this leads to a periodic structure of the spectrum. For instance, if we take an N-level model for the molecule, the Floquet spectrum will consist of a group of N eigenvalues that are repeated at a distance $k\omega$ for all $k \in \mathbb{Z}$ — that is, an infinite number of times. This periodic structure can be called Floquet zones, or Brillouin zones in analogy to a similar property in crystals. Although the Floquet Hamiltonian has an infinite number of eigenvalues and eigenvectors, once N are known, all the others can be constructed trivially. As the examples in Sections VI show, the energies of two different Brillouin zones can overlap and even lead to resonances that strongly couple the zones, leading to nontrivial physical effects.

In the quasiperiodic case of two or several incommensurate frequencies the Floquet eigenvalues cover the real line densely, and the overlap between Brillouin zones is much more intricate.

This structure of the eigenvectors and eigenvalues of Floquet Hamiltonians can be understood by considering an alternative interpretation of the Floquet eigenvalue problem.

We look for a unitary transformation $C(x, \boldsymbol{\theta})\colon \mathscr{H} \to \mathscr{H}$ (with $\boldsymbol{\theta}$ interpreted as a parameter) such that the semiclassical Schrödinger equation is transformed into an equation with a time-independent Hamiltonian B—that is, such that

$$U^B(t, t_0; \boldsymbol{\theta}) = C(\boldsymbol{\Theta}(t))^{-1} U(t, t_0; \boldsymbol{\theta}) C(\boldsymbol{\Theta}(t_0)) \tag{A16}$$

satisfies

$$i\hbar\frac{\partial}{\partial t}U^B(t,t_0;\boldsymbol{\theta}) = BU^B(t,t_0;\boldsymbol{\theta}) \tag{A17}$$

where $B = B(x)$ is a constant operator (i.e., independent of t and $\boldsymbol{\theta}$), acting on $\mathscr{H}$. If such a transformation exists, then

$$U(t,t_0;\boldsymbol{\theta}) = C(\boldsymbol{\Theta}(t))e^{-iB(t-t_0)/\hbar}C(\boldsymbol{\Theta}(t_0))^{-1} \tag{A18}$$

and B can be expressed in terms of $C(\boldsymbol{\Theta}(t))$ and $H(\boldsymbol{\Theta}(t))$:

$$B = C(\boldsymbol{\Theta}(t))^{-1}H(\boldsymbol{\Theta}(t))C(\boldsymbol{\Theta}(t)) - i\hbar C(\boldsymbol{\Theta}(t))^{-1}\frac{\partial C(\boldsymbol{\Theta}(t))}{\partial\boldsymbol{\theta}}\cdot\frac{d\boldsymbol{\Theta}(t)}{dt} \tag{A19}$$

Acting with $\mathscr{T}_{-\boldsymbol{\omega}t}$ from the left on (A18) and with $\mathscr{T}_{\boldsymbol{\omega}t_0}$ from the right, one obtains that C induces a unitary transformation of the Floquet Hamiltonian in the enlarged space $\mathscr{K}$:

$$\begin{aligned}\mathscr{T}_{-\boldsymbol{\omega}t}U(t,t_0;\boldsymbol{\theta})\mathscr{T}_{\boldsymbol{\omega}t_0} &\equiv e^{-iK(t-t_0)/\hbar}\\ &= C(\boldsymbol{\theta})\mathscr{T}_{-\boldsymbol{\omega}t}e^{-iB(t-t_0)/\hbar}\mathscr{T}_{\boldsymbol{\omega}t_0}C(\boldsymbol{\theta})^{-1}\\ &= C(\boldsymbol{\theta})e^{-(i/\hbar)(t-t_0)(-i\hbar\boldsymbol{\omega}\cdot\partial/\partial\boldsymbol{\theta}+B)}C(\boldsymbol{\theta})^{-1}\end{aligned} \tag{A20}$$

that is,

$$K = C(\boldsymbol{\theta})\left(-i\hbar\boldsymbol{\omega}\cdot\frac{\partial}{\partial\boldsymbol{\theta}}+B\right)C(\boldsymbol{\theta})^{-1} \tag{A21}$$

Hence, the determination of the eigenelements of K in $\mathscr{K}$ is reduced to the determination of those of B in $\mathscr{H}$. When such a transformation $C(\boldsymbol{\theta})$ can be found, the operator B is called the *dressed Hamiltonian*. Although it acts only on the molecular Hilbert space $\mathscr{H}$, it contains the information on the photons, which "dress" the molecule. The transformation $C(\boldsymbol{\theta})$ can be interpreted as a change of representation. We remark that the transformation $C(x,\boldsymbol{\theta})$, and thus the dressed Hamiltonian B, is clearly not unique since $C(x,\boldsymbol{\theta})$ can be composed with any unitary transformation that acts inside $\mathscr{H}$.

We consider only the situation in which B has a purely discrete spectrum: $B\phi_m^B = \lambda_m^B\phi_m^B$. Since $-i\hbar\boldsymbol{\omega}\frac{\partial}{\partial\boldsymbol{\theta}}$ commutes with B, and its eigenelements are $\hbar\mathbf{k}\cdot\boldsymbol{\omega}$ and $e^{i\mathbf{k}\cdot\boldsymbol{\theta}}$, $\mathbf{k}\in\mathbb{Z}^M$, we can conclude that the eigenvalues and eigenvectors of K have the general structure given in Eqs. (A15) and (A14). Since the sets of functions $\{\phi_m^B\}$ and $\{e^{i\mathbf{k}\cdot\boldsymbol{\theta}}\}$ are complete orthonormal bases of their respective spaces $\mathscr{H}$ and $\mathscr{L}_2(\mathbf{T}^d, d\boldsymbol{\theta}/2\pi)$, and since C is unitary, we conclude that $\{\psi_{m,\mathbf{k}}\}$ forms a complete basis of $\mathscr{K}$.

In order to arrive at the eigenvalue equation for K, we remark that since B is Hermitian in $\mathscr{H}$ there is a unitary transformation T that diagonalizes it in a reference base $\{|f_m\rangle\}$ of $\mathscr{H}$ (which, e.g., in the case of a two-level model can be represented in coordinates by $\{(1,0),(0,1)\}$):

$$T^\dagger BT = D = \sum_m \lambda_m^B |f_m\rangle\langle f_m| \tag{A22}$$

T can be thought of as a matrix whose columns are the components of the eigenvectors of B expressed in the basis $\{f_m\}$:

$$T = \sum_m |\phi_m^B\rangle\langle f_m| \tag{A23}$$

which allows us to write

$$\psi_{m,\mathbf{k}}(x,\boldsymbol{\theta}) = C(x,\boldsymbol{\theta})(T\otimes \mathbb{1}_{\mathscr{L}})|f_m \otimes e^{i\mathbf{k}\cdot\boldsymbol{\theta}}\rangle \tag{A24}$$

Inserting (A22) into (A21), we obtain

$$K = C(\boldsymbol{\theta})T\left(-i\hbar\boldsymbol{\omega}\cdot\frac{\partial}{\partial\boldsymbol{\theta}} + D\right)T^\dagger C(\boldsymbol{\theta})^{-1} \tag{A25}$$

or

$$KC(\boldsymbol{\theta})T = C(\boldsymbol{\theta})T\left(-i\hbar\boldsymbol{\omega}\cdot\frac{\partial}{\partial\boldsymbol{\theta}} + D\right) \tag{A26}$$

Applying both sides of this operator relation to the elements of the basis $f_m \otimes e^{i\mathbf{k}\cdot\boldsymbol{\theta}}$ and using the fact that

$$\left(-i\hbar\boldsymbol{\omega}\cdot\frac{\partial}{\partial\boldsymbol{\theta}} + D\right)|f_m \otimes e^{i\mathbf{k}\cdot\boldsymbol{\theta}}\rangle = (\lambda_m^B + \hbar\mathbf{k}\cdot\boldsymbol{\omega})|f_m \otimes e^{i\mathbf{k}\cdot\boldsymbol{\theta}}\rangle \tag{A27}$$

we arrive at the eigenvalue equation for K:

$$K\psi_{m,\mathbf{k}}(x,\boldsymbol{\theta}) = \lambda_{m,\mathbf{k}}\psi_{m,\mathbf{k}}(x,\boldsymbol{\theta}) \tag{A28}$$

Remark: Systems for which such a transformation $C(x,\boldsymbol{\theta})$ exists are called reducible. Due to the Floquet theorem, this is always the case for time-periodic

Hamiltonians (for finite- or infinite-dimensional $\mathscr{H}$). However, in the case of several incommensurate frequencies (quasi-periodic Hamiltonian), reducibility is not always satisfied, even for finite-dimensional $\mathscr{H}$ [19,33–36]. We remark that for finite-dimensional $\mathscr{H}$, reducibility is equivalent to the property of K having no continuous spectrum [19].

3. Relation Between Eigenvectors and Diagonalization Transformations

The preceding arguments are an adaptation of the following elementary relations of finite-dimensional linear algebra, which we illustrate with two-dimensional Hermitian matrices: We will use the notation S, A, D for the operators and $\tilde{S}, \tilde{A}, \tilde{D}$ for the corresponding matrix representations in a reference orthonormal basis $\{|e_1\rangle, |e_2\rangle\}$. We will use $|\psi\rangle$ for the intrinsic vectors, and we will use $\tilde{\psi}$ for their column vectors of components with respect to the reference basis. Accordingly, $\tilde{e_1} = \binom{1}{0}$ and $\tilde{e_2} = \binom{0}{1}$. We consider a Hermitian operator A represented by the matrix $\tilde{A}$. This operator can be diagonalized with respect to the reference basis by a unitary transformation S represented by a matrix $\tilde{S}$: $\tilde{S}^{\dagger}\tilde{A}\tilde{S} = \tilde{D}$, where $\tilde{D} = \mathrm{diag}(\lambda_1, \lambda_2)$. The eigenvalue equations can be written in matrix form as

$$\tilde{A}\tilde{S} = \tilde{S}\tilde{D} \tag{A29}$$

The components of the normalized eigenvectors $\tilde{\psi}_1 = \binom{u_1}{v_1}$, $\tilde{\psi}_2 = \binom{u_2}{v_2}$ can be identified as the column vectors of the matrix

$$\tilde{S} = \begin{pmatrix} u_1 & u_2 \\ v_1 & v_2 \end{pmatrix} \equiv (\mathrm{column}(\tilde{\psi}_1)\ \mathrm{column}(\tilde{\psi}_2)) \tag{A30}$$

since

$$\tilde{A}\tilde{S} = (\mathrm{column}(\tilde{A}\tilde{\psi}_1)\ \mathrm{column}(\tilde{A}\tilde{\psi}_2)) \tag{A31}$$

and

$$\tilde{S}\tilde{D} = (\mathrm{column}(\tilde{\psi}_1\lambda_1)\ \mathrm{column}(\tilde{\psi}_2\lambda_2)) \tag{A32}$$

Equation (A29) is equivalent to $\tilde{A}\tilde{\psi}_j = \lambda_j\tilde{\psi}_j$, $j = 1, 2$.

In order to extend these relations to operators in an infinite-dimensional Hilbert space, we first write them in terms of operators S, A, D instead of in terms of their matrix representations $\tilde{S}, \tilde{A}, \tilde{D}$. The diagonalization formula is

$$S^{\dagger}AS = D \tag{A33}$$

where D is an operator that is diagonal in the reference basis; that is, $\langle e_i|D|e_j\rangle = \delta_{ij}\lambda_j$. The diagonalization equation (A33) is equivalent to the eigenvalue equation written in operator form:

$$AS = SD \tag{A34}$$

The relation with the eigenvectors was made in the matrix representation by the statement that the column vectors of the matrix S that diagonalizes A are the components $\tilde{\psi}_j$ of the eigenvectors with respect to the reference basis. Extracting the jth column vector of a matrix $\tilde{S}$ is done by letting the matrix act on the coordinates $\tilde{e}_j$ of the jth basis element: $\tilde{S}\tilde{e}_j = j$th column of $\tilde{S}$. In terms of operators, this translates into the statement that the jth eigenvector can be expressed as

$$|\psi_j\rangle = S|e_j\rangle \tag{A35}$$

since, indeed

$$AS|e_j\rangle = SD|e_j\rangle = S\lambda_j|e_j\rangle = \lambda_j S|e_j\rangle \tag{A36}$$

The operator S that diagonalizes A in the basis $\{|e_1\rangle, |e_2\rangle\}$ can be written as

$$S = \sum_j |\psi_j\rangle\langle e_j| \tag{A37}$$

We remark that in this formulation the choice of the reference basis is fixed but arbitrary. We use this formulation in the discussion of the adiabatic theorem in Appendix C.

APPENDIX B: COHERENT STATES IN THE FLOQUET REPRESENTATION

In this Appendix we show that the coherent states are represented in Floquet theory by a generalized function $\Phi_{\theta_0}(\theta)$, which is real and depends on $\theta - \theta_0$, where $\theta_0 \in \mathbb{S}^1$ is a fixed angle, and

$$(\Phi_{\theta_0}(\theta))^2 = 2\pi\delta(\theta - \theta_0) \tag{B1}$$

This can be obtained as follows. The photon field coherent states are eigenvectors of the annihilation operator

$$a|\alpha\rangle = \alpha|\alpha\rangle, \qquad \alpha = |\alpha|e^{-i\theta_0} \tag{B2}$$

In the usual Fock number state representation they are given, up to a phase factor, by

$$|\alpha\rangle = e^{-|\alpha|^2/2} \sum_{n=0}^{\infty} \frac{\alpha^n}{\sqrt{n!}} |n\rangle \tag{B3}$$

In the phase representation they can be written as

$$\begin{aligned} \Phi_{\theta_0}^{(\bar{n})}(\theta) &= e^{i\zeta} e^{-|\alpha|^2/2} \sum_{n=0}^{\infty} \frac{\alpha^n}{\sqrt{n!}} e^{i(n-\bar{n})\theta} \\ &= e^{-|\alpha|^2/2} \sum_{n=0}^{\infty} \frac{|\alpha|^n}{\sqrt{n!}} e^{i(n-\bar{n})(\theta-\theta_0)} \end{aligned} \tag{B4}$$

(where ζ is an arbitrary constant phase that we have chosen as $\zeta = \bar{n}\theta_0$). In order to obtain the representation of coherent states in Floquet theory, we have to take $|\alpha| = \sqrt{\bar{n}}$, since the average photon number in a coherent state is given by $|\alpha|^2$, and then apply the limit $\bar{n} \to \infty$.

This can be rigorously done using directly the representation (B4), as was shown in Ref. 9. Here we discuss an alternative construction, which is formal but gives a useful intuition. We use an approximate expression of the coherent states for large $\bar{n}$, obtained in [8], by developing

$$a_{\bar{n},\theta} = \sqrt{\bar{n}}\, e^{-i\theta} \sqrt{1 - \frac{1}{\bar{n}} i \frac{\partial}{\partial \theta}} \underset{\bar{n}\to\infty}{\rightsquigarrow} \sqrt{\bar{n}}\, e^{-i\theta} \left(1 - \frac{1}{2\bar{n}} i \frac{\partial}{\partial \theta}\right) \tag{B5}$$

This leads to the following asymptotic expression [8] for the normalized coherent state corresponding to $\alpha = \sqrt{\bar{n}}\, e^{-i\theta_0}$, obtained as a solution of $e^{-i\theta}(1 - i/(2\bar{n})\partial/\partial\theta)\, \Phi_{\theta_0}^{(\bar{n})} = e^{-i\theta_0} \Phi_{\theta_0}^{(\bar{n})}$:

$$\Phi_{\theta_0}^{(\bar{n})} \underset{\bar{n}\to\infty}{\rightsquigarrow} \frac{1}{\nu} \exp\{-2\bar{n}[1 - \cos(\theta - \theta_0) - i(\sin(\theta - \theta_0) - (\theta - \theta_0))]\} \tag{B6}$$

where the normalization constant is

$$\nu^2 = e^{-4\bar{n}} I_0(4\bar{n}) \tag{B7}$$

with I_0 a Bessel function, which behaves asymptotically as

$$I_0(4\bar{n}) = \int_0^{2\pi} \frac{d\theta}{2\pi} \exp(4\bar{n}\cos\theta) \underset{\bar{n}\to\infty}{\rightsquigarrow} \frac{e^{4\bar{n}}}{(8\pi\bar{n})^{1/2}} \tag{B8}$$

Therefore

$$|\Phi_{\theta_0}^{(\bar{n})}(\theta)|^2 \underset{\bar{n}\to\infty}{\rightsquigarrow} (8\pi\bar{n})^{1/2} \exp\{-4\bar{n}[1-\cos(\theta-\theta_0)]\} \tag{B9}$$

Noticing that the function $\exp\{-4\bar{n}[1-\cos(\theta-\theta_0)]\}$ behaves like $\exp\{-2\bar{n}(\theta-\theta_0)^2\}$ for $\bar{n}\to\infty$, we get

$$|\Phi_{\theta_0}^{(\bar{n})}(\theta)|^2 \underset{\bar{n}\to\infty}{\rightsquigarrow} 2\pi\delta(\theta-\theta_0) \tag{B10}$$

where $\delta(\theta-\theta_0)$ is the usual Dirac delta function.

We remark that since the phase term in (B4) [or in (B6)] is odd in $\theta-\theta_0$, we obtain that $\Phi_{\theta_0}^{(\bar{n})}(\theta)\to\Phi_{\theta_0}(\theta)$ with $\Phi_{\theta_0}(\theta)$ real and

$$(\Phi_{\theta_0}^{(\bar{n})}(\theta))^2 \underset{\bar{n}\to\infty}{\longrightarrow} 2\pi\delta(\theta-\theta_0) \tag{B11}$$

Furthermore, using the well-known properties of the expectation values of N^m on coherent states, we obtain

$$\langle\Phi_{\theta_0}^{(\bar{n})}(\theta)| - i\frac{\partial}{\partial\theta}|\Phi_{\theta_0}^{(\bar{n})}(\theta)\rangle_{\mathscr{L}} = 0, \qquad \text{for all } \bar{n} \tag{B12}$$

$$\langle\Phi_{\theta_0}^{(\bar{n})}(\theta)|(-i)^m\frac{\partial^m}{\partial\theta^m}|\Phi_{\theta_0}^{(\bar{n})}(\theta)\rangle_{\mathscr{L}} \underset{\bar{n}\to\infty}{\longrightarrow} \infty, \qquad m\geq 2 \tag{B13}$$

The subscripts in the scalar product symbols ($\langle\ |\ \rangle_{\mathscr{L}}$) indicate on which space they act. Thus we conclude that in Floquet theory the photon coherent states are represented by the "square root of a δ-function," which we denote by $\Phi_{\theta_0}(\theta)=(2\pi)^{1/2}\delta_{1/2}(\theta-\theta_0)$. Since we will be interested in expectation values, only $|\Phi_{\theta_0}|^2$ will appear in our calculations. The formal calculus rules for $\delta_{1/2}(\theta-\theta_0)$ are given in Ref. 9.

APPENDIX C. THE ADIABATIC THEOREM FOR FLOQUET HAMILTONIANS

In this Appendix we sketch an argument that leads to the adiabatic theorem for an N-level system with a Floquet Hamiltonian denoted $K^{\mathbf{r}}$ which generates the dressed Schrödinger equation [Eq. (229) of Section IV].

$$i\hbar\frac{\partial\psi(\theta,t)}{\partial t} = K^{\mathbf{r}(s)}\psi(\theta,t) \tag{C1}$$

We have to show that in the adiabatic limit up to corrections of order $\mathcal{O}(1/\tau)$, the evolution is approximated by

$$\psi(\theta, t) \simeq \sum_{m\in\mathcal{S}} c_m \exp[i\delta_m^{\mathbf{r}(s)}(t)]\psi_m^{\mathbf{r}(s)}(\theta) \tag{C2}$$

where the c_m are complex numbers determined by the initial condition

$$c_m = \langle \exp[i\delta_m^{\mathbf{r}(s_0)}(t_0)]\psi_m^{\mathbf{r}(s_0)}(\theta)|\psi(\theta, t_0)\rangle \tag{C3}$$

Let $\{\psi_m^{\mathbf{r}(s)}\}$ be an orthonormal basis of instantaneous eigenvectors of $K^{\mathbf{r}(s)}$, which we assume to be sufficiently smooth as a function of s. We define the unitary operator $\mathsf{S}^{\mathbf{r}(s)} := \sum_m |\psi_m^{\mathbf{r}(s)}\rangle\langle\psi_m^{\mathbf{r}(s_0)}|$, where s_0 is the initial time. This operator transforms the Floquet Hamiltonian by

$$\mathsf{D}^{\mathbf{r}(s)} := (\mathsf{S}^{\mathbf{r}(s)})^\dagger K^{\mathbf{r}(s)}\mathsf{S}^{\mathbf{r}(s)} \tag{C4}$$

into an operator $\mathsf{D}^{\mathbf{r}(s)}$, which for all s is diagonal in the basis taken at s_0, $\{\psi_m^{\mathbf{r}(s_0)}\}$. Defining transformed states by

$$\tilde{\psi}(\theta, s) = (\mathsf{S}^{\mathbf{r}(s)})^\dagger \psi(\theta, \tau s) \tag{C5}$$

the Schrödinger equation (C1) can be rewritten as

$$\frac{i\hbar}{\tau}\frac{\partial\tilde{\psi}(\theta, s)}{\partial s} = \left[\mathsf{D}^{\mathbf{r}(s)} - \frac{i\hbar}{\tau}(\mathsf{S}^{\mathbf{r}(s)})^\dagger \frac{\partial \mathsf{S}^{\mathbf{r}(s)}}{\partial s}\right]\tilde{\psi}(\theta, s) \tag{C6}$$

The last term on the right-hand side of Eq. (C6) characterizes the *nonadiabatic couplings* between the instantaneous Floquet states (off-diagonal terms, of the form $|\psi_\ell^{\mathbf{r}(s_0)}\rangle\langle\psi_\ell^{\mathbf{r}(s)}|\dot{\mathbf{r}}\cdot\nabla_{\mathbf{r}}\psi_m^{\mathbf{r}(s)}\rangle_{\mathcal{K}}\langle\psi_m^{\mathbf{r}(s_0)}|$, $\ell \neq m$) and the Berry phase (real diagonal terms). In the adiabatic limit $\tau \to \infty$, one can neglect the nonadiabatic couplings—that is, the nondiagonal terms that are of order $\mathcal{O}(1/\tau)$:

$$\frac{i\hbar}{\tau}\frac{\partial\tilde{\psi}(\theta, s)}{\partial s} \simeq \left[\mathsf{D}^{\mathbf{r}(s)} - \frac{i\hbar}{\tau}\mathrm{diag}_{(s_0)}\left((\mathsf{S}^{\mathbf{r}(s)})^\dagger \frac{\partial \mathsf{S}^{\mathbf{r}(s)}}{\partial s}\right)\right]\tilde{\psi}(\theta, s) \tag{C7}$$

where $\mathrm{diag}_{(s_0)}$ denotes the diagonal part with respect to the initial time basis $\{\psi_m^{\mathbf{r}(s_0)}\}$. Developing $\psi(\theta, t)$ at an initial time $t_0 = \tau s_0$ in the eigenvector basis of $\mathsf{K}^{\mathbf{r}(s_0)}$, spanning the subspace $\mathcal{S}$

$$\psi(\theta, \tau s_0) = \sum_{m\in\mathcal{S}} c_m \psi_m^{\mathbf{r}(s_0)}(\theta) \tag{C8}$$

the equation is easily solved and one recovers Eq. (233). We remark that in many applications the initial time s_0 is taken before the rise of the laser pulse. In this case, since the interaction is off, the initial basis coincides with the eigenvectors of the free molecule multiplied by those of the field. The operator $\mathsf{D}^{\mathbf{r}(s)}$ of Eq. (C4) can then be written as

$$\mathsf{D}^{\mathbf{r}(s)} = -i\hbar\omega\frac{\partial}{\partial\theta} + \mathsf{d}^{\mathbf{r}(s)} \tag{C9}$$

where $\mathsf{d}^{\mathbf{r}(s)}$ is an operator in $\mathscr{H}$ that is diagonal in the basis of the eigenvectors of the free molecule.

Acknowledgments

We thank O. Atabek, K. Bergmann, C. Dion, O. Faucher, A. Joye, A. Keller, B. Lavorel, R. Marquardt, F. Monti, N. Sangouard, B. W. Shore, S. Thomas, R. G. Unanyan, M. Amniat-Talab, N. V. Vitanov, and L. P. Yatsenko for many fruitful discussions.

References

1. J. H. Shirley, *Phys. Rev.* **138**, B979 (1965).
2. H. Sambe, *Phys. Rev. A* **7**, 2203 (1973).
3. J. Howland, *Math. Ann.* **207**, 315 (1974).
4. J. Bellissard, Stability and instability in quantum mechanics, in *Trends and Developments in the Eighties*, S. Albeverio and P. Blanchard, eds., World Scientific, Singapore, 1985.
5. S. I. Chu, *Adv. At. Mol. Phys.* **21**, 197 (1985).
6. S. I. Chu, *Adv. Chem. Phys.* **73**, 739 (1987).
7. I. Bialynicki-Birula and Z. Bialynicka-Birula, *Phys. Rev. A* **14**, 1101 (1976).
8. I. Bialynicki-Birula and C. L. Van, *Acta Physica Polonica A* **57**, 599 (1980).
9. S. Guérin, F. Monti, J.-M. Dupont, and H. R. Jauslin, *J. Phys. A* **30**, 7193 (1997).
10. C. Cohen-Tannoudji, J. Dupont-Roc, and G. Grynberg, *Atom–Photon Interactions*, John Wiley & Sons, New York, 1992.
11. G. Compagno, R. Passante, and F. Persico, *Atom–Field Interactions and Dressed Atoms*, Cambridge University Press, Cambridge, England, 1995.
12. P. Carruthers and M. M. Nieto, *Rev. Mod. Phys.* **40**, 411 (1968).
13. J. M. Lévy-Leblond, *Ann. Phys.* **101**, 319 (1976).
14. P. A. M. Dirac, *The Principles of Quantum Mechanics*, 4th ed., Clarendon Press, London, 1958.
15. J. Von Neumann, *Mathematische Grundlagen der Quantenmechanik*, Springer, Berlin, 1932.
16. G. A. Raggio and S. Zivi, *J. Math. Phys.* **26**, 2529 (1985).
17. T. S. Ho and S. I. Chu, *J. Phys. B: At. Mol. Phys.* **17**, 2101 (1984).
18. H. R. Jauslin and J. L. Lebowitz, *Chaos* **1**, 114 (1991).
19. P. Blekher, H. R. Jauslin, and J. L. Lebowitz, *J. Stat. Phys.* **68**, 271 (1992).
20. H. R. Jauslin, in *II Granada Lectures in Computational Physics*, P. L. Garrido and J. Marro, eds., World Scientific, Singapore, 1993.

21. W. Scherer, *J. Phys. A* **27**, 8331 (1994).
22. W. Scherer, *J. Phys. A* **30**, 2825 (1997).
23. W. Scherer, *J. Math. Phys.* **39**, 2597 (1998).
24. W. Scherer, *Phys. Lett. A* **233**, 1 (1997).
25. W. Scherer, *Phys. Rev. Lett.* **74**, 1495 (1995).
26. C. Chandre and H. R. Jauslin, *J. Math. Phys.* **39**, 5856 (1998).
27. C. Chandre and H. R. Jauslin, *Phys. Rep.* **365**, 1 (2002).
28. J. H. Van Vleck, *Phys. Rev.* **33**, 467 (1929).
29. H. R. Jauslin, S. Guérin, and S. Thomas, *Physica A* **279**, 432 (2000).
30. H. Primas, *Rev. Mod. Phys.* **35**, 710 (1963).
31. V. G. Tyuterev and V. I. Perevalov, *Chem. Phys. Lett.* **74**, 494 (1980).
32. A. Keller, C. M. Dion, and O. Atabek, *Phys. Rev. A* **61**, 023409 (2000).
33. L. H. Eliasson, *Commun. Math. Phys.* **146**, 447 (1992).
34. L. H. Eliasson, Ergodic skew systems on $\mathbb{T}^d \times SO(3, \mathbb{R})$, preprint, ETH, Zürich, 1991.
35. R. Krikorian, *C. R. Acad. Sci. Paris Ser. I* **321**, 1039 (1995).
36. R. Krikorian, *Ergodic Th. Dynam. Syst.* **19**, 61 (1999).
37. H. R. Jauslin, Small divisors in driven quantum systems, in *Stochasticity and Quantum Chaos*, Z. Haba, W. Cegla, L. Jakóbczyk, eds., Kluwer Publishers, Hingham, MA, 1995.
38. S. Guérin, R. Unanyan, L. Yatsenko, and H. R. Jauslin, *Optics Express* **4**, 84 (1999).
39. B. W. Shore, *The Theory of Coherent Atomic Excitation*, John Wiley & Sons, New York, 1990.
40. L. Allen and J. H. Eberly, *Optical Resonance and Two-Level Atoms*, Dover, New York, 1987.
41. B. Friedrich and D. Herschbach, *Phys. Rev. Lett.* **74**, 4623 (1995).
42. S. Guérin, L. P. Yatsenko, H. R. Jauslin, O. Faucher, and B. Lavorel, *Phys. Rev. Lett.* **88**, 22601 (2002).
43. C. Dion, Ph.D. thesis, Université de Paris-Sud and Université de Sherbrooke, 1999.
44. M. V. Berry, *Proc. R. Soc. London A* **392**, 45 (1984).
45. J. E. Avron and A. Elgart, *Phys. Rev. A* **58** 4300 (1998).
46. J. E. Avron and A. Elgart, *Commun. Math. Phys.* **203**, 445 (1999).
47. K. Drese and M. Holthaus, *Eur. Phys. J. D* **5**, 119 (1999).
48. A. M. Dykhne, *Sov. Phys. JETP* **14**, 941 (1962).
49. J. P. Davis and P. Pechukas, *J. Chem. Phys.* **64**, 3129 (1976).
50. J.-T. Hwang and P. Pechukas, *J. Chem. Phys.* **67**, 4640 (1977).
51. A. Joye, H. Kuntz, and C.-Ed. Pfister, *Ann. Phys.* **208**, 299 (1991).
52. A. Joye, G. Mileti, and C.-Ed. Pfister, *Phys. Rev. A* **44**, 4280 (1991).
53. M. V. Berry, *Proc. R. Soc. London A* **429**, 61 (1990).
54. M. V. Berry, *Proc. R. Soc. London A* **414**, 31 (1987).
55. A. Joye and C.-E. Pfister, *J. Math. Phys.* **34**, 454 (1993).
56. A. Joye, *J. Phys. A* **26**, 6517 (1993).
57. K. Drese and M. Holthaus, *Eur. Phys. J. D* **3**, 73 (1998).
58. S. Guérin, S. Thomas, and H. R. Jauslin, *Phys. Rev. A* **65**, 023409 (2002).
59. P. R. Berman, L. Yan, K.-H. Chiam, and R. Sung, *Phys. Rev. A* **57**, 79 (1998).
60. M. Holthaus and B. Just, *Phys. Rev. A* **49**, 1950 (1994).

61. M. V. Korolkov, J. Manz, and G. K. Paramonov, *Chem. Phys.* **217**, 341 (1997).
62. S. Guérin and H. R. Jauslin, *Phys. Rev. A* **55**, 1262 (1997).
63. L. P. Yatsenko, N. V. Vitanov, B. W. Shore, T. Rickes, and K. Bergmann, *Opt. Commun.* **204**, 413 (2002).
64. L. D. Landau, *Phys. Z. Sowjetunion* **2**, 46 (1932).
65. C. Zener, *Proc. R. Soc. London A* **137**, 696 (1932).
66. H. P. Breuer and M. Holthaus, *Phys. Lett. A* **140**, 507 (1989).
66a. C. E. Carrol and F. T. Hioe, *J. Phys. A* **19**, 2061 (1986).
67. S. Guérin, L. P. Yatsenko, and H. R. Jauslin, *Phys. Rev. A* **63**, R031403 (2001).
68. L. P. Yatsenko, S. Guérin, and H. R. Jauslin, *Phys. Rev. A* **65**, 043407 (2002).
69. N. V. Vitanov, M. Fleischhauer, B. W. Shore, and K. Bergmann, *Adv. At. Mol. Opt. Phys.* **46**, 55, (2001).
70. S. Chelkowski and A. D. Bandrauk, *J. Raman Spectrosc.* **28**, 459 (1997).
71. L. P. Yatsenko, B. W. Shore, T. Halfmann, K. Bergmann, and A. Vardi, *Phys. Rev. A* **60**, R4237 (1999).
72. T. Rickes, L. P. Yatsenko, S. Steuerwald, T. Halfmann, B. W. Shore, N. V. Vitanov, and K. Bergmann, *J. Chem. Phys.* **113**, 534 (2000).
73. M. V. Danileiko, V. I. Romanenko, and L. P. Yatsenko, *Opt. Commun.* **109**, 462 (1994).
74. S. Guérin, L. P. Yatsenko, T. Halfmann, B. W. Shore, and K. Bergmann, *Phys. Rev. A* **58**, 4691 (1998).
75. D. Grischkowsky, M. M. T. Loy, and P. F. Liao, *Phys. Rev. A* **12**, 2514 (1975).
76. N. V. Vitanov, K.-A. Suominen, and B. W. Shore, *J. Phys. B* **32**, 4535 (1999).
77. M. Holthaus, *Phys. Rev. Lett.* **69**, 1596 (1992).
78. M. Grifoni and P. Hänggi, *Phys. Rep.* **304**, 229 (1998).
79. S. Guérin, *Phys. Rev. A* **56**, 1458 (1997).
80. V. I. Romanenko and L. P. Yatsenko, *JETP* **90**, 407 (2000).
81. R. G. Unanyan, N. V. Vitanov, and K. Bergmann, *Phys. Rev. Lett.* **87**, 137902 (2001).
82. S. Guérin, R. G. Unanyan, L. P. Yatsenko, and H. R. Jauslin, *Phys. Rev. A* **66**, 032311 (2002).
83. R. G. Unanyan, S. Guérin, and H. R. Jauslin, *Phys. Rev. A* **62**, 043407 (2000).
84. R. G. Unanyan, S. Guérin, B. W. Shore, and K. Bergmann, *Eur. Phys. J. D* **8**, 443 (2000).
85. S. Guérin, H. R. Jauslin, and R. Unanyan, Multiphoton processes in lambda three-level systems, in *Multiphoton Processes 1999*, L. F. DiMauro, R. F. Freeman, and K. C. Kulander, eds., AIP Conference Proceedings, Vol. 525, p. 571, 2000.

RECENT ADVANCES IN THE THEORY OF VIBRATION–ROTATION HAMILTONIANS

JANNE PESONEN and LAURI HALONEN

Laboratory of Physical Chemistry, University of Helsinki, Helsinki, Finland

CONTENTS

I. Introduction
II. Classical Mechanics
  A. Hamiltonian from Lagrangian
    1. Lagrangian
    2. Hamiltonian
  B. Hamiltonian via Scalar Chain Rules
  C. Hamiltonian via Measuring Vectors
    1. Infinitesimal Approach
    2. Geometric Algebra Approach
III. Canonical Quantization
  A. Unconstrained Quantization
  B. Constrained Quantization
IV. Volume Elements
V. Conclusion
Appendix A: Rudiments of Geometric Algebra
  1. Sums and Products
    a. Addition of Vectors
    b. Multiplication of Vectors
    c. Generalization of Multiplication
    d. Magnitude
    e. Expansion Rules
  2. Basis Representation for Three-Dimensional Geometric Algebra
  3. Geometry
    a. Projections
    b. Reflections
    c. Rotations
    d. Angles
    e. Spherical Trigonometry

Advances in Chemical Physics, Volume 125, Edited by I. Prigogine and Stuart A. Rice.
ISBN 0-471-21452-3.

I. INTRODUCTION

Vibration–rotation theory has its origin in the birth of quantum mechanics when it became possible to develop necessary theoretical tools to understand internal and rotational motion of polyatomic molecules. The first systematic formulations of this theory by Wilson and Dennison and their co-workers were based on the rectilinear normal coordinate concept [1]. The expansion of the exact kinetic energy terms combined with the use of perturbation theory provided a powerful tool in understanding spectroscopic observations [2]. A major contribution in this approach was made in the 1960s when Watson transformed the exact kinetic energy operator of both nonlinear and linear molecules into a simple form [3,4]. However, in the 1970s it became apparent that there is need for approaches based on curvilinear internal coordinates instead of rectilinear ones. Indeed, an exact and correct vibrational kinetic energy operator expressed in terms of internal bond coordinates for triatomic molecules was published in the review by Carney, Sprandel, and Kern [5]. This operator including also rotational and Coriolis terms was later used, for example, in variational calculations for water [6]. A good agreement with experimental values was obtained. Another development was the discovery that excited stretching states in hydrogen-containing symmetrical molecules were well-described by the local mode model which in contrast to the standard vibration–rotation theory was modeled in terms of bond displacement coordinates (for reviews see Refs. 7 and 8). The signal from these observations was clear: It was time to develop a general theory based on nonrectilinear coordinates in order to cope with larger than triatomic semirigid molecules.

Curvilinear internal bond coordinates versus rectilinear normal coordinates [9,10] is, of course, not the only choice to be made. There is a larger selection of coordinates to choose from: Radau, Jacobi, hyperspherical, and so on, coordinates; see, for example, Refs. 11–14, and the review by Bačić and Light [15] (and references therein). In addition to the rovibrational states of semirigid molecules, these can be used for different types of problems; for example, for systems where molecular bonds are broken and formed, chemical reactions occur, and so on. It is clear that both kinetic energy operators and potential energy surfaces are needed for all these systems. We only deal with kinetic energy operators in this review. There are currently a variety of methods available which are used to derive them. One of them is based on the Lagrangian, which is transformed to the Hamiltonian form [16–21]. The bottleneck in this approach is the analytical inversion of matrices (if exact analytic kinetic energy operators are needed). Another approach is to start from the Cartesian representation of the kinetic energy and then use the usual chain rule to make the coordinate transformation [22–28]. This leads to complicated intermediate mathematical formulas that can become tedious (or at least impractical) to derive by hand for larger than three atomic molecules in general coordinates and body axes. The use of symbolic algebra programs has helped in this respect and has thus advanced the field [23]. The Hamiltonian formulation is largely based on the work by Wilson, which originally treated harmonic vibrations in polyatomic molecules [1]. Wilson's infinitesimal analysis to construct kinetic energy coefficients has made it possible to form exact Hamiltonians at least for semirigid molecules. The rotational motion remained a bottleneck in this approach for several years. However, Lukka's brilliant idea to employ infinitesimal rotational coordinates solved the problem, although the extension of this method may become tedious for larger than triatomic molecules and for arbitrary axis systems [29]. Lukka's initial approach is based on Wilson's type infinitesimal analysis, but an algebraic version also exists [30,31]. The quantization of the classical Hamiltonian can be achieved by considering the transformation of the Cartesian Laplacian operator to the coordinate system of interest and using the appropriate volume element of integration.

It seems that from the existing literature it is somewhat difficult to obtain a general picture how the exact quantum mechanical kinetic energy operators are formed. The use of a branch of mathematics called geometric algebra by Hestenes [32–35] has changed the view about how to go beyond triatomic molecules in curvilinear internal bond coordinates and how these problems should be addressed [36–41]. This powerful method provides a general algebraic approach to form these operators for any molecular size and for any chosen coordinate system. But it is not just a generalization; it also provides a new picture of how some of the details should be understood. It has also given a solution to some previous problems: For example, it provides a method of how to form kinetic energy operators for systems where the coordinates used are not explicit functions of nuclear positions [37], and it offers the first practical way to calculate volume elements for integration [39]. However, there is a price to be paid in order to be able to understand this approach. One has to master the field of geometric algebra, where unfamiliar operations (at least to most scientists) such as addition of numbers to vectors and dividing a vector by another vector can be performed. We try to help the reader by providing an appendix with a summary of mathematical details. We have put emphasis on calculus because geometric algebra is particularly powerful in differentiating vectors directly without the need of going into component representations. This review contains a summary of this new approach to construct molecular Hamiltonians. Examples are given in terms of curvilinear bond coordinates, but the treatment is general and not restricted to any particular choice of the internal coordinates. In addition, in explaining what has been known before, new results and insights are also given. There are some aspects that we do not cover. These include possible singularities in the Hamiltonian [42], the choice between different molecular axis systems, where interaction between vibration–rotation has been minimized [43], including Eckart axes [24,25,44–46] (where the interaction has been minimized at a reference configuration), and the gauge-invariant representations of the vibration–rotation Hamiltonians [47]. We also restrict the treatment to molecules, which are nonlinear at the reference configuration. As is common, we work within the Born–Oppenheimer approximation [48,49]. We hope that this contribution will help research in the field not only in theory but also in practice.

II. CLASSICAL MECHANICS

It is possible to formulate the classical laws of motion in several ways. Newton's equations are taught in every basic course of classical mechanics. However, especially in the presence of constraint forces, the equations of motion can often be presented in a simpler form by using either Lagrangian or Hamiltonian formalism. In short, in the Newtonian approach, an N-point particle system is described by specifying the position $\mathbf{x}_\alpha = \mathbf{x}_\alpha(t)$ of each particle α as a function of time. The positions are found by solving the equations of motion,

$$m_\alpha \ddot{\mathbf{x}}_\alpha = -\boldsymbol{\nabla}_\alpha V + \mathbf{f}_\alpha + \mathbf{N}_\alpha \tag{1}$$

where m_α is the mass of the particle α, an overdot signifies differentiation with respect to t, V is the potential for the conservative forces, $\mathbf{f}_\alpha$ is the force function for nonconservative forces acting on the particle α, and $\mathbf{N}_\alpha = \sum_i \lambda_i \boldsymbol{\nabla}_\alpha \phi_i$ is the resultant force of the P constraints $\{\phi_1, \phi_2, \ldots, \phi_P\}$ acting on the particle α (λ_i is a proportionality constant) [50].

A. Hamiltonian from Lagrangian

1. *Lagrangian*

The state of a system is specified by $A = 3N - P$ generalized coordinates $\{q_1, q_2, \ldots, q_A\}$ in the Lagrangian approach. The coordinates q_i can be found as a function of time by solving Lagrange's equations [50,51]

$$\frac{d}{dt}\frac{\partial}{\partial \dot{q}_i}L - \frac{\partial}{\partial q_i}L = F_i \tag{2}$$

where $F_i = F_i(q_1, q_2, \ldots, q_A; \dot{q}_1, \dot{q}_2, \ldots, \dot{q}_A; t) = \sum_\alpha \mathbf{f}_\alpha \cdot (\partial \mathbf{x}_\alpha / \partial q_i)$ is the q_i component of the generalized force. The constraint forces do not appear explicitly in the Lagrangian equations. The Lagrangian L is given by

$$L = T - V = \frac{1}{2}\sum_{ij} g_{q_i q_j} \dot{q}_i \dot{q}_j + \sum_i b_i \dot{q}_i + c - V \tag{3}$$

where the covariant metric tensor $g_{q_i q_j}$ can be formed as the sum of the inner products

$$g_{q_i q_j} = \sum_\alpha^N m_\alpha \frac{\partial \mathbf{x}_\alpha}{\partial q_i} \cdot \frac{\partial \mathbf{x}_\alpha}{\partial q_j} \tag{4}$$

and

$$b_i = \sum_\alpha^N \frac{\partial \mathbf{x}_\alpha}{\partial q_i} \cdot \frac{\partial \mathbf{x}_\alpha}{\partial t} \tag{5}$$

$$c = \sum_\alpha^N \frac{\partial \mathbf{x}_\alpha}{\partial t} \cdot \frac{\partial \mathbf{x}_\alpha}{\partial t} \tag{6}$$

For the conservative systems, such as free molecules, or molecules subject to time-independent constraints, $c = \sum_i b_i \dot{q}_i = 0$ [50]. The tangent $\partial \mathbf{x}_\alpha / \partial q_i$ is denoted by $\mathbf{e}_{q_i}^{(\alpha)}$ from now on.

The shape, rotational, and translational coordinates are used to properly account for the different types of molecular motion. The word "molecule" refers to any N-point particle system moving under the influence of the potential V, which is a function of the shape coordinates only. Furthermore, the molecules may rotate and translate freely, but their shape may be subject to P constraints. We call them free molecules, because the constraints in shape can be thought to follow from the potential of the molecule. For example, a bond becomes rigid, if the bond stretching force constant tends to infinite. However, it is simple to generalize the treatment given here to include the rotational and translational constraints, if needed.

The position of the molecule can be parameterized by three Cartesian coordinates $X_i = \mathbf{X} \cdot \mathbf{u}_i$ (where $\{\mathbf{u}_1, \mathbf{u}_2, \mathbf{u}_3\}$ are three orthonormal space-fixed vectors) of the center of mass of the molecule

$$\mathbf{X} = \sum_\alpha^N \frac{m_\alpha \mathbf{x}_\alpha}{M} \tag{7}$$

where $M = \sum_{\alpha}^{N} m_\alpha$ is the mass of the molecule. By writing each nuclear position $\mathbf{x}_\alpha$ as

$$\mathbf{x}_\alpha = \mathbf{y}_\alpha + \mathbf{X} = \mathbf{y}_\alpha + X_1\mathbf{u}_1 + X_2\mathbf{u}_2 + X_3\mathbf{u}_3 \tag{8}$$

it is seen that the tangents associated with the center-of-mass coordinates are given by

$$\mathbf{e}_{X_i}^{(\alpha)} = \frac{\partial \mathbf{x}_\alpha}{\partial X_i} = \mathbf{u}_i \tag{9}$$

The shape of the molecule can be described by $A_\mathrm{V} = A - 6$ shape coordinates $\{s_1, s_2, \ldots, s_{A_\mathrm{V}}\}$. Typical examples of shape coordinates used in the molecular spectroscopy are the interparticle distance

$$r_{\alpha\beta} = |\mathbf{r}_{\alpha\beta}| = |\mathbf{x}_\beta - \mathbf{x}_\alpha| \tag{10}$$

the angle $\theta_{\beta\alpha\gamma}$ between the interparticle distance vectors $\mathbf{r}_{\alpha\beta}$ and $\mathbf{r}_{\alpha\gamma}$ (the bond angle)

$$\theta_{\beta\alpha\gamma} = \arccos(\mathbf{u}_{\mathbf{r}_{\alpha\beta}} \cdot \mathbf{u}_{\mathbf{r}_{\alpha\gamma}}) \tag{11}$$

and the dihedral angle (sometimes called the torsional angle)

$$\tau_{\gamma\alpha\beta\kappa} = \arccos\left(\frac{\mathbf{u}_{\mathbf{r}_{\alpha\gamma}} \wedge \mathbf{u}_{\mathbf{r}_{\alpha\beta}}}{|\mathbf{u}_{\mathbf{r}_{\alpha\gamma}} \wedge \mathbf{u}_{\mathbf{r}_{\alpha\beta}}|} \cdot \frac{\mathbf{u}_{\mathbf{r}_{\alpha\beta}} \wedge \mathbf{u}_{\mathbf{r}_{\beta\kappa}}}{|\mathbf{u}_{\mathbf{r}_{\alpha\beta}} \wedge \mathbf{u}_{\mathbf{r}_{\beta\kappa}}|}\right) \tag{12}$$

between the planes $\mathbf{u}_{\mathbf{r}_{\alpha\gamma}} \wedge \mathbf{u}_{\mathbf{r}_{\alpha\beta}}$ and $\mathbf{u}_{\mathbf{r}_{\alpha\beta}} \wedge \mathbf{u}_{\mathbf{r}_{\beta\kappa}}$. If both sides of Eq. (7) are differentiated with respect to the shape coordinate s_i, the relation

$$\sum_{\alpha}^{N} m_\alpha \mathbf{e}_{s_i}^{(\alpha)} = 0 \tag{13}$$

follows, because the center of mass $\mathbf{X}$ is independent of the shape coordinates. Thus, the change in shape does not translate the molecule.

The orientation of a nonlinear molecule can be described by three Euler angles $\{\phi, \theta, \chi\}$, because it takes two angles to describe the orientation of any body-fixed vector and takes one angle to describe the orientation of the body about that vector. The Euler angles relate the orientation of an orthonormal molecule-fixed axis system $\{\mathbf{u}'_1, \mathbf{u}'_2, \mathbf{u}'_3\}$ to some standard orthonormal space-fixed frame $\{\mathbf{u}_1, \mathbf{u}_2, \mathbf{u}_3\}$ (see Fig. 1 and Eq. (A73) in Appendix A, Section 3.c).

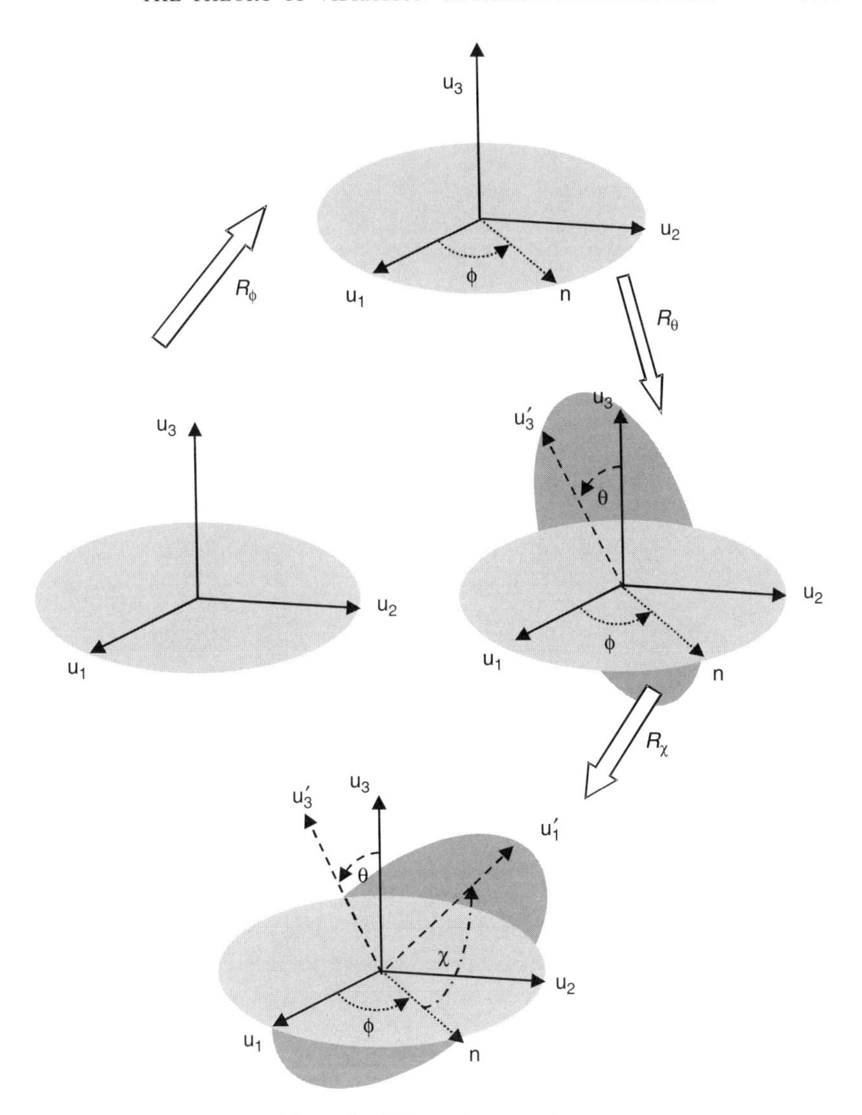

Figure 1. Euler angle convention.

Each choice of the body-axes specifies a reference orientation, in which the body frame coincides with the space-fixed frame. Let us choose one of the body axes, say $\mathbf{u}'_3$, to be in the direction of

$$\mathbf{r} = \sum_{\alpha} c_{\alpha} \mathbf{x}_{\alpha} \tag{14}$$

.

where the coefficients c_α are known functions of nuclear positions $\mathbf{x}_1, \mathbf{x}_2, \ldots$. Thus,

$$\mathbf{u}'_3 = \frac{\mathbf{r}}{|\mathbf{r}|} \tag{15}$$

By choosing another direction as

$$\mathbf{s} = \sum_\alpha d_\alpha \mathbf{x}_\alpha \tag{16}$$

where the coefficients d_α are also known functions of nuclear positions, and $\mathbf{s}$ is not collinear with $\mathbf{r}$ (i.e. $\mathbf{r} \times \mathbf{s} \neq 0$), one can define the body-fixed axis $\mathbf{u}'_2$ as

$$\mathbf{u}'_2 = \frac{\mathbf{r} \times \mathbf{s}}{|\mathbf{r} \times \mathbf{s}|} \tag{17}$$

and the body-fixed axis $\mathbf{u}'_1$ is given by

$$\mathbf{u}'_1 = \mathbf{u}'_2 \times \mathbf{u}'_3 \tag{18}$$

The coefficients c_α and d_α fulfill the relations

$$\sum_\alpha c_\alpha = 0 \tag{19}$$

$$\sum_\alpha d_\alpha = 0 \tag{20}$$

in order to preserve the translational invariance of the body-fixed frame. For a given shape, the change in the orientation of the molecule is identical to the change in the orientation of the body axes, but it is independent of any particular choice of the body axes. However, if the molecule deforms (i.e., the initial and the final shape differ), it makes no sense to ask "how much has the molecule rotated," because the answer would depend on the choice of the body axes [47]. Because the positions $\mathbf{y}_\alpha$ are related to the body-frame positions $\mathbf{y}'_\alpha$ as [compare with Eq. (A73), which is written for the basis vectors]

$$\begin{aligned}\mathbf{y}_\alpha &= R^\dagger \mathbf{y}'_\alpha R = e^{-\chi i\mathbf{u}'_3/2} e^{-\theta i\mathbf{n}/2} e^{-\phi i\mathbf{u}_3/2} \mathbf{y}'_\alpha e^{\phi i\mathbf{u}_3/2} e^{\theta i\mathbf{n}/2} e^{\chi i\mathbf{u}'_3/2} \\ &= e^{-i\phi\mathbf{u}_3/2} e^{-i\theta\mathbf{u}_1/2} e^{-i\chi\mathbf{u}_3/2} \mathbf{y}'_\alpha e^{i\chi\mathbf{u}_3/2} e^{i\theta\mathbf{u}_1/2} e^{i\phi\mathbf{u}_3/2}\end{aligned} \tag{21}$$

(where $\mathbf{n} = \mathbf{u}_3 \times \mathbf{u}'_3/|\mathbf{u}_3 \times \mathbf{u}'_3|$ is the line of nodes), the derivatives of nuclear positions with respect to Euler angles are given by [see Eqs. (A8) and (A32)]

$$\mathbf{e}_\phi^{(\alpha)} = \frac{\partial \mathbf{x}_\alpha}{\partial \phi} = \frac{\partial \mathbf{y}_\alpha}{\partial \phi} = \frac{i}{2}(-\mathbf{u}_3 R^\dagger \mathbf{y}'_\alpha R + R^\dagger \mathbf{y}'_\alpha R\, \mathbf{u}_3) = \mathbf{u}_3 \times \mathbf{y}_\alpha \tag{22}$$

$$\mathbf{e}_\theta^{(\alpha)} = \frac{\partial \mathbf{x}_\alpha}{\partial \theta} = \mathbf{n} \times \mathbf{y}_\alpha \tag{23}$$

$$\mathbf{e}_\chi^{(\alpha)} = \frac{\partial \mathbf{x}_\alpha}{\partial \chi} = \mathbf{u}'_3 \times \mathbf{y}_\alpha \tag{24}$$

In deriving Eq. (23), we have used the identity $R_\phi R_\phi^\dagger = 1$ and Eq. (A74). Because $\sum_\alpha m_\alpha \mathbf{y}_\alpha = 0$, it follows that

$$\sum_\alpha^N m_\alpha \mathbf{e}_\phi^{(\alpha)} = \sum_\alpha^N m_\alpha \mathbf{e}_\theta^{(\alpha)} = \sum_\alpha^N m_\alpha \mathbf{e}_\chi^{(\alpha)} = 0 \tag{25}$$

as one suspects, because rotation a free molecule does not change its center of mass.

Because of the translational invariance of the shape coordinates and Euler angles [Eqs. (13) and (25)], it follows from Eqs. (4) and (9) that translation is separated from the vibrational and rotational degrees of freedom; that is, the matrix $[g_{ij}]$ is partitioned into an internal block $\mathrm{g}_{\mathrm{int}}$ (depending only on the shape coordinates and Euler angles) of the size $(A-3)\times(A-3)$ and to a translational block $\mathrm{g}_{\mathrm{transl}}$ of the size 3×3 (depending only on the center-of-mass coordinates) as

$$[g_{ij}] = \begin{bmatrix} \mathrm{g}_{\mathrm{int}} & 0 \\ 0^{\mathrm{T}} & \mathrm{g}_{\mathrm{transl}} \end{bmatrix} \tag{26}$$

where 0 represents an $(A-3)\times 3$ block of zeros and 0^{T} represents a $3\times(A-3)$ block of zeros. The Lagrangian of a free molecule can be written as

$$L = \frac{1}{2}\left[\sum_{i=1}^{3}\left(M\dot{X}_i^2 + \sum_{j=1}^{A_{\mathrm{V}}} 2\dot{B}_i g_{B_i s_j}\dot{s}_j + \sum_{j=1}^{3}\dot{B}_i g_{B_i B_j}\dot{B}_j\right) + \sum_{i,j=1}^{A_{\mathrm{V}}}\dot{s}_i g_{s_i s_j}\dot{s}_j\right] - V \tag{27}$$

where $B_1 = \phi$, $B_2 = \theta$, and $B_3 = \chi$.

It is common to express the rotation of the molecule in terms of the body-frame components $\omega_i' = \boldsymbol{\omega}'\cdot\mathbf{u}_i = \boldsymbol{\omega}\cdot\mathbf{u}_i'$ of the rotational velocity (see Ref. 50, pages 306–316)

$$\boldsymbol{\omega} = 2i\dot{R}^\dagger R = -2iR^\dagger\dot{R} \tag{28}$$

instead of the time derivatives of the Euler angles $\{\dot\phi, \dot\theta, \dot\chi\}$. The quantity ω_i' is an example of *quasi-velocity*, because it is not obtained as a time derivative of some scalar coordinate. The transformations from $\{\omega_1', \omega_2', \omega_3'\}$ to $\{\dot\phi, \dot\theta, \dot\chi\}$, and vice versa, are easy because the rotational velocity $\boldsymbol{\omega}$ is related to the time derivatives of the Euler angles as (see Ref. 50, page 315)

$$\boldsymbol{\omega} = \dot\phi\mathbf{u}_3 + \dot\theta\mathbf{n} + \dot\chi\mathbf{u}_3' \tag{29}$$

We demonstrate the elegance of geometric algebra by deriving Eq. (29) as follows:

Example 1. *The rotor R can be parameterized by $R = e^{i\chi\mathbf{u}_3/2}e^{i\theta\mathbf{u}_1/2}e^{i\phi\mathbf{u}_3/2} = Q_\chi Q_\theta R_\phi$ (see Appendix A, Section 3.c), so*

$$-2i\dot{R} = \dot{\chi}\mathbf{u}_3 Q_\chi Q_\theta R_\phi + \dot{\theta} Q_\chi Q_\theta \mathbf{u}_1 R_\phi + \dot{\phi} Q_\chi Q_\theta R_\phi \mathbf{u}_3 \tag{30}$$

Because $Q_\theta = R_\phi R_\theta R_\phi^\dagger$ [Eq. (A81)] and $\mathbf{n} = R_\phi^\dagger \mathbf{u}_1 R_\phi$ [Eq. (A74)], the equation

$$-2i\dot{R} = \dot{\chi}\mathbf{u}_3 R + \dot{\theta} Q_\chi R_\phi \mathbf{n} R_\theta + \dot{\phi} R \mathbf{u}_3 \tag{31}$$

follows. Furthermore, because $Q_\chi = RR_\phi^\dagger Q_\theta^\dagger = RR_\theta^\dagger R_\phi^\dagger$, this reads as

$$-2i\dot{R} = \dot{\chi}\mathbf{u}_3 R + \dot{\theta} R\mathbf{n} + \dot{\phi} R\mathbf{u}_3 \tag{32}$$

When both sides of Eq. (32) are multiplied from the left by $R^\dagger$, Eq. (29) follows.

In what follows, we relate the components of the rotational velocity to the time derivatives of the Euler angles and vice versa. First, we express the nodal line vectors $\{\mathbf{n}_1 = \mathbf{u}_3, \mathbf{n}_2 = \mathbf{n}, \mathbf{n}_3 = \mathbf{u}'_3\}$ in Eq. (29) in terms of the body-axes $\{\mathbf{u}'_1, \mathbf{u}'_2, \mathbf{u}'_3\}$ as

$$\begin{bmatrix} \mathbf{u}_3 \\ \mathbf{n} \\ \mathbf{u}'_3 \end{bmatrix} = \Omega \begin{bmatrix} \mathbf{u}'_1 \\ \mathbf{u}'_2 \\ \mathbf{u}'_3 \end{bmatrix} \tag{33}$$

where the elements $[\Omega]_{ij} = \mathbf{n}_i \cdot \mathbf{u}'_j$ of the matrix Ω are given by

$$\Omega = \begin{bmatrix} \sin\theta\sin\chi & \sin\theta\cos\chi & \cos\theta \\ \cos\chi & -\sin\chi & 0 \\ 0 & 0 & 1 \end{bmatrix} \tag{34}$$

Then, we obtain ω'_i by taking the inner product of both sides of Eq. (29) with the body axis $\mathbf{u}'_i$. The result is

$$\begin{bmatrix} \omega'_1 \\ \omega'_2 \\ \omega'_3 \end{bmatrix} = \Omega^T \begin{bmatrix} \dot{\phi} \\ \dot{\theta} \\ \dot{\chi} \end{bmatrix} = \begin{bmatrix} \sin\theta\sin\chi & \cos\chi & 0 \\ \sin\theta\cos\chi & -\sin\chi & 0 \\ \cos\theta & 0 & 1 \end{bmatrix} \begin{bmatrix} \dot{\phi} \\ \dot{\theta} \\ \dot{\chi} \end{bmatrix} \tag{35}$$

where Ω^T is the transpose of Ω. The time derivatives of the Euler angles are given in terms of the body-frame components ω_i' by inverting the relations in Eq. (35). This produces

$$\begin{bmatrix} \dot{\phi} \\ \dot{\theta} \\ \dot{\chi} \end{bmatrix} = (\Omega^T)^{-1} \begin{bmatrix} \omega_1' \\ \omega_2' \\ \omega_3' \end{bmatrix} = \begin{bmatrix} \frac{\sin\chi}{\sin\theta} & \frac{\cos\chi}{\sin\theta} & 0 \\ \cos\chi & -\sin\chi & 0 \\ -\frac{\sin\chi\cos\theta}{\sin\theta} & -\frac{\cos\chi\cos\theta}{\sin\theta} & 1 \end{bmatrix} \begin{bmatrix} \omega_1' \\ \omega_2' \\ \omega_3' \end{bmatrix} \tag{36}$$

The *covariant measuring vectors*[1] $\mathbf{e}_{\omega_i'}^{(\alpha)}$ associated with the nucleus α and the components ω_i' can be obtained, when Eq. (36) is substituted to Eq. (27). This produces

$$L = \frac{1}{2}\left[\sum_{i,k=1}^{3}\left(M\dot{X}_i^2 + \sum_{j=1}^{A_V} 2\omega_k'\Omega_{ki}^{-1} g_{B_i s_j}\dot{s}_j + \sum_{j,l=1}^{3}\omega_k'\Omega_{ki}^{-1} g_{B_i B_j}\Omega_{lj}^{-1}\omega_l'\right) + \sum_{i,j=1}^{A_V}\dot{s}_i g_{s_i s_j}\dot{s}_j\right] - V \tag{37}$$

(because $[(\Omega^T)^{-1}]_{ik} = \Omega_{ki}^{-1}$). By collecting the terms proportional to ω_k', we may define the covariant measuring vectors as

$$\mathbf{e}_{\omega_i'}^{(\alpha)} = \sum_{j=1}^{3}\Omega_{ij}^{-1}\mathbf{e}_{B_j}^{(\alpha)} \tag{38}$$

Stating explicitly,

$$\mathbf{e}_{\omega_1'}^{(\alpha)} = \left(\frac{\sin\chi}{\sin\theta}\mathbf{u}_3 + \cos\chi\mathbf{n} - \frac{\cos\theta\sin\chi}{\sin\theta}\mathbf{u}_3'\right) \times \mathbf{y}_\alpha = \mathbf{u}_1' \times \mathbf{y}_\alpha \tag{39}$$

$$\mathbf{e}_{\omega_2'}^{(\alpha)} = \left(\frac{\cos\chi}{\sin\theta}\mathbf{u}_3 - \sin\chi\mathbf{n} - \frac{\cos\theta\cos\chi}{\sin\theta}\mathbf{u}_3'\right) \times \mathbf{y}_\alpha = \mathbf{u}_2' \times \mathbf{y}_\alpha \tag{40}$$

$$\mathbf{e}_{\omega_3'}^{(\alpha)} = \mathbf{u}_3' \times \mathbf{y}_\alpha \tag{41}$$

[1]We use the term measuring vector because the vector $\mathbf{e}_{\omega_i'}^{(\alpha)}$ is not obtained by differentiating the position $\mathbf{x}_\alpha$ with respect to some coordinate q_i (i.e., it is not a tangent vector to any coordinate q_i). However, because the vectors $\mathbf{e}_{\omega_i'}^{(\alpha)}$ can be written as a linear combinations of the tangent vectors, they can be used to relate the change in the position $\mathbf{x}_\alpha$ to the change in coordinates (which is given by $d\mathbf{x}_\alpha = \sum_i \frac{\partial \mathbf{x}_\alpha}{\partial q_i} dq_i$); that is, they *measure* the change. From now on, we collectively denote both the tangents $\frac{\partial \mathbf{x}_\alpha}{\partial q_i}$ and $\mathbf{e}_{\omega_i'}^{(\alpha)}$ as the *covariant* measuring vectors (to emulate the terminology of classical tensor analysis [52]).

where Eqs. (22)–(24) have been employed. These equations are remarkably similar to those that relate $\mathbf{e}_{B_i}^{(\alpha)}$ to the positions $\mathbf{y}_\alpha$ [Eqs. (22)–(24)]. The measuring vector associated with ω_3' is the same as the measuring vector associated with χ, that is, $\mathbf{e}_{\omega_3'}^{(\alpha)} = \mathbf{e}_\chi^{(\alpha)}$. The measuring vector $\mathbf{e}_{\omega_1'}^{(\alpha)}$ is obtained from the measuring vector $\mathbf{e}_\phi^{(\alpha)}$ by replacing the space-fixed axis $\mathbf{u}_3$ by the body axis $\mathbf{u}_1'$ (and vice versa). Similarly, the measuring vector $\mathbf{e}_{\omega_2'}^{(\alpha)}$ is obtained from the measuring vector $\mathbf{e}_\theta^{(\alpha)}$ by replacing the nodal line vector $\mathbf{n}$ by the body axis $\mathbf{u}_2'$ (and vice versa). Thus, following the notation of Ref. 17, the internal part of the covariant metric tensor, g_{int}, can be written as

$$\mathsf{g}_{\text{int}} = \begin{bmatrix} \mathsf{S} & \tilde{\mathsf{C}}^T \\ \tilde{\mathsf{C}} & \tilde{\mathsf{I}} \end{bmatrix} \tag{42}$$

where the sub-blocks are given in terms of measuring vectors as

$$\mathsf{S}_{ij} = g_{s_i s_j} = \sum_\alpha^N m_\alpha \mathbf{e}_{s_i}^{(\alpha)} \cdot \mathbf{e}_{s_j}^{(\alpha)} \tag{43}$$

$$\tilde{\mathsf{C}}_{ij} = g_{s_i \omega_j'} = \sum_\alpha^N m_\alpha \mathbf{e}_{s_i}^{(\alpha)} \cdot \mathbf{e}_{\omega_j'}^{(\alpha)} \tag{44}$$

$$\tilde{\mathsf{I}}_{ij} = g_{\omega_i' \omega_j'} = \sum_\alpha^N m_\alpha \mathbf{e}_{\omega_i'}^{(\alpha)} \cdot \mathbf{e}_{\omega_j'}^{(\alpha)} \tag{45}$$

(see Ref. 17 for a different way of obtaining these elements). The Lagrangian becomes

$$L = \frac{1}{2}\left[\sum_{i=1}^{3}\left(M\dot{X}_i^2 + \sum_{j=1}^{A_V} 2\omega_i' g_{\omega_i' s_j}\dot{s}_j + \sum_{j=1}^{3}\omega_i' g_{\omega_i'\omega_j'}\omega_j'\right) + \sum_{i,j=1}^{A_V}\dot{s}_i g_{s_i s_j}\dot{s}_j\right] - V \tag{46}$$

The Lagrangian in standard shape coordinates can be formed in the following way. We obtain the tangents by differentiating the linear combinations

$$\mathbf{z}_\alpha = \sum_\beta c_{\alpha\beta}\mathbf{x}_\beta \tag{47}$$

of the nuclear positions $\mathbf{x}_\alpha$ as

$$\mathbf{e}_{s_i}^{(\alpha)} = \sum_\beta c_{\alpha\beta}^{-1}\mathbf{e}_{s_i}^{(\mathbf{z}_\beta)} \tag{48}$$

where $c^{-1}_{\alpha\beta}$ is the element of the inverse $\mathbf{c}^{-1}$ of the matrix $[\mathbf{c}]_{\alpha\beta} = c_{\alpha\beta}$, and

$$\mathbf{e}^{(\mathbf{z}_\beta)}_{s_i} = \frac{\partial \mathbf{z}_\beta}{\partial s_i} \tag{49}$$

is the derivative of the vector $\mathbf{z}_\beta$ with respect to the coordinate s_i. The coefficients $c_{\alpha\beta}$ do not depend on the nuclear positions and the vectors $\mathbf{z}_1, \mathbf{z}_2, \ldots, \mathbf{z}_{N-1}$ are translationally invariant (i.e., $\sum_\beta c_{\alpha\beta} = 0$ for $\alpha = 1, 2, \ldots, N-1$). The vector $\mathbf{z}_N$ is the position of the center of mass $\mathbf{X}$. The vectors $\mathbf{z}_1, \mathbf{z}_2, \ldots, \mathbf{z}_{N-1}$ can be the bond vectors $\mathbf{r}_{\alpha\beta} = \mathbf{x}_\beta - \mathbf{x}_\alpha$, if standard spectroscopical shape coordinates (such as bond lengths and bond angles) are used. In what follows, we assume for simplicity that the shape coordinates include lengths $z_\alpha = |\mathbf{z}_\alpha|$, angles $\theta_{\alpha\beta} = \arccos(\mathbf{u}_{\mathbf{z}_\alpha} \cdot \mathbf{u}_{\mathbf{z}_\beta})$ between vectors $\mathbf{z}_\alpha$ and $\mathbf{z}_\beta$, and the dihedral angles $\tau_{\alpha\beta\gamma} = \arccos(\mathbf{i}_{\alpha\beta} \cdot \mathbf{i}_{\beta\gamma})$ between planes $\mathbf{i}_{\alpha\beta} = \mathbf{z}_\alpha \wedge \mathbf{z}_\beta / |\mathbf{z}_\alpha \wedge \mathbf{z}_\beta|$ and $\mathbf{i}_{\beta\gamma} = \mathbf{z}_\beta \wedge \mathbf{z}_\gamma / |\mathbf{z}_\beta \wedge \mathbf{z}_\gamma|$. They are a complete set of shape coordinates (i.e., distinct configurations are mapped uniquely; see Ref. 53). The relations between angles and dihedral angles as given in Appendix A, Section 3.e can be used to obtain the tangents in some other set of shape coordinates, if needed. The derivatives $\mathbf{e}^{(\mathbf{z}_\beta)}_{s_i}$ can be found by considering the following parameterization in the body frame, where $\boldsymbol{\mu}'_3$ is in the direction of $\mathbf{z}_1$:

1. A vector $\mathbf{z}_1$ can be obtained from the unit vector $\mathbf{u}_3$ by the rotation R and the dilation by z_1. It can be written parametrically as

$$\mathbf{z}_1 = z_1 R^\dagger(\phi, \theta, \chi)\mathbf{u}_3 R(\phi, \theta, \chi) \tag{50}$$

where we obtain

$$\frac{\partial \mathbf{z}_1}{\partial z_1} = \mathbf{u}_{\mathbf{z}_1} \tag{51}$$

All the other vibrational derivatives $\partial \mathbf{z}_1 / \partial s_i$ are taken to be zero.

2. A vector $\mathbf{z}_2$ can be obtained by rotating the unit vector $\mathbf{u}_{\mathbf{z}_1}$ in the plane $\mathbf{i}_{12}(\theta, \phi, \chi) = \mathbf{u}_{\mathbf{z}_1} \wedge \mathbf{u}_{\mathbf{z}_2} / |\mathbf{u}_{\mathbf{z}_1} \wedge \mathbf{u}_{\mathbf{z}_2}| = \mathbf{u}_{\mathbf{z}_1} \wedge \mathbf{u}_{\mathbf{z}_2} / \sin\theta_{12}$ (whose orientation can be parameterized by three Euler angles, as indicated) by the angle θ_{12} and by multiplying it by the factor z_2. The vector $\mathbf{z}_2$ can be written parametrically as

$$\mathbf{z}_2(z_2, \theta_{12}, \phi, \theta, \chi) = z_2 e^{-\theta_{12}\mathbf{i}_{12}/2} \mathbf{u}_{\mathbf{z}_1} e^{\theta_{12}\mathbf{i}_{12}/2} = z_2 \mathbf{u}_{\mathbf{z}_1} e^{\theta_{12}\mathbf{i}_{12}} \tag{52}$$

where we have used $e^{-\theta_{12}\mathbf{i}_{12}/2}\mathbf{u}_{\mathbf{z}_1} = \mathbf{u}_{\mathbf{z}_1} e^{\theta_{12}\mathbf{i}_{12}/2}$ in the second equality ($\mathbf{u}_{\mathbf{z}_1}\mathbf{i}_{12} = \mathbf{u}_{\mathbf{z}_1} \cdot \mathbf{i}_{12} = -\mathbf{i}_{12}\mathbf{u}_{\mathbf{z}_1}$, because $\mathbf{u}_{\mathbf{z}_1}$ is in the plane $\mathbf{i}_{12}$). Thus, we obtain

$$\frac{\partial \mathbf{z}_2}{\partial z_2} = \mathbf{u}_{\mathbf{z}_2} \tag{53}$$

$$\frac{\partial \mathbf{z}_2}{\partial \theta_{12}} = z_2 \frac{\partial \mathbf{u}_{\mathbf{z}_1} e^{\theta_{12}\mathbf{i}_{12}}}{\partial \theta_{12}} = \mathbf{z}_2 \cdot \mathbf{i}_{12} = \frac{z_2(\cos\theta_{12}\mathbf{u}_{\mathbf{z}_2} - \mathbf{u}_{\mathbf{z}_1})}{\sin\theta_{12}} \tag{54}$$

[we have replaced the geometric product $\mathbf{z}_2\mathbf{i}_{12}$ by the inner product $\mathbf{z}_2 \cdot \mathbf{i}_{12}$, because the outer product $\mathbf{z}_2 \wedge \mathbf{i}_{12}$ vanishes. The last equality follows from the expansion rule of Eq. (A31)]. All the other vibrational derivatives $\partial \mathbf{z}_2/\partial s_i$ are taken to be zero.

3. Other vectors $\mathbf{z}_3, \mathbf{z}_4, \ldots, \mathbf{z}_{N-1}$ can be parameterized with respect to two previously defined vectors $\mathbf{z}_\beta$ and $\mathbf{z}_\gamma$. Suitable coordinates are the length z_α, the angle $\theta_{\alpha\beta}$, and the torsional angle $\tau_{\alpha\beta\gamma} = \arccos(\mathbf{i}_{\alpha\beta} \cdot \mathbf{i}_{\beta\gamma})$ between the planes $\mathbf{i}_{\alpha\beta} = \mathbf{z}_\alpha \wedge \mathbf{z}_\beta / |\mathbf{z}_\alpha \wedge \mathbf{z}_\beta|$ and $\mathbf{i}_{\beta\gamma} = \mathbf{z}_\beta \wedge \mathbf{z}_\gamma / |\mathbf{z}_\beta \wedge \mathbf{z}_\gamma|$ (see Fig. 2).

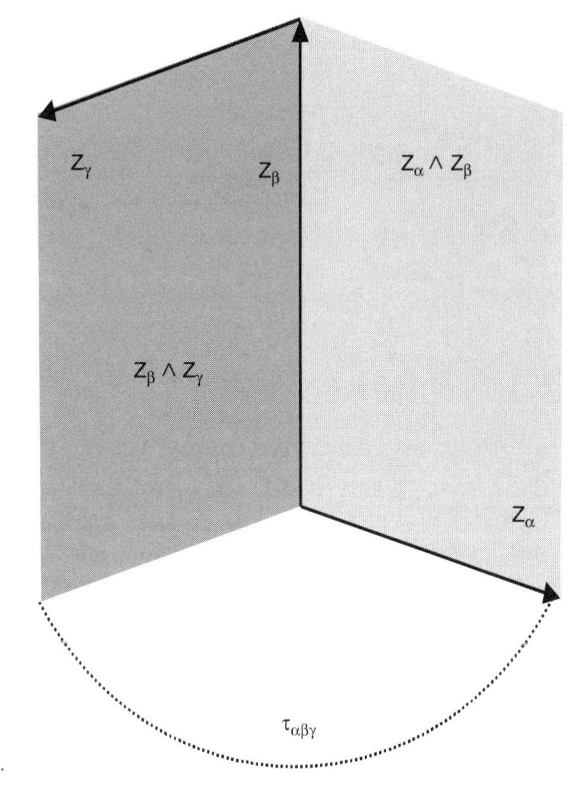

Figure 2. Dihedral angle $\tau_{\alpha\beta\gamma}$.

The vector $\mathbf{z}_\alpha$ is obtained from the unit vector $\mathbf{u}_{\mathbf{z}_\beta}$ by rotating it in the plane $\mathbf{i}_{\beta\gamma}$ by an angle $\theta_{\alpha\beta}$, followed by another rotation about the axis $\mathbf{u}_{\mathbf{z}_\beta}$ by an dihedral angle $\tau_{\alpha\beta\gamma}$, and a dilation by z_α. It can be written parametrically as

$$
\begin{aligned}
&\mathbf{z}_\alpha(z_\alpha, \theta_{\alpha\beta}, \tau_{\alpha\beta\gamma}, \phi, \theta, \chi) \\
&\quad = z_\alpha e^{-\tau_{\alpha\beta\gamma} i \mathbf{u}_{\mathbf{z}_\beta}/2} e^{-\theta_{\alpha\beta}\mathbf{i}_{\beta\gamma}/2} \mathbf{u}_{\mathbf{z}_\beta} e^{\theta_{\alpha\beta}\mathbf{i}_{\beta\gamma}/2} e^{\tau_{\alpha\beta\gamma} i \mathbf{u}_{\mathbf{z}_\beta}/2} \\
&\quad = z_\alpha e^{-\tau_{\alpha\beta\gamma} i \mathbf{u}_{\mathbf{z}_\beta}/2} \mathbf{u}_{\mathbf{z}_\beta} e^{\theta_{\alpha\beta}\mathbf{i}_{\beta\gamma}} e^{\tau_{\alpha\beta\gamma} i \mathbf{u}_{\mathbf{z}_\beta}/2} \\
&\quad = z_\alpha\{[\cos\theta_{\alpha\beta} + \sin\theta_{\alpha\beta}((1-\cos\tau_{\alpha\beta\gamma})\cot\theta_{\beta\gamma} - \tan\theta_{\beta\gamma})]\mathbf{u}_{\mathbf{z}_\beta} \\
&\qquad + \frac{\sin\theta_{\alpha\beta}}{\sin\theta_{\beta\gamma}}(\cos\tau_{\alpha\beta\gamma}\mathbf{u}_{\mathbf{z}_\gamma} + \sin\tau_{\alpha\beta\gamma}\mathbf{u}_{\mathbf{z}_\beta} \times \mathbf{u}_{\mathbf{z}_\gamma})\}
\end{aligned}
\tag{55}
$$

where we obtain

$$
\frac{\partial \mathbf{z}_\alpha}{\partial z_\alpha} = \mathbf{u}_{\mathbf{z}_\alpha} \tag{56}
$$

$$
\begin{aligned}
\frac{\partial \mathbf{z}_\alpha}{\partial \theta_{\alpha\beta}} &= z_\alpha\{[-\sin\theta_{\alpha\beta} + \cos\theta_{\alpha\beta}((1-\cos\tau_{\alpha\beta\gamma})\cot\theta_{\beta\gamma} - \tan\theta_{\beta\gamma})]\mathbf{u}_{\mathbf{z}_\beta} \\
&\quad + \frac{\cos\theta_{\alpha\beta}}{\sin\theta_{\beta\gamma}}(\cos\tau_{\alpha\beta\gamma}\mathbf{u}_{\mathbf{z}_\gamma} + \sin\tau_{\alpha\beta\gamma}\mathbf{u}_{\mathbf{z}_\beta} \times \mathbf{u}_{\mathbf{z}_\gamma})\}
\end{aligned}
\tag{57}
$$

$$
\frac{\partial \mathbf{z}_\alpha}{\partial \tau_{\alpha\beta\gamma}} = \frac{i}{2}(-\mathbf{u}_{\mathbf{z}_\beta}\mathbf{z}_\alpha + \mathbf{z}_\alpha\mathbf{u}_{\mathbf{z}_\beta}) = \mathbf{u}_{\mathbf{z}_\beta} \times \mathbf{z}_\alpha \tag{58}
$$

All the other vibrational derivatives $\partial\mathbf{z}_\alpha/\partial s_i$ $(\alpha = 3, 4, \ldots, N-1)$ are taken to be zero, except the derivatives with respect to the valence angles and dihedral angles, which are used to parameterize the vectors $\mathbf{z}_\beta$ and $\mathbf{z}_\gamma$. If these coordinates are included to the shape coordinate set $\{s_1, s_2, \ldots, s_{A_V}\}$, the non-zero derivatives $\partial\mathbf{z}_\alpha/\partial s_i$ can be found by substituting the derivatives $\partial\mathbf{u}_{\mathbf{z}_\beta}/\partial s_i$ and $\partial\mathbf{u}_{\mathbf{z}_\gamma}/\partial s_i$ derived in the previous steps to

$$
\begin{aligned}
\frac{\partial \mathbf{z}_\alpha}{\partial s_i} &= z_\alpha\left\{[\cos\theta_{\alpha\beta} + \sin\theta_{\alpha\beta}((1-\cos\tau_{\alpha\beta\gamma})\cot\theta_{\beta\gamma} - \tan\theta_{\beta\gamma})]\frac{\partial\mathbf{u}_{\mathbf{z}_\beta}}{\partial s_i}\right. \\
&\quad \left. + \frac{\sin\theta_{\alpha\beta}}{\sin\theta_{\beta\gamma}}\left[\cos\tau_{\alpha\beta\gamma}\frac{\partial\mathbf{u}_{\mathbf{z}_\gamma}}{\partial s_i} + \sin\tau_{\alpha\beta\gamma}\left(\frac{\partial\mathbf{u}_{\mathbf{z}_\beta}}{\partial s_i} \times \mathbf{u}_{\mathbf{z}_\gamma} + \mathbf{u}_{\mathbf{z}_\beta} \times \frac{\partial\mathbf{u}_{\mathbf{z}_\gamma}}{\partial s_i}\right)\right]\right\}
\end{aligned}
\tag{59}
$$

We may as well use other coordinates to parameterize $\mathbf{z}_\alpha$. As an example, we consider the parameterization by the length z_α, the angle $\theta_{\alpha\beta}$, and the angle $\phi_{\alpha\beta\gamma}$ between the vector $\mathbf{z}_\alpha$ and the plane $\mathbf{z}_\beta \wedge \mathbf{z}_\gamma$ (see Fig. 3):

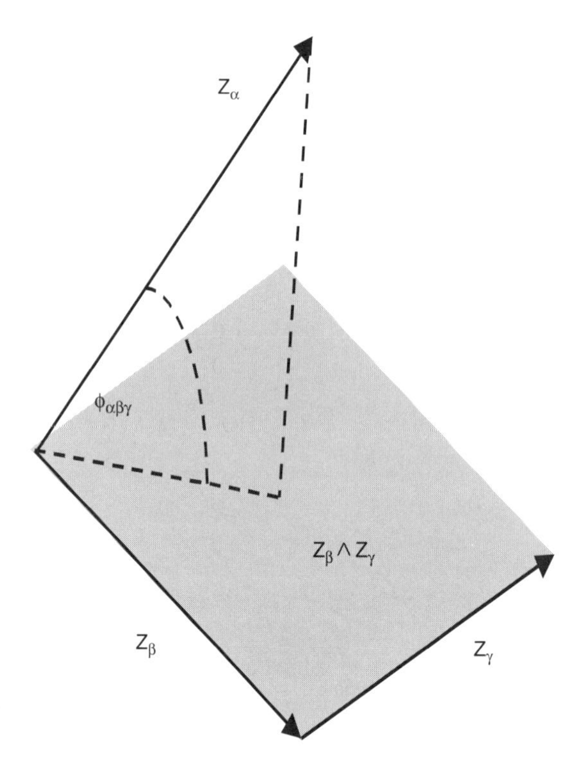

Figure 3. Angle between vector and plane $\phi_{\alpha\beta\gamma}$.

Example 3. *We find the tangent* $\partial \mathbf{z}_\alpha / \partial \phi_{\alpha\beta\gamma}$ *from the tangent* $\partial \mathbf{z}_\alpha / \partial \tau_{\alpha\beta\gamma}$ *[Eq. (58)] by performing the coordinate transformation from* $\{z_\alpha, \theta_{\alpha\beta}, \tau_{\alpha\beta\gamma}\}$ *to* $\{z_\alpha, \theta_{\alpha\beta}, \phi_{\alpha\beta\gamma}\}$. *Because both coordinate sets include* $\{z_\alpha, \theta_{\alpha\beta}\}$, *it suffices to consider* $\phi_{\alpha\beta\gamma}$ *as function of* $\tau_{\alpha\beta\gamma}$ *only; that is, we can write*

$$\sin \phi_{\alpha\beta\gamma} = \sin \theta_{\alpha\beta} \sin \tau_{\alpha\beta\gamma} = f(\tau_{\alpha\beta\gamma}) \tag{60}$$

(see Appendix A, Section 3.e). The partial derivation of the equation given above w.r.t. $\phi_{\alpha\beta\gamma}$ *produces*

$$\cos \phi_{\alpha\beta\gamma} = \sin \theta_{\alpha\beta} \cos \tau_{\alpha\beta\gamma} \frac{\partial \tau_{\alpha\beta\gamma}}{\partial \phi_{\alpha\beta\gamma}} \tag{61}$$

or

$$\frac{\partial \tau_{\alpha\beta\gamma}}{\partial \phi_{\alpha\beta\gamma}} = \frac{\cos \phi_{\alpha\beta\gamma}}{\sin \theta_{\alpha\beta} \cos \tau_{\alpha\beta\gamma}} = \frac{\cos \phi_{\alpha\beta\gamma}}{\sin \theta_{\alpha\beta} \cos[\arcsin(\sin \phi_{\alpha\beta\gamma} / \sin \theta_{\alpha\beta})]} \tag{62}$$

and we have

$$\frac{\partial \mathbf{z}_\alpha}{\partial \phi_{\alpha\beta\gamma}} = \frac{\partial \mathbf{z}_\alpha}{\partial \tau_{\alpha\beta\gamma}} \frac{\partial \tau_{\alpha\beta\gamma}}{\partial \phi_{\alpha\beta\gamma}} = \frac{\mathbf{u}_{\mathbf{z}_\beta} \times \mathbf{z}_\alpha \cos \phi_{\alpha\beta\gamma}}{\sin \theta_{\alpha\beta} \cos[\arcsin(\sin \phi_{\alpha\beta\gamma}/\sin \theta_{\alpha\beta})]} \tag{63}$$

Although the covariant rotational measuring vectors $\mathbf{e}_{\omega_i'}^{(\alpha)}$ can be obtained from Eqs. (39)–(41), it is advantageous to use the vectors

$$\mathbf{e}_{\omega_i'}^{(\mathbf{z}_\alpha)} = \mathbf{u}_i' \times \mathbf{z}_\alpha \tag{64}$$

to obtain the measuring vectors [analogously to Eq. (48)] as

$$\mathbf{e}_{\omega_i'}^{(\alpha)} = \sum_\beta c_{\alpha\beta}^{-1} \mathbf{e}_{\omega_i'}^{(\mathbf{z}_\beta)} \tag{65}$$

We present a fully worked-out example to make this concrete:

Example 4. *We derive the Lagrangian for a triatomic molecule in the bond-z frame and in the shape coordinates* $\{r_{31}, r_{32}, \theta_{132}\}$. *The suitable linear combinations* $\{\mathbf{z}_1, \mathbf{z}_2, \mathbf{z}_3\}$ *are now the bond vectors* $\mathbf{r}_{31} = \mathbf{x}_1 - \mathbf{x}_3$ *and* $\mathbf{r}_{32} = \mathbf{x}_2 - \mathbf{x}_3$ *and the position of the center of mass* $\mathbf{X} = (m_1\mathbf{x}_1 + m_2\mathbf{x}_2 + m_3\mathbf{x}_3)/M$ *(where* $M = m_1 + m_2 + m_3$*); that is, now*

$$\begin{bmatrix} \mathbf{z}_1 \\ \mathbf{z}_2 \\ \mathbf{z}_3 \end{bmatrix} = \begin{bmatrix} 1 & 0 & -1 \\ 0 & 1 & -1 \\ \frac{m_1}{M} & \frac{m_2}{M} & \frac{m_3}{M} \end{bmatrix} \begin{bmatrix} \mathbf{x}_1 \\ \mathbf{x}_2 \\ \mathbf{x}_3 \end{bmatrix} \tag{66}$$

By inverting the above relations, we obtain

$$\begin{bmatrix} \mathbf{x}_1 \\ \mathbf{x}_2 \\ \mathbf{x}_3 \end{bmatrix} = \begin{bmatrix} \frac{m_3+m_2}{M} & -\frac{m_2}{M} & 1 \\ -\frac{m_1}{M} & \frac{m_3+m_1}{M} & 1 \\ -\frac{m_1}{M} & -\frac{m_2}{M} & 1 \end{bmatrix} \begin{bmatrix} \mathbf{z}_1 \\ \mathbf{z}_2 \\ \mathbf{z}_3 \end{bmatrix} \tag{67}$$

By using the tangents in Eqs. (51), (53), and (54), we obtain the following derivatives $\partial \mathbf{z}_\alpha / \partial s_i$ *differing from zero:*

$$\frac{\partial \mathbf{r}_{31}}{\partial r_{31}} = \mathbf{u}_{\mathbf{r}_{31}} \tag{68}$$

$$\frac{\partial \mathbf{r}_{32}}{\partial r_{32}} = \mathbf{u}_{\mathbf{r}_{32}} \tag{69}$$

$$\frac{\partial \mathbf{r}_{32}}{\partial \theta_{132}} = \frac{r_{32}(\cos \theta_{132} \mathbf{u}_{\mathbf{r}_{32}} - \mathbf{u}_{\mathbf{r}_{31}})}{\sin \theta_{132}} \tag{70}$$

and we finally obtain the vibrational tangents $\mathbf{e}_{s_i}^{(\alpha)}$ *via Eqs. (48) and (67). For example,*

$$\mathbf{e}_{r_{31}}^{(1)} = \frac{\partial \mathbf{x}_1}{\partial r_{31}} = \frac{(m_2 + m_3)\mathbf{u}_{\mathbf{r}_{31}}}{M} \tag{71}$$

$$\mathbf{e}_{r_{31}}^{(2)} = \frac{\partial \mathbf{x}_2}{\partial r_{31}} = -\frac{m_1 \mathbf{u}_{\mathbf{r}_{31}}}{M} \tag{72}$$

$$\mathbf{e}_{r_{31}}^{(3)} = \frac{\partial \mathbf{x}_3}{\partial r_{31}} = -\frac{m_1 \mathbf{u}_{\mathbf{r}_{31}}}{M} \tag{73}$$

The directions $\mathbf{r}$ *and* $\mathbf{s}$ *in Eqs. (14) and (16) are in the chosen bond-z frame* $\mathbf{r} = \mathbf{r}_{31}$ *and* $\mathbf{s} = \mathbf{r}_{32}$, *so the body axes are*

$$\mathbf{u}'_3 = \mathbf{u}_{\mathbf{r}_{31}} \tag{74}$$

$$\mathbf{u}'_2 = \frac{\mathbf{u}_{\mathbf{r}_{31}} \times \mathbf{u}_{\mathbf{r}_{32}}}{\sin\theta_{132}} \tag{75}$$

$$\mathbf{u}'_1 = \mathbf{u}'_2 \times \mathbf{u}'_3 = \frac{\mathbf{u}_{\mathbf{r}_{32}} - \cos\theta_{132}\mathbf{u}_{\mathbf{r}_{31}}}{\sin\theta_{132}} \tag{76}$$

[where we have used the triple product in Eq. (A34) to obtain the latter equality in Eq. (76)] and we obtain the covariant rotational measuring vectors associated with the bond vectors by the direct substitution of Eq. (64) as

$$\mathbf{e}_{\omega'_1}^{(\mathbf{r}_{31})} = \mathbf{u}'_1 \times \mathbf{r}_{31} = \frac{\mathbf{u}_{\mathbf{r}_{32}} \times \mathbf{r}_{31}}{\sin\theta_{132}} = -r_{31}\mathbf{u}'_2 \tag{77}$$

$$\mathbf{e}_{\omega'_2}^{(\mathbf{r}_{31})} = \mathbf{u}'_2 \times \mathbf{r}_{31} = r_{31}\mathbf{u}'_1 \tag{78}$$

$$\mathbf{e}_{\omega'_3}^{(\mathbf{r}_{31})} = \mathbf{u}'_3 \times \mathbf{r}_{31} = 0 \tag{79}$$

and

$$\mathbf{e}_{\omega'_1}^{(\mathbf{r}_{32})} = \mathbf{u}'_1 \times \mathbf{r}_{32} = -\frac{\cos\theta_{132}\mathbf{u}_{\mathbf{r}_{31}} \times \mathbf{r}_{32}}{\sin\theta_{132}} = -r_{32}\cos\theta_{132}\mathbf{u}'_2 \tag{80}$$

$$\mathbf{e}_{\omega'_2}^{(\mathbf{r}_{32})} = \mathbf{u}'_2 \times \mathbf{r}_{32} = \frac{r_{32}(\cos\theta_{132}\mathbf{u}_{\mathbf{r}_{32}} - \mathbf{u}_{\mathbf{r}_{31}})}{\sin\theta_{132}} \tag{81}$$

$$\mathbf{e}_{\omega'_3}^{(\mathbf{r}_{32})} = \mathbf{u}'_3 \times \mathbf{r}_{32} = r_{32}\sin\theta_{132}\mathbf{u}'_2 \tag{82}$$

where we finally obtain via Eqs. (65) and (67) the rotational tangents $\mathbf{e}_{\omega'_i}^{(\alpha)}$. *For example,*

$$\mathbf{e}_{\omega'_3}^{(3)} = \frac{m_2 r_{32}\mathbf{u}'_2}{M} \tag{83}$$

The metric tensor elements (differing from zero) in the Lagrangian [Eq. (46)] are obtained from the tangents as

$$g_{r_{31}r_{31}} = \frac{m_1(m_2+m_3)}{M} \tag{84}$$

$$g_{r_{31}r_{32}} = -\frac{\cos\theta_{132}m_1m_2}{M} \tag{85}$$

$$g_{r_{31}\theta_{132}} = \frac{\sin\theta_{132}r_{32}m_1m_2}{M} \tag{86}$$

$$g_{r_{31}\omega'_2} = \frac{r_{32}\sin\theta_{132}m_1m_2}{M} \tag{87}$$

$$g_{r_{32}r_{32}} = \frac{m_2(m_1+m_3)}{M} \tag{88}$$

$$g_{r_{32}\omega'_2} = -\frac{r_{31}\sin\theta_{132}m_1m_2}{M} \tag{89}$$

$$g_{\theta_{132}\theta_{132}} = \frac{r_{32}^2m_2(m_1+m_3)}{M} \tag{90}$$

$$g_{\theta_{132}\omega'_2} = \frac{m_2r_{32}[r_{32}m_3+m_1(r_{32}-r_{31}\cos\theta_{132})]}{M} \tag{91}$$

$$g_{\omega'_1\omega'_1} = \frac{r_{32}^2\cos^2\theta_{132}m_2m_3+m_1[r_{31}^2m_3+m_2(r_{31}-r_{32}\cos\theta_{132})^2]}{M} \tag{92}$$

$$g_{\omega'_1\omega'_3} = -\frac{m_2r_{32}\sin\theta_{132}[m_3r_{32}\cos\theta_{132}+m_1(-r_{31}+r_{32}\cos\theta_{132})]}{M} \tag{93}$$

$$g_{\omega'_2\omega'_2} = \frac{m_2m_3r_{32}^2+m_1[m_3r_{31}^2+m_2(r_{31}^2-2r_{31}r_{32}\cos\theta_{132}+r_{32}^2)]}{M} \tag{94}$$

$$g_{\omega'_3\omega'_3} = \frac{r_{32}^2\sin^2\theta_{132}m_2(m_1+m_3)}{M} \tag{95}$$

2. *Hamiltonian*

The state of a system is represented by a single trajectory in a $2A$-dimensional phase space $\{q_1, q_2, \ldots, q_A;\ p_1, p_2, \ldots, p_A\}$ in the Hamiltonian formalism. The generalized coordinates q_i and the generalized momenta

$$p_i = \frac{\partial L}{\partial \dot{q}_i} = \sum_j g_{q_iq_j}\dot{q}_j + b_i \tag{96}$$

are regarded as independent variables, and they can be solved as explicit functions of time from the Hamilton's canonical equations [51,54]

$$\frac{\partial H}{\partial p_i} = \dot{q}_i \tag{97}$$

$$\frac{\partial H}{\partial q_i} = -\dot{p}_i \tag{98}$$

where the Hamiltonian H is given by

$$H = -L + \sum_i^A p_i \dot{q}_i \tag{99}$$

The quantity $b_i = 0$ for conservative systems in Eq. (96), and it is possible to solve

$$\dot{q}_i = \sum_j^A g^{(q_i q_j)} p_j \tag{100}$$

where the contravariant metric tensor $g^{(q_i q_j)}$ is the inverse of the covariant metric tensor $g_{q_i q_j}$, that is,

$$\sum_k^A g^{(q_i q_k)} g_{q_k q_j} = \delta_{ij} \tag{101}$$

where Kronecker's delta δ_{ij} is one, if $i = j$, and zero otherwise. By direct substitution of Eqs. (3) and (100) to Eq. (99), the Hamiltonian of a conservative system is seen to be equal to the total energy,

$$H = T + V = \frac{1}{2} \sum_{ij}^A p_i g^{(q_i q_j)} p_j + V \tag{102}$$

The body-frame components $l'_i = \mathbf{u}'_i \cdot \mathbf{l} = \mathbf{u}_i \cdot \mathbf{l}'$ of the internal angular momentum (Ref. 50, page 339)

$$\mathbf{l} = \sum_\alpha^N m_\alpha \mathbf{y}_\alpha \times \dot{\mathbf{y}}_\alpha = \sum_\alpha^N m_\alpha \mathbf{y}_\alpha \times (\boldsymbol{\omega} \times \mathbf{y}_\alpha + R^\dagger \dot{\mathbf{y}}'_\alpha R) \tag{103}$$

can be extracted by differentiating the Lagrangian with respect to the quasi-velocity ω_i' as

$$l_i' = \frac{\partial L}{\partial \omega_i'} = \sum_{j=1}^{3} g_{\omega_i'\omega_j'}\omega_j' + \sum_{j=1}^{A_V} g_{\omega_i' s_j}\dot{s}_j \tag{104}$$

The first Hamiltonian equation reads as

$$\frac{\partial H}{\partial l_i'} = \omega_i' \tag{105}$$

in terms of ω_i' and l_i'. It is possible to write the analogue of the second Hamiltonian equation in terms of the quasi-coordinates and the time derivative of the quasi-momenta. The interested reader is referred to Ref. 54. The first term on the right-hand side of Eq. (104),

$$j_i' = \sum_{j=1}^{3} g_{\omega_i'\omega_j'}\omega_j' \tag{106}$$

is often referred as the rotational contribution, and the second term,

$$k_i' = \sum_{j=1}^{A_V} g_{\omega_i s_j}\dot{s}_j \tag{107}$$

is often referred to as the vibrational contribution to the body-frame component l_i' of the internal angular momentum $\mathbf{l}$. However, as pointed out in Ref. 47, this decomposition is not gauge-invariant (it depends on the choice of the body frame), so one should not address physical significance to it. The momentum conjugated to the shape coordinate s_i is given by

$$p_{s_i} = \frac{\partial L}{\partial \dot{s}_i} = \sum_{j=1}^{A_V} g_{s_i s_j}\dot{s}_j + \sum_{j=1}^{3} g_{s_i\omega_j'}\omega_j' \tag{108}$$

and the momentum of the center of mass is given by

$$p_{X_i} = \frac{\partial L}{\partial \dot{X}_i} = M\dot{X}_i$$

Thus, the Hamiltonian reads as

$$H = \frac{1}{2}\left[\sum_{i=1}^{3}\left(\frac{p_{X_i}^2}{M} + \sum_{j=1}^{3} l'_i g^{(l'_i l'_j)} l'_j + 2\sum_{j=1}^{A_V} l'_i g^{(l'_i s_j)} p_{s_j}\right) + \sum_{i,j=1}^{A_V} p_{s_i} g^{(s_i s_j)} p_{s_j}\right] + V \tag{109}$$

where the metric tensor elements $g^{(l'_i l'_j)}$, $g^{(l'_i s_j)}$, and $g^{(s_i s_j)}$ are obtained from the inverse of the covariant metric tensor as

$$\left[\begin{array}{ccc|ccc} g^{(s_1 s_1)} & \cdots & g^{(s_1 s_{A_V})} & g^{(s_1 l'_1)} & g^{(s_1 l'_2)} & g^{(s_1 l'_3)} \\ & \ddots & \vdots & \vdots & \vdots & \vdots \\ & & g^{(s_{A_V} s_{A_V})} & g^{(s_{A_V} l'_1)} & g^{(s_{A_V} l'_2)} & g^{(s_{A_V} l'_3)} \\ \hline & & & g^{(l'_1 l'_1)} & g^{(l'_1 l'_2)} & g^{(l'_1 l'_3)} \\ & & & & g^{(l'_2 l'_2)} & g^{(l'_2 l'_3)} \\ & & & & & g^{(l'_3 l'_3)} \end{array}\right]$$

$$= \left[\begin{array}{ccc|ccc} g_{s_1 s_1} & \cdots & g_{s_1 s_{A_V}} & g_{s_1 \omega'_1} & g_{s_1 \omega'_2} & g_{s_1 \omega'_3} \\ & \ddots & \vdots & \vdots & \vdots & \vdots \\ & & g_{s_{A_V} s_{A_V}} & g_{s_{A_V} \omega'_1} & g_{s_{A_V} \omega'_2} & g_{s_{A_V} \omega'_3} \\ \hline & & & g_{\omega'_1 \omega'_1} & g_{\omega'_1 \omega'_2} & g_{\omega'_1 \omega'_3} \\ & & & & g_{\omega'_2 \omega'_2} & g_{\omega'_2 \omega'_3} \\ & & & & & g_{\omega'_3 \omega'_3} \end{array}\right]^{-1} \tag{110}$$

We have indicated only upper triangles, because the matrices are symmetric. We have also separated the "rotational," "Coriolis," and "shape" blocks by the border lines, although this categorization has no true physical significance. The Hamiltonian in Eq. (109) can be represented in various other ways. The interested reader is referred to Ref. 47.

Much of the practically oriented literature utilizing the Lagrangian is devoted for finding suitable factorizations of $[g_{\text{int}}]$, which allows one to avoid the full inversion of $[g_{ij}]$ to obtain $[g^{(ij)}]$ (see Refs. 17 and 18). For example, the matrix g_{int} in Eq. (42) can be represented as a product

$$\mathsf{g}_{\text{int}} = \begin{bmatrix} 1 & 0 \\ 0 & \Omega \end{bmatrix} \begin{bmatrix} \mathsf{S} & \mathsf{C}^T \\ \mathsf{C} & \mathsf{I} \end{bmatrix} \begin{bmatrix} 1 & 0 \\ 0 & \Omega^T \end{bmatrix}$$

where $\mathsf{1}$ is an $(A-3)\times(A-3)$ diagonal unit block, $\mathsf{I}=\Omega^{-1}\tilde{\mathsf{I}}(\Omega^T)^{-1}$ and $\mathsf{C}=\Omega^{-1}\tilde{\mathsf{C}}$. Consequently, the contravariant metric tensor $[g^{(ij)}]$, which appears in the Hamiltonian of Eq. (109), can be obtained (at least in principle, if not in practice) as in Ref. 17

$$[g^{(ij)}]=\begin{bmatrix}\mathsf{1} & \mathsf{0}\\ \mathsf{0} & (\Omega^T)^{-1}\end{bmatrix}\mathsf{J}^{-1}\begin{bmatrix}\mathsf{1} & \mathsf{0}\\ \mathsf{0} & \Omega^{-1}\end{bmatrix} \tag{111}$$

where J^{-1} is an $(A-3)\times(A-3)$ matrix

$$\mathsf{J}^{-1}=\begin{bmatrix}\mathsf{1} & -\mathsf{S}^{-1}\mathsf{C}^T\\ \mathsf{0} & \mathsf{1}\end{bmatrix}\begin{bmatrix}\mathsf{S}^{-1} & \mathsf{J}^{-1}\\ \mathsf{0} & (\mathsf{I}^*)^{-1}\end{bmatrix}\begin{bmatrix}\mathsf{1} & \mathsf{0}\\ -\mathsf{C}\mathsf{S}^{-1} & \mathsf{1}\end{bmatrix} \tag{112}$$

and $\mathsf{I}^*=\mathsf{I}-\mathsf{C}\mathsf{S}^{-1}\mathsf{C}^T$. The largest matrix one needs to invert is the $(A-6)\times(A-6)$-dimensional matrix S, instead of the whole $(A-3)\times(A-3)$-dimensional metric tensor. The elements of S^{-1} are known for all standard shape coordinates (such as bond lengths and bond angles) of a free molecule not subject to any constraints [1,36]. Thus, in the absence of constraints, it suffices to invert only the 3×3-dimensional matrix I^* to obtain the Hamiltonian in the standard shape coordinates. Unfortunately, the calculations may remain laborious even after this simplification. The Hamiltonians in standard internal coordinates (such as bond lengths and bond angles) have been obtained by this approach for triatomic molecules [18]. The Hamiltonians have been formed up to six-atomic molecules in principal-axis hyperspherical coordinates [19–21].

B. Hamiltonian via Scalar Chain Rules

We give a brief overlook to the derivation of the vibration–rotation Hamiltonian with the coordinate chain rules of differentiation (e.g., Refs. 22–28). For example, in Ref. 23, $g^{(q_iq_j)}$ is written as

$$g^{(q_iq_j)}=\sum_{\alpha}^{N}\frac{1}{m_\alpha}\sum_{r}^{3}\frac{\partial q_i}{\partial x_{\alpha_r}}\frac{\partial q_j}{\partial x_{\alpha_r}} \tag{113}$$

where the variables $x_{\alpha_i}=\mathbf{x}_\alpha\cdot\mathbf{u}_i$ are the Cartesian coordinates of the position vectors $\mathbf{x}_\alpha$ in some orthonormal basis $\{\mathbf{u}_1,\mathbf{u}_2,\mathbf{u}_3\}$. The correctness of Eq. (113) can be verified by direct substitution

$$\begin{aligned}g^{(q_iq_k)}g_{q_kq_j}&=\sum_{k}^{3N}\sum_{\alpha\beta}^{N}\frac{m_\beta}{m_\alpha}\sum_{rs}^{3}\frac{\partial q_i}{\partial x_{\alpha_r}}\frac{\partial q_k}{\partial x_{\alpha_r}}\frac{\partial x_{\beta_s}}{\partial q_k}\frac{\partial x_{\beta_s}}{\partial q_j}=\sum_{\alpha\beta}^{N}\frac{m_\beta}{m_\alpha}\sum_{rs}^{3}\frac{\partial q_i}{\partial x_{\alpha_r}}\frac{\partial x_{\beta_s}}{\partial x_{\alpha_r}}\frac{\partial x_{\beta_s}}{\partial q_j}\\&=\sum_{\alpha}^{N}\sum_{r}^{3}\frac{\partial q_i}{\partial x_{\alpha_r}}\frac{\partial x_{\alpha_r}}{\partial q_j}=\frac{\partial q_i}{\partial q_j}=\delta_{ij}\end{aligned} \tag{114}$$

It is possible to reduce the amount of work needed in applying the chain rule with a suitable choice of Cartesian coordinates, shape coordinates, and body frames (see Refs. 26–28). For example, in Ref. 26, the chain rule analogous to Eq. (113), namely

$$g^{(q_i q_j)} = \left[\sum_{\alpha}^{N-1} \frac{c_{i\alpha} c_{j\alpha}}{m_\alpha} - \frac{1}{M}\left(\sum_{\alpha}^{N-1} c_{i\alpha}\right)\left(\sum_{\alpha}^{N-1} c_{j\alpha}\right)\right] \sum_{rs}^{3} \frac{\partial q_i}{\partial z_{\alpha_r}} \frac{\partial q_j}{\partial z_{\beta_s}} \tag{115}$$

involves $3(N-1)$ coordinates $z_{\alpha_i} = \mathbf{z}_\alpha \cdot \mathbf{u}_i$ of some $N-1$ translationally invariant linear combinations $\mathbf{z}_\alpha = \sum_{\beta}^{N-1} c_{\alpha\beta} \mathbf{y}_\beta$ of the nuclear positions $\mathbf{y}_\alpha$, such as the Jacobi or Radau vectors ($\mathbf{z}_N$ is the center of mass $\mathbf{X}$). The directions $\mathbf{r}$ and $\mathbf{s}$, which determine the body frame, are chosen as $\mathbf{r} = \mathbf{z}_1$ and $\mathbf{s} = \mathbf{z}_2$. The internal geometry of the molecule is described by $N-1$ lengths $z_1, z_2, \ldots, z_{N-1}$, $N-2$ angles $\theta_{1\alpha}$ between the vectors $\mathbf{z}_1$ and $\mathbf{z}_\alpha$ (where $\alpha = 2, 3, \ldots, N-1$), and $N-3$ dihedral angles $\tau_{\alpha 12}$ between the planes $\mathbf{z}_1 \wedge \mathbf{z}_2$ and $\mathbf{z}_1 \wedge \mathbf{z}_\alpha$ (where $\alpha = 3, 4, \ldots, N-1$).

It is tacitly assumed that all degrees of freedom are included in Eqs. (113) and (115). Thus, the formulas are valid only in the absence of constraints. The unconstrained contravariant metric tensor (denoted now for clarity by $h^{(q_i q_j)}$) can be used to obtain the $A \times A$-dimensional contravariant metric tensor in the presence of $P = 3N - A$ rigid constraints as [1,55]

$$g^{(ij)} = h^{(ij)} - \sum_{r,s=A+1}^{3N} h^{(ir)} x_{r-A,\, s-A} h^{(sj)} \tag{116}$$

where the active coordinates are indexed by $i, j = 1, 2, \ldots, A$, and x_{rs} is an element of the matrix

$$\mathsf{X} = \begin{bmatrix} h^{(A+1,A+1)} & h^{(A+1,A+2)} & \cdots & h^{(A+1,3N)} \\ & h^{(A+2,A+2)} & \cdots & h^{(A+2,3N)} \\ & & \ddots & \vdots \\ & & & h^{(3N,3N)} \end{bmatrix}^{-1} \tag{117}$$

Only the upper triangle is shown, because the matrix is symmetric.

C. Hamiltonian via Measuring Vectors

An inspection of Eq. (113) shows that in the absence of constraints and in the special case when the last term in Eq. (116) is zero, it is possible to write

$$g^{(q_i q_j)} = \sum_{\alpha}^{N} \frac{1}{m_\alpha} \mathbf{e}_\alpha^{(q_i)} \cdot \mathbf{e}_\alpha^{(q_j)} \tag{118}$$

where

$$\mathbf{e}_{\alpha}^{(q_i)} = \boldsymbol{\nabla}_{\alpha} q_i \tag{119}$$

is the vector derivative of the coordinate q_i with respect to $\mathbf{x}_{\alpha}$. We refer these quantities as contravariant measuring vectors (the same way we denote the tangents $\mathbf{e}_{q_i}^{(\alpha)} = \partial \mathbf{x}_{\alpha}/\partial q_i$ as covariant measuring vectors). The values of $\mathbf{e}_{\alpha}^{(q_i)}$ at the reference configuration $\{s_1, s_2, \ldots, s_{3N-6}\} = \{s_1^{(e)}, s_2^{(e)}, \ldots, s_{3N-6}^{(e)}\}$ are often (especially in the older literature) denoted as the "s vectors," although sometimes no such distinction in notations is made. The contravariant measuring vectors are reciprocal to the covariant measuring vectors, that is,

$$\sum_{\alpha}^{N} \mathbf{e}_{\alpha}^{(q_i)} \cdot \mathbf{e}_{q_j}^{(\alpha)} = \delta_{ij} \tag{120}$$

which follows from the general chain rule

$$\frac{\partial q_i}{\partial t} = \sum_{\alpha}^{N} (\boldsymbol{\nabla}_{\alpha} q_i) \cdot \frac{\partial \mathbf{x}_{\alpha}}{\partial t} \tag{121}$$

by setting $t = q_j$. Equation (121) also relates the rate of change in the coordinate $q_i(\mathbf{x}_{\alpha})$ for any given rate of the change $d\mathbf{x}_{\alpha}/dt$ of the nuclear position $\mathbf{x}_{\alpha}$ (which in turn justifies the use of the term measuring vector). The reciprocality condition in Eq. (120) is another way of writing Eq. (114).

The measuring vectors $\mathbf{e}_{\alpha}^{(l_i')}$ associated with the ith body-frame component of the total angular momentum can be obtained as described in the following sections. As seen from Eq. (110), the contravariant metric tensor, where the "rotational" part has been expressed in the components of the internal angular momentum, is inverse to the covariant metric tensor, where the "rotational" part has been given in terms of the components of the rotational velocity. Thus, the vectors $\mathbf{e}_{\alpha}^{(l_i')}$ are reciprocal to the vectors $\mathbf{e}_{\omega_j'}^{(\alpha)}$; that is, they obey Eq. (120):

$$\sum_{\alpha}^{N} \mathbf{e}_{\alpha}^{(l_i')} \cdot \mathbf{e}_{\omega_j'}^{(\alpha)} = \sum_{\alpha}^{N} \sum_{k}^{3} \Omega_{jk}^{-1} \mathbf{e}_{\alpha}^{(l_i')} \cdot \mathbf{e}_{B_k}^{(\alpha)} = \delta_{ij} \tag{122}$$

where we used Eqs. (38) to express $\mathbf{e}_{\omega_j'}^{(\alpha)}$ as linear combination of $\mathbf{e}_{B_k}^{(\alpha)}$. By inspecting the above equation, the sum of the dot products $\mathbf{e}_{\alpha}^{(l_i')} \cdot \mathbf{e}_{B_k}^{(\alpha)}$ (over the particle index α) is

$$\sum_{\alpha}^{N} \mathbf{e}_{\alpha}^{(l_i')} \cdot \mathbf{e}_{B_k}^{(\alpha)} = \Omega_{ki} \tag{123}$$

If we express $\mathbf{e}_\alpha^{(l'_i)}$ as linear combinations of $\mathbf{e}_\alpha^{(B_j)}$—that is, if we substitute $\mathbf{e}_\alpha^{(l'_i)} = \sum_l f_{li}\mathbf{e}_\alpha^{(B_l)}$ into Eq. (122)—we find

$$\sum_\alpha^N \sum_{kl}^3 \Omega_{jk}^{-1} f_{li}\mathbf{e}_\alpha^{(B_l)} \cdot \mathbf{e}_{B_k}^{(\alpha)} = \sum_\alpha^N \sum_k^3 \Omega_{jk}^{-1} f_{ki}\mathbf{e}_\alpha^{(B_k)} \cdot \mathbf{e}_{B_k}^{(\alpha)} = \sum_k^3 \Omega_{jk}^{-1} f_{ki} = \delta_{ij} \quad (124)$$

where we may solve the unknown coefficients as $f_{ki} = \Omega_{ki}$. Thus, the rotational measuring vector $\mathbf{e}_\alpha^{(l'_i)}$ associated with the nucleus α and the ith body-frame component of the internal angular momentum is given by

$$\mathbf{e}_\alpha^{(l'_i)} = \sum_j^3 \Omega_{ji}\mathbf{e}_\alpha^{(B_j)} \quad (125)$$

The reciprocality condition in Eq. (120) can be used to deduce some general properties of the measuring vectors associated with the shape coordinates and Euler angles. For example, because the measuring vector $\mathbf{e}_{X_i}^{(\alpha)} = \mathbf{u}_i$ is the same for any nucleus α, it follows that

$$\sum_\alpha^N \mathbf{e}_\alpha^{(s_i)} = \sum_\alpha^N \mathbf{e}_\alpha^{(B_i)} = 0 \quad (126)$$

for any s_i and B_i. Furthermore, because the vectors $\mathbf{e}_{\omega'_i}^{(\alpha)}$ and $\mathbf{e}_\alpha^{(s_i)}$ are reciprocal by the same argument as given before Eq. (122), the substitution of $\mathbf{e}_{\omega'_k}^{(\alpha)} = \mathbf{u}'_k \times \mathbf{y}_\alpha$ to Eq. (120) allows us to extract

$$\sum_\alpha^N \mathbf{y}_\alpha \times \mathbf{e}_\alpha^{(s_i)} = \sum_\alpha^N \mathbf{x}_\alpha \times \mathbf{e}_\alpha^{(s_i)} = 0 \quad (127)$$

where we have used $\mathbf{a} \cdot \mathbf{b} \times \mathbf{c} = \mathbf{a} \times \mathbf{b} \cdot \mathbf{c}$ to exchange the cross and the dot in the triple box product, and the second form follows from the application of Eq. (126). These are known relations from the literature (see, e.g., Ref. 10), although they have been derived before in an indirect way. Some other general properties of the measuring vectors (which to our knowledge have not been presented before) are derived in Appendix B, Section 1. We present different approaches in obtaining the contravariant measuring vectors in the following sections.

1. Infinitesimal Approach

One can use infinitesimal nuclear displacements to determine the contravariant measuring vectors. Infinitesimals[2] were commonly utilized during the first two centuries of the calculus, but they fell into disuse as algebraic entities in the mathematical literature during the latter half of the nineteenth century, when the concept of limit took a rigorous and final form. However, they did not completely vanish. Physicists and engineers did not abandon their use in deriving correct answers to physical problems. Indeed, it was customary for a long time to obtain the contravariant vibrational measuring vectors by infinitesimal reasoning [1] (called Wilson's s-vector method after its inventor E. B. Wilson). The rotational analogue of Wilson's s-vector method [1], namely Lukka's s-vector method [29], was the *first* technique to obtain the rotational measuring vectors. Smooth infinitesimal calculus has been axiomatized recently [56].

Shape Coordinates. As shown in Appendix A, Section 4.a,

$$\frac{dq_i}{da_\alpha} = \mathbf{u}_{d\mathbf{a}_\alpha} \cdot \boldsymbol{\nabla}_\alpha q_i = \mathbf{u}_{d\mathbf{a}_\alpha} \cdot \mathbf{e}_\alpha^{(q_i)} \tag{128}$$

where $da_\alpha = |d\mathbf{a}_\alpha|$ is the length of an infinitesimal nuclear displacement $d\mathbf{a}_\alpha = \mathbf{x}_\alpha(q_i^{(e)} + dq_i) - \mathbf{x}_\alpha(q_i^{(e)})$, $dq_i = q_i(\mathbf{x}_\alpha^{(e)} + d\mathbf{a}_\alpha) - q_i(\mathbf{x}_\alpha^{(e)})$ is the (infinitesimal) change in the coordinate q_i caused by the infinitesimal nuclear displacement, and $\mathbf{u}_{d\mathbf{a}_\alpha} = d\mathbf{a}_\alpha / da_\alpha$ is the unit vector in the direction of the displacement. If the nucleus α is displaced to the direction of the greatest change (i.e., if the direction of $\mathbf{u}_{d\mathbf{a}_\alpha}$ coincides with the direction of $\boldsymbol{\nabla}_\alpha q_i$), we may replace the inner product in Eq. (128) by the geometric product and solve

$$\boldsymbol{\nabla}_\alpha q_i = \mathbf{u}_\alpha^{(q_i)} \frac{dq_i}{da_\alpha} \tag{129}$$

where $\mathbf{u}_\alpha^{(q_i)} = \mathbf{e}_\alpha^{(q_i)} / |\mathbf{e}_\alpha^{(q_i)}|$. As an example [1], we have the following:

Example 5. *The most efficient way to increase the distance* $r_{\alpha\beta} = |\mathbf{r}_{\alpha\beta}| = |\mathbf{x}_\beta - \mathbf{x}_\alpha|$ *is to displace the nucleus* β *to the direction of* $\mathbf{r}_{\alpha\beta}$ *(by intuition). Thus, the direction of the gradient* $\boldsymbol{\nabla}_\alpha r_{\alpha\beta}$ *is the unit vector* $\mathbf{u}_{\mathbf{r}_{\alpha\beta}}$. *The change in* $r_{\alpha\beta}$

[2] By the word "infinitesimals" we refer to the nilpotent quantities, whose second or higher powers are zero. Some of their algebraic properties are quite different from the the properties of the finite quantities [56]. For example, $\epsilon\xi = 0$ does not necessarily imply $\epsilon = 0$ or $\xi = 0$, if ϵ and ξ are infinitesimals. Also, the infinitesimals do not possess inverses; that is, ϵ^{-1} is not defined.

produced by the displacement of the nucleus β *to the direction of* $\mathbf{u}_{\mathbf{r}_{\alpha\beta}}$ *is equal to the magnitude of the displacement, that is,* $da_\beta = dr_{\alpha\beta}$. *Thus, we obtain*

$$\nabla_\beta r_{\alpha\beta} = \mathbf{u}_{\mathbf{r}_{\alpha\beta}} \tag{130}$$

In a similar manner, we can derive

$$\nabla_\alpha r_{\alpha\beta} = \mathbf{u}_{\mathbf{r}_{\beta\alpha}} = -\mathbf{u}_{\mathbf{r}_{\alpha\beta}} \tag{131}$$

Example 6. *The most efficient way to increase the angle* $\theta_{\beta\alpha\gamma}$ *between the vectors* $\mathbf{r}_{\alpha\beta}$ *and* $\mathbf{r}_{\alpha\gamma}$ *is (by intuition) to displace the nucleus* β *to the direction perpendicular to* $\mathbf{r}_{\alpha\beta}$ *(outwards) in the plane* $\mathbf{u}_{\mathbf{r}_{\alpha\beta}} \wedge \mathbf{u}_{\mathbf{r}_{\alpha\gamma}}$. *This is the direction of the gradient* $\mathbf{u}_\beta^{(\theta_{\beta\alpha\gamma})} = \nabla_\beta\theta_{\beta\alpha\gamma}/|\nabla_\beta\theta_{\beta\alpha\gamma}|$, *and in terms of bond vectors it can be obtained as the opposite to the unit vector to the direction of the rejection of* $\mathbf{u}_{\mathbf{r}_{\alpha\gamma}}$ *from* $\mathbf{u}_{\mathbf{r}_{\alpha\beta}}$ *as (see Appendix A, Section 3.a)*

$$\mathbf{u}_\beta^{(\theta_{\beta\alpha\gamma})} = -\frac{\hat{P}^{\perp}_{\mathbf{u}_{\mathbf{r}_{\alpha\beta}}}(\mathbf{u}_{\mathbf{r}_{\alpha\gamma}})}{|\hat{P}^{\perp}_{\mathbf{u}_{\mathbf{r}_{\alpha\beta}}}(\mathbf{u}_{\mathbf{r}_{\alpha\gamma}})|} = -\frac{(\mathbf{u}_{\mathbf{r}_{\alpha\gamma}} \wedge \mathbf{u}_{\mathbf{r}_{\alpha\beta}}) \cdot \mathbf{u}_{\mathbf{r}_{\alpha\beta}}}{|(\mathbf{u}_{\mathbf{r}_{\alpha\gamma}} \wedge \mathbf{u}_{\mathbf{r}_{\alpha\beta}}) \cdot \mathbf{u}_{\mathbf{r}_{\alpha\beta}}|} = \frac{\mathbf{u}_{\mathbf{r}_{\alpha\beta}}\cos\theta_{\beta\alpha\gamma} - \mathbf{u}_{\mathbf{r}_{\alpha\gamma}}}{\sin\theta_{\beta\alpha\gamma}} \tag{132}$$

[where we use the expansion rule in Eq. (A31)]. The displacement da_β *of the nucleus* β *to the direction* $\mathbf{u}_\beta^{(\theta_{\beta\alpha\gamma})}$ *produces a change* $d\theta_{\beta\alpha\gamma}$ *in* $\theta_{\beta\alpha\gamma}$, *where*

$$\sin(d\theta_{\beta\alpha\gamma}) = \frac{da_\beta}{r_{\alpha\beta}} \approx d\theta_{\beta\alpha\gamma} \tag{133}$$

The latter equality is exact in the case of an infinitesimal displacement. Hence,

$$\nabla_\beta\theta_{\beta\alpha\gamma} = \mathbf{u}_\beta^{(\theta_{\beta\alpha\gamma})}\frac{d\theta_{\beta\alpha\gamma}}{da_\beta} = \frac{\mathbf{u}_{\mathbf{r}_{\alpha\beta}}\cos\theta_{\beta\alpha\gamma} - \mathbf{u}_{\mathbf{r}_{\alpha\gamma}}}{r_{\alpha\beta}\sin\theta_{\beta\alpha\gamma}} \tag{134}$$

Unfortunately, the correctness of the result depends on the configuration the nucleus is displaced about. The interested reader should look at Ref. 36, Appendix B2, where Wilson's method is applied to the out-of-plane bending in the planar configuration producing an incomplete result.

Rotational Degrees of Freedom. Lukka's approach in obtaining the rotational measuring vectors via infinitesimal deduction has gained popularity recently (see, e.g., Refs. 30 and 31). In what follows, we generalize the results in Ref. 29

to apply the method to an *arbitrary* body frame. In the limit of an infinitesimal rotation, we may write

$$\mathbf{e}_\alpha^{(I_1')} = \boldsymbol{\nabla}_\alpha \theta \tag{135}$$

$$\mathbf{e}_\alpha^{(I_2')} = \theta \boldsymbol{\nabla}_\alpha \phi - \chi \boldsymbol{\nabla}_\alpha \theta \tag{136}$$

$$\mathbf{e}_\alpha^{(I_3')} = \boldsymbol{\nabla}_\alpha \phi + \boldsymbol{\nabla}_\alpha \chi \tag{137}$$

To obtain these expressions, take the Taylor series expansions of the elements Ω_{ji} in Eq. (125) and neglect higher than the first-order terms. If we can calculate the gradients of the Euler angles as

$$\boldsymbol{\nabla}_\alpha B_i = \mathbf{u}_\alpha^{(B_i)} \frac{dB_i}{da_\alpha} \tag{138}$$

where $\mathbf{u}_\alpha^{(B_i)} = \mathbf{e}_\alpha^{(B_i)} / |\mathbf{e}_\alpha^{(B_i)}|$, we can use Eqs. (135)–(137) to find the rotational measuring vectors. The infinitesimal change dB_i in the Euler angle B_i caused by the infinitesimal displacement $d\mathbf{a}_\alpha$ of the nucleus α can be obtained from the directional derivative [Eq. (128)]

$$dB_i = (d\mathbf{a}_\alpha) \cdot \boldsymbol{\nabla}_\alpha B_i = da_\alpha \mathbf{u}_{da_\alpha} \cdot \boldsymbol{\nabla}_\alpha B_i \tag{139}$$

As an example, we evaluate the gradient of θ to obtain $\mathbf{e}_\alpha^{(I_1')}$ for those body frames, where the coefficients c_β do *not* depend on the nuclear positions—that is, for which $\boldsymbol{\nabla}_\alpha c_\beta = 0$ for any α and β:

Example 7. *Generally, the directional derivative* $\mathbf{a}_\alpha \cdot \boldsymbol{\nabla}_\alpha \mathbf{u}_3'$ *can be evaluated as*

$$\begin{aligned}\mathbf{a}_\alpha \cdot \boldsymbol{\nabla}_\alpha \mathbf{u}_3' &= \mathbf{a}_\alpha \cdot \boldsymbol{\nabla}_\alpha \frac{\mathbf{r}}{r} = \frac{\mathbf{a}_\alpha \cdot \boldsymbol{\nabla}_\alpha \mathbf{r}}{r} + \left(\mathbf{a}_\alpha \cdot \boldsymbol{\nabla}_\alpha \frac{1}{r}\right)\mathbf{r} = \frac{\mathbf{a}_\alpha \cdot \boldsymbol{\nabla}_\alpha \mathbf{r}}{r} - \frac{\mathbf{a}_\alpha \cdot \boldsymbol{\nabla}_\alpha r}{r^2}\mathbf{r} \\ &= \frac{\mathbf{a}_\alpha \cdot \boldsymbol{\nabla}_\alpha \mathbf{r}}{r} - \frac{\mathbf{r} \cdot (\mathbf{a}_\alpha \cdot \boldsymbol{\nabla}_\alpha \mathbf{r})}{r^3}\mathbf{r} = \frac{\left[c_\alpha \mathbf{a}_\alpha + \sum_\beta (\mathbf{a}_\alpha \cdot \boldsymbol{\nabla}_\alpha c_\beta)\mathbf{x}_\beta\right]}{r} \\ &\quad - \frac{\mathbf{r} \cdot \left[c_\alpha \mathbf{a}_\alpha + \sum_\beta (\mathbf{a}_\alpha \cdot \boldsymbol{\nabla}_\alpha c_\beta)\mathbf{x}_\beta\right]}{r^3}\mathbf{r}\end{aligned} \tag{140}$$

where $r = |\mathbf{r}|$. *Because* $\cos\theta = \mathbf{u}_3' \cdot \mathbf{u}_3$, *we obtain the directional derivative as*

$$\begin{aligned}\mathbf{a}_\alpha \cdot \boldsymbol{\nabla}_\alpha \theta = &- \frac{\left[c_\alpha \mathbf{a}_\alpha + \sum_\beta (\mathbf{a}_\alpha \cdot \boldsymbol{\nabla}_\alpha c_\beta)\mathbf{x}_\beta\right] \cdot \mathbf{u}_3}{r \sin\theta} \\ &+ \frac{\mathbf{r} \cdot \left[c_\alpha \mathbf{a}_\alpha + \sum_\beta (\mathbf{a}_\alpha \cdot \boldsymbol{\nabla}_\alpha c_\beta)\mathbf{x}_\beta\right]}{r^3 \sin\theta}\mathbf{r} \cdot \mathbf{u}_3\end{aligned} \tag{141}$$

Now we calculate the gradients for the body frames with $\nabla_\alpha c_\beta = 0$ for any α and β. In this case, the directional derivative $\mathbf{a} \cdot \nabla_\alpha \theta$ reads as follows:

$$\mathbf{a}_\alpha \cdot \nabla_\alpha \theta = \frac{c_\alpha(\mathbf{u}'_3 \cdot \mathbf{a}_\alpha \mathbf{u}'_3 \cdot \mathbf{u}_3 - \mathbf{a}_\alpha \cdot \mathbf{u}_3)}{r \sin \theta} \approx \frac{c_\alpha(\mathbf{u}'_3 \cdot \mathbf{a}_\alpha \mathbf{u}'_3 \cdot \mathbf{u}_3 - \mathbf{a}_\alpha \cdot \mathbf{u}_3)}{r\theta} \tag{142}$$

where the latter equality is exact in the infinitesimal limit (i.e., when we substitute $\mathbf{a}_\alpha$ by $d\mathbf{a}_\alpha$). By Eq. (139), the latter equality gives the infinitesimal change $d\theta$ in the Euler angle θ caused by the infinitesimal displacement $d\mathbf{a}_\alpha$ of the nucleus α, that is,

$$d\theta = \frac{c_\alpha(\mathbf{u}'_3 \cdot d\mathbf{a}_\alpha \mathbf{u}'_3 \cdot \mathbf{u}_3 - d\mathbf{a}_\alpha \cdot \mathbf{u}_3)}{r\theta} \tag{143}$$

The fastest way to increase the Euler angle θ (i.e., the direction of $\nabla_\alpha \theta$) is (by intuition) to displace the nucleus α to the direction orthogonal to the body axis $\mathbf{u}'_3$ (i.e. $d\mathbf{a}_\alpha \cdot \mathbf{u}'_3 = 0$). Thus, by substituting

$$\mathbf{u}_3 = \cos\theta \mathbf{u}'_3 + \sin\theta(\cos\chi \mathbf{u}'_2 + \sin\chi \mathbf{u}'_1) \approx \mathbf{u}'_3 + \theta \mathbf{u}'_2 \tag{144}$$

(where the latter equality is exact in the infinitesimal limit) and using Eq. (143) in Eq. (139), we may solve

$$\nabla_\alpha \theta = -\frac{c_\alpha \mathbf{u}'_2}{r} \tag{145}$$

It is also possible to obtain the measuring vectors in body frames where the coefficients c_α and d_α depend on the nuclear positions. This can be tedious for an arbitrary body frame: The labor needed increases fast when the coefficients become complicated functions of nuclear positions. The risk of missing parts of the gradients also increases due to the choice of the particular reference configuration used to evaluate $\nabla_\alpha c_\beta$ and $\nabla_\alpha d_\beta$. We do not pursue this approach any further, because we have algebraic tools in our possession for calculating the rotational measuring vectors.

2. *Geometric Algebra Approach*

Geometric algebra approach offers some advantages over other methods presented in the literature. First of all, atomic position vectors themselves are manipulated instead of their components, and hence all expressions are simple at each stage of derivation. This is not the case when Cartesian components and back substitutions are used to obtain contravariant measuring vectors [57]. As a

TABLE I
Some Vector Derivatives[a]

F	$\nabla_\beta F$	$\nabla_\alpha F$
$\mathbf{r}_{\alpha\beta}$	3	-3
$\mathbf{r}_{\alpha\beta}\mathbf{a}$	$3\mathbf{a}$	$-3\mathbf{a}$
$\mathbf{a}\mathbf{r}_{\alpha\beta}$	$-\mathbf{a}$	$\mathbf{a}$
$\mathbf{a}\cdot\mathbf{r}_{\alpha\beta}$	$\mathbf{a}$	$-\mathbf{a}$
$\mathbf{a}\wedge\mathbf{r}_{\alpha\beta}$	$-2\mathbf{r}_{\alpha\beta}$	$2\mathbf{r}_{\alpha\beta}$
$r_{\alpha\beta}^k$	$kr_{\alpha\beta}^{k-2}\mathbf{r}_{\alpha\beta}$	$-kr_{\alpha\beta}^{k-2}\mathbf{r}_{\alpha\beta}$
$r_{\alpha\beta}^k\mathbf{r}_{\alpha\beta}$	$(k+3)r_{\alpha\beta}^k$	$-(k+3)r_{\alpha\beta}^k$
$(\mathbf{r}_{\alpha\beta}\times\mathbf{a})\cdot\mathbf{b}$	$\mathbf{a}\times\mathbf{b}$	$-\mathbf{a}\times\mathbf{b}$
$\mathbf{r}_{\alpha\beta}A_{\bar{p}}$	$3A_{\bar{p}}$	$-3A_{\bar{p}}$
$\mathbf{r}_{\alpha\beta}\wedge A_{\bar{p}}$	$(3-p)A_{\bar{p}}$	$-(3-p)A_{\bar{p}}$
$\mathbf{r}_{\alpha\beta}\cdot A_{\bar{p}}$	$pA_{\bar{p}}$	$-pA_{\bar{p}}$
$\mathbf{r}_{\alpha\beta}\cdot\mathbf{r}_{\alpha\gamma}$	$\mathbf{r}_{\alpha\gamma}$	$-(\mathbf{r}_{\alpha\beta}+\mathbf{r}_{\alpha\gamma})$

[a]The $\mathbf{a}$ and $\mathbf{b}$ are vectors, and $A_{\bar{p}} = \mathbf{a}_1\wedge\cdots\wedge\mathbf{a}_p$ is a p blade ($p = 1, 2$, and 3) independent of a vectors $\mathbf{x}_\alpha$ and $\mathbf{x}_\beta$. Furthermore, $\mathbf{r}_{\alpha\beta} = \mathbf{x}_\beta - \mathbf{x}_\alpha$, and $k = 0, \pm1, \pm2, \pm3, \ldots$.

simple example, the measuring vectors for the Cartesian coordinates of the center of the mass are given by the direct vectorial differentiation as

$$\mathbf{e}_\alpha^{(X_i)} = \nabla_\alpha X_i = \sum_{\beta=1}^{N} \nabla_\alpha \frac{m_\beta \mathbf{u}_i\cdot\mathbf{x}_\beta}{M} = \frac{m_\alpha}{M}\mathbf{u}_i \tag{146}$$

(where we have used the rule $\nabla_\alpha \mathbf{a}\cdot\mathbf{x}_\alpha = \mathbf{a}$ valid for any vector $\mathbf{a}$ independent of $\mathbf{x}_\alpha$; see Eq. (A129) and Table I).

Shape Coordinates. The gradients of any explicitly defined curvilinear shape coordinate can be found by the methods of geometric algebra. The results given in Table I are useful in this context. We solve the gradients of a bond angle $\theta_{\beta\alpha\gamma}$ by direct vectorial differentiation as a simple example:

Example 8. *By writing the definition of the bond angle* $\theta_{\beta\alpha\gamma}$ *as*

$$\mathbf{r}_{\alpha\beta}\cdot\mathbf{r}_{\alpha\gamma} = r_{\alpha\beta}r_{\alpha\gamma}\cos\theta_{\beta\alpha\gamma} \tag{147}$$

we may differentiate the left-hand side using the product rule as

$$\nabla_\beta(\mathbf{r}_{\alpha\beta}\cdot\mathbf{r}_{\alpha\gamma}) = \mathbf{r}_{\alpha\gamma} \tag{148}$$

and the right-hand side as

$$\begin{aligned}\nabla_\beta(r_{\alpha\beta}r_{\alpha\gamma}\cos\theta_{\beta\alpha\gamma}) &= (\nabla_\beta r_{\alpha\beta})r_{\alpha\gamma}\cos\theta_{\beta\alpha\gamma} + r_{\alpha\beta}r_{\alpha\gamma}\nabla_\beta(\cos\theta_{\beta\alpha\gamma}) \\ &= \mathbf{u}_{\mathbf{r}_{\alpha\beta}}r_{\alpha\gamma}\cos\theta_{\beta\alpha\gamma} - r_{\alpha\beta}r_{\alpha\gamma}\sin\theta_{\beta\alpha\gamma}\nabla_\beta\theta_{\beta\alpha\gamma}\end{aligned} \tag{149}$$

where we may solve

$$\nabla_\beta\theta_{\beta\alpha\gamma} = \frac{\mathbf{u}_{\mathbf{r}_{\alpha\beta}}\cos\theta_{\beta\alpha\gamma} - \mathbf{u}_{\mathbf{r}_{\alpha\gamma}}}{r_{\alpha\beta}\sin\theta_{\beta\alpha\gamma}} \tag{150}$$

Similarly,

$$\nabla_\gamma\theta_{\beta\alpha\gamma} = \frac{\mathbf{u}_{\mathbf{r}_{\alpha\gamma}}\cos\theta_{\beta\alpha\gamma} - \mathbf{u}_{\mathbf{r}_{\alpha\beta}}}{r_{\alpha\gamma}\sin\theta_{\beta\alpha\gamma}} \tag{151}$$

and by the translational invariance [Eq. (126)] we obtain

$$\nabla_\alpha\theta_{\beta\alpha\gamma} = -(\nabla_\beta\theta_{\beta\alpha\gamma} + \nabla_\gamma\theta_{\beta\alpha\gamma}) \tag{152}$$

The next example is somewhat more interesting:

Example 9. *A new ammonia inversion coordinate S_2 is defined in Ref. 58 as*

$$S_2(A_2'') = \pm\frac{1}{3^{1/4}}(2\pi - \theta_{23} - \theta_{13} - \theta_{12})^{1/2} \tag{153}$$

where θ_{ij} is the angle between the bond vectors $\mathbf{x}_{H_i} - \mathbf{x}_N$ and $\mathbf{x}_{H_j} - \mathbf{x}_N$, and the plus sign is for the right-handed and the minus sign for the left-handed configuration. (The other symmetrized displacement coordinates $S_1(A_1')$, $S_{3a}(E')$, $S_{3b}(E')$, $S_{4a}(E')$, and $S_{4b}(E')$ are defined as linear combinations of the bond stretching and bond angle displacements in the usual way [59]. The symmetry labels refer to the D_{3h} point group.) It is unnecessary to express the sign factor explicitly, because the coordinate gradients can be calculated from the square S_2^2 of the inversion coordinate with the chain rule of Eq. (A131). As a simple example, the vector derivative of S_2 with respect to the position $\mathbf{x}_{H_1}$ of the proton H_1 can be solved from

$$2S_2\nabla_{H_1}S_2 = -\frac{\nabla_{H_1}\theta_{13} + \nabla_{H_1}\theta_{12}}{3^{1/2}} \tag{154}$$

by substituting the known vector derivatives $\nabla_{H_1}\theta_{12}$ and $\nabla_{H_1}\theta_{13}$ [which are obtained from Eq. (150)]. The result can be expressed in terms of the symmetry *coordinates by using inverse coordinate relations, such as*

$$\theta_{13} - \frac{2\pi}{3} = -\frac{1}{\sqrt{3}}S_2^2 - \frac{1}{\sqrt{6}}S_{4a} + \frac{1}{\sqrt{2}}S_{4b} \tag{155}$$

The interested reader should consult Ref. 36 for other explicitly defined curvilinear shape coordinates. There are two ways to read off coordinate gradients $\boldsymbol{\nabla}_\beta s_i$ for the standard internal coordinates from the vector derivatives $\boldsymbol{\nabla}_{\mathbf{z}_\alpha} s_i$ in Table II.

TABLE II
Vector Derivatives of the Internal Coordinates

q_i	Definition	$\boldsymbol{\nabla}_{\mathbf{z}_\alpha} q_i$
z_α	$\lvert\mathbf{z}_\alpha\rvert$	$\mathbf{u}_{\mathbf{z}_\alpha}$
$z_{\alpha\beta}$	$\lvert\mathbf{z}_\beta - \mathbf{z}_\alpha\rvert$	$-\mathbf{u}_{\mathbf{z}_{\alpha\beta}}$
$z_{\beta\alpha}$	$\lvert\mathbf{z}_\alpha - \mathbf{z}_\beta\rvert$	$\mathbf{u}_{\mathbf{z}_{\alpha\beta}}$
$t_{\alpha\beta}$	$\mathbf{z}_\alpha \cdot \dfrac{\mathbf{z}_\gamma \wedge \mathbf{z}_\kappa}{\mathbf{z}_\beta \wedge \mathbf{z}_\gamma \wedge \mathbf{z}_\kappa}$	$\dfrac{\mathbf{z}_\gamma \wedge \mathbf{z}_\kappa}{\mathbf{z}_\beta \wedge \mathbf{z}_\gamma \wedge \mathbf{z}_\kappa}$
$\theta_{\alpha\beta}$	$\arccos(\mathbf{u}_{\mathbf{z}_\alpha} \cdot \mathbf{u}_{\mathbf{z}_\beta})$	$\dfrac{\mathbf{u}_{\mathbf{z}_\alpha}\cos\theta_{\alpha\beta} - \mathbf{u}_{\mathbf{z}_\beta}}{z_\alpha \sin\theta_{\alpha\beta}}$
$\phi_{\alpha\beta\gamma}$	$\arcsin\left(-i\dfrac{\mathbf{z}_\alpha \wedge \mathbf{z}_\beta \wedge \mathbf{z}_\gamma}{z_\alpha z_\beta z_\gamma \sin\theta_{\beta\gamma}}\right)$	$\dfrac{1}{z_\alpha \cos\phi_{\alpha\beta\gamma}}\left(\dfrac{\mathbf{u}_{\mathbf{z}_\beta} \times \mathbf{u}_{\mathbf{z}_\gamma}}{\sin\theta_{\beta\gamma}} - \mathbf{u}_{\mathbf{z}_\alpha}\sin\phi_{\alpha\beta\gamma}\right)$
$\phi_{\beta\alpha\gamma}$	$\arcsin\left(-i\dfrac{\mathbf{z}_\beta \wedge \mathbf{z}_\alpha \wedge \mathbf{z}_\gamma}{z_\alpha z_\beta z_\gamma \sin\theta_{\alpha\gamma}}\right)$	$\dfrac{1}{z_\alpha \cos\phi_{\beta\alpha\gamma}}\left(\dfrac{\mathbf{u}_{\mathbf{z}_\gamma} \times \mathbf{u}_{\mathbf{z}_\beta}}{\sin\theta_{\alpha\gamma}} - \sin\phi_{\beta\alpha\gamma}\dfrac{\mathbf{u}_{\mathbf{z}_\alpha} - \mathbf{u}_{\mathbf{z}_\gamma}\cos\theta_{\alpha\gamma}}{\sin^2\theta_{\alpha\gamma}}\right)$
$\tau_{\alpha\beta\gamma}$	$\arccos\left(\dfrac{\mathbf{u}_{\mathbf{z}_\alpha} \wedge \mathbf{u}_{\mathbf{z}_\beta}}{\lvert\mathbf{u}_{\mathbf{z}_\alpha} \wedge \mathbf{u}_{\mathbf{z}_\beta}\rvert} \cdot \dfrac{\mathbf{u}_{\mathbf{z}_\beta} \wedge \mathbf{u}_{\mathbf{z}_\gamma}}{\lvert\mathbf{u}_{\mathbf{z}_\beta} \wedge \mathbf{u}_{\mathbf{z}_\gamma}\rvert}\right)$	$\dfrac{1}{z_\alpha}\left(\dfrac{\cot\theta_{\beta\gamma}}{\sin\theta_{\alpha\beta}\sin\tau_{\alpha\beta\gamma}} - \dfrac{\cot\theta_{\alpha\beta}\cot\tau_{\alpha\beta\gamma}}{\sin\theta_{\alpha\beta}}\right)\mathbf{u}_{\mathbf{z}_\beta} + \dfrac{\cot\tau_{\alpha\beta\gamma}}{z_\alpha}(1+\cot^2\theta_{\alpha\beta})\mathbf{u}_{\mathbf{z}_\alpha} - \dfrac{1}{z_\alpha \sin\theta_{\alpha\beta}\sin\theta_{\beta\gamma}\sin\tau_{\alpha\beta\gamma}}\mathbf{u}_{\mathbf{z}_\gamma}$
$\tau_{\beta\alpha\gamma}$	$\arccos\left(\dfrac{\mathbf{u}_{\mathbf{z}_\beta} \wedge \mathbf{u}_{\mathbf{z}_\alpha}}{\lvert\mathbf{u}_{\mathbf{z}_\beta} \wedge \mathbf{u}_{\mathbf{z}_\alpha}\rvert} \cdot \dfrac{\mathbf{u}_{\mathbf{z}_\alpha} \wedge \mathbf{u}_{\mathbf{z}_\gamma}}{\lvert\mathbf{u}_{\mathbf{z}_\alpha} \wedge \mathbf{u}_{\mathbf{z}_\gamma}\rvert}\right)$	$\frac{1}{z_\alpha}\Big[\cot\tau_{\beta\alpha\gamma}(\cot^2\theta_{\alpha\beta} + \cot^2\theta_{\alpha\gamma} + 2) - \dfrac{2\cos\theta_{\beta\gamma}}{\sin\theta_{\alpha\beta}\sin\theta_{\alpha\gamma}\sin\tau_{\beta\alpha\gamma}}\Big]\mathbf{u}_{\mathbf{z}_\alpha} + \dfrac{1}{z_\alpha}\left[\dfrac{\cos\theta_{\alpha\gamma}}{\sin\theta_{\alpha\beta}\sin\theta_{\alpha\gamma}\sin\tau_{\beta\alpha\gamma}} - \dfrac{\cot\tau_{\beta\alpha\gamma}\cos\theta_{\alpha\beta}}{\sin^2\theta_{\alpha\beta}}\right]\mathbf{u}_{\mathbf{z}_\beta} + \dfrac{1}{z_\alpha}\left[\dfrac{\cos\theta_{\alpha\beta}}{\sin\theta_{\alpha\beta}\sin\theta_{\alpha\gamma}\sin\tau_{\beta\alpha\gamma}} - \dfrac{\cot\tau_{\beta\alpha\gamma}\cos\theta_{\alpha\gamma}}{\sin^2\theta_{\alpha\gamma}})\right]\mathbf{u}_{\mathbf{z}_\gamma}$
$\varphi_{\alpha\beta\gamma}$	$-i\dfrac{\mathbf{z}_\alpha \wedge \mathbf{z}_\beta \wedge \mathbf{z}_\gamma}{z_\alpha z_\beta z_\gamma}$	$\dfrac{\mathbf{z}_\beta \times \mathbf{z}_\gamma}{z_\alpha z_\beta z_\gamma} - \varphi_{\alpha\beta\gamma}\dfrac{\mathbf{z}_\alpha}{z_\alpha^2}$

1. One can utilize the independency of $\mathbf{x}_\alpha$ form $\mathbf{x}_\beta$. It is possible to regard the transformation from $\boldsymbol{\nabla}_\beta$ to $\boldsymbol{\nabla}_{\mathbf{r}_{\alpha\beta}}$ as a shift of origin. However, the vector derivative is invariant under the shift of origin (by a simple chain rule, $\boldsymbol{\nabla}_\beta$ has the same form in the Cartesian coordinates $x_{\beta_i} = \mathbf{x}_\beta \cdot \mathbf{u}_i$ and $x'_{\beta_i} = (\mathbf{x}_\beta - \mathbf{a}) \cdot \mathbf{u}_i = x_{\beta_i} - a_i$, where $\mathbf{a}$ is a vector independent of $\mathbf{x}_\beta$). Thus, we may write $\boldsymbol{\nabla}_\beta s_i = \boldsymbol{\nabla}_{\mathbf{r}_{\alpha\beta}} s_i$ and we obtain the coordinate gradients by substituting the vectors $\mathbf{z}_\alpha$ by the appropriate bond vectors $\mathbf{r}_{\alpha\beta}$.
2. More generally, the gradients $\boldsymbol{\nabla}_\alpha s_i$ can be obtained from the vector derivatives $\boldsymbol{\nabla}_{\mathbf{z}_\alpha} s_i$ by utilizing the relation

$$\boldsymbol{\nabla}_\alpha = \sum_\beta c_{\beta\alpha} \boldsymbol{\nabla}_{\mathbf{z}_\beta} \tag{156}$$

This equality can be derived from Eq. (47):

Example 10. *Because $z_{\alpha_i} = \sum_\beta c_{\alpha\beta} x_{\beta_i}$, using the simple chain rule it is seen that*

$$\boldsymbol{\nabla}_\alpha = \sum_i^3 \mathbf{u}_i \frac{\partial}{\partial x_{\alpha_i}} = \sum_{ij}^3 \mathbf{u}_i \sum_\beta \frac{\partial z_{\beta_j}}{\partial x_{\alpha_i}} \frac{\partial}{\partial z_{\beta_j}} = \sum_{ij}^3 \mathbf{u}_i \sum_{\beta\gamma} c_{\beta\gamma} \frac{\partial x_{\gamma_j}}{\partial x_{\alpha_i}} \frac{\partial}{\partial z_{\beta_j}} \tag{157}$$

where Eq. (156) follows due to the relation $\partial x_{\gamma_j} / \partial x_{\alpha_i} = \delta_{\alpha\gamma}\delta_{ij}$.

Alternatively, the gradients $\boldsymbol{\nabla}_\alpha s_i$ can be obtained from the variation of the coordinate s_i along the path of a particle α. In practice, we (formally) differentiate the coordinate s_i with respect to a scalar parameter t, write the result in the form $\dot{s}_i = \sum_\alpha^N \mathbf{f}_\alpha \cdot \dot{\mathbf{x}}_\alpha$, and apply Eq. (121) to pick the gradient as $\boldsymbol{\nabla}_\alpha s_i = \mathbf{f}_\alpha$ (see Ref. 37). As an example, we have the following:

Example 11. *When both sides of the equation*

$$r_{\alpha\beta}^2 = \mathbf{r}_{\alpha\beta} \cdot \mathbf{r}_{\alpha\beta} \tag{158}$$

are differentiated with respect to some scalar parameter t, the result

$$2 r_{\alpha\beta} \dot{r}_{\alpha\beta} = 2\mathbf{r}_{\alpha\beta} \cdot \dot{\mathbf{r}}_{\alpha\beta} \tag{159}$$

follows. After the terms are rearranged, one obtains

$$\dot{r}_{\alpha\beta} = \frac{\mathbf{r}_{\alpha\beta} \cdot \dot{\mathbf{r}}_{\alpha\beta}}{r_{\alpha\beta}} = \frac{\mathbf{r}_{\alpha\beta} \cdot (\dot{\mathbf{x}}_\beta - \dot{\mathbf{x}}_\alpha)}{r_{\alpha\beta}} \tag{160}$$

where the coordinate gradients are easily picked with the help of Eq. (121) as

$$\nabla_\beta r_{\alpha\beta} = \frac{\mathbf{r}_{\alpha\beta}}{r_{\alpha\beta}} = -\nabla_\alpha r_{\alpha\beta} \tag{161}$$

This approach is particularly useful, when the coordinates s_i (and the derivatives $\dot{s}_i$) are defined as *implicit* functions of the nuclear positions. The eigenvalues λ_1, λ_2, and λ_3 of the moment tensor

$$\mathfrak{M}(\boldsymbol{\omega}) = \sum_\alpha m_\alpha \mathbf{y}_\alpha \mathbf{y}_\alpha \cdot \boldsymbol{\omega} \tag{162}$$

provide an example of these kind of coordinates. They have been used to describe some interesting features of reactive scattering involving three- and four-particle systems (such as atom and diatom reactions); see, e.g. [13]. The gradients $\nabla_\alpha \lambda_i$ can be obtained from the variation of the coordinate λ_i along the path of a particle α more easily than by the direct vectorial differentiation. In short,

$$\dot{\lambda}_i = \sum_\alpha^N 2m_\alpha \mathbf{y}_\alpha \cdot \mathbf{u}_i'\mathbf{u}_i' \cdot \left(\dot{\mathbf{x}}_\alpha - \sum_\beta^N \frac{m_\beta \dot{\mathbf{x}}_\beta}{M}\right) = \sum_\alpha^N \dot{\mathbf{x}}_\alpha \cdot \nabla_\alpha \lambda_i \tag{163}$$

where we may extract the gradients as

$$\nabla_\alpha \lambda_i = 2m_\alpha \mathbf{y}_\alpha \cdot \mathbf{u}_i'\mathbf{u}_i' \tag{164}$$

(see Ref. 37 for details).

Rotational Degrees of Freedom. We can evaluate the gradients of the Euler angles from the directional cosines. For example, we can obtain $\nabla_\alpha \theta$ from Eq. (141) as

$$\nabla_\alpha \theta = \nabla_{\mathbf{a}_\alpha} \mathbf{a}_\alpha \cdot \nabla_\alpha \theta = \frac{c_\alpha(\mathbf{u}_3' \cos\theta - \mathbf{u}_3) + \sum_\beta (\nabla_\alpha c_\beta)(\mathbf{x}_\beta \cdot \mathbf{u}_3' \cos\theta - \mathbf{x}_\beta \cdot \mathbf{u}_3)}{r \sin\theta} \tag{165}$$

[see Appendix A, Section 4.b, Eq. (A134)]. If we express this as the function of the body axes and substitute the result and the gradients $\nabla_\alpha \phi$ and $\nabla_\alpha \chi$ to Eq. (125), we obtain the measuring vectors $\mathbf{e}_\alpha^{(l_i')}$. It should be emphasized that the resulting rotational measuring vectors do not depend on the Euler angles. One should suspect that there is an easier *algebraic* way of obtaining the rotational measuring vectors without the intervention of the Euler angles (or any other rotational parameters, which cancel out in the final result). We have already seen that the rotational measuring vectors can be deduced via infinitesimal displacements in a manner that does not use extensively the Euler angles in the course of derivation. Thus, we shall not pursue the direct vectorial differentiation of the Euler angles any further.

It turns out (as a result of a derivation described shortly at the end of this section) that the rotational measuring vectors are given by

$$\mathbf{e}_{\alpha}^{(l'_i)} = \boldsymbol{\nabla}_{\mathbf{a}}[(\mathbf{a} \cdot \boldsymbol{\nabla}_{\alpha}\mathbf{u}'_j) \cdot \mathbf{u}'_k] \tag{166}$$

The target of differentiation is implied by the parentheses, and the indices i, j, and k are in cyclic order. We may express the rotational measuring vectors as explicit functions of the nuclear positions. As an example, we obtain from Eq. (140)

$$\begin{aligned} \mathbf{e}_{\alpha}^{(l'_2)} &= \boldsymbol{\nabla}_{\mathbf{a}}[(\mathbf{a} \cdot \boldsymbol{\nabla}_{\alpha}\mathbf{u}'_3) \cdot \mathbf{u}'_1] = \boldsymbol{\nabla}_{\mathbf{a}} \frac{c_{\alpha}\mathbf{a} \cdot \mathbf{u}'_1 + \sum_{\beta}(\mathbf{a} \cdot \boldsymbol{\nabla}_{\alpha}c_{\beta})\mathbf{x}_{\beta} \cdot \mathbf{u}'_1}{r} \\ &= \frac{c_{\alpha}\mathbf{u}'_1 + \sum_{\beta}\mathbf{x}_{\beta} \cdot \mathbf{u}'_1 \boldsymbol{\nabla}_{\alpha}c_{\beta}}{r} \end{aligned} \tag{167}$$

By a similar derivation, the other rotational measuring vectors are found to be

$$\mathbf{e}_{\alpha}^{(l'_1)} = -\frac{c_{\alpha}\mathbf{u}'_2 + \sum_{\beta}\mathbf{x}_{\beta} \cdot \mathbf{u}'_2 \boldsymbol{\nabla}_{\alpha}c_{\beta}}{r} \tag{168}$$

$$\begin{aligned} \mathbf{e}_{\alpha}^{(l'_3)} &= -\frac{1}{|\mathbf{r} \times \mathbf{s}|}[\mathbf{s} \cdot \mathbf{u}'_3(c_{\alpha}\mathbf{u}'_2 + \sum_{\beta}\mathbf{x}_{\beta} \cdot \mathbf{u}'_2 \boldsymbol{\nabla}_{\alpha}c_{\beta}) \\ &\quad - r(d_{\alpha}\mathbf{u}'_2 + \sum_{\beta}\mathbf{x}_{\beta} \cdot \mathbf{u}'_2 \boldsymbol{\nabla}_{\alpha}d_{\beta})] \end{aligned} \tag{169}$$

The gradients $\boldsymbol{\nabla}_{\alpha}c_{\beta}$ and $\boldsymbol{\nabla}_{\alpha}d_{\beta}$ depend on the explicit formulas of coefficients c_{β} and d_{β}. These coefficients can be functions of the bond lengths, bond angles, and so on. In any case, the gradients can be calculated straightforwardly by the methods of geometric algebra. As an example, we have the following:

Example 12. *Let us choose the directions* $\mathbf{r}$ *and* $\mathbf{s}$ *as*

$$\mathbf{r} = \mathbf{x}_1 + \mathbf{x}_2 + \mathbf{x}_3 + \cdots + \mathbf{x}_{N-1} - (N-1)\mathbf{x}_N \tag{170}$$

$$\mathbf{s} = \mathbf{x}_1 - \mathbf{x}_2 \tag{171}$$

Then, the nonzero coefficients are $c_{\alpha} = 1$ *for* $1 \leq \alpha < N$, $c_N = -(N-1)$, $d_1 = 1$, *and* $d_2 = -1$ *and all the gradients of the coefficients are zero (i.e.,* $\boldsymbol{\nabla}_{\alpha}c_{\beta} = \boldsymbol{\nabla}_{\alpha}d_{\beta} = 0$ *for all* α *and* β*). By a direct substitution to Eqs. (167)–(169),* *the rotational measuring vectors* $\mathbf{e}_\alpha^{(l'_1)}$ *are seen to be*

$$\mathbf{e}_\alpha^{(l'_1)} = -\frac{\mathbf{u}'_2}{r} \quad \text{for} \quad 1 \leq \alpha < N \tag{172}$$

$$\mathbf{e}_N^{(l'_1)} = (N-1)\frac{\mathbf{u}'_2}{r} \tag{173}$$

Similarly, the measuring vectors $\mathbf{e}_\alpha^{(l'_2)}$ *are*

$$\mathbf{e}_\alpha^{(l'_2)} = \frac{\mathbf{u}'_1}{r} \quad \text{for} \quad 1 \leq \alpha < N \tag{174}$$

$$\mathbf{e}_N^{(l'_2)} = -(N-1)\frac{\mathbf{u}'_1}{r} \tag{175}$$

and the measuring vectors $\mathbf{e}_\alpha^{(l'_3)}$ *are*

$$\mathbf{e}_1^{(l'_3)} = -\frac{(\mathbf{s}\cdot\mathbf{u}'_3 - r)\mathbf{u}'_2}{|\mathbf{r}\times\mathbf{s}|} \tag{176}$$

$$\mathbf{e}_2^{(l'_3)} = -\frac{(\mathbf{s}\cdot\mathbf{u}'_3 + r)\mathbf{u}'_2}{|\mathbf{r}\times\mathbf{s}|} \tag{177}$$

$$\mathbf{e}_\alpha^{(l'_3)} = -\frac{\mathbf{s}\cdot\mathbf{u}'_3\mathbf{u}'_2}{|\mathbf{r}\times\mathbf{s}|} \quad \text{for} \quad 3 \leq \alpha < N \tag{178}$$

$$\mathbf{e}_N^{(l'_3)} = (N-1)\frac{\mathbf{s}\cdot\mathbf{u}'_3\mathbf{u}'_2}{|\mathbf{r}\times\mathbf{s}|} \tag{179}$$

In what follows, we describe the derivation of Eq. (166). The gradients of the Euler angles can be related to the contravariant rotational measuring vectors via the *mobile velocity* $\boldsymbol{\omega}^{(\alpha)}(\mathbf{a})$ of the body frame. The mobile velocity can be obtained from [32,34]

$$\mathbf{a}\cdot\boldsymbol{\nabla}_\alpha\mathbf{u}'_j = \boldsymbol{\omega}^{(\alpha)}(\mathbf{a}) \times \mathbf{u}'_j \tag{180}$$

This resembles the equation expressing the time dependency of the orientation of a unit vector in term of its rotational velocity (it can be obtained by replacing in Eq. (180) the directional derivative $\mathbf{a}\cdot\boldsymbol{\nabla}_\alpha$ with the time derivative d/dt, and the mobile velocity by the rotational velocity). The mobile velocity $\boldsymbol{\omega}^{(\alpha)}(\mathbf{a})$ can be solved from Eq. (180) in terms of the body axes as

$$\boldsymbol{\omega}^{(\alpha)}(\mathbf{a}) = -\frac{i}{2}\sum_j^3 \mathbf{u}'_j\mathbf{a}\cdot\boldsymbol{\nabla}_\alpha\mathbf{u}'_j = \frac{1}{2}\sum_j^3 \mathbf{u}'_j \times (\mathbf{a}\cdot\boldsymbol{\nabla}_\alpha\mathbf{u}'_j) \tag{181}$$

where we have been able to replace $-i\mathbf{u}_j'\mathbf{a}\cdot\nabla_\alpha$ by $\mathbf{u}_j'\times(\mathbf{a}\cdot\nabla_\alpha\mathbf{u}_j')$ in the second equality due to the orthogonality of $\mathbf{u}_j'$ to $\mathbf{a}\cdot\nabla_\alpha\mathbf{u}_j'$, which follows from

$$\mathbf{a}\cdot\nabla_\alpha 1 = 0 = \mathbf{a}\cdot\nabla_\alpha(\mathbf{u}_j'\cdot\mathbf{u}_j') = 2(\mathbf{a}\cdot\nabla_\alpha\mathbf{u}_j')\cdot\mathbf{u}_j' \tag{182}$$

Equation (181) is derived here for the interested reader:

Example 13. *Because generally (see, e.g., Ref. 32 and also Appendix B, Section 1)* $\sum_j^3 \mathbf{u}_j' A_{\bar{r}}\mathbf{u}_j' = (-1)^r(3-2r)A_{\bar{r}}$ *(where* $A_{\bar{r}}$ *is an r-blade), one can write*

$$\sum_{j=1}^{3}\mathbf{u}_j'[i\boldsymbol{\omega}^{(\alpha)}(\mathbf{a})]\mathbf{u}_j' = -i\boldsymbol{\omega}^{(\alpha)}(\mathbf{a}) \tag{183}$$

(note that $i\boldsymbol{\omega}^{(\alpha)}(\mathbf{a})$ *is a bivector, so* $r=2$*). On the other hand,*

$$\boldsymbol{\omega}^{(\alpha)}(\mathbf{a})\times\mathbf{u}_j' = \mathbf{u}_j'\cdot[i\boldsymbol{\omega}^{(\alpha)}(\mathbf{a})] = \frac{1}{2}[\mathbf{u}_j' i\boldsymbol{\omega}^{(\alpha)}(\mathbf{a}) - i\boldsymbol{\omega}^{(\alpha)}(\mathbf{a})\mathbf{u}_j'] \tag{184}$$

where Eq. (A33) is used in the first equality, and the last form follows from the direct definition $\mathbf{a}\cdot\mathbf{B} = \frac{1}{2}(\mathbf{aB}-\mathbf{Ba})$ *valid for any vector* $\mathbf{a}$ *and bivector* $\mathbf{B}$*. Thus, if the both sides of Eq. (184) are multiplied on the left with* $\mathbf{u}_j'$ *and summed over j, we obtain [by using Eq. (183) and* $\sum_j^3 \mathbf{u}_j'\mathbf{u}_j' = 3$*]*

$$\sum_{j=1}^{3}\mathbf{u}_j'\boldsymbol{\omega}^{(\alpha)}(\mathbf{a})\times\mathbf{u}_j' = \frac{1}{2}\sum_{j=1}^{3}[\mathbf{u}_j'\mathbf{u}_j' i\boldsymbol{\omega}^{(\alpha)}(\mathbf{a}) - \mathbf{u}_j'(i\boldsymbol{\omega}^{(\alpha)}(\mathbf{a}))\mathbf{u}_j'] = 2i\boldsymbol{\omega}^{(\alpha)}(\mathbf{a}) \tag{185}$$

where Eq. (181) follows.

The mobile velocity $\boldsymbol{\omega}^{(\alpha)}(\mathbf{a})$ can be related to the Euler angles by the relation

$$\boldsymbol{\omega}^{(\alpha)}(\mathbf{a}) = \mathbf{u}_3\mathbf{a}\cdot\nabla_\alpha\phi + \mathbf{n}\mathbf{a}\cdot\nabla_\alpha\theta + \mathbf{u}_3'\mathbf{a}\cdot\nabla_\alpha\chi \tag{186}$$

which is similar to Eq. (29) and can be obtained by a similar derivation. The directional derivative of the Euler angles can be picked by dotting both sides of Eq. (186) with the reciprocal nodal line vectors (see page 262 in Ref. 50 for reciprocal frames)

$$\mathbf{n}^{(i)} = \frac{\mathbf{n}_j\times\mathbf{n}_k}{\mathbf{n}_1\cdot\mathbf{n}_2\times\mathbf{n}_3} \tag{187}$$

(where i,j,k are in cyclic order, and $\mathbf{n}_1=\mathbf{u}_3, \mathbf{n}_2=\mathbf{n}=\mathbf{u}_3\times\mathbf{u}_3'/|\mathbf{u}_3\times\mathbf{u}_3'|$, $\mathbf{n}_3=\mathbf{u}_3'$ as before), that is,

$$\mathbf{a}\cdot\nabla_\alpha B_i = \mathbf{n}^{(i)}\cdot\boldsymbol{\omega}^{(\alpha)}(\mathbf{a}) \tag{188}$$

and the gradient of the Euler angle is obtained via the chain rule [Eq. (A134)] as

$$\nabla_\alpha B_i = \nabla_{\mathbf{a}}\mathbf{a} \cdot \nabla_\alpha B_i = \nabla_{\mathbf{a}}\mathbf{n}^{(i)} \cdot \boldsymbol{\omega}^{(\alpha)}(\mathbf{a}) \tag{189}$$

If we substitute Eq. (189) to Eq. (125), we obtain

$$\mathbf{e}_\alpha^{(l'_i)} = \nabla_{\mathbf{a}} \sum_j^3 \Omega_{ji}\mathbf{n}^{(j)} \cdot \boldsymbol{\omega}^{(\alpha)}(\mathbf{a}) = \nabla_{\mathbf{a}} \sum_j^3 \mathbf{u}'_i \cdot \mathbf{n}_j\mathbf{n}^{(j)} \cdot \boldsymbol{\omega}^{(\alpha)}(\mathbf{a}) \tag{190}$$

where we used the definition of Ω_{ji} [given in Eq. (34)] to obtain the last equality. Because $\sum_j \mathbf{n}_j\mathbf{n}^{(j)} \cdot \boldsymbol{\omega}^{(\alpha)}(\mathbf{a})$ is the representation of the vector $\boldsymbol{\omega}^{(\alpha)}(\mathbf{a})$ in the nonorthogonal frame $\{\mathbf{n}_1, \mathbf{n}_2, \mathbf{n}_3\}$ (see, e.g., Ref. 34), we obtain

$$\mathbf{e}_\alpha^{(l_i)} = \nabla_{\mathbf{a}}\mathbf{u}'_i \cdot \boldsymbol{\omega}^{(\alpha)}(\mathbf{a}) \tag{191}$$

By inserting the expression of $\boldsymbol{\omega}^{(\alpha)}(\mathbf{a})$ in Eq. (181), the above equation can be written as

$$\mathbf{e}_\alpha^{(l'_i)} = \frac{1}{2}\sum_j^3 \nabla_{\mathbf{a}}\mathbf{u}'_i \cdot [\mathbf{u}'_j \times (\mathbf{a} \cdot \nabla_\alpha \mathbf{u}'_j)] = \frac{1}{2}\sum_j^3 \nabla_{\mathbf{a}}(\mathbf{u}'_i \times \mathbf{u}'_j) \cdot (\mathbf{a} \cdot \nabla_\alpha \mathbf{u}'_j) \tag{192}$$

where we may extract the result in Eq. (166) by using

$$\mathbf{u}'_i \times \mathbf{u}'_i = 0 \tag{193}$$

$$\mathbf{u}'_i \times \mathbf{u}'_j = \mathbf{u}'_k \tag{194}$$

(where the indices i, j, k are in cyclic order), and

$$\mathbf{u}'_i \cdot (\mathbf{a} \cdot \nabla_\alpha \mathbf{u}'_j) = -(\mathbf{a} \cdot \nabla_\alpha \mathbf{u}'_i) \cdot \mathbf{u}'_j \tag{195}$$

(which can be obtained by taking the directional derivative of $\mathbf{u}'_i \cdot \mathbf{u}'_j = \delta_{ij}$). As an example, consider the case $i = 1$. Then, by the direct substitution to Eq. (192), we obtain

$$\begin{aligned}\mathbf{e}_\alpha^{(l'_1)} &= \frac{1}{2}\nabla_{\mathbf{a}}[(\mathbf{u}'_1 \times \mathbf{u}'_2) \cdot (\mathbf{a} \cdot \nabla_\alpha \mathbf{u}'_2) + (\mathbf{u}'_1 \times \mathbf{u}'_3) \cdot (\mathbf{a} \cdot \nabla_\alpha \mathbf{u}'_3)] \\ &= \frac{1}{2}\nabla_{\mathbf{a}}[\mathbf{u}'_3 \cdot (\mathbf{a} \cdot \nabla_\alpha \mathbf{u}'_2) - \mathbf{u}'_2 \cdot (\mathbf{a} \cdot \nabla_\alpha \mathbf{u}'_3)]\end{aligned} \tag{196}$$

By using Eq. (195), we may simplify this to

$$\mathbf{e}_\alpha^{(l'_1)} = \boldsymbol{\nabla}_\mathbf{a}[(\mathbf{a} \cdot \boldsymbol{\nabla}_\alpha \mathbf{u}'_2) \cdot \mathbf{u}'_3] \tag{197}$$

The measuring vectors $\mathbf{e}_\alpha^{(l'_2)}$ and $\mathbf{e}_\alpha^{(l'_3)}$ are obtained by a similar derivation.

Example 14. *We form the Hamiltonian for a bent triatomic molecule using the bond lengths r_{31}, r_{32} and the bond angle θ_{132} as coordinates and using the bond-z body frame defined in Eqs. (74)–(76). The coefficients c_α and d_α differing from zero are in this case $c_3 = 1$, $c_1 = -1$, $d_3 = 1$, and $d_2 = -1$. The measuring vectors (differing from zero) for the bond lengths are given as*

$$\mathbf{e}_1^{(r_{31})} = \mathbf{u}_{\mathbf{r}_{31}} = -\mathbf{e}_3^{(r_{31})} \tag{198}$$

$$\mathbf{e}_2^{(r_{32})} = \mathbf{u}_{\mathbf{r}_{32}} = -\mathbf{e}_3^{(r_{32})} \tag{199}$$

The measuring vectors for the bond angle θ_{132} are obtained by substituting the indices $1, 2, 3$ to Eqs. (150)–(152). For example,

$$\mathbf{e}_1^{(\theta_{132})} = \frac{\mathbf{u}_{\mathbf{r}_{31}} \cos\theta_{132} - \mathbf{u}_{\mathbf{r}_{32}}}{r_{31} \sin\theta_{132}} \tag{200}$$

The rotational measuring vectors are obtained by the substitution of Eqs. (167)–(169). The measuring vectors different from zero are

$$\mathbf{e}_1^{(l'_1)} = -\frac{\mathbf{u}'_2}{r_{31}} \tag{201}$$

$$\mathbf{e}_3^{(l'_1)} = \frac{\mathbf{u}'_2}{r_{31}} \tag{202}$$

$$\mathbf{e}_1^{(l'_2)} = \frac{\mathbf{u}'_1}{r_{31}} \tag{203}$$

$$\mathbf{e}_3^{(l'_2)} = -\frac{\mathbf{u}'_1}{r_{31}} \tag{204}$$

$$\mathbf{e}_1^{(l'_3)} = -\frac{\cos\theta_{132} \mathbf{u}'_2}{r_{31} \sin\theta_{132}} \tag{205}$$

$$\mathbf{e}_2^{(l'_3)} = \frac{\mathbf{u}'_2}{r_{32} \sin\theta_{132}} \tag{206}$$

$$\mathbf{e}_3^{(l'_3)} = -\frac{(-r_{32}\cos\theta_{132} + r_{31})\mathbf{u}'_2}{r_{31} r_{32} \sin\theta_{132}} \tag{207}$$

The nonzero elements of the contravariant metric tensor are given by the substitution of Eq. (118) as

$$g^{(r_{31}r_{31})} = \frac{1}{m_1} + \frac{1}{m_3} \tag{208}$$

$$g^{(r_{31}r_{32})} = \frac{\cos\theta_{132}}{m_3} \tag{209}$$

$$g^{(r_{31}\theta_{132})} = -\frac{\sin\theta_{132}}{m_3 r_{32}} \tag{210}$$

$$g^{(r_{32}r_{32})} = \frac{1}{m_2} + \frac{1}{m_3} \tag{211}$$

$$g^{(r_{32}\theta_{132})} = -\frac{\sin\theta_{132}}{m_3 r_{31}} \tag{212}$$

$$g^{(r_{32}l'_2)} = \frac{\sin\theta_{132}}{m_3 r_{31}} \tag{213}$$

$$g^{(\theta_{132}\theta_{132})} = \frac{1}{m_1 r_{31}^2} + \frac{1}{m_2 r_{32}^2} + \frac{1}{m_3}\left(\frac{1}{r_{31}^2} + \frac{1}{r_{32}^2} - \frac{2\cos\theta_{132}}{r_{31}r_{32}}\right) \tag{214}$$

$$g^{(\theta_{132}l'_2)} = -\frac{1}{m_1 r_{31}^2} + \frac{\cos\theta_{132} r_{31} - r_{32}}{m_3 r_{31}^2 r_{32}} \tag{215}$$

$$g^{(l'_1 l'_1)} = \frac{m_1 + m_3}{m_1 m_3 r_{31}^2} \tag{216}$$

$$g^{(l'_1 l'_3)} = \frac{1}{r_{31}^2}\left[\frac{\cot\theta_{132}}{m_1} + \frac{\csc\theta_{132}}{m_3 r_{32}}(-r_{31} + r_{32}\cos\theta_{132})\right] \tag{217}$$

$$g^{(l'_2 l'_2)} = \frac{m_1 + m_3}{m_1 m_3 r_{31}^2} \tag{218}$$

$$g^{(l'_3 l'_3)} = \frac{m_2 m_3 r_{32}^2 \cot^2\theta_{132} + m_1 \csc^2\theta_{132}[m_3 r_{31}^2 + m_2(r_{31} - r_{32}\cos\theta_{132})^2]}{m_1 m_2 m_3 r_{31}^2 r_{32}^2} \tag{219}$$

III. CANONICAL QUANTIZATION

A. Unconstrained Quantization

The quantum mechanical kinetic energy operator is given by

$$\hat{T} = -\frac{\hbar^2}{2}\sum_{\alpha}^{N}\frac{1}{m_\alpha}\boldsymbol{\nabla}_\alpha^2 \tag{220}$$

in the absence of constraints—that is, for $A = 3N$ or $P = 0$. By utilizing the formal analogy of the metric tensors appearing in Lagrangians and Hamiltonians to the metric tensors appearing in the Riemannian differential geometry, we may represent $\hat{T}$ in terms of the coordinates $\{q_1, q_2, \ldots, q_{3N}\} = \{s_1, s_2, \ldots, s_{3N-6}, \phi, \theta, \chi, X_1, X_2, X_3\}$ as

$$\hat{T} = -\frac{\hbar^2}{2}\sum_{ij}^{3N}\left(\frac{\partial}{\partial q_i} + \frac{1}{J}\frac{\partial J}{\partial q_i}\right) g^{(q_iq_j)}\frac{\partial}{\partial q_j} \tag{221}$$

where $J = |\det g_{q_iq_j}|^{1/2}\prod_{\alpha}^{N} m_\alpha^{-3/2} = |\det g^{(q_iq_j)}|^{-1/2}\prod_{\alpha}^{N} m_\alpha^{-3/2}$ is the Jacobian for the coordinate transformation $\{x_{1_1}, x_{1_2}, x_{1_3}, \ldots, x_{N_1}, x_{N_2}, x_{N_3}\} \rightarrow \{s_1, s_2, \ldots, s_{3N-6}, \phi, \theta, \chi, X_1, X_2, X_3\}$. The factor $\prod_{\alpha}^{N} m_\alpha^{-3/2}$ is needed to eliminate the mass dependency of $\det g^{(q_iq_j)}$. Equation (221) is identical, up to the constant factor $-\hbar^2/2$, to the coordinate representation of an N-dimensional Laplacian operator in Euclidean space [52]. A direct evaluation of the determinant $\det g_{q_iq_j}$ by conventional means is tedious, because the elements of the metric tensor can be complicated functions of the shape coordinates. A better way of obtaining J is presented in Section IV. Occasionally (e.g., Ref. 17), the quantity $-i\hbar\partial/\partial q_j$ is denoted as the generalized momentum operator $\hat{p}_j$, and $-i\hbar[\partial/\partial q_j + (\partial J/\partial q_j)/J]$ as the "adjoint" momentum operator $\hat{p}_j^\dagger$, because by this definition, $\int J(\hat{p}_j\Psi)^*\Phi\, dq_1dq_2\ldots dq_{3N} = \int J\Psi^*\hat{p}_j^\dagger\Phi dq_1dq_2\ldots dq_{3N}$, if the scalar functions Φ and Ψ vanish at the integration limits (the asterisk implies the complex conjugation). All integrations are performed using the volume element $d\tau = Jdq_1dq_2\ldots dq_{3N}$. If instead one wishes to integrate employing the volume element $d\tau_w = wdq_1dq_2\ldots dq_{3N}$, the corresponding kinetic energy operator $\hat{T}_w$ is given as [17]

$$\hat{T}_w = J^{1/2}w^{-1/2}\hat{T}w^{1/2}J^{-1/2} \tag{222}$$

in terms of the kinetic energy operator $\hat{T}$ of Eq. (221). However, if constraints are imposed to the molecule, the representation is more complicated, and it has been a subject of controversy (see Section III.B).

In terms of the shape coordinates, the Euler angles, and the Cartesian coordinates of the center of mass, the kinetic energy operator can be written as the sum

$$\hat{T} = \hat{T}^{(\mathrm{int})} + \hat{T}^{(\mathrm{transl})} \tag{223}$$

where the translational part is

$$\hat{T}^{(\mathrm{transl})} = -\frac{\hbar^2}{2M}\sum_{i=1}^{3}\frac{\partial^2}{\partial X_i^2} \tag{224}$$

and the internal part is

$$\hat{T}^{(\text{int})} = -\frac{\hbar^2}{2}\left[\sum_{i=1}^{3N-6}\left(\frac{\partial}{\partial s_i} + J^{-1}\frac{\partial J}{\partial s_i}\right)\left(\sum_{j=1}^{3N-6} g^{(s_i s_j)}\frac{\partial}{\partial s_j} + \sum_{j=1}^{3} g^{(s_i B_j)}\frac{\partial}{\partial B_j}\right)\right.$$
$$\left. + \sum_{i=1}^{3}\left(\frac{\partial}{\partial B_i} + J^{-1}\frac{\partial J}{\partial B_i}\right)\left(\sum_{j=1}^{3N-6} g^{(B_i s_j)}\frac{\partial}{\partial s_j} + \sum_{j=1}^{3} g^{(B_i B_j)}\frac{\partial}{\partial B_j}\right)\right] \tag{225}$$

The internal part is usually written in terms of the body-frame components $\hat{l}'_i = \mathbf{u}'_i \cdot \hat{\mathbf{l}}$ of the dual $\hat{\mathbf{l}} = -i\hat{\mathbf{L}}$ of the internal angular momentum operator

$$\hat{\mathbf{L}} = -i\hbar\sum_{\alpha}^{N-1}\mathbf{y}_\alpha \times \boldsymbol{\nabla}_{\mathbf{y}_\alpha} \tag{226}$$

rather than in terms of the Euler angle partial derivative operators $\{\partial/\partial\phi, \partial/\partial\theta, \partial/\partial\chi\}$ as

$$\hbar\left(\boldsymbol{\nabla}_\alpha\phi\frac{\partial}{\partial\phi} + \boldsymbol{\nabla}_\alpha\theta\frac{\partial}{\partial\theta} + \boldsymbol{\nabla}_\alpha\chi\frac{\partial}{\partial\chi}\right) = \sum_{i}^{3}\mathbf{e}_\alpha^{(l'_i)}\hat{l}'_i \tag{227}$$

in the Schrödinger equation. In order to use the formalism of geometric algebra consistently, the angular momentum operator $\hat{\mathbf{L}}$ is treated as a bivector; that is, we assume that i in Eq. (226) is the unit trivector—in contrast to conventional treatments, where no such interpretation is assigned to i. The internal part of the kinetic energy operator reads in terms of the body-frame components of the internal angular momentum as

$$\hat{T}^{(\text{int})} = -\frac{\hbar^2}{2}\left[\sum_{i=1}^{3N-6}\left(\frac{\partial}{\partial s_i} + J^{-1}\frac{\partial J}{\partial s_i}\right)\left(\sum_{j=1}^{3N-6} g^{(s_i s_j)}\frac{\partial}{\partial s_j} + \frac{1}{\hbar}\sum_{j=1}^{3} g^{(s_i l'_j)}\hat{l}'_j\right)\right.$$
$$\left. + \frac{1}{\hbar}\sum_{i=1}^{3}\hat{l}'_i\left(\sum_{j=1}^{3N-6} g^{(l'_i s_j)}\frac{\partial}{\partial s_j} + \frac{1}{\hbar}\sum_{j=1}^{3} g^{(l'_i l'_j)}\hat{l}'_j\right)\right] \tag{228}$$

Example 15. *The Jacobian for a triatomic molecule in bond-angle coordinates, and in the bond-z frame is [see Eq. (260)]*

$$J = m_1^{-3/2} m_2^{-3/2} m_3^{-3/2} \sin\theta r_{31}^2 r_{32}^2 \sin\theta_{132} \tag{229}$$

The kinetic energy operator is given by

$$\hat{T}^{(\mathrm{int})} = \frac{1}{2}\sum_{i,j=1}^{3}[\hat{p}_{s_i}^{\dagger} g^{(s_i s_j)} \hat{p}_{s_j} - i(\hat{p}_{s_i}^{\dagger} g^{(s_i l'_j)} \hat{l}'_j + \hat{l}'_i g^{(l'_i s_j)} \hat{p}_{s_j}) - \hat{l}'_i g^{(l'_i l'_j)} \hat{l}'_j] \tag{230}$$

The factor $-i$ in front of the "Coriolis" and the minus sign in front of the "rotational" terms arise from the interpretation of i as the unit trivector. Of course, they can be set to 1 (as in conventional treatments), if no such interpretation is wanted. The momentum operators are given by

$$\hat{p}_{r_{31}} = -i\hbar\frac{\partial}{\partial r_{31}} \tag{231}$$

$$\hat{p}_{r_{32}} = -i\hbar\frac{\partial}{\partial r_{32}} \tag{232}$$

$$\hat{p}_{\theta_{132}} = -i\hbar\frac{\partial}{\partial \theta_{132}} \tag{233}$$

the "adjoint" momentum operators by

$$\hat{p}_{r_{31}}^{\dagger} = -i\hbar\left(\frac{\partial}{\partial r_{31}} + J^{-1}\frac{\partial J}{\partial r_{31}}\right) = -i\hbar\left(\frac{\partial}{\partial r_{31}} + \frac{2}{r_{31}}\right) \tag{234}$$

$$\hat{p}_{r_{32}}^{\dagger} = -i\hbar\left(\frac{\partial}{\partial r_{32}} + J^{-1}\frac{\partial J}{\partial r_{32}}\right) = -i\hbar\left(\frac{\partial}{\partial r_{32}} + \frac{2}{r_{32}}\right) \tag{235}$$

$$\hat{p}_{\theta_{132}}^{\dagger} = -i\hbar\left(\frac{\partial}{\partial \theta_{132}} + J^{-1}\frac{\partial J}{\partial \theta_{132}}\right) = -i\hbar\left(\frac{\partial}{\partial \theta_{132}} + \frac{\cos\theta_{132}}{\sin\theta_{132}}\right) \tag{236}$$

and the contravariant metric tensor elements $g^{(s_i s_j)}$, $g^{(s_i l'_j)}$, and $g^{(l'_i l'_j)}$ are derived in Example 14.

B. Constrained Quantization

The quantization of constrained many-body systems has been of interest since the birth of the modern quantum theory (see, e.g., Refs. 60–66 and references therein). This problem is often referred as the quantization in Riemannian or curved space by the analogy to the differential geometry of surfaces [62]. Because generalized momentas are differential operators, it is more difficult to obtain the quantum Hamiltonian in the presence of constraints compared to the classical case (see the end of Section II.B). There are many ways to approach the problem. In some references, the kinetic energy operator is sought as the coordinate representation of an A-dimensional tangential Laplacian operator embedded in a $3N$-dimensional configuration space. In other approaches, the classical Hamiltonian is quantized, or the classical Lagrangian is transformed to a quantum mechanical Hamiltonian by using some sets of correspondence rules, although most prescriptions and correspondence rules do not produce correct quantum mechanical equivalents of the classical quantities [67–69].

It is accepted generally that the kinetic energy operator for conservative systems can be written in terms of the active coordinates $\{q_1, q_2, \ldots, q_A\}$ as

$$\hat{T} = -\frac{\hbar^2}{2}\sum_{ij}^{A}(\det g_{q_iq_j})^{-1/2}\frac{\partial}{\partial q_i}g^{(q_iq_j)}(\det g_{q_iq_j})^{1/2}\frac{\partial}{\partial q_j} + U \tag{237}$$

where $g_{q_iq_j}$ is the $(A \times A)$-dimensional covariant metric tensor and $g^{(q_iq_j)}$ is its inverse. Unfortunately, different references disagree on the form of the extra term U. For example, in Ref. 63 it assumed that

$$U = 0 \tag{238}$$

On the other hand, in Ref. 60 it is suggested that

$$U = \frac{\hbar^2}{4}\kappa \tag{239}$$

where κ is the intrinsic scalar curvature, and in Ref. 62 it is suggested that

$$U = \frac{\hbar^2}{8}\sum_{i,j=1}^{A}\sum_{r,s=A+1}^{3N}\Gamma^i_{ir}\Gamma^j_{js}h^{(q_rq_s)} \tag{240}$$

where Γ^i_{ir} are the Christoffel symbols of the second kind [52] in the $3N$-dimensional space. These kinds of differences have turned the subject of constrained quantization to a source of controversy.

IV. VOLUME ELEMENTS

In this section, we present a simple method for obtaining the Jacobian J for the coordinate transformation $\{x_{1_1}, x_{1_2}, x_{1_3}, \ldots, x_{N_1}, x_{N_2}, x_{N_3}\} \rightarrow \{s_1, s_2, \ldots, s_{3N-6}, \phi, \theta, \chi, X_1, X_2, X_3\}$. It was shown in Ref. 39 that the volume element of integration $d\tau = J dq_1 dq_2 \ldots dq_{3N-6} d\phi d\theta d\chi dX_1 dX_2 dX_3$ can be written as a product of N volume elements, each associated with a separate set of three coordinates as

$$d\tau = \frac{|d^3\mathbf{z}_1||d^3\mathbf{z}_2|\ldots|d^3\mathbf{z}_{N-1}|dX_1dX_2dX_3}{|\det[\mathbf{c}]|^3} \tag{241}$$

The vectors $\mathbf{z}_\alpha$ are linear combinations of the vectors $\mathbf{x}_\alpha$ [Eq. (47)]. They are chosen so that $\mathbf{z}_1$ can be parameterized by two Euler angles ϕ and θ, and the shape coordinate s_1, $\mathbf{z}_2$ can be parameterized by the Euler angle χ and two shape coordinates s_2 and s_3, and the vectors $\mathbf{z}_3, \mathbf{z}_4, \ldots, \mathbf{z}_{N-1}$ can be parameterized by three shape coordinate each (up to an orientation, which does not affect the volume element). The quantity det $[\mathsf{c}]$ is the determinant of the matrix $[c_{\alpha\beta}]$ [see Eq. (47)]. The individual volume elements were shown to be

$$|d^3\mathbf{z}_1| = \frac{d\phi d\theta ds_1}{|\mathbf{e}_{\mathbf{z}_1}^{(\phi)} \wedge \mathbf{e}_{\mathbf{z}_1}^{(\theta)} \wedge \mathbf{e}_{\mathbf{z}_1}^{(s_1)}|} \tag{242}$$

$$|d^3\mathbf{z}_2| = \frac{d\chi ds_2 ds_3}{|\mathbf{e}_{\mathbf{z}_2}^{(\chi)} \wedge \mathbf{e}_{\mathbf{z}_2}^{(s_2)} \wedge \mathbf{e}_{\mathbf{z}_2}^{(s_3)}|} \tag{243}$$

$$|d^3\mathbf{z}_\alpha| = \frac{ds_i ds_j ds_k}{|\mathbf{e}_{\mathbf{z}_\alpha}^{(s_i)} \wedge \mathbf{e}_{\mathbf{z}_\alpha}^{(s_j)} \wedge \mathbf{e}_{\mathbf{z}_\alpha}^{(s_k)}|} \qquad \text{for } \alpha = 3, 4, \ldots, N-1 \tag{244}$$

where

$$\mathbf{e}_{\mathbf{z}_\alpha}^{(B_j)} = \sum_i^3 \Omega_{ij}^{-1} \mathbf{e}_{\mathbf{z}_\alpha}^{(l'_i)} = \sum_i^3 \Omega_{ij}^{-1} \boldsymbol{\nabla}_{\mathbf{a}}[(\mathbf{a} \cdot \boldsymbol{\nabla}_{\mathbf{z}_\alpha} \mathbf{u}'_j) \cdot \mathbf{u}'_k] \tag{245}$$

and $\mathbf{e}_{\mathbf{z}_\alpha}^{(s_i)} = \boldsymbol{\nabla}_{\mathbf{z}_\alpha} s_i$, as before. However, we could use the contravariant measuring vectors $\mathbf{e}_{\mathbf{z}_\alpha}^{(l'_i)}$ instead of $\mathbf{e}_{\mathbf{z}_\alpha}^{(B_i)}$ in the volume-element calculation, because

$$g' = \sin^2\theta g \tag{246}$$

where $g = \det g^{(q_i q_j)}$ is the determinant of the contravariant metric tensor with the rotational degrees of freedom represented in terms of the Euler angles, and $g' = \det g^{(ij)}$ is the determinant of the contravariant metric tensor with the rotational degrees of freedom represented in terms of the body-frame components of the total angular momentum ($\det[\Omega]^{-1} = -1/\sin\theta$). Thus, we may as well write

$$|d^3\mathbf{z}_1| = \frac{\sin\theta d\phi d\theta ds_1}{|\mathbf{e}_{\mathbf{z}_1}^{(l'_1)} \wedge \mathbf{e}_{\mathbf{z}_1}^{(l'_2)} \wedge \mathbf{e}_{\mathbf{z}_1}^{(s_1)}|} \tag{247}$$

$$|d^3\mathbf{z}_2| = \frac{d\chi ds_2 ds_3}{|\mathbf{e}_{\mathbf{z}_2}^{(l'_3)} \wedge \mathbf{e}_{\mathbf{z}_2}^{(s_2)} \wedge \mathbf{e}_{\mathbf{z}_2}^{(s_3)}|} \tag{248}$$

The volume element is independent of the choice of the body axes, as explicitly proven in Ref. 39. The body frame used in the volume-element calculations does not need to be the body frame used to represent the vibration–rotation Hamiltonian. If the shape coordinates employed include $z_1 = |\mathbf{z}_1|$, z_2, and $\theta_{12} = \arccos(\mathbf{u}_{\mathbf{z}_1} \cdot \mathbf{u}_{\mathbf{z}_2})$, one may choose the "bond-z" frame, where $\mathbf{r} = \mathbf{z}_1$ and $\mathbf{s} = \mathbf{z}_2$ in Eqs. (14) and (16). Then,

$$\mathbf{e}_{\mathbf{z}_1}^{(l'_1)} = -\frac{\mathbf{u}'_2}{z_1} \tag{249}$$

$$\mathbf{e}_{\mathbf{z}_1}^{(l'_2)} = \frac{\mathbf{u}'_1}{z_1} \tag{250}$$

$$\mathbf{e}_{\mathbf{z}_2}^{(l'_3)} = \frac{z_1 \mathbf{u}'_2}{|\mathbf{z}_1 \times \mathbf{z}_2|} = \frac{\mathbf{u}_{\mathbf{z}_1} \times \mathbf{u}_{\mathbf{z}_2}}{z_2 \sin^2 \theta_{12}} \tag{251}$$

and we have

$$|d^3\mathbf{z}_1| = \frac{z_1^2 \sin\theta d\phi d\theta ds_1}{|\mathbf{u}'_1 \wedge \mathbf{u}'_2 \wedge \mathbf{e}_{\mathbf{z}_1}^{(s_1)}|} = \frac{z_1^2 \sin\theta d\phi d\theta ds_1}{|\mathbf{u}_{\mathbf{z}_1} \cdot \mathbf{e}_{\mathbf{z}_1}^{(s_1)}|} \tag{252}$$

$$|d^3\mathbf{z}_2| = \frac{z_2 \sin^2 \theta_{12} d\chi ds_2 ds_3}{|(\mathbf{u}_{\mathbf{z}_1} \times \mathbf{u}_{\mathbf{z}_2}) \wedge \mathbf{e}_{\mathbf{z}_2}^{(s_2)} \wedge \mathbf{e}_{\mathbf{z}_2}^{(s_3)}|} \tag{253}$$

where the last equality in Eq. (252) follows from $\mathbf{u}'_1 \wedge \mathbf{u}'_2 \wedge \mathbf{e}_{\mathbf{z}_1}^{(s_1)} = -(i\mathbf{u}'_1 \times \mathbf{u}'_2) \wedge \mathbf{e}_{\mathbf{z}_1}^{(s_1)} = -i(\mathbf{u}'_1 \times \mathbf{u}'_2) \cdot \mathbf{e}_{\mathbf{z}_1}^{(s_1)}$ [see Eqs. (A32) and (A33)]. This is the result of Ref. 39, derived a bit more tediously with $\mathbf{e}_{\mathbf{z}_\alpha}^{(B_j)}$. Any constant factor in $d\tau$ (such as $1/|\det[\mathsf{c}]|^3$) can be thought to be absorbed in the normalization of the wave functions Ψ_i (generally, $\int \Psi_i^* \Psi_i \, d\tau = 1$).

Example 16. *Consider a parameterization of* $\mathbf{z}_1$ in terms of ϕ, θ, *and its length* $z_1 \equiv s_1$. *From Table II, we obtain the vector derivative*

$$\mathbf{e}_{\mathbf{z}_1}^{(z_1)} = \boldsymbol{\nabla}_{\mathbf{z}_1} z_1 = \mathbf{u}_{\mathbf{z}_1} \tag{254}$$

Thus, by the direct substitution of Eq. (252),

$$|d^3\mathbf{z}_1| = \frac{z_1^2 \sin\theta \, d\phi d\theta ds_1}{|\mathbf{u}_{\mathbf{z}_1} \cdot \mathbf{u}_{\mathbf{z}_1}|} = z_1^2 \sin\theta \, d\phi d\theta dz_1 \tag{255}$$

If $\mathbf{z}_2$ *is parameterized in terms of* χ, *its length* $z_2 = s_2$, *and the angle* $\theta_{12} = \arccos(\mathbf{u}_{\mathbf{z}_1} \cdot \mathbf{u}_{\mathbf{z}_2}) = s_3$, *then we obtain*

$$\mathbf{e}_{\mathbf{z}_2}^{(z_2)} = \boldsymbol{\nabla}_{\mathbf{z}_2} z_2 = \mathbf{u}_{\mathbf{z}_2} \tag{256}$$

$$\mathbf{e}_{\mathbf{z}_2}^{(\theta_{12})} = \boldsymbol{\nabla}_{\mathbf{z}_2} \theta_{12} = \frac{\mathbf{u}_{\mathbf{z}_2} \cos\theta_{12} - \mathbf{u}_{\mathbf{z}_1}}{z_2 \sin\theta_{12}} \tag{257}$$

(see Table II) and the volume element $|d^3\mathbf{z}_2|$ is

$$|d^3\mathbf{z}_2| = \frac{z_2^2 \sin^3\theta_{12}\, d\chi ds_2 ds_3}{|(\mathbf{u}_{\mathbf{z}_1} \times \mathbf{u}_{\mathbf{z}_2}) \wedge \mathbf{u}_{\mathbf{z}_2} \wedge (\mathbf{u}_{\mathbf{z}_2} \cos\theta_{12} - \mathbf{u}_{\mathbf{z}_1})|}$$
$$= \frac{z_2^2 \sin^3\theta_{12}\, d\chi dz_2 d\theta_{12}}{|(\mathbf{u}_{\mathbf{z}_1} \times \mathbf{u}_{\mathbf{z}_2}) \wedge \mathbf{u}_{\mathbf{z}_2} \wedge \mathbf{u}_{\mathbf{z}_1}|} = z_2^2 \sin\theta_{12}\, d\chi dz_2 d\theta_{12} \tag{258}$$

(where the last equality follows, because $|(\mathbf{u}_{\mathbf{z}_1} \times \mathbf{u}_{\mathbf{z}_2}) \wedge \mathbf{u}_{\mathbf{z}_2} \wedge \mathbf{u}_{\mathbf{z}_1}| = |(\mathbf{u}_{\mathbf{z}_1} \times \mathbf{u}_{\mathbf{z}_2})(\mathbf{u}_{\mathbf{z}_2} \wedge \mathbf{u}_{\mathbf{z}_1})| = |\mathbf{u}_{\mathbf{z}_1} \times \mathbf{u}_{\mathbf{z}_2}||\mathbf{u}_{\mathbf{z}_2} \wedge \mathbf{u}_{\mathbf{z}_1}| = \sin^2\theta_{12}$), and we obtain the full volume element

$$d\tau = \frac{|d^3\mathbf{z}_1||d^3\mathbf{z}_2| dX_1 dX_2 dX_3}{|\det[\mathsf{c}]|^3} = \frac{z_1^2 z_2^2 \sin\theta_{12}\, dz_1 dz_2 d\theta_{12} \sin\theta d\phi d\theta d\chi dX_1 dX_2 dX_3}{|\det[\mathsf{c}]|^3} \tag{259}$$

In particular, if $\mathbf{z}_1 = \mathbf{r}_{31}$ and $\mathbf{z}_2 = \mathbf{r}_{32}$ (in which case $|\det[\mathsf{c}]|^3 = 1$), the volume element for a bent triatomic molecule in bond-angle coordinates is obtained as

$$d\tau = |d^3\mathbf{r}_{31}||d^3\mathbf{r}_{32}| dX_1 dX_2 dX_3$$
$$= \sin\theta r_{31}^2 r_{32}^2 \sin\theta_{132}\, dr_{31} dr_{32} d\theta_{132} dX_1 dX_2 dX_3 d\phi d\theta d\chi \tag{260}$$

Example 17. *Let us consider the volume element of integration for a tetra-atomic molecule in $\{z_1, z_2, z_3, \theta_{12}, \theta_{23}, \phi_{321}\}$, where ϕ_{321} is the angle between the vector $\mathbf{z}_3$ and the plane $\mathbf{z}_2 \wedge \mathbf{z}_1$ (i.e., Wilson's definition for the out-of-bending coordinate [1]). Now, because*

$$\mathbf{e}_{\mathbf{z}_3}^{(z_3)} = \mathbf{u}_{\mathbf{z}_3} \tag{261}$$

$$\mathbf{e}_{\mathbf{z}_3}^{(\theta_{23})} = \frac{\mathbf{u}_{\mathbf{z}_3} \cos\theta_{23} - \mathbf{u}_{\mathbf{z}_2}}{z_3 \sin\theta_{23}} \tag{262}$$

$$\mathbf{e}_{\mathbf{z}_3}^{(\phi_{321})} = \frac{1}{z_3 \cos\phi_{321}} \left(\frac{\mathbf{u}_{\mathbf{z}_2} \times \mathbf{u}_{\mathbf{z}_1}}{\sin\theta_{12}} - \mathbf{u}_{\mathbf{z}_3} \sin\phi_{321} \right) \tag{263}$$

(see Table II), the volume element $|d^3\mathbf{z}_3|$ is given by Eq. (244) in $\{z_3, \theta_{23}, \phi_{321}\}$ as

$$|d^3\mathbf{z}_3| = \frac{z_3^2 \sin\theta_{23} \sin\theta_{12} |\cos\phi_{321}| dz_3 d\theta_{23} d\phi_{321}}{|\mathbf{u}_{\mathbf{z}_3} \wedge \mathbf{u}_{\mathbf{z}_2} \wedge (\mathbf{u}_{\mathbf{z}_2} \times \mathbf{u}_{\mathbf{z}_1})|}$$
$$= \frac{z_3^2 \sin\theta_{23} \sin\theta_{12} |\cos\phi_{321}|}{|\cos\theta_{13} - \cos\theta_{23} \cos\theta_{12}|} dz_3 d\theta_{23} d\phi_{321} \tag{264}$$

where the last equality follows from

$$\mathbf{u}_{\mathbf{z}_\alpha} \wedge \mathbf{u}_{\mathbf{z}_\beta} \wedge (\mathbf{u}_{\mathbf{z}_\beta} \times \mathbf{u}_{\mathbf{z}_\gamma}) = -i\mathbf{u}_{\mathbf{z}_\alpha} \cdot (\mathbf{u}_{\mathbf{z}_\gamma} - \cos\theta_{\beta\gamma}\mathbf{u}_{\mathbf{z}_\beta}) \tag{265}$$

Thus, by multiplying the volume element of Eq. (259) with $|d^3\mathbf{z}_3|$, *the full volume element is obtained as*

$$d\tau = \frac{z_1^2 z_2^2 z_3^2 \sin^2\theta_{12} \sin\theta_{23} |\cos\phi_{321}| dz_1 dz_2 dz_3 d\theta_{12} d\theta_{23} d\phi_{321} \sin\theta \, d\phi d\theta d\chi dX_1 dX_2 dX_3}{|\cos\theta_{13} - \cos\theta_{23}\cos\theta_{12}||\det[\mathbf{c}]|^3} \tag{266}$$

In particular, if $\mathbf{z}_1 = \mathbf{r}_{41}$, $\mathbf{z}_2 = \mathbf{r}_{42}$, *and* $\mathbf{z}_3 = \mathbf{r}_{43}$ *(in which case* $|\det[\mathbf{c}]|^3 = 1$*), we obtain the volume-element in bond lengths* $\{r_{41}, r_{42}, r_{43}\}$, *bond angles* $\{\theta_{142}, \theta_{143}\}$, *and the angle between the bond* $\mathbf{r}_{43}$ *and the plane* $\mathbf{r}_{41} \wedge \mathbf{r}_{42}$. *These coordinates are suitable for formaldehyde type of molecules.*

The interested reader should look at Ref. 39 for more examples of the volume-element calculations. It is straightforward to derive all the known volume elements (such as the volume element for the polyspherical coordinates in Ref. 53) by the present method.

V. CONCLUSION

Geometric algebra is shown to produce molecular vibration–rotation kinetic energy operators in a simple, coherent, and general fashion. This branch of mathematics has been used before in some other related areas [70–73] but for the first time we have shown it to be powerful in vibration–rotation theory. Both the classical Lagrangians and Hamiltonians are formed in a simpler way than before, and the quantization is done by known rules. Curvilinear internal bond and angle displacement coordinates are used as an example in the case of shape coordinates, although our approach is completely general and can be used for any chosen shape coordinate system. In the case of the rotational motion, any molecular axis system can be employed. In addition, new insight is given with regard to the well-known infinitesimal approach to form both Wilson's vibrational and Lukka's rotational measuring vectors (s vectors). In conclusion, geometric algebra is clearly more powerful in vibration–rotation theory than ordinary vector algebra, and we strongly recommend its use in future work.

APPENDIX A: RUDIMENTS OF GEOMETRIC ALGEBRA

To put it briefly, geometric algebra [32–35] is an extension of the real number system to incorporate the geometric concept of direction; that is, it is a system of directed numbers. Geometric algebra integrates the well-established branches of complex analysis, matrix, vector, and tensor algebras [32], and the lesser-known simplicial calculus [74], calculus of differential forms [32,75], and the quaternion and spinor algebras [32,50] into a coherent mathematical language, which retains the advantages of each of these subalgebras, but also possesses powerful new capabilities. Unlike the other algebraic systems, it integrates the projective geometry fully into its formalism [76,77].

As any algebra, geometric algebra is completely characterized by a set of axioms. However, the choice of axioms is governed by geometric considerations. This enriches the algebra: Besides scalars and vectors, there are other objects as well. The interpretation of some familiar objects differs from the conventional one. For example, it turns out that in many physical applications, complex numbers should *not* be regarded as scalars, and there exist more than one type of unit imaginaries—that is, unit quantities with square -1. In geometric algebra, one can also perform mathematical operations that are forbidden, say, in vector algebra. One can, for example, add a scalar to a vector or divide a vector by a vector.

1. Sums and Products

We present the sum and product rules of geometric algebra in this section. There is only one kind of addition, but in contrast there are several products. However, *geometric product* is in a special position, because all other products (such as the inner and cross products) can be derived from it.

a. Addition of Vectors

The sum $\mathbf{a} + \mathbf{b}$ of the vectors $\mathbf{a}$ and $\mathbf{b}$ is found by joining the head of the vector $\mathbf{a}$ to the tail of the vector $\mathbf{b}$ (see Fig. A1). Since this parallelogram rule can be found on any book on vector analysis, it does not need to be discussed here, apart to mention that it is associative, distributive, and commutative.

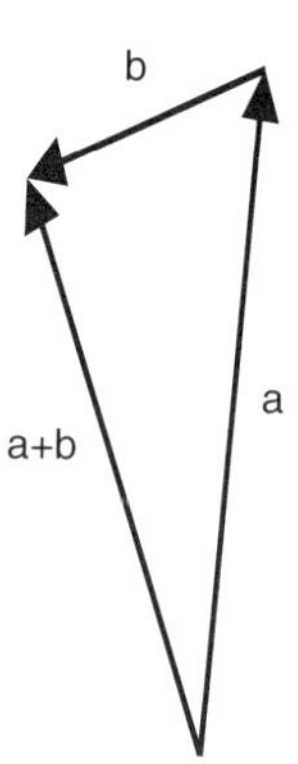

Figure A1. Vector additon.

b. Multiplication of Vectors

The product **ab** of two arbitrary vectors **a** and **b** should specify their relative orientation. Furthermore, it should be both distributive and associative, that is,

$$\mathbf{a}(\mathbf{b}+\mathbf{c}) = \mathbf{ab} + \mathbf{ac} \tag{A1}$$

$$\mathbf{abc} = \mathbf{a}(\mathbf{bc}) = (\mathbf{ab})\mathbf{c} \tag{A2}$$

The square of the vector **a** should be equal to the square of its length, that is,

$$\mathbf{a}^2 = |\mathbf{a}|^2 \geq 0 \tag{A3}$$

It follows that the square of the sum of two vectors is

$$(\mathbf{a}+\mathbf{b})^2 = |\mathbf{a}|^2 + |\mathbf{b}|^2 + \mathbf{ab} + \mathbf{ba} \tag{A4}$$

By the Pythagorean theorem, one can write

$$|\mathbf{a}+\mathbf{b}|^2 = |\mathbf{a}|^2 + |\mathbf{b}|^2 + 2\mathbf{a}\cdot\mathbf{b} \tag{A5}$$

where $\mathbf{a}\cdot\mathbf{b}$ denotes the inner product (the "dot" product) of **a** and **b**. Geometrically, the inner product $\mathbf{a}\cdot\mathbf{b}$ is equal to the projection of **a** onto **b** multiplied by the length of **b**, and it is familiar from the ordinary vector analysis. One can use Eqs. (A4) and (A5) to *define* the dot product $\mathbf{a}\cdot\mathbf{b}$ of any two vectors **a** and **b** in terms of the yet unknown product **ab** as

$$\mathbf{a}\cdot\mathbf{b} = \frac{\mathbf{ab}+\mathbf{ba}}{2} \tag{A6}$$

Note that it is not assumed that the product **ab** would be commutative. On the contrary: If **a** is perpendicular to **b**, then $\mathbf{a}\cdot\mathbf{b} = 0$ and it follows that for any two perpendicular vectors we have $\mathbf{ab} = -\mathbf{ba}$. These properties can be combined by defining a geometric product for arbitrary vectors **a** and **b** as

$$\mathbf{ab} = \mathbf{a}\cdot\mathbf{b} + \mathbf{a}\wedge\mathbf{b} \tag{A7}$$

where

$$\mathbf{a}\wedge\mathbf{b} = \frac{\mathbf{ab}-\mathbf{ba}}{2} = -\mathbf{b}\wedge\mathbf{a} \tag{A8}$$

is the antisymmetric part of the geometric product.

The entity $\mathbf{a} \wedge \mathbf{b}$ cannot be a scalar, because it anticommutes with the vector $\mathbf{a}$:

$$\begin{aligned}\mathbf{a}(\mathbf{a} \wedge \mathbf{b}) &= \mathbf{a}\frac{\mathbf{ab} - \mathbf{ba}}{2} = \frac{|\mathbf{a}|^2\mathbf{b} - \mathbf{aba}}{2} = \frac{\mathbf{b}|\mathbf{a}|^2 - \mathbf{aba}}{2} = \frac{\mathbf{ba}^2 - \mathbf{aba}}{2} \\ &= (\mathbf{b} \wedge \mathbf{a})\mathbf{a} = -(\mathbf{a} \wedge \mathbf{b})\mathbf{a} \end{aligned} \tag{A9}$$

Nor is $\mathbf{a} \wedge \mathbf{b}$ a vector, because its square is *negative*, as seen by

$$\begin{aligned}(\mathbf{a} \wedge \mathbf{b})^2 &= \left(\frac{\mathbf{ab} - \mathbf{ba}}{2}\right)^2 = \frac{(\mathbf{ab})^2 - 2|\mathbf{a}|^2|\mathbf{b}|^2 + (\mathbf{ba})^2}{4} \\ &= \frac{(\mathbf{a} \cdot \mathbf{b} + \mathbf{a} \wedge \mathbf{b})^2 - 2|\mathbf{a}|^2|\mathbf{b}|^2 + (\mathbf{a} \cdot \mathbf{b} - \mathbf{a} \wedge \mathbf{b})^2}{4} \\ &= \frac{(\mathbf{a} \wedge \mathbf{b})^2 + (\mathbf{a} \cdot \mathbf{b})^2 - |\mathbf{a}|^2|\mathbf{b}|^2}{2} \end{aligned} \tag{A10}$$

(because $(\mathbf{a} \cdot \mathbf{b})^2 \leq |\mathbf{a}|^2|\mathbf{b}|^2$, where the equality holds only for $\mathbf{a}$ which is collinear with $\mathbf{b}$), and its direction does not change when its vector factors $\mathbf{a}$ and $\mathbf{b}$ are both multiplied by -1. It is a *bivector*, a new kind of entity. It can be pictured as an oriented parallelogram with sides $\mathbf{a}$ and $\mathbf{b}$ (see Fig. A2). Note,

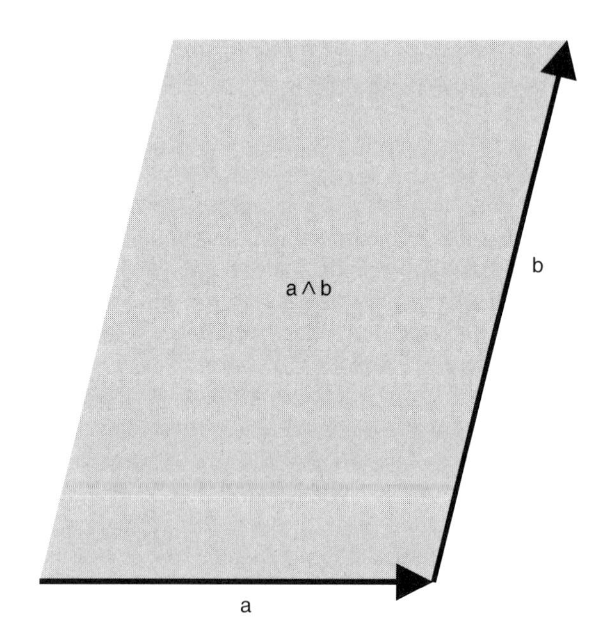

Figure A2. Bivector $\mathbf{a} \wedge \mathbf{b}$.

however, that the same bivector could as well be pictured as any other planar object with the same orientation and area: The particular shape is unimportant. If the bivector is multiplied by the scalar λ, its area is dilated by the factor $|\lambda|$. The orientation also changes to opposite, if the scalar is negative. Any bivector formed as an outer product of two unit vectors $\mathbf{u}$ and $\mathbf{v}$ can be written as

$$\mathbf{u} \wedge \mathbf{v} = \mathbf{i} \sin\theta \tag{A11}$$

where $\mathbf{i}$ is the unit bivector of the $\mathbf{u} \wedge \mathbf{v}$-plane [i.e., $|\mathbf{i}| = 1$, see Eq. (A28) for the general definition of the magnitude of a multivector]. By denoting the angle between $\mathbf{a}$ and $\mathbf{b}$ as θ, Eq. (A7) can be written as

$$\mathbf{ab} = ab \cos\theta + \mathbf{i}ab \sin\theta \tag{A12}$$

where $\mathbf{i} = \mathbf{a} \wedge \mathbf{b}/|\mathbf{a} \wedge \mathbf{b}|$ is the unit bivector in the $\mathbf{a} \wedge \mathbf{b}$-plane.

c. *Generalization of Multiplication*

A natural way to generalize the outer product between a k-blade $A_{\bar{k}} = \mathbf{a}_1 \wedge \mathbf{a}_2 \wedge \cdots \wedge \mathbf{a}_k$ and an l-blade $B_{\bar{l}} = \mathbf{b}_1 \wedge \mathbf{b}_2 \wedge \cdots \wedge \mathbf{b}_l$ is to choose it as the $k + l$ part of the geometric product $A_{\bar{k}}B_{\bar{l}}$, that is,

$$A_{\bar{k}} \wedge B_{\bar{l}} = \langle A_{\bar{k}}B_{\bar{l}}\rangle_{\overline{k+l}} \tag{A13}$$

where $\langle A_{\bar{k}}B_{\bar{l}}\rangle_{\bar{m}}$ denotes the m-blade part of $A_{\bar{k}}B_{\bar{l}}$. Because the outer product of any two vectors is antisymmetric, the outer product $B_{\bar{l}} \wedge A_{\bar{k}}$ differs from $A_{\bar{k}} \wedge B_{\bar{l}}$ at most by sign, that is,

$$A_{\bar{k}} \wedge B_{\bar{l}} = (-1)^{kl} B_{\bar{l}} \wedge A_{\bar{k}} \tag{A14}$$

The outer product is associative; for example,

$$\mathbf{a} \wedge \mathbf{b} \wedge \mathbf{c} = (\mathbf{a} \wedge \mathbf{b}) \wedge \mathbf{c} = \mathbf{a} \wedge (\mathbf{b} \wedge \mathbf{c}) \tag{A15}$$

The generalization of the inner product is more demanding. The grade of the inner product between $A_{\bar{k}}$ and $B_{\bar{l}}$ should be the difference of the grades of $A_{\bar{k}}$ and $B_{\bar{l}}$. An extension of the inner product is defined in Ref. 32 as

$$A_{\bar{k}} \cdot B_{\bar{l}} = \langle A_{\bar{k}}B_{\bar{l}}\rangle_{\overline{|k-l|}} \qquad \text{if } k, l > 0 \tag{A16}$$

$$A_{\bar{k}} \cdot B_{\bar{l}} = 0 \qquad \text{if } k = 0 \quad \text{or} \quad l = 0 \tag{A17}$$

where the grade of $A_{\bar{k}} \cdot B_{\bar{l}}$ is $|k - l|$ with the exception for the scalars in Eq. (A17). This dot product is either symmetric or antisymmetric; that is,

$$A_{\bar{k}} \cdot B_{\bar{l}} = (-1)^{k(l-1)} B_{\bar{l}} \cdot A_{\bar{k}} \tag{A18}$$

It suffices to use solely this dot product in our work, although there are other generalizations of the inner products, which are more efficient in the formal mathematical derivations and proofs (see Ref. 78 for a careful discussion). Analogously to Eq. (A7), we can write the geometric product for a vector $\mathbf{x}$ and an arbitrary multivector A as

$$\mathbf{x}A = \mathbf{x} \cdot A + \mathbf{x} \wedge A \tag{A19}$$

$$A\mathbf{x} = A \cdot \mathbf{x} + A \wedge \mathbf{x} \tag{A20}$$

where we may use $A \cdot \mathbf{x} = -\mathbf{x} \cdot \breve{A}$ and $A \wedge \mathbf{x} = \mathbf{x} \wedge \breve{A}$ to solve

$$\mathbf{x} \cdot A = \frac{1}{2}(\mathbf{x}A - \breve{A}\mathbf{x}) \tag{A21}$$

$$A \cdot \mathbf{x} = \frac{1}{2}(A\mathbf{x} - \mathbf{x}\breve{A}) \tag{A22}$$

and

$$\mathbf{x} \wedge A = \frac{1}{2}(\mathbf{x}A + \breve{A}\mathbf{x}) \tag{A23}$$

$$A \wedge \mathbf{x} = \frac{1}{2}(A\mathbf{x} + \mathbf{x}\breve{A}) \tag{A24}$$

where the accent above A implies the reversion of the sign for odd blades; that is,

$$\breve{A}_{\bar{k}} = (-1)^k A_{\bar{k}} \tag{A25}$$

This operation is known as the grade involution.

The reversion of a multivector is a common operation besides the grade involution. For any k-blade $A_{\bar{k}}$, its reverse $A_{\bar{k}}^{\dagger}$ is given as

$$A_{\bar{k}}^{\dagger} = (\mathbf{a}_1 \wedge \mathbf{a}_2 \wedge \cdots \wedge \mathbf{a}_k)^{\dagger} = \mathbf{a}_k \wedge \mathbf{a}_{k-1} \wedge \cdots \wedge \mathbf{a}_1 = (-1)^{k(k-1)/2} A_{\bar{k}} \tag{A26}$$

Generally, the geometric product of two blades $A_{\bar{k}}$ and $B_{\bar{l}}$ is *not* related by the formula analogous to Eq. (A7), if both $k, l > 1$. The geometric product $A_{\bar{k}}B_{\bar{l}}$ of two blades $A_{\bar{k}}$ and $B_{\bar{l}}$ results in terms of intermediate grade from $|k-l|$ to $k+l$ in the steps of two; that is,

$$A_{\bar{k}}B_{\bar{l}} = \sum_{m=0}^{(k+l-|k-l|)/2} \langle A_{\bar{k}}B_{\bar{l}} \rangle_{\overline{|k-l|+2m}} \tag{A27}$$

d. *Magnitude*

The square of the magnitude of any multivector A is given by the scalar part of the product of A with its reverse; that is,

$$|A|^2 = \langle A^\dagger A\rangle_{\bar{0}} = \sum_{k=0} |\langle A\rangle_{\bar{k}}|^2 \tag{A28}$$

e. *Expansion Rules*

The dot product of two p-blades can be expressed as a linear combination of the contractions of $(p-1)$-blades as [32]

$$\begin{aligned}&(\mathbf{a}_p \wedge \cdots \wedge \mathbf{a}_1)\cdot(\mathbf{b}_1 \wedge \cdots \wedge \mathbf{b}_p)\\ &= \sum_{k=1}^{p}(-1)^{k+1}(\mathbf{a}_1\cdot\mathbf{b}_k)(\mathbf{a}_p \wedge \cdots \wedge \mathbf{a}_2)\cdot(\mathbf{b}_1 \wedge \cdots \wedge \breve{\mathbf{b}}_k \wedge \cdots \wedge \mathbf{b}_p)\end{aligned} \tag{A29}$$

where $\breve{\mathbf{b}}_k$ means that the vector $\mathbf{b}_k$ is omitted from the product. This expansion rule may be used repeatedly to reduce the dot product of p-blades to a linear combination of dot products of vectors.

The expansion rule for the dot product of a vector $\mathbf{a}$ and a p-blade $\mathbf{b}_1 \wedge \cdots \wedge \mathbf{b}_p$ is given by

$$\mathbf{a}\cdot(\mathbf{b}_1 \wedge \cdots \wedge \mathbf{b}_p) = \sum_{k=1}^{p}(-1)^{k+1}\mathbf{a}\cdot\mathbf{b}_k(\mathbf{b}_1 \wedge \cdots \wedge \breve{\mathbf{b}}_k \wedge \cdots \wedge \mathbf{b}_p) \tag{A30}$$

Especially, the dot product of a vector $\mathbf{a}$ and a bivector $\mathbf{b}\wedge\mathbf{c}$ is given by

$$\mathbf{a}\cdot(\mathbf{b}\wedge\mathbf{c}) = \mathbf{a}\cdot\mathbf{b}\mathbf{c} - \mathbf{a}\cdot\mathbf{c}\mathbf{b} \tag{A31}$$

2. Basis Representation for Three-Dimensional Geometric Algebra

The outer product $\mathbf{a}_1 \wedge \mathbf{a}_2 \wedge \cdots \wedge \mathbf{a}_k$ is zero for $k > 3$ in the three-dimensional space, and any trivector can be expressed as a multiple of a unit trivector i. As implied by its name, the unit trivector i is of the unit magnitude; that is, $i^\dagger i = 1 = |i|$. On the other hand, $i^2 = -1$. Furthermore, in the three-dimensional space, the unit trivector commutes with all other elements of the algebra. The trivector $\mathbf{a}\wedge\mathbf{b}\wedge\mathbf{c}$ can be pictured as an oriented parallelepiped with sides $\mathbf{a}$, $\mathbf{b}$, and $\mathbf{c}$ (see Fig. A3). The vector cross product $\mathbf{a}\times\mathbf{b}$ is related to the bivector $\mathbf{a}\wedge\mathbf{b}$ as

$$\mathbf{a}\times\mathbf{b} = -i(\mathbf{a}\wedge\mathbf{b}) \tag{A32}$$

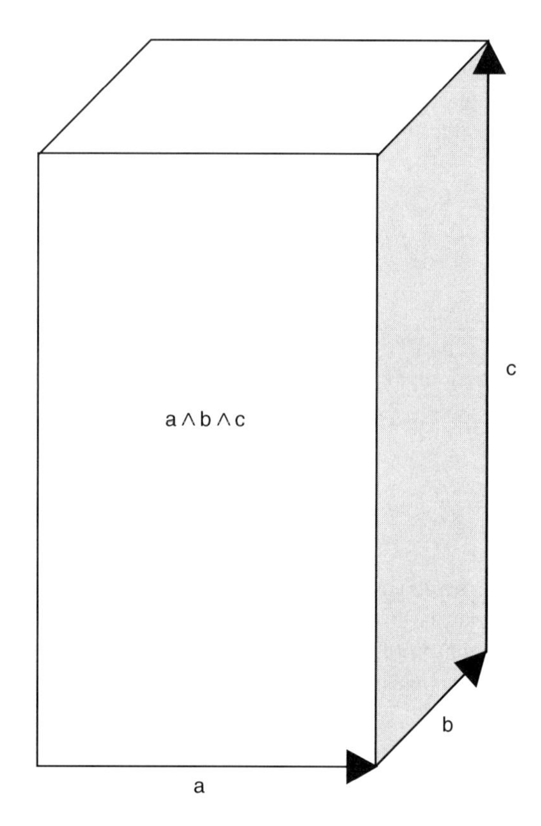

Figure A3. Trivector $\mathbf{a} \wedge \mathbf{b} \wedge \mathbf{c}$.

where $\mathbf{a} \times \mathbf{b}$ is a vector perpendicular to the plane $\mathbf{a} \wedge \mathbf{b}$ (see Fig. A4). Also, the unit trivector can be used to switch between the wedge and the dot as

$$i(\mathbf{a} \wedge \mathbf{b}) = \mathbf{a} \cdot (i\mathbf{b}) = \mathbf{b} \times \mathbf{a} \tag{A33}$$

where by using Eq. (A31) we may obtain the triple cross product as

$$\mathbf{a} \times (\mathbf{b} \times \mathbf{c}) = -\mathbf{a} \cdot (\mathbf{b} \wedge \mathbf{c}) = \mathbf{a} \cdot \mathbf{c}\mathbf{b} - \mathbf{a} \cdot \mathbf{b}\mathbf{c} \tag{A34}$$

The grade of any blade in a three-dimensional space is equal or smaller than 3. Thus, any multivector A in a three-dimensional space can be written as

$$A = \alpha + i\beta + \mathbf{a} + i\mathbf{b} \tag{A35}$$

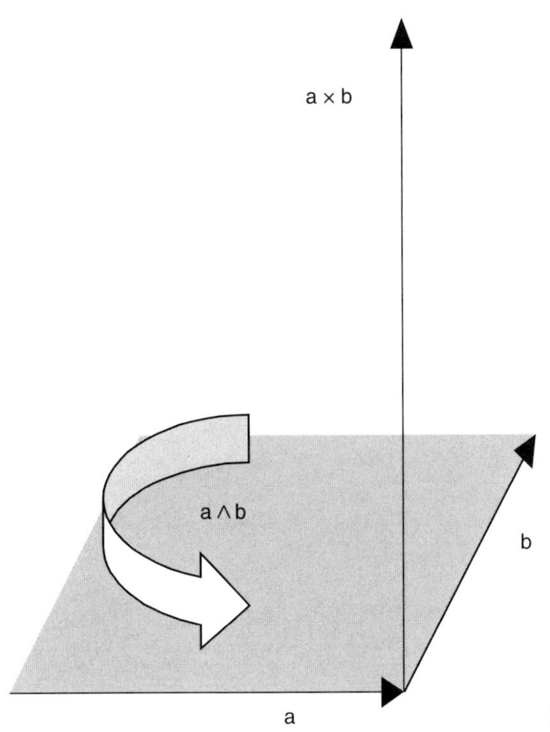

Figure A4. Cross product $\mathbf{a} \times \mathbf{b}$.

where $\alpha = \langle A \rangle_{\bar{0}}$ is the scalar part of A, $\mathbf{a} = \langle A \rangle_{\bar{1}}$ is the vector part of A, $i\mathbf{b} = \langle A \rangle_{\bar{2}}$ is the bivector part of A, and $i\beta = \langle A \rangle_{\bar{3}}$ is the trivector part of A.

There are only three linearly independent orthonormal vectors $\{\mathbf{u}_1, \mathbf{u}_2, \mathbf{u}_3\}$ and three linearly independent bivectors $\{\mathbf{i}_1, \mathbf{i}_2, \mathbf{i}_3\}$. These bivectors can be represented in terms of $\{\mathbf{u}_1, \mathbf{u}_2, \mathbf{u}_3\}$ as

$$\mathbf{i}_1 = \mathbf{u}_2\mathbf{u}_3 = i\mathbf{u}_1 \tag{A36}$$

$$\mathbf{i}_2 = \mathbf{u}_3\mathbf{u}_1 = i\mathbf{u}_2 \tag{A37}$$

$$\mathbf{i}_3 = \mathbf{u}_1\mathbf{u}_2 = i\mathbf{u}_3 \tag{A38}$$

where the set $\{\mathbf{i}_1, \mathbf{i}_2, \mathbf{i}_3\}$ is right-handed. Any bivector $\mathbf{B}$ can be expanded in this bivector basis as

$$\mathbf{B} = B_1\mathbf{i}_1 + B_2\mathbf{i}_2 + B_3\mathbf{i}_3 \tag{A39}$$

where $B_i = \mathbf{B} \cdot \mathbf{i}_i^{\dagger}$ is the scalar component of the bivector $\mathbf{B}$ in $\mathbf{i}_i$. By including the remaining basis elements, namely the scalar unit 1 and the unit trivector i, we may present the multiplication table of the elements of a three-dimensional geometric algebra as

	1	$\mathbf{u}_1$	$\mathbf{u}_2$	$\mathbf{u}_3$	$\mathbf{i}_1$	$\mathbf{i}_2$	$\mathbf{i}_3$	i
1	1	$\mathbf{u}_1$	$\mathbf{u}_2$	$\mathbf{u}_3$	$\mathbf{i}_1$	$\mathbf{i}_2$	$\mathbf{i}_3$	i
$\mathbf{u}_1$	$\mathbf{u}_1$	1	$\mathbf{i}_3$	$-\mathbf{i}_2$	i	$-\mathbf{u}_3$	$\mathbf{u}_2$	$\mathbf{i}_1$
$\mathbf{u}_2$	$\mathbf{u}_2$	$-\mathbf{i}_3$	1	$\mathbf{i}_1$	$\mathbf{u}_3$	i	$-\mathbf{u}_1$	$\mathbf{i}_2$
$\mathbf{u}_3$	$\mathbf{u}_3$	$\mathbf{i}_2$	$-\mathbf{i}_1$	1	$-\mathbf{u}_2$	$\mathbf{u}_1$	i	$\mathbf{i}_3$
$\mathbf{i}_1$	$\mathbf{i}_1$	i	$-\mathbf{u}_3$	$\mathbf{u}_2$	-1	$-\mathbf{i}_3$	$\mathbf{i}_2$	$-\mathbf{u}_1$
$\mathbf{i}_2$	$\mathbf{i}_2$	$\mathbf{u}_3$	i	$-\mathbf{u}_1$	$\mathbf{i}_3$	-1	$-\mathbf{i}_1$	$-\mathbf{u}_2$
$\mathbf{i}_3$	$\mathbf{i}_3$	$-\mathbf{u}_2$	$\mathbf{u}_1$	i	$-\mathbf{i}_2$	$\mathbf{i}_1$	-1	$-\mathbf{u}_3$
i	i	$\mathbf{i}_1$	$\mathbf{i}_2$	$\mathbf{i}_3$	$-\mathbf{u}_1$	$-\mathbf{u}_2$	$-\mathbf{u}_3$	-1

where the geometric product is taken in the order a column times a row. It is also seen that

$$\mathbf{i}_1\mathbf{i}_2\mathbf{i}_3 = -i\mathbf{u}_1\mathbf{u}_2\mathbf{u}_3 = 1 \tag{A40}$$

3. Geometry

In the geometric algebra, each geometrical point is represented by a vector, and any geometric quantity can be described in terms of its intrinsic properties alone, without introducing any external coordinate frames. An unlimited number of *geometrical* relations can be extracted by the simple algebraic manipulation of the rules given above. As an elementary example, we derive "the law of sines" familiar from trigonometry:

Example A1. *Let us represent the sides of a triangle as the vectors* **a**, **b**, *and* **c** *(as in Fig. A1), and let us denote the angle between* **a** *and* **b** *by C, the angle between* **b** *and* **c** *by A, and the angle between* **c** *and* **a** *by B. Then,*

$$\mathbf{a} + \mathbf{b} = \mathbf{c} \tag{A41}$$

If both sides in Eq. (A41) are wedged by **a**, **b**, *and* **c** *in turn, we obtain*

$$\mathbf{a} \wedge \mathbf{b} = \mathbf{c} \wedge \mathbf{b} = \mathbf{a} \wedge \mathbf{c} \tag{A42}$$

where the "law of sines"

$$ab \sin C = cb \sin A = ac \sin B \tag{A43}$$

or

$$\frac{\sin C}{c} = \frac{\sin A}{a} = \frac{\sin B}{b} \tag{A44}$$

follows from the magnitudes of the bivectors.

a. Projections

Any vector **a** can be decomposed to the components parallel ($\hat{P}^{||}_{\mathbf{b}}(\mathbf{a})$) and orthogonal ($\hat{P}^{\perp}_{\mathbf{b}}(\mathbf{a})$) to some given vector **b** by simply multiplying it by $\mathbf{bb}^{-1}$. This results in

$$\mathbf{abb}^{-1} = \frac{\mathbf{abb}}{b^2} = \frac{(\mathbf{ab})\mathbf{b}}{b^2} = \frac{1}{b^2}(\mathbf{a}\cdot\mathbf{b} + \mathbf{a}\wedge\mathbf{b})\mathbf{b} = \hat{P}^{||}_{\mathbf{b}}(\mathbf{a}) + \hat{P}^{\perp}_{\mathbf{b}}(\mathbf{a}) \tag{A45}$$

where $\hat{P}^{||}_{\mathbf{b}}(\mathbf{a}) = \mathbf{a}\cdot\mathbf{bb}/b^2$ is the parallel, and $\hat{P}^{\perp}_{\mathbf{b}}(\mathbf{a}) = \mathbf{a}\wedge\mathbf{bb}/b^2$ is the perpendicular component or the rejection of **a** from **b** (see Fig. A5). Similarly,

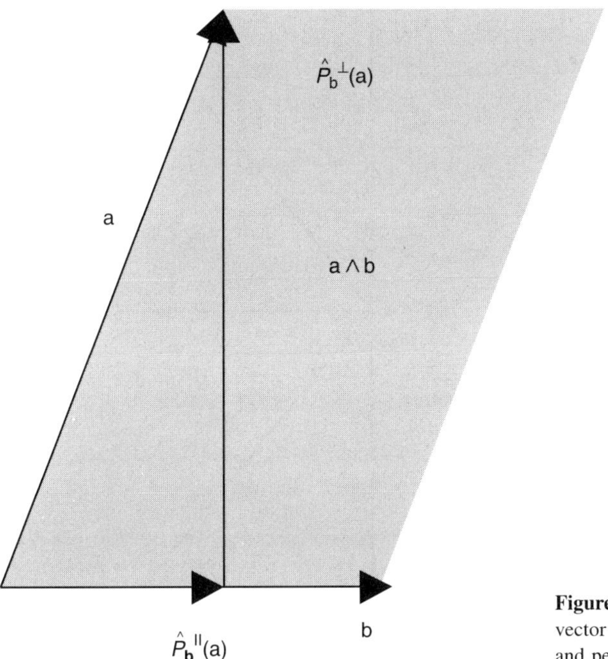

Figure A5. Decomposition of vector **a** to components along and perpendicular to vector **b**.

any vector **a** can be decomposed to the components parallel ($\hat{P}^{||}_{\mathbf{A}}(\mathbf{a})$) and orthogonal ($\hat{P}^{\perp}_{\mathbf{A}}(\mathbf{a})$) to some given plane $\mathbf{A} = \mathbf{b} \wedge \mathbf{c}$ as

$$\hat{P}^{||}_{\mathbf{A}}(\mathbf{a}) = \mathbf{a} \cdot \mathbf{A}\mathbf{A}^{-1} \tag{A46}$$

$$\hat{P}^{\perp}_{\mathbf{A}}(\mathbf{a}) = \mathbf{a} \wedge \mathbf{A}\mathbf{A}^{-1} \tag{A47}$$

b. Reflections

A vector **a** can be reflected along a unit vector **u** to **a**′ by

$$\mathbf{a}' = -\mathbf{u}\mathbf{a}\mathbf{u} \tag{A48}$$

(see Fig. A6).

c. Rotations

A vector **a** can be rotated in the plane $\mathbf{i} = \mathbf{u} \wedge \mathbf{w}/|\mathbf{u} \wedge \mathbf{w}|$ through the bivector angle $\mathbf{B} = B\mathbf{i}$ ($B = |\mathbf{B}| \geq 0$ is the magnitude of the rotation) spanned by the unit

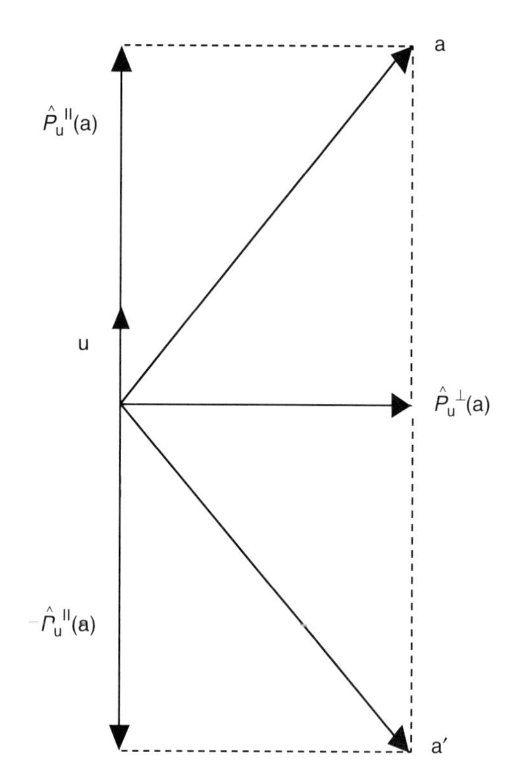

Figure A6. Reflection of vector **a** along vector **u**.

vectors **u** and **w** by reflecting it twice, first along the unit vector **u** and then along the unit vector **w** as

$$\mathbf{a}' = \mathbf{wuauw} \tag{A49}$$

The product $\mathbf{uw} = \mathbf{u} \cdot \mathbf{w} + \mathbf{u} \wedge \mathbf{w}$ is a spinor; that is, it is a sum of a scalar and a bivector. It can be written exponentially as

$$\mathbf{uw} = e^{\mathbf{i}B/2} = \cos\frac{B}{2} + \mathbf{i}\sin\frac{B}{2} = R(\mathbf{B}) \tag{A50}$$

(see Fig. A7). Because the unit plane of rotation **i** can be expressed as the dual $\mathbf{i} = i\mathbf{v}$ of the unit normal **v** to the rotation plane (i.e., the rotation axis), it is customary to write the rotor $R(\mathbf{B})$ as

$$R = \alpha + i\boldsymbol{\beta} \tag{A51}$$

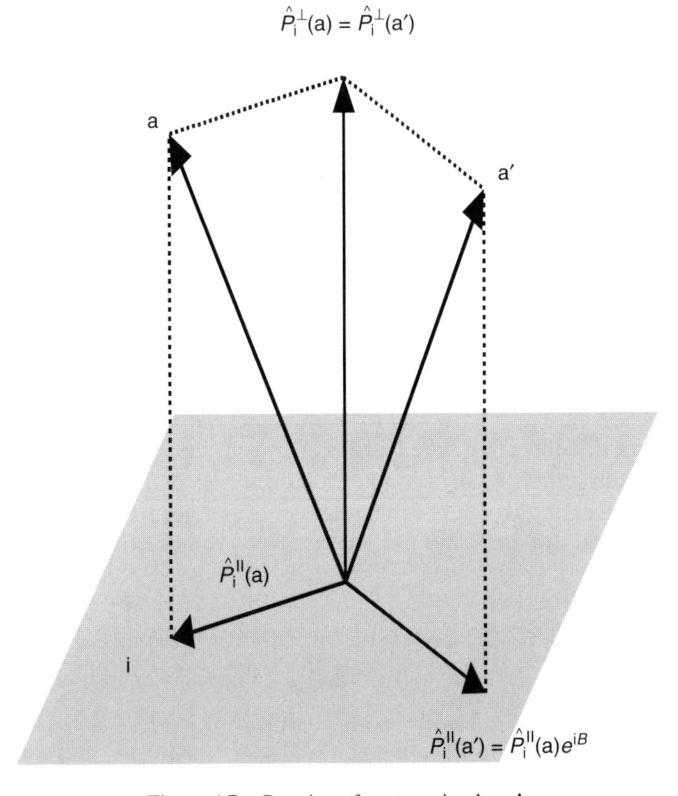

Figure A7. Rotation of vector **a** in plane **i**.

where the Euler scalar is given by

$$\alpha = \cos\frac{B}{2} \tag{A52}$$

and the Euler vector is expressed as

$$\boldsymbol{\beta} = \mathbf{v}\sin\frac{B}{2} \tag{A53}$$

Any multivector M (vector, bivector, etc., or any of their combination) can be rotated by the formula similar to Eq. (A49). If M' is the multivector M rotated through a bivector angle $\mathbf{B}$, then M' is given by sandwiching the multivector M between exponentials of the rotation plane $\mathbf{B}$ [50]:

$$M' = e^{-\mathbf{B}/2} M e^{\mathbf{B}/2} \tag{A54}$$

Such a simple expression does not exist in the ordinary vector algebra, where supplementary algebraic structures in the form of rotation matrices are needed. Geometric algebra offers an effective way of describing rotations. For example, the spinor $e^{\mathbf{B}/2}$ describing the net rotation of two successive rotations, first $e^{\mathbf{B}_1/2}$, then $e^{\mathbf{B}_2/2}$, is found by multiplying

$$e^{\mathbf{B}/2} = e^{\mathbf{B}_1/2} e^{\mathbf{B}_2/2} \tag{A55}$$

In many respects, the rotor $e^{\mathbf{B}/2}$ behaves like the ordinary exponential function $e^{B/2}$ of a scalar variable B. For example,

$$e^{-\mathbf{B}/2} e^{\mathbf{B}/2} = 1 = e^{\mathbf{B}/2} e^{-\mathbf{B}/2} \tag{A56}$$

Or, if the rotor $R(\mathbf{B})$ is multiplied n times by itself, the result is simply the same rotor with the argument bivector angle $\mathbf{B}$ multiplied by n, that is, $R^n(\mathbf{B}) = R(n\mathbf{B})$. However, the product of two rotors does *not* generally commute, that is,

$$e^{\mathbf{B}_1/2} e^{\mathbf{B}_2/2} \neq e^{\mathbf{B}_2/2} e^{\mathbf{B}_1/2} \tag{A57}$$

and the net rotation $\mathbf{B}$ produced by two successive partial rotations $R(\mathbf{B}_1)R(\mathbf{B}_2)$ is *not* generally given as the sum of partial rotation planes, that is,

$$e^{\mathbf{B}_1/2} e^{\mathbf{B}_2/2} \neq e^{\mathbf{B}_1/2 + \mathbf{B}_2/2} \tag{A58}$$

unless the both partial rotations share the same rotation axis.

The computationally convenient way to find the compound rotor $R = R_1 R_2$ in terms of the rotors R_1 and R_2 is to combine the two rotors in the Euler form as in Eq. (A50). The Euler parameters for the net rotation R are obtained from the Euler parameters of $\alpha_k, \boldsymbol{\beta}_k$ of $R_k = R(\mathbf{B}_k)$ $(k = 1, 2)$ by a simple substitution

$$\begin{aligned} R &= \alpha + i\boldsymbol{\beta} = (\alpha_1 + i\boldsymbol{\beta}_1)(\alpha_2 + i\boldsymbol{\beta}_2) \\ &= \alpha_1\alpha_2 - \boldsymbol{\beta}_1 \cdot \boldsymbol{\beta}_2 + i(\alpha_1\boldsymbol{\beta}_2 + \alpha_2\boldsymbol{\beta}_1 + i\boldsymbol{\beta}_1 \wedge \boldsymbol{\beta}_2) \end{aligned}$$

where the scalar and vector parts can be separated as

$$\alpha = \alpha_1\alpha_2 - \boldsymbol{\beta}_1 \cdot \boldsymbol{\beta}_2 \tag{A59}$$

$$\boldsymbol{\beta} = \alpha_1\boldsymbol{\beta}_2 + \alpha_2\boldsymbol{\beta}_1 + \boldsymbol{\beta}_2 \times \boldsymbol{\beta}_1 \tag{A60}$$

In terms of the Euler parameters, the rotated vector $\mathbf{a}'$ is given by

$$\mathbf{a}' = (\alpha - i\boldsymbol{\beta})\mathbf{a}(\alpha + i\boldsymbol{\beta}) = (2\alpha^2 - 1)\mathbf{a} + 2\alpha\boldsymbol{\beta} \times \mathbf{a} + 2\mathbf{a} \cdot \boldsymbol{\beta}\boldsymbol{\beta} \tag{A61}$$

Example A2. *A vector is rotated first by* $R_1 = e^{i(2\pi/3)(\mathbf{u}_1 + \mathbf{u}_2/2)}$, *then by* $R_2 = e^{i(\pi/10)(\mathbf{u}_1 + \mathbf{u}_3)}$, *where* $\{\mathbf{u}_1, \mathbf{u}_2, \mathbf{u}_3\}$ *is some right-handed set of orthonormal vectors. We may pick the Euler scalar and vector by changing the exponential to the form*

$$R_i = e^{iB_i\mathbf{v}_i/2}$$

where $\mathbf{v}_i$ *is the ith unit rotation axis. In the present case,*

$$\mathbf{v_1} = \frac{2}{\sqrt{5}}(\mathbf{u}_1 + \mathbf{u}_2/2) \tag{A62}$$

$$\mathbf{v_2} = \frac{1}{\sqrt{2}}(\mathbf{u}_1 + \mathbf{u}_3) \tag{A63}$$

and the Euler scalars are

$$\alpha_1 = \cos\left(\frac{\pi\sqrt{5}}{3}\right) \tag{A64}$$

$$\alpha_2 = \cos\left(\frac{\pi}{5\sqrt{2}}\right) \tag{A65}$$

and the Euler vectors are

$$\boldsymbol{\beta}_1 = \sin\left(\frac{\pi\sqrt{5}}{3}\right)\mathbf{v}_1 \tag{A66}$$

$$\boldsymbol{\beta}_2 = \sin\left(\frac{\pi}{5\sqrt{2}}\right)\mathbf{v}_2 \tag{A67}$$

Because

$$\boldsymbol{\beta}_1 \cdot \boldsymbol{\beta}_2 = \sqrt{\frac{2}{5}}\sin\left(\frac{\pi\sqrt{5}}{3}\right)\sin\left(\frac{\pi}{5\sqrt{2}}\right) \tag{A68}$$

$$\boldsymbol{\beta}_2 \times \boldsymbol{\beta}_1 = \sqrt{\frac{2}{5}}\sin\left(\frac{\pi\sqrt{5}}{3}\right)\sin\left(\frac{\pi}{5\sqrt{2}}\right)\left(-\frac{\mathbf{u}_1}{2} + \mathbf{u}_2 + \frac{\mathbf{u}_3}{2}\right) \tag{A69}$$

the Euler parameters for the compound rotor $R = R_1R_2$ are given approximately as

$$\alpha = -0.82408 \tag{A70}$$

$$\begin{aligned}\boldsymbol{\beta} &= 0.27007\mathbf{u}_1 + 0.48467\mathbf{u}_2 - 0.11425\mathbf{u}_3 \\ &= 0.56648\mathbf{v}\end{aligned} \tag{A71}$$

where the vector

$$\mathbf{v} = 0.47676\mathbf{u}_1 + 0.85558\mathbf{u}_2 - 0.20168\mathbf{u}_3 \tag{A72}$$

is the unit rotation axis for the compound rotation.

Euler Angles. In the three-dimensional space, the rotor R can be parameterized by three scalar parameters. The Euler angles ϕ, θ, and χ are popular among physicists and astronomers. The laboratory-fixed frame $\{\mathbf{u}_1, \mathbf{u}_2, \mathbf{u}_3\}$ can be made to coincide with the body-fixed frame $\{\mathbf{u}'_1, \mathbf{u}'_2, \mathbf{u}'_3\}$ by three successive rotations

$$\mathbf{u}'_i = R^\dagger_\chi R^\dagger_\theta R^\dagger_\phi \mathbf{u}_i R_\phi R_\theta R_\chi \tag{A73}$$

The rotation is composed as follows (see Fig. 1):

1. First, $\mathbf{u}_i$ is rotated by $R_\phi = e^{i\phi\mathbf{u}_3/2}$ about the laboratory-fixed axis $\mathbf{u}_3$.

2. Second, the resulting vector is rotated by $R_\theta = e^{i\theta\mathbf{n}/2}$ about the line of nodes

$$\mathbf{n} = R_\phi^\dagger \mathbf{u}_1 R_\phi = \mathbf{u}_1 e^{i\phi\mathbf{u}_3} = \frac{\mathbf{u}_3 \times \mathbf{u}_3'}{|\mathbf{u}_3 \times \mathbf{u}_3'|} \tag{A74}$$

3. Finally, the resulting vector is rotated by $R_\chi = e^{i\chi\mathbf{u}_3'/2}$ about the body-fixed axis $\mathbf{u}_3'$.

However, this *hybrid* interpretation is not unique, but the same rotation is achieved as a product of rotations about the *laboratory-fixed* axes $\mathbf{u}_3$ and $\mathbf{u}_1$ as

$$\mathbf{u}_i' = R_\phi^\dagger Q_\theta^\dagger Q_\chi^\dagger \mathbf{u}_i Q_\chi Q_\theta R_\phi \tag{A75}$$

where the rotors Q_χ and Q_θ are defined as

$$Q_\chi = e^{i\chi\mathbf{u}_3/2} \tag{A76}$$

$$Q_\theta = e^{i\theta\mathbf{u}_1/2} \tag{A77}$$

The same rotation is also achieved as a product of rotations about the *body-fixed* axes $\mathbf{u}_3'$ and $\mathbf{u}_1'$ as

$$\mathbf{u}_i' = R_\phi'^\dagger Q_\theta'^\dagger R_\chi^\dagger \mathbf{u}_i R_\chi Q_\theta' R_\phi' \tag{A78}$$

where the rotors R_ϕ' and Q_θ' are defined as

$$R_\phi' = e^{i\phi\mathbf{u}_3'/2} \tag{A79}$$

$$Q_\theta' = e^{i\theta\mathbf{u}_1'/2} \tag{A80}$$

One can find relations between the rotors by the direct multiplication. As an example,

$$R_\phi^\dagger Q_\theta R_\phi = R_\phi^\dagger \left(\cos\frac{\theta}{2} + i\sin\frac{\theta}{2}\mathbf{u}_1\right) R_\phi = \cos\frac{\theta}{2} + i\sin\frac{\theta}{2}\mathbf{n} = e^{i\theta\mathbf{n}/2} = R_\theta \tag{A81}$$

d. *Angles*

The angle between a vector $\mathbf{z}_\alpha$ and a plane $\mathbf{z}_\beta \wedge \mathbf{z}_\gamma$ is given by [50, page 662]

$$\sin\phi_{\alpha\beta\gamma} = -i\mathbf{u}_{\mathbf{z}_\alpha} \wedge \mathbf{i}_{\beta\gamma} \tag{A82}$$

where $\mathbf{u}_{\mathbf{z}_\alpha}$ denotes a unit vector in the direction of $\mathbf{z}_\alpha$, and $\mathbf{i}_{\beta\gamma}$ denotes the unit bivector in the plane $\mathbf{z}_\beta \wedge \mathbf{z}_\gamma$. This is part of a more fundamental relation

$$\mathbf{u}_{\mathbf{z}_\alpha}\mathbf{i}_{\beta\gamma} = \mathbf{u}_{\phi_{\alpha\beta\gamma}} e^{i\phi_{\alpha\beta\gamma}} \tag{A83}$$

where $\mathbf{u}_{\phi_{\alpha\beta\gamma}} = \boldsymbol{\phi}_{\alpha\beta\gamma}/\phi_{\alpha\beta\gamma}$ is the unit axis of rotation, which moves the vector $\mathbf{z}_\alpha$ to the plane $\mathbf{i}_{\beta\gamma}$.

The dihedral angle $\tau_{\alpha\beta\gamma}$ between planes $\mathbf{z}_\alpha \wedge \mathbf{z}_\beta$ and $\mathbf{z}_\beta \wedge \mathbf{z}_\gamma$ is given by

$$\tau_{\alpha\beta\gamma} = \arccos(\mathbf{i}_{\alpha\beta} \cdot \mathbf{i}_{\beta\gamma}) \tag{A84}$$

(note that the orientation of the plane $\mathbf{i}_{\alpha\beta}$ is opposite of the orientation of $\mathbf{i}_{\beta\gamma}$, when the two planes coincide) The product of the bivectors $\mathbf{i}_{\alpha\beta}$ and $\mathbf{i}_{\beta\gamma}$ can be written in terms of the dihedral angle $\tau_{\alpha\beta\gamma}$ as

$$\mathbf{i}_{\alpha\beta}\mathbf{i}_{\beta\gamma} = \cos\tau_{\alpha\beta\gamma} - i\mathbf{u}_{\mathbf{z}_\beta}\sin\tau_{\alpha\beta\gamma} \tag{A85}$$

e. *Spherical Trigonometry*

Let $\mathbf{u}_{\mathbf{z}_\alpha}$, $\mathbf{u}_{\mathbf{z}_\beta}$, and $\mathbf{u}_{\mathbf{z}_\gamma}$ be three unit vectors, let $\theta_{\alpha\beta}$ be the angle between $\mathbf{u}_{\mathbf{z}_\alpha}$ and $\mathbf{u}_{\mathbf{z}_\beta}$, let $\theta_{\beta\gamma}$ be the angle between $\mathbf{u}_{\mathbf{z}_\beta}$ and $\mathbf{u}_{\mathbf{z}_\gamma}$, and let $\theta_{\gamma\alpha}$ be the angle between $\mathbf{u}_{\mathbf{z}_\gamma}$ and $\mathbf{u}_{\mathbf{z}_\alpha}$. Furthermore, let $\tau_{\alpha\beta\gamma}$ be the dihedral angle between the planes $\mathbf{u}_{\mathbf{z}_\alpha} \wedge \mathbf{u}_{\mathbf{z}_\beta}$ and $\mathbf{u}_{\mathbf{z}_\beta} \wedge \mathbf{u}_{\mathbf{z}_\gamma}$, let $\tau_{\beta\alpha\gamma}$ be the dihedral angle between the planes $\mathbf{u}_{\mathbf{z}_\beta} \wedge \mathbf{u}_{\mathbf{z}_\alpha}$ and $\mathbf{u}_{\mathbf{z}_\alpha} \wedge \mathbf{u}_{\mathbf{z}_\gamma}$, and let $\tau_{\alpha\gamma\beta}$ be the dihedral angle between the planes $\mathbf{u}_{\mathbf{z}_\alpha} \wedge \mathbf{u}_{\mathbf{z}_\gamma}$ and $\mathbf{u}_{\mathbf{z}_\gamma} \wedge \mathbf{u}_{\mathbf{z}_\beta}$.

By taking the geometric products of the unit vectors $\mathbf{u}_{\mathbf{z}_\alpha}$, $\mathbf{u}_{\mathbf{z}_\beta}$, and $\mathbf{u}_{\mathbf{z}_\gamma}$ as

$$(\mathbf{u}_{\mathbf{z}_\alpha}\mathbf{u}_{\mathbf{z}_\beta})(\mathbf{u}_{\mathbf{z}_\beta}\mathbf{u}_{\mathbf{z}_\gamma})(\mathbf{u}_{\mathbf{z}_\gamma}\mathbf{u}_{\mathbf{z}_\alpha}) = 1 \tag{A86}$$

we obtain

$$e^{\mathbf{i}_{\alpha\beta}}e^{\mathbf{i}_{\beta\gamma}}e^{\mathbf{i}_{\gamma\alpha}} = 1 \tag{A87}$$

where many (scalar) relations of spherical trigonometry can be read off. For example, by solving

$$e^{-\mathbf{i}_{\gamma\alpha}} = e^{\mathbf{i}_{\alpha\beta}}e^{\mathbf{i}_{\beta\gamma}} \tag{A88}$$

and rewriting the above equality in Euler form, we get

$$\begin{aligned}\cos\theta_{\gamma\alpha} - \mathbf{i}_{\gamma\alpha}\sin\theta_{\gamma\alpha} &= (\cos\theta_{\alpha\beta} + \mathbf{i}_{\alpha\beta}\sin\theta_{\alpha\beta})(\cos\theta_{\beta\gamma} + \mathbf{i}_{\beta\gamma}\sin\theta_{\beta\gamma})\\ &= \cos\theta_{\alpha\beta}\cos\theta_{\beta\gamma} + \mathbf{i}_{\alpha\beta}\mathbf{i}_{\beta\gamma}\sin\theta_{\alpha\beta}\sin\theta_{\beta\gamma}\\ &\quad + \mathbf{i}_{\beta\gamma}\sin\theta_{\beta\gamma}\cos\theta_{\alpha\beta} + \mathbf{i}_{\alpha\beta}\cos\theta_{\beta\gamma}\sin\theta_{\alpha\beta}\end{aligned} \tag{A89}$$

By Eq. (A85), this reads as

$$\cos\theta_{\gamma\alpha} - \mathbf{i}_{\gamma\alpha}\sin\theta_{\gamma\alpha} = \cos\theta_{\alpha\beta}\cos\theta_{\beta\gamma} + \cos\tau_{\alpha\beta\gamma}\sin\theta_{\alpha\beta}\sin\theta_{\beta\gamma} + \mathbf{i}_{\beta\gamma}\sin\theta_{\beta\gamma}\cos\theta_{\alpha\beta} + \mathbf{i}_{\alpha\beta}\cos\theta_{\beta\gamma}\sin\theta_{\alpha\beta} - i\mathbf{u}_{\mathbf{z}_\beta}\sin\tau_{\alpha\beta\gamma}\sin\theta_{\alpha\beta}\sin\theta_{\beta\gamma} \qquad \text{(A90)}$$

where we may extract from the scalar part a standard relation of spherical trigonometry:

$$\cos\theta_{\gamma\alpha} = \cos\theta_{\alpha\beta}\cos\theta_{\beta\gamma} + \cos\tau_{\alpha\beta\gamma}\sin\theta_{\alpha\beta}\sin\theta_{\beta\gamma} \qquad \text{(A91)}$$

Other similar relations are easily found by cyclically permuting the indices α, β, γ. From the bivector part we obtain

$$\mathbf{i}_{\gamma\alpha}\sin\theta_{\gamma\alpha} = i\mathbf{u}_{\mathbf{z}_\beta}\sin\tau_{\alpha\beta\gamma}\sin\theta_{\alpha\beta}\sin\theta_{\beta\gamma} - \mathbf{i}_{\beta\gamma}\sin\theta_{\beta\gamma}\cos\theta_{\alpha\beta} - \mathbf{i}_{\alpha\beta}\cos\theta_{\beta\gamma}\sin\theta_{\alpha\beta} \qquad \text{(A92)}$$

which has no counter part in the conventional spherical trigonometry. By using Laplace's expansion rule in Eq. (A29), one obtains

$$|\mathbf{u}_{\mathbf{z}_\alpha} \wedge \mathbf{u}_{\mathbf{z}_\beta} \wedge \mathbf{u}_{\mathbf{z}_\gamma}|^2 = 1 - \cos^2\theta_{\alpha\beta} - \cos^2\theta_{\beta\gamma} - \cos^2\theta_{\gamma\alpha} + 2\cos\theta_{\alpha\beta}\cos\theta_{\beta\gamma}\cos\theta_{\gamma\alpha} \qquad \text{(A93)}$$

One can use dihedral angles $\tau_{\alpha\beta\gamma}$ to relate

$$\frac{\mathbf{u}_{\mathbf{z}_\alpha} \wedge \mathbf{u}_{\mathbf{z}_\beta} \wedge \mathbf{u}_{\mathbf{z}_\gamma}}{i} = \sin\theta_{\beta\gamma}\sin\theta_{\alpha\gamma}\sin\tau_{\alpha\gamma\beta} = \sin\theta_{\beta\gamma}\sin\theta_{\alpha\beta}\sin\tau_{\alpha\beta\gamma} = \sin\theta_{\alpha\gamma}\sin\theta_{\alpha\beta}\sin\tau_{\beta\alpha\gamma} \qquad \text{(A94)}$$

and one can use angles $\phi_{\alpha\beta\gamma}$ between vectors and planes to obtain

$$\frac{\mathbf{u}_{\mathbf{z}_\alpha} \wedge \mathbf{u}_{\mathbf{z}_\beta} \wedge \mathbf{u}_{\mathbf{z}_\gamma}}{i} = \sin\phi_{\alpha\beta\gamma}\sin\theta_{\beta\gamma} = \sin\phi_{\beta\gamma\alpha}\sin\theta_{\gamma\alpha} = \sin\phi_{\gamma\alpha\beta}\sin\theta_{\alpha\beta} \qquad \text{(A95)}$$

To find more on the use of geometric algebra in spherical trigonometry, the reader should consult Ref. 50, pages 661–667.

4. Geometric Calculus

The machinery of geometric algebra makes it possible to differentiate and integrate functions of vector variables in a coordinate-free manner. The conventionally separated concepts of the gradient, divergence, and curl are obtainable from a single vector derivative. First, we introduce the concept of directional derivative [32,34,71].

a. Directional Derivative

The directional derivative of any multivector $F(\mathbf{x}_\alpha)$ to the direction of a vector $\mathbf{a}$ is defined [50] as

$$\mathbf{a} \cdot \nabla_\alpha F(\mathbf{x}_\alpha) = \lim_{\delta \to 0} \frac{F(\mathbf{x}_\alpha + \delta \mathbf{a}) - F(\mathbf{x}_\alpha)}{\delta} = \frac{d}{d\delta} F(\mathbf{x}_\alpha + \delta \mathbf{a})|_{\delta \to 0} \tag{A96}$$

where we use the subscript α in the vector variable $\mathbf{x}$ to emphasize that these results are applicable in the case of several vector variables $\mathbf{x}_1, \mathbf{x}_2, \ldots$. The operator ∇_α in Eq. (A96) is the vector derivative operator

$$\nabla_\alpha = \mathbf{u}_1 \frac{\partial}{\partial x_{\alpha_1}} + \mathbf{u}_2 \frac{\partial}{\partial x_{\alpha_2}} + \mathbf{u}_3 \frac{\partial}{\partial x_{\alpha_3}} \tag{A97}$$

Once the most basic directional derivative,

$$\mathbf{a} \cdot \nabla_\alpha \mathbf{x}_\alpha = \frac{d}{d\delta} (\mathbf{x}_\alpha + \delta \mathbf{a}) \bigg|_{\delta \to 0} = \mathbf{a} \tag{A98}$$

is known, one can use a few simple rules to obtain the directional derivatives of more complicated functions without the need to use Eq. (A96). For example, with the help of the distributive rule

$$\mathbf{a} \cdot \nabla_\alpha (F + G) = \mathbf{a} \cdot \nabla_\alpha F + \mathbf{a} \cdot \nabla_\alpha G \tag{A99}$$

one can calculate the directional derivative of the internal position $\mathbf{y}_\alpha = \mathbf{x}_\alpha - \mathbf{X}$ (where $\mathbf{X} = \sum_\beta \frac{m_\beta}{M} \mathbf{x}\beta$) as

$$\mathbf{a} \cdot \nabla_\alpha \mathbf{y}_\alpha = \mathbf{a} - \sum_\beta \frac{m_\beta}{M} \mathbf{a} \cdot \nabla_\alpha \mathbf{x}_\beta = \mathbf{a}\left(1 - \frac{m_\alpha}{M}\right) \tag{A100}$$

because the positions $\mathbf{x}_1$, $\mathbf{x}_2, \ldots$ are independent. Similarly, the directional derivative of the internal position $\mathbf{y}_\beta = \mathbf{x}_\beta - \mathbf{X}$ ($\alpha \neq \beta$) is given by

$$\mathbf{a} \cdot \nabla_\alpha \mathbf{y}_\beta = -\frac{m_\alpha}{M} \mathbf{a} \tag{A101}$$

The product rule

$$\mathbf{a} \cdot \nabla_\alpha (FG) = (\mathbf{a} \cdot \nabla_\alpha F) G + F \mathbf{a} \cdot \nabla_\alpha G \tag{A102}$$

is valid for any multivectors F and G. It can be used to evaluate the directional derivative of the product FG.

Example A3. *If* $\mathbf{b}$ *is a vector independent of* $\mathbf{x}_\alpha$, *then*

$$\mathbf{a} \cdot \boldsymbol{\nabla}_\alpha(\mathbf{b}\mathbf{x}_\alpha) = (\mathbf{a} \cdot \boldsymbol{\nabla}_\alpha \mathbf{b})\mathbf{x}_\alpha + \mathbf{b}\mathbf{a} \cdot \boldsymbol{\nabla}_\alpha \mathbf{x}_\alpha = \mathbf{b}\mathbf{a} \tag{A103}$$

Furthermore, for any vectors $\mathbf{f}$ *and* $\mathbf{g}$, *the following rules are valid:*

$$\mathbf{a} \cdot \boldsymbol{\nabla}_\alpha(\mathbf{f} \cdot \mathbf{g}) = (\mathbf{a} \cdot \boldsymbol{\nabla}_\alpha \mathbf{f}) \cdot \mathbf{g} + \mathbf{f} \cdot (\mathbf{a} \cdot \boldsymbol{\nabla}_\alpha \mathbf{g}) \tag{A104}$$

$$\mathbf{a} \cdot \boldsymbol{\nabla}_\alpha(\mathbf{f} \wedge \mathbf{g}) = (\mathbf{a} \cdot \boldsymbol{\nabla}_\alpha \mathbf{f}) \wedge \mathbf{g} + \mathbf{f} \wedge (\mathbf{a} \cdot \boldsymbol{\nabla}_\alpha \mathbf{g}) \tag{A105}$$

$$\mathbf{a} \cdot \boldsymbol{\nabla}_\alpha(\mathbf{f} \times \mathbf{g}) = (\mathbf{a} \cdot \boldsymbol{\nabla}_\alpha \mathbf{f}) \times \mathbf{g} + \mathbf{f} \times (\mathbf{a} \cdot \boldsymbol{\nabla}_\alpha \mathbf{g}) \tag{A106}$$

Example A4. *The directional derivative of* $\mathbf{x}_\alpha^2$ *is given by the product rule in Eq. (A104) as*

$$\mathbf{a} \cdot \boldsymbol{\nabla}_\alpha \mathbf{x}_\alpha^2 = \mathbf{a} \cdot \boldsymbol{\nabla}_\alpha(\mathbf{x}_\alpha \cdot \mathbf{x}_\alpha) = 2\mathbf{a} \cdot \mathbf{x}_\alpha \tag{A107}$$

Also, if $F(\mathbf{x}_\alpha) = F(\lambda(\mathbf{x}_\alpha))$ is an arbitrary multivector valued compound function of the scalar valued function $\lambda(\mathbf{x}_\alpha)$, then

$$\mathbf{a} \cdot \boldsymbol{\nabla}_\alpha F(\lambda(\mathbf{x}_\alpha)) = \frac{dF}{d\lambda} \mathbf{a} \cdot \boldsymbol{\nabla}_\alpha \lambda(\mathbf{x}_\alpha) \tag{A108}$$

Example A5. *By the chain rule in Eq. (A108),*

$$\mathbf{a} \cdot \boldsymbol{\nabla}_\alpha x_\alpha^2 = 2x_\alpha \mathbf{a} \cdot \boldsymbol{\nabla}_\alpha x_\alpha \tag{A109}$$

But $\mathbf{a} \cdot \boldsymbol{\nabla}_\alpha \mathbf{x}_\alpha^2 = \mathbf{a} \cdot \boldsymbol{\nabla}_\alpha x_\alpha^2$, *so by equating Eqs. (A107) and (A109),*

$$\mathbf{a} \cdot \boldsymbol{\nabla}_\alpha x_\alpha = \frac{\mathbf{a} \cdot \mathbf{x}_\alpha}{x_\alpha} \tag{A110}$$

follows.

More generally, if $f = |\mathbf{f}|$, one can write

$$\mathbf{a} \cdot \boldsymbol{\nabla}_\alpha f = \frac{\mathbf{f} \cdot (\mathbf{a} \cdot \boldsymbol{\nabla}_\alpha \mathbf{f})}{f} \tag{A111}$$

for any vector $\mathbf{f}$. This kind of *implicit differentiation* is generally a useful way to obtain the directional derivatives of the functions of F when the directional derivative of F is known (F may be any multivector). As an example, because

$$F^{-1}F = 1 \Leftrightarrow (\mathbf{a} \cdot \boldsymbol{\nabla}_\alpha F^{-1})F + F^{-1}\mathbf{a} \cdot \boldsymbol{\nabla}_\alpha F = 0 \tag{A112}$$

the directional derivative of F^{-1} is[3]

$$\mathbf{a} \cdot \boldsymbol{\nabla}_\alpha F^{-1} = -F^{-1}(\mathbf{a} \cdot \boldsymbol{\nabla}_\alpha F)F^{-1} \tag{A113}$$

Occasionally the gradient $\boldsymbol{\nabla}_\alpha f$ is known for some *scalar*-valued function $f = f(\mathbf{x}_\alpha)$. Then, the directional derivative of f to the direction of $\mathbf{a}$ is found by taking the inner product between the gradient of f and $\mathbf{a}$, that is,

$$\mathbf{a} \cdot \boldsymbol{\nabla}_\alpha f = \mathbf{a} \cdot (\boldsymbol{\nabla}_\alpha f) \tag{A114}$$

Note, however, that this does *not* hold generally for vector- or bivector-valued functions.

Variation of a Function Along the Path. If the paths traversed by the vectors $\mathbf{x}_1, \mathbf{x}_2, \ldots$ can be parameterized by a scalar parameter t, the net variation of $F \equiv F(\mathbf{x}_1, \mathbf{x}_2, \ldots)$ along the paths $\mathbf{x}_1(t), \mathbf{x}_2(t), \ldots$ is given by

$$\frac{dF}{dt} = \sum_{\alpha}^{N} \frac{d\mathbf{x}_\alpha(t)}{dt} \cdot \boldsymbol{\nabla}_\alpha F \tag{A115}$$

Differential. The change in F caused by an infinitesimal displacement $d\mathbf{a}_\alpha$ (i.e., for $d\mathbf{a}_\alpha$, whose second- or higher-order powers vanish) is given by

$$F(\ldots, \mathbf{x}_\alpha + d\mathbf{a}_\alpha, \ldots) - F(\ldots, \mathbf{x}_\alpha, \ldots) = dF(\ldots, d\mathbf{a}_\alpha, \ldots) \tag{A116}$$

where the differential dF is the directional derivative of F evaluated in the infinitesimal displacement $d\mathbf{a}_\alpha$. Because

$$dF(\ldots, d\mathbf{a}_\alpha, \ldots) = d\mathbf{a}_\alpha \cdot \boldsymbol{\nabla}_\alpha F \tag{A117}$$

[3]Not all multivectors have inverses. For example, $A = (1 + \mathbf{u}_1)/2$ does not possess an inverse (note that $A^2 = A$).

we may write

$$\frac{dF}{da_\alpha} = \mathbf{u}_{d\mathbf{a}_\alpha} \cdot \nabla_\alpha F \tag{A118}$$

Taylor Expansion. In the region of convergence, the value of a continuous function F at the point $\mathbf{x}_1 + \mathbf{a}_1, \mathbf{x}_2 + \mathbf{a}_2, \ldots$ can be obtained from its value at the point $\mathbf{x}_1, \mathbf{x}_2, \ldots$ as

$$\begin{aligned} F(\mathbf{x}_1 + \mathbf{a}_1, \mathbf{x}_2 + \mathbf{a}_2, \ldots) &= F(\mathbf{x}_1, \mathbf{x}_2, \ldots) + \mathbf{a}_1 \cdot \nabla_1 F(\mathbf{x}_1, \mathbf{x}_2, \ldots) + \mathbf{a}_2 \cdot \nabla_2 F(\mathbf{x}_1, \mathbf{x}_2, \ldots) + \cdots \\ &\quad + (\mathbf{a}_1 \cdot \nabla_1)(\mathbf{a}_2 \cdot \nabla_2) F(\mathbf{x}_1, \mathbf{x}_2, \ldots) + \cdots \\ &= \prod_{\alpha=1}^{N} \sum_{k_\alpha=0}^{\infty} \frac{(\mathbf{a}_\alpha \cdot \nabla_\alpha)^{k_\alpha}}{k_\alpha!} F(\mathbf{x}_1, \mathbf{x}_2, \ldots) \end{aligned} \tag{A119}$$

where it is understood that $(\mathbf{a}_\alpha \cdot \nabla_\alpha)^0 = 1$.

b. *Vectorial Differentiation*

Conventionally, the vector derivative $\nabla_\alpha F$ of a function $F(\mathbf{x}_\alpha)$ of a vector variable $\mathbf{x}_\alpha$ is defined only for scalar values functions F; and in order to calculate it, one expresses the vector derivative operator ∇_α in some coordinates as

$$\nabla_\alpha = \sum_i (\nabla_\alpha q_i) \frac{\partial}{\partial q_i} \tag{A120}$$

If F is a vector—that is, if $F = \mathbf{f}(\mathbf{x}_\alpha)$—its divergence and curl are defined as

$$\mathrm{div}_\alpha\, \mathbf{f} = \nabla_\alpha \cdot \mathbf{f} \tag{A121}$$

$$\mathrm{curl}_\alpha\, \mathbf{f} = \nabla_\alpha \times \mathbf{f} \tag{A122}$$

By using the definition of the geometric product, we can write

$$\nabla_\alpha \mathbf{f} = \nabla_\alpha \cdot \mathbf{f} + \nabla_\alpha \wedge \mathbf{f} = \nabla_\alpha \cdot \mathbf{f} + i \nabla_\alpha \times \mathbf{f} \tag{A123}$$

so the divergence is just the scalar part, and curl is the dual of the bivector part of the vector derivative of $\mathbf{f}$. Because the last form is restricted to the three-dimensional space only, it is more appropriate to regard the curl as the bivector part of the vector derivative. The vector derivative $\nabla_\alpha F$ is defined for all elements F, not just for scalars and vectors—that is, generally

$$\nabla_\alpha F = \nabla_\alpha \cdot F + \nabla_\alpha \wedge F \tag{A124}$$

It follows from Eq. (A124) that the vector derivative operator changes the grade of the object it operates on by ± 1. For example, the vector derivative of the scalar $\lambda(\mathbf{x}_\alpha)$ is a vector (because $\mathbf{a} \cdot \lambda \equiv 0$ for any scalar λ, so $\mathbf{a}\lambda = \mathbf{a} \wedge \lambda$), and the vector derivative of the vector $\mathbf{f}(\mathbf{x}_\alpha)$ is a scalar plus a bivector. The differentiation with respect to the vector variable $\mathbf{x}_\alpha$ greatly resembles the differentiation with respect to some scalar variable x_α. For example, the vector differentiation is distributive,

$$\nabla_\alpha (F + G) = \nabla_\alpha F + \nabla_\alpha G \tag{A125}$$

for any F and G. If $\lambda = \lambda(\mathbf{x}_\alpha)$ is a scalar-valued function, then

$$\nabla_\alpha (\lambda G) = (\nabla_\alpha \lambda) G + \lambda \nabla_\alpha G \tag{A126}$$

However, the vector derivative operator does not commute with multivectors, and the product rule must be written as

$$\nabla_\alpha (FG) = \dot{\nabla}_\alpha \dot{F} G + \grave{\nabla}_\alpha F \grave{G} \tag{A127}$$

in the general case, where the target of differentiation is implicated by the accents.

We have not yet discussed how to find the vector derivative $\nabla_\alpha F$ in practice. It suffices to use the simple vector derivatives

$$\nabla_\alpha \mathbf{x}_\alpha = \sum_{k,j}^{3} \mathbf{u}_k \mathbf{u}_j \frac{\partial x_{\alpha_j}}{\partial x_{\alpha_k}} = \sum_{k}^{3} \frac{\partial x_{\alpha_k}}{\partial x_{\alpha_k}} = 3 \tag{A128}$$

$$\nabla_\alpha \mathbf{a} \cdot \mathbf{x}_\alpha = \sum_{k}^{3} \mathbf{u}_k \left(\mathbf{a} \cdot \frac{\partial \mathbf{x}_\alpha}{\partial x_{\alpha_k}} \right) = \sum_{k}^{3} \mathbf{u}_k \mathbf{a} \cdot \mathbf{u}_k = \mathbf{a} = \nabla_\alpha \mathbf{x}_\alpha \cdot \mathbf{a} \tag{A129}$$

(where $\{\mathbf{u}_1, \mathbf{u}_2, \mathbf{u}_3\}$ is some orthonormal frame, $x_{\alpha_k} = \mathbf{u}_k \cdot \mathbf{x}_\alpha$, and $\mathbf{a}$ is any vector independent of $\mathbf{x}_\alpha$) and to combine them with the product and chain rules allowing the evaluation of the vector derivative of any function.

Example A6. *By the product rule and Eq. (A129),*

$$\nabla_\alpha x_\alpha^2 = \nabla_\alpha \mathbf{x}_\alpha \cdot \mathbf{x}_\alpha = \dot{\nabla}_\alpha \dot{\mathbf{x}}_\alpha \cdot \mathbf{x}_\alpha + \dot{\nabla}_\alpha \mathbf{x}_\alpha \cdot \dot{\mathbf{x}}_\alpha = 2\mathbf{x}_\alpha \tag{A130}$$

(where $x_\alpha = |\mathbf{x}_\alpha|$).

Some other rules are also useful. For example, if $F(\lambda(\mathbf{x}_\alpha))$ is a multivector function of the scalar argument $\lambda(\mathbf{x})$, then

$$\boldsymbol{\nabla}_\alpha F(\lambda(\mathbf{x}_\alpha)) = (\boldsymbol{\nabla}_\alpha \lambda(\mathbf{x}_\alpha))\frac{\partial F}{\partial \lambda} \tag{A131}$$

The use of this rule is illustrated in the next example.

Example A7. *The left-hand side of Eq. (A130) can be written as*

$$\boldsymbol{\nabla}_\alpha x_\alpha^2 = 2x_\alpha \boldsymbol{\nabla}_\alpha x_\alpha \tag{A132}$$

where by Eq. (A130) the derivative $\boldsymbol{\nabla}_\alpha x_\alpha$ *can be solved as*

$$\boldsymbol{\nabla}_\alpha x_\alpha = \frac{\mathbf{x}_\alpha}{x_\alpha} \tag{A133}$$

The vector derivative $\boldsymbol{\nabla}_\alpha F$ can be obtained from the directional derivative $\mathbf{a} \cdot \boldsymbol{\nabla}_\alpha F$ as

$$\boldsymbol{\nabla}_\alpha F = \boldsymbol{\nabla}_\mathbf{a} \mathbf{a} \cdot \boldsymbol{\nabla}_\alpha F \tag{A134}$$

The right-hand side in Eq. (A134) may appear at first sight peculiar to the reader unfamiliar with geometric algebra, because such an expression does not exist in ordinary vector algebra. Eq. (A134) can be derived by substituting $A = \boldsymbol{\nabla}_\alpha$ to the general rule $\boldsymbol{\nabla}_\mathbf{a}\mathbf{a} \cdot A = A$ (where A is any multivector independent of $\mathbf{a}$). As a hopefully useful example,

Example A8. *By Example A4, the directional derivative of* $\mathbf{x}_\alpha^2$ *is* $\mathbf{a} \cdot \boldsymbol{\nabla}_\alpha \mathbf{x}_\alpha^2 = 2\mathbf{a} \cdot \mathbf{x}_\alpha$. *Thus, by Eq. (A134),*

$$\boldsymbol{\nabla}_\alpha x_\alpha^2 = \boldsymbol{\nabla}_\mathbf{a} \mathbf{a} \cdot \boldsymbol{\nabla}_\alpha x_\alpha^2 = \boldsymbol{\nabla}_\mathbf{a}(2\mathbf{a} \cdot \mathbf{x}_\alpha) = 2\mathbf{x}_\alpha \tag{A135}$$

which agrees with Eq. (A130).

If $F(\mathbf{x}_\alpha)$ is a scalar-valued function, the vector derivative $\boldsymbol{\nabla}_\alpha F$ can be extracted from the variation of F along the path $\mathbf{x}_\alpha(t)$ in Eq. (A115). This method has the advantage of reducing the calculation of the vector derivative of a scalar-valued function to ordinary scalar differentiation. As an example,

Example A9. *Because*

$$\frac{dx_\alpha^2}{dt} = 2x_\alpha \frac{dx_\alpha}{dt} = \frac{d(\mathbf{x}_\alpha \cdot \mathbf{x}_\alpha)}{dt} = 2\mathbf{x}_\alpha \cdot \frac{d\mathbf{x}_\alpha}{dt} \tag{A136}$$

or

$$\frac{dx_\alpha}{dt} = \frac{\mathbf{x}_\alpha}{x_\alpha} \cdot \frac{d\mathbf{x}_\alpha}{dt} \tag{A137}$$

the gradient $\nabla_\alpha x_\alpha$ *can be picked by Eq. (A115) as*

$$\nabla_\alpha x_\alpha = \frac{\mathbf{x}_\alpha}{x_\alpha} \tag{A138}$$

The results in Table I can be proved by using the above rules.

APPENDIX B: MISCELLANEOUS RESULTS

Some miscellaneous results are derived in the following subsections.

1. Relations of the Co- and Contravariant Measuring Vectors

It is worth the trouble to relate the properties of the covariant measuring vectors to those of the contravariant measuring vectors. We shall consider only molecules, which are not subject to constraints. By using the coordinate representation of ∇_α, it follows that

$$\nabla_\alpha(A_{\bar{r}}\mathbf{x}_\beta) = \delta_{\alpha\beta} \sum_i^{3N} \mathbf{e}_\alpha^{(q_i)} A_{\bar{r}} \frac{\partial \mathbf{x}_\beta}{\partial q_i} = \delta_{\alpha\beta} \sum_i^{3N} \mathbf{e}_\alpha^{(q_i)} A_{\bar{r}} \mathbf{e}_{q_i}^{(\beta)} \tag{B1}$$

where $A_{\bar{r}}$ is an r-blade independent of $\mathbf{x}_\alpha$, and $r = 0, 1, 2, 3$. But because

$$A_{\bar{r}}\mathbf{x}_\beta = A_{\bar{r}} \cdot \mathbf{x}_\beta + A_{\bar{r}} \wedge \mathbf{x}_\beta = (-1)^{r+1}\mathbf{x}_\beta \cdot A_{\bar{r}} + (-1)^r \mathbf{x}_\beta \wedge A_{\bar{r}} \tag{B2}$$

and

$$\nabla_\alpha \mathbf{x}_\beta \cdot A_{\bar{r}} = \delta_{\alpha\beta} r A_{\bar{r}} \tag{B3}$$

$$\nabla_\alpha \mathbf{x}_\beta \wedge A_{\bar{r}} = \delta_{\alpha\beta}(3 - r) A_{\bar{r}} \tag{B4}$$

it follows that

$$\sum_{i}^{3N} \mathbf{e}_{\alpha}^{(q_i)} A_{\bar{r}} \mathbf{e}_{q_i}^{(\beta)} = \delta_{\alpha\beta}(-1)^r(3-2r)A_{\bar{r}} \tag{B5}$$

where some special relations can be read off. For example, by setting $A_{\bar{r}} = 1$ (so $r = 0$), the identity

$$\sum_{i}^{3N} \mathbf{e}_{\alpha}^{(q_i)} \mathbf{e}_{q_i}^{(\beta)} = 3\delta_{\alpha\beta} \tag{B6}$$

follows. By decomposing this into a sum of inner and outer products, equations

$$\sum_{i}^{3N} \mathbf{e}_{\alpha}^{(q_i)} \cdot \mathbf{e}_{q_i}^{(\beta)} = 3\delta_{\alpha\beta} \tag{B7}$$

and

$$\sum_{i}^{3N} \mathbf{e}_{\alpha}^{(q_i)} \times \mathbf{e}_{q_i}^{(\beta)} = 0 \tag{B8}$$

follow.

2. Some Properties of Shape Coordinates

In general, if the nuclei α are displaced by the amount $\mathbf{a}_\alpha$ ($\alpha = 1, 2, \ldots$), the value of the shape coordinate s_i changes by

$$\triangle s_i = s_i(\mathbf{x}_1^{(e)} + \mathbf{a}_1, \mathbf{x}_2^{(e)} + \mathbf{a}_2, \ldots) - s_i(\mathbf{x}_1^{(e)}, \mathbf{x}_2^{(e)}, \ldots) \tag{B9}$$

where $\mathbf{x}_\alpha^{(e)}$ is the value of $\mathbf{x}_\alpha$ at the reference configuration. The direction of the measuring vector $\mathbf{e}_\alpha^{(s_i)}$ also changes, if the coordinate s_i is curvilinear. To be more precise, for the displacement $\mathbf{a}_\alpha$ of the nucleus α, the unit vector $\mathbf{u}_\alpha^{(s_i)} = \mathbf{e}_\alpha^{(s_i)}/|\mathbf{e}_\alpha^{(s_i)}|$ changes its direction at a rate [32,34]

$$\mathbf{a}_\alpha \cdot \boldsymbol{\nabla}_\alpha \mathbf{u}_\alpha^{(s_i)} = \boldsymbol{\omega}_\alpha^{(s_i)}(\mathbf{a}_\alpha) \times \mathbf{u}_\alpha^{(s_i)} \tag{B10}$$

where $\boldsymbol{\omega}_\alpha^{(s_i)}(\mathbf{a}_\alpha)$ is a vector evaluated at the displacement $\mathbf{a}_\alpha$. Because the component of $\boldsymbol{\omega}_\alpha^{(s_i)}(\mathbf{a}_\alpha)$ to the direction of $\mathbf{u}_\alpha^{(s_i)}$ plays no role in the change of the direction of $\mathbf{u}_\alpha^{(s_i)}$, we may replace Eq. (B10) by

$$\mathbf{a}_\alpha \cdot \boldsymbol{\nabla}_\alpha \mathbf{u}_\alpha^{(s_i)} = \boldsymbol{\omega}_\alpha^{(s_i)\perp}(\mathbf{a}_\alpha) \times \mathbf{u}_\alpha^{(s_i)} = -i\boldsymbol{\omega}_\alpha^{(s_i)\perp}(\mathbf{a}_\alpha)\mathbf{u}_\alpha^{(s_i)} \tag{B11}$$

where $\boldsymbol{\omega}_\alpha^{(s_i)\perp}$ is the part of $\boldsymbol{\omega}_\alpha^{(s_i)}$ perpendicular to $\mathbf{u}_\alpha^{(s_i)}$. It can be solved as

$$\boldsymbol{\omega}_\alpha^{(s_i)\perp}(\mathbf{a}_\alpha) = i(\mathbf{a}_\alpha \cdot \boldsymbol{\nabla}_\alpha \mathbf{u}_\alpha^{(s_i)})\mathbf{u}_\alpha^{(s_i)} = \mathbf{u}_\alpha^{(s_i)} \times (\mathbf{a}_\alpha \cdot \boldsymbol{\nabla}_\alpha \mathbf{u}_\alpha^{(s_i)}) \tag{B12}$$

where in the last term we have used Eq. (182) to replace the geometric product by the cross product. We may regard the following as the *definition* of the curvilinearity: The coordinate s_i is curvilinear if

$$\boldsymbol{\omega}_\alpha^{(s_i)\perp}(\mathbf{a}_\alpha) \neq 0 \tag{B13}$$

for an arbitrary displacement $\mathbf{a}_\alpha$.

Example B1. *Because the directional derivative $\mathbf{a}_\beta \cdot \boldsymbol{\nabla}_\beta \mathbf{u}_{\mathbf{r}_{\alpha\beta}}$ of the measuring vector $\mathbf{e}_\beta^{(r_{\alpha\beta})} = \boldsymbol{\nabla}_\beta r_{\alpha\beta} = \mathbf{u}_{\mathbf{r}_{\alpha\beta}} = \mathbf{u}_\beta^{(r_{\alpha\beta})}$ is*

$$\mathbf{a}_\beta \cdot \boldsymbol{\nabla}_\beta \mathbf{u}_{\mathbf{r}_{\alpha\beta}} = \frac{\mathbf{r}_{\alpha\beta}\mathbf{r}_{\alpha\beta} \wedge \mathbf{a}_\beta}{r_{\alpha\beta}} \tag{B14}$$

it follows that

$$\boldsymbol{\omega}_\beta^{(r_{\alpha\beta})\perp}(\mathbf{a}_\beta) = -i\frac{\mathbf{r}_{\alpha\beta} \wedge \mathbf{a}_\beta \mathbf{r}_{\alpha\beta}}{r_{\alpha\beta}} \mathbf{u}_{\mathbf{r}_{\alpha\beta}} = \mathbf{r}_{\alpha\beta} \times \mathbf{a}_\beta \tag{B15}$$

which, for an arbitrary displacement $\mathbf{a}_\beta$, differs from zero. Thus, the internuclear distance coordinate $r_{\alpha\beta}$ is curvilinear.

In particular, the vector $\boldsymbol{\omega}_\alpha^{(s_i)\perp}(\mathbf{a}_\alpha/a_\alpha)$ gives us the rate of change of the direction of the unit measuring vector $\mathbf{u}_\alpha^{(s_i)}$ to the direction of the nuclear displacement $\mathbf{a}_\alpha/a_\alpha$, that is,

$$\frac{d\mathbf{u}_\alpha^{(s_i)}}{da_\alpha} = \frac{\mathbf{a}_\alpha}{a_\alpha} \cdot \boldsymbol{\nabla}_\alpha \mathbf{u}_\alpha^{(s_i)} = \boldsymbol{\omega}_\alpha^{(s_i)\perp}\left(\frac{\mathbf{a}_\alpha}{a_\alpha}\right) \times \mathbf{u}_\alpha^{(s_i)} \tag{B16}$$

The rectilinear shape coordinates are defined by the condition

$$\boldsymbol{\omega}_\alpha^{(s_i)\perp}(\mathbf{a}_\alpha) = 0 \tag{B17}$$

that is, the unit measuring vectors $\mathbf{u}_\alpha^{(s_i)}$ are transported parallelly along the displacement $\mathbf{a}_\alpha$. An example of the rectilinear shape coordinate is the first-order approximation $\triangle\tilde{s}_i$ of the curvilinear displacement coordinate $\triangle s_i$. This rectilinear internal displacement coordinate is *defined* by the first term

$$\triangle\tilde{s}_i = \sum_{\alpha}^{N} \mathbf{a}_\alpha \cdot \boldsymbol{\nabla}_\alpha\, s_i\big|_{\mathbf{s}^{(e)}} \tag{B18}$$

in the Taylor expansion of s_i about the reference configuration $\mathbf{s}^{(e)} = \{s_1^{(e)}, s_2^{(e)}, \ldots, s_{3N-6}^{(e)}\}$ [see Eq. (A118)]. Thus,

$$\mathbf{e}_\alpha^{(\tilde{s}_i)} = \boldsymbol{\nabla}_\alpha s_i\big|_{\mathbf{s}^{(e)}} \tag{B19}$$

which satisfies Eq. (B17). In the past, these coordinates were used almost exclusively in the theoretical molecular spectroscopy (see, e.g., Ref. 1). In the limit of infinitesimal deformations, $\triangle\tilde{s}_i$ equals the change $\triangle s_i$. Of course, $\triangle\tilde{s}_i$ is also well-defined for finite displacements, but it differs, in general, from $\triangle s_i$.

Acknowledgments

We would like to thank Vincenzo Aquilanti, Andrea Beddoni, Tucker Carrington, Jr., Robert Littlejohn, and Brian Sutcliffe for useful discussions concerning our work. We are also grateful to EU (contract number HPRN-CT-1999-00005) and the Academy of Finland for financial support.

References

1. E. B. Wilson, J. C. Decius, and P. C. Cross, *Molecular Vibrations*, Dover, New York, 1980 (republication of the original 1955 edition).
2. H. H. Nielsen, The vibration–rotation energies of molecules and their spectra in the infra-red, in *Handbuch der Physik*, part XXXVII/1, Springer-Verlag, Berlin, 1959, p. 173–313.
3. J. K. G. Watson, Simplification of the molecular vibration–rotation Hamiltonian. *Mol. Phys.* **15**, 479–490 (1968).
4. J. K. G. Watson, The vibration–rotation Hamiltonian of linear molecules. *Mol. Phys.* **19**, 465–487 (1970).
5. G. D. Carney, L. L. Sprandel, and C. W. Kern, Variational approaches to vibration–rotation spectroscopy for polyatomic molecules. *Adv. Chem. Phys.* **37**, 305–379 (1978).
6. S. Carter and N. C. Handy, A theoretical determination of the rovibrational energy levels of the water molecule. *J. Chem. Phys.* **87**, 4294–4301 (1987).
7. M. S. Child and L. Halonen, Overtone frequencies and intensities in the local mode picture. *Adv. Chem. Phys.* **57**, 1–58 (1984).
8. L. Halonen, Local mode vibrations in polyatomic molecules. *Adv. Chem. Phys.* **104**, 41–179 (1998).
9. A. R. Hoy, G. Strey, and I. M. Mills, Anharmonic force constant calculations. *Mol. Phys.* **24**, 1265–1290 (1972).
10. G. O. Sørensen, A new approach to the Hamiltonian of nonrigid molecules. *Topics Curr. Chem.* **82**, 97–175 (1979).
11. V. Aquilanti and S. Cavalli, Coordinates for molecular dynamics: Orthogonal local systems. *J. Chem. Phys.* **85**, 1355–1361 (1986).

12. V. Aquilanti and S. Cavalli, Hyperspherical coordinates for molecular dynamics by the methods of trees and the mapping of potential energy surfaces for triatomic systems. *J. Chem. Phys.* **85**, 1362–1375 (1986).
13. R. G. Littlejohn and M. Reinsch, Internal or shape coordinates in the N-body problem. *Phys. Rev. A* **52**, 2035–2051 (1995).
14. V. Aquilanti, A. Beddoni, S. Cavalli, A. Lombard, and R. G. Littlejohn, Collective hyperspherical coordinates for polyatomic molecules and clusters. *Mol. Phys.* **98**, 1763–1770 (2000).
15. Z. Bačić and J. C. Light, Theoretical methods for rovibrational states of floppy molecules, *Annu. Rev. Phys. Chem.* **40**, 469–498 (1989).
16. X. Chapuisat, A. Nauts, and G. Durand, A method to obtain the Eckart Hamiltonian and the equations of motion of a highly deformable polyatomic system in terms of generalized coordinates. *Chem. Phys.* **56**, 91–105 (1981).
17. A. Nauts and X. Chapuisat, Momentum, quasi-momentum and Hamiltonian operators in terms of arbitrary curvilinear coordinates, with special emphasis on molecular Hamiltonians. *Mol. Phys.* **55**, 1287–1318 (1985).
18. X. Chapuisat, A. Nauts, and J.-P. Brunet, Exact quantum molecular Hamiltonians. Part 1. Application to the dynamics of three particles. *Mol. Phys.* **72**, 1–31 (1991).
19. X. Chapuisat, Principal-axis hyperspherical description of N-particle systems: Classical treatment. *Phys. Rev. A* **44**, 1328–1351 (1991).
20. X. Chapuisat and A. Nauts, Principal-axis hyperspherical description of N-particle systems: Quantum mechanical treatment. *Phys. Rev. A* **45**, 4277–4292 (1992).
21. E. Baloitcha and M. N. Hounkonnou, Principal-axis hyperspherical description of six particle systems: Quantum-mechanical treatment. *J. Phys. B* **32**, 4823–4837 (1999).
22. B. T. Sutcliffe, Some problems in defining a nuclear motion Hamiltonian and their relation to the problem of molecular shape, in *Current Aspects of Quantum Chemistry*, R. Carbo, ed., Elsevier, Amsterdam, 1982, p. 99–125.
23. N. C. Handy, The derivation of vibration-rotation kinetic energy operators, in internal coordinates. *Mol. Phys.* **61**, 207–223 (1987).
24. H. Wei and T. Carrington, Jr., The triatomic Eckart-frame kinetic energy operator in bond coordinates. *J. Chem. Phys.* **107**, 9493–9501 (1997).
25. H. Wei and T. Carrington, Jr., An exact Eckart-embedded kinetic energy operator in Radau coordinates for triatomic molecules. *Chem. Phys. Lett.* **287**, 289–300 (1998).
26. M. Mladenovic, Rovibrational Hamiltonians for general polyatomic molecules in spherical polar parameterization I. Orthogonal representations. *J. Chem. Phys.* **112**, 1070–1081 (2000).
27. M. Mladenovic, Rovibrational Hamiltonians for general polyatomic molecules in spherical polar parameterization II. Nonorthogonal representations of internal molecular geometry. *J. Chem. Phys.* **112**, 1081–1095 (2000).
28. X. G. Wang and T. Carrington, Jr., A simple method for deriving kinetic energy operators. *J. Chem. Phys.* **113**, 7097–7101 (2000).
29. T. J. Lukka, A simple method for the derivation of exact quantum-mechanical vibration–rotation Hamiltonians in terms of internal coordinates. *J. Chem. Phys.* **102**, 3945–3955 (1995).
30. S. M. Colwell and N. C. Handy, The derivation of vibration–rotation kinetic energy operators in internal coordinates II. *Mol. Phys.* **92**, 317–330 (1997).
31. J. R. Alvarez-Collado, On derivation of curvilinear ro-vibrational quantum kinetic energy operator for polyatomic molecules. *J. Mol. Struct. (Theochem.)* **433**, 69–81 (1998).

32. D. Hestenes and G. Sobczyk, *Clifford Algebra to Geometric Calculus*, Reidel, Dordrecht, 1984.

33. D. Hestenes, Synopsis of geometric algebra (http://modelingnts.la.asu.edu/pdf/NFMPchapt1.pdf, 1998).

34. D. Hestenes, Geometric calculus (http://modelingnts.la.asu.edu/pdf/NFMPchapt2.pdf, 1998).

35. P. Lounesto, *Clifford Algebras and Spinors*, Cambridge University Press, Cambridge, England, 2001.

36. J. Pesonen, Vibrational coordinates and their gradients: A geometric algebra approach. *J. Chem. Phys.* **112**, 3121–3132 (2000).

37. J. Pesonen, Gradients of vibrational coordinates from the variation of coordinates along the path of a particle. *J. Chem. Phys.* **115**, 4402–4403 (2001).

38. J. Pesonen, Vibration–rotation kinetic energy operators: A geometric algebra approach. *J. Chem. Phys.* **114**, 10598–10607 (2001).

39. J. Pesonen and L. Halonen, Volume-elements of integration: A geometric algebra approach. *J. Chem. Phys.* **116**, 1825–1833 (2002).

40. J. Pesonen, *Application of geometric algebra to theoretical molecular spectroscopy* (http://ethesis.helsinki.fi/julkaisut/mat/kemia/vk/pesonen/). PhD thesis, Helsinki, 2001.

41. J. Pesonen, Exact kinetic energy operators for polyatomic molecules, in *Applications of Geometric Algebra in Computer Science and Engineering,* L. Dorst, C. Doran, and J. Lasenby, eds., Birkhäuser, Boston, 2002, p. 261–270.

42. R. G. Littlejohn, K. A. Mitchell, V. Aquilanti, and S. Cavalli, Body frames and frame singularities for three-atom systems. *Phys. Rev. A* **58**, 3705–3717 (1998).

43. T. J. Lukka and E. Kauppi, Seminumerical contact transformation: From internal coordinate rovibrational Hamiltonian to effective rotational Hamiltonians. Framework of the method. *J. Chem. Phys.* **103**, 6586–6596 (1995).

44. J. Louck and H. W. Galbraith, Eckart vectors, Eckart frames, and polyatomic molecules. *Rev. Mod. Phys.* **48**, 69–106 (1976).

45. K. L. Mardis and E. L. Sibert III, Derivation of rotation–vibration Hamiltonians that satisfy the Casimir condition. *J. Chem. Phys.* **106**, 6618–6621 (1997).

46. H. Wei and T. Carrington, Jr., Explicit expressions for triatomic Eckart frames in Jacobi, Radau, and bond coordinates. *J. Chem. Phys.* **107**, 2813–2818 (1997).

47. R. G. Littlejohn and M. Reinsch, Gauge fields in the separation of rotations and internal motions in the *N*-body problem. *Rev. Mod. Phys.* **69**, 213–275 (1997).

48. B. T. Sutcliffe, The idea of a potential energy surface. *J. Mol. Struct. (Theochem.)* **341**, 217–235 (1994).

49. B. K. Kendrick, C. A. Mead, and D. G. Truhlar, Properties of nonadiabatic couplings and the generalized Born–Oppenheimer approximation. *Chem. Phys.* **277**, 31–41 (2002).

50. D. Hestenes, *New Foundations for Classical Mechanics*, 2nd. ed., Kluwer Academic Publishers, Dordrecht, 1999.

51. H. Goldstein, C. Poole, and J. Safko, *Classical Mechanics*, 3rd. ed., Addison-Wesley, San Francisco, 2002.

52. G. B. Arfken and H. J. Weber, *Mathematical Methods for Physicists*, 4th. ed., *Academic Press, San Diego, 1995.*

53. J. H. Frederick and C. Woywood, General formulation of the vibrational kinetic energy operator in internal bond-angle coordinates. *J. Chem. Phys.* **111**, 7255–7271 (1999).

54. H. C. Corben and P. Stehle, *Classical Mechanics*, John Wiley & Sons, New York, 1950.
55. X.-G. Wang and T. Carrington, Jr., Six-dimensional variational calculation of the bending energy levels of HF trimer and DF trimer. *J. Chem. Phys.* **115**, 9781–9796 (2001).
56. J. L. Bell, *A Primer of Infinitesimal Analysis*, Cambridge University Press, Cambridge, England, 1998.
57. S. Califano, *Vibrational States*, John Wiley & Sons, London, 1976.
58. J. Pesonen, A. Miani, and L. Halonen, Curvilinear internal bond coordinate Hamiltonian for ammonia I. Application to a CCSD(T) bidimensional potential energy surface. *J. Chem. Phys.* **115**, 1243–1250 (2001).
59. F. A. Cotton, *Chemical Applications of Group Theory*, John Wiley & Sons, New York, 1971.
60. T. Kimura, T. Ohtani, and R. Sugarno, On the consistency between Lagrangian and Hamiltonian formalisms in quantum mechanics III. *Prog. Theor. Phys.* **48**, 1395–1407 (1972).
61. R. C. T. da Costa, Constraints in quantum mechanics. *Phys. Rev. A* **25**, 2893–2900 (1982).
62. N. Ogawa, N. Chepilko, and A. Kobushkin, Quantum mechanics in Riemannian manifold II. *Prog. Theor. Phys.* **85**, 1189–1201 (1991).
63. J. E. Hadder and J. H. Frederick, Molecular Hamiltonian for highly constrained model systems. *J. Chem. Phys.* **97**, 3500–3520 (1992).
64. L. Kaplan, N. T. Maitra, and J. Heller, Quantizing constrained systems. *Phys. Rev. A* **56**, 2592–2599 (1997).
65. X. Chapuisat and A. Nauts, A general property of quantum mechanical Hamiltonian for constrained systems. *Mol. Phys.* **91**, 47–57 (1997).
66. K. E. Mitchell, Gauge fields and extrapotentials in constrained quantum systems. *Phys. Rev. A.* **63**, 04211-21–042112-20 (2001).
67. J. R. Shewell, On the formation of quantum mechanical operators. *Am. J. Phys.* **27**, 16–20 (1959).
68. G. R. Gruber, Comments on the correspondence principles of quantum mechanical operators. *Found. Phys.* **4**, 19–22 (1974).
69. G. R. Gruber, Quantization in generalized coordinates III—Lagrangian formulation. *Int. J. Theor. Phys.* **7**, 253–257 (1973).
70. T. Havel and I. Najfeld, Applications of geometric algebra to the theory of molecular conformation. Part 1. The optimum alignment problem. *J. Mol. Struct. (Theochem.)* **308**, 241–262 (1994).
71. T. Havel and I. Najfeld, Applications of geometric algebra to the theory of molecular conformation. 2. The local deformation problem. *J. Mol. Struct. (Theochem.)* **136**, 175–189 (1995).
72. D. Hestenes, Point groups and space groups in geometric algebra (`http://modelingnts.la.asu.edu/pdf/crystalsymmetry.pdf`), in *Applications of Geometric Algebra in Computer Science and Engineering*, L. Dorst, C. Doran, and J. Lasenby, eds., Birkhäuser, Boston, 2002, p. 3–34.
73. D. Hestenes and E. D. Fasse, Modeling elastically coupled rigid bodies with geometric algebra (`http://modelingnts.la.asu.edu/pdf/ElasticModeling.pdf`), in *Applications of Geometric Algebra in Computer Science and Engineering*, L. Dorst, C. Doran, and J. Lasenby, eds., Birkhäuser, Boston, 2002, p. 197–212.
74. G. Sobczyk, Simplicial calculus with geometric algebra (`http://modelingnts.la.asu.edu/pdf/SIMP_CAL.pdf`), in *Clifford Algebras and Their Applications in Mathematical Physics*, A. Micali, R. Boudet, and J. Helmstetter, eds., Kluwer, Dordrecht, 1992, p. 279–292.

75. D. Hestenes, Differential forms in geometric calculus (http://modelingnts.la.asu.edu/pdf/DIF_FORM.pdf), in *Clifford Algebras and Their Applications in Mathematical Physics*, F. Brackx et al., eds., Kluwer, Dordrecht, 1993, p. 269–285.

76. D. Hestenes and R. Ziegler, Projective geometry with Clifford algebra (http://modelingnts.la.asu.edu/pdf/PGwithCA.pdf). *Acta Appl. Math.* **23**, 25–63 (1991).

77. D. Hestenes, H. Li, and A. Rockwood, New algebraic tools for computational geometry (http://modelingnts.la.asu.edu/pdf/, 1999).

78. L. Dorst, The inner products of geometric algebra, in *Applications of Geometric Algebra in Computer Science and Engineering*, L. Dorst, C. Doran, and J. Lasenby, eds., Birkhäuser, Boston, 2002, p. 35–46.

PROTON TRANSFER AND COHERENT PHENOMENA IN MOLECULAR STRUCTURES WITH HYDROGEN BONDS

V. V. KRASNOHOLOVETS, P. M. TOMCHUK, and S. P. LUKYANETS

Department of Theoretical Physics, Institute of Physics, National Academy of Sciences, Kyïv, Ukraine

CONTENTS

I. Specific Physical Effects in Structures with Hydrogen Bonds
II. Quantum Mechanical Descriptions of Hydrogen-Bonded Systems
  A. Hydrogen Bond Vibrations
  B. Proton Transfer Incorporating Acoustic Phonons
  C. Proton Polaron
  D. Proton Ordering
  E. Bending Vibrations of Hydrogen Bond
  F. Tunneling Transition and Coupled Protons
III. Transport Properties of Hydrogen-Bonded Complexes: Polarons and Polaritons in Systems with Hydrogen Bonds
  A. Orientational-Tunneling Model of One-Dimensional Molecular System
  B. Proton Ordering Model
  C. Proton Conductivity at the Superionic Phase Transitions
  D. Polaronic Conductivity Along a Hydrogen-Bonded Chain
  E. The Anharmonicity Influence
  F. Influence of Coulomb Correlations and the Electric Field Local Heterogeneities on Proton Conductivity
  G. External Influences on the Proton Conductivity
  H. Vibration Fluctuations of a Resonance Integral in the Polaron Problem
  I. An Example of Superionic Conductivity: The $NH_4IO_3 \cdot 2NHIO_3$ Crystal
  K. Polariton Effect in Crystals with Symmetric $O \cdots H \cdots O$ Hydrogen Bonds
IV. Bacteriorhodopsin Considered from the Microscopic Physics Standpoint
  A. Active Site of Bacteriorhodopsin and the Proton Path
  B. Light-Excited Retinal and Evolution of Excitations in the Retinal
  C. Proton Ejection

Advances in Chemical Physics, Volume 125, Edited by I. Prigogine and Stuart A. Rice.
ISBN 0-471-21452-3.

V. Mesomorphic Transformations and Proton Subsystem Dynamics in Alkyl- and Alkoxybenzoic Acids
A. Molecular Associates
B. Rearrangement of Hydrogen Bonds: Mechanism of Open Associates Formation
VI. Quantum Coherent Phenomena in Structures with Hydrogen Bonds
A. Mesoscopic Quantum Coherence and Tunneling in Small Magnetic Grains and Ordered Molecules
B. Two Possible Mechanisms of Coherent Tunneling of the Repolarization of Hydrogen-Bonded Chain
C. Can Coherent Tunneling of Heavy Particles Be More Probable than That of Light Particles? The Role of Proton–Phonon Coupling
VII. Unusual Properties of Aqueous Systems
A. Organization and Thermodynamic Features of Degassed Aqueous Systems
1. Experimental Results
2. Thermodynamics
3. Organization of Water System
B. Determination of Water Structure by Pulsed Nuclear Magnetic Resonance (NMR) Technique
C. Water-Dependent Switching in Continuous Metal Films
VIII. Clustering in Molecular Systems
A. Clusterization as Deduced from the Most General Statistical Mechanical Approach
B. Cluster Formation in Solid Phase of Alkyl- and Alkoxybenzoic Acids
C. Clustering of H_2O Molecules in Water
IX. Summary
Appendix A: Stretching and Bending Energies as Functions of $R_{O\cdots O}$
Appendix B: A Possible Mechanism of Sonoluminescence
Appendix C: Diagonalization of Phonon Variables
Appendix D: Proton Bifurcation and the Phonon Mixture
Acknowledgments
References

I. SPECIFIC PHYSICAL EFFECTS IN STRUCTURES WITH HYDROGEN BONDS

Hydrogen bonds and the motion of protons in hydrogen-bonded networks exhibit a great number of interesting physical effects, which, in turn, bring profound theoretical studies into being.

The most reliable information on the behavior of hydrogen bonds and proton motion in a hydrogen bond and through hydrogen-containing compounds yields X-ray diffraction, Raman and infrared (IR) spectra, nuclear magnetic resonance (NMR), absorption electron spectra, microwave and submicrowave spectroscopy, quasi-elastic neutron scattering (QENS), and rotational spectroscopy of supersonically expanded jets. The arrival of intense neutron sources has had a new dramatic impact on hydrogen bonding studies in solids [1]: Neutron diffraction and vibrational spectroscopy can also be performed with inelastic neutron scattering (INS) techniques. Besides, a very remarkable result has recently been obtained by Isaacs et al. [2]: Studying high-momentum transfer inelastic (Compton) X-ray scattering of the hydrogen bond in ice I_h, they have revealed the Compton profile anisotropy that has reasonably been interpreted as the first direct experimental evidence for the substantial covalent character of the hydrogen bonds. The results obtained in Ref. 2 indeed demonstrate a high sensitivity of Compton scattering at the investigation of the phase of the electronic wave function. Thus, a pure classical (electrostatic) bonding model turns out to be inconsistent to describe the dynamics of the hydrogen bond, and that is why the role of quantum mechanical description of the behavior of the hydrogen bond is growing.

By means of the IR spectroscopy, Zundel [3–6] could show that various homoconjugated hydrogen bonds demonstrate a large proton polarizability. He found that in the case when a system of hydrogen bonds is completely structurally symmetrical, the whole system shows a large proton polarizability due to the strong coupling of proton motions.

Inelastic neutron scattering studies have shown [7–11] that many hydrogen-bonded crystals [potassium carbonate (i.e., $KHCO_3$), various polyanilines, $Ca(OH)_2$, and others] are characterized by the proton dynamics that is very decoupled from the backbone lattice.

The availability of the coherent proton tunneling and cooperative proton tunneling and transfer of four protons in hydrogen bonds of benzoic acids crystals (including dye doped) have been revealed by Trommsdorff and collaborators [12,13] by reading the electron spectra of the crystals.

Describing different techniques employed at the investigation of the motion of protons in hydrogen bonds, Fillaux [11] in particular notes that optical techniques are very sensitive to the hybridization state of valence electrons, which are largely unknown, and quantum calculations are not yet able to provide reliable values. On the contrary, the neutron scattering process is entirely attributable to nuclear interactions, and cross sections are independent of chemical bonding. It is interesting that the inelastic neutron scattering spectrum of the $KHCO_3$ cannot be described by conventional harmonic force fields because its protons are almost totally decoupled from surrounding heavy atoms, and it is also interesting that the dynamics of the protons are rather specified by localized modes in a fixed frame [11]. Similar results have been obtained for a whole series of molecular crystals, and this means that the description of the dynamics of molecular crystals based on the normal modes formalism should be reconsidered. Fillaux [14] has suggested that such a peculiar dynamics of protons might be resulted from quantum mechanics of a nonlocal sort.

At the same time, Trommsdorff [15] advocates low-temperature spectroscopy combined with single-molecule spectroscopy, calling them an exraordinal powerful tool to study the structure and dynamics in condensed phases. He intimates that high-resolution optical spectroscopic methods make it possible to directly investigate some aspects of proton transfer reactions in which the proton displacement takes place by tunneling [15,16]. The observation of intramolecular coherent proton tunneling has allowed the detailed theoretical examination and qualitative numerical calculations for such compounds as porphyrine, namalonaldehyde, topolone, and so on [15]. Intermolecular proton tunneling occurs in a great number of systems; as a rule, researchers talk about incoherent tunneling, which implies the direct influence of the thermal bath [15,17].

There is no energy difference in the symmetric double-well potential, and this gives rise to the direct proton tunneling, however, any asymmetry of the double-well potential creates the energy difference between vibrational levels. In this case one can infer that the tunneling motion should incorporate the vibrating energy of surrounding atoms. Indeed, it seems reasonable that when the proton moves, surrounding atoms should adjust to new equilibrium positions and hence the potential energy, or potential energy surface will govern the proton motion; in the general case it should incorporate the interaction with surrounding atoms including the rest protons and outside fields [18,19]. Theoretical consideration conducted in Refs. 20–22 [see also Refs. 23 (review paper) and 24] showed that the tunneling motion could occur when the tunneling motion of a proton is associated with phonons.

In particular, recent results [25] obtained by quasi-elastic neutron scattering are consistent with the fast localized motion of hydrogen associated with phonons within the hexagons formed by interstitial $g(Hf_2Mo_2)$ sites. Besides, the researchers [25] has also revealed the slower jump process corresponding to hopping of hydrogen from one hexagon to another, which has been proven by nuclear magnetic resonance measurements. The ratio of the characteristic hopping rates for the two jump processes is $\sim 10^3$ at 300 K.

A very peculiar dynamics has been revealed in the $Ca(OH)_2$ crystal by means of inelastic neutron scattering technique [26]. It has been found that anharmonic terms must be included, which mix the vibrational states of the OH and lattice modes. In particularly, the lattice modes have successfully been represented as the superposition of oxygen and proton synchronous oscillators, and it appears that the proton bending mode E_u is strongly coupled to the lattice modes. The contribution of the proton harmonic wave functions has been taken as the zero-order approximation.

Thus it seems that the whole variety of hydrogen-containing compounds might be subdivided to three conditionally independent groups. The first group embraces compounds that feature coherent proton tunneling, the second one covers compositions in which incoherent tunneling occurs, and the third one is intermediate between the first and second groups.

Other problems are associated with the motion of an excess proton through the network of hydrogen bonds (e.g., ice) or along a hydrogen-bonded chain.

Such a motion in many aspects is very different from the discussed above. We can mention here an early review paper by Morowitz [27] on biological systems considered as proton semiconductors, and we can also mention experimental work [28] on the first evidence for high conduction of protons along the interface between water and a polar lipid monolayer. A number of hydrogen-containing solids, such as lithium hydrazium sulfate $Li(N_2H_5)SO_4$, tri-ammonium hydrogen disulfate $(NH_4)_3H(SO_4)_2$, and others, exhibit the protonic conductivity three orders of magnitude greater along the *c* axis than in the perpendicular directions [29,30] (this is because of very long hydrogen-bonded chains, which are parallel to *c* axis and spread along the whole crystal [31,32]). Hydrogen-bonded chains are also present in many biological objects, in proton-conducting membrane proteins in particular [1,33–37]. The titled crystals together with other ones—for example, $(NH_4)_3H(SeO_4)_2$ [38–42] and $NH_4IO_3{\cdot}2HIO_3$ [43–48]—belong to superionic conductors due to abnormal high proton conductivity.

Modern inelastic neutron scattering technique has made it possible to discover free protons in solids [49]; that is, free protons have been found in manganese dioxides, coals, graphite nitric acid intercalation compounds, polypyrolles and polyanilines, and β-alumina. Perhaps the said compounds may be called protonic conductors as well, though in the solids the density of free protons is very small and the distribution of proton kinetic momentum is hidden by the zero-point oscillations of the host matrix [49].

In liquid phase, hydrogen atoms (or hydrogen bonds) display other peculiarities. For instance, the hydrogen bond structure of liquid water and alcohols is exemplified by the existence of three different species, as it follows from the infrared overtone spectra [50]: $OH_{\text{nonbounded}}$, $OH_{\text{weak-cooperat.}}$, and $OH_{\text{strong-cooperat}}$. By increasing the temperature, the content of strong cooperative hydrogen bonds decreases and a similar number of weak cooperative hydrogen bonds increases. Thus the difference spectra indicate identically the necessity to differ between three types of hydrogen-bonded interacting OH groups. In the next sections, we touch some other subtle structural characteristics of water.

Based on the nuclear magnetic resonance results, Huyskens [51,52] (see also Ref. 53) has corrected the ratio of a definite number of hydrogen bonds for a given number of molecules, $r_{\text{f}} = (1 - N_{\text{H-bond}}/N_{\text{molecules}})$, deduced for alcohol by Luck [54–56]. Huyskens has shown that the hydrogen bonds perpetually jump from one partner to another; in other words, molecules are completely inserted in a chain, acting at the same time in hydrogen bonding as proton donor and proton acceptor.

The arrangement of hydrogen bonds in hydrogen-containing compounds, which takes place in condensed phases (liquid, liquid crystal, and solid) due to the phase transition, is often accompanied by the appearance of open associates with their following polymerization, or clustering as it happens in alkoxycyanobipheniles. A sophisticated theoretical analysis of the mechanism of dissociation of the cyclic dimmers accompanied by formation of opened chain-like associates and monomers has recently been proposed by Krasnoholovets and co-workers [57–62].

One more interesting aspect that we would like to mention has recently been emphasized by Gavrilov and Mukina [63]: Common to all protonated crystals with hydrogen bonds is the existence of distinguishing ranges of temperature, which are typical for crystals with weak hydrogen bonds (100 ± 10 K) and those with strong hydrogen bonds (210 ± 10, 260 ± 10, 380 ± 10 K).

In the next section we describe the most general approaches that are successfully employed at the theoretical study of the hydrogen behavior in compounds with hydrogen bonds. Then in the following sections we will show how some concrete problems are solved using methods developed in quantum mechanics, quantum field theory of solids, and statistical mechanics.

II. QUANTUM MECHANICAL DESCRIPTION OF HYDROGEN-BONDED SYSTEMS

A. Hydrogen Bond Vibrations

First consideration for the motion of the hydrogen atom in double-well potentials were proposed by Kovner and Chuenkov [64] and Hadzi [65]. However, the detailed analysis of the separability of the X—H stretching vibrations (with energy about 3000 cm^{-1}) from the hydrogen bond vibrations (with energy around 100 cm^{-1}) in the linear triatomic X—H $\cdots$ Y system was carried out by Marechal and Witkowski [66].

Their attempt [66], rather successful, was aimed to fully explain the experimental infrared spectra. The normal stretching coordinates representing the X—H motion and the X—H $\cdots$ Y motion were designated as q and Q, respectively (Fig. 1). The Hamiltonian describing such a system was written as

$$H(q,Q) = \frac{P^2}{2M} + \frac{p^2}{2m_{\text{H}}} + U(q,Q) = \frac{P^2}{2M} + H_1(q,Q) \tag{1}$$

Here P and p are the conjugate momenta of the coordinates Q and q, respectively; M and m_{H} are the masses of points described by the coordinates Q and q,

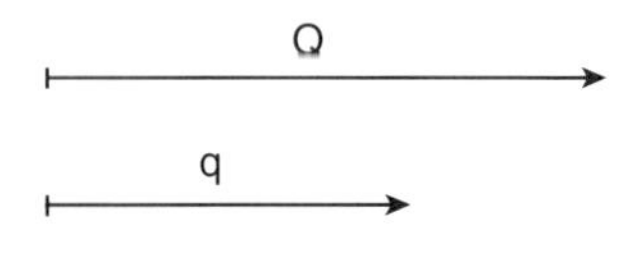

Figure 1. Q and q are normal coordinates.

respectively. $U(q, Q)$ is the potential energy for this motion. Using the fact that the oscillatory motion of the q coordinate is several tens faster than the oscillatory motion of the Q coordinate, the eigenfunctions Ψ of the Hamiltonian $H(q, Q)$ was expanded as

$$\Psi = \sum_n \alpha_n(Q)\phi_n(q, Q) \tag{2}$$

Thus $\phi_n(q, Q)$ is the wave function for the q motion when Q is fixed. The corresponding characteristic equation is

$$H\phi_n(q, Q) = \mathscr{E}_n(Q)\phi_n(q) \tag{3}$$

The wave function $\alpha_n(Q)$ for the Q motion is the characteristic of the equation

$$\begin{aligned} &\left[\frac{P^2}{2M} + \mathscr{E}_n(Q) + \langle\varphi_n(q, Q)|\frac{P^2}{2M}|\phi_n(q, Q)\rangle_q\right]\alpha_n(Q) + \sum_{n,n_1} \\ &\quad \times \left[\langle\phi_n(q, Q)|\mathbf{P}|\phi_{n_1}(q, Q)\rangle_q \frac{\mathbf{P}}{M} + \langle\phi_n(q, Q)|\frac{P^2}{2M}|\phi_{n_1}(q, Q)\rangle_q\right] \\ &\quad \times \alpha_n(Q) = E\alpha_n(Q) \end{aligned} \tag{4}$$

The function ϕ_n is reasonably approximated as a harmonic-oscillator function whose force constant k and equilibrium position q_0 depend on Q. Such an approximation signifies that the potential energy in expression (1) can be written as

$$U(q, Q) = \frac{1}{2}k(Q)[q - q_0(Q)]^2 + U'(Q) \tag{5}$$

Thus the frequency of the q motion is

$$\omega_{\text{H}}(Q) = \sqrt{k(Q)/m_{\text{H}}} \tag{6}$$

Here $\omega_{\text{H}}(Q)$ is equal to the X—H stretching mode when the hydrogen bonding is still not switched. This allows the calculation of the eigenvalues [66]

$$\begin{aligned} \mathscr{E}_n(Q) &= \langle\varphi_n(q, Q)|H_1|\phi_n(q, Q)\rangle_q \\ &= \left(n + \frac{1}{2}\right)\hbar\omega_{\text{H}}(Q) + \frac{1}{2}m_{\text{H}}\omega_{\text{H}}^2 q_0^2(Q) + U'(Q) \end{aligned} \tag{7}$$

$$\langle\varphi_n(q, Q)|\frac{P^2}{2M}|\phi_n(q, Q)\rangle_q = \left(n + \frac{1}{2}\right)\frac{\hbar^2}{M}\left(\frac{d\omega}{dQ}\right)^2(1 - 4\mu) \tag{8}$$

where

$$\lambda(Q) = -q_0(Q)[m_{\mathrm{H}}\omega_{\mathrm{H}}(Q_0)/2\hbar]^{1/2}$$
$$\mu(Q) = [\omega_{\mathrm{H}}(Q_0) - \omega_{\mathrm{H}}(Q)]/2[\omega_{\mathrm{H}}(Q_0) + \omega_{\mathrm{H}}(Q)] \tag{9}$$

The other matrix elements involved in Eq. (4) are obtained [66] by using a Lippinkott–Schröder potential [67]. Using expressions (7) and (8), the following Hamiltonian can be derived:

$$H^{(n)} = \frac{P^2}{2M} + \left(n + \frac{1}{2}\right)\left[\hbar\omega_{\mathrm{H}}(Q) + \frac{\hbar^2}{M}\left(\frac{d\lambda}{dQ}\right)^2(1 - 4\mu)\right] + \frac{1}{2}m_{\mathrm{H}}\omega_{\mathrm{H}}^2(Q)q_0^2(Q) + U'(Q) \tag{10}$$

The two main consequences follow from the Hamiltonian (10): (i) The difference between the Hamiltonians of the first excited state ($n = 1$) and the ground state ($n = 0$) for the q motion is

$$H^{(1)} - H^{(0)} \cong \text{const} + B_{\mathrm{H}}Q, \quad B_{\mathrm{H}} = \hbar\left(\frac{d\omega_{\mathrm{H}}}{dQ}\right)_{Q_0} \tag{11}$$

(ii) The isotope effect is characterized by the substitution $B_{\mathrm{D}} = B_{\mathrm{H}}/\sqrt{2}$.

Many carboxylic acids in gases and condensed states form dimers connected by a couple of hydrogen bonds (Fig. 2). The said acids are often treated as model systems for the study of spectra of hydrogen bonds [66], intermolecular

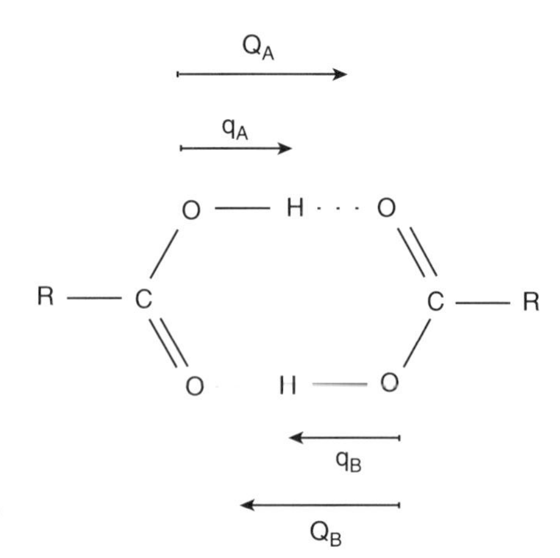

Figure 2. The cyclic planar carboxylic acid dimer.

hydrogen bonding and proton reactions [20–23], and rearrangement of hydrogen bonds [57–60].

The total Hamiltonian of a cyclic dimmer can be written as

$$H_{\text{total}} = H_A + H_B + H_{\text{int}} \tag{12}$$

where $H_{A(B)}$ is the Hamiltonian (1). The individual wave functions for the O—H stretching modes of bond A and B are taken as the basis wave functions. In the basis state the appropriate wave function is

$$\phi_0^l = \beta_0^l(Q_A, Q_B)\phi_0(q_A, Q_A)\phi_0(q_B, Q_B) \tag{13}$$

The corresponding effective Hamiltonian is

$$H_0 = \frac{P_A^2}{2M} + \frac{P_B^2}{2M} + \mathscr{E}_0(Q_A) + \mathscr{E}_0(Q_B) \tag{14}$$

The wave functions that specify a single excitation in the group O—H are

$$\begin{aligned}\phi_1^l = {} & \beta_1^l(Q_A, Q_B) \quad \phi_1(q_A, Q_A) \quad \phi_0(q_B, Q_B) \\ & + \beta_1^{l'}(Q_A, Q_B) \quad \phi_0(q_A, Q_A) \quad \phi_1(q_B, Q_B)\end{aligned} \tag{15}$$

Functions (15) split the effective Hamiltonian to

$$H^{\pm} = \frac{P_A^2}{2M} + \frac{P_B^2}{2M} + \mathscr{E}_1(Q_A) + \mathscr{E}_2(Q_B) \pm H_{\text{int}}(Q_A, Q_B)C_2 \tag{16}$$

where C_2 is the symmetry operator.

The transition probability from the l'th level of H_0 is defined as

$$D_l^{l'(\pm)} = \text{const}|\langle \beta_0^{l'}(Q_A, Q_B)|1 \mp C_2|\beta_l^{\pm}(Q_A, Q_B)\rangle_{Q_A+Q_B, Q_A-Q_B}|^2 \tag{17}$$

$D_l^{l'(\pm)}$ was computed [66] by a variational method; in the first approximation

$$D_{u,v}^{u',v'(\pm)} = \Gamma_{u,u'}^2 \left(\frac{B_{\text{H}}}{\sqrt{2M\hbar\omega_{\text{OO}}}} \right) \cdot (C_v^{v'(\pm)})^2 \tag{18}$$

Here l stands for the two indices u and v and $\Gamma_{u,u'}$ is the Franck–Condon factor. The calculated transition probability (18) was very successfully applied for the description of the experimental spectra for CD_3COOH and CD_3COOD [66].

The dependence of stretching vibrations of X—H on the length of the hydrogen bond is studied very well both theoretically and experimentally (see, e.g., Refs. 68–70). However, the reasons of the bending vibrations of hydrogen bonds still remained unclosed so far. In Section II.E we will discuss a possible mechanism of such a behavior of the hydrogen bond.

B. Proton Transfer Incorporating Acoustic Phonons

The interconversion of the two tautomer forms of dimers (Fig. 2) by a concerted two proton transfer is governed by a double-well potential. In condensed phases the two possible tautomers are identical. Skinner and Trommsdorff [21] treated a model of crystalline benzoic acids in which the dynamics of each proton pair is uncorrelated with other pairs. Their model was based on a single double-minimum potential coupled to a thermal bath—that is, crystal vibrations. It was believed that in a condensed phase the crystal field breaks the symmetry of the two wells (Fig. 3). If Fig. 3 indeed represents the real situation when the proton transfer from the left well to the right one should take place at the participation of vibrations of the crystal; that is, phonons of some sort should activate the proton transfer. We would like to emphasize that this is the conventional viewpoint, which is widely employed by physicists.

For instance, a new method for dealing with phonon modes in path integrals was proposed by Sethna [71,72]. Using the path integrals and an instanton

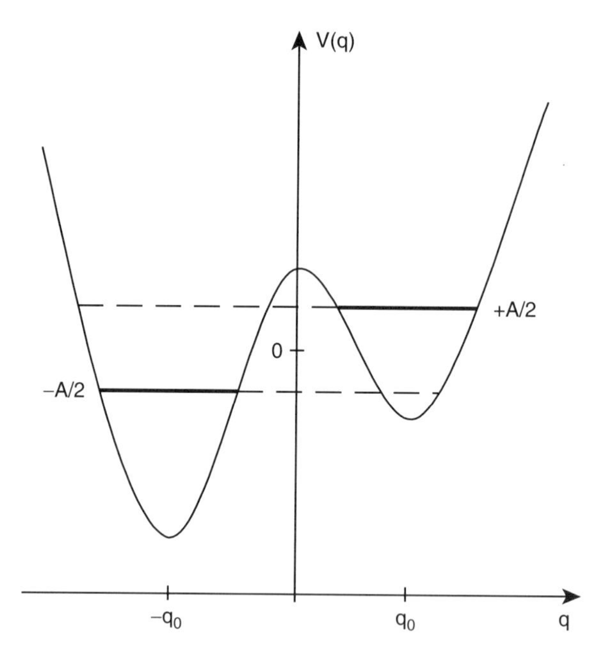

Figure 3. Asymmetric double-well potential $v(q)$. The minima are located at $\pm q_0$; the energy of the localized states $|\beta\rangle$ and $|\alpha\rangle$ are, respectively, $\pm A/2$.

calculation, he could construct a detailed theory of the tunneling event in the presence of phonons. The influence of dissipation on quantum coherence of a quantum mechanical particle that moves in a symmetric double-well potential and interacts with the environment by a phenomenological friction coefficient was considered by Bray and Moore [73]. Those methods are very interesting because they allow the quantitative calculation of rate constant which characterizes the process of tunneling in a double-well potential.

Using the mentioned approach [71–73], Skinner and Trommsdorff [21] started from the Hamiltonian

$$H = H_0 + H_{\rm ph} + H' \tag{19}$$

Here

$$H_0 = T_q + v(q) \tag{20}$$

is the Hamiltonian for the proton coordinate q, and T_q and $v(q)$ are the kinetic energy and potential energy, respectively;

$$H_{\rm ph} = T_Q + V(Q) \tag{21}$$

is the Hamiltonian of the crystalline coordinate Q, and T_Q and $V(Q)$ are the kinetic energy and potential energy, respectively;

$$H' = V(q, Q) \tag{22}$$

is the interaction potential energy.

The potential energy for proton coordinate written in the linear approximation is

$$v(q) = v_0(q) + \tilde{A}q \tag{23}$$

where $v_0(q)$ is a symmetric potential with minima at $\pm q_0$. For the symmetric case when $A = 0$, there are two lowest states $|\psi_1^0\rangle$ and $|\psi_2^0\rangle$ with an energy difference $E_2^0 - E_1^0 = J > 0$, which is also called the tunneling splitting. Setting $E_1^0 = -J/2$, one can choose new states $|\alpha\rangle = \frac{1}{2}(|\psi_1^0\rangle - |\psi_2^0\rangle)$ and $|\beta\rangle = \frac{1}{2}(|\psi_1^0\rangle + |\psi_2^0\rangle)$. The wave functions $\psi_\alpha(q) = \langle q \mid \alpha\rangle$ and $\psi_\beta(q) = \langle q \mid \beta\rangle$ are localized in the left and right wells, respectively, besides $\langle\alpha|q|\beta\rangle = 0$.

In such a presentation

$$H_0 = \left\| \begin{matrix} -A/2 & -J/2 \\ -J/2 & A/2 \end{matrix} \right\| \tag{24}$$

where $A \cong 2\tilde{A}q_0$ is the energy asymmetry between the localized levels (Fig. 3). The eigenvalues corresponding to the Hamiltonian (24) are

$$E_1 = -\frac{1}{2}\sqrt{A^2 + J^2}, \qquad E_2 = -E_1 \tag{25}$$

In the framework of the model, the difference $E_2 - E_1$ is within the acoustic phonon bandwidth, and that is why Skinner and Trommsdorff [21] reasonably assumed that the longitudinal acoustic modes could influence the tunneling proton transfer. The interaction potential (22) was chosen in the form of the deformation potential approximation (see Ref. 74)

$$\begin{aligned} H' &\cong \frac{1}{2}a\delta\rho(\mathbf{R}) \\ \delta\rho(\mathbf{R}) &= -\hbar\sum_{\mathbf{k}}\sqrt{\frac{\omega_{\mathbf{k}}\omega_{\mathrm{D}}}{N}}(\hat{b}_{\mathbf{k}}e^{i\mathbf{kR}} + \hat{b}^{+}_{\mathbf{k}}e^{-i\mathbf{kR}}) \end{aligned} \tag{26}$$

Here a is a dimensionless constant, $\delta\rho(\mathbf{R})$ is the density fluctuation of the medium at the position $\mathbf{R}$ (the center of symmetry of the benzoic acid dimer), ω_{D} is the Debye frequency, and N is the number of acoustic modes, $\omega_{\mathbf{k}} = v_{\mathrm{sound}}|\mathbf{k}|$, $\hat{b}^{+}_{\mathbf{k}}(\hat{b}_{\mathbf{k}})$ is the Bose operator of creation (annihilation of a acoustic phonon with the wave vector $\mathbf{k}$). In the localized representation we have

$$H_0 = \begin{Vmatrix} -\delta A/2 & 0 \\ 0 & \delta A/2 \end{Vmatrix} \tag{27}$$

where $\delta A = a\delta\rho(\mathbf{R})$.

The second-order perturbation theory makes it possible to calculate the rate constant for the transition from the eigenstate $|\psi_m\rangle$ to $|\psi_l\rangle$ [21]:

$$k_{lm} = \frac{1}{\hbar^2}\int_{-\infty}^{\infty}\langle H'^{+}_{lm}(t)H'_{lm}\rangle e^{i\omega_{ml}t}\,dt \tag{28}$$

where

$$\begin{aligned} &\omega_{ml} = (E_m - E_l)/\hbar, && H'(t) = e^{iH_{\mathrm{ph}}t/\hbar}H'e^{-iH_{\mathrm{ph}}t/\hbar} \\ &\langle\ldots\rangle = \frac{\mathrm{Tr}\{e^{-H_{\mathrm{ph}}/k_{\mathrm{B}}T}\ldots\}}{\mathrm{Tr}e^{-H_{\mathrm{ph}}/k_{\mathrm{B}}T}}, && H_{\mathrm{ph}} = \sum_k \hbar\omega_k\left(\hat{b}^{+}_k\hat{b}_k + \frac{1}{2}\right) \end{aligned} \tag{29}$$

For the chosen model ($l, m = 1, 2$), Skinner and Trommsdorff [21] calculated the rate constant

$$k_{21} = k_{21}^0 \cdot (n(\omega_{21}) + 1) \tag{30}$$

$$k_{21}^0 = k\omega_{21}, \qquad k = \frac{3\pi}{2} a^2 \left(\frac{J}{\hbar\omega_D} \right)^2, \qquad n(\omega) = \frac{1}{e^{\hbar\omega/k_B T} - 1} \tag{31}$$

Rate constants k_{21} and k_{12} satisfy the detailed balance condition

$$k_{21}/k_{12} = e^{-\hbar\omega_{21}/k_B T} \tag{32}$$

The fundamental relaxation rate is

$$\frac{1}{\tau} = k_{12} + k_{11} = k_{21}^0 \coth \frac{\hbar\omega_{21}}{2k_B T} \tag{33}$$

In Ref. 21 it was also considered the model of two interacting double wells coupled to a thermal bath. The problem was simplified by the consideration of two dimers separated by the dye molecule when the distance between the dimmers was about 1 nm. In this case the interaction Hamiltonian was approximated by

$$H' = \delta A \cdot (|\phi_4\rangle\langle\phi_4| - |\phi_1\rangle\langle\phi_1|) \tag{34}$$

where $|\phi_i\rangle$ ($i = 1, \ldots, 4$) are the basis states for the Hamiltonian H_0, which is the four-row matrix that includes parameters A and J determined above and a new one, $B \propto 2v_0 q_0^2$. In the case when the inequality $J \ll |A \pm B|$ is not satisfied, H_0 can be diagonalized exactly [21]:

$$H_0 = \begin{Vmatrix} -A & -J/\sqrt{2} & 0 & 0 \\ -J/2 & -B & -J/\sqrt{2} & 0 \\ 0 & -J/\sqrt{2} & A & 0 \\ 0 & 0 & 0 & -B \end{Vmatrix} \tag{35}$$

The calculation of the rate constant for the model described by the Hamiltonians (35) and (34) results in [21]

$$k_{ij} = k_{ij}^0 \cdot (n(\omega_{ij}) + 1) \tag{36}$$

$$k_{ij} = k_{ij}^0 n(\omega_{ij}), \; k_{ij}^0 = k\omega_{ij} \left(\frac{2\hbar\omega_{ij}}{J} \right)^2 [M_{3i}M_{3j} - M_{1i}M_{1j}]^2 \tag{37}$$

where M_{ij} are matrix elements for the eigenstates $|\psi_j\rangle$, which are functions of $|\phi_j\rangle$.

The rate constant (36) and (37) as a function of temperature correlated well with the experimental data obtained for the carboxylic acid protons of crystalline perprotobenzoic acid and ring-deuterobenzoic acids by nuclear magnetic resonance T_1 [75] and inelastic neutron scattering (for an analysis of the experiment see Refs. 76 and 77). It should be noted that some of the major parameters of the model (for instance, J) allowed the direct determination by fluorescence line narrowing technique.

The single-phonon absorption/emission is dominant at the low temperature limit when $A \gg k_B T$ and $a \gg 1$ (a is the dimensionless constant). When temperature increases, new factors begin to influence the proton transfer, namely, multiphoton processes and the tunneling through intermediate excited states.

In particular, in Ref. 78 the proton transfer dynamics of hydrogen bonds in the vicinity of the guest molecules has been studied by using a field-cyclic nuclear magnetic relaxometry (spin-lattice relaxometry). It has been revealed that the spin lattice relaxation incorporates the two major members:

$$T_1^{-1}(\omega) = T_1^{-1}(\omega)|_{\mathrm{p}} + T_1^{-1}(\omega)|_{\mathrm{g}} \tag{38}$$

The first member on the right in expression (38) is the relaxation rate caused by dimmers of the pure material. The second member arises from dynamics of the benzoic acid dimers contained within the range of the thiondigo quest molecule; $T_1^{-1}(\omega)|_g$ is stipulated by modulation of proton dipolar interaction induced by the proton transfer in the hydrogen bonds.

A very new methodology called the perturbative instanton approach has recently been developed by Benderskii et al. [79–84] for the description of tunneling spilling, which manifests itself in vibrational spectra of hydrogen containing nonrigid molecules. The mentioned studies allow the detailed consideration of resonances in tunneling splittings at the transverse frequency Ω close to the energy difference D between potential wells. Taking into account the asymmetry of the double-well potential, they chose the potential in the form (for simplicity we treat here only one-dimensional case) [82]

$$V = \frac{m\Omega^2 a_0^2}{8}(1 - X^2)^2 + \frac{3D}{4}X\left(1 - \frac{X^2}{3}\right) \tag{39}$$

The dimensionless asymmetry D has been assumed to be of the order of the semiclassical parameter γ^{-1} where

$$\gamma = \frac{m\Omega a_0^2}{2\hbar} \tag{40}$$

The corresponding Schrödinger equation is

$$\frac{d^2\psi}{dX^2} + \left[4\gamma\left(\varepsilon - \frac{3}{2}aX\left(1 - \frac{X^2}{3}\right)\right) - \gamma^2(1 - X^2)^2\right]\psi = 0 \tag{41}$$

which includes the dimensionless parameters

$$\varepsilon = \frac{E}{\hbar\Omega}, \qquad a = \frac{D}{2\hbar\Omega} \tag{42}$$

The solutions to Eq. (41) are [83]

$$\psi_0^L = D_{\nu_L}(-2\sqrt{\omega_L\gamma}(1 + X)), \quad \psi_0^R = D_{\nu_R}(-2\sqrt{\omega_R\gamma}(1 - X)) \tag{43}$$

where

$$\omega_{L,D} = \left(1 \pm \frac{3a}{2\gamma}\right)^{1/2}, \qquad \varepsilon + a = \omega_L\left(\nu_L + \frac{1}{2}\right), \qquad \varepsilon - a = \omega_R\left(\nu_R + \frac{1}{2}\right) \tag{44}$$

and $D_\nu(x)$ are parabolic cylinder functions. Then solutions (43) and (44) make it possible to obtain the equation for the instanton quantization [83], which, if the asymmetry is small ($a \ll 1$), is reduced to the rules

$$\nu_L = n + \chi_{nn}^-, \qquad \nu_R = n - 2a + \chi_{nn}^+ \tag{45}$$

Here n is the integer and

$$\chi_{nn}^{\pm} = a \pm \sqrt{a^2 + (\chi_{nn}^0)^2} \tag{46}$$

where the value

$$\chi_{nn}^0 = \frac{1}{\sqrt{2\pi}} \frac{2^{4n+2}\gamma^{n+1/2}}{n!} e^{-4/3\gamma} \tag{47}$$

written in the dimensionless energy units is the tunneling splitting in the symmetric potential. The appropriate spectrum of energy eigenvalues is

$$\varepsilon_n^{L(R)} = n + \frac{1}{2} \mp \sqrt{a^2 + (\chi_{nn}^0)^2} \tag{48}$$

The further complicated studies [83,84] have solved the quantum problem within the perturbative instanton approach generalized for excited states simulated above the barrier and for anharmonic transverse vibrations. The parameters of many-dimensional torsion–vibration Hamiltonians of some molecules (H_2O_2, malonaldehyde, etc.) have been derived. It has been shown that the torsion motion and bending vibrations are responsible for vibration-assisted tunneling and for significant dependence of the tunneling splittings on quantum numbers of transverse vibrations. The dependence of tunneling splittings on isotope effects of H/D and 13C/12C has also been calculated.

We have cited evidence supporting the proton transfer along hydrogen bonds; however, there are investigations carried out on other species, which bring forward an argument that the simple hydrogen-bonded dimer is favored over the proton transfer. For instance, the title of a review article by Legon [85] includes the following words: "Hydrogen bonding versus proton transfer." Based on so-called fast-mixing technique and pulsed-nozzle, Fourier-transform microwave spectroscopy (i.e., rotational spectroscopy of supersonic jets) Legon [85] has examined rotational spectra and spectroscopic constants of heterodimers in solid particles in gas mixtures in the series ($R_{3-n}H_nN \cdots HX$). Three possibilities have been analyzed, namely, the hydrogen-bonded form, a form with partial proton transfer, and a form with complete proton transfer. The conclusion has been drawn that the species $H_3N \cdots HX$ and $H_3P \cdots HX$ (where X = F, Cl, Br, and I) can all be described as the simple hydrogen-bonded type, without the need to invoke an appreciable extent of proton transfer (thus the double-well potential of the hydrogen bond is symmetric). And only the series $(CH_3)_3N \cdots HX$ (where X = F, Cl, Br, and I) has shown that the progressive weakening of the HX bond with respect to the dissociation products H^+ and X^- favors the ion pair.

C. Proton Polaron

In the case of the proton transfer, which takes place in the double-well potential, the potential energy (or potential energy surface) governs the proton motion and generally it incorporates the interaction with surrounding atoms including the rest protons and outside fields. Diffusion and mobility of hydrogen atoms and protons in compounds with hydrogen bonds allow the consideration very similar to the one described above. Indeed, since in compounds protons are characterized by localized wave functions, the motion of protons through the crystal should include the tunneling matrix element, or the tunneling (resonance) integral J and the possible interaction with the environment—that is, the thermal bath. By definition, the resonance integral is

$$J = \int \psi^*(\mathbf{r}_1) H \psi(\mathbf{r}_2)\, dV \tag{49}$$

where $\psi(\mathbf{r_i})$ is the proton wave function of a proton in the site described by the radius vector $\mathbf{r_i}$. The wave function $\psi(\mathbf{r_i})$ is compared to a ground state and H is the Hamiltonian, which includes the kinetic and potential energies of the proton.

The localized proton interacting with the corresponding ion (or atom) gives rise to its displacement from the equilibrium position, which in turn should lead to the lowering the potential energy of the atom. The problem is known as the displaced harmonic oscillator (see, e.g., Ref. 86). The classical function of Hamilton for the oscillator is

$$H = \frac{p^2}{2m} + \frac{m}{2}\omega^2 q^2 \tag{50}$$

The Schrödinger equation for the oscillator is

$$\left(-\frac{\hbar^2}{2m}\frac{d^2}{dq^2} + \frac{m}{2}\omega^2 q^2\right)\psi(q) = E\psi(q) \tag{51}$$

or in the dimensionless presentation

$$\frac{\hbar\omega}{2}\left(-\frac{d^2}{d\xi^2} + \xi^2\right)\psi(\xi) = E\psi(\xi) \tag{52}$$

where the dimensionless coordinate ξ is determined by the relation

$$q = \sqrt{\frac{\hbar}{m\omega}}\xi \tag{53}$$

One can introduce the Bose operators of creation (annihilation) of one oscillation, $\hat{b}^+(\hat{b})$:

$$\xi = \frac{1}{\sqrt{2}}(\hat{b} + \hat{b}^+), \qquad \frac{d}{d\xi} = \frac{1}{\sqrt{2}}(\hat{b} - \hat{b}^+) \tag{54}$$

and then the Schrödinger equation (52) is transformed to

$$\hbar\omega\,\hat{b}^+\hat{b}\psi = \left(E - \frac{1}{2}\right)\psi \tag{55}$$

and the operators satisfy the commutation relation

$$\hat{b}\hat{b}^+ - \hat{b}^+\hat{b} = 1 \tag{56}$$

Now let the oscillator is subjected to the action of an outside force $\sqrt{2}\gamma\hbar\omega$, which is not time-dependent. Then Eq. (52) becomes

$$\frac{\hbar\omega}{2}\left(-\frac{d^2}{d\xi^2}+\xi^2-\sqrt{2}\gamma\xi\right)\psi(\xi)=E\psi(\xi) \tag{57}$$

and in terms of operators $\hat{b}^+$ and $\hat{b}$, Schrödinger equation (55) changes to

$$[\hat{b}^+\hat{b}-\gamma(\hat{b}^+ +\hat{b})]\psi=\left[E/(\hbar\omega)-\frac{1}{2}\right]\psi \tag{58}$$

The solution of Eq. (58) is very simple. Equation (58) is satisfied by the operators

$$\hat{b}=\hat{\hat{b}}+\gamma, \qquad \hat{b}^+=\hat{\hat{b}}^+ +\gamma \tag{59}$$

and instead of Eq. (58) we acquire

$$\hat{\hat{b}}^+\hat{\hat{b}}\varphi=\left[E/(\hbar\omega)-\gamma^2-\frac{1}{2}\right]\varphi \tag{60}$$

where the new operators $\hat{\hat{b}}^+$ and $\hat{\hat{b}}$ obey the same commutation relation (56).

The displacement $-\gamma^2\hbar\omega$ of the potential energy of the oscillator directly indicates the possible strong interaction of a particle, which induced the displacement, with the vibrating lattice. First of all the charged particle should interact with polar modes—that is, optical phonons of the crystal studied. The detailed theory of the particle behavior resulted in the developing the small polaron theory for electrons was elaborated by Holstein [87] and Firsov [88]. The first theoretical research concerning the protonic small polaron was carried out later by Flynn and Stoneham [20], Fischer et al. [89], Roberts et al. [90], Klinger and Azizyan [91,92], and Tonks and Silver [93]. In the following sections we will widely use the small polaron model, applying it for the study of proton transport and rearrangements in systems with hydrogen bonds.

D. Proton Ordering

The necessity to analyze the phase transition in KH_2PO_4 crystals allowed Blinc [94–96] (see also Refs. 97 and 98) to construct the special pseudospin formalism describing the proton subsystem in many hydrogen-containing compounds.

So far we have dealt with the one-particle approximation, and the long-range proton interaction has not been taken into account. Experimental examination of ferroelectrics with hydrogen—that is, the KDP crystal (KH_2PO_4) and others (CsH_2AsO_4, $NH_4H_2AsO_4$, etc.)—showed that their physical properties are essentially caused by the proton subsytem (see, e.g., Blinc and Žecš [96]). The appearance of the soft mode in such crystals is associated with the parameter of proton cooperation, or proton ordering. The corresponding theoretical model is the following. Protons being strongly localized along hydrogen bonds undergo anharmonic vibrations near equilibrium positions in the potential with two wells. The energy of the proton subsystem is presented in the form [96,98]

$$H_{\text{proton}} = \sum_{l} H_1(l) + \sum_{\substack{j,l \\ (l>j)}} H_2(jl) + \cdots \tag{61}$$

Here $H_1(l)$ corresponds to the part of the energy that depends on the configuration of the lth proton, and $H_2(jl)$ is the part of the energy that depends on the pair configuration of the lth and the jth protons. It is assumed that protons tunnel between two equilibrium positions (hence the double-well potential should be symmetric). In the presentation of second quantization in which the one-particle Hamiltonian $H_1(l)$ is diagonalized, the total Hamiltonian is written as

$$H_{\text{proton}} = \sum_{\alpha} E_{\alpha} \hat{a}^{+}_{\alpha,l} \hat{a}_{\alpha,l} + \sum_{\alpha,\beta,\gamma,\delta} \mathrm{v}^{jl}_{\alpha\beta\gamma\delta} \hat{a}^{+}_{\alpha,l} \hat{a}_{\beta,l} \hat{a}^{+}_{\gamma,j} \hat{a}_{\delta,j} \tag{62}$$

where $\hat{a}^{+}_{\alpha,l}(\hat{a}_{\alpha,l})$ is the Fermi or Bose operator of the creation (annihilation) of a proton/deuteron in the lth position in the one-particle quantum state α where $\alpha = +, -$ (note that the ground state of a hydrogen atom in the potential with two minima is doublet).

The corresponding eigenfunctions (see Fig. 4)

$$\psi_{+} = \frac{1}{\sqrt{2}}(\varphi_L + \varphi_R) \tag{63}$$

$$\psi_{-} = \frac{1}{\sqrt{2}}(\varphi_L - \varphi_R) \tag{64}$$

are symmetric and asymmetric linear combinations of wave functions localized in the left (φ_L) and the right (φ_R) equilibrium positions. The condition of the presence of a hydrogen atom in the hydrogen bond is expressed as

$$\hat{a}^{+}_{+,l} \hat{a}_{+,l} + \hat{a}^{+}_{-,l} \hat{a}_{-,l} = 1 \tag{65}$$

The operators $\hat{a}^{+}_{\alpha,i}(\hat{a}_{\alpha,i})$ make it possible to introduce the so-called pseudospin formalism that features the effective "spin-1/2" operators

$$S^x_l = \frac{1}{2}(\hat{a}^{+}_{+,l} \hat{a}^{i}_{+,l} - \hat{a}^{+}_{-,l} \hat{a}_{-,l}) \tag{66}$$

$$S^y_l = \frac{1}{2}(\hat{a}^{+}_{+,l} \hat{a}_{-,l} - \hat{a}^{+}_{-,l} \hat{a}_{+,l}) \tag{67}$$

$$S^z_l = \frac{1}{2}(\hat{a}^{+}_{-,l} \hat{a}_{-,l} + \hat{a}^{+}_{+,l} \hat{a}_{+,l}) \tag{68}$$

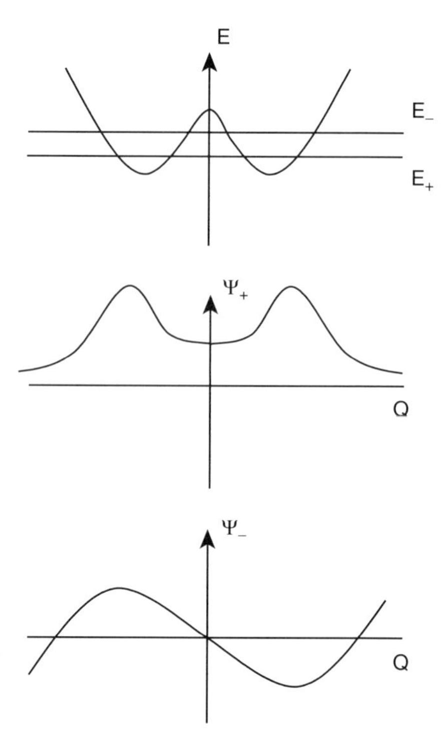

Figure 4. The basis state and proper functions for one-particle potential with two minima.

The pseudospin operators obey the known rules

$$[S_j^x, S_l^y]_- = i\delta_{jl}S_l^z;\ [S_j^y, S_l^z]_- = i\delta_{jl}S_l^x;\ [S_j^z, S_l^x]_- = i\delta_{jl}S_j^y \tag{69}$$

The operators of creation (annihilation) of a particle in symmetric quantum states, $\hat{a}_{+,l}^+(\hat{a}_{+,l})$, and in asymmetric ones, $\hat{a}_{-,l}^+(\hat{a}_{-,l})$, can be presented through the operators of creation and annihilation of a particle in the left and the right equilibrium positions in the hydrogen bond. Then one can express the pseudospin operators in the form

$$S_l^x = \frac{1}{2}(\hat{a}_{L,l}^+\hat{a}_{R,l} + \hat{a}_{R,l}^+\hat{a}_{L,l}) \tag{70}$$

$$S_l^y = \frac{1}{2}(\hat{a}_{L,l}^+\hat{a}_{R,l} - \hat{a}_{R,l}^+\hat{a}_{R,l}) \tag{71}$$

$$S_l^z = \frac{1}{2}(\hat{a}_{L,l}^+\hat{a}_{L,l} - \hat{a}_{R,l}^+\hat{a}_{R,l}) \tag{72}$$

It can be seen from expressions (70)–(72) that the z component of the pseudospin determines the operator of dipole momentum, the y component is the operator of tunneling current, and the x component is the operator of proton tunneling. In other words, S^z characterizes the difference between the populations of the left and the right equilibrium positions and S^x characterizes the difference between the populations of symmetric and asymmetric states.

Thus the Hamiltonian of the proton subsystem becomes

$$H_{\text{proton}} \cong -\hbar\Omega \sum_l S_l^x - \frac{1}{2} \sum_{j,l} \mathcal{J}_{jl} S_j^z S_l^z \tag{73}$$

where the tunneling integral and the energy of proton interaction are, respectively,

$$\hbar\Omega \cong E_- - E_+ \tag{74}$$

$$\mathcal{J}_{jl} = -4\nu_{++--}^{jl} \tag{75}$$

In the mentioned approximation we have

$$H_{\text{proton}} \cong -\hbar\Omega \sum_l S_l^x - \frac{1}{2} \sum_{j,l} \mathcal{J}_{jl} \langle S_j^z \rangle S_l^z \tag{76}$$

The parameter $\sum_l \mathcal{J}_{jl} \langle S_l^z \rangle$ contains information on the collective proton coupling, and the order parameter $\langle S_l^z \rangle$ points to the degree of proton ordering.

The pseudospin methodology is widely used not only for the description of hydrogen containing ferro- and antiferroelectrics, but also for the study of many other systems with hydrogen bonds. In particular, the pseudospin methodology was applied by Silbey and Trommsdorff [99] for examining the influence of two-phonon process on the rate constant of molecular compounds. In the next sections we will also employ the pseudospin formalism for the investigation of some problems where protons are exemplified by the cooperative behavior.

E. Bending Vibrations of Hydrogen Bond

Perhaps the first theoretical model describing the three-dimensional pattern that makes it possible to explain correlations between dependences of stretching and bending vibrations of hydrogen bonds has been constructed by Rozhkov et al. [100]. Their research was based on a work by Isaacs et al. [101], who showed that hydrogen forms the covalent bond with two oxygen atoms. In the framework of adiabatic approximation, Rozhkov et al. [100] chose the model potential of the oxygen–hydrogen interaction in the form of the Morse potential:

$$V(\mathbf{r}) = D(e^{-2a(r-r_0)} - 2e^{-a(r-r_0)}) \tag{77}$$

where D, a, and r_0 are parameters of the potential. In a recent work by Isaacs et al. [101], it has been shown that the hydrogen atom forms the covalent bond with two oxygen atoms. Because of the model for the construction of the potential of hydrogen bond U, one more oxygen atom spaced at R along the x axis (i.e. the O···O line) has been added. Therefore

$$U(\mathbf{r}) = V(\boldsymbol{\rho}, X + R/2) + V(\boldsymbol{\rho}, X - R/2) \tag{78}$$

where X is the coordinate of the proton reckoned from the center of the hydrogen bond and $\boldsymbol{\rho}$ is the radius vector of hydrogen in the yz plane; that is, $\boldsymbol{\rho} = (X, Y)$.

Let us rewrite expression (2) explicitly:

$$U(\mathbf{r}) = D[e^{-2a(r_- - r_0)} + e^{-2a(r_+ - r_0)} - 2e^{-a(r_- - r_0)} - 2e^{-a(r_+ - r_0)}] \tag{79}$$

where $r_\pm = \{(X \pm R/2)^2 + y^2 + z^2\}^{1/2}$. Coordinates of the minimum of the potential (3) are

$$X_0 = \pm\frac{1}{a}\operatorname{arccot}(e^{(R-R_c)/2}) \tag{80}$$

where $y_0 = z_0 = 0$ and $R_c = (2r_0 + a^{-1}\ln 2)$ is the minimum length of the hydrogen bond at which the double-well potential is transformed to the one-dimensional one (note that the coordinates are defined from the condition grad U).

Having explained experimental dependences of frequencies of the stretching and bending vibrations on the length of hydrogen bond, Rozhkov et al. [100] have proposed to include one more parameter $b \neq a$ to the potential (77). They employed the method developed by Tanaka [69], which allowed the passage from the dynamics of a proton in an isolated hydrogen bond to the thermodynamics of a system of hydrogen bonds. The new parameter b becomes responsible for the curvature of the potential (77) in transversal directions. In so doing, ar is replaced for $[a^2X^2 + b^2(y^2 + z^2)]^{1/2}$ in expression (77).

Using the perturbation theory developed in Ref. 102, one can choose the Hamiltonian of the three-dimensional oscillator with frequencies ω_ν and ω_δ (stretching and bending, respectively) and a new equilibrium position x, which are vibration parameters of the free energy. In this case the free energy per hydrogen can be represented as

$$F = \langle U_0 \rangle + F_0 - \langle H_0 \rangle_0 \tag{81}$$

$$F_0 = -k_B T \ln\left(2\sinh\frac{\hbar\omega_\nu}{k_B T}\right) - 2k_B T \ln\left(2\sinh\frac{\hbar\omega_\delta}{k_B T}\right) \tag{82}$$

where

$$\langle \ldots \rangle_0 = \frac{\mathrm{Tr}\{\ldots e^{-H_0/k_B T}\}}{\mathrm{Tr}\, e^{-H_0/k_B T}} \tag{83}$$

For the calculation of $\langle U_0 \rangle$, one can put $\mathbf{r} = \mathbf{r}_0 + \mathbf{u}$, where $\mathbf{r}_0$ is the equilibrium position and $\mathbf{u}$ is the displacement. In the harmonic approximation we obtain

$$\langle e^{i\mathbf{k}\mathbf{u}} \rangle_0 = e^{-\frac{1}{2}\sum_i \langle k_i \rangle^2 \langle u_i \rangle^2} \tag{84}$$

Then the average value of a function $f(\mathbf{r})$ is

$$\langle f(\mathbf{r}) \rangle_0 = \frac{\int d\xi f(\mathbf{r}_0 + \xi)\exp\left(-\frac{1}{2}\sum_i \xi_i^2/\sigma_i^2\right)}{(2\pi)^{3/2}\sigma_x\sigma_y\sigma_z} \tag{85}$$

where $\sigma_i^2 = \langle u_i \rangle_0^2$ and $i = x, y, z$.

Setting $\langle U(\mathbf{r}) \rangle_0 = 2DW$, we obtain the dimensionless average potential

$$\begin{aligned} W \cong\ & e^{-2\xi_0+2\lambda}\,\frac{(\xi^2 - \chi^2)\cosh 2\chi + 2\mu(\xi\cosh 2\chi - \chi\sinh\chi)}{(\xi + 2\mu)^2 - \chi^2} \\ & -2e^{-\xi_0+\lambda/2}\,\frac{(\xi^2 - \chi^2)\cosh 2\chi + \mu(\xi\cosh\chi - \chi\sinh\chi)}{(\xi + \mu)^2 - \chi^2} \end{aligned} \tag{86}$$

where $\xi = aR/2$, $\xi_0 = \xi - ar_0$, $\chi = ax$, $\lambda = a^2\sigma_\nu^2$, $\mu = b^2\sigma_\delta$, $\sigma_\nu \equiv \sigma_x$, and $\sigma_\delta = \sigma_y = \sigma_z$. In the harmonic approximation we obtain

$$\sigma_\nu^2 = \frac{\hbar\coth(\hbar\omega_\nu/k_B T)}{2m\omega_\nu} \tag{87}$$

$$\sigma_\delta^2 = \frac{\hbar\coth(\hbar\omega_\delta/k_B T)}{2m\omega_\nu} \tag{88}$$

where m is the proton/deuteron mass. Allowing for expressions (81)–(83) and (87) and (88), one can show that the minimization of the free energy, $\partial F/\partial\omega_{\nu,\delta}$, is reduced to the equations

$$\omega_\nu^2 = 2\Omega^2\, \partial W/\partial\lambda \tag{89}$$

$$\omega_\delta^2 = 2\Omega^2\, \partial W/\partial\mu \tag{90}$$

where

$$\Omega^2 = 2a^2 D/m \tag{91}$$

Equations (13) and (14) along with the equation for determining the localization of hydrogen in the hydrogen bond,

$$\partial W/\partial x = 0 \tag{92}$$

makes it possible to define the needed parameters.

Numerical calculations have been performed [100] using the simplex method [103]. Then calculated curves $\omega_\nu(R)$ and $\omega_\delta(R)$ and experimental dependences of $\omega_{\nu,\delta}$ as functions of R (see Ref. 70) have been superimposed for comparison. We would like to note that the agreement is indeed remarkable. The increase of the $\omega_\delta(R)$ observed with the shortened length of hydrogen bond has accounted for the increment of the rigidity of hydrogen bond in transverse directions. The strong fall of the curve $\omega_\nu(R)$, which takes place in the same region of R, is explained by the transformation of the potential (78) in the vicinity of the equilibrium position of the proton to the one-dimensional potential $U = a^4x^4D/4$ [104].

Thus the model of the three-dimensional hydrogen bond presented in Ref. 100 should be considered very realistic. The interaction between hydrogen bonds (or hydrogen-bonded chains), which is taken into account in the model, plays a key role in the behavior of the $\omega_\nu(R)$ and $\omega_\delta(R)$, especially in the range of small R that is typical for the strong hydrogen bond.

Another approach of O—H—O hydrogen bond dynamics is discussed in Appendix A; the approach also accurately reproduces the well-known dependence of the O—H stretching mode on the hydrogen bond length.

F. Tunneling Transition and Coupled Protons

Quantum effects and strong interactions with vibrating surrounding atoms complicate the detailed study of proton transfer in the hydrogen bond AH $\cdots$ B. Owing to the small mass, quantum tunneling of the proton plays an important role at a symmetric double-well potential.

The study of the time scale of proton transfer has shown [110] that the transfer is specified by the residence time in a given well or by the frequency of excursions to other well. Besides, the proton oscillates within the local potential minimum of a given well with the period of 10^{-15} to 10^{-13} s, which is much shorter than the residence time. The thermal motion of the molecules should modulate the potential over a period of 10^{-13} to 10^{-11} s. Fillaux et al. [110] note that the vibration assisted tunneling mechanism [21] has emerged from extensive studies (mostly of carboxylic acid dimers) with nuclear magnetic resonance and quasi-elastic neutron scattering [111–114], which probe a time scale of the order of 10^{-9} s. This scale is significantly longer than that of proton dynamics. This means that the two mentioned methods provide only the information averaged over many excursions of the proton between the two wells, and quantum effects are then not observed directly.

Vibrational spectroscopy measures atomic oscillations practically on the scale as the scale of proton dynamics, 10^{-15} to 10^{-12} s. Fillaux et al. [110] note that optical spectroscopies, infrared and Raman, have disadvantages for the study of proton transfer that preclude a complete characterization of the potential. (However, the infrared and Raman techniques are useful to observe temperature effects; inelastic neutron spectra are best observed at low temperature.) As mentioned in Ref. 110, the main difficulties arise from the non-specific sensitivity for proton vibrations and the lack of a rigorous theoretical framework for the interpretation of the observed intensities.

Inelastic neutron scattering spectroscopy is characterized by completely different intensities because the neutron scattering process is entirely attributable to nuclear interactions [110]: Each atom features its nuclear cross section, which is independent of its chemical bonding. Then the intensity for any transition is simply related to the atomic displacements scaled by scattering cross sections. And because the cross section of the proton is about one order of magnitude greater than that for any other atom, the method is able to record details of quantum dynamics of proton transfer.

As a rule, vibration dynamics of atoms and molecules is treated in the framework of harmonic force fields. In the model, eigenvalues are the normal frequencies, and eigenvectors determine displacements of atoms for each normal mode (see, e.g. Ref. 115) and the eigenvectors are related to the band intensities. However, transition moment operators are largely unknown for optical spectra, and therefore theoretical descriptions are remaining questionable. The inelastic neutron scattering technique allows the calculation of band intensities, because they are proportional to the mean-square amplitudes of atomic displacements scaled by nuclear cross sections [116–118]. This specified force fields and makes it possible to extract modes involving large proton displacements while contributions from other atoms can be ignored.

The inelastic neutron scattering spectra of potassium hydrogen carbonate ($KHCO_3$), some molecular crystals, and polymers have shown [8–11,110] that these compounds can be regarded as crystals of protons so weakly coupled to surrounding atoms that the framework of the atoms and ions can be virtually ignored. This allows one to support Fillaux's [119] quasi-symmetric double minimum potential along the proton stretching mode coordinate. In this case, proton transfer is associated with the pure tunneling transition, and the "phonon assistance" of proton tunneling is unnecessary. Thus proton tunneling is purely a local dynamics.

What is the reason for decoupling the proton dynamics from the crystal lattice revealed in a great number of compounds? Fillaux [119] has investigated the reason for this decoupling, supposing that the proton dynamics could be

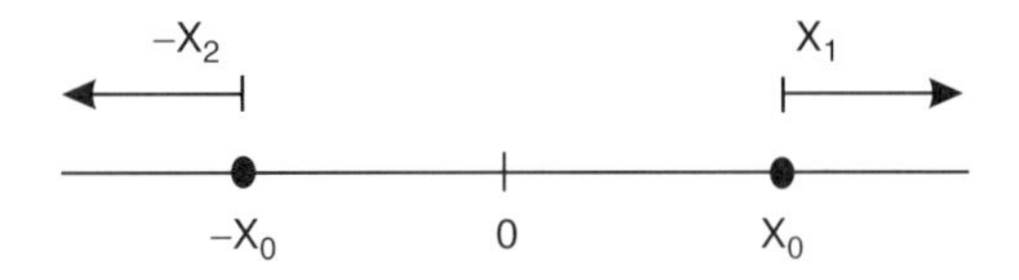

Figure 5. Two identical harmonic oscillators.

caused by the spin correlation for indistinguishable fermions according to the Pauli exclusion principle. He has considered the dynamics of a centrosymmetric pair of protons, assuming that the protons can be represented with identical harmonic oscillators with mass m moving along collinear coordinates x_1 and x_2 and coupled to one another (Fig. 5). The Hamiltonian was chosen in the form

$$H = \frac{1}{2m}(P_1^2 + P_2^2) + \frac{1}{2}m\omega_{0x}^2 \times [(x_1 - X_0)^2 + (x_2 + X_0)^2 + 2\gamma_x(x_1 - x_2)^2] \tag{93}$$

Here P_1 and P_2 are the momenta of the two oscillating masses; the harmonic frequency of the uncoupled oscillators at equilibrium positions $\pm X_0$ is ω_{0x}. The coupling potential proportional to γ_x depends only on the distance between equilibrium positions of the two oscillating masses at $\pm X_0' = X_0/(1 + 4\gamma_x)$.

A system of coupled oscillators can be expressed as linear oscillations of normal modes. The symmetric (x_s) and asymmetric (x_a) displacements of the two particles represent normal coordinates. The coordinates are defined with accuracy to an arbitrary factor, but frequencies and wave functions are independent of it. Normalized normal coordinates corresponding to an effective mass m are the following:

$$x_a = \frac{1}{\sqrt{2}}(x_1 + x_2), \qquad x_s = \frac{1}{\sqrt{2}}(x_1 - x_2) \tag{94}$$

$$P_a = \frac{1}{\sqrt{2}}(P_1 + P_2), \qquad P_s = \frac{1}{\sqrt{2}}(P_1 - P_2) \tag{95}$$

Substituting expressions (94) and (95) into the Hamiltonian (93), we get the Hamiltonian that includes separated normal modes, that is,

$$H = H_a + H_s + m\omega_{0x}^2 X_0^2 \frac{4\gamma_x}{1 + 4\gamma_x} \tag{96}$$

where

$$H_a = \frac{P_a^2}{2m} + \frac{1}{2}m\omega_a^2 x_a^2 \tag{97}$$

$$H_s = \frac{P_s^2}{2m} + \frac{1}{2}m\omega_s^2(x_s - \sqrt{2}X_0')^2 \tag{98}$$

Here, in expressions (97) and (98), the normal frequencies are

$$\omega_a = \omega_{0x}, \qquad \omega_s = \omega_{0x}\sqrt{1+4\gamma_x} \tag{99}$$

The corresponding eigenfunctions and eigenvalues are (see also Ref. 120)

$$\psi_{\rm ln} = \psi_l(x_a)\cdot\psi_{sn}(x_s - \sqrt{2}X_0') \tag{100}$$

$$E_{\rm ln} = \left(al + \frac{1}{2}\right)\hbar\omega_a + \left(sn + \frac{1}{2}\right)\hbar\omega_s \tag{101}$$

Spins of the two protons are correlated in a singlet ($s = 0$) and a triplet ($s = 1$) state. The spatial part of the proton wave function, with respect to particle permutation, is symmetrical in the singlet state and antisymmetrical in the triplet state. Explicitly, the wave functions are

$$\Theta_{0\pm}(x_1, x_2) = \frac{1}{\sqrt{2}}\psi_{s0}(x_a)[\psi_{s0}(x_s - \sqrt{2}X_0') \pm \psi_{a0}(x_s + \sqrt{2}X_0')] \tag{102}$$

If we drop the spin interaction and the tunneling resonance integral for two protons in a dimer, the energy of splitting for the singlet and triplet states become ignored. Hence the ground state remains degenerate similarly to the bosons.

If the proton oscillations are not coupled to the dimer oscillations, the dynamics can be modeled by symmetric and antisymmetric normal coordinates for proton modes (x_a, x_s) and for dimer modes (X_a, X_s). Then the total wave function is

$$\Xi_0(x_1, x_2;\ X_1, X_2) = \psi_{a0}(x_a)\psi_{s0}(x_s - \sqrt{2}X_0')\varphi_{a0}(x_a)\varphi_{s0}(x_s - \sqrt{2}X_0') \tag{103}$$

If the nuclear spins are taken into account, the total vibrational wave function according to the Pauli principle becomes similar to function (103):

$$\begin{aligned}\Xi_{0\pm}(x_1, x_2;\ X_1, X_2) = \frac{1}{\sqrt{2}}&\psi_{a0}(x_a)[\psi_{s0}(x_s - \sqrt{2}X_0') \\ &\pm \psi_{s0}(x_s - \sqrt{2}X_0')]\varphi_{a0}(x_a)\varphi_{s0}(x_s - \sqrt{2}X_0'')\end{aligned} \tag{104}$$

In the case when the proton and dimer modes are coupled, the normal coordinates become linear combinations of the proton and dimer's atoms coordinates. Fermions and bosons are not distinguished, and we have a conflict with the Pauli principle: The two protons are not distinguished in the ground state.

Thus we can consider the two types of dynamics. The first one is usual, which is met in vibrational spectroscopy; in this case, normal coordinates represent the mixture of fermion and boson displacements. The second one is based on the Pauli principle, namely, that there is no mixing of the normal coordinates for fermions and bosons and therefore the total wave function is factored:

$$\psi(x_i; X_i) = \psi(x_i)\Phi(X_i) \tag{105}$$

Thus Fillaux [119], assuming that the spin interaction is extremely weak, has argued that the factorization of the wave function must be regarded as a spin-related selection rule. The rule applies to pairs of coupling terms and to the mean distance between their equilibrium positions. This is treated as the fundamental justification of the localized proton modes introduced empirically to elucidate his previous observations [7–9].

The excited states, however, have no degeneracy, and that is why in this case the dynamics can be represented by conventional normal modes that can include the total number of atomic coordinates. The normal coordinates may be different in the ground and excited states; and because of that, the calculation of the scattering function can be more complicated with the analysis based on the normal mode approach.

The spin-related section rule can be proved by elastic neutron scattering measurements. In order to establish the specific fingerprint of the spin correlation, the scattering functions for the linear harmonic oscillator, for the double-well minimum function, and for pairs of coupled oscillators have been calculated in Ref. 119.

Considering the harmonic potential, Fillaux chooses the wave functions for an isolated harmonic oscillator along the x coordinate in the form [120]

$$\psi_n(x) = \left(\frac{a_x^2}{\pi}\right)^{1/4} \frac{1}{\sqrt{2^n n!}} H_n(a_x x) e^{-\frac{1}{2}a_x^2 x^2} \tag{106}$$

with

$$a_x^2 = \frac{m\omega_{0x}}{\hbar} = \frac{1}{2u_{0x}^2} \tag{107}$$

H_n is the Hermite polynomial of degree n, m is the oscillator mass, and u_{0x}^2 is the mean-square amplitude in the ground state. The scattering process, which is very important for the studies by the neutron techniques, is characterized by the scattering function

$$S(\mathbf{Q}, \omega) = |\langle \psi_{\text{fin}}(\mathbf{r})|\exp(i\mathbf{Q}\mathbf{r})|\psi_{\text{in}}(\mathbf{r})\rangle|^2 \delta(E_{\text{in,fin}} - \hbar\omega) \tag{108}$$

where $\psi_{\text{in}}(\mathbf{r})$ and $\psi_{\text{fin}}(\mathbf{r})$ are the proton wave functions in the initial and final states respectively, $E_{\text{in,fin}}$ is the energy of the transition, and $\hbar\omega$ is the neutron energy transfer.

In the case of the harmonic oscillator, the elastic scattering function has the form

$$S(\mathbf{Q}, \omega) = \exp(-Q_x^2, u_{0x}^2)\delta(\omega) \tag{109}$$

A single particle in a symmetrical two-well potential is presented by the symmetric (0^+) and antisymmetric (0^-) substates. The difference between two corresponding energies $E_{0^-} - E_{0^+} = \hbar\omega_0$ represents the tunneling splitting. In the case of the high potential barrier, the corresponding wave functions are symmetrical and antisymmetrical combinations of the harmonic wave functions (106) centered at the minima of two wells, $\pm X_0$:

$$\psi_{0^\pm}(x) = \frac{1}{\sqrt{2[1 \pm \exp(-a_x^2 X_0^2)]}}[\psi_0(x - X_0) \pm \psi_0(x + X_0)] \tag{110}$$

The appropriate scattering functions, which characterize symmetric and antisymmetric combinations of harmonic wave functions, are [119]

$$S(Q_x, \omega)_{0^+0^+} = \frac{1}{1 + \exp(-X_0^2/2u_{0x}^2)} \times \left| \left[\cos(Q_x X_0) + \exp\left(-\frac{X_0^2}{2u_{0x}^2}\right)\right] \exp\left(-\frac{Q_x^2 u_{0x}^2}{2}\right) \right|^2 \delta(\omega) \tag{111}$$

$$S(Q_x, \omega)_{0^\mp 0^\mp} = \frac{1}{1 - \exp(-X_0^2/2u_{0x}^2)} \times \left| i \sin(Q_x X_0) \exp\left(-\frac{Q_x^2 u_{0x}^2}{2}\right) \right|^2 \delta(\omega_0 \pm \omega) \tag{112}$$

$$S(Q_x, \omega)_{0^-0^-} = \frac{1}{1 - \exp(-X_0^2/2u_{0x}^2)} \times \left| \left[\cos(Q_x X_0) - \exp\left(-\frac{X_0^2}{2u_{0x}^2}\right)\right] \exp\left(-\frac{Q_x^2 u_{0x}^2}{2}\right) \right|^2 \delta(\omega) \tag{113}$$

Here, the periodic terms are caused by interferences between waves scattered by the same proton at the two sites, and the exponent terms are stipulated by the overlapping the wave functions centered at $\pm X_0$.

Functions $S_{0^+0^+}$ and $S_{0^-0^-}$ are analogous to the optical fringes; functions $S_{0^+0^-}$ and $S_{0^-0^+}$ are associated with the tunneling transition at energy $\hbar\omega_{0^-}$.

The elastic incoherent scattering functions defined in expression (100) yields [119]

$$S(Q_x, \omega) = 2\exp\left[-Q_x^2\left(\frac{u_{0x}^2}{2\sqrt{1+4\gamma_x}} + \frac{u_{0x}^2}{2}\right)\right]\delta(\omega) \tag{114}$$

It is easily seen that the Gaussian profile in expression (114) is practically the same as for the harmonic oscillator (109). Consequently, isolated oscillators and coupled pairs of bosons are virtually are not distinguished.

Now let us look at the coupled fermions. Fillaux considers, as an example, the hydrogen molecule and argues that a similar pattern should take place for the oscillators in $KHCO_3$. The vibrational ground-state wave functions are

$$\Theta_{0^\pm}(x_1, x_2) = \frac{1}{\sqrt{2}}[\psi_{s0}(x_1 - x_2 - R) \pm \psi_{s0}(x_1 - x_2 + R)] \tag{115}$$

where R_0 is the bond length. The part of the scattering function depending on the spatial coordinates is equal to

$$S(Q_x, \omega) = |\langle\Theta_{0^\pm}(x_1, x_2)|e^{iQ_x x_1} \pm e^{iQ_x x_2}|\Theta_{0^\pm}(x_1, x_2)\rangle|^2\delta(\omega) \tag{116}$$

Thus in the case of the coupled fermions when the singlet and tripled states belong to the same molecular species, the corresponding scattering functions are

$$\begin{aligned} S(Q_x, \omega)_{0^+0^+} &= 2\cos^2(Q_x X_0')\left[\cos(Q_x X_0') + \exp\left(-\frac{X_0'^2}{2u_{0x}^2}\right)\right]^2 \\ &\quad \times \exp\left[-Q_x^2\left(\frac{u_{0x}^2}{2\sqrt{1+4\gamma_x}} + \frac{u_{0x}^2}{2}\right)\right]\delta(\omega) \end{aligned} \tag{117}$$

$$S(Q_x, \omega)_{0^\pm 0^\mp} = 2\sin^4(Q_x X_0')^2 \times \exp\left[-Q_x^2\left(\frac{u_{0x}^2}{2\sqrt{1+4\gamma_x}} + \frac{u_{0x}^2}{2}\right)\right]\delta(\omega) \tag{118}$$

$$\begin{aligned} S(Q_x, \omega)_{0^-0^-} &= 2\cos^2(Q_x X_0')\left[\cos(Q_x X_0') - \exp\left(-\frac{X_0'^2}{2u_{0x}^2}\right)\right]^2 \\ &\quad \times \exp\left[-Q_x^2\left(\frac{u_{0x}^2}{2\sqrt{1+4\gamma_x}} + \frac{u_{0x}^2}{2}\right)\right]\delta(\omega) \end{aligned} \tag{119}$$

As can be seen from expressions (117)–(119), quantum interference leads to the modulation of the Gaussian profiles by the $\cos^2(Q_x, X_0')$ and $\sin^4(Q_x, X_0')$ terms.

Thus the dynamics of the coupled oscillators is a function of their quantum nature. In the case of bosons, the dynamics is described by usual spatial symmetric and antisymmetric normal modes; but in the case of fermions, spin–spin correlations result in a singlet and triplet state. In crystalline ground state, owing to the Pauli principle the proton dynamics is the dynamics of fermions and it is virtually decoupled from the dynamics of bosons—that is, O, C, N, and so on, atoms.

The further study of pairs of coupled oscillators, both theoretical and experimental, has been performed on the potassium hydrogen carbonate crystal, $KHCO_3$, and its deuterated analog, $KDCO_3$, by Ikeda and Fillaux [121]. It has been shown in their research that (i) the proton dynamics in fact is amenable to the harmonic oscillator and in the ground state is largely decoupled from the lattice; (ii) classical normal coordinates apply only to coupled pairs of bosons; (iii) coupled pairs of fermions must be regarded as singlet and triplet states, and the spin correlation for indistinguishable protons gives rise to quantum interference; and (iv) deutron and lattice dynamics are correlates.

An interesting study concerning the tunneling of protons has been conducted by Willison [122], who has treated the phenomenon of sonoluminescence in liquid water just as caused by the proton tunneling between oxygens of nearest water molecules (see Appendix B).

III. TRANSPORT PROPERTIES OF HYDROGEN-BONDED COMPLEXES: POLARONS AND POLARITONS IN SYSTEMS WITH HYDROGEN BONDS

A. Orientational-Tunneling Model of One-Dimensional Molecular System

Proton transport along a hydrogen-bonded chain strongly depends on structural peculiarities of the chain. In Ref. 135 the description of proton transport has been based on the proton transfer process that involves the Grotthus mechanism with the further reorientation caused by "tumbling" (vibrational and/or librational) of the neighboring ionic group. In other words, the first step of the transport process consists of ionic defect motion, and the next one is accompanied by occurrence of the so-called orientational defect. Their existence is accepted in models considering proton transport in ice and the orientational relaxation rates of icy substances. Perhaps the first experimental study of the orientational (Bjerrum) defect charge was carried out by Hubman [136] based on the measuring the dielectric constant of ammonia-doped ice. In I_h ice, the energy of the Bjerrum defects was determined [137] by relaxation of the surrounding neighbors when the defects were forming.

The model proposed by Stasyuk et al. [135] describes the chain of hydrogen bonds connecting by ionic groups. The model is a development of their previous "pseudo-spins reduced basis model" [138], which took into account only the motion of a proton along the hydrogen bond. Reference 135 discusses the orientational degrees of freedom, which make it possible to include rotations of covalent bonds connecting protons with ionic groups. The quasi-one-dimensional chain with L- and D-Bjerrum proton defects (for a discussion about the defects see, e.g., Ref. 139) as well as the chain with only one proton have been considered. The energy levels, polarization, and susceptibility of the proton subsystem that features a small number of hydrogen bonds have also been investigated. It has been shown that the thermodynamic properties of such molecular systems change strongly with the change of orientational motion frequency and external electric field strength.

Stasyuk et al.'s model is the following. Let the molecular system AH$\cdots$CH...CH$\cdots$B contain a chain of N hydrogen bonds between the ions or ionic groups C (Fig. 6). A and B denote ions or ionic groups placed on the ends of the chain. Stasyuk et al. [135] proceed from the concept implying the availability of a double-well potential for each bond. The ionic defect represents the motion of a proton within a single hydrogen bond between the two heavy ions. The orientational defects allow a proton to move along a chain. In the approach, there are only two lowest proton states in a bond and only two different orientational positions for the covalent bond. The starting Hamiltonian presented in the second quantization form is based on wave functions ψ_a and ψ_b of the lowest proton states in the left (a) and right (b) minima of the double-well potential:

$$H = H_A + \sum_{l=1}^{N-1} H_l + H_B + H_{\text{tun}} + H_{\text{rot}} + H_{\text{C}} + H_{\mathscr{E}} \tag{120}$$

The short-range interaction's part of the Hamiltonian includes three terms:

$$\begin{aligned} H_A &= \varepsilon_A(1 - n_{1,a}) + w_A n_{1,a} \\ H_l &= w'(1 - n_{l,b})(1 - n_{l+1,a}) + w n_{l,b} n_{l+1,a} + \varepsilon(1 - n_{k,b}) n_{l+1,a} \\ &\quad + \varepsilon(1 - n_{l,b}) n_{l+1,a} + \varepsilon n_{l,b}(1 - n_{l+1,a}) \\ H_B &= \varepsilon_B(1 - n_{N,b}) + w_B n_{N,b} \end{aligned} \tag{121}$$

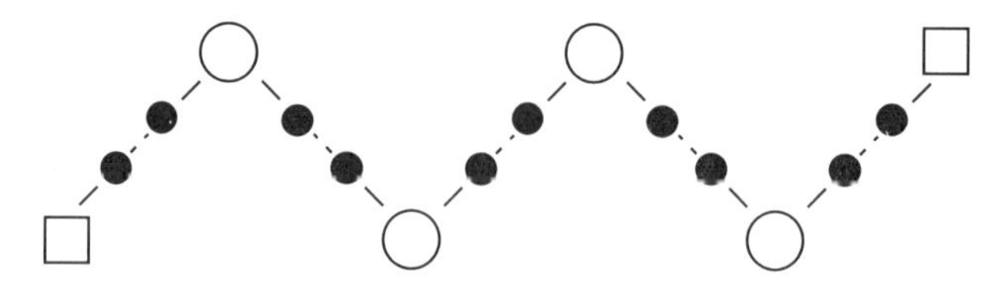

Figure 6. Scheme of the hydrogen-bonded chain. The arrows show the direction of the motion of a proton. (From Ref. 136.)

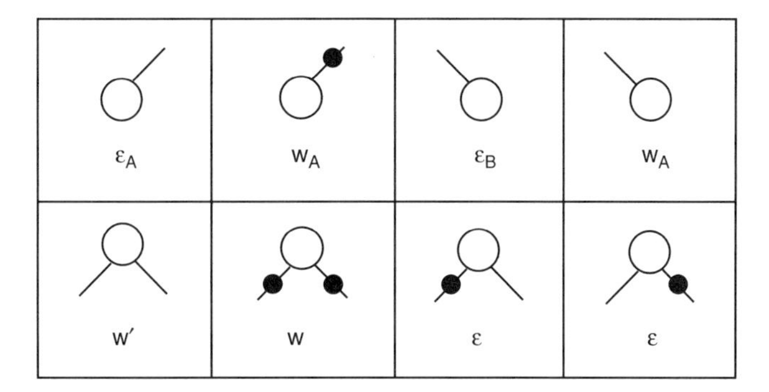

Figure 7. Proton configurations near ionic groups and the proton energies. (From Ref. 136.)

where $\varepsilon, \varepsilon_{A,B}, w, w', w_{A,B}$ are the energies of proton configurations in the potential minima nearest to the ionic groups (Fig. 7); $n_{l,\mu}$ are the occupation numbers of protons in the positions $\mu = a, b$ in the bond $k(k = 1, \ldots, N)$.

The tunneling energy H_{tun} is presented in the form

$$H_{\text{tun}} = \Omega_A(\hat{a}^+_{1,a}\hat{a}_{1,b} + \hat{a}^+_{1,b}\hat{a}_{1,a}) + \Omega_B(\hat{a}^+_{N,a}\hat{a}_{N,b} + \hat{a}^+_{N,b}\hat{a}_{N,a}) + \Omega_0 \sum_{l=2}^{N-1} (\hat{a}^+_{l,a}\hat{a}_{l,b} + \hat{a}^+_{l,b}\hat{a}_{l,a}) \quad (122)$$

where $\hat{a}^+_{l,\mu}(\hat{a}_{l,\mu})$ are the proton creation (annihilation) operators—that is, Fermi operators. The tunneling frequencies of the outer and inner bonds are Ω_A, Ω_B, and Ω_0, respectively. The effect associated with the orientational motion is described by the term

$$H_{\text{rot}} = \Omega_{\text{rot}} \sum_{l=1}^{N-1} (\hat{a}^+_{l,b}\hat{a}_{l+1,a} + \hat{a}^+_{l+1,a}\hat{a}_{l,b}) \quad (123)$$

where Ω_{rot} is characterized by the pseudo-tunneling effect. The term

$$H_{\text{C}} = U \sum_{l=1}^{N-1} n_{l,a} n_{l,b} + V \sum_{l=1}^{N} (1 - n_{l,a})(1 - n_{l,b}) \quad (124)$$

corresponds to the energy of Coulomb repulsion between two protons in the hydrogen bond, the first term, and between free electron pairs without protons, the second term (this is so-called lone pairs [138,140]). The last term in expression (120)

$$H_{\mathscr{E}} = e\mathscr{E} \sum_{l=1}^{N} \sum_{\mu=\{a,b\}} R_{l,\mu} n_{l,\mu} \quad (125)$$

describes the interaction between a proton with charge e and the external electric field $\mathscr{E}$; $R_{l,\mu}$ is the distance between the left edge of the chain and the corresponding potential well (l, μ).

In the case of a symmetric system, some parameters are reduced to the following: $\Omega_A = \Omega_B = \Omega$, $\varepsilon_A = \varepsilon_B = \tilde{\varepsilon}$, and $w_A = w_B = \tilde{w}$.

There exist 2^{2N} many-particle states in a chain with N bonds, namely, $|r\rangle = |n_{1,a}, \ldots, n_{N,b}\rangle$, where $n_{k,\mu} = \{0, 1\}$ is the number of protons in the position $\mu = \{a, b\}$ in the lth bond. In this case the Hamiltonian expressed in terms of the Hubbard operators $X^{r,r'} = |r\rangle\langle r'|$ acting on the basis $|r\rangle$ decomposed into $(2N + 1)$ pairs that correspond to the various values of the proton number $n = \sum_{l,\mu} n_{l,\mu} = 0, 1, \ldots, 2N$:

$$H = H^0 \oplus H^1 \oplus \cdots \oplus H^{2N} \tag{126}$$

The eigenvalues of these Hamiltonians give the energy spectrum of the system studied. Short-range proton correlations and a strong Coulomb interaction between protons in a bond could result in the equilibrium distribution of protons without high-energy proton configurations [138]. Similarly, the Hubbard-type interaction can also be excluded. Let us write, for example, the basis for $|r\rangle$ for $N = 2$:

$$\begin{aligned}
c_{0,a} &= X^{1,2} + X^{3,6} + X^{4,7} + X^{5,8} + X^{9,12} + X^{10,13} + X^{11,14} + X^{15,16} \\
c_{1,a} &= X^{1,4} - X^{2,7} - X^{3,9} + X^{5,11} + X^{6,12} - X^{8,14} - X^{10,15} + X^{13,16} \\
c_{0,b} &= X^{1,3} - X^{2,6} + X^{4,9} + X^{5,10} - X^{7,12} - X^{8,13} + X^{11,15} - X^{14,16} \\
c_{1,b} &= X^{1,5} - X^{2,8} - X^{3,10} - X^{4,11} + X^{6,13} + X^{7,14} + X^{9,15} - X^{12,16}
\end{aligned} \tag{127}$$

The Hamiltonian decomposes into five terms:

$$H = H_2^0 \oplus H_2^1 \oplus H_2^2 \oplus H_2^3 \oplus H_2^4 \tag{128}$$

where

$$H_2^0 = a_2^0 \cdot \hat{1} \tag{129}$$

$$\begin{aligned}
H_2^1 = {} & ((b - a) + \mathscr{E}(M_a + M_{ab}))X^{2,2} + M_a\mathscr{E}(X^{3,3} - X^{4,4}) \\
& + ((b - a) - \mathscr{E}(M_a + M_{ab}))X^{5,5} + \Omega(X^{2,3} + X^{3,2}) \\
& + \Omega(X^{4,5} + X^{5,4}) + \Omega_{\text{rot}}(X^{3,4} + X^{4,3}) + a_2^1 \cdot \hat{1}
\end{aligned} \tag{130}$$

$$H_2^2 = [U + V + \mathscr{E}(2M_a + M_{ab})]X^{6,6} + M_{ab}(X^{7,7} - X^{10,10})$$
$$+ (b - a)X^{8,8} + [J - (b - a)]X^{9,9} + \Omega(X^{8,10} + X^{5,4})$$
$$+ \Omega(X^{9,10} + X^{10,9}) + \Omega_{\text{rot}}(X^{6,7} + X^{7,6})$$
$$+ \Omega_{\text{rot}}(X^{10,11} + X^{11,10}) + a_2^2 \cdot \hat{1} \qquad (131)$$

$$H_2^3 = [U + J - (b - a) + \mathscr{E}(M_a + M_{ab})]X^{12,12} + (U + \mathscr{E}M_a)X^{13,13}$$
$$+ (U - \mathscr{E}M_a)X^{14,14} + [U + J - (b - a) - \mathscr{E}(M_a + M_{ab})]X^{15,15}$$
$$+ \Omega(X^{12,13} + X^{13,12}) + \Omega(X^{14,15} + X^{15,14})$$
$$+ \Omega_{\text{rot}}(X^{13,14} + X^{14,13}) + a_2^3 \cdot \hat{1} \qquad (132)$$

$$H_2^4 = (2U + J)X^{16,16} + a_2^4 \cdot \hat{1}$$

Here $(b - a) = (\tilde{w} - \tilde{\varepsilon}) - (\varepsilon - w')$ is the difference between energies of proton configurations at the first and last ionic groups (A and B) and an internal potential well (C). The value $a_2^2 = \tilde{w} + \tilde{\varepsilon} + \varepsilon$ determines the energy of domain walls at the ends of the chain (due to the fact that the boundary of the chain is characterized by other parameters than the internal part). The other parameters a_i^j are the following:

$$a_2^0 = a_2^2 - (\tilde{w} - \tilde{\varepsilon}) - (\varepsilon - w')$$
$$a_2^1 = a_2^2 - (\tilde{w} - \tilde{\varepsilon})$$
$$a_2^3 = a_2^2 + (\tilde{w} - \tilde{\varepsilon})$$
$$a_2^4 = a_2^2 + (\tilde{w} - \tilde{\varepsilon}) + (\varepsilon - w')$$

The parameter $J = w + w' - 2\varepsilon$ is the effective short-range interaction between the protons near an ionic group; $M_a = -eR_a$, $M_{ab} = -eR_{ab}$, where R_a is the distance between the ionic group and the neighboring potential well and R_{ab} is the distance between two neighboring wells of the double minimum potential. Note that for a chain with any finite number of N, the corresponding Hamiltonian can be written in the same way.

The equation for eigenvalues and eigenvectors of the Hamiltonian (126)

$$\sum_{r'} \langle r|H|r'\rangle u_{rr'} = \lambda_\mu u_{r\mu} \qquad (134)$$

allows one to obtain the energy spectrum of the system.

The dielectric susceptibility χ of proton subsystem can be expressed via the Green function

$$\chi(\omega) = -\frac{2\pi}{\hbar}\langle\langle P \mid P\rangle\rangle_{\omega} \tag{135}$$

where $P = -e\sum_{l,\mu} R_{l,\mu} n_{l,\mu}$ is the operator of electric dipole momentum of protons in the chain. The Hamiltonian represented on the basis $|\tilde{r}\rangle = \sum_{r} u_{rr'}|r'\rangle$ has the diagonal form

$$\tilde{H} = \sum_{p} \lambda_p \tilde{X}^{p,p} \tag{136}$$

where the operator $\tilde{X}^{p,p}$ is determined by the unitary transformation of Hubbard operators $X^{p,p}$,

$$X^{r,r'} = \sum_{\mu,\nu} u^{*}_{r\mu}\tilde{X}^{\mu,\nu}u_{\nu r'} \tag{137}$$

The equation of motion of the operator $\tilde{X}^{\mu,\nu}$ is the following:

$$i\hbar\frac{\partial}{\partial t}\tilde{X}^{\mu,\nu} = [\tilde{X}^{\mu,\nu}, \tilde{H}]_t = (\lambda_\nu - \lambda_\mu)\tilde{X}^{\mu,\nu} = \lambda_{\nu\mu}\tilde{X}^{\mu,\nu} \tag{138}$$

This allows the determination of the susceptibility of the proton subsystem,

$$\chi(\omega) = \sum_{\mu<\nu} 2\tilde{P}^{2}_{\mu\nu}\lambda_{\nu\mu}\frac{\langle\tilde{X}^{\mu\mu} - \tilde{X}^{\nu\nu}\rangle}{(\hbar\omega)^2 - \lambda^2_{\nu\mu}} \tag{139}$$

where the operator of dipole moment constructed on the basis $|r'\rangle$ is equal to

$$\tilde{P}_{\mu\nu} = \sum_{l,l'} u_{\mu l}\langle l|P|l'\rangle u^{*}_{l'\nu} \tag{140}$$

and the average occupation number of state $|\mu\rangle$ is defined as

$$\langle X^{\mu\mu}\rangle = \frac{e^{-\beta\lambda_\mu}}{\sum_\nu e^{-\beta\lambda_\nu}} \tag{141}$$

The energy spectrum and the susceptibility, for instance, for a short chain with the number of ions $N = 2$ and two protons ($n = 2$) in the two hydrogen bonds have been calculated in Ref. 135. In this case the system studied is described by the Hamiltonian (131) and the corresponding eigenvalues are

$$\begin{aligned} \lambda_{1,2} &= \frac{1}{2}(U + V \pm q_-) + \mathscr{E}(M_a + M_{ab}) + a_2^2 \\ \lambda_3 &= (b - a) + a_2^2 \\ \lambda_4 &= J - (b - a) + a_2^2 \\ \lambda_{5,6} &= \frac{1}{2}(U + V \pm q_+) - \mathscr{E}(M_a + M_{ab}) + a_2^2 \end{aligned} \tag{142}$$

where

$$q_{\pm} = \sqrt{(U + V \mp 2\mathscr{E}M_a)^2 + 4\Omega_{\text{rot}}^2} \tag{143}$$

The examination of the spectrum (142) as a function of Ω_{rot} that features the proton's orientational motion and at $\mathscr{E} = 0$ shows that the behavior of the system is specified by two regimes. At $\Omega_{\text{rot}} < \Omega_{\text{rot}}^*$, where $\Omega_{\text{rot}}^* = \{(b-a)^2 - (U+V)\,(b-a)\}^{1/2}$, the ground state corresponds to the location of the domain wall in the center of the chain when the protons are found in potential wells near the outer ionic groups A and B. At $\Omega_{\text{rot}} > \Omega_{\text{rot}}^*$ one of the protons is localized at an internal ion. The appropriate static susceptibility at low temperature becomes

$$\chi = \begin{cases} 0, & \Omega_{\text{rot}} < \Omega_{\text{rot}}^* \\ 2\frac{M_a^2}{q}\left(1 - \frac{U^2}{q^2}\right), & \Omega_{\text{rot}} > \Omega_{\text{rot}}^* \end{cases} \tag{144}$$

Numerical calculations performed for the system $N = n = 2$, $N = n = 3$, and $N = n = 4$ have shown [135] that χ as a function of Ω_{rot} is smoothed in the vicinity of Ω_{rot}^* and that the value of Ω_{rot}^* shifts to the low frequency region with the increase of N.

The statistical sum for the considering system ($N = n = 2$) is equal to

$$\begin{aligned} Z &= e^{-(b-a)/k_{\text{B}}T} + e^{-(J-(b-a))/k_{\text{B}}T} + 2e^{-(U+V)/2k_{\text{B}}T} \\ &\times \left(e^{-\mathscr{E}(M_a+M_{ab})/k_{\text{B}}T} \cosh\frac{q_-}{2k_{\text{B}}T} + e^{\mathscr{E}(M_a+M_{ab})/k_{\text{B}}T} \cosh\frac{q_+}{2k_{\text{B}}T} \right) \end{aligned} \tag{145}$$

The average dipole moment is obtained from expression (145)

$$\langle P \rangle = -\frac{\partial F}{\partial \mathscr{E}} \tag{146}$$

where the free energy is given by

$$F = -\frac{\ln Z}{k_B T} \tag{147}$$

Explicitly,

$$\langle P \rangle = -2\frac{e^{-(U+V)/2k_B T}}{Z}\{e^{-\mathscr{E}(M_a+M_{ab})/k_B T}\xi_- - e^{\mathscr{E}(M_a+M_{ab})/k_B T}\xi_+\} \tag{148}$$

$$\xi_\pm = (M_a + M_{ab})\cosh\frac{q_\pm}{2k_B T} + M_a \frac{U + V \mp 2\mathscr{E}M_a}{q_\pm}\sinh\frac{q_\pm}{2k_B T} \tag{149}$$

Expressions from (145) to (149) allows one to obtain the analytical expression for the static susceptibility,

$$\begin{aligned}\chi(\mathscr{E}) = &\frac{e^{-(U+V)/2k_B T}}{Z^2}\{e^{-\mathscr{E}(M_a+M_{ab})/k_B T}\xi_- - e^{\mathscr{E}(M_a+M_{ab})/k_B T}\xi_+\}^2 \\ &+ \frac{e^{-(U+V)/2k_B T}}{Z}\{e^{-\mathscr{E}(M_a+M_{ab})/k_B T}\eta_- - e^{\mathscr{E}(M_a+M_{ab})/k_B T}\eta_+\}\end{aligned} \tag{150}$$

where

$$\begin{aligned}\eta_\pm = &\frac{2}{k_B T}(M_a + M_{ab})^2 - M_a^2\left(\frac{U + V \mp 2\mathscr{E}M_a}{q_\pm}\right)^2 \cosh\frac{q_\pm}{2k_B T} \\ &- 4\frac{M_a^2}{q_\pm}\left[1 - \left(\frac{U + V \mp 2\mathscr{E}M_a}{q_\pm}\right)^2\right]\sinh\frac{q_\pm}{2k_B T}\end{aligned} \tag{151}$$

Expressions (142), (148), and (150) allow the consideration of the energy spectrum, the dipole moment, and the susceptibility on the field strength $\mathscr{E}$. A numerical inspection of the spectrum performed in Ref. 135 has shown that the ground state is reconstructed at $\mathscr{E} = \mathscr{E}_2 = (U + V)/2M_a$. The transfer of all the protons along the field to an appropriate edge of the chain induces D defects at this edge, and consequently L defects appear at the opposite edge. This changes the inclination of the dipole moment $\langle P \rangle$ as a function of $\mathscr{E}$ just at $\mathscr{E} = \mathscr{E}_2$. An anomaly of susceptibility $\chi(\mathscr{E})$ appears at $\mathscr{E} = \mathscr{E}_2$ as well. Besides, it has been concluded that the motion of protons along the chain under an electric field is very sensitive to the value of Ω_{rot}. In the general case, the peculiar value $\mathscr{E}_2$ depends on the number of ions in the chain [135],

$$\mathscr{E}_2 = \frac{U + V}{2(N - 1)M_a} \tag{152}$$

The appearance of high-energy proton configurations, D and L defects, has occurred at $\mathscr{E} > \mathscr{E}_2$, which is caused by the increasing the dipole moment of the proton subsystem with the gain of the number of hydrogen bonds.

The motion of L defects has been treated in detail in the case $N = 3$ and $n = N - 1$. The energy spectrum as a function of $\Omega_{\rm rot}$ has been obtained analytically at $\mathscr{E} = 0$ and $\Omega = 0$. Two peculiarities have been found in the spectrum at the parameters $\Omega_{\rm rot}^{(1)} = -(b - a)$ and $\Omega_{\rm rot}^{(2)} = -(b - a)/(\sqrt{2} - 1)$. The rearrangement of the spectrum influences the behavior of the susceptibility: $\chi(\Omega_{\rm rot})$ decreases with $\Omega_{\rm rot} > \Omega_{\rm rot}^{(2)}$, and this takes place at the arbitrary number of hydrogen bonds. With the increase of the number of hydrogen bonds, the distance between $\Omega_{\rm rot}^{(1)}$ and $\Omega_{\rm rot}^{(2)}$ becomes smaller. Applying an external field leads to the proton transfer along the field and the migration of an L defect in the opposite direction. Thus an L defect is localized at the left edge of the chain, and all the protons occupy the right potential wells in the remaining $(N - 1)$ double-well potentials. In other words, the protons are ordered and therefore the left domain wall should appear between the first bond with an L defect and the second left bond. Meanwhile the right domain wall is formed at the right edge; that is, a D defect is created at the right edge of the chain. By the estimation [135], D and L defects may be realized at value $\mathscr{E}$, which decreases with the growing the number of hydrogen bonds.

As an example of an object with more than one L defect, the chain with N hydrogen bonds and only one proton has been studied [135]. They have considered the energy spectrum as a function of the field $\mathscr{E}$ at the fitted parameters $(b - a), \Omega, \Omega_0, M_a, M_{ab}$ and various values of $\Omega_{\rm rot}$. The susceptibility falls within two regions that are specified by $\Omega_{\rm rot} < -(b - a)$ and $\Omega_{\rm rot} > -(b - a)$. The $(N - 1)$ energy levels form the energy band with the increase of N. A peculiarity revealed in the behavior of $\langle P \rangle$ and χ at $\Omega_{\rm rot} > -(b - a)$ has been related to the motion of a delocalized proton from the inner hydrogen bond toward the outer bond along the applied field.

B. Proton Ordering Model

A number of hydrogen-containing compounds, such as lithium hydrazium sulfate $Li(N_2H_5)SO_4$ (LiHzS), tri-ammonium hydrogen disulfate $(NH_4)_3H(SO_4)_2$ (TAHS), and others, exhibit an anisotropy of the protonic electrical conductivity σ at temperatures $T \geq 300$ K: along the c axis the σ_c is about 10^3 larger than in the perpendicular directions [29,30]. This effect is stipulated by long hydrogen-bonded chains that spread along the c axis along the whole crystal. Thus the relatively large σ_c is caused by the motion of protons along hydrogen-bonded chains. The large proton mobility of ice is explained by the net of hydrogen bonds [140]. The comparatively high proton conductivity was observed also in proton-conducting proteins [141–143], which also connected with proton transfer along the hydrogen-bonded chains.

Having understood the microscopic mechanism of the motion of protons in the hydrogen-bonded network and along the hydrogen-bonded chain, we should know first the behavior of the proton subsystem in the aforementioned compounds. The compounds are characterized by the complicated structure of the primitive cell, and many of them possess typical ferroelectric properties caused by the presence of hydrogen bonds. Experimental studies (see, e.g., Refs. 144 and 145) specify a significant role of the proton subsystem in proton superionic phase transitions in a group of crystals $M_3H(XO_4)_2$ (where $M = K$, Rb, Cs, MH_4; $X = Se$, S), which indicates the prime importance of the consideration of a hydrogen bond network and its rearrangement at the number of phase transitions of the crystals [42]. For instance, the structural X-ray studies [146,147] revealed the following phase transition sequence in the $(NH_4)_3H(SeO_4)_2$ crystal [42]: superionic phase I (with trigonal symmetry $R\bar{3}m$), superionic phase II (space group $R\bar{3}$), ferroelastic phase III (triclinic symmetry $C\bar{1}$ or $P\bar{1}$, with small deviations from monoclinic), ferroelastic phase IV (monoclinic symmetry Cc) with the transition temperatures 332 K, 302 K, 275 K, 181 K, and 101 K, respectively.

In a series of works by Salejda and Dzhavadov [148] and Stasyuk et al. [38–42], a microscopic description of the phase transitions in $(NH_4)_3H(SeO_4)_2$ has been proposed. The results obtained have been compared with those obtained from the Landau phenomenological theory. The original Hamiltonian has been presented in a bilinear form

$$H_{\text{int}} = \frac{1}{2} \sum_{m,m';f,f'} \Phi_{ff'}(mm') n_{mf} n_{m'f'} \tag{153}$$

where $n_{mf} = 0$ or 1 is the occupation number of a proton in position f of unit cell m; $\Phi_{ff'}(mm')$ denotes the energy of pair interaction. The Hamiltonian of the proton subsystem includes, in addition to expression (153), the sum of single-particle energies

$$H_0 = \sum_{m,f} (E - \mu) n_{mf} \tag{154}$$

The chemical potential μ of the proton subsystem at the given number of protons is determined from condition

$$\sum_{m;f} \bar{n}_{mf} = \bar{n} \tag{155}$$

where n is the number of protons per unit cell, which for $(NH_4)_3 \cdot H(SeO_4)_2$-type crystals one can put $\bar{n} = 1$. In the mean-field approximation

$$H_{\text{MF}} = U_0 + \sum_{m;f} \gamma_f(m) n_{mf} + \sum_{m;f} (E - \mu) n_{mf} \tag{156}$$

where

$$U_0 = -\frac{1}{2} \sum_{m,m';f,f'} \Phi_{ff'}(mm')\bar{n}_{mf}\bar{n}_{m'f'} \tag{157}$$

we obtain

$$\gamma_f(m) = \sum_{m';f'} \Phi_{ff'}(mm')\bar{n}_{m'f'} \tag{158}$$

The statistical sum Z and thermodynamic potential Ω for the system with the Hamiltonian (156) are defined as

$$Z = \mathrm{Tr}\, e^{-H_{\mathrm{MF}}/k_{\mathrm{B}}T} = \prod_{m;f} \sum_{n_f=\{0,1\}} e^{-(\gamma_f(m)+E-\mu)n_m/k_{\mathrm{B}}T} e^{-U_0/k_{\mathrm{B}}T} \tag{159}$$

$$\Omega = -k_{\mathrm{B}}T \ln Z = U_0 - k_{\mathrm{B}}T \sum_{m;f} \ln(1 + e^{-(\gamma_f(m)+E-\mu)/k_{\mathrm{B}}T}) \tag{160}$$

Thus, the mean number of protons per unit cell obeys the Fermi–Dirac statistics

$$\bar{n}_{mf} = (e^{(\gamma_f(m)+E-\mu)/k_{\mathrm{B}}T} + 1)^{-1} \tag{161}$$

Let us now represent the mean number of protons $\bar{n}_{mf}$ as follows:

$$\bar{n}_{mf} = \frac{1}{3}\bar{n} + \delta\bar{n}_{mf} \tag{162}$$

where $\delta\bar{n}_{mf}$ is the deviation from the mean proton occupation number. The three components can be represented in the form [38]

$$\begin{aligned} \delta\bar{n}_{m1} &= \frac{1}{\sqrt{2}} u e^{i\mathbf{k}_3\mathbf{R}_m} + \frac{1}{\sqrt{6}}\nu \\ \delta\bar{n}_{m2} &= -\frac{1}{\sqrt{2}} u e^{i\mathbf{k}_3\mathbf{R}_m} + \frac{1}{\sqrt{6}}\nu \\ \delta\bar{n}_{m3} &= -\frac{2}{\sqrt{6}}\nu \end{aligned} \tag{163}$$

The presentation (163) makes it possible to obtain the observed occupations of proton positions ($\bar{n}_{m1} = 1$, $\bar{n}_{m2} = 0$, $\bar{n}_{m3} = 0$ or $\bar{n}_{m1} = 0, \bar{n}_{m2} = 1, \bar{n}_{m3} = 0$) for the saturation values of order parameters u and v: $u = \pm 1/\sqrt{2}$ and $v = 1/\sqrt{6}$, where

$$u = \pm \frac{1}{\sqrt{2}} \langle n_{m1} - n_{m2} \rangle, \qquad v = \frac{1}{\sqrt{6}} \langle n_{m1} + n_{m2} - 2n_{m3} \rangle \tag{164}$$

The resulting orderings are induced by the irreducible representation $A_g(\tau_1)$ of the wave vector group $G_{\mathbf{k}_i}$ as well as by the representation $E_g(\tau_3 + \tau_5)$ of the point group $G_{\mathbf{k}_7=0}$. Indeed, in a general case the deviation can be written as

$$\begin{aligned}
\delta \bar{n}_1 &= \frac{1}{\sqrt{2}} [u_3 e^{i\mathbf{k}_3 \mathbf{R}_m} - u_2 e^{i\mathbf{k}_2 \mathbf{R}_m}] + \frac{1}{3}(\eta + \zeta) \\
\delta \bar{n}_2 &= \frac{1}{\sqrt{2}} [u_1 e^{i\mathbf{k}_1 \mathbf{R}_m} - u_3 e^{i\mathbf{k}_3 \mathbf{R}_m}] + \frac{1}{3}(\varepsilon^2 \eta + \varepsilon \zeta) \\
\delta \bar{n}_3 &= \frac{1}{\sqrt{2}} [u_2 e^{i\mathbf{k}_2 \mathbf{R}_m} - u_1 e^{i\mathbf{k}_1 \mathbf{R}_m}] + \frac{1}{3}(\varepsilon \eta + \varepsilon^2 \zeta)
\end{aligned} \tag{165}$$

where $\varepsilon = e^{i2\pi/3}$; the variables

$$\begin{aligned}
\eta &= \bar{n}_1 + \varepsilon \bar{n}_2 + \varepsilon^2 \bar{n}_3 \\
\zeta &= \bar{n}_1 + \varepsilon^2 \bar{n}_2 + \varepsilon \bar{n}_3
\end{aligned} \tag{166}$$

are transformed according to the E_g representation under point group operations. Setting

$$\eta = |\eta| e^{i\psi}, \qquad \zeta = |\eta| e^{-i\psi} \tag{167}$$

we obtain

$$\begin{aligned}
\eta + \zeta &= 2|\eta| \cos \psi \\
\varepsilon^2 \eta + \varepsilon \zeta &= 2|\eta| \cos\left(\psi - \frac{2\pi}{3}\right) \\
\varepsilon \eta + \varepsilon^2 \zeta &= 2|\eta| \cos\left(\psi - \frac{4\pi}{3}\right)
\end{aligned} \tag{168}$$

The choice $|\eta| = \sqrt{3/2}v$, $\psi = \{\frac{1}{3}\pi;\ \pi;\ \frac{1}{3}\pi\}$ corresponds to the orientation states with vectors $\mathbf{k}_3 - \frac{1}{2}\mathbf{b}_3$, $\mathbf{k}_1 = \frac{1}{2}\mathbf{b}_1$, and $\mathbf{k}_2 = \frac{1}{2}\mathbf{b}_2$, respectively

The same proton occupation in each unit cell occurs in phase IV. Hence the change in symmetry of the proton distribution in phase IV, in comparison with phase II, may be formally connected with one of irreducible representations of group $G_{\mathbf{k}=0}$, if only the proton subsystem is taken into account.

For the basis $(n_1,\ n_2,\ n_3)$, only a two-dimensional irreducible representation E_g becomes suitable. In this case the expressions for the deviations $\delta\mathbf{n}_f$ can be derived from Eqs. (165) with restriction to uniform terms. The values $|\eta| = 1$ and $\psi = 0,\ \frac{2}{3}\pi,\ \frac{2}{3}\pi$ correspond to three orientation states with $\mathbf{n}_{mf} = (1,0,0)$, $(0,1,0)$, $(0,0,1)$, respectively. Thus, Eqs. (165) for the deviations of proton occupation numbers include the cases of all three phases (II, III, and IV). The representation of $\delta\mathbf{n}_f$ in the form of Eqs. (163) allows one to describe not only the ordering to an orientation state with vector $\mathbf{k}_3$ of phase III (at $u \neq 0$ and $v \neq 0$), but also the ordering with $n_{mf} = (0,0,1)$ for phase IV (with $u = 0$, $v \neq 0$ and saturation value $v = -2/\sqrt{6}$).

Equations (163) allow one to derive equations for the proton mean occupation numbers and thermodynamic functions in the mean-field approximation. Substituting variables $\delta\bar{n}_{mi}$ where $i = 1,\ 2,\ 3$ into Eqs. (157)–(159), we get

$$\begin{aligned}\gamma_1(m) &= \gamma_0 + av + be^{i\mathbf{k}_3\mathbf{R}_m}u\\ \gamma_2(m) &= \gamma_0 + av - be^{i\mathbf{k}_3\mathbf{R}_m}u\\ \gamma_3(m) &= \gamma_0 - 2av\end{aligned} \tag{169}$$

Here

$$\begin{aligned}\gamma_0 &= \frac{1}{3}\bar{n}\sum_{f'}\Phi_{ff'}(0)\\ a &= \frac{1}{\sqrt{6}}[\varphi_{11}(0) - \varphi_{12}(0)]\\ b &= \frac{1}{\sqrt{2}}[\varphi_{11}(\mathbf{k}_3) - \varphi_{12}(\mathbf{k}_3)]\end{aligned} \tag{170}$$

where the following designation is used:

$$\varphi_{ff'}(\mathbf{k}) = \sum_{m'}\Phi_{ff'}(mm')\,e^{i\mathbf{k}\cdot(\mathbf{R}_m - \mathbf{R}_{m'})} \tag{171}$$

which is the Fourier transform of the proton interaction matrix. Then instead of expression (157) we have

$$U_0 = -\frac{N}{2}\gamma_0 - \frac{N}{2}\sqrt{6}av^2 - \frac{N}{2}\sqrt{2}bu^2 \tag{172}$$

The symbols u and v here play the role of proton order parameters. Then the thermodynamic equilibrium conditions $\partial\Omega/\partial\bar{n}_{mf} = 0$ and $\partial F/\partial\bar{n}_{mf} = 0$ are rewritten with respect to the parameters u and ν as follows:

$$\frac{\partial}{\partial u}\left(\frac{1}{N}F\right) = 0, \qquad \frac{\partial}{\partial \nu}\left(\frac{1}{N}F\right) = 0 \tag{173}$$

With the explicit form of the free energy F as a function of order parameters u and ν,

$$\begin{aligned}\frac{1}{N}F = \frac{1}{N}U_0 - k_{\mathrm{B}}T\bigg[\ln\left(1+\frac{1}{y}e^{-(a\nu+bu)/k_{\mathrm{B}}T}\right) \\ + \ln\left(1+\frac{1}{y}e^{-(a\nu-bu)/k_{\mathrm{B}}T}\right) + \ln\left(1+\frac{1}{y}e^{2a\nu/k_{\mathrm{B}}T}\right)\bigg] + \mu\bar{n}\end{aligned} \tag{174}$$

where $y = e^{(E-\mu+\gamma_0)/k_{\mathrm{B}}T}$, one derives the following equations for u and ν:

$$\begin{aligned} u &= \frac{1}{\sqrt{2}}\left[\frac{1}{ye^{(a\nu+bu)/k_{\mathrm{B}}T}+1} - \frac{1}{ye^{(a\nu-bu)/k_{\mathrm{B}}T}+1}\right] \\ \nu &= \frac{1}{\sqrt{26}}\left[\frac{1}{ye^{(a\nu+bu)/k_{\mathrm{B}}T}+1} - \frac{1}{ye^{(a\nu-bu)/k_{\mathrm{B}}T}+1} - \frac{Z}{ye^{-2a\nu/k_{\mathrm{B}}T}+1}\right] \end{aligned} \tag{175}$$

The parameter y is deduced from the equation

$$\frac{1}{ye^{(a\nu+bu)/k_{\mathrm{B}}T}+1} - \frac{1}{ye^{(a\nu-bu)/k_{\mathrm{B}}T}+1} - \frac{Z}{ye^{-2a\nu/k_{\mathrm{B}}T}+1} = \bar{n} \tag{176}$$

which follows from Eq. (161). For the sake of simplicity, let $\bar{n} = 1$. The solution to Eqs. (175) and (176) and the analysis of extrema of the function F has been performed numerically [38]. It has been found that a first-order transition to phase IV takes place at $b = \sqrt{3}\,a$ with $u = 0$ and $\nu \neq 0$. At $b > \sqrt{3}a$ a transition to the ordered state corresponding to phase III with $u \neq 0$ and $\nu \neq 0$ occurs. Tricritical takes place at $(b/a)_{\mathrm{c}} = 2.07$ and $T_{\mathrm{c}} = 0.65a$ when the transition to phase III changes from first to second. At $b/a = (b/a)^{*} = \sqrt{3}$ all three absolute minima of free energy coexist (at $T < T_{\mathrm{c}}$) having equal depth.

C. Proton Conductivity at the Superionic Phase Transitions

The rearrangement of the hydrogen-bonded network at the superionic phase transitions studied in the framework of the lattice-gas-type model, which has been described in the previous subsection, makes it possible to evaluate the main aspects of proton conductivity in the mentioned compounds.

Let us treat the proton dynamics starting from the modified Hamiltonian (153), that is [149,150],

$$H_{\text{int}} = \frac{1}{2} \sum_{m,m';f,f'} \Phi_{ff'}(mm') n_{mf} n_{m'f'} - \mu \sum_{m;f} n_{mf} \tag{177}$$

where $n_{mf} = \{0, 1\}$ is the proton occupation number for position f in the primitive unit cell with the coordinate $\mathbf{R}_m$, $\Phi_{ff'}(mm')$ is the energy of proton interactions, and μ denotes the chemical potential that determines the average proton concentration.

The formation of a hydrogen bond induces the deformation of the XO_4 group [151–153], which leads to the shortening of the distance between the vertex oxygens $O(2)'$ and $O(2)''$. In its turns, this localizes the proton, resulting in the typical polaron effect described in Section II.C. Thus the appearance of the polaron means the increase of the activation energy for the bond breaking and the hopping of the proton to another localized position in the lattice.

The potential energy of the oxygen subsystem in the harmonic approximation is given by expression

$$\phi = \phi_0 + \sum_{m,m'} \sum_{k,k'} \sum_{\alpha,\beta} \phi_{\alpha\beta}(mk; m'k') u_\alpha(mk) u_\beta(m'k') \tag{178}$$

where $k = 1, 2$ is the sublattice number of the mth unit cell and the force constants

$$\phi_{\alpha\beta}(mk; m'k') = \left. \frac{\partial^2}{\partial u_\alpha(mk) \partial u_\beta(m'k')} \right|_0 \tag{179}$$

The matrices $\phi_{\alpha\beta}(mk; m'k')$ for the $M_3H(X0_4)_2$ class of crystals are represented as follows [149]:

$$\begin{aligned}
\phi(m1; m + \mathbf{a}_1, 2) &= \begin{pmatrix} a & c \\ c & a + 2/\sqrt{3} \end{pmatrix} \\
\phi(m1; m + \mathbf{a}_2, 2) &= \begin{pmatrix} a & -c \\ -c & a + 2/\sqrt{3} \end{pmatrix} \\
\phi(m1; m + \mathbf{a}_3, 2) &= \begin{pmatrix} a + \sqrt{3}c & c \\ c & a - 2/\sqrt{3} \end{pmatrix} \\
\phi(mk; mk) &= \begin{pmatrix} h & 0 \\ 0 & h \end{pmatrix}
\end{aligned} \tag{180}$$

If we consider only displacements of the O(2) oxygens in (x,y)-plane, we gain the dynamic matrix

$$D(\mathbf{k}) = \begin{pmatrix} D^{11} & D^{12} \\ D^{21} & D^{22} \end{pmatrix} \tag{181}$$

where

$$D^{12} = \begin{pmatrix} a(e^{i\mathbf{k}\mathbf{a}_1} + e^{i\mathbf{k}\mathbf{a}_2}) + (a+\sqrt{3}c)e^{i\mathbf{k}\mathbf{a}_3} & c(e^{i\mathbf{k}\mathbf{a}_1} - e^{i\mathbf{k}\mathbf{a}_2}) \\ c(e^{i\mathbf{k}\mathbf{a}_1} - e^{i\mathbf{k}\mathbf{a}_2}) & (a+2\sqrt{3}c)(e^{i\mathbf{k}\mathbf{a}_1} + e^{i\mathbf{k}\mathbf{a}_2}) \\ & +(a-\sqrt{3}c)e^{i\mathbf{k}\mathbf{a}_3} \end{pmatrix}$$

$$D^{11}(\mathbf{k}) = D^{22}(\mathbf{k}) = \begin{pmatrix} h & 0 \\ 0 & h \end{pmatrix}, \qquad D^{21}(\mathbf{k}) = D^{12}(-\mathbf{k}) \tag{182}$$

Thus the determination of the normal vibration modes is reduced to the evaluation of the vibration frequencies and polarization vectors of the matrix $D(\mathbf{k})$. For instance, in the case $\mathbf{k} = 0$ we have

$$\omega_{1/3}(0) = h + (3a + \sqrt{3}c), \qquad \omega_{2/4}(0) = h - (3a + \sqrt{3}c) \tag{183}$$

$$u_1 = \frac{1}{\sqrt{2}}\begin{pmatrix} 1 \\ 0 \\ 1 \\ 0 \end{pmatrix}, \qquad u_2 = \frac{1}{\sqrt{2}}\begin{pmatrix} 0 \\ -1 \\ 0 \\ 1 \end{pmatrix}$$
$$u_3 = \frac{1}{\sqrt{2}}\begin{pmatrix} 0 \\ 1 \\ 0 \\ 1 \end{pmatrix}, \qquad u_4 = \frac{1}{\sqrt{2}}\begin{pmatrix} -1 \\ 0 \\ 1 \\ 0 \end{pmatrix} \tag{184}$$

The solutions (183) and (184) point to the existence of the two different types (in- and antiphase) of the oxygen vibrations with different frequencies.

The change of the proton potential on the hydrogen bond caused by the antiphase vibrations of the oxygens results in the shortening of the bond length. Hence the modes $j = 2$ and $j = 4$ are treated as having the coordinates of polarization vectors $\mathbf{u}_2$ and $\mathbf{u}_4$ approximated by their values at $\mathbf{k} = 0$. Thus the interaction of the protons with the oxygen vibrations can be represented in the second quantization form

$$H_{\text{pr-ph}} = \sum_{m;f}\sum_{\mathbf{k},j} \tau_{mf}(\mathbf{k}j)(\hat{b}_{\mathbf{k}j} + \hat{b}^{+}_{\mathbf{k}j})n_{mf} \tag{185}$$

where $\hat{b}^+_{\mathbf{k}j}(\hat{b}_{\mathbf{k}j})$ is the Bose operator of phonon creation (annihilation) in the jth branch with the wave vector $\mathbf{k}$. The coefficients $\tau_{mf}(\mathbf{k}j)$ are given in Ref. 150. The vibration energy of the oxygen subsystem is

$$H_{\text{ph}} = \sum_{\mathbf{k},j} \hbar\omega_j(\mathbf{k})\hat{b}^+_{\mathbf{k}j}\hat{b}_{\mathbf{k}j} \tag{186}$$

Neglecting the intrabond proton potential, the proton transport process is considered as the dynamical breaking and formation of the hydrogen bonds connected to the HXO_4 ionic group rotations. This signifies that the tunneling integral is presented as follows [149]:

$$H_{\text{tun}} = \Omega_{\text{rot}} \sum_{m;f\neq f'} \{\hat{a}^+_{mf}\hat{a}_{mf'} + \hat{a}^+_{m+a_f-a_{f'},f}\hat{a}_{mf'}\} + h.c. \tag{187}$$

Here $\hat{a}^+_{mf}(\hat{a}_{mf})$ is the operator of proton creation (annihilation) in a position that is characterized by indices m and f defined above. Ω_{rot} is the transfer integral that describes the interbond proton hopping as a quantum tunneling. Once again we would like to emphasize that the parameter Ω_{rot} does not characterized the direct overlapping of the wave functions of protons located at the nearest sites. Ω_{rot} is associated with the rotational motion of ion groups, which in the first approximation one can consider as a proton "tunneling" of some sort.

Applying the canonical transformation $\tilde{H} = e^{iS}He^{-iS}$ used in the theory of small polaron (see, e.g., Ref. 88), we obtain instead of the Hamiltonians (185)–(187)

$$\begin{aligned}\tilde{H} = -\tilde{\mu}\sum_{m,f} n_{mf} + \frac{1}{2}\sum_{m,m';f,f'} \tilde{\Phi}_{ff'}(mm')n_{mf}n_{m'f'} \\ + \sum_{\mathbf{k},j} \hbar\omega_j(\mathbf{k})\hat{b}^+_{\mathbf{k}j}\hat{b}_{\mathbf{k}j} + \tilde{H}_{\text{tun}}\end{aligned} \tag{188}$$

$$\begin{aligned}\tilde{H}_{\text{tun}} = \Omega_{\text{rot}} \sum_{m;f\neq f'} \{\hat{a}^+_{mf}\hat{a}_{mf'}X_{ff'}(mm) \\ + \hat{a}^+_{m+a_f-a_{f'},f}\hat{a}_{mf'}X_{ff'}(m+a_f-a_{f'},m)\} + h.c.\end{aligned} \tag{189}$$

Here the band narrowing factor is

$$\begin{aligned}X_{ff'}(mm') = e^{-\sum_{\mathbf{k},j}\Delta\tau_{ff'}(mm';\mathbf{k}j)(\hat{b}_{\mathbf{k}j}+\hat{b}^+_{\mathbf{k}j})/\hbar\omega_j(\mathbf{k})} \\ \Delta\tau_{ff'}(mm';\mathbf{k}j) = \tau_{mf}(\mathbf{k}j) - \tau_{m'f'}(\mathbf{k}j)\end{aligned} \tag{190}$$

the proton chemical potential is

$$\tilde{\mu} = \mu + \sum_{\mathbf{k},j} \frac{|\tau_{mf}(\mathbf{k}j)|^2}{\hbar\omega_j(\mathbf{k})} \tag{191}$$

and the proton energy renormalized due to the polaron shift is

$$\tilde{\Phi}_{ff'}(mm') = \Phi_{ff'}(mm') - 2\sum_{\mathbf{k},j} \frac{\tau_{mf}(\mathbf{k}j)\tau_{m'f'}(-\mathbf{k}j)}{\hbar\omega_j(\mathbf{k})} \tag{192}$$

Proton conductivity of the systems studied has been considered in the framework of the Kubo linear response theory [154], i.e., the conductivity has been written as

$$\sigma(\omega, T) = \frac{1}{\mathcal{V}} \int_0^\infty dt e^{i(\omega+i\varepsilon)t} \int_0^\infty d\lambda \langle J(t - i\hbar\lambda)J(0)\rangle \tag{193}$$

here $\mathcal{V}$ is the effective volume of the crystal studied, the proton current is given by

$$J = \frac{e}{i\hbar}[H, x] = \frac{e\Omega_{\text{rot}}}{i\hbar} \sum_m \sum_{f \neq f'} \mathbf{R}_{ff'} [\hat{a}^+_{mf} \hat{a}_{mf'} X_{ff'}(mm) \\ + \hat{a}^+_{m+a_f-a_{f'},f} \hat{a}_{mf'} X_{ff'}(m + a_f - a_{f'}, m)] + h.c. \tag{194}$$

where $x = \sum_{m,f} n_{mf}\mathbf{r}_{mf}$ is the proton polarization operator and the vector connected the centers of hydrogen bonds is equal to $\mathbf{R}_{ff'} = \mathbf{R}_{ff'}(mm) = \mathbf{r}_{mf} - \mathbf{r}_{mf'}$.

Using the procedure with the deformation of integration counter in the complex plane proposed in Ref. 155, Pavlenko and Stasyuk [149,150] have arrived at the following result:

$$\sigma(\omega) = \frac{e^2\Omega^2_{\text{rot}}}{\hbar^2} \frac{2\sqrt{\pi}}{\mathcal{V}} \frac{\sinh(\hbar\omega/2k_B T)}{\hbar\omega/2} e^{-5E_0/(12k_B T)} \\ \times \tilde{\tau} \sum_{f,f'} \mathbf{R}^2_{ff'} \bar{n}_f(1 - \bar{n}_{f'}) e^{\hbar\alpha_{ff'}/(2k_B T)} e^{-\tau^2(\omega+\alpha_{ff'})^2} \tag{195}$$

where $\alpha_{ff'} = \Phi_0(1 - \bar{n}_f + \bar{n}_{f'})/2\hbar$ and

$$\Phi_0 = \Phi_{12}(m, m + \mathbf{a}_1 - \mathbf{a}_2) = \Phi_{12}(m, m + \mathbf{a}_1 - \mathbf{a}_3) = \Phi_{12}(m, m + \mathbf{a}_1 - \mathbf{a}_3)$$

are the interaction energy matrix elements presented in the approximation of nearest neighbors.

The activation energy has the form

$$E_{\rm a}^{ff'} = \frac{5}{12}E_0 - \frac{1}{4}\Phi_0 \cdot (1 - \bar{n}_f + \bar{n}_{f'}) + \frac{3}{80}\Phi_0^2 \cdot (1 - \bar{n}_f + \bar{n}_{f'})^2 / E_0 \qquad (196)$$

which includes the contribution from different transfer processes. The first term in expression (196) is the typical polaronic part, and the next two terms are caused by the interproton interaction and proton ordering. $E_{\rm a}^{ff'}$ changes with the temperature T as the proton ordering $\bar{n}_f$ is a function of T.

The temperature dependence of the different phonon-activated transfer processes is defined by the redistribution of proton occupancies $\bar{n}_j$ of three sublattices $(j = 1, 2, 3)$. For example, in the web-superionic phase the hydrogen-bonded network is disordered, and that is why the system is exemplified by only one activation energy, $E_{\rm a}^0 = \frac{5}{12}E_0 - \frac{1}{4}\Phi_0 + \frac{3}{80}\Phi_0^2/E_0$.

At low temperature, in phase III the saturation is realized (or $\bar{n}_2 = 1$ and $\bar{n}_1 = \bar{n}_3 = 0$); in this case, $E_{\rm a}^{23} = \frac{5}{12}E_0 - \frac{1}{4}\Phi_0 + \frac{3}{80}\Phi_0^2/E_0$ and $E_{\rm a}^{12} = E_{\rm a}^{13} = \frac{5}{12}E_0$, which typically holds for the strong polaron effect. In phase IV, the ordering features the components $\bar{n}_3 = 1$ and $\bar{n}_1 = \bar{n}_2 = 0$, and then the activation energy is $E_{\rm a}^{31} = E_{\rm a}^{32} = \frac{5}{12}E_0$. Therefore, in ordered phases, the activation energy is always higher than in the superionic phase that agrees with the experiments [155,156].

A numerical calculation of the conductivity (195) has been performed in Refs. 149 and 150 at different sets of parameters. The energy of activation estimated in Ref. 149 varied from 0.288 to 0.32×10^{-19} J, while the experimentally obtained value was approximately $0.48 \cdot 10^{-19}$ J. The better fit to the experimental data as mentioned in Ref. 149 could be possible if the additional short-range proton correlations would be taken into account.

Short-range proton correlations have been introduced in the starting Hamiltonian in Ref. 150; that is, the Hubbard operators describing the energy of the hydrogen bond configuration with two possible protons have entered a new term in the total Hamiltonian of the system studied, namely,

$$J \sum_{m,f} (X_{mf}^{ab} + X_{mf}^{ba}) \qquad (197)$$

where J is the usual resonance integral that has been added to the term that described the reorientational "tunneling" motion of protons (187). $X_{mf}^{ab} = |p\rangle_{mf}\langle q|_{mf}$ are the projection Hubbard operators, constructed on states

$$\begin{aligned} |0\rangle_{mf} &= |00\rangle_{mf} \qquad \text{(there is no hydrogen bond)} \\ |a\rangle_{mf} &= |10\rangle_{mf} \qquad \text{(proton is in the left well)} \\ |b\rangle_{mf} &= |01\rangle_{mf} \qquad \text{(proton is in the right well)} \end{aligned} \qquad (198)$$

In particular, this allows the representation of the operators of occupation of hydrogen bonds through the Hubbard operators

$$n_{mf} = X_{mf}^{aa} + X_{mf}^{bb} \tag{199}$$

The numerical estimate [150] of the proton conductivity (195) derived in the framework of the model described above is in good agreement with the experimental results. This is the direct theoretical justification of the experimentalists' remark [152,153,157] that the reorientation of ion groups should assist the proton transport in some superionic conductors.

D. Polaronic Conductivity Along a Hydrogen-Bonded Chain

In Section A, we have already considered the approach that includes the movement of two types of faults: ionic state (positive or/and negative) and Bjerrum faults. It has been suggested (see also, e.g., Refs. 158–160) that the drift of the ionic state (i.e., strictly speaking, an excess proton or proton hole) along the ordering chain changes the polarization of each site and therefore changes the polarization of the whole chain.

Thus, for the motion of the next state along the chain it is necessary to repolarize the chain into its initial state, which can be achieved by Bjerrum-fault transfer. The semiphenomenological theory of proton transfer along the hydrogen-bonded chain of the ice-like structure, developed in Ref. 161, includes the influence of longitudinal acoustic vibrations of the chain sites on the proton subsystem. Reference 162, in which the dynamics of the ionic state formation in the hydrogen-bonded chains is considered, resembles roughly Ref. 161.

One more approach based on the small polaron concept regarding the description of the motion of protons along a hydrogen-bonded chain, which allows the detailed calculation of the proton electric conductivity, was proposed in Refs. 163–165. In Ref. 163 the ionic state transfer along the weak hydrogen-bonded chain was treated. In that paper, expressions for the proton current density are derived in the small polaron model with allowance for the strong ionic coupling with intrasite vibrations of the nearest-neighbor A—H groups, $\nu(\text{AH}) \approx 10^{14}$ Hz (the weak coupling slightly decreases the A—H group longitudinal vibrations frequency, it is characterized by a large equilibrium distance $\text{A} \cdots \text{A}$, $0.29\ \text{nm} \leq r_{\text{A}\ldots\text{A}} \leq 0.34\ \text{nm}$ [166], and, therefore, the proton is strongly localized in the potential well near one of the A atoms). However, as was mentioned in Ref. 167 (see also Ref. 3), the necessary reason for the proton transfer in the hydrogen-bonded chain was the large proton polarizability of hydrogen bonds. This polarizability is typical for the strong hydrogen bonds, which are specified by Ref. 167: (i) a considerable drop of the longitudinal vibrations ν (AH) from 6×10^{13} Hz to 3×10^{12} Hz in comparison with the previous mentioned 10^{14} Hz; (ii) a short equilibrium distance A...A, $0.25\ \text{nm} \leq r_{\text{A...A}} \leq 0.28\ \text{nm}$; and (iii) a small height of the energy barrier between two wells, $E_w \ll h\nu_0$ ($\nu_0 \sim 10^{14}$ Hz, the proton frequency in an isolated well). The consequence of the above-mentioned facts is the proton delocalization and the appearance of optical polarization vibrations of it in the hydrogen bond with frequency $\nu = (3.6 \text{ to } 6) \times 10^{12}\ \text{Hz} \approx k_{\text{B}}T/h$, where $T \approx 300$ K. These conditions have been well known for many years (see, e.g., Ref. 168). Besides, numerical calculations by Scheiner [168] show that in a chain with the number of bonds exceeding four, the influence of marginal effects upon the excess proton located on the middle molecule can almost be neglected.

Let us suppose that the strong hydrogen-bonded chain is absolutely ordered either by a strong external electric field $\mathscr{E}$ or by the asymmetrical arrangement of side bonds (Fig. 8). Then we assume that thermally activated structural defects in a hydrogen-bonded chain (like those in ice) are practically excluded (in the similar structure system—that is, ice—the concentration of Bjerrum faults is 5×10^{-7} per molecule H_2O, and the concentration of ions is 10^{-12} per molecule H_2O [159]).

Let us consider the migration of a charge carrier—that is, of an ionic state (an excess proton or a proton hole)—assuming that the chain does not change the polarization after a charge carrier has passed along it (Fig. 9). This is true if, firstly, the charge carrier lifetime τ_0 on the lth site considerably exceeds the time $\delta\tau$ required for hopping from the neighboring $(l-1)$th site, and, secondly, the initial polarization recovers on the $(l-1)$th site for times less than or of the order of τ_0. Such an approach assumes that the Bjerrum fault, which follows the ionic state, is removed by the repolarization of the $(l-1)$th site before an ionic state can pass to the next site. This Bjerrum-fault turning can be caused by electrical momenta of the neighboring ordered A–H chain groups. If we treat the case of a strong longitudinal field $\mathscr{E}$ (for instance, $\mathscr{E} = 10^7$ to 10^8 V/m), the

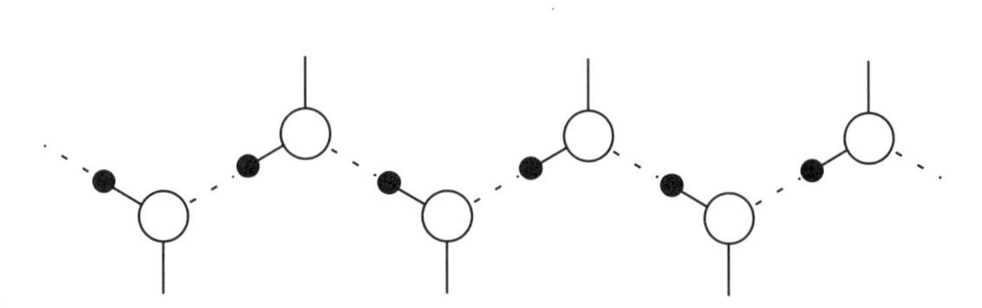

Figure 8. Chain of hydrogen bonds in longitudinal electric field $\mathscr{E}$; ○ is atom A, where A = O, N, F; ● is H; ----- is hydrogen bond; —— is covalent bond.

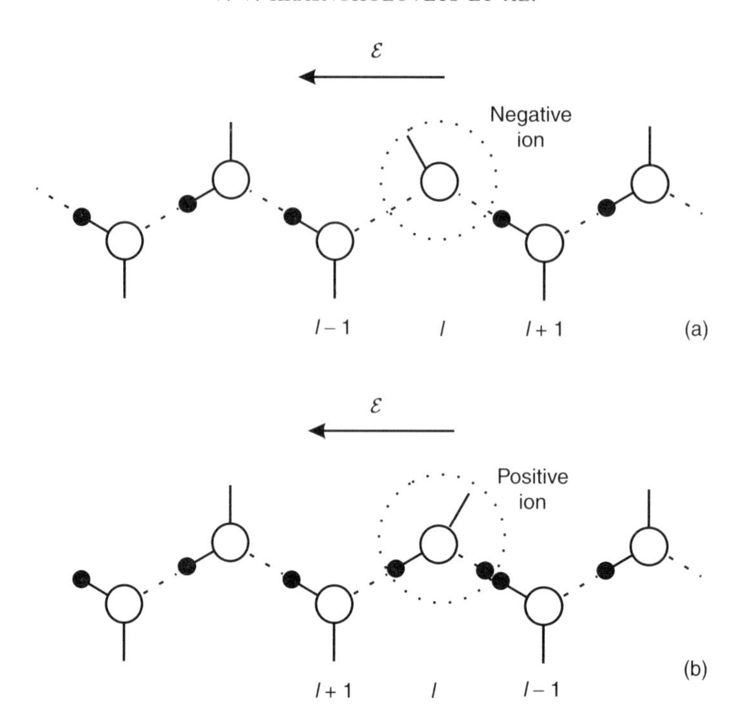

Figure 9. (a) Ionic state (−) moves to the right; this corresponds to the motion of the chain protons to the left. (b) Ionic state (+) moves to the left; this corresponds to the movement of chain protons to the left, too.

initial polarization of the A—H···A complex broken in the neighborhood of the ion recovers both under the influence of the large total polarization of other ordered chain links and under the influence of the field itself. The fact that the total polarization of the ordered chains is large in the electric field is a consequence of the strong polarizability of hydrogen bonds. The latter, if we neglect the coupling of separate O—H···O complexes, was shown in earlier works by Zundel and collaborators [169,170]. They demonstrated that in field $\mathscr{E} = 10^7$ to 10^8 V/m the strong hydrogen bond polarizability is of the order of 10^{-28} m^3. This value exceeds the polarizability of the isolated $(H_3O)^+$ ion and water molecules by 10^2 times.

Thus in view of the aforementioned reasons, for times $t \geq \tau_0$ we can ignore the local changes of the chain polarization while an ionic state moves along it, and imagine the latter as is shown in Fig. 9.

In the small concentration of charge carriers the hydrogen-bonded chain Hamiltonian can be written as

$$H = H_0 + H_1 + H_{\text{tun}} \tag{200}$$

$$\begin{aligned} H_0 &= \sum_l E\hat{a}_l^+\hat{a}_l + \sum_q \hbar\omega_0(q)\left(\hat{b}_q^+\hat{b}_q + \frac{1}{2}\right) \\ &\quad - \sum_{l,q} \hbar\omega_0(q)\hat{a}_l^+\hat{a}_l[u_l(q)\hat{b}_q^+ + u_l^*(q)\hat{b}_q] \end{aligned} \tag{201}$$

$$H_{\text{tun}} = \sum_{l,m} J(R_m)\hat{a}_{l+m}^+\hat{a}_l \tag{202}$$

where E is the quasi-particle energy in the site; $\hat{a}_l^+(\hat{a}_l)$ is the charge carrier creation (annihilation) Fermi operator in the lth site; $\hat{b}_q^+(\hat{b}_q)$ is the optical phonon creation (annihilation) Bose operator with momentum q and frequency $\omega_0(q)$; the tunneling Hamiltonian operator (202) is taken as perturbance; $J(R_m)$ is the wave-function overlap integral of charge carriers responsible for the carrier intersite hopping;

$$u_l(q) = N^{-1/2}e^{iqR_l}[u_1e^{-iqg} + u_0 + u_2e^{iqg}]$$

is a dimensionless quantity characterizing the displacement of protons in the nearest-neighbor $(l-1)$th and $(l+1)$th sites (as well as that of the lth site, in the case when the carrier is an excess proton) due to polarization of carrier interaction with the proton vibrations on the hydrogen bonds to the left (to the right) of it; u_0 is a constant responsible for the carrier interaction with the lth site proton (in case when the charge carrier is a proton hole, then $u_0 \equiv 0$); g is the chain constant; $R_l = lg$, where l is the site index; and N is the number of chain sites.

The operator H_2 introduced in expression (200) corresponds to a possible optical and acoustical phonon interaction in the chain (this is because AH groups of which the chain is constructed would be treated as external, and then the entire compound governs the behavior of the A atoms embedded in the chain; see also the next subsection). In the adiabatic approximation, it is expressed in the following way:

$$\begin{aligned} H_2 &= \sum_k \hbar\omega_{\text{ac}}(k)\left(\hat{B}_k^+\hat{B}_k + \frac{1}{2}\right) \\ &\quad + \hbar\chi\sum_{l;k,q} n_l[u_l(q)\hat{b}_q^+ + u_l^*(q)\hat{b}_q][\upsilon_l(k)\hat{B}_k^+ + \upsilon_l^*(k)\hat{B}_k] \end{aligned} \tag{203}$$

where the second term describes the reverse influence of displaced protons ($u_l(q)$ value) of the $(l-1)$th and $(l+1)$th sites upon the ion located in the lth site which has induced these displacements. The interaction results in the attraction of the ion itself to the $(l+1)$th site proton (Fig. 9). In expression (203), $\hat{B}_k^+(\hat{B}_k)$ stands for an acoustical phonon creation (annihilation) Bose operator with momentum k and frequency $\omega_{ac}(k)$; $\upsilon_l(k)$ is the dimensionless site vibration amplitude; χ is the constant of possible interaction of acoustic and optical phonons with charged fault (in s^{-1}); and $n_l = \langle \hat{a}_l^+ \hat{a}_l \rangle$ is the number of carriers in the chain (below we put $\langle n_l \rangle = n = \text{const}$).

The Hamiltonian (203) is diagonalized in the operators $\hat{b}_q$, $\hat{B}_k$ by the Bogolyubov–Tyablikov canonical transformation [171,172]. The diagonalization results are (see Appendix C)

$$H = \tilde{H}_0 + H_{\text{tun}} \tag{204}$$

$$\tilde{H}_0 = \sum_l E\hat{a}_l^+\hat{a}_l + \sum_{l=1}^{2}\sum_q \hbar\Omega(q)\left(\hat{b}_l^+(q)\hat{b}_l(q) + \frac{1}{2}\right) - \sum_{\alpha=1}^{2}\sum_{l;q} \hbar\Omega(q)\hat{a}_l^+\hat{a}_l[\tilde{u}_{\alpha l}(q)\hat{b}_\alpha^+(q) + \tilde{u}_{\alpha l}^*(q)\hat{b}_\alpha(q)] \tag{205}$$

Here

$$\Omega_{1,2}(q) = \left\{\frac{1}{2}(\omega_0^2(q) + \omega_{ac}^2(q)) \pm \frac{1}{2}\left[(\omega_0^2(q) - \omega_{ac}^2(q))^2 + 4\chi^2 n^2 |u_q|^2 |\upsilon_q|^2 \omega_0^2(q)\omega_{ac}^2(q)\right]^{1/2}\right\}^{1/2} \tag{206}$$

$$\tilde{u}_{\alpha l}(q) = 2u_l(q)\begin{cases} \dfrac{\omega_0^2(q)}{\Omega_1(q)[\Omega_1(q) + \omega_0(q)]} & (\alpha = 1) \\[2ex] \dfrac{\Omega_2(q) - \omega_{ac}(q)}{\chi n |u_q||\upsilon_q|} & (\alpha = 2) \end{cases} \tag{207}$$

where u_q and υ_q are defined by $u_l(q)$ and $\upsilon_l(q)$, respectively, divided by $N^{-1/2}\exp(iqR_l)$.

Let us note that with the variation of χ from zero to the maximally permissible value $\chi_{\max} = [\omega_0(q)\omega_{ac}(q)]^{1/2}/2n|u_q||\upsilon_q|$, the frequencies $\Omega_{1,2}(q)$ change within

$$\omega_0(q) < \Omega_1(q) < \sqrt{\omega_0^2(q) + \omega_{ac}^2(q)}, \qquad \omega_{ac}(q) > \Omega_2(q) > 0 \tag{208}$$

$\chi_{\max}$ corresponds to $\Omega_2(q) = 0$.

In the Hamiltonian (205) the operator $\hat{b}_1^+(q)(\hat{b}_1(q))$ describes the creation the creation (annihilation) of an optical phonon with frequency $\Omega_1(q)$ and takes into account the polarizing influence of the charged fault on proton subsystem. The operator $\hat{b}_2^+(q)(\hat{b}_2(q))$ corresponds to creation (annihilation) of an acoustic phonon with frequency $\Omega_2(q)$ relative to new equilibrium site positions; in this case the renormalized acoustic phonons become also polarizational ones. This is taken into account in the last term of the Hamiltonian $\tilde{H}_0$ describing the ionic state interaction with the new optical and acoustical vibrations of the chain.

Let us now consider the proton current in the hydrogen-bonded chain studied. With the use of the unitary operator

$$S = \sum_{\alpha=1}^{2} \sum_{l;q} \hat{a}_l^+ \hat{a}_l [\tilde{u}_{\alpha l}(q) \hat{b}_\alpha^+(q) - \tilde{u}_{\alpha l}^*(q) \hat{b}_\alpha(q)] \tag{209}$$

we perform the canonical transformation $\bar{H} = e^{-S} \tilde{H} e^{S}$, which is diagonalized with an accuracy to terms of fourth order in the Fermi operator:

$$\bar{H}_0 = \sum_l \bar{E} \hat{a}_l^+ \hat{a}_l + \sum_{\alpha=1}^{2} \sum_q \hbar \Omega_\alpha(q) \left[\hat{b}_\alpha^+(q) \hat{b}_\alpha(q) + \frac{1}{2} \right] \tag{210}$$

where

$$\bar{E} = \bar{E}_l = E - \sum_{\alpha=1}^{2} \sum_q \Omega_\alpha(q) |\tilde{u}_{\alpha l}(q)|^2 \tag{211}$$

(the second term on the right-hand side is the so-called polaron shift). The operator $\bar{H}_0$ is site-diagonal and characterizes stationary states of the system, and it resembles the small polaron Hamiltonian with allowance for $\bar{H}_1 = e^{-S} H_1 e^{S}$ (the Hamiltonian H_1 is assumed to be a perturbation). But, since $\bar{H}_0$ includes in addition to polarizational also optical renormalized phonons, operator (204)—or, more exactly $\bar{H}_0 + \bar{H}_1$—would be more properly called Hamiltonian for a polaron–condenson whose binding energy is defined by (211).

Let us calculate the proton current density of I due to hopping (at room temperature). On calculating I, we apply the method based on the usage of the statistical operator, proposed earlier by Hattori [173] for another type of systems. Setting $J(R_m) \ll k_{\mathrm{B}}T$ where $k_{\mathrm{B}}T \approx 300$ K, with accuracy to terms of order $J^2(R_m)$ we have for the current density

$$I = \mathrm{Tr}(\rho_{\mathscr{E}} j) \tag{212}$$

The current density operator is

$$j_{\mathscr{E}} = \frac{e}{\mathscr{V} i\hbar}\left[\sum_l R_l \hat{a}_l^+ a_l, H_1\right] = \frac{e}{i\hbar}\sum_{l,m} J(R_m)(R_{l+m} - R_l)\hat{a}_{l+m}^+ a_l \tag{213}$$

where $\mathscr{V}$ is an effective volume occupied by a carrier in the chain and the operator $\rho_{\mathscr{E}}$ is the density matrix (or statistical operator) correction due to the interaction with the external field $\mathscr{E}$:

$$\rho_{\mathscr{E}} = -\frac{i}{\hbar}\int_{-\infty}^{t} d\tau \exp\left[-\frac{i}{\hbar}(t-\tau)(H + H_{\mathscr{E}})\right][H_{\mathscr{E}}, \rho_1] \times \exp\left[\frac{i}{\hbar}(t-\tau)(H + H_{\mathscr{E}})\right] \tag{214}$$

Here the field Hamiltonian is

$$H_{\mathscr{E}} = -e\mathscr{E}\sum_l R_l \hat{a}_l^+ \hat{a}_l \tag{215}$$

and the correction to the statistical operator is

$$\rho_1 = -\frac{e^{-\tilde{H}_0/k_B T}}{\mathrm{Tr} e^{-\tilde{H}_0/k_B T}}\int_0^{1/k_B T} d\lambda e^{\lambda\tilde{H}_0} H_{\mathrm{tun}} e^{-\lambda\tilde{H}_0} \tag{216}$$

Expression (212) is reduced to

$$I = \frac{\mathscr{E}}{\mathscr{V}\,\mathrm{Tr} e^{-\tilde{H}_0/k_B T}}\int_0^{\infty} d\tau \int_0^{1/k_B T} d\lambda \mathrm{Tr}\{e^{-\tilde{H}_0/k_B T} e^{\tilde{H}_0\lambda} j(-\tau) e^{-\tilde{H}_0\lambda} j\} \tag{217}$$

$$j(-\tau) = \exp\left[-\frac{i}{\hbar}\tau(\tilde{H}_0 + H_{\mathscr{E}})\right] j \exp\left[\frac{i}{\hbar}\tau(\tilde{H}_0 + H_{\mathscr{E}})\right] \tag{218}$$

Using the invariability of the trace, we perform the canonical transformation in expression (217) by means of quantum mechanical thermal averaging over Fermi operators with an accuracy to terms linear in the carrier concentration, and then averaging over the phonon bath, we obtain [165]

$$I = \mathscr{E} e^2 g^2 J^2 n h^{-2}\int_0^{\infty} d\tau \int_0^{1/k_B T} d\lambda \exp\left(\frac{i}{\hbar}\tau e\mathscr{E} g\right)\exp\left(-2\sum_l S_{T,\alpha}\right) \times \left\{\exp\sum_{\alpha;q}|\Delta^{(\alpha)}(q)|^2 \mathrm{cosech}\frac{\hbar\Omega_\alpha(q)}{2k_B T}\cos\left[\left(\tau - \frac{i\hbar}{2k_B T} + i\hbar\lambda\right)\Omega_\alpha(q)\right]\right\} \tag{219}$$

where the constant of charge carrier (i.e., proton) coupling with optical ($\alpha = 1$) and acoustical ($\alpha = 2$) phonons is

$$S_{T,\alpha} = \frac{1}{2}\sum_{q} |\Delta^{(\alpha)}(q)|^2 \coth\frac{\hbar\Omega_\alpha(q)}{2k_\mathrm{B}T} \tag{220}$$

$$|\Delta^{(\alpha)}(q)|^2 = 4|u_q|^2\frac{2}{N}(1-\cos qg)\begin{cases} \dfrac{\omega_0^4(q)}{\Omega_1^2(q)[\Omega_1(q)+\omega_0(q)]^2} & (\alpha = 1) \\ \dfrac{[\Omega_2(q)-\omega_\mathrm{ac}(q)]^2}{\chi^2 n^2 |u_q|^2 |v_q|^2} & (\alpha = 2) \end{cases} \tag{221}$$

In view of the comparatively large mass, a proton drifting along the hydrogen-bonded system resembles a classical particle; therefore in order to obtain expression (219), we have performed the following simplifications. The assumption mentioned above is that the overlap integral is $J(R_m) \ll k_\mathrm{B}T$, and for $\delta E = E - \bar{E} > \hbar\Omega_1(q) \geq \hbar\omega_0(q) \sim k_\mathrm{B}T$ the inequality $J(R_m) \ll \delta E$ is valid. This shows that on drifting along the chain a proton hops only to neighboring sites. Under these conditions the overlap integral is nonzero only between nearest neighbors, $J(g) = J = \text{const}$. Besides, in expression (217) after averaging over Fermi operators, terms with $R_{l+m} - R_l = g$ are nonzero in the sums over l and m. Let us also note that the assumed condition $J \ll k_\mathrm{B}T$ enables us to take into account the field $\mathscr{E}$ to practically any order in expressions (217) and (219).

In the small polaron theory the coupling constant S_T is much greater than 1, which allows the calculation of integrals in expression (219) [165]

$$\begin{aligned} I = {} & 2^{3/2}\pi^{1/2} egnJ^2\hbar^{-2} e^{-\frac{E_\mathrm{a}}{k_\mathrm{B}T}} \left[\sum_{\alpha;q} |\Delta^{(\alpha)}(q)|^2 \Omega_l^2(q) \operatorname{cosech}\frac{\hbar\Omega_\alpha(q)}{2k_\mathrm{B}T}\right]^{-1/2} \\ & \times \sinh\frac{eg\mathscr{E}}{2k_\mathrm{B}T} \exp\left(-\frac{(eg\mathscr{E})^2}{\sum_{\alpha;q} |\Delta^{(\alpha)}(q)|^2 \hbar^2 \Omega_l^2(q) \operatorname{cosech}\frac{\hbar\Omega_\alpha(q)}{2k_\mathrm{B}T}}\right) \end{aligned} \tag{222}$$

where n is the polaron concentration and the activation energy of polaron hops is

$$E_\mathrm{a} = \frac{1}{k_\mathrm{B}T}\sum_{\alpha;q} |\Delta^{(\alpha)}(q)|^2 \tanh\frac{\hbar\Omega_\alpha(q)}{4k_\mathrm{B}T} \tag{223}$$

An analysis conducted in Ref. 165 has shown that the phonon–phonon interaction that is described by the constant χ does not virtually influence $S_{T,1}$. However, the interaction leads to a considerable coupling of a polaron with acoustic phonons, and therefore the value of $S_{T,2}$ is able to reach $S_{T,1}$ (in the case when the concentration of protons is not very small).

A hydrogen-bonded chain with no temperature-activated free charge carriers is a dielectric. But charge carriers can be injected by the external electric field. Such a situation corresponds to the case of space-charge-limited proton current. A number of materials, including ice crystals—that is, a system with a hydrogen-bonded network—exhibit stationary proton injection currents [174]. Thus the space-charge-limited proton current can be found in hydrogen-bonded chains. To calculate the current, we start from the following equations [175]:

$$I = en\mu(\mathscr{E})\mathscr{E} \tag{224}$$

$$\frac{d\mathscr{E}}{dx} = \frac{en}{\varepsilon_0 \varepsilon} \tag{225}$$

where $\mu(\mathscr{E})$ is the carrier mobility, ε is dielectric constant of the system studied, and ε_0 is the dielectric constant of free space. In Eqs. (224) and (225) we usually ignore the diffusion current, because the connection of the current with the local field is the same as that with the field average.

Let us calculate the space-charge-limited density of proton current $I \cdot \mu(\mathscr{E})$ needed for this purpose is obtained from the right-hand side of formula (222) divided by $en\mathscr{E}$. Putting en from Eq. (224) into Eq. (225), we arrive at the differential equation

$$\mu(\mathscr{E}')\mathscr{E}'\, d\mathscr{E}' = \frac{I}{\varepsilon_0 \varepsilon}\, dx \tag{226}$$

Here $0 \le \mathscr{E}' \le \mathscr{E}$, where $\mathscr{E}$ is the applied field; $0 \le x \le L$, where L is the screening length of the charge in the chain. The solution to Eq. (226) with account for expression (223) is [165]

$$I = \frac{2^{5/2}\pi^{1/2}\varepsilon_0}{\hbar^2 e} \frac{\varepsilon k_{\mathrm{B}} T J^2}{L} \left(\cosh \frac{e\mathscr{E} g}{2k_{\mathrm{B}}T} - 1 \right) \times \frac{\exp(-E_{\mathrm{a}}/k_{\mathrm{B}}T)}{\left[\sum_{\alpha;q} |\Delta^{(\alpha)}(q)|^2 \Omega_\alpha^2(q) \operatorname{cosech} \frac{\hbar\Omega_\alpha(q)}{2k_{\mathrm{B}}T} \right]^{1/2}} \tag{227}$$

The electrical conductivity σ is easy defined by expression (217) or (221) for I because, by definition, $I = \sigma(\mathscr{E})\mathscr{E}$.

Experimental values of conductivity σ_c at room temperature in the $Li(N_2H_5)SO_4$ and $(NH_4)_3H(SO_4)_2$ crystals (i.e., conductivity along hydrogen-bonded chains) are equal to $0.23 \times 10^{-8}\ \Omega^{-1}\,\mathrm{cm}^{-1}$ [29] and $5 \times 10^{-8}\ \Omega^{-1}\,\mathrm{cm}^{-1}$ [30], respectively. Choosing reasonable values of three major parameters J, S_T, and Ω, we obtain [165] the same values of σ by derived expressions (223) and (227). Moreover, expressions (223) and (227) correlate well with the measured proton conductivity of proton-conducting membrane proteins (see Refs. 33 and 34).

E. The Anharmonicity Influence

In complex compounds including proton-conducting proteins the hydrogen-bonded chain is formed by means of AH groups, the position of which is rather rigidly fixed by the side chemical bonds. Therefore the distances between the sites (AH groups) and, thus, the chain rigidity appear to be functions of external parameters—that is, thermostat functions. In other words, if we neglect a comparatively weak influence of the light H atoms upon the vibrations of A atoms, then the A atom participation in acoustic vibrations of the molecular chain will be entirely environmental, with A atoms being rigidly connected to the environment by the side chemical bonds. Therefore, these acoustic vibrations appear to be external with respect to the chain. Owing to this fact, a bilinear connection of the acoustic vibrations with the longitudinal optical polarizational phonons (the chain proton vibrations on the hydrogen bonds) responsible for the proton formation appears. On calculating the proton conductivity, all these peculiarities of a hydrogen-bonded chain have been taken into account.

Apart from the bilinear connection between acoustic and optical vibrations, the usual phonon anharmonicity takes place in the chain. The acoustic vibrations modulate the distances between A atoms and thus influence the frequency of proton vibrations on hydrogen bonds. The proton vibration frequency can be presented as

$$\omega(q) \to \omega_q + \frac{\partial \omega}{\partial U_q(r)} \delta U_q(r) \tag{228}$$

where $U_q(r)$ is the coordinate of the normal acoustic vibrations of the chain. The calculation of the second member in expression (228) results in an addendum of the type

$$H' = \sum_q (\kappa_q \hat{B}_q^+ + \kappa_q^* \hat{B}_q) \hat{b}_q^+ \hat{b}_q \tag{229}$$

appearing in the Hamiltonian that describes the motion of an excess proton (or proton hole) along a hydrogen-bonded chain; here κ_q is the anharmonicity constant, $\hat{B}_q^+ (\hat{B}_q)$ is the acoustic phonon creation (annihilation) operator, and $\hat{b}_q^+ (\hat{b}_q)$ is the analogous optical phonon operator.

In Ref. 176 the estimation of anharmonicity on the proton conductivity, neglecting the bilinear phonon–phonon interaction, has been calculated. The proton current density is obtained from expression (219); in this formula we should now formally put $\nu \to 1$, $|\Delta^{(\nu)}(q)| \to |\Delta_q|$, $S_{T1} \to S_T$, and $\Omega_1(q) \to \omega_q [1 + (\kappa_q \hat{B}_q^+ + \kappa_q^* \hat{B}_q)]$, and the thermal averaging after Gibbs should be additionally performed:

$$\frac{e^{-\frac{1}{k_B T}\sum_q \hbar\Omega_q \left(\hat{B}_q^+ \hat{B}_q + \frac{1}{2}\right)}}{\mathrm{Tr}\, e^{-\frac{1}{k_B T}\sum_q \hbar\Omega_q \left(\hat{B}_q^+ \hat{B}_q + \frac{1}{2}\right)}} \tag{230}$$

where $\hbar\Omega_q$ is the energy of the thermal vibration quantum—that is, the lattice acoustic phonon energy. As a result, we gain

$$\begin{aligned} I = {} & 2^{3/2}\pi^{1/2}\hbar^{-2} engJ^2 \sinh\frac{e\mathscr{E}g}{2k_B T} \\ & \times e^{-E_a/k_B T}\left[\sum_q |\Delta(q)|^2 \omega_q^2 \operatorname{cosech}\frac{\hbar\omega_q}{2k_B T}\right]^{-1/2} \\ & \times \exp\left\{\sum_q |\kappa_q|^2 |\Delta_q|^2 \left(\frac{\hbar\omega_q}{2k_B T}\right)^2 \operatorname{cosech}^4 \frac{\hbar\omega_q}{2k_B T}\right. \\ & \left. \times \left(1 - \operatorname{cosech}\frac{\hbar\omega_q}{2k_B T}\right)^2 \coth\frac{\hbar\Omega_q}{2k_B T}\right\} \end{aligned} \tag{231}$$

Expression (231) is correct at $\mathscr{E} < 10^8$ V/m and $|\kappa_q| \ll 1$. Setting $\kappa_q = 0$, we arrive at the current density derived in the pure small polaron theory [88].

If we put $|\kappa_q| = 0.1$, $|\Delta_q| = 20$–30, and $\omega_q = 2\pi \cdot 6 \times 10^{12}\ \mathrm{s}^{-1}$ and T is around room temperature, we can estimate the last term exponential factor in expression (231) by

$$\exp\left\{\alpha \coth\frac{\hbar\Omega}{2k_B T}\right\} \tag{232}$$

where α varies from 10^{-3} to 10^{-2} and Ω is a typical frequency of the acoustic phonons conditioning the anharmonicity. Since the charge carriers (protons or proton holes) in a hydrogen-bonded chain should be described by strongly localized functions (see also Kittel [177]), they should also interact with the shortest wave vibrations. Let us take as the Debye value $\Omega_{\text{Debye}} = (1 \text{ to } 5)\times 10^{12}\ \mathrm{s}^{-1}$, then we can assume that $\Omega = (0.01 \text{ to } 0.1)\Omega_{\text{Debye}}$. Substituting these values into expression (232), we get a rough estimation of the last exponential factor in I (231): $e^{0.2}$ to e^1.

Thus, the adduced estimation shows that under the indicated values of parameters the anharmonicity can appreciably influence the proton current in compounds with hydrogen-bonded chains.

F. Influence of Coulomb Correlations and the Electric Field Local Heterogeneities on Proton Conductivity

Nonlinear transfer and transport phenomena in hydrogen-bonded chains can be caused not only by a strong coupling of moving protons with the chain sites, but also by the relative large concentration of charge carriers in the chain. The model proton polaron Hamiltonian, which takes into account proton–proton correlation in the chain, can be written as [178]

$$H = H_0 + H_{\text{tun}} \tag{233}$$

$$\begin{aligned} H_0 = {} & \sum_l E\hat{a}_l^+\hat{a}_l + \sum_q \hbar\omega_q\left(\hat{b}_q^+\hat{b}_q + \frac{1}{2}\right) \\ & - \sum_q \hbar\omega_q \hat{a}_l^+\hat{a}_l[u_l(q)\hat{b}_q^+ + u_l^*(q)\hat{b}_q] + \sum_{l,m} U_{lm}\hat{a}_l^+\hat{a}_l\hat{a}_m^+\hat{a}_m \end{aligned} \tag{234}$$

$$H_{\text{tun}} = \sum_l J(mg)\hat{a}_{l+m}^+\hat{a}_l + h.c. \tag{235}$$

Here once again $\hat{a}_l^+(\hat{a}_l)$ is the Fermi operator of charge carrier creation (annihilation) in the lth site, E is the quasi-particle (i.e., polaron) energy in the site, $\hat{b}_q^+(\hat{b}_q)$ is the optical phonon creation (annihilation) Bose operator with the momentum q and the frequency ω_q, U_{lm} is the energy of the Coulomb repulsion among the carriers sitting in the l and m sites, operator (235) is the tunnel Hamiltonian assumed to be perturbation, $J(mg)$ is the resonance overlap integral, $u_l(q)$ is the dimensionless value characterizing the displacement of protons in the $(l-1)$th and $(l+1)$th sites by the carrier located in the lth site, and $R_l = lg$, where l is the site ordinal index and g is the chain constant.

It should be noted that the Coulomb correlations among the carriers on one site should have been taken into account in the Hamiltonian H in the general case, yet we assume that the energy of such interaction is too large and thus the Hamiltonian in Eqs. (233)–(235) describes the carrier migration along the lower Hubbard band (about Hubbard correlations see, e.g., Ref. 173).

To derive the current density, we apply the method employed in the previous subsection. Thus we proceed from the formula [cf. with Eq. (212)]

$$I = \text{Tr}(\rho_{\text{tot}} j) \simeq \text{Tr}(\rho_{\mathscr{E}} j) \tag{236}$$

where ρ_{tot} is the system total statistical operator and j is the operator of current density. The operator ρ_{tot} satisfies the equation of motion

$$i\hbar\frac{\partial\rho_{\text{tot}}}{\partial t} = i\hbar\frac{\partial\rho_{\mathscr{E}}}{\partial t} = [H_{\text{tot}}, \rho] \tag{237}$$

where

$$H_{\mathrm{tot}} = H + H_{\mathscr{E}}, \qquad H_{\mathscr{E}} = H_{\mathscr{E}}^{(0)} + H_{\mathscr{E}}^{(1)} \tag{238}$$

$$\rho_{\mathrm{tot}} = \rho + \rho_{\mathscr{E}} \tag{239}$$

Here the operator $H_{\mathscr{E}}$ describes the interaction with the field $\mathscr{E}$ and

$$\begin{aligned} H_{\mathscr{E}}^{(0)} &= -e\mathscr{E} \sum_{l} R_l \hat{a}_l^{+} \hat{a}_l \\ H_{\mathscr{E}}^{(1)} &= -e\mathscr{E} \sum_{l,m} \mathscr{L}(mg) \hat{a}_{l+m}^{+} \hat{a}_l \end{aligned} \tag{240}$$

$\rho \simeq \mathrm{const}$ is the statistical operator of the system studied, when the field is absent, in linear approximation $\rho = \rho_0 + \rho_1$:

$$\rho_0 = \frac{e^{-H_0/k_{\mathrm{B}}T}}{\mathrm{Tr}\, e^{-H_0/k_{\mathrm{B}}T}} \tag{241}$$

$$\rho_1 = -\frac{e^{-H_0/k_{\mathrm{B}}T}}{\mathrm{Tr}\, e^{-H_0/k_{\mathrm{B}}T}} \int_0^{1/k_{\mathrm{B}}T} d\lambda e^{\lambda H_0} H_{\mathrm{tun}} e^{-\lambda H_0} \tag{242}$$

$\rho_{\mathscr{E}}$ is the correlation to ρ due to the interaction $H_{\mathscr{E}}$ [$\rho_{\mathscr{E}}$ is directly derived from Eq. (237)]:

$$\rho_{\mathscr{E}} = -\frac{i}{\hbar} \int_{-\infty}^{t} d\tau e^{-i\frac{t-\tau}{\hbar} H_{\mathrm{tot}}} [H_{\mathscr{E}}, \rho] e^{i\frac{t-\tau}{\hbar} H_{\mathrm{tot}}} \tag{243}$$

In the case where hopping motion is involved, the density operator j is written as

$$j = \frac{e}{\mathscr{V} i\hbar} [H_{\mathrm{tun}}, \Re] \tag{244}$$

where $\mathscr{V}$ is an effective volume occupied by a carrier in the chain and (in the general case) the coordinate operator is expressed as

$$\Re = \sum_{l} R_l \hat{a}_l^{+} \hat{a}_l + \sum_{l,m} \mathscr{L}(mg) \hat{a}_{l+m}^{+} \hat{a}_l \tag{245}$$

It should be noted that in expression (245) and, therefore, in expressions (240) we have neglected the coordinate dependence of the phonon subsystem (i.e., indirect transitions have been dropped). Then, with allowance for expressions (235) and

(245) we obtain the following from commutation (244) for j:

$$j = j^{(0)} + j^{(1)} \tag{246}$$

$$j^{(0)} = \frac{e}{\mathcal{V} i\hbar} \sum_{l,m} J(mg)(R_{l+m} - R_l)\hat{a}^{+}_{l+m}\hat{a}_l \tag{247}$$

$$j^{(1)} = \frac{e}{\mathcal{V} i\hbar} \sum_{l,m,n} [J(ng)\mathscr{L}((m-n)g) - J((m-n)g)\mathscr{L}(ng)]\hat{a}^{+}_{l+m}\hat{a}_l \tag{248}$$

In the Hamiltonian H_{tot} contained in expression (243), we neglect the operator H_1; such an approximation corresponds to the calculation of I with the accuracy of $J^2(g)$. Besides, in H_{tot} in expression (243) the operator $H^{(1)}_{\mathscr{E}}$ can also be neglected; in this case the criterion that it be small can be derived as follows. Let us formally expand in expression (243) $\exp[\pm i(t-\tau)(H_0 + H^{(0)}_{\mathscr{E}} + H^{(1)}_{\mathscr{E}})/\hbar]$ in a series in the operator $H^{(1)}_{\mathscr{E}}$ with restriction to the first members of the expansion. Then the corrections proportional to $J^2(mg)e\mathscr{E}\mathscr{L}(g)$ will appear in the current density. Hence as we neglect the members proportional to $J^3(g)$, acquire the following restriction on the parameters of the operator $H^{(1)}_{\mathscr{E}}$:

$$e\mathscr{E}\mathscr{L}(g) \leq J(g) \tag{249}$$

Note that the function $\mathscr{L}(g)$ characterizes the site neighborhood in which the meaning of the electric field is different from the applied field $\mathscr{E}$; that is, just $\mathscr{L}(g)$ is responsible for the electric field local heterogeneity.

With allowance for the above-mentioned facts, the current density I can be presented as the sum of three terms:

$$I = I_0 + I_1 + I_2 \tag{250}$$

where

$$I_0 = \text{Tr}(\rho^{(0)}_{\mathscr{E}} j^{(0)}) \tag{251}$$

$$I_1 = \text{Tr}(\rho^{(1)}_{\mathscr{E}} j^{(0)}) \tag{252}$$

$$I_2 = \text{Tr}(\rho^{(0)}_{\mathscr{E}} j^{(1)}) \tag{253}$$

$$\rho^{(0,1)}_{\mathscr{E}} = -\frac{i}{\hbar}\int_{-\infty}^{t} d\tau e^{-i\frac{t-\tau}{\hbar}(H_0 + H^{(0)}_{\mathscr{E}})}[H^{(0,1)}_{\mathscr{E}}, \rho_{(1,0)}]e^{i\frac{t-\tau}{\hbar}(H_0 + H^{(0)}_{\mathscr{E}})} \tag{254}$$

By means of the unitary operator

$$S = \sum_{l;q} \hat{a}^{+}_l \hat{a}_l [u_l(q)\hat{b}^{+}_q - u^{*}_l(q)\hat{b}_q] \tag{255}$$

we perform the canonical transformation $\bar{H}_0 = e^{-S} H_0 e^S$, diagonalizing the Hamiltonian with respect to the phonon variables

$$\bar{H}_0 = \sum_l \bar{E} \hat{a}_l^+ \hat{a}_l + \sum_q \hbar\omega_q \left(\hat{b}_q^+ \hat{b}_q + \frac{1}{2} \right) + \sum_{l,m} \bar{U}_{lm} \hat{a}_l^+ \hat{a}_l \hat{a}_m^+ \hat{a}_m \tag{256}$$

Here $\bar{E}$ is the carrier coupling energy on the site including the polaron shift, and $\tilde{U}_{lm}$ is the energy including not only the Coulomb correlations but also the correlations according for phonon exchange,

$$\bar{E} = E_l = E - \sum_q \hbar\omega_q |u_l(q)|^2 \tag{257}$$

$$\bar{U}_{lm} = \bar{U}_{ml} = U_{lm} - \sum_q \hbar\omega_q \mathrm{Re}[u_l(q) u_l^*(q)] \tag{258}$$

According to expressions (227) and (254), we get for the current density I_0 [178]

$$I_0 = \frac{\mathscr{E}}{\mathscr{V}\,\mathrm{Tr}\, e^{-H_0/k_B T}} \mathrm{Tr}\Big\{ e^{-H_0/k_B T} \int_0^\infty d\tau e^{-i\frac{\tau}{\hbar}(H_0 + H_{\mathscr{E}}^{(0)})}$$
$$\times \int_0^{1/k_B T} d\lambda e^{\lambda H_0} j^{(0)} e^{-\lambda H_0} e^{-i\frac{\tau}{\hbar}(H_0 + H_{\mathscr{E}}^{(0)})} j^{(0)} \Big\} \tag{259}$$

Calculating the integrals and performing thermal averaging over the phonon subsystem in expression (259), we obtain [178]

$$I_0 \simeq 2^{3/2} \pi^{1/2} \hbar^{-2} n g J^2 \sinh \frac{e\mathscr{E} g}{2k_B T}$$
$$\times e^{-E_a/k_B T} \left[\sum_q |\Delta(q)|^2 \omega_q^2 \,\mathrm{cosech}\, \frac{\hbar\omega_q}{2k_B T} \right]^{-1/2}$$
$$\times e^{\bar{U}/2k_B T} \exp\left(-\frac{(eg\mathscr{E} - \bar{U})^2}{2\sum_q |\Delta(q)|^2 \hbar^2 \omega_q^2 \,\mathrm{cosech}\, \frac{\hbar\omega_q}{2k_B T}} \right) \tag{260}$$

Here n is concentration of polarons and E_a is the activation energy [see expression (223)].

Making use of the explicit form of the commutator $[H_{\mathscr{E}}^{(1)}, \rho_0]$, we gain for the current density I_1 instead of (252) and (254)

$$I_1 = -\frac{i}{\hbar} \int_{-\infty}^0 d\tau \frac{1}{\mathscr{V}\,\mathrm{Tr} e^{-H_0/k_B T}} \mathrm{Tr}\Big\{ e^{i\frac{\tau}{\hbar}(H_0 + H_{\mathscr{E}}^{(0)})}$$
$$\times \left[-e\mathscr{E} \sum_{l,m} \mathscr{L}(mg) \hat{a}_{l+m}^+ \hat{a}_l, e^{-H_0/k_B T} \right] e^{-i\frac{\tau}{\hbar}(H_0 + H_{\mathscr{E}}^{(0)})} j^{(0)} \Big\} \tag{261}$$

The calculation of the right-hand side of (261) is analogous to that in the case of I_0. The final expression is [178]

$$I_1 \simeq \sqrt{2\pi}\hbar^{-2} ne\mathscr{E}\mathscr{L}gJe^{-E_a/k_B T}\left[\sum_q |\Delta(q)|^2 \omega_q^2 \operatorname{cosech} \frac{\hbar\omega_q}{2k_B T}\right]^{-1/2}$$
$$\times\, e^{\bar{U}/k_B T}(1 - e^{(3\bar{U} - e\mathscr{E}g)/2k_B T})\exp\left(-\frac{(eg\mathscr{E} - \bar{U})^2}{2\sum_q |\Delta(q)|^2 \hbar^2 \omega_q^2 \operatorname{cosech} \frac{\hbar\omega_q}{2k_B T}}\right) \quad (262)$$

Since the theory, which is developing, is applied to strongly localized heavy charge carriers (i.e., protons), we have performed the calculations in the limiting transition $l \to 1$, reflecting the fact that protons hop only to nearest sites. In this approximation, the contribution of I_2 to the total current density can be neglected [178]. Thus, the total current is

$$I_{\text{tot}} = I_0 + I_1 \quad (263)$$

where I_0 and I_1 are given by expressions (260) and (262), respectively.

Let us calculate the space-charge-limited density of proton current (263). In this case the polaron potential energy changes from site to site within the chain and, being due to the field, remains smaller than the value of the Coulomb repulsion between the carriers, U (the latter obstructs the filling of a chain by carriers by definition). The value of the Coulomb correlation energy is limited by the inequalities $J \ll \bar{U} < E - \bar{E}$, where $J \ll k_B T$ and the polaron shift energy $E - \bar{E}$ can be approximately identical to the fault (ionic state) activation energy within a chain, its value being of the order of 10–15 $k_B T$ [29,158,160]. Since $\mathscr{L}$ describes the neighborhood into which the carrier arrives after its hopping from the nearest neighbor at distances between the sites smaller than g, the inequality

$$e\mathscr{E}g < \bar{U} \quad (264)$$

should be considered as the criterion of validity of the small carrier concentration approximation. At such values of the field $\mathscr{E}$ the contribution of Coulomb correlations to the Hamiltonian (229) may be dropped [formally it corresponds to $\bar{U} = 0$ in expressions for current (260) and (262)].

To derive the space-charge-limited density of proton current that takes into account the local heterogeneities, which is given by the parameter $\mathscr{L}$, we proceed from Eq. (225) and the equation

$$I = I_0 + I_1 = en[\mu_0(\mathscr{E}) + \mu_1(\mathscr{E})]\mathscr{E} \quad (265)$$

where $\mu_0(\mathscr{E})$ is the carrier mobility without allowance for local changes of the field; $\mu_1(\mathscr{E})$ is a correction to $\mu_0(\mathscr{E})$ with allowance for local changes of the field. With $\mathscr{E} \leq 10^8$ V/cm the mobilities are [178]

$$\mu_0(\mathscr{E}) = 2J\mathscr{E}^{-1}\sinh(e\mathscr{E}g/2k_BT)M \tag{266}$$

$$\mu_1(\mathscr{E}) = e\mathscr{L}\exp(-e\mathscr{E}g/2k_BT)[1 - \exp(-e\mathscr{E}g/k_BT)]M \tag{267}$$

where

$$M = \sqrt{2\pi}\hbar^{-2}gJ\exp(-E_a/k_BT)\left[\sum_q |\Delta(q)|^2\omega_q^2 \operatorname{cosech}\frac{\hbar\omega_q}{2k_BT}\right]^{-1/2} \tag{268}$$

Solving equations (225) and (265), we derive the needed current density. It is convenient to write the solution in terms of the effective electroconductivity

$$\begin{aligned}\sigma(\mathscr{E}) &= \sigma_0(\mathscr{E}) + \sigma_1(\mathscr{E}) \\ &= \frac{2^{5/2}\pi^{1/2}\varepsilon_0\varepsilon k_BT}{\hbar^2 eL}\frac{\exp(-E_a/k_BT)}{\left[\sum_q |\Delta(q)|^2\omega_q^2\operatorname{cosech}\frac{\hbar\omega_q}{2k_BT}\right]^{1/2}} \\ &\quad \times\left(\frac{\cosh(e\mathscr{E}g/2k_BT)}{\mathscr{E}} + \frac{k_BT}{J}\frac{\mathscr{L}}{g}F(\mathscr{E})\right)\end{aligned} \tag{269}$$

where

$$F(\mathscr{E}) = 1 - \left(1 + \frac{e\mathscr{E}g}{2k_BT}\right)e^{-\frac{e\mathscr{E}g}{2k_BT}} - \frac{1}{9}\left[1 - \left(1 + \frac{3e\mathscr{E}g}{2k_BT}\right)e^{-\frac{3e\mathscr{E}g}{2k_BT}}\right] \tag{270}$$

An analysis of the results obtained shows [178] that the correction $\sigma_1(\mathscr{E})$ in (269) reaches the value of $\sigma_0(\mathscr{E})$ at the field $\mathscr{E} > 10^6$ V/m and at the ratio $\mathscr{L}/g = 0.001$ to 0.05.

Thus, considering the transport phenomenon in a hydrogen-bonded chain in an electric field $\mathscr{E} > 10^6$ V/m, the calculation of the local field changes from the value of the applied field becomes essential if $\mathscr{L}/g \geq 0.01$ (however, $\mathscr{L}/g \ll 1$). The Coulomb correlations must be accounted for at large values of the field, but are not likely to considerably change the general trend of the curve $\sigma(\mathscr{E})$, yet influence quantitative estimates of the values $\sigma_0(\mathscr{E})$ and $\sigma_1(\mathscr{E})$.

G. External Influences on the Proton Conductivity

Let us consider following Ref. 179 the character and value of the change of the hydrogen-bonded chain proton conductivity under the influence of light (two cases can be treated: absorption by a polaron and absorption by a chain) and ultrasound.

As is known from electron small polaron theory [88], transitions of charge carriers from site to site are possible not only due to phonon activation but owing to the action of light (so-called intraband absorption). With this regard the adiabatic transition of a carrier from the ground state of the lth site to the ground state of the $(l+1)$th site occurs through an interval of excited states. Assume that an excited virtual state is described by function $|l'\rangle$. The matrix element of such a process, corresponding to second order of perturbation theory, can be written as

$$\frac{\langle l|e\tilde{\mathscr{E}}(t)\cdot\mathbf{r}|l'\rangle\langle l'|e\tilde{\mathscr{E}}(t)\cdot\mathbf{r}|l+1\rangle}{E'-E} = \frac{(\mathbf{d}\cdot\tilde{\mathscr{E}}(t))^2}{\Xi}\hat{a}_{l+1}^{+}\hat{a}_l \tag{271}$$

Here $e\tilde{\mathscr{E}}(t)\cdot\mathbf{r}$ is the potential energy of a charge carrier, $\tilde{\mathscr{E}}(t)=\tilde{\mathscr{E}}_0\cos\Omega t$ is the variable electric field (in dipole approach), $\mathbf{r}$ is the charge carrier coordinate in a unit cell between the lth site and the $(l+1)$th site, $\Xi = E' - E$ is the difference between the energies of the carrier in excited and ground states, $\mathbf{d}$ is the dipole transition moment, and $\hat{a}_l^{+}$ $(\hat{a}_l)$ is the creation (annihilation) Fermi operator of a charge carrier in the lth site.

With regard to these radiation transitions, the Hamiltonian of the system available in the direct external electric field can be written in the small carrier concentration approximation as

$$H = H_0 + H_{\text{tun}} + H_{\mathscr{E}} + H_{\tilde{\mathscr{E}}} \tag{272}$$

$$H_0 = \sum_l E\hat{a}_l^{+}\hat{a}_l + \sum_q \hbar\omega_q\left(\hat{b}_q^{+}\hat{b}_q + \frac{1}{2}\right) - \sum_q \hbar\omega_q\hat{a}_l^{+}\hat{a}_l[u_l(q)\hat{b}_q^{+} + u_l^{*}(q)\hat{b}_q] \tag{273}$$

$$H_{\text{tun}} = \sum_l J\hat{a}_{l+1}^{+}\hat{a}_l + h.c. \tag{274}$$

$$H_{\mathscr{E}} = -e\sum_l R_l\hat{a}_l^{+}\hat{a}_l \tag{275}$$

$$H_{\tilde{\mathscr{E}}} = \sum_l \frac{(\mathbf{d}\cdot\tilde{\mathscr{E}}_0)^2}{2\Xi}\hat{a}_{l+1}^{+}\hat{a}_l + h.c. \tag{276}$$

Here $H_0 + H_{\text{tun}}$ is the small polaron Hamiltonian, $H_{\mathscr{E}}$ describes the correlation between the carriers and the applied field $\mathscr{E}$, $H_{\tilde{\mathscr{E}}}$ is the operator describing intraband transitions conditioned by light absorption, E is the carrier coupling energy in the site, J is the resonance integral, and $R_l = lg$, where l is the site index and g the chain constant.

Let us calculate the correction to the proton polaron direct current density conditioned by light-induced transitions between the sites. This photocurrent calculation is analogous to that of the correction to the drift activation current carried out in the previous subsection [the analogy lies in the fact that operator (276) is similar to the correction to the Hamiltonian if the electric field is taken into account, with the correction being nondiagonal on the operator of the coordinate; see expression (236)].

The current density sought for is obtained according to

$$I_{\tilde{\mathscr{E}}} = [\mathrm{Tr}(\rho_{\mathscr{E}}^{(1)} j^{(2)}) + \mathrm{Tr}\,(\rho_{\mathscr{E}}^{(2)} j^{(1)})] \tag{277}$$

Here

$$\rho_{\mathscr{E}}^{(1,2)} = -\frac{i}{\hbar}\int_{-\infty}^{0} d\tau e^{-i\frac{t-\tau}{\hbar}(H_0+H_{\mathscr{E}})}[H_{\mathscr{E}}, \rho_{\mathrm{tun}}^{(1,2)}]e^{i\frac{t-\tau}{\hbar}(H_0+H_{\mathscr{E}})} \tag{278}$$

is the corresponding correction to the statistic operator of the system conditioned by the correlation (276), where

$$\rho_{\mathrm{tun}}^{(1,2)} = -\frac{e^{-H_0/k_{\mathrm{B}}T}}{\mathrm{Tr}\,e^{-H_0/k_{\mathrm{B}}T}}\int_{0}^{1/k_{\mathrm{B}}T} d\lambda e^{\lambda H_0} H_{\mathrm{tun},\tilde{\mathscr{E}}}\, e^{-\lambda H_0} \tag{279}$$

and

$$j^{(1,2)} = \frac{e}{\mathscr{V} i\hbar}\left[\sum_{l} R_l \hat{a}_l^{+} \hat{a}_l, H_{\mathrm{tun},\tilde{\mathscr{E}}}\right] \tag{280}$$

is the current density operator.

The calculation of the density current (277) is very similar to that carried out in the previous subsection. The final expression is

$$\begin{aligned} I_{\tilde{\mathscr{E}}} &= \sqrt{2\pi}\hbar^{-2} engJ \frac{(\mathbf{d}\cdot\tilde{\mathscr{E}}_0)^2}{2\Xi} \sinh\frac{e\mathscr{E}g}{2k_{\mathrm{B}}T} \\ &\times \frac{\exp(-E_{\mathrm{a}}/k_{\mathrm{B}}T)}{\left[\sum_q |\Delta(q)|^2 \omega_q^2 \,\mathrm{cosech}\,\frac{\hbar\omega_q}{2k_{\mathrm{B}}T}\right]^{-1/2}} \end{aligned} \tag{281}$$

where n is the carrier concentration in the chain, E_{a} is the polaron hopping activation energy [see expression (223)], and $|\Delta(q)|^2$ is the coupling function.

The resulting proton current density (which includes the drift activation current), considered in the two previous subsections and photocurrent (281) is

$$I_{\tilde{\mathscr{E}}} = \sqrt{2\pi}\hbar^{-2} engJ\left(2J + \frac{(\mathbf{d}\cdot\tilde{\mathscr{E}}_0)^2}{2\Xi}\right)\sinh\frac{e\mathscr{E}g}{2k_B T} \times \frac{\exp(-E_a/k_B T)}{\left[\sum_q |\Delta(q)|^2 \omega_q^2 \mathrm{cosech}\frac{\hbar\omega_q}{2k_B T}\right]^{-1/2}} \tag{282}$$

If the dipole transition moment $\mathbf{d}$ is comparatively large and the value $(\mathbf{d}\cdot\tilde{\mathscr{E}}_0)^2/2\Xi$ is of the order of J (recall that $J \sim (10^{-4}$ to $10^{-2})k_B T$, where $T \simeq 300$ K), then expression (282) gives a considerable increase of the current. With this regard the stationary photocurrent for $\mathscr{E} < 10^8$ V/m below room temperature is linear with the constant field $\mathscr{E}$ and quadratic with the amplitude of the variable field $\tilde{\mathscr{E}}_0^2$.

The maximum $|\mathbf{d}|$ is reached in the region of the polaron resonance absorption with radiation frequency $\Omega \approx 4E_a/\hbar$, where the width of the absorption curve is proportional to $\sqrt{E_a k_B T}$ [88,89]. According to data [165,180,181], $E_a = 3$–$20\ k_B T$ for systems with hydrogen-bonded chains. Therefore, the frequency interval $\Omega = 2\pi \times 1$–6×10^{14} s^{-1} should correspond to the proton polaron absorption in a hydrogen-bonded chain.

If we consider proton vibrations of a chain on hydrogen bonds as vibrations of individual oscillators with the typical frequency $\omega_0 \approx 2\pi \times 5 \times 10^{12}$ s^{-1}, then under the influence of infrared radiation with frequency ω_0, resonance absorption should be observed. A wide absorption band in the crystal LiN_2H_5 SO_4 with maximum in the region ω_0 has really been observed experimentally [180]. Strongly correlated vibrations of protons of the hydrogen-bonded chain, which result in a large proton polarizability of the chain, were observed by Zundel and collaborator [see, e.g., Ref. 6 (review article); this effect is considered in the following subsections.

The power absorbed by protons vibrating on hydrogen bonds in a chain can be considered as that absorbed by N oscillators, the cooperative effect being neglected [182]

$$\mathscr{P} = N\frac{\frac{e^2}{m}\tilde{\mathscr{E}}_0^2\eta\tilde{\omega}^2\cos\vartheta}{(\tilde{\omega}^2 - \omega_0^2)^2 + 4\eta^2\tilde{\omega}^2} \tag{283}$$

Here e is the elementary charge, m is the proton mass, $\tilde{\mathscr{E}}_0$ is the amplitude of the electromagnetic wave with frequency $\tilde{\omega}$, ϑ is the angle between the dipole axis (i.e., the hydrogen-bonded chain) and the vector $\tilde{\mathscr{E}}_0$, and η is the absorption coefficient (the half-width of the absorption band). The absorbed energy can cause the heating of chain protons and, therefore, the increase of optical phonon temperature: $T \rightarrow T^* > T$. Since in the first approximation the phonon heating is proportional to the temperature difference—that is, $P = C(T^* - T)$ [183], where C is a constant—then assuming $\mathscr{P} = P$ with resonance ($\tilde{\omega} = \omega_0$), we get

$$T^* = T + \alpha(\omega_0)\tilde{\mathscr{E}}_0^2 \tag{284}$$

where $\alpha(\omega_0) = (4m\eta C)^{-1}Ne^2\cos\vartheta$.

With the above-mentioned parameter values of a hydrogen-bonded chain, with regard to anharmonicity (231) and heating (284) we obtain for the current density instead of (283)

$$\begin{aligned} I = {} & 2^{3/2}\pi^{1/2}\hbar^{-2}engJ^2 \sinh\frac{e\mathscr{E}g}{2k_{\mathrm{B}}T^*} \\ & \times e^{-E_{\mathrm{a}}/k_{\mathrm{B}}T}\left[\sum_q |\Delta(q)|^2\omega_q^2 \operatorname{cosech}\frac{\hbar\omega_q}{2k_{\mathrm{B}}T^*}\right]^{-1/2} \\ & \times \exp\left\{\sum_q |\kappa_q|^2|\Delta_q|^2\left(\frac{\hbar\omega_q}{2k_{\mathrm{B}}T^*}\right)^2 \operatorname{cosech}^4\frac{\hbar\omega_q}{2k_{\mathrm{B}}T^*}\right. \\ & \times \left.\left(1 - \operatorname{cosech}\frac{\hbar\omega_q}{2k_{\mathrm{B}}T^*}\right)^2 \coth\frac{\hbar\Omega_q}{2k_{\mathrm{B}}T}\right\} \end{aligned} \tag{285}$$

Here T is the temperature of acoustic phonons (thermostat), T^* is the temperature of optical phonons (284), the anharmonicity constant $|\kappa_q|$ is much less than 1, and Ω_q is the frequency of acoustic phonons; it is possible to assume $\Omega_q = (10^{-2}$ to $10^{-1})\Omega_{\mathrm{Debye}}$. The coupling between the optical and acoustic phonons is strongest near Ω_{Debye}; and because of this, for sufficiently large anharmonicity, $|\kappa_q| \geq 10^{-2}$ even at $T = T^*$, the last exponential multiplier can be approximated by the exponent below, with the dispersion being neglected:

$$\exp\left\{2|\kappa|^2|\Delta|^2 \frac{(10^{-2} \text{ to } 10^{-1})\hbar\Omega_{\mathrm{Debye}}}{\hbar\omega_0}\left(\frac{T^*}{T}\right)^2\right\} \tag{286}$$

Hence it is clear that an increase of the temperature of the optical phonons T^* activating the charge carrier hopping leads to an increase of the proton current density (285). Such a stimulation mechanism remains working until the heating (284) is small in comparison with the polaron shift energy $\approx |\Delta|^2\hbar\omega_0$ (the coupling function $|\Delta|^2 \approx 20$).

Let us now treat the influence of ultrasound (hypersound) on the proton mobility in the hydrogen-bonded chain. The impact of ultrasound on free electrons weakly connected with matrix vibrations (e.g., metals and nonpolar semiconductors) has been studied in detail (see, e.g., Ref. 184). The effect of ultrasound on the diffusion of atoms in a solid has also been studied in detail (see, e.g., Ref. 185).

While considering the influence of sound on the mobility of protons, their peculiar position should be taken into account, because in a solid they are specified both by the quantum features (the availability of the overlap integral between the nearest-neighbor sites in crystal with hydrogen bonds) and by classical ones (a large mass and hence a usual diffusion in metals and nonpolar semiconductors). The peculiar position occupied by the charge carriers in a hydrogen-bonded chain enables us to point out a specific mechanism of the proton conductivity stimulation by ultrasound.

The overlap integral J securing the carrier tunneling through the barrier between the two neighboring polaron wells in the quasi-classical approximation has the following appearance:

$$J \simeq E \exp\left\{-\frac{2}{\hbar}\int_0^a dx\sqrt{2m^*[V(x) - K]}\right\} \tag{287}$$

where a is the potential barrier width, m^* is the effective mass of a carrier in a chain, $V(x)$ is the potential energy, and K is the kinetic energy of a carrier. Since the proton mass is much larger than that of the electron, the absolute value of the exponential function is not small (compare with Ref. 186). The probability of tunneling between polaron wells depends on the barrier width, which can be modulated in turn by an external ultrasound in this case.

The sites of hydrogen-bonded chains are rigidly fixed by the side chemical bonds in various compounds. That is why the distances between the sites and, therefore, the chain rigidity are functions of external parameters related to it—that is, the thermostat functions. In that event, longitudinal vibrations of the chain sites imposed by the matrix lead to periodical oscillations of the barrier width a between the neighboring polaron wells. If the barrier becomes thinner—that is, if $a \to a - \delta a$—the overlap integral

$$J \to J \exp\left(\frac{2}{\hbar}\langle R\rangle\delta a\right) \tag{288}$$

where $\langle R \rangle$ is the average meaning of the integrand $\sqrt{2m^*[V(x) - K]}$ in expression (287). The displacement of a can be quantized on the acoustic phonon operators: $\delta a \rightarrow \sum_q (\alpha_q \hat{B}_q^+ + \alpha_q^* \hat{B}_q)$. Then, performing quantum mechanical thermal averaging in expression (288) we have

$$\tilde{J} = J\exp\left\{\frac{\langle R \rangle}{\hbar} \sum_q |\alpha_q|^2 \coth \frac{\hbar \Omega_q}{2k_B T}\right\} \tag{289}$$

where Ω_q is the acoustic phonon frequency with the wave number q.

On switching on ultrasound with the wave vector directed along the chain, the vibration amplitudes of framework atoms will be already defined in the chain not by temperature of the medium but by the ultrasound source intensity. In this case we have instead of expression (289)

$$\tilde{J}(A, \Omega_{\text{sound}}) = J\exp\left\{\frac{\langle R \rangle}{\hbar} \delta a \left(1 + |\zeta(\Omega_{\text{sound}})|^2 \frac{A}{\delta a}\right)\right\} \tag{290}$$

where $|\zeta(\Omega_{\text{sound}})|^2$ is a constant characterizing the interaction between ultrasound with frequency Ω_{sound} and the chain framework vibrations, and A is the ultrasound amplitude. The results obtained, (289) and (290), also agree qualitatively with the behavior of the proton tunnel frequency in the complex O—H$\cdots$O of the KH_2PO_4 crystal considered in Refs. 187 and 188 when the pressure and temperature were taken as external factors.

Since the value δa is under the exponential sign, even a small correction to it should lead to a considerable change of the overlap integral. In fact, it is clearly seen from expression (290) that the correction $|\zeta(\Omega_{\text{sound}})|^2 A$ caused by ultrasound to δa brings about a considerable increase of the vibration amplitude of chain framework atoms. This amplifies the modulation of the distance between the neighboring polaron wells. Moreover, since the expression for the proton current density is proportional to J^2 when external effects are not available [see, for instance, expressions (222) or (231)], the replacement of this value for $\tilde{J}^2(A, \Omega_{\text{sound}})$ including an exponential dependence on the ultrasound amplitude should result in an appreciable exponential growth of the current (at least, by several times).

Of course, expression (290) is correct only in cases when tunneling of a quasi-particle (i.e., proton polaron), through the barrier takes place for times shorter than the period of the barrier oscillations. The amplitude of ultrasound vibrations in this situation should not exceed some critical A_c under which the warning-up of the chain is essential (in particular, the critical power of ultrasound for biotissues varies from 10^{-2} to 10^{-1} W/cm^2).

H. Vibration Fluctuations of a Resonance Integral in the Polaron Problem

In the small polaron model a charge carrier that is strongly connected with a site locally deforms the crystal lattice and a consequence of the deformation is the induction of the interaction of the carrier with polarized optical phonons. For the most part, the impact of the phonon subsystem is reduced to the renormalization of the coupling energy. Considering the phonon subsystem in terms of second quantization makes it possible to take into account some peculiarities associated with external factors such as temperature, electromagnetic field, and ultrasound. Phonons are able to modulate the resonance overlap integral, which results into the effective thinning of the barrier between two polaron wells. In turn it leads to a significant increase of the current density (see previous sections and also Ref. 189).

A small proton polaron is different in some aspects from the electron polaron; that is, the hydrogen atom is able to participate in the lattice vibrations in principle (in any case it is allowable for excited states; see Section II.F), but the electron cannot. This means that one more mechanism of phonon influence on the proton polaron is quite feasible. That is, phonon fluctuations would directly influence wave functions of the protons and thereby contribute to the overlapping of their wave functions. In other words, phonons can directly increase the overlap integral in concept. Such an approach allows one to describe the proton transfer without using the concept of transfer from site to site through an intermediate state.

For instance, a similar approach was developed by Kagan and Klinger [190]. They studied the fluctuating diffusion through barrier (coherent) and over barrier (incoherent) of ^{3}He atoms in the crystal of ^{4}He atoms. In other words, the tunnel integral was treated as a matrix element constructed not only on wave functions $y_{\mathbf{l}}$ and $y_{\mathbf{m}}$ of diffusing ^{3}He atoms (**l** and **m** are the radius vectors of the lattice sites between which the quantum diffusion occurs), but also on wave functions $j_{\mathbf{l}n}$ and $j_{\mathbf{m}n}$ of vibrating motion of the crystal where n is the number of occupation of an excited state. The intersite diffusion of particles, which took into account diagonal phonon transitions, was treated in detail in Ref. 190, but the method of calculation did not allow them to investigate the contribution of off-diagonal phonon transfers.

Another approach has been proposed in Ref. 191. The approach is based on the model of small polaron and makes it possible to extend the range of the model. In particular, it includes the influence of vibration wave functions on the tunnel integral and provides a way of the estimation of diagonal and off-diagonal phonon transfers on the proton polaron mobility. It turns out that the one phonon approximation is able significantly to contribute to the proton mobility. Therefore, we will further deal with matrix elements constructed on the phonon function $\varphi(\mathbf{q}) = \hat{b}_{\mathbf{q}}^{+}\varphi_{\mathbf{q}}(0)$ and will examine only the first excited state with $\nu = 1$.

In the approximation of small concentration of carriers we can write the polaron Hamiltonian as follows:

$$H = H_0 + H_{\text{tun}} + H_{\mathscr{E}} \tag{291}$$

$$H_0 = \sum_{\mathbf{l}} E\hat{a}_{\mathbf{l}}^{+}\hat{a}_{\mathbf{l}} + \sum_{\mathbf{q}} \hbar\omega_{\mathbf{q}}\left(\hat{b}_{\mathbf{q}}^{+}\hat{b}_{\mathbf{q}} + \frac{1}{2}\right) - \sum_{\mathbf{q}} \hbar\omega_{\mathbf{q}}\hat{a}_{\mathbf{l}}^{+}\hat{a}_{\mathbf{l}}[u_{\mathbf{l}}(\mathbf{q})\hat{b}_{\mathbf{q}}^{+} + u_{\mathbf{l}}^{*}(\mathbf{q})\hat{b}_{\mathbf{q}}] \tag{292}$$

$$H_{\text{tun}} = \sum_{\mathbf{l},\mathbf{m}}\sum_{\mathbf{k},\mathbf{q}} \hat{b}_{\mathbf{k}}^{+}\varphi_{\mathbf{k}}^{*}(0)J_{\mathbf{lm}}\hat{a}_{\mathbf{l}}^{+}\hat{a}_{\mathbf{m}}\hat{b}_{\mathbf{q}}\varphi_{\mathbf{q}}(0) + h.c. \tag{293}$$

Here E is the energy of connection of a carrier with the $\mathbf{l}$th site; $u_{\mathbf{l}}(\mathbf{q}) = ue^{i\mathbf{q}\cdot\mathbf{l}}$, where u is the dimensionless constant, which characterizes the degree of local deformation of the lattice by a carrier. The tunnel Hamiltonian (293) includes in addition the overlapping the phonon functions $\hat{b}_{\mathbf{q}}\varphi_{\mathbf{q}}(0)$ and $\hat{b}_{\mathbf{k}}^{+}\varphi_{\mathbf{k}}(0)$ as we conjecture that the proton polaron engages in the lattice vibration as well (hence $J_{\mathbf{lm}}$ becomes a function of the momentum $\mathbf{q}$ and/or $\mathbf{k}$, too). In the present form, Eq. (293), the total number of phonons is not kept; however, we imply that the phonons are absorbed and emitted by polarons and thus their total number remains invariable. Let us introduce the designation

$$V_{\substack{\mathbf{lm}\\ \mathbf{kq}}} = \varphi_{\mathbf{k}}^{*}(0)J_{\mathbf{lm}}\varphi_{\mathbf{q}}(0) \tag{294}$$

Then the tunnel Hamiltonian (293) changes to

$$H_{\text{tun}} = \sum_{\mathbf{l},\mathbf{m}}\sum_{\mathbf{k},\mathbf{q}} V_{\substack{\mathbf{lm}\\ \mathbf{kq}}}\hat{a}_{\mathbf{l}}^{+}\hat{a}_{\mathbf{m}}\hat{b}_{\mathbf{q}}^{+}\hat{b}_{\mathbf{q}} + h.c. \tag{295}$$

The Hamiltonian $H_{\mathscr{E}}$ in expression (291) has the form

$$H_{\mathscr{E}} = -e\sum_{\mathbf{l}} \mathscr{E}\cdot\mathbf{l}\hat{a}_{\mathbf{l}}^{+}a_{\mathbf{l}} \tag{296}$$

and describes the behavior of the field $\mathscr{E}$ on a proton polaron located in the $\mathbf{l}$th site.

The diagonalization of the Hamiltonian H_0 (292) by phonon variables is realized by the canonical transformation

$$\bar{H}_0 = e^{-S}H_0e^{S}, \qquad S = \sum_{\mathbf{l};\mathbf{q}} \hat{a}_{\mathbf{l}}^{+}\hat{a}_{\mathbf{l}}[u_{\mathbf{l}}(\mathbf{q})\hat{b}_{\mathbf{q}}^{+} - u_{\mathbf{l}}^{*}(\mathbf{q})\hat{b}_{\mathbf{q}}] \tag{297}$$

Under the transformation (297) the Hamiltonian (1) is transformed to

$$\bar{H} = \bar{H}_0 + \bar{H}_{\text{tun}} + H_{\mathscr{E}} \tag{298}$$

$$\bar{H}_0 = \sum_{\mathbf{l}} \left(E - \sum_{\mathbf{q}} |u|^2 \hbar\omega_{\mathbf{q}} \right) \hat{a}_{\mathbf{l}}^{+} \hat{a}_{\mathbf{l}} + \sum_{\mathbf{q}} \hbar\omega_{\mathbf{q}} \left(\hat{b}_{\mathbf{q}}^{+} \hat{b}_{\mathbf{q}} + \frac{1}{2} \right) \tag{299}$$

$$\bar{H}_{\text{tun}} = \sum_{\mathbf{l,m}} \sum_{\mathbf{k,q}} V_{\substack{\mathbf{lm}\\ \mathbf{kq}}} \hat{a}_{\mathbf{l}}^{+} \hat{a}_{\mathbf{m}} \hat{\Phi}_{\mathbf{lm}}(0) e^{-S} \hat{b}_{\mathbf{q}}^{+} \hat{b}_{\mathbf{q}} e^{S} + h.c. \tag{300}$$

where

$$\hat{\Phi}_{\mathbf{lm}}(0) = \exp\left\{ \sum_{\mathbf{q}} \hat{a}_{\mathbf{l}}^{+} \hat{a}_{\mathbf{m}} [\Delta_{\mathbf{lm}}(\mathbf{q}) \hat{b}_{\mathbf{q}}^{+} - \Delta_{\mathbf{lm}}^{*}(\mathbf{q}) \hat{b}_{\mathbf{q}}] \right\} \tag{301}$$

$$\Delta_{\mathbf{lm}}(\mathbf{q}) = u_{\mathbf{l}}(\mathbf{q}) - u_{\mathbf{m}}(\mathbf{q}) \tag{302}$$

With regard for the rules of transformation of the operator functions $\hat{b}_{\mathbf{q}}^{+}$ and $\hat{b}_{\mathbf{q}}$ [88,192], the Hamiltonian $\bar{H}_{\text{tun}}$ can be presented in the form

$$\bar{H}_{\text{tun}} = \sum_{\mathbf{l,m}} \sum_{\mathbf{k,q}} V_{\substack{\mathbf{lm}\\ \mathbf{kq}}} \hat{a}_{\mathbf{l}}^{+} \hat{a}_{\mathbf{m}} \hat{M}_{\substack{\mathbf{lm}\\ \mathbf{kq}}} \hat{\Phi}_{\mathbf{lm}}(0) + h.c. \tag{303}$$

where the following designation has been introduced:

$$\hat{M}_{\substack{\mathbf{lm}\\ \mathbf{kq}}} = -\frac{\partial^2}{\partial \Delta_{\mathbf{lm}}(\mathbf{q}) \partial \Delta_{\mathbf{lm}}^{*}(\mathbf{k})} - \sum_{\mathbf{l}'} u_{\mathbf{l}'}^{*}(\mathbf{q}) \frac{\partial}{\partial \Delta_{\mathbf{lm}}^{*}(\mathbf{q})}$$
$$+ \sum_{\mathbf{m}'} u_{\mathbf{m}'}^{*}(\mathbf{q}) \frac{\partial}{\partial \Delta_{\mathbf{lm}'}(\mathbf{q})} + \sum_{\mathbf{l}',\mathbf{m}'} u_{\mathbf{l}'}^{*}(\mathbf{q}) u_{\mathbf{m}'}(\mathbf{k}) \tag{304}$$

Let us now calculate the current density of proton polarons by setting $|V_{\substack{\mathbf{lm}\\ \mathbf{kq}}}| \ll \hbar\omega_{\mathbf{q}},\ E,\ k_{\text{B}}T$. We start from the expression

$$I = \text{Tr}(\rho_{\text{tun}}, j) \tag{305}$$

where the correction to the statistical operator of the system studied caused by the operator $\bar{H}_{\text{tun}}$ (303) has the form

$$\rho_{\text{tun}} = -\frac{i}{\hbar} \int_{-\infty}^{t} d\tau e^{-i(t-\tau)\bar{H}_0/\hbar} [\bar{H}_{\text{tun}}, \rho_0] e^{i(t-\tau)\bar{H}_0/\hbar} \tag{306}$$

the undisturbed statistical operator is equal to

$$\rho_0 = \frac{e^{-\bar{H}_0/k_B T}}{\mathrm{Tr}\, e^{-\bar{H}_0/k_B T}} \tag{307}$$

and the operator of current density is

$$j = \frac{e}{\mathscr{V} i\hbar}\left[\bar{H}_{\mathrm{tun}}, \sum_{\mathbf{l}} \mathbf{l}\hat{a}_{\mathbf{l}}^{+}\hat{a}_{\mathbf{l}}\right] \tag{308}$$

($\mathscr{V}$ is the effective volume of the crystal which contains one polaron). Substituting expressions from (306) to (308), the current density (305) can be written as

$$\mathbf{I} = \frac{1}{\hbar^2 \mathrm{Tr}\, e^{-\bar{H}_0/k_B T}} \int_0^{\infty} d\tau\, \mathrm{Tr}\left\{\left\{e^{-\bar{H}_0/k_B T}\bar{H}_{\mathrm{tun}}\left(-\frac{i}{\hbar}\tau\right)\left[j\left(-\frac{1}{k_B T}\right) - j\right]\right\}\right. \tag{309}$$

where the following designations of the operators are introduced:

$$\bar{H}_{\mathrm{tun}}\left(-\frac{i}{\hbar}\tau\right) = e^{-\frac{i}{\hbar}\tau(\bar{H}_0 + H_{\mathscr{E}})}\bar{H}_{\mathrm{tun}} e^{\frac{i}{\hbar}\tau(\bar{H}_0 + H_{\mathscr{E}})} \tag{310}$$

$$j\left(-\frac{1}{k_B T}\right) = e^{-\frac{1}{k_B T}(\bar{H}_0 + H_{\mathscr{E}})} j e^{\frac{1}{k_B T}(\bar{H}_0 + H_{\mathscr{E}})} \tag{311}$$

Then the following transformations can be performed in expression (309):

$$\begin{aligned}
\mathbf{I} &= \frac{1}{\mathscr{V}\hbar^2\, \mathrm{Tr}\, e^{-\bar{H}_0/k_B T}} \int_0^{\infty} d\tau\, \mathrm{Tr}\left\{e^{-\bar{H}_0/k_B T}. \sum_{\mathbf{l},\mathbf{m},\mathbf{l}',\mathbf{m}'}\ \sum_{\mathbf{k},\mathbf{q},\mathbf{k}',\mathbf{q}'} V_{\substack{\mathbf{lm}\\ \mathbf{kq}}} V_{\substack{\mathbf{l}'\mathbf{m}'\\ \mathbf{k}'\mathbf{q}'}} (\mathbf{l}-\mathbf{m})\right. \\
&\times \hat{a}_{\mathbf{l}}^{+}\hat{a}_{\mathbf{m}}\hat{a}_{\mathbf{l}'}^{+}\hat{a}_{\mathbf{m}'}\left[\hat{M}_{\substack{\mathbf{lm}\\ \mathbf{kq}}}\hat{\Phi}_{\mathbf{lm}}\left(-\frac{i}{\hbar}\tau\right)\hat{M}_{\substack{\mathbf{l}'\mathbf{m}'\\ \mathbf{k}'\mathbf{q}'}}\hat{\Phi}_{\mathbf{l}'\mathbf{m}'}\left(-\frac{1}{k_B T}\right)e^{-e\mathscr{E}\cdot(\mathbf{l}-\mathbf{m})/2k_B T}\right. \\
&\left.\left. - e^{-5e\mathscr{E}\cdot(\mathbf{l}-\mathbf{m})/2k_B T}\hat{M}_{\substack{\mathbf{lm}\\ \mathbf{kq}}}\hat{\Phi}_{\mathbf{lm}}\left(-\frac{i}{\hbar}\tau\right)\hat{M}_{\substack{\mathbf{l}'\mathbf{m}'\\ \mathbf{k}'\mathbf{q}'}}\hat{\Phi}_{\mathbf{l}'\mathbf{m}'}(0)\right]\right\} \\
&= \frac{1}{\hbar^2\, \mathrm{Tr}\, e^{-\bar{H}_0/k_B T}} \sum_{\mathbf{l},\mathbf{m}}\sum_{\mathbf{k},\mathbf{q}} e^{-\bar{H}_0/k_B T}\frac{\hat{a}_{\mathbf{l}}^{+}\hat{a}_{\mathbf{l}}}{\mathscr{V}}(\mathbf{l}-\mathbf{m})|V_{\substack{\mathbf{lm}\\ \mathbf{kq}1}}|^2(\hat{M}_{\substack{\mathbf{lm}\\ \mathbf{kq}}})^2 \\
&\times \int_0^{\infty} d\tau\left\{\hat{\Phi}_{\mathbf{lm}}\left(-\frac{i}{\hbar}\tau\right)\hat{\Phi}_{\mathbf{ml}}\left(-\frac{1}{k_B T}\right)e^{-e\mathscr{E}\cdot(\mathbf{l}-\mathbf{m})/2k_B T}\right. \\
&\left. - \hat{\Phi}_{\mathbf{lm}}\left(-\frac{i}{\hbar}\tau\right)\hat{\Phi}_{\mathbf{ml}}(0)e^{-5e\mathscr{E}\cdot(\mathbf{l}-\mathbf{m})/2k_B T}\right\}
\end{aligned} \tag{312}$$

[here the structure of the operator $\hat{\Phi}_{\mathbf{lm}}(x)$ is defined by formulas (310) and (311)]. Preserving in expression (312) only members approximated by nearest neighbors, we then carry out the thermal averaging and integrate over τ. As a result we obtain

$$\mathbf{I} = \frac{\boldsymbol{\mathscr{E}}}{|\boldsymbol{\mathscr{E}}|}\frac{2^{3/2}\pi^{1/2}ng}{\hbar^2}\sinh\frac{e\mathscr{E}g}{k_BT}\sum_{\mathbf{k},\mathbf{q}}|V_{\mathbf{kq}}|^2\hat{M}^2_{\mathbf{qk}} \times \frac{\exp(-E_a/k_BT)}{[\sum_{\mathbf{q}'}|\Delta(\mathbf{q}')|^2\omega^2_{\mathbf{q}'}\operatorname{cosech}\hbar\omega_{\mathbf{q}'}]^{1/2}} \tag{313}$$

where the activation energy is equal to

$$E_a = k_BT\sum_{\mathbf{q}'}|\Delta(\mathbf{q}')|^2\tanh(\hbar\omega_{\mathbf{q}'}/4k_BT) - 3e\mathscr{E}g/2 \tag{314}$$

Here the following designations are introduced: n is the concentration of polarons, g is the lattice constant, $V_{\substack{\mathbf{lm}\\ \mathbf{kq}}} \to V_{\substack{\mathbf{ll+g}\\ \mathbf{kq}}} = V_{\mathbf{kq}}$,

$$\Delta_{\mathbf{lm}}(\mathbf{q}) \to \Delta_{\mathbf{l+g}}(\mathbf{q}) = \Delta(\mathbf{q}), \text{ and } \hat{M}_{\substack{\mathbf{lm}\\ \mathbf{qk}}} \to \hat{M}_{\substack{\mathbf{ll+g}\\ \mathbf{qk}}} = \hat{M}_{\mathbf{qk}}$$

Let us analyze the result obtained in Eq. (313) neglecting the dependence of values on wave vectors $\mathbf{q}$ and $\mathbf{k}$. The elements of operator $\hat{M}^2$ can be written in agreement with expression (304) as below:

$$\hat{M}^2 \propto \left\{\frac{\partial^4}{\partial|\Delta|^4},\quad \pm|u|^2\frac{\partial^2}{\partial|\Delta|^2},\quad u\frac{\partial^3}{\partial|\Delta|^3},\quad |u|^4\right\} \tag{315}$$

On the other hand, in expression (313) the structure of the term (as a function of parameter $|\Delta|$), which falls under the influence of the operator $\hat{M}^2$, has the form

$$f(|\Delta|) \propto \frac{1}{|\Delta|}\exp(-C|\Delta|^2), \qquad 0 < C < 1 \tag{316}$$

Taking into account the inequalities $\hbar\omega < k_BT$ and $|u|^2 \approx |\Delta|^2 \gg 1$, we can compare all the terms of the series $\hat{M}^2 f(|\Delta|)$. It is easy to see that the term proportional to $|u|^4$ is maximal in expression (315); the other ones are one to several orders of magnitude smaller. Thus, we can retain only one, with the biggest term proportional to $|u|^4$ in the series (315). Then setting $|u|^2 \approx |\Delta|^2$ and assuming that $e\mathscr{E}g < k_BT$ (it is correct until $\mathscr{E} \leq 10^7$ V/m and for moderate temperatures, see Ref. 88 for details), we finally obtain instead of expression (313)

$$\boldsymbol{I} = \frac{\boldsymbol{\mathcal{E}}}{|\boldsymbol{\mathcal{E}}|}\frac{2^{3/2}\pi^{1/2}ng}{\hbar^2}\sinh\frac{e\mathcal{E}g}{k_{\rm B}T}|V|^2|\Delta|^4 \times \frac{\exp(-E_{\rm a}/k_{\rm B}T)}{[|\Delta|^2\omega^2 {\rm cosech}(\hbar\omega/2k_{\rm B}T)]^{1/2}} \tag{317}$$

Expression (317) for the current density is distinguished from that derived in the standard model of small polaron [see, e.g., formula (227)] only by multiplier $|\Delta|^4$. However, in the theory of small polaron is assumed that $|\Delta|^2$ significantly exceeds the unit. Consequently, the result obtained makes it possible to spread the framework of the small polaron model very considerably. The inclusion of phonon transitions is culminated in the renormalization of the overlap integral,

$$J \to V = J|\Delta|^2 \tag{318}$$

Therefore, in cases when the value of J is too small, the proton conductivity can still be very substantial due to the fluctuating absorption and radiation of phonons by charged carriers. This allows also the application of the approach developed above to compounds in which hopping of heavy carriers occurs by comparatively distant structural groups (for example, 0.4 to 0.8 nm, which takes place in some superionic crystals).

I. An Example of Superionic Conductivity: The $NH_4IO_3 \cdot 2NHIO_3$ Crystal

The crystal of ammonium triiodate was investigated by different methods: X-ray diffraction analysis [193], neutronography [194], polarization optical techniques [195], differential scanning calorimetry and vibrational spectroscopy [196], and nuclear quadrupole resonance [197]. In Refs. 193 and 195 it was established that at a temperature above $T_0 = 213$ K, this is a superionic crystal with a conductivity of the protonic type ($\sigma \simeq 10^{-5}$ Ω^{-1}cm^{-1} at 300 K [193]). Transition to the superionic state is accompanied by anomalies in permittivity and conductivity, characteristic of phase transition of the second kind [195], but the polarization-optical changes indicate that at $T = T_0$ the symmetry of the crystal does not change. Neutronographic investigation [194] revealed the presence of a system of bifurcated hydrogen contacts. In view of this fact, transition to the superionic state was attributed to disordering of the protonic subsystem (at $T > T_0$). Raman spectroscopy studies [196,198] demonstrated that ordering of the protonic subsystem actually occurs at a much lower temperature T_c, which lies in the range of about 120–125 K and is accompanied by doubling of the unit cell volume.

Indeed, the results obtained in Refs. 196 and 198 have shown that the phase diagram of ammonium triiodate $NH_4IO_3 \cdot 2HIO_3$ crystal is essentially more complicated than was assumed previously: In addition to the second-order transition to the high proton conductivity state at the temperature $T_c = 213$ K, the differential scanning calorimetry examinations has revealed two transitions at the temperatures $T_1 = 365.6$ K (first-order transition) and at $T_c^1 = 365.6$ K (second-order transition). Starting from general theoretical views, Puchkovska and Tarnavski [196] have proposed the symmetry of crystal phases which agrees with the results of investigations of the vibrational spectra of the crystal. Apparently the new phase 3 existing in the narrow temperature range from $T_c^1 = 365.6$ to $T_c^2 = 211.6$ K is a metastable phase, and therefore it was not found previously in the dielectric and conductivity measurements [195]. Metastable phases have been revealed also in other proton conductors. For instance, in Ref. 199, transitions between stable and metastable phases in the ammonium hydrogen selenates NH_4HSeO_4 and ND_4SeO_4 were discussed.

The finding of several phase transitions in triiodate ammonium allows the examination of another proton conductor, the $KIO_3 \cdot 2HIO_3$ crystal, which at room temperature has the same structure as triiodate ammonium and whose dielectric properties and temperature dependencies are similar to the latter [200,201].

Having understood the mechanism of the motion of protons in the $NH_4IO_3 \cdot 2HIO_3$ crystal, the investigation of the far-infrared absorption spectrum (in the region from 60 to 400 cm^{-1}) and the Raman spectrum (in the region from 600 to 800 cm^{-1}) has been performed [46]. The temperature behavior of the vibration spectra has been studied in the range from 77 to 300 K. The Raman spectra have been obtained from a polycrystalline samples on a DFS-24 (LOMO) spectrometer. The fragments of these spectra discussed herein are illustrated in Fig. 10. Figure 11 shows the IR spectra recorded by a FIS-3 spectrometer (Hitachi, Japan). The temperature of the sample was stabilized with the help of an "Utrex" K-23 thermostabilization system (Institute of Physics, Kyiv) with an accuracy of ± 1 K.

Measurements performed revealed anomalous behavior of (a) the Raman-active mode with a frequency of ≈ 756 cm^{-1} (Fig. 10) and (b) an infrared-active mode with a frequency of ≈ 99 cm^{-1} (Fig. 11). The intensities of these modes in the superionic phase are virtually constant, but at $T < T_0$ the first of them increases in accordance with the law $|(T - T_0)/T_0|^{1/2}$, whereas the second of these modes decreases in accordance with the same law.

The conductivity of the crystal in Arrhenius coordinates is represented by a straight line, and the activation energy is evaluated from the conductivity plot as a function of temperature, $E_a \simeq 46$ kJ/mol [193]. The activation energy, assessed from broadening of the PMR band in accordance with the Waugh–Fedin relation, proved to be equal to $E_a^{(2)} \simeq 33.1$ kJ/mol [193,202]. Investigation of

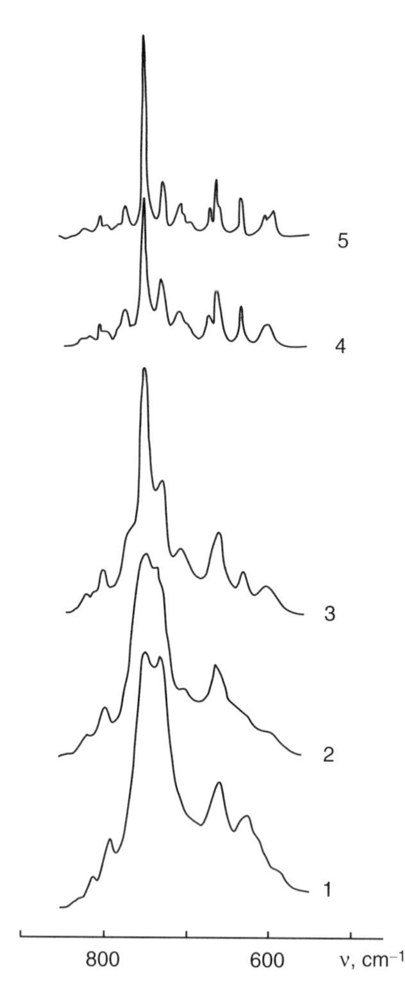

Figure 10. Raman spectrum of the $NH_4IO_3 \cdot 2HIO_3$ crystal in the region of stretching vibrations of the I–O group as a function of temperature: 297 (1), 216 (2), 185 (3), 125 (4), 77 K (6). (From Ref. 46.)

the spin-lattice relaxation gives an activation energy of $E_a^{(1)} \simeq 8.8$ kJ/mol (at $T > T_0$) [202], and the value $E_a^{(1)}$ determined by this procedure characterizes exclusively the hopping mobility of the protons. Such an appreciable spread of activation energy values measured by different methods suggests that the conductivity in ammonium triiodate is governed by two different mechanisms with the activation energies $E_a^{(1)}$ and $E_a^{(2)}$ $(E_a^{(1)} + E_a^{(2)} \simeq E_a)$. Therefore, it is natural to suppose that anomalous modes with frequencies of 99 and 756 cm^{-1} found in Ref. 46 are directly connected with the two competing protonic conductivity mechanisms.

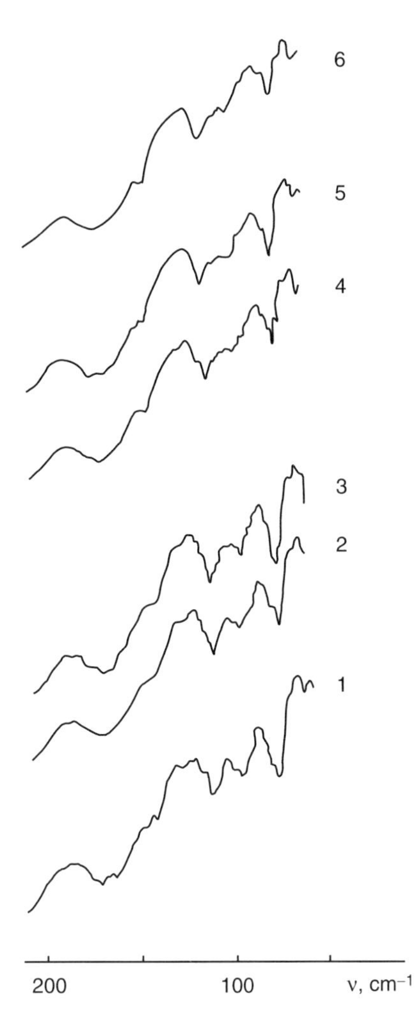

Figure 11. Infrared absorption spectrum of the $NH_4IO_3 \cdot 2HIO_3$ crystal in the region of the lattice vibrations as a function of temperature: 132 (1), 172 (2), 182 (3), 227 (4), 245 (5), 286 K (6). (From Ref. 46.)

The presence of the polar mode oriented in the direction of maximum conductivity gives grounds to postulate the feasibility of the polaronic mechanism of motion of the protons in this crystal. As was shown in the previous subsection (see also Refs. 57,58, and 191) the proton can traverse a comparatively large distance between the nearest sites of the protonic sublattice (0.4 to 0.8 nm) with the participation of vibrational quanta, that is, phonons; the virtual absorption of such a quantum can appreciably increase the resonance integral of overlapping of the wave functions of the proton on the nearest sites; see expression (318).

As we pointed out above, not only the low-frequency mode (99 cm^{-1}), but also the high-frequency mode (756 cm^{-1}), is involved in the protonic conductivity. The criterion for the low-frequency polaronic conductivity is the inequality $\hbar\omega < 2k_{\text{B}}T$, where ω is the frequency of the phonons; $\omega = 99\ \text{cm}^{-1}$ falls under the inequality at $T > 50$ K. At the inequality $\hbar\omega > 2k_{\text{B}}T$, which holds true for the mode with a frequency of 756 cm^{-1}, the standard model of the mobility of the small is not valid; therefore, this case should be treated separately.

Let us apply expression (317) for the calculation of the proton current density associated with the hopping mobility of protons. Neglecting the dispersion in expression (317) and assuming that the applied electric field $\mathscr{E}$ is small, we can write the hopping current density as follows:

$$I = \frac{2^{3/2}\pi^{1/2}e^2ng^2}{\hbar^2 k_{\text{B}}T}|V|^2|\Delta^{(1)}|^4 \times \frac{\exp(-E_{\text{a}}/k_{\text{B}}T)}{[|\Delta^{(1)}|^2\omega_1^2 \text{cosech}(\hbar\omega_1/2k_{\text{B}}T)]^{1/2}} \tag{319}$$

where the activation energy is (314)

$$E_{\text{a}}^{(1)} \simeq k_{\text{B}}T|\Delta^{(1)}|^2 \tanh(\hbar\omega_1/4k_{\text{B}}T) \tag{320}$$

Here $|\Delta^{(1)}|^2$ is the coupling constant, which ties protons in the polar low-frequency mode ($\omega_1 = 99\ \text{cm}^{-1}$); n is the proton concentration, and g the lattice constant.

As is well known from the small polaron theory [88], in the region of moderate temperatures the contribution made by the band mechanism to the polaronic current is considerably smaller than the contribution made by the hopping conductivity. However, this result was obtained in approximation of low-frequency phonons, when $\hbar\omega < 2k_{\text{B}}T$. At the same time, the polaronic conductivity can be considered for the opposite case as well (in any case, for heavy carriers, such as protons) [163] when the energy of the phonons exceeds the thermal energy k_BT. Then, it follows that the ratio of the hopping and band conductivities satisfies the inequality in favor of the band mechanism $(I_{\text{band}} \gg I_{\text{hop}})|_{\hbar\Omega \gg k_{\text{B}}T}$. Consequently, it is reasonable to calculate the density of the band polaronic current I_{band}, brought about by mode with a frequency $\omega_2/2\pi = 756\ \text{cm}^{-1}$, and to compare the result obtained with that for the hopping current I_{hop} (319) where the role of the activating system is played by the low-frequency mode with $\omega_1/2\pi = 99\ \text{cm}^{-1}$.

It is convenient to start from the canonical form of the polaron Hamiltonian $\bar{H} = e^{-S} H e^{S}$ written in the momentum presentation

$$\bar{H} = \bar{H}_0 + \bar{H}_{\text{tun}} \tag{321}$$

$$\bar{H}_0 = \sum_{\mathbf{k}} E(\mathbf{k}) \hat{a}^+_{\mathbf{k}} \hat{a}_{\mathbf{k}} + \sum_{\mathbf{q}} \hbar\omega_{2\mathbf{q}} \left(\hat{b}^+_{\mathbf{q}} b_{\mathbf{q}} + \frac{1}{2} \right) \tag{322}$$

$$\bar{H}_{\text{tun}} = \sum_{\mathbf{k},\mathbf{k}'} \hat{a}^+_{\mathbf{k}'} \hat{a}_{\mathbf{k}} \frac{1}{N} \sum_{\mathbf{l},\mathbf{m}} V_{\substack{\mathbf{lm} \\ \mathbf{kk}'}} \hat{\Phi}_{\mathbf{lm}} e^{i(\mathbf{k}-\mathbf{k}')\mathbf{l} - i\,\mathbf{k}'\mathbf{m}'} \tag{323}$$

Here

$$E(\mathbf{k}) = -E_{\text{p}} + 2\langle V \rangle e^{-S_T} \cos \mathbf{kg} \tag{324}$$

is the energy of band polaron where the first term

$$E_{\text{p}} = \sum_{\mathbf{q}} |u_{\mathbf{q}}|^2 \hbar\omega_{2\mathbf{q}} \tag{325}$$

is the polaron shift—that is, the energy gap that separates the narrow polaron band [the second term in Eq. (324)] from the valence zone. The phonon operator is expressed as

$$\hat{\Phi}_{\mathbf{lm}} \equiv \hat{\Phi}_{\mathbf{lm}}(0) = \exp\left\{ \sum_{\mathbf{q}} [\Delta^{(2)}_{\mathbf{lm}}(\mathbf{q}) \hat{b}^+_{\mathbf{q}} - \Delta^{(2)^*}_{\mathbf{lm}}(\mathbf{q}) \hat{b}_{\mathbf{q}}] \right\} - e^{-S_T} \tag{326}$$

$$\Delta^{(2)}_{\mathbf{lm}}(\mathbf{q}) = u_{\mathbf{l}}(\mathbf{q}) - u_{\mathbf{m}}(\mathbf{q}) \tag{327}$$

and $u_{\mathbf{l}}(\mathbf{q}) = ue^{i\mathbf{lq}}$, where u is the dimensionless constant ($|u|^2 \gg 1$) that characterizes the degree of local deformation of the crystal by the carrier with respect to its interaction with the high-frequency mode, $\omega_2/2\pi = 756\ \text{cm}^{-1}$. The matrix element $V_{\substack{\mathbf{lm} \\ \mathbf{kk}'}}$ in the tunnel Hamiltonian (323) has been defined in the previous subsection, expression (294). The value $\langle V \rangle$ in expression (324) is the thermal average of the aforementioned matrix element (we do not define concretely the phonon mode which leads to the increase of the resonance integral because as it will be seen from the following consideration the parameter $\langle V \rangle$ will drop out of the net result). $\hat{a}^+_{\mathbf{k}}(\hat{a}_{\mathbf{k}})$ is the Fermi operator of creation (annihilation) of a phonon with the wave vector $\mathbf{k}$; the constant of proton–phonon coupling is [see Eq. (220)]

$$S_T = \frac{1}{2} \sum_{\mathbf{q}} |\Delta^{(2)}(\mathbf{q})|^2 \coth \frac{\hbar\omega_{2\mathbf{q}}}{2k_{\text{B}}T} \tag{328}$$

By definition [88] (see also Ref. 164), the operator of density of the band current is equal to

$$j_{\text{band}} = \frac{e}{\mathcal{V} i\hbar}\left[i\sum_{\mathbf{k},\mathbf{k}'}\left(\frac{\partial}{\partial \mathbf{k}}\delta_{\mathbf{k},0}\right)\hat{a}^{+}_{\mathbf{k}+\mathbf{k}'}\hat{a}_{\mathbf{k}'},\ \bar{H}_0\right]$$

$$= \frac{e}{\mathcal{V}\hbar}\sum_{\mathbf{k},\mathbf{k}'}\hat{a}^{+}_{\mathbf{k}}\hat{a}_{\mathbf{k}}\frac{\partial}{\partial \mathbf{k}}E(\mathbf{k}) \tag{329}$$

Performing here the thermal averaging, we obtain the following equation for the density of band current:

$$I_{\text{band}} = \frac{e}{\mathcal{V}\hbar}\sum_{\mathbf{k},\mathbf{k}'}\langle\hat{a}^{+}_{\mathbf{k}}\hat{a}_{\mathbf{k}}\rangle\frac{\partial}{\partial \mathbf{k}}E(\mathbf{k}) \tag{330}$$

where the brackets $\langle\ldots\rangle$ just mean the thermal averaging. The average value $\langle\hat{a}^{+}_{\mathbf{k}}\hat{a}_{\mathbf{k}}\rangle$ of the operator of number of quasi-particles in the polaron band is defined from the kinetic equation

$$\left(i\hbar\frac{\partial}{\partial t} + ie\mathscr{E}\right)\langle\hat{a}^{+}_{\mathbf{k}}\hat{a}_{\mathbf{k}}\rangle = -\langle[\bar{H}_{\text{tun}}, \hat{a}^{+}_{\mathbf{k}}\hat{a}_{\mathbf{k}}]\rangle \tag{331}$$

When we uncouple the commutator on the left-hand side of Eq. (331), nondiagonal corrections to the operator $\hat{a}^{+}_{\mathbf{k}}\hat{a}_{\mathbf{k}}$ appear. A correction $\hat{a}^{+}_{\mathbf{k}'}\hat{a}_{\mathbf{k}}$ caused by the electric field $\mathscr{E}$ and the interaction with phonons can be defined from the equation of motion

$$\left(i\hbar\frac{\partial}{\partial t} + E(\mathbf{k}') - E(\mathbf{k})\right)\hat{a}^{+}_{\mathbf{k}'}\hat{a}_{\mathbf{k}} + [H_{\mathscr{E}}, \hat{a}^{+}_{\mathbf{k}'}\hat{a}_{\mathbf{k}}] = [\bar{H}_{\text{tun}}, \hat{a}^{+}_{\mathbf{k}'}\hat{a}_{\mathbf{k}}] \tag{332}$$

where the field operator $\bar{H}_{\mathscr{E}}$ (296) written in **k**-presentation has the form

$$\bar{H}_{\mathscr{E}} = ie\mathscr{E}\sum_{\mathbf{k}'}\left(\frac{\partial}{\partial \mathbf{k}}\delta_{\mathbf{k},0}\right)\hat{a}^{+}_{\mathbf{k}+\mathbf{k}'}\hat{a}_{\mathbf{k}'} \tag{333}$$

With regard to Eq. (332) we derive from Eq. (331) the kinetic equation for the function $f(\mathbf{k}, t) = \langle\hat{a}^{+}_{\mathbf{k}}\hat{a}_{\mathbf{k}}\rangle$ which describes the quasi-particle distribution,

$$\frac{\partial f}{\partial t} + \frac{e\mathscr{E}}{\hbar}\frac{\partial f}{\partial \mathbf{k}} = \sum_{\mathbf{k}'}[f(\mathbf{k}, t) - f(\mathbf{k}', t)]W_{\mathscr{E}}(\mathbf{k}', \mathbf{k}) \tag{334}$$

where $W_{\mathscr{E}}(\mathbf{k}', \mathbf{k})$ is the transitions probability that is given by the expression

$$W_{\mathscr{E}}(\mathbf{k}', \mathbf{k}) = 2\langle V \rangle \hbar^{-2} \int_0^\infty d\tau \langle \hat{\Phi}_{\mathbf{g}}(\tau) \hat{\Phi}_{\mathbf{g}}(0) \rangle \times \exp\left\{ -e + i\hbar^{-1} \int_0^\tau d\tau' [E(\mathbf{k}' - e\mathscr{E}\mathbf{g}/\hbar) - E(\mathbf{k} - e\mathscr{E}\mathbf{g}/\hbar)] \right\} \quad (335)$$

Let us represent the distribution function in the form

$$f(\mathbf{k}, t) \simeq f^{(0)}(\mathbf{k}) + f^{(1)}(\mathbf{k}, t) \quad (336)$$

where $f^{(0)}$ is the unperturbed distribution function and $f^{(1)}$ is a correction to it. Substituting f from expression (336) into Eq. (334), we get

$$f^{(1)}(\mathbf{k}, t) \simeq \frac{1}{\sum_{\mathbf{k}'} W_{\mathscr{E}}(\mathbf{k}', \mathbf{k})} \frac{e}{\hbar} E(\mathbf{k}) \frac{\partial f^{(0)}}{\partial \mathbf{k}} \quad (337)$$

Rewriting expression (330) in terms of the result (337), we obtain

$$I_{\text{band}} = \frac{e^2}{\hbar} \sum_{\mathbf{k}} \left(\sum_{\mathbf{k}'} W_{\mathscr{E}}(\mathbf{k}', \mathbf{k}) \right)^{-1} \frac{\partial f^{(0)}(\mathbf{k})}{\partial \mathbf{k}} \frac{\partial E(\mathbf{k})}{\partial \mathbf{k}} \quad (338)$$

First of all, let us calculate the value $W_{\mathscr{E}}(\mathbf{k}', \mathbf{k})$ (335). Allowance for the narrowness of the polaron band means that we can put $E(\mathbf{k}' - e\mathscr{E}\mathbf{g}/\hbar) \simeq E(\mathbf{k} - e\mathscr{E}\mathbf{g}/\hbar)$. This assumption reduces the calculation of the integral to the integration over τ carried out in expression (309). The final result is

$$W_{\mathscr{E}}(\mathbf{k}', \mathbf{k}) \simeq \frac{2\pi^{1/2} \langle V \rangle^2}{\hbar^2 S_T^{1/2} \omega_2} e^{-S_T} e^{-e\mathscr{E}g/\hbar\omega_2} \quad (339)$$

Substituting $W_{\mathscr{E}}(\mathbf{k}', \mathbf{k})$ from (339) to expression (338), setting the Boltzmann distribution for the population of polarons in the band, $f^{(0)}(\mathbf{k}) = \exp(-E(\mathbf{k})/k_B T)$, and not taking the dispersion into account, we finally arrive at the density of band proton current:

$$I_{\text{band}} = \frac{e^2 g^2 n \mathscr{E} \omega_2 |\Delta^{(2)}|^2 \exp(-|\Delta^{(2)}|^2/2)}{2^{1/2} \pi^{1/2} k_B T} \exp(-|\Delta^{(2)}|^2 \hbar\omega_2 / k_B T) \quad (340)$$

(when deriving expression (340), we have taken into account that in the range from low to room temperature $\coth(\hbar\omega_2/2k_B T) \simeq 1$ and that the inequality $e\mathscr{E}g/\hbar\omega_2 \ll 1$ holds right up to $\mathscr{E} \sim 10^9$ V/m).

The experimental protonic conductivity versus temperature dependence curves $\sigma(T)$ are described in Refs. 193 and 195 by the simple Arrhenius law

$$\sigma(T) = \frac{\text{const}}{T} \exp\left(-\frac{E_a}{k_B T}\right)$$

in which the parameters const and E_a are determined empirically irrespective of the microscopic characteristics of the crystal. In our present consideration we derive formulas for polaronic conductivity $\sigma_{hop}(T) = I_{hop}/\mathscr{E}$ and $\sigma_{band}(T) = I_{band}/\mathscr{E}$ [where I_{hop} and I_{band} imply expressions (319) and (340) respectively], proceeding from the notion of small proton polaron. The expressions for $\sigma_{hop}(T)$and $\sigma_{band}(T)$ will now can be compared with the experimental dependence $\sigma(T)$.

In expressions (319) and (340) the lattice constant g equals 0.8 nm, the concentration of bifurcated protons (between two hydrogen bonds of the cell) per unit volume, n, is approximately 4×10^{27} m^{-3}, and the frequencies $\omega_1 = 2\pi \times 99$ cm^{-1} and $\omega_2 = 2\pi \times 756$ cm^{-1} are expressed in s^{-1}. The coupling constant $|\Delta^{(1)}|^2$ connected the charge carrier with the vibrations of the crystal characterized by ω_1 can easily be found by comparing the calculated value of the activation energy (320) with the experimental value $E_a^{(1)} \simeq 8.8$ kJ/mol; from the equation thus formed we get $|\Delta^{(1)}|^2 \simeq 25$. Analogously, by comparing the exponential term $|\Delta^{(2)}|^2\hbar\omega_2$ from expression (340) with the experimental value of the activation energy $E_a^{(2)} \simeq 33.1$ kJ/mol, we obtain the coupling constant of the carrier with the mode of collective vibrations of the IO_3 pyramids: $|\Delta^{(2)}|^2 \simeq 4$. Only one parameter of the theory—namely, the tunnel integral to the second power $|V|^2$—remains indeterminate.

Figure 12 shows the experimental curve for conductivity $\sigma(T)$ in the direction of maximum mobility of the protons [193] and the plots of the band and hopping polaron conductivity versus T dependence curves ($\sigma_{hop}(T)$ and $\sigma_{band}(T)$, respectively), constructed in conformity with Eqs. (319) and (340); the tunnel integral is assumed to be equal to $|V| = 10^{-26}$ J. As can be seen from Fig. 12, for the adopted parameter values, based on the experimental values of the frequencies of optically active modes $\omega_1 = 2\pi \times 99$ cm^{-1} and $\omega_2 = 2\pi \times 756$ cm^{-1} and on the activation energies $E_a^{(1)} \simeq 8.8$ kJ/mol and $E_a^{(2)} \simeq 33.1$ kJ/mol, the calculated total conductivity $\sigma_{hop}(T) + \sigma_{band}(T)$ is qualitatively consistent with the experimental value of the conductivity $\sigma(T)$ in the region $T > T_0 = 213$ K. Furthermore, from Fig. 12 it seen that in the region $T > T_0$ the band polaron conductivity prevails (the activation of carriers to the band is effected by the mode with the frequency of 756 cm^{-1}), whereas the hopping polaron conductivity activated by the mode with the frequency of 99 cm^{-1} should prevail in the region of $T < T_0$.

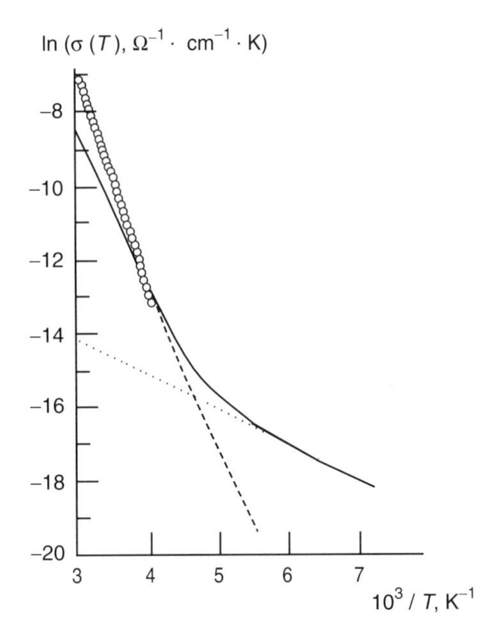

Figure 12. Conductivity of the $NH_4IO_3 \cdot 2HIO_3$ crystal as a function of the reciprocal temperature: the experimental dependence (○) in the direction of maximum conductivity of the crystal; the dependence charts of hopping conductivity (······) and of band conductivity (-----), plotted in accordance with Eqs. (319) and (340), respectively. The continuous curve is a plot of the total conductivity $\sigma_{\text{hop}} + \sigma_{\text{band}}$ as a function of the reciprocal temperature T^{-1}. (From Ref. 46.)

Thus, the polaronic mechanism of the superionic conductivity in the crystal of ammonium triiodate, which we propose in this chapter, is yet another, novel conception in the range of the known [149,150,203,204] modernized models of the hopping and continuous diffusion.

At the same time it is interesting to understand why the intracellular 756 cm^{-1} mode influences the proton conductivity. As we mentioned above, the polarized optical 99-cm^{-1} mode activates the proton mobility in the range $T_c < T < T_0$, where $T_c = 120\,\text{K}$ and $T_0 = 213\,\text{K}$. However, the intracellular 756-cm^{-1} mode is not polarized; nevertheless, it is responsible for the proton mobility for $T > T_0$. With $T > T_c = 120$ K, these two modes demonstrate an anomalous temperature behavior and the intracellular mode begins to intensify [47]. It is the intensification of the cellular mode with T, which leads to its strong coupling with charge carriers in the crystal studied. A detailed theory of the mixture of the two modes is posed in Appendix D.

K. Polariton Effect in Crystals with Symmetric O···H···O Hydrogen Bonds

Up to now there is still no complete information on the mechanisms of ν(XH) stretching vibration band broadening in the infrared spectra of hydrogen-bonded systems. Among these systems the most intriguing seem to be those with strong hydrogen bonds, when the effect is fairly pronounced and increases up to continuous absorption of several thousands wavenumbers. According to Hadzi [205], "With hydrogen bonds stronger than, for instance, those in carboxylic dimers the absorption pattern in the region above 1600 cm^{-1} is very peculiar. Instead of one, more or less structured, X—H stretching band there appear several, most often three bands.... This type of spectrum has been designated (i) and the bands *A*, *B*, *C* may be taken as characteristic of a class of very strong hydrogen bonds." So far there is a large amount of experimental data on the spectroscopic behavior of strong hydrogen bonds (the $R_{\mathrm{O\cdots O}}$ distance ranging from 0.254 to less than 0.248 nm) [206–210], and the results obtained are related mainly to liquids and solids. The origin of the *ABC* structure could be explained by anharmonic coupling between the stretching high frequency X—H $\cdots$ X and low frequency X $\cdots$ X motions, called the strong-coupling theory [66], which we discussed in Section II.A. The substructure of shape of the *ABC* region can also be described by anharmonic resonance interactions with other intramolecular vibrations and Fermi resonance [211].

The hydrogen bond potential for the systems with strong hydrogen bonds is expected to be a single- or two-minima potential with comparably deep minima and low barriers, so that the proton may be considered localized in the deeper well. Furthermore, with hydrogen bond strength increasing, the redistribution of the relative intensity of thee bands is observed (the intensity of the band *A* decreases, whereas that of the *B* and particularly *C* band increases) along with the band *A* lowering in frequency [209]. Thus, for the very short hydrogen bonds the ν(XH) stretching band developed into "very strong and broad absorption culminating somewhere between 600 and 1200 cm^{-1}" [66] with sharp low-frequency edge. Furthermore, according to Hadzi [205], these spectra classified as (ii) have led to later diffraction studies. Several models have been proposed [211,212] for the description of the continuos absorption, though an explanation of such spectroscopic behavior of strong symmetrical hydrogen bonds is still questioned. It is likely that different mechanisms contribute more or less to the proton behavior in solids with strong hydrogen bonds that we propose could contribute to the knowledge of the ν(XH) shaping mechanisms under various conditions.

Let us consider molecular excitations in a system with strong hydrogen bonds, which is under the light radiation according to Ref. 213.

The existence of hydrogen bonds with a high polarizability often leads to the formation of a wide extended band in the infrared absorption spectra. This is especially clearly manifested in compounds with strong hydrogen bonds when the distance between oxygen atoms $R_{\mathrm{O\cdots O}} \leq 0.25$ nm. Hence, the motion of a proton located at the center of a O $\cdots$ H $\cdots$ O bridge will cause strong modulation of the cell dipole moment. When incident light is applied to a crystal, those polarized proton oscillations will be coupled with falling photons forming polaritons. In such a case, high-frequency intermolecular vibrations will interact just with polaritons.

High-frequency intermolecular vibrations and their interaction with optical phonons are described by the Hamiltonian [214,215]

$$H_0 = \sum_n E_n^f(0)\hat{a}_n^{f^+}\hat{a}_n^f + \frac{1}{\sqrt{N}}\sum_{n,\mathbf{k}} \hbar\chi_{\mathbf{k}}^{nf}\hat{a}_n^{f^+}\hat{a}_n^f(\hat{b}_{\mathbf{k}} + \hat{b}_{-\mathbf{k}}^+)$$
$$+ \sum_{\mathbf{k}} \hbar\Omega(\mathbf{k})\hat{b}_{\mathbf{k}}^+\hat{b}_{\mathbf{k}} \tag{341}$$

Here $E_n^f(0)$ refers to the fth excited state of a free molecule in the crystal; $\hat{a}_n^f + (\hat{a}_n^f)$ is the Bose operator of creation (annihilation) of an intramolecular vibrational excitation in the nth molecule; $\hbar\Omega(\mathbf{k})$ refers to the energy of an optical phonon with the wave vector $\mathbf{k}$ connected with proton oscillations in the O $\cdots$ H $\cdots$ O bridge; $\hat{b}_{\mathbf{k}}^+(\hat{b}_{\mathbf{k}})$ is the Bose operator of phonon creation (annihilation); and $\hbar\chi_{\mathbf{k}}^{nf}$ is the coupling energy between the molecular excitation and phonons.

The most important feature of the present model refers to the strong interaction between the incident photons and molecular excitations. Such a model seems to be one more mechanism that explains the appearance of the extended wing in the ν(OH) infrared absorption.

The Hamiltonian of photon–phonon interaction has the form [171]

$$H = \frac{1}{2}\sum_{\mathbf{k}} H_{\mathbf{k}} \tag{342}$$

$$H_{\mathbf{k}} = \hbar\omega(\mathbf{k})(\hat{\alpha}_{\mathbf{k}}^+\hat{\alpha}_{\mathbf{k}} + \hat{\alpha}_{-\mathbf{k}}^+\hat{\alpha}_{-\mathbf{k}}) + \hbar\Omega_l(\hat{b}_{\mathbf{k}}^+\hat{b}_{\mathbf{k}} + \hat{b}_{-\mathbf{k}}^+\hat{b}_{-\mathbf{k}})$$
$$- D_{\mathbf{k}}[(\hat{\alpha}_{\mathbf{k}}^+ - \hat{\alpha}_{-\mathbf{k}})(\hat{b}_{\mathbf{k}}^+ + \hat{b}_{-\mathbf{k}}) + (\hat{\alpha}_{-\mathbf{k}}^+ - \hat{\alpha}_{\mathbf{k}})(\hat{b}_{-\mathbf{k}}^+ + \hat{b}_{\mathbf{k}})] \tag{343}$$

Here the interaction $\hbar\omega_{\mathbf{k}}$ is the energy of an incident photon with the wave vector $\mathbf{k}$, and $\hat{\alpha}_{\mathbf{k}}^+(\hat{\alpha}_{\mathbf{k}})$ the Bose operator of creation (annihilation) of the photon; the interaction matrix is

$$D_{\mathbf{k}} = -D_{\mathbf{k}}^* = -\frac{i}{2}\sqrt{\frac{\Omega_l\omega(\mathbf{k})(\varepsilon_0 - \varepsilon_\infty)}{\varepsilon_0\varepsilon_\infty}} \tag{344}$$

and neglecting the dispersion we can assume that in expressions (343) and (344) $\Omega(\mathbf{k}) = \Omega_l$ (hereafter Ω_l and Ω_t are well-known threshold frequencies for longitudinal and transversal optical modes, respectively; see, e.g., Ref. 171). The diagonalization of the Hamiltonian (343) is done by transition to new operators [171]

$$\hat{\beta}_{\mu\mathbf{k}} = u_{\mu_1}\hat{\alpha}_{\mathbf{k}} + u_{\mu_2}\hat{b}_{\mathbf{k}} - \nu_{\mu_1}\hat{\alpha}_{-\mathbf{k}}^+ - \nu_{\mu_2}\hat{b}_{-\mathbf{k}}^+ \tag{345}$$

where u_{μ_j} and v_{μ_j} are four real functions that should be determined ($\mu = 1, 2$; $j = 1, 2$).

The diagonalized Hamiltonian has the form

$$H_{\mathbf{k}} = \sum_{\mu=1}^{2} \hbar\omega_{\mu}(\mathbf{k})\hat{\beta}_{\mu\mathbf{k}}^{+}\hat{\beta}_{\mu\mathbf{k}} \tag{346}$$

The operator $\hat{\beta}_{\mu\mathbf{k}}$ satisfies the Bose commutation condition

$$[\hat{\beta}_{\mu\mathbf{k}}, \hat{\beta}_{\mu'\mathbf{k}}^{+}] = \delta_{\mu\mu'} \tag{347}$$

The following equalities are valid:

$$v_{\mu_1} = \frac{\omega_{\mu}(\mathbf{k}) - \omega(\mathbf{k})}{\omega_{\mu}(\mathbf{k}) + \omega(\mathbf{k})} u_{\mu_1}, \qquad v_{\mu_2} = \frac{\Omega_l - \omega(\mathbf{k})}{\Omega_l + \omega(\mathbf{k})} u_{\mu_2} \tag{348}$$

allowing one to eliminate v_{μ_1} and v_{μ_2}. These functions satisfy the equations

$$\sum_{s=1}^{2} (u_{\mu s}u_{\eta s} - v_{\mu s}v_{\eta s}) = \delta_{\mu\eta} \tag{349}$$

Equations (348) and (349) lead to the following equations for quantities $u_{\mu s}$ and $v_{\mu s}$:

$$u_{11}u_{21}\left[1 - \frac{(\omega_1 - \omega(\mathbf{k}))(\omega_2 - \omega(\mathbf{k}))}{(\omega_1 + \omega(\mathbf{k}))(\omega_2 + \omega(\mathbf{k}))}\right] + u_{12}u_{22}\left[1 - \frac{(\Omega_l - \omega_1)(\Omega_l - \omega_2)}{(\Omega_l + \omega_1)(\Omega_l + \omega_2)}\right] = 0 \tag{350}$$

$$u_{11}^2\left[1 - \left(\frac{\omega_1 - \omega(\mathbf{k})}{\omega_1 + \omega(\mathbf{k})}\right)^2\right] + u_{12}^2\left[1 - \left(\frac{\Omega_l - \omega_1}{\Omega_l + \omega_1}\right)^2\right] = 1 \tag{351}$$

$$u_{21}^2\left[1 - \left(\frac{\omega_2 - \omega(\mathbf{k})}{\omega_2 + \omega(\mathbf{k})}\right)^2\right] + u_{22}^2\left[1 - \left(\frac{\Omega_l - \omega_2}{\Omega_l + \omega_2}\right)^2\right] = 1 \tag{352}$$

One of the four quantities $u_{\mu s}$ and $v_{\mu s}$ is the free parameter; let $u_{12} = 1$. Then solving Eqs. (350)–(352), one can rewrite the phonon operator $(\hat{b}_{\mathbf{k}}^{+} + \hat{b}_{-\mathbf{k}})$ in the interaction term of the Hamiltonian (341) via polariton operators $\hat{\beta}_{1,2\mathbf{k}}$

$$
\begin{aligned}
(\hat{b}^{+}_{\mathbf{k}} + \hat{b}_{-\mathbf{k}}) = \frac{\Omega_l + \omega_1}{2\omega_1 u_{12}} \Bigg\{ & (\hat{\beta}_{1\mathbf{k}} + \hat{\beta}^{+}_{1,-\mathbf{k}}) \\
& - \left[\frac{u_{11}}{u_{12}} \frac{\omega(\mathbf{k})(\Omega_l + \omega_1)}{\omega_1(\omega_1 + \omega(\mathbf{k}))} - \frac{u_{21}}{u_{22}} \frac{\omega(\mathbf{k})(\Omega_l + \omega_2)}{\omega_2(\omega_2 + \omega(\mathbf{k}))} \right] \\
& \times \left[\frac{u_{11}}{u_{12}} \frac{\omega(\mathbf{k})(\Omega_l + \omega_1)}{\omega_1(\omega_1 + \omega(\mathbf{k}))} (\hat{\beta}_{1\mathbf{k}} + \hat{\beta}^{+}_{1,-\mathbf{k}}) \right. \\
& \left. - \frac{u_{11}}{u_{22}} \frac{\omega(\mathbf{k})(\Omega_l + \omega_2)}{\omega_2(\omega_1 + \omega(\mathbf{k}))} (\hat{\beta}_{2\mathbf{k}} + \hat{\beta}^{+}_{2,-\mathbf{k}}) \right] \Bigg\}
\end{aligned}
\tag{353}
$$

Here $\omega(\mathbf{k}) = ck$ and the frequencies of polariton branches are

$$
\omega^2_{1,2}(\mathbf{k}) = \frac{1}{2\varepsilon_\infty} (\varepsilon_0 \Omega_t^2 + c^2k^2 \pm [(\varepsilon_0\Omega_t^2 + c^2k^2)^2 - 4\varepsilon_\infty c^2 \Omega_t^2 k^2]^{1/2}) \tag{354}
$$

At small k these two solutions are reduced to

$$
\omega_1^2(\mathbf{k}) = \frac{c^2k^2}{\varepsilon_\infty} \tag{355}
$$

$$
\omega_2^2(\mathbf{k}) = \Omega_t^2 \frac{\varepsilon_0}{\varepsilon_\infty} + \frac{c^2k^2}{\varepsilon_\infty} = \Omega_l^2 \frac{c^2k^2}{\varepsilon_\infty} \tag{356}
$$

In approximation $ck \gg \Omega_l,\ \Omega_t$, Eq. (353) becomes

$$
(\hat{b}^{+}_{\mathbf{k}} + \hat{b}_{-\mathbf{k}}) = \frac{\Omega_l - \Omega_t}{8\sqrt{\Omega_t \Omega_l}} \sqrt{ck} (\hat{\beta}_{2\mathbf{k}} + \hat{\beta}^{+}_{2,-\mathbf{k}}) \tag{357}
$$

When obtaining Eq. (357), we let $\Omega_l > \Omega_t$ at least by several times. Hence, instead of the exciton–phonon Hamiltonian (341), one can turn to the exciton–polariton Hamiltonian:

$$
\begin{aligned}
H_0 = & \sum_n E_n^f(0) \hat{a}_n^{f^+} \hat{a}_n^f + \frac{1}{\sqrt{N}} \sum_{n,\mathbf{k}} \kappa_{\mathbf{q}} \hbar\omega_2(\mathbf{k}) \hat{a}_n^{f^+} \hat{a}_n^f (\hat{\beta}_{2\mathbf{k}} + \hat{\beta}^{+}_{2,-\mathbf{k}}) \\
& + \sum_{\mathbf{k}} \hbar\omega_2(\mathbf{k}) \hat{\beta}^{+}_{2\mathbf{k}} \hat{\beta}_{2\mathbf{k}}
\end{aligned}
\tag{358}
$$

where the effective coupling function is

$$
\kappa_{\mathbf{k}} = \tilde{\chi}^{nf}_{\mathbf{k}} \frac{\Omega_l - \Omega_t}{8\sqrt{\Omega_t \Omega_l}} \sqrt{ck} \tag{359}
$$

and $\tilde{\chi}_{\mathbf{k}}^{nf}$ is the dimensionless function of exciton–phonon interaction. The Hamiltonian (18) is diagonalized over $\hat{\beta}_{2\mathbf{k}}$ by means of the well-known transformation [214]

$$\hat{a}_n^f = \hat{A}_n^f e^{-\hat{\sigma}_n^f} \tag{360}$$

$$\hat{\beta}_{2\mathbf{k}} = \hat{B}_{\mathbf{k}} - \sum_n \hat{A}_n^{f^+} \hat{A}_n^f \frac{1}{\sqrt{N}} \kappa_{\mathbf{k}}^* \tag{361}$$

$$\hat{\sigma}_n^f = \frac{1}{\sqrt{N}} \sum_{\mathbf{k}} (\kappa_{\mathbf{k}}^* \hat{B}_{\mathbf{k}}^+ - \kappa_{\mathbf{k}} \hat{B}_{\mathbf{k}}) \tag{362}$$

where $\hat{\sigma}_n^{f^+} = -\hat{\sigma}_n^f$. Consequently, the Hamiltonian is transformed to

$$H = \sum_n \varepsilon_n^f \hat{A}_n^{f^+} \hat{A}_n^f + \sum_{\mathbf{k}} \hbar\omega_2(\mathbf{k}) \hat{B}_{\mathbf{k}}^+ \hat{B}_{\mathbf{k}} \tag{363}$$

$$\varepsilon_n^f = E_n^f(0) - \frac{1}{N} \sum_{\mathbf{k}} |\kappa_{\mathbf{k}}|^2 \hbar\omega_2(\mathbf{k}) \tag{364}$$

Having analyzed infrared absorption spectra, let us calculate the absorption coefficient $K(\omega)$ (see, e.g., Refs. 215 and 216):

$$G_n^f(t) = \langle\langle \hat{a}_n^f(t), \hat{a}_n^{f^+}(0) \rangle\rangle \tag{365}$$

The time-dependent operators $\hat{A}_n^f$ and $\hat{B}_{\mathbf{k}}$ are as follows:

$$\hat{A}_n^f(t) = \hat{A}_n^f e^{-i\varepsilon_n^f t}, \qquad \hat{B}_{\mathbf{k}}(t) = \hat{B}_{\mathbf{k}} e^{-i\omega_2(\mathbf{k})t} \tag{366}$$

Equation (365) changes to [211,214]

$$G_n^f(t) = -i\theta(t) \langle \exp[-\hat{\sigma}_n^f(t)] \exp[\hat{\sigma}_n^f(0)] \rangle_0 \exp(-i\varepsilon_n^f t) \tag{367}$$

where

$$\theta(t) = \begin{cases} 0, & \text{if} \quad t < 0 \\ 1, & \text{if} \quad t \geq 0 \end{cases} \tag{368}$$

The correlation function from Eq. (367) is reduced to

$$\langle \exp[-\hat{\sigma}_n^f(t)] \exp[\hat{\sigma}_n^f(0)] \rangle_0 = [g_n^f(t)] \tag{369}$$

where the coupling function is expressed as

$$g_n^f(t) = \frac{1}{N} \sum_{\mathbf{k}} |\kappa_{\mathbf{k}}|^2 \{ (n_{\mathbf{k}} + 1) e^{-i\omega_2(\mathbf{k})t} + n_{\mathbf{k}} e^{i\omega_2(\mathbf{k})t} - (2n_{\mathbf{k}} + 1) \} \tag{370}$$

Here $n_{\mathbf{k}}$ is the quantity of polarons and is defined by an external light source. In our case the intensity of the source is small, of the order of 10 μW. Therefore one would expect that the value of $n_{\mathbf{k}}$ is near unity of smaller.

The overall result for the Fourier component of the Green function, Eq. (367), is

$$G_n^f(\omega) = -i \int_t^{\infty} dt \exp[i(\omega - \varepsilon_n^f + i\gamma)t + g_n^f(t)] \tag{371}$$

where γ is a positive parameter that describes the adiabatic switching of exciton–polaron interaction. Furthermore, one should take into account that the coupling constant is small, $\tilde{\chi} \ll 1$. Hence, the function $g_n^f(t)$ in Eq. (370) can be expanded into a Taylor series; and for the absorption coefficient, one can gain (only two terms of expanding are saved)

$$K(\omega) \propto \operatorname{Im} i \int_t^{\infty} dt\, e^{i(\omega - \varepsilon_n^f + i\gamma)t} \left[1 + \tilde{\chi}^2 \left(\left(\frac{\Omega_l - \Omega_t}{8\sqrt{\Omega_t \Omega_l}} \right)^2 \sum_{\mathbf{k}} (\sqrt{ck})^2 e^{-ickt} + \cdots \right) \right] \tag{372}$$

Here at the expansion we have put $n_{\mathbf{k}} = 1$, and $\tilde{\chi}_{\mathbf{k}}$ is taken in the step form: $\tilde{\chi}_{\mathbf{k}} = \tilde{\chi}$ if $k_0 \le k \le k_{\max}$ and $\tilde{\chi}_{\mathbf{k}} = 0$ for all other k; that is, we believe that the exciton–phonon interaction is effective only in the region between phonon wavenumbers $k_0 = 50\ \mathrm{cm}^{-1}$ and $k_{\max} = 200\ \mathrm{cm}^{-1}$. Note that we have also made an extrapolation of the function $\kappa_{\mathbf{k}}$, expression (359), from large values of k to small values.

Integrating over t, we gain from Eq. (372)

$$K(\omega) = K_0(\omega) + K_1(\omega) \tag{373}$$

$$K_0(\omega) \propto \frac{\gamma}{(\omega - \varepsilon_n^f)^2 + \gamma^2} \tag{374}$$

$$K_1(\omega) \propto \frac{\tilde{\chi}^2 (\Omega_l - \Omega_t)^2}{64 \Omega_l^2 \Omega_t} \gamma \sum_{\mathbf{k}} \frac{ck}{(\omega - \varepsilon_n^f - ck)^2 + \gamma^2} \tag{375}$$

Let us make a transfer in Eq. (375) from summation over **k** to integration:

$$\sum_{\mathbf{k}} \frac{ck}{(\omega - \varepsilon_n^f - ck)^2 + \gamma^2} = I(\omega) = \frac{2\pi}{ck_0} 2 \int_1^{k_{\max}/k_0} \frac{x\,dx}{(\Xi - x)^2 + (\gamma/ck_0)^2} \tag{376}$$

where k_0 and $k_{\max}$ are effective lower and upper values of the wavenumber; moreover, in Eq. (376) the following parameter is introduced:

$$\Xi = \frac{\omega - \varepsilon_n^f}{ck_0} \tag{377}$$

The integral $I(\omega)$ in expression (376) can easily be taken:

$$I(\omega) = \frac{1}{2}\ln\left|\frac{(ck_{\max})^2 - 2(\omega - \varepsilon_n^f)ck_{\max} + (\omega - \varepsilon_n^f)^2 + \gamma^2}{(ck_0)^2 - 2(\omega - \varepsilon_n^f)ck_0 + (\omega - \varepsilon_n^f)^2 + \gamma^2}\right| + \frac{\omega - \varepsilon_n^f}{\gamma}\left(\arctan\frac{ck_{\max} - (\omega - \varepsilon_n^f)}{\gamma} - \arctan\frac{ck_0 - (\omega - \varepsilon_n^f)}{\gamma}\right) \qquad (378)$$

Thus, the correction to the absorption coefficient $K_0(\omega)$ (373) takes the form

$$K_1(\omega) \propto \frac{\pi}{16}\tilde{\chi}^2 \frac{(\Omega_l - \Omega_t)^2}{\Omega_l^2 \Omega_t} \gamma I(\omega) \qquad (379)$$

Utilizing expressions (374) and (379), we can calculate the absorption coefficient (373) as a function of the following parameters: the constant of exciton–phonon coupling $\tilde{\chi}$, the longitudinal phonon frequency Ω_l (that is, the proton stretching frequency in the hydrogen bridge $O \cdots H \cdots O$), the transversal phonon frequency Ω_t (that is, the bending proton frequency in the bond $O \cdots H \cdots O$), the effective boundary values of the polariton branch ck_0 and $ck_{\max}$, and the damping constant γ. Numerical calculations of the absorption coefficient $K(\omega)$ have been performed while varying all of these parameters [213]. The spectra calculated have shown very significant deviations of the ν(OH) band from the Lorentzian profile (i.e., when $\tilde{\chi} = 0$).

In order to check the validity of the suggested approach, we have applied theoretical results to the OH stretching band-shape modeling for the α-modification of the acid potassium iodate $KIO_3 \cdot HIO_3$ crystal (the spectra of the said crystal have been recorded in Ref. 217). The α-modification of the $KIO_3 \cdot HIO_3$ crystal is known to exhibit ferroelectric behavior [218]. The structure is monoclinic with space group $P2_1/c$ in the paraelectric phase, and is assumed to change to $P2_1$ below the Curie temperature $T_c = 223$ K. The principal feature of the crystal structure of α-$KIO_3 \cdot HIO_3$, as revealed by inelastic neutron scattering [219], is a short $O–H \cdots O$ hydrogen bond ($O \cdots O$ distance about 0.249 nm) linking two iodate ions. In the paraelectric phase, the symmetric $O \cdots H \cdots O$ hydrogen bond takes place).

In Fig. 13 the experimental spectrum of the α-$KIO_3 \cdot HIO_3$ crystal in the ν(OH) region and the calculated spectrum [by formulas (373), (374), and (379)] are presented. It is easy to see that the theoretical curve gives a fairly good fitting of the experimental spectrum. The negligible deviation of the experimental and calculated spectra observed in the middle part of the figure may be explained by the overlapping of the δ(OH) absorption band in this region of the spectrum.

Thus the inclusion of the polariton part into the absorption coefficient of a solid with strong symmetric hydrogen bonds predicts that instead of the narrow

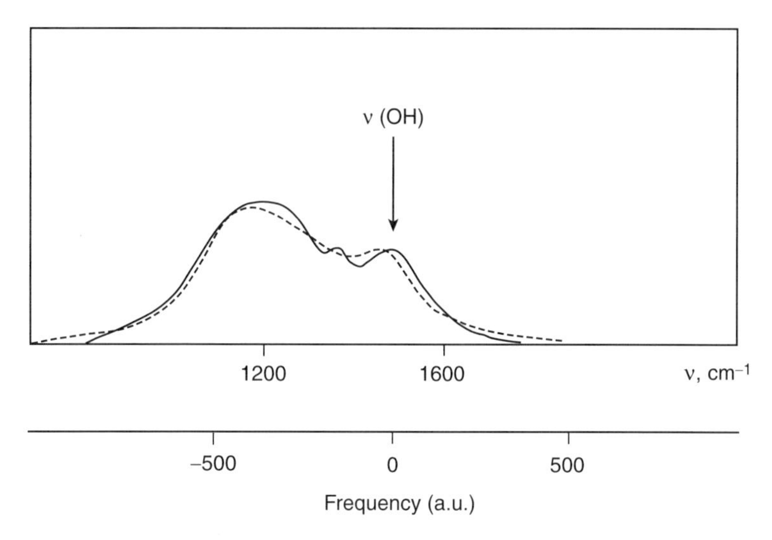

Figure 13. Infrared absorption spectrum of the α-$KH(IO_3)_2$ crystal in the ν(OH) stretching region (——) and the numerically calculated spectrum (-----). The values of the fitting parameters (in cm^{-1}) are: $\Omega_l = 200$, $\Omega_t = 50$, $ck_0 = 10$, $ck_{\max} = 500$, $\gamma = 100$. The dimensionless coupling constant $\tilde{\chi} = 0.008$. (From Ref. 214.)

high-frequency OH stretching band, a very strong broad absorption should appear in the frequency region below 1200 cm^{-1}. The theory described can be applied to the analysis of any real crystal that includes strong symmetric hydrogen bonds.

IV. BACTERIORHODOPSIN CONSIDERED FROM THE MICROSCOPIC PHYSICS STANDPOINT

A. Active Site of Bacteriorhodopsin and the Proton Path

Bacteriorhodopsin is a membrane protein, which is present in the purple membrane of *Halobacterium halobium*. Bacteriorhodopsin is an energy-transducing molecule operating as a light-driven molecular generator of proton current: It transports protons from the cytoplasmic to the extracellular side of the membrane [220]. Upon light excitation, the light-adapted form of bacteriorhodopsin undergoes a cycle of chemical transformations including at least five intermediates, K, L, M, N, and O (and the ground state, bR), with absorption bands covering almost all the visible range [221,222].

Henderson et al. [223] presented a detailed pattern of the structure of bacteriorhodopsin using high-resolution cryoelectron microscopy. Using X-ray and neutron diffraction techniques, Dencher et al. [224–227] could decode the secondary and tertiary structure of bacteriorhodopsin during the photocycle. Nevertheless, we should emphasize that the resolution is still restricted by several tenths of a nanometer. The most important amino acid residues of bacteriorhodopsin molecule participating in a protein–chromophore interaction have been determined by methods of genetic engineering (see, e.g., Refs. 228–230).

It is now firmly established that the primary light-induced processes include a very fast shift of electron density along the protonated Schiff base retinal and its all-*trans*–13-*cis* isomerization, leading to the formation of the first stable intermdiate K [231] which stores about one-third of the absorbed light energy [232]. Resonance Raman and Fourier transform-infrared (FT-IR) spectroscopies have provided significant information about the nature of the alternations of the retinal and of neighboring amino acids residues constituting an active site during the bacteriorhodopsin photocycle. Various intermediates have been stabilized, and FT-IR difference spectra of the intermediates have been taken (see, e.g., Refs. 6, 222, 230, and 233). The key changes involve Tyr 185, Asp 85, Asp 96, Asp 212, Glu 204, and very plausible Tyr 57.

Different aspects of bacteriorhodopsin functioning have been studied early [35–37, 223, 233–249]. However, as a rule, the researchers have described the mechanism of proton transport phenomenologically, though the analysis of the results taken from the FT-IR difference spectra yielded detailed descriptions of the proton state in a proton pathway in all of the above-mentioned intermediates (see, e.g., Zundel [6]).

Below we propose a microscopic model of the light-induced proton transport in bacteriorhodopsin, which is based on the active site model proposed in Refs. 6, 35–37, 233, and 248. Namely, we shall analyze the influence of the charge separation in the excited chromophore on a set of proton absorption bands covering almost all the visible range [221,222].

Henderson et al. [223] presented a detailed pattern of the structure of bacteriorhodopsin using high-resolution cryoelectron microscopy. Using X-ray and neutron diffraction techniques, Dencher et al. [224–227] could decode the secondary and tertiary structure of bacteriorhodopsin during the photocycle. Nevertheless, we should emphasize that the resolution still shows transitions in the active site (protonation of counterions, deprotonation of Schiff base, and reprotonation of counterions), leading to a metastable state of the protein.

Since bacteriorhodopsin is a chromoprotein complex, a change in the interaction between the Schiff base of retinal-chromophore and the protein-opsin, taking place in the active site, plays a central role in light energy conversion and membrane color regulation. Opsin consists of a single polypeptide chain of 248 amino acids residues. It forms seven transmembrane segments, A–G, having α-helical secondary structure. The tertiary structure of these transmembrane segments determined by electron microscopy [223] is

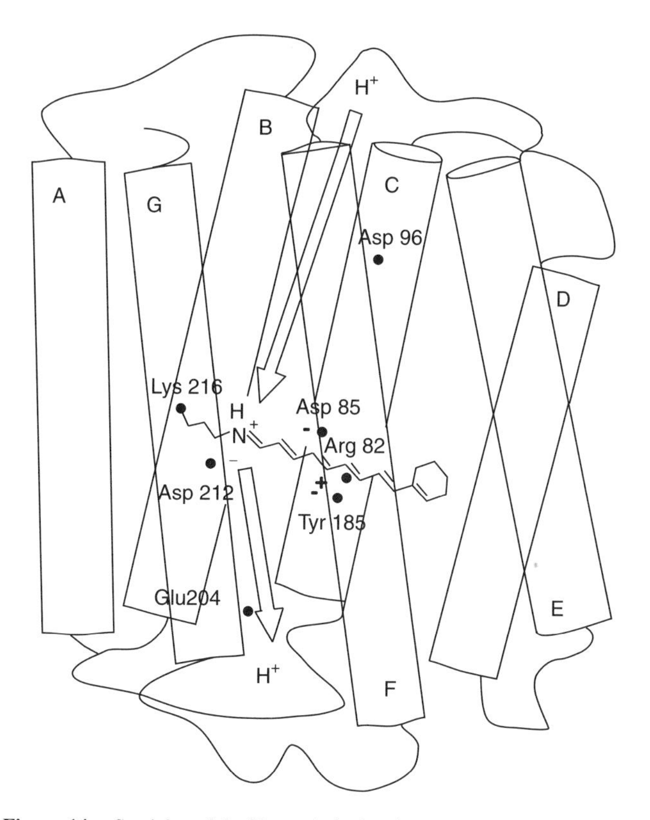

Figure 14. Spatial model of bacteriorhodopsin structure. (From Ref. 37.)

shown schematically in Fig. 14. The retinal covalently bound via the protonated Schiff base to the amino acid residue Lys 216 [220] in the middle of the G segment is also shown in Fig. 14. The retinal long axis is inclined at an angle of about 20° relative to membrane plane [250]. The β-ionone ring of the retinal points away from the cytoplasmic side of the membane [251]. Two orientations of the N–H bond of the retinal are possible [238,246]: The N–H bond may point toward either the cytoplasmic or the extracellular side of the membrane. Experimental data on photocycle kinetics [238] show that both these orientations may occur in the native membrane. Henceforth, we will consider only one orientation state, shown in Fig. 14.

Figure 14 also shows the amino acid residues playing a key role in the proton transport: Asp 85, Asp 212, and Tyr 185 with negatively charged terminal groups that are counterions with respect to the positively charged protonated Schiff base; Asp 85 and Asp 96 are inner proton acceptor and donor, respectively; Arg 82 favors the deprotonation of the retinal Schiff base. The data on light-induced FT-IR difference spectroscopy indicate that aspartic and tyrosine acid terminal groups undergo the largest changes during the bacteriorhodopsin photocycle [6,252,253]. Zundel and co-workers [6,233,241, 254] constantly noticed that the proton pathway includes tyrosines and structural water. And at the end of the pathway, Glu 204 is present [6,230,255].

The principal structure of the outlet proton channel is now considered as being composed of water molecules and the terminal —OH group of Asp 212, Glu 204 and, possibly, tyrosine(s). In contrast, the inlet channel has hydrophobic side chains of amino acid residues, except for the polar side chain of Asp 96 which is positioned about 0.1 nm from the Schiff base and about 0.6 nm from the cytoplasmic end of the C segment [224]. One or two water molecules close to the Schiff base [256–258] perhaps play a structure forming role and promote the proton dynamics.

The light-induced isomerization of the retinal shifts the protonated Schiff base into a new environment. A subsequent charge separation triggers the rearrangement of the active site, resulting in determination of the connection of the proton with the retinal Schiff base down to the critical value when the Schiff's base proton is transferred to Asp 85.

The measurements carried out by light-induced FT-IR difference spectroscopy indicate that the bR $\rightarrow$ K transition is accompanied by the protonation of Tyr 185 ($-O^- \rightarrow -OH$) [259–261] and the creation of new H bonds [233,6]. These bonds are formed between the protonated Schiff base and the terminal group of Asp 85, as well as between the terminal groups of Asp 212 and Tyr 185. The protonation of Tyr 185 can easily be explained using the results of quantum chemical calculations of the charge distribution in the intermediate K [262]. In accordance with these results, the negative charge is increased near the C_{14} of 13-*cis*-retinal, which is located close to the terminal group of Tyr 185 (Fig. 15). This suggests that the negative charge favors H^+ transfer from the terminal group $-NH_3^+$ of Arg 82 to that of Tyr 185 via the side chain of Asp 212, because the negative charge of its terminal group is partially neutralized by the positive charge concentrated near C_{15} and the side chain of Lys 216 (Fig. 15b).

FT-IR spectrometric measurements [252,263] show that Asp 96 undergoes deprotonation ($-COOH \rightarrow -COO^-$) during the K $\rightarrow$ L transition. From our point of view, this process can be understood if one considers the positive charge at the C_{15} of 13-*cis*-retinal and the side chain of Lys 216 (Fig. 16). It is obviously favors the deprotonation process for the terminal group of Asp 96. In turn, the appearance of the negative charge distant from the active side can promote changes in the protein structure (leading, in particular, to a variation in geometry and strength of the H bonds [245]), which have been observed experimentally [264–269].

The protonated Schiff base of all-*trans*-retinal in the ground state has a pK_a greater than 13 [270], which provides its strong connection with the retinal by

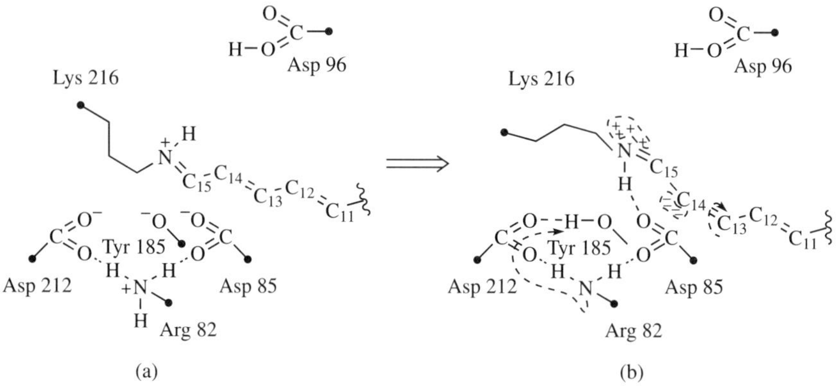

Figure 15. Scheme of the active site changes at the transition from the ground state bR (a) to the intermediate K (b). (From Ref. 37.)

means of an electrostatic attractive interaction between the positive protonated Schiff base and negative counterions. Charge separation in the active site taking place after the all-*trans*–13-*cis* isomerization of the retinal makes the connection of the proton with the Schiff base weaker. This naturally results in a decrease in the pK_a value for the Schiff's base proton.

At the L → M transition, two protons neutralize the counterions [233]: One proton transfers from the Schiff base to the terminal group of Asp 85 [271,272], and another transfers from the terminal group of Tyr 185 to that of Asp 212 [273]. Note that this process follows the deprotonation of Asp 96, including an additional negative charge on the terminal group –COOH of Asp 85 and Asp 212. This is the mechanism of additional decrease of the Schiff's base

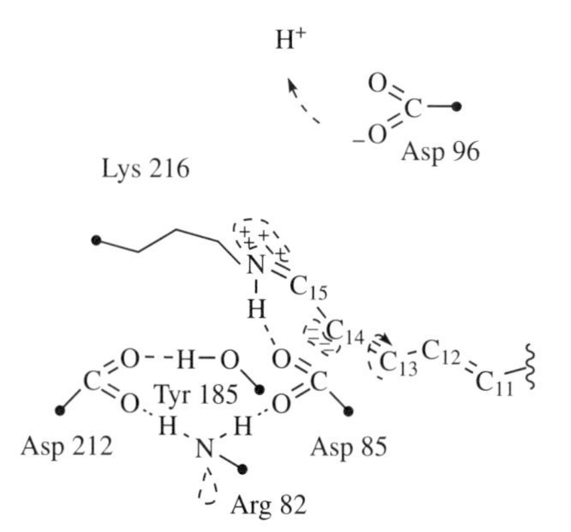

Figure 16. Scheme of the active site for the intermediate L. (From. Ref. 37.)

proton pK_a down to the critical value ($\approx$ 3 [274]), at which the proton transfers from the Schiff base to the terminal group of Asp 85. This process, in turn, decreases the charge separation in the active site and thereby increases the acceptor property of Asp 212. The proton from Tyr 185 is then captured by this acceptor (Fig. 17a).

In the next stage of the L $\rightarrow$ M transition the terminal group of Asp 96 is reprotonated. This is followed by deprotonation of the Schiff base, which becomes less positive than that in the ground state of the retinal. Hence, the inductive interaction between the Schiff base and Asp 96 decreases; as a result, Asp 96 is able to resume its initial state (—COOH).

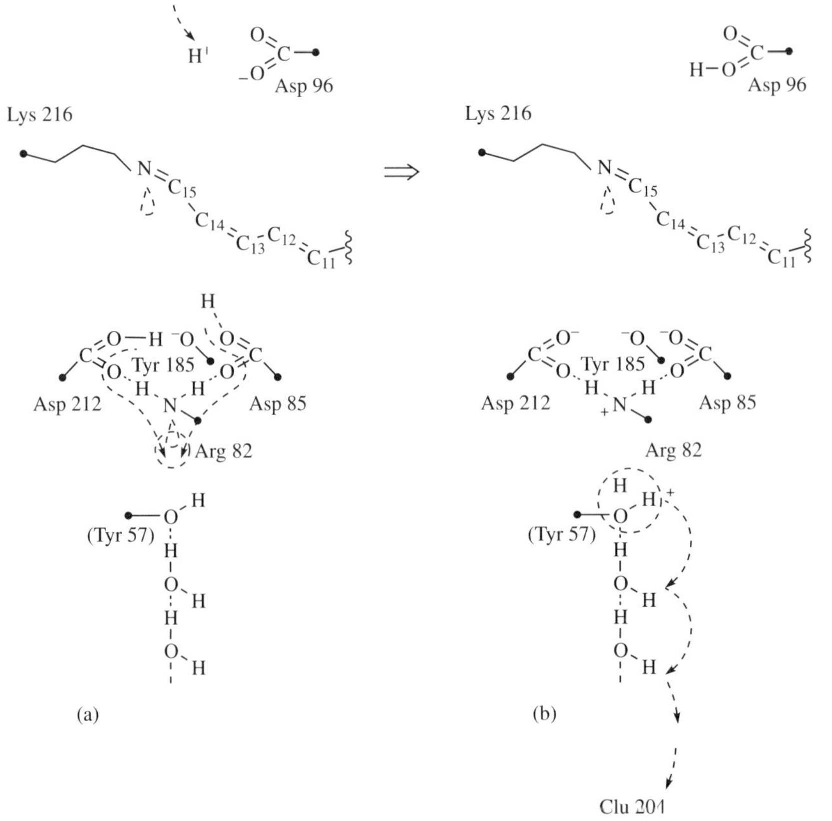

Figure 17. Scheme of the active site changes during the L $\rightarrow$ M transition. (a) Formation of the hydrogen-bonded chain at the moment before the reprotonation of Arg 82. (b) Reconstruction of the active site (see Figure 15a) and the excess proton hopping motion along the hydrogen-bonded chain. (From Ref. 37.)

The dominant effect in the proton transport is due to changes in protein structure (see also Refs. 258, 275, and 276). These structural changes lead to the formation of a hydrogen bond between the amino acid residues (Asp 85–Arg 82–Asp 212), holding a proton of the Schiff base, and the outlet proton channel to the terminal group $-NH_2$ of Arg 82 (Fig. 17a). This is supported by data from Ref. 233. The most possible candidate for the formation of this hydrogen bond is terminal group of Tyr 57.

B. Light-Excited Retinal and Evolution of Excitations in the Retinal

After light with energy $\hbar\omega \simeq 213$ kJ/mol (or $85k_BT$, where $T = 300$ K) has been absorbed by retinal in bacteriorhodopsin, the retinal transfers to the excited state ($^1B_u \leftarrow S_0$ transition [238]) and electron density is shifted along the retinal polyene chain toward the nitrogen atom over a few femtoseconds. This provides a large change in the retinal dipole momentum $\Delta\mu \approx 13.5$ D, which has been confirmed using second harmonic generation and two-photon absorption experiments [277,278].

Let us treat the process of retinal excitation ($\pi \rightarrow \pi^*$ electron transition) and its subsequent relaxation into the vibronic subsystem.

One of the main elements of the retinal molecule is a polyene chain that is long enough. The absorption band of such a chain lies in visible range (see, e.g., Ref. 279); therefore just the polyene chain is responsible for the absorption by $\lambda_{max} = 570$ nm. The spectra of polyenes are characterized by a vibrational structure of the main absorption band. In bacteriorhodopsin retinal, this structure is on the band of absorption with $\lambda_{max} = 419$ nm [280]. Therefore, the main, more intensive band of observing the vibrational structure on the main absorption band is attributed probably to the stronger interaction of the excited π-electronic (π^*-electronic) cloud in bacteriorhodopsin with a polar environment (in this case the energy of the π^*-electronic cloud is a little bit higher than in the previous case, and therefore the energy is nearer to energies of the protein polar groups' vibration). These facts indicate that when considering the kinetics of the π-electronic cloud, we must take into account the connection of the cloud with vibration of the conjugate chain nuclei. Experiments on the isolated retinal have enabled us to conclude [35] that a *cis–trans* distribution occurs on vibration-excited levels. And the effective intersystem transition at these levels plays a substantial role.

Now we will discuss the characteristics of the vibrational and π-electronic subsystems of the chain of coupled C=C bonds [35]. An isolated C=C group undergoes high-frequency valence (intramolecular) vibration, $\Omega = 2\pi \times 5 \times 10^{13}$ s^{-1}. The groups in the polyene chain are interconnected by σ bonds (the vibrational frequency of the carbon atoms in the isolated C–C group equals about $2\pi \times 2 \times 10^{13}$ s^{-1}), hence the virtual vibrational quantum interchange is substantial. By such a strong bond the spectrum of intramolecular vibrations of the retinal polyene chain is formed in an energetic band with a width $2\hbar\Omega$. On the other hand, in the polyene chain the absolute meaning of the resonance integral of the electron overlap in the C=C bond may be evaluated by the value $\hbar M = (4\text{–}6) \times 10^{-19}$J (therefore, $M = 2\pi \times (6\text{–}9) \times 10^{14}\ \text{s}^{-1}$), and the resonance integral of the π-electron overlap between two nearest groups C=C can be evaluated by the value $0.7\hbar M$. The magnitude of the resonance integral of π electrons in the lowest excited states should be at least as much as one mentioned above. Consequently, the width of the π^*-electronic cloud energetic band (excitonic band) can be assumed to be equal to $2\hbar M$; the vibrational band comes out to $2\hbar\Omega$, and the inequality $2\hbar M \gg 2\hbar\Omega$ holds true.

The electronic excitation in the retinal is coherent [35], and this means that the interaction of excited electrons with retinal's vibrations is small. That is, the energy of exciton–vibration interaction $\hbar\chi$ is much smaller than the excitonic and vibrational bands, $\hbar\chi \lll 2\hbar\Omega \lll 2\hbar M$.

The Hamiltonian of intramolecular collective vibrations of the quasi-one-dimensional system of coupled C=C bonds can be written as

$$H_{\text{vib}} = \sum_{l} \hbar\Omega \hat{B}_l^+ \hat{B}_l + \sum_{l,m} \hbar\delta\Omega_{l-m} \hat{B}_l^+ \hat{B}_m \tag{380}$$

where Ω is a characteristic vibrational frequency of carbon atoms (sites) in polyene chain, $\hat{B}_l^+(\hat{B}_l)$ is the Bose creation (annihilation) operator of an vibrational quantum in the lth site, and $\delta\Omega_{l-m}$ is an exchange frequency of the virtual vibrational quantum between the lth and the mth sites.

To diagonalize the Hamiltonian (380), let us go from the site representation to the momentum representation of the creation (annihilation) operators of vibrations through the transform $\hat{B}_l = N^{-1/2} \sum_q \hat{B}_q \exp(-iqR_l)$. Here N is a number of carbon atoms in the chain; $R_l = lg_0$, where g_0 is a distance between the nearest C atoms; we suppose $g_0 = r_{\text{C=C}} \approx r_{\text{C–C}}$—that is, $g_0 = 0.135$ to 0.146 nm. In the momentum representation the Hamiltonian

$$H_{\text{vib}} = \sum_{q} \hbar\Omega_q \left(\hat{B}_q^+ \hat{B}_q + \frac{1}{2} \right) \tag{381}$$

Here $\hat{B}_q^+(\hat{B}_q)$ is the Bose creation (annihilation) operator of the collective intensive vibrational quantum with an energy of $\hbar\Omega_q$ and quasi-momentum of $\hbar q$.

The excitonic Hamiltonian

$$H_{\text{exc}} = \sum_{k} \hbar\varepsilon_k \hat{A}_k^+ \hat{A}_k \tag{382}$$

where $\hat{A}_k^+(\hat{A}_k)$ is the Fermi creation (annihilation) operator of a coherent exciton in the band having the energy $\hbar\varepsilon_k$ and quasi-momentum $\hbar k$.

In the case of the coherent vibronic state excitations, we shall take into account a change of the frequency of the collective vibrations in the polyene chain caused by the exciton excitation:

$$\Omega_q \to \Omega_q + \frac{\partial \Omega_q}{\partial N_q} \delta N_q$$

Next, we pass to the operators of the exciton-state occupation numbers $\delta N_q \to \hat{N}_q = \hat{A}_q^+ \hat{A}_q$ and replace $\partial \Omega_q / \partial N_q \to \Delta \Omega_q$.

The total Hamiltonian of the system under consideration assumes as

$$H = H_0 + H_1 \tag{383}$$

$$H_0 = \sum_q H_{0q} = \sum_q \left\{ \hbar \varepsilon_q \hat{A}_q^+ \hat{A}_q + \hbar \Omega_q \left(\hat{B}_q^+ \hat{B}_q + \frac{1}{2} \right) \right.$$
$$\left. + \hbar \Delta \Omega_q \hat{A}_q^+ \hat{A}_q \left(\hat{B}_q^+ \hat{B}_q + \frac{1}{2} \right) \right\} \tag{384}$$

$$H_1 = \sum_{k,q} \left\{ \hbar \chi_q \hat{B}_q \hat{A}_{k-q}^+ \hat{A}_k + \hbar \chi_q^* \hat{B}_q^+ \hat{A}_k^+ \hat{A}_{k-q} \right\} \tag{385}$$

If we neglect the exciton–vibration interaction (i.e., the operator H_1), the wave functions of collective vibrations and of coherent excitonic states of the retinal molecule (they are specified by quasi-momentum $\hbar q$ and are the proper functions of the Hamiltonian H_{0q}) form an orthonormal basis

$$|q; f_q, \mu_q\rangle = \frac{1}{\sqrt{\mu_q!}} (\hat{B}_q^+)^{\mu_q} (\hat{A}_q^+)^{f_q} |q; 0, 0\rangle \tag{386}$$

where $f_q = 0, 1$; $\mu_q = 0, 1, \ldots$; $|q; 0, 0\rangle$ is the vector of a ground state and describes the system without quasi-particles.

Let us consider the kinetics of the excitonic excitation in the retinal using the method of nonequilibrium statistical operator [281]. We denote the statistical operator of the system by $\rho(t)$ and shall study the evolution of its following matrix elements:

$$\begin{aligned} \mathcal{N}_{qq}(t) &= \mathrm{Tr}\{\rho(t) \hat{B}_q^+ \hat{B}_q\} \\ \rho_{kk}(t) &= \mathrm{Tr}\{\rho(t) \hat{A}_k^+ \hat{A}_k\} \\ P_{\substack{kk \\ qq}}(t) &= \mathrm{Tr}\{\rho(t) \hat{A}_k^+ \hat{A}_k \hat{B}_q^+ \hat{B}_q\} \end{aligned} \tag{387}$$

Here $\mathcal{N}_{qq}(t)$ is a nonequilibrium function of the excitations distribution in vibrational subsystem at the moment of time t; matrix elements $\rho_{kk}(t)$ determine a probability of existence at the moment of t of the π-electronic cloud in an excited state with a quasi-momentum $\hbar k$ and with an energy $\hbar\varepsilon_k$; matrix elements $P_{\substack{kk\\qq}}(t)$ describe a coupling between the π^*-electronic cloud and the vibrational subsystem.

The equation of evolution of the matrix elements $\mathcal{N}_{qq}(t)$ has the form

$$i\hbar\frac{\partial\mathcal{N}_{qq}(t)}{\partial t} = -\mathrm{Tr}\{\rho^{(0)}(t)[H_0, \hat{B}_q^+\hat{B}_q]\} - \frac{i}{\hbar}\lim_{\eta\to 0}\int_{-\infty}^{t} d\tau e^{\eta\tau}\,\mathrm{Tr}\{\rho^{(0)}(t)[H_1(\tau), [H_1, \hat{B}_q^+\hat{B}_q]]\} \tag{388}$$

Similar equations are derived for $\rho_{kk}(t)$ and $P_{\substack{kk\\qq}}(t)$. In Eq. (389), $\rho^{(0)}(t)$ is statistical operator of the excitonic vibrational system (retinal) in thermostat (protein) in the absence of interaction H_1; the operator $H_1(\tau)$ equals $\exp(i\tau H_0/\hbar)H_1\exp(-i\tau H_0/\hbar)$. From Eq. (388) and from similar equations for $\rho_{kk}(t)$ and $P_{\substack{kk\\qq}}(t)$ with regard to expressions (383) to (385), we derive the following system of the differential coupling equations:

$$\frac{\partial}{\partial t}\mathcal{N}_{qq}(t) = \sum_k \upsilon_{kq}(t)\{\rho_{kk}(t) + P_{\substack{k-q,k-q\\q,q}}(t) - P_{\substack{kk\\qq}}(t)\} \tag{389}$$

$$\frac{\partial}{\partial t}\rho_{kk}(t) = \sum_k \upsilon_{kq}(t)\Big\{-\rho_{kk}(t) + \rho_{k-q,k-q}(t) - 2P_{\substack{kk\\qq}}(t) + P_{\substack{k-q,k-q\\q,q}}(t) + P_{\substack{k+q,k+q\\q,q}}(t)\Big\} \tag{390}$$

$$\frac{\partial}{\partial t}P_{\substack{kk\\qq}}(t) = \upsilon_{kq}(t)\Big\{\rho_{k-q,k-q}(t) + 3[P_{\substack{k-q,k-q\\q,q}}(t) - P_{\substack{kk\\qq}}(t)]\Big\} \tag{391}$$

In the equations derived, the terms containing products of four and more Fermi operators $\hat{A}_q$ and $\hat{A}_q^+$ are omitted. Besides, when deriving the equations, in $H_1(\tau)$ the uncoupling of the operator products have been made:

$$2\hat{A}_q^+\hat{A}_q\hat{B}_q^+\hat{B}_q|_\tau \to \langle\hat{A}_q^+\hat{A}_q\rangle_\tau\hat{B}_q^+\hat{B}_q|_\tau + \langle\hat{B}_q^+\hat{B}_q\rangle_\tau\hat{A}_q^+\hat{A}_q|_\tau = \rho_{qq}(\tau)\hat{B}_q^+\hat{B}_q|_\tau + \mathcal{N}_{qq}(\tau)\hat{A}_q^+\hat{A}_q|_\tau \tag{392}$$

It is admissible if

$$\rho_{qq}(\tau)|\Delta\Omega_q| < \varepsilon_q, \Omega_q, \qquad \mathcal{N}_{qq}(\tau)|\Delta\Omega_q| < \varepsilon_q, \Omega_q \tag{393}$$

In Eqs. (389)–(392) the function of a momentary exciton–vibration interaction is

$$\upsilon_{kq}(t) = |\chi_q|^2\frac{\tilde{\Omega}_q(t)}{[\tilde{\varepsilon}_{k-q}(t) - \tilde{\varepsilon}_k(t)]^2 - \tilde{\Omega}_q^2(t)} \tag{394}$$

where

$$\tilde{\varepsilon}_k(t) = \varepsilon_k + \Delta\Omega_k\left[\mathcal{N}_{kk}(\tau) + \frac{1}{2}\right], \qquad \tilde{\Omega}_q(t) = \Omega_q + \Delta\Omega_q\rho_{qq}(t) \tag{395}$$

The inclusion of the term $\hbar\Delta\Omega_q\hat{A}_q^+\hat{A}_q\hat{B}_q^+\hat{B}_q$ in the Hamiltonian (384) makes it possible to avoid a singularity in expression (394) at certain values of k and q. Note that in expression (394), $v_{kq}(t) > 0$ when $q\pi/2g_0$. A choice of the function of exciton–vibration interaction as

$$\chi_q = \begin{cases} \chi > 0, & q \geq \pi/2g_0 \\ 0, & q \leq \pi/2g_0 \end{cases} \tag{396}$$

satisfies the demands mentioned above.

The initial conditions for equations (389)–(391) are

$$\rho_{kk}(t)\big|_{t=0} = \delta_{k0}, \qquad \mathcal{N}_{qq}(t)\big|_{t=0} = P_{\substack{kk\\qq}}(t)\big|_{t=0} = 0 \tag{397}$$

Now let us pass to the Laplace transformation over time variable t in Eqs. (390)–(391):

$$\rho_{kk}^L(p) = \int_0^\infty e^{-pt}\rho_{kk}(t)\,dt, \qquad P_{\substack{kk\\qq}}^L(p) = \int_0^\infty e^{-pt}P_{\substack{kk\\qq}}(t)\,dt \tag{398}$$

For this purpose we should multiply both sides in Eqs. (390) and (391) by e^{-pt} and integrate over t from 0 to ∞. As a result, we get the equations

$$\begin{aligned} p\rho_{kk}^L(p) - \rho_{kk}(t)\big|_{t=0} = \sum_q v_{kq}\big\{ &-\rho_{kk}^L(p) + \rho_{k-q,k-q}^L(p) \\ &- 2P_{\substack{kk\\qq}}^L(p) + P_{\substack{k-q,k-q\\q,q}}^L(p) + P_{\substack{k+q,k+q\\q,q}}^L(p)\big\} \end{aligned} \tag{399}$$

$$P_{\substack{kk\\qq}}^L(p) - P_{\substack{kk\\qq}}(t)\big|_{t=0} = v_{kq}\big\{\rho_{k-q,\,k-q}^L(p) + 3[P_{\substack{k-q,\,k-q\\q,\,q}}^L(p) - P_{\substack{kk\\qq}}^L(p)]\big\} \tag{400}$$

Here the neglecting of the dependence of v_{kq} on t occurs owing to relationships (393) and (395). With the help of Eq. (400), correlators $P_{\substack{kk\\qq}}^L(p)$ in Eq. (399) can be eliminated, and after that the inverse Laplace transform can be made. This brings us to equation

$$\frac{\partial}{\partial t}\rho_{kk}(t) = -\sum_q \tilde{v}_{kq}\{\rho_{kk}(t) + \rho_{k-q,k-q}(t)\} \tag{401}$$

where $\tilde{v}_{kq} = v_{kq}(1 + 1/9)^{-1}$; below we set $\tilde{v}_{kq} \simeq v_{kq} \simeq \bar{v}$.

Let us substitute now the difference $[P_{\substack{k-q,k-q\\ q,q}}(t) - P_{\substack{kk\\ qq}}(t)]$ from Eq. (391) into Eq. (389) and then perform uncoupling $P_{\substack{kk\\ qq}}(t) \to \rho_{kk}(t)\mathcal{N}_{qq}(t)$. The term $(\partial/\partial t)\rho_{kk}(t)$ on the left-hand side of Eq. (390) can now be replaced by the right-hand side of Eq. (401). Then for the function $R(t) = \sum_k \rho_{kk}(t)$ and for the characteristic function $\mathcal{N}(t) = \mathcal{N}_{qq}(t)|_{\max}$ we obtain the following equations:

$$\frac{\partial}{\partial t}R(t) = -2N\bar{\upsilon}R(t) \tag{402}$$

$$\frac{\partial}{\partial t}\mathcal{N}(t) = \frac{1}{N}[2N\bar{\upsilon}t + \ln(3 - e^{-2N\bar{\upsilon}t})](3 - e^{-2N\bar{\upsilon}t}) \tag{403}$$

In the retinal the number of sites, N, is 12 and $\bar{\upsilon} \ll \Omega$; and because of $\Omega = 2\pi \times 5 \times 10^{13}$ s^{-1} and $M \geq 2\pi \times 5 \times 10^{14}$ s^{-1}, it is possible to set $\upsilon \simeq 2 \times 10^{13}$ s^{-1}. The characteristic time of vibration relaxation of the electronic excitation has an order 10^{-13} to 10^{-12} s [282]. In this case for the light-excited retinal in bacteriorhodopsin, according to experimental data [283] we have $t = t_{\mathrm{rel}} = 7 \times 10^{-13}$ s^{-1}. Thus, according to Eq. (402) $R(t_{\mathrm{rel}}) = 0$. Then from Eq. (403) we derive the distribution function, which is not in equilibrium, of collective intramolecular vibrations of the retinal at the moment of t_{rel}:

$$\mathcal{N}(t_{\mathrm{rel}}) \simeq \frac{2}{3}\bar{\upsilon}t_{\mathrm{rel}} \simeq 10 \tag{404}$$

The distribution function (404) is much greater than the Planck distribution function

$$\frac{1}{e^{\hbar\Omega/k_{\mathrm{B}}T} - 1} \simeq e^{-\hbar\Omega/k_{\mathrm{B}}T} \ll 1 \tag{405}$$

which characterized vibrations of sites in retinal before the photon absorption (we recall that $\hbar\Omega \simeq 85\, k_{\mathrm{B}}T,\ T = 300$ K). The function $\mathcal{N}(t)$ can be written formally in the form of the Planck function written for more high effective temperature T^*, $[\exp(\hbar\Omega/k_{\mathrm{B}}T^*) - 1]^{-1}$. Then at the moment t_{rel} the effective temperature is expressed as

$$T^* = \frac{\hbar\Omega}{k_{\mathrm{B}}\ln(11/10)} \simeq 80\,T \tag{406}$$

Thus the energy passed into the vibrational subsystem is equal to $\hbar\Omega\mathcal{N}(t_{\mathrm{rel}}) \simeq 80\, k_{\mathrm{B}}T$; that is, virtually all the total energy of the absorbed photon ($85\, k_{\mathrm{B}}T$) has passed on to the vibrational subsystem by the time of $t - t_{\mathrm{rel}}$.

Clearly, the excited π^*-electron subsystem should introduce a significant charge distribution in polyene chain of the retinal. It is the charge separation that plays an important role in the destabilization of the excited retinal. In fact, the initial all-*trans* conformation of the retinal is stabilized by an attractive electrostatic interaction between the protonated Schiff base and nearby negatively charged residues (counterions). In contrast, the electronic excitation of the retinal generating a negative charge near the retinal Schiff base destabilizes the all-*trans* conformation; as a consequence, fast torsional vibrations can be initiated [284–286]. The frequency of these vibrations can estimated from a simple scheme of repulsive electrostatic interaction between the π^*-electron cloud, having an effective charge of e^*, and the counterions, with a general of e. We can get an evaluation of the energy of this interaction by the expression

$$W = \frac{ee^*}{4\pi\varepsilon_0\varepsilon r} \tag{407}$$

where $e^* \approx 0.1e$ [262], $r \approx 0.5$ nm, and $\varepsilon \approx 3$. Thus the vibrational frequency is $\nu = W/h \approx 10^{13}\ \mathrm{s}^{-1}$, which is consistent with the data of Ref. 238. Hence we can conclude that the π^* electron relaxes via the induced torsional vibrations of the retinal. The intensity of these vibrations is increased, as is readily seen from the consideration given above; and, finally, picosecond isomerization from all-*trans* to the 13-*cis* conformation of the retinal takes place.

The light-induced 13-*cis*-retinal of bacteriorhodopsin features a strained state with separated charges: An additional dipole almost perpendicular to the retinal long axis arises after *trans–cis* isomerization (see Fig. 15b). Thus, part of the absorbed photon energy is stored. The energy difference between the 13-*cis* and all-*trans* conformations of the retinal amounts to $\Delta U \approx 65$ kJ/mol [287].

C. Proton Ejection

A broad structureless absorption band in the infrared spectrum (1700–2800 cm^{-1}) observed for the intermediate L by Zundel and co-workers [6,238,288] was unambiguously interpreted as proton collective fluctuations occurring in the outlet proton channel. In this channel, charges are shifted within less than a picosecond [250,289]; and an uninterrupted hydrogen bond, which features a large proton polarizability [6], is formed. Thus the outlet channel in the intermediate L makes up the hydrogen-bonded chain whose polarization periodically changes: $\mathrm{OH}\cdots\mathrm{OH}\cdots\mathrm{OH}\ldots \rightleftarrows \mathrm{HO}\cdots\mathrm{HO}\cdots\mathrm{HO}\ldots$ As we mentioned in Ref. 37, such collective proton vibrations in the hydrogen-bonded chain could be considered as longitudinal polarized optical phonons with a frequency of more than 10^{12} Hz.

Figure 17a demonstrates the state of the hydrogen-bonded chain (possible terminal group of Tyr 57 followed by water molecules) and that of the active site before an excess proton enters the outlet channel. At this moment, the terminal group $-\mathrm{NH}_2$ of Arg 82 appears to be connected with two other protons of the terminal groups of Asp 85 and Asp 212. These two protons destabilize the active site because they both pretend to occupy (indirectly) the vacancy in the terminal group of Arg 82. In the next step, these protons occupy the vacancy, and the initial states of Asp 85, Asp 212, and Arg 82 residues are resumed. However, one of the transferred protons becomes coupled with Arg 82 while another proton is in excess. This can be regarded as a short-lived ion state ($\tau_0 \approx 10^{-6}$ s; see, e.g., Ref. 34), located at the site of the hydrogen-bonded chain neighboring Arg 82 (Fig. 5b). It is obvious that the excess proton strongly interacts with the collective fluctuations of the hydrogen bonds or in other terms with polarized optical phonons. This enables us to apply the concept of the proton polaron, which has been discussed in the previous sections, to the problem of proton transport in bacteriorhodopsin, because during the L → M transition the excess proton finds itself in a strongly polarized outlet channel.

The polaron state may be created on any site of the hydrogen-bonded chain along which the excess proton moves. In this state, the charge carrier is coupled with a chain site for a time τ_0, leading to a local deformation of the chain. Figure 18 illustrates the excess proton hopping motion along the outlet hydrogen-bonded chain as a set of polaron wells. The positively charged terminal group $-NH_3^+$ of Arg 82 has to set the direction of the excess proton motion from the active site to the extracellular surface of the membrane. Hence, this proton drifts under the electrostatic field of the group $-NH_3^+$.

We can write the Hamiltonian of the excess proton in the hydrogen-bonded chain in the form

$$H = H_0 + H_{\text{tun}} + H_{\mathscr{E}} \tag{408}$$

$$\begin{aligned} H_0 = & \sum_l E\hat{a}_l^+\hat{a}_l + \sum_q \hbar\omega_q\left(\hat{b}_q^+\hat{b}_q + \frac{1}{2}\right) \\ & - \sum_l \hat{a}_l^+\hat{a}_l\hbar\omega_q[u_l(q)\hat{b}_q^+ + u_l^*(q)\hat{b}_q] \end{aligned} \tag{409}$$

$$H_{\text{tun}} = \sum_l J\hat{a}_{l+1}^+\hat{a}_l + h.c. \tag{410}$$

$$H_{\mathscr{E}} = -\mathscr{E}\sum_l R_l\hat{a}_l^+\hat{a}_l \tag{411}$$

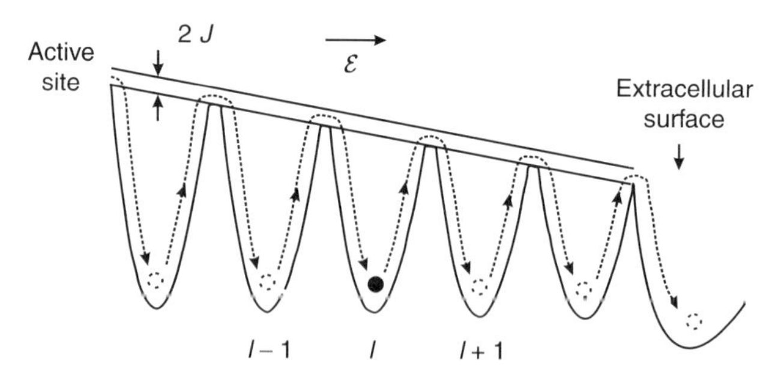

Figure 18. Hopping motion of the excess proton in the system of polaron wells. $2J$ is width of the narrow polaron band; $\mathscr{E}$ is the pulling electric field. (From Ref. 37.)

where H_0 is the basic small polaron Hamiltonian containing energy operators for the excess charge carrier [first term in expression (409)], polarized vibrations of the hydrogen-bonded chain [that is, optical phonons; second term in expression (409)], and a local deformation induced by the excess proton [third term in expression (409)]; H_{tun} is the tunnel Hamiltonian describing the proton motion from site to site, and $H_{\mathscr{E}}$ is the electric field Hamiltonian depicting the change in the proton potential energy along the chain. In expression (409), E is the coupling energy of the excess proton with the lth site of the chain; $\hat{a}_l^+ (\hat{a}_l)$ is the Fermi operator of creation (annihilation) of the charge carrier in the lth site; and $\hat{b}_q^+ (\hat{b}_q)$ is the Bose operator of creation (annihilation) of the optical phonon with the cycle frequency ω_q and wavenumber q; $u_l(q) = u \exp(iql)$, where u is the dimensionless constant responsible for the coupling between the excess proton and the lth site; J is the resonance integral that characterizes the overlapping of wave functions of the excess proton in neighboring sites; and $R_l = lg$, where l is the site number in the chain and g the chain constant.

The calculation of the proton current is reduced to the method described in Section III. The density of the hopping proton current is given by

$$I = \text{Tr}(\rho_{\mathscr{E}} j) \tag{412}$$

where $\mathscr{V}$ is the effective volume of the chain and the statistical operator $\rho_{\mathscr{E}}$ and the operator of current density are determined as follows:

$$j = \frac{e}{\mathscr{V} i\hbar} [H_{\mathscr{E}}, H_{\text{tun}}] \tag{413}$$

$$\rho_{\mathscr{E}} = \frac{i}{\hbar} \int_{-\infty}^{t} d\tau e^{-\frac{i}{\hbar}(t-\tau)(H_0 + H_{\mathscr{E}})} [H_{\mathscr{E}}, \rho_{\text{tun}}] e^{\frac{i}{\hbar}(t-\tau)(H_0 + H_{\mathscr{E}})} \tag{414}$$

where

$$\rho_1 = \rho_0 \int_0^{1/k_B T} d\lambda e^{\lambda H_0} H_1 e^{-\lambda H_0}, \qquad \rho_0 = \frac{e^{-H_0/k_B T}}{\text{Tr}\, e^{-H_0/k_B T}} \tag{415}$$

After calculations and leaving out the phonon dispersion, we obtain the following expression for the density current of the outlet proton channel of bacteriorhodopsin [37]:

$$I = 2^{2/3} \pi^{1/3} \frac{eg}{\mathscr{V} \hbar^2 \omega} \sinh \frac{e\mathscr{E} g}{2k_B T} \frac{\exp(-E_a/k_B T)}{[2u^2 \operatorname{cosech}(\hbar\omega/k_B T)]^{1/2}} \tag{416}$$

where the activation energy is expressed as

$$E_a = 2k_B T u^2 \tanh(\hbar\omega/4k_B T) \tag{417}$$

Let us present the current density I on the left-hand side of expression (416) in the form $I = e/At_{\text{eject}}$, where A is the area of cross section of the chain and t_{eject} is the time of proton drift along the chain. Since $\mathscr{V} = AL$, where L is the chain length, we can derive from expression (417) the expression for the time drift of excess proton in the chain under consideration:

$$t_{\text{eject}} = 2^{-3/2}\pi^{-1/2}\frac{L}{g}\frac{\hbar^2\omega}{J^2}e^{E_a/k_BT}\frac{[2u^2\,\text{cosech}\,(\hbar\omega/k_BT)]^{1/2}}{\sinh(e\mathscr{E}g/2k_BT)} \tag{418}$$

All the parameters in expression (418) can be readily estimated. In the proton polaron concept there are only two free parameters: the coupling function, written here as u^2, and the resonance integral J. Their orders of magnitude are known from the general model of small polarons. Here we can set $u^2 = 15$ and $J = 0.9 \times 10^{-2}k_BT$. The phonon frequency can be put equal to $\omega = 2\pi \times 6 \times 10^{12}\ \text{s}^{-1}$, as we discussed above; and in this case $\hbar\omega = k_BT$, where $T = 300$ K. Then the activation energy E_a equals $7.35k_BT$. The excess proton energy in the field of the active site effective charge $e\mathscr{E}g$ has been chosen to be equal to 10^7 V/m, which is typical for living cells. Finally, setting the number of the chain sites $L/g = 5$, we gain from expression (418)

$$t_{\text{eject}} \approx 80\ \mu\text{s} \tag{419}$$

This estimation is consistent with experimental data [290] according to which the time of proton ejection equals 76 μs. In Ref. 230 it is pointed out that at pH 7, the proton appears at the surface with a time constant of about 80 μs. Thus, the theory presented herein is in good agreement with the available experiments.

V. MESOMORPHIC TRANSFORMATIONS AND PROTON SUBSYSTEM DYNAMICS IN ALKYL- AND ALKOXYBENZOIC ACIDS

A. Molecular Associates

Mesomorphism of alkyl- and alkoxybenzene acids, which can be either smectic or nematic types, was explained [291] by aggregation of their molecules via intermolecular hydrogen bonds in cyclic dimers (hereafter we will use abbreviations *n*ABA and *n*AOBA for alkyl- and alkoxybenzene acids, respectively, where *n* determines the number of C atoms in the alkyl radical). Experimental investigations of the dynamics of hydrogen bonds in the vicinity of the phase transitions have been performed in a series of works (see, e.g., Ref. 292). It was shown early [291,293] that in the infrared spectrum new bands appear with heating of the solid-state phase well in advance of the melt point.

This is testimony to a partial dissociation of cyclic dimers with the formation of open associates and monomers. In Refs. 57, 58, 60 the quantitative estimations of the relative number of different building blocks as functions of temperature and phase of the sample have been made. For reference, the infrared spectra of *n*-carbonic and perftoralkylbenzoic acids, which do not possess mesomorphic properties, have been studied as well.

The absorption infrared spectra of *n*ABA ($n = 1$–9) and *n*AOBA ($n = 3$–12) have been measured in the spectral range from 100 to 4000 cm^{-1}(± 3 cm^{-1}) and in the temperature interval from 100 to 550 K (± 1 K) [57,58,60]. The assignment of the spectral infrared bands to the vibrations of atoms of the dimer ring (Fig. 2) has been made based on calculations of frequencies and shapes of normal coordinates performed for dimers of homologies of carbon acids [292]. Vibrations of the parareplaced benzoic acid and its alkyl radical have been interpreted on the basis of calculations carried out in Refs. 294 and 295, respectively. The maxima recorded at 2900 cm^{-1} (ν_{OH} band), 1692 cm^{-1} ($\nu_{C=O}$ band), and 940 cm^{-1} (ρ_{OH} band) for all phases of the ABA and AOBA are direct evidence for the connection of molecules ABA and AOBA via hydrogen bonds.

In the crystal state ($T = 300$ K), molecules of AOBA are united into centrosymmetrical dimers by means of a pair of hydrogen bonds ($R_{O\cdots O} =$ 0.2629 nm). For short homologies ($n \leq 5$) the packing of molecules in the crystal is caused by the interaction between the rings of benzoic acids (Fig. 19a). The layer packing of molecules with more long alkyl radicals ($n \geq 6$) stems from the interaction between methylene chains aligned parallel to each other (Fig. 19b). The distance between benzoic acids from the nearest layers equals 0.45 nm. The packing of molecules in the ABA crystal has a similar nature.

The enthalpy of hydrogen bond in cyclic dimers (Fig. 20a) in the ABA and AOBA is equal to -35.3 ± 0.8 kJ/mol per hydrogen bond at $T = 300$ K. For the quantitative estimation of the enthalpy the Iogansen method [296] has been applied, which is based on the analysis of the shift of the frequency of twisting vibration ρ_{OH} and that of the gravity center of the ν_{OH} band of associates with relation to the corresponding shifts for monomers. The determination of frequencies $\rho_{OH}^{(mon)}$ and $\nu_{OH}^{(mon)}$ of ABA and AOBA monomers, which are needed for the calculation of the enthalpy, has been performed by the infrared spectra of ABA and AOBA measured in the solution of CCl_4 and the gaseous state in the temperature range from 540 to 550 K. The frequencies prove to be 612 and 3550 cm^{-1}, respectively, practically for all the acids studied.

As it follows from the infrared spectra, only cyclic dimers are present in the crystal state on cooling to 100 K. In this state, the bond energy of the hydrogen bond has been determined to be 36 ± 1 kJ/mol at 100 K and 34 ± 1 kJ/mol at 300 K. The crystals, on being heated, exhibit the appearance of a new band (920–925 cm^{-1}) well before the transition to a metaphase. The intensity of the band increases with *T*, and the maximum of the band slightly moves to low

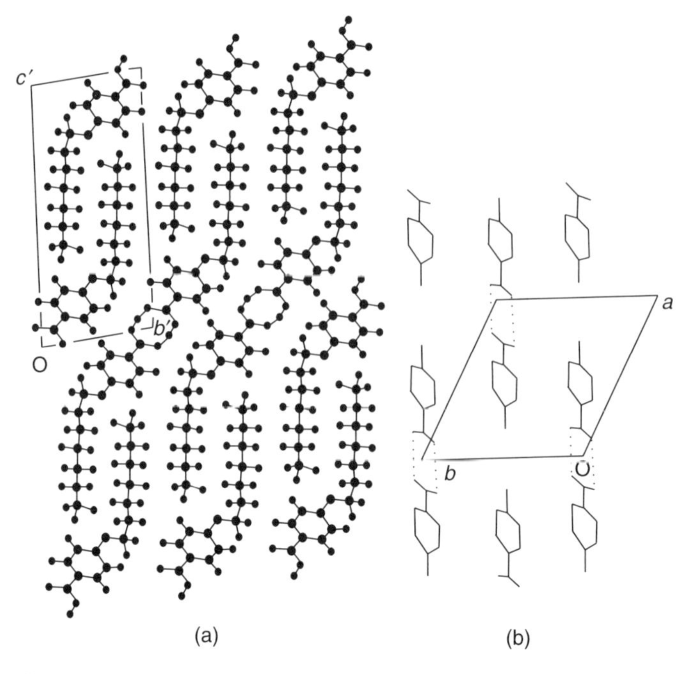

Figure 19. (a) Molecular stacking in the crystal state of the 7AOBA in the projection on the *a* axis. (b) The crystal structure of toluil acid in the projection on the *c* axis. (From Ref. 57.)

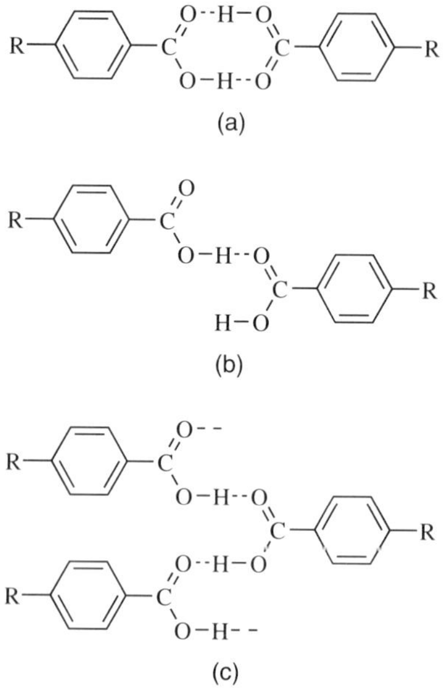

Figure 20. Possible types of hydrogen-bonded associates: (a) Cyclic dimer, (b) open dimer, and (c) chain associate. (From Ref. 58.)

frequencies. Coincidentally with this band, one more band with the maximum at 1710 cm^{-1}, a wide shoulder (3300 cm^{-1}), and new bands in the range from 500 to 700 cm^{-1} appear as well. Taking into account peculiarities of the crystal structure of the 7AOBA and low intensities of the bands caused by the absorption of unbound hydroxyl group OH (612 and 3550 cm^{-1}), it is necessary to suggest that three or more molecules take part in the formation of such associates (Fig. 20c). An appraisal of the enthalpy of hydrogen bond of the formed chain associates by the Iogansen's frequency rule [296] yields magnitude -31.5 ± 0.8 kJ/mol. Near the phase transition from the solid crystal to the liquid crystal, the quantity of open associates accounts for 30% of the total number of *n*AOBA (*n*ABA) molecules; a small part of monomers is available as well.

Heating over the melting temperature is accompanied by amplification of the intensity of bands resulted from monomers (612, 1730, and 3550 cm^{-1}) and decay of the intensity of bands stemming from cyclic dimers (940, 1690, and 2900 cm^{-1}). That is, with heating, cyclic dimers dissociate into monomers; the quantity of open associates holds constant in the framework of the mesophase (Fig. 21). These clusters (i.e., open associates), which feature a certain order

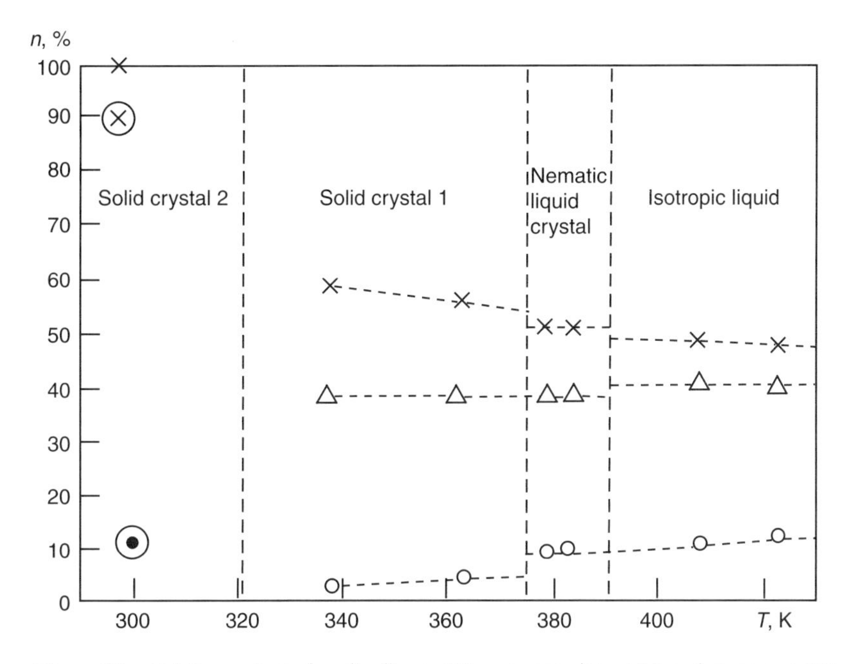

Figure 21. Relative content of cyclic dimers (×), open associates (△), and monomers (○) in different phase states of the 7AOBA; (⊗, ⊙) are the data for dilution of the 7AOBA in CCl_4. The vertical dashed lines correspond to the phase transition temperatures. (From Ref. 57.)

parameter, set conditions for a smectic layer type of the mesophase. In particular, electron paramagnetic resonance spectroscopy also points to the formation of polymer hydrogen-bonded associates in the mesophase of 9AOBA [297].

Cyclic dimers, open associates, and monomers exist also in the liquid state; however, with temperature the dynamic equilibrium is shifted to monomers. Note that a more simple pattern of the molecular structure is set with dissolving of the acids studied in CCl_4: Only cyclic dimers and monomers are found at equilibrium, and the quantity of monomers increases with dilution of the solution or with its heating; open associates do not arise. The enthalpy of hydrogen bond of dimer molecules in the solution is approximately the same as in the crystal at the same temperature.

What is the reason for the transformation of hydrogen bonds revealed at the phase transitions in the homology series studied? Bernal's hypothesis [298] indicates that it is the character of packing of molecules in the crystal state which determines the possibility of mesophase formation and its type. In the case of nAOBA-type benzoic acids with short alkyl radicals ($n \leq 5$), the ordering of molecules ("piles") is determined by the interaction between benzoic rings. If the radicals are longer, the intermolecular interaction between methylene chains fixes the layer packing of cyclic dimers (Fig. 20). The conjugation of π electrons of benzoic acids with those of carbonyl groups dictates the flat structure of the cyclic dimer core. With melting, the liquid crystal is formed: in the first case nematic and in the second case smectic. The distance between benzoic rings of neighboring dimers is 0.45 nm. In similar manner, the nABA crystals are constructed [57,58,60].

B. Rearrangement of Hydrogen Bonds: Mechanism of Open Associates Formation

Let us consider the process of disruption of a pair of hydrogen bonds, which have connected two nABA (or nAOBA) molecules into a dimer, and the following formation of one crossed hydrogen bond combining two nABA molecules which previously have been unconnected. The problem can be reduced to the model below. Let two hydrogen bonds from two nearest dimer rings is characterized by two-well potentials (Figure 20a) whose parameters are given (see Section II). It would seem that the dynamics of the crystal net of hydrogen bonds, which are marked by two-well potentials, can be investigated in principle in the framework of the Ising pseudospin model (see Section II.D). However, in our case there is reliable experimental evidence of the origination of "crossed" hydrogen bonds that characterize the creation of open associates discussed in the previous subsection.

It is reasonable to assume that at the phase transition "solid crystal 1 $\rightarrow$ solid crystal 2" that is close to the phase transition "solid crystal $\rightarrow$ liquid crystal" the intradimer hydrogen bonds are in a predissociate state. This state can be unstable with respect to the proton transfer from the dimer to a new spatial position that corresponds to a "crossed" hydrogen bond. Such a transfer can be activated by two types of intradimer vibrations of the proton [292]: (i) $\rho(\mathrm{O}\cdots\mathrm{H}) =$ 30–50 cm^{-1} (in plane of the dimer ring), which is transversal to the $\mathrm{O}\cdots\mathrm{O}$ line of the dimer; (ii) $\alpha_{\perp}(\mathrm{O}\cdots\mathrm{H}) = 30$–$50\ \mathrm{cm}^{-1}$ (normal to the dimer plane).

Let us employ now the small polaron model for the consideration of proton transfer from the intradimer state to the interdimer one (Fig. 20a and Fig. 20b, respectively) [57]. The initial Hamiltonian can be written in the form

$$H = H_0 + H_{\text{tun}} \tag{420}$$

$$\begin{aligned} H_0 = {} & \sum_{l=1}^{2} E_l \hat{a}_l^+ \hat{a}_l + \sum_{\alpha;\mathbf{q}} \hbar\omega_{\alpha\mathbf{q}} \left(\hat{b}_{\alpha\mathbf{q}}^+ \hat{b}_{\alpha\mathbf{q}} + \frac{1}{2} \right) \\ & - \sum_{\alpha=1}^{2} \hat{a}_l^+ \hat{a}_l \sum_{\alpha;\mathbf{q}} \hbar\omega_{\alpha\mathbf{q}} [u_\alpha(\mathbf{q}) \hat{b}_{\alpha\mathbf{q}}^+ + u_\alpha^*(\mathbf{q}) \hat{b}_{\alpha\mathbf{q}}] \end{aligned} \tag{421}$$

$$H_{\text{tun}} = V_{12}(\hat{a}_1^+ \hat{a}_2 + \hat{a}_2^+ \hat{a}_1) \tag{422}$$

Here H_0 is the Hamiltonian that includes the bond energy of a proton in the intradimer ($E_1 < 0$) and interdimer ($E_2 < 0$) hydrogen bonds (we suppose that $|E_2| > |E_1|$), the energy of phonons, and the interaction of the proton with the lattice phonons. H_{tun} is the tunnel Hamiltonian that provides for proton transfer between two types of the hydrogen bonds. $\hat{a}_l^+(\hat{a}_l)$ is the Fermi operator of creation (annihilation) of hydrogen atom in the intradimer ($l = 1$) and interdimer ($l = 2$) hydrogen bond, respectively; $\hat{b}_{\alpha\mathbf{q}}^+(\hat{b}_{\alpha\mathbf{q}})$ is the Bose operator of creation (annihilation) of a lattice polarized optical phonon, which belongs to the lth branch, with the energy $\hbar\omega_{\alpha\mathbf{q}}$ and the wave vector $\mathbf{q}$; $u_l(\mathbf{q}) = u_l \exp(i\mathbf{q}\cdot\mathbf{l})$, where u_l is the dimensionless value that characterizes the displacement of the pair of oxygens $\mathrm{O}\cdots\mathrm{O}$, which form the hydrogen bond, from their initial equilibrium positions due to the localization of the hydrogen atom H between them (u_1 refers to the intradimer hydrogen bond, and u_2 refers to the interdimer hydrogen bond). The matrix element from expression (422) is determined as

$$V_{12} = \int d\mathbf{r} \langle \Phi | \psi_1^* W \psi_2 | \Phi \rangle \tag{423}$$

where W is the potential that specifies two possible positions (1 and 2) of the hydrogen atom, intradimer and interdimer, respectively (Fig. 22); ψ_1 and ψ_2 are wave functions of the hydrogen atom, which are localized at the aforementioned position;

$$|\Phi\rangle = \hat{B}^+ \Phi_0, \qquad \langle \Phi | = \hat{B} \Phi_0 \tag{424}$$

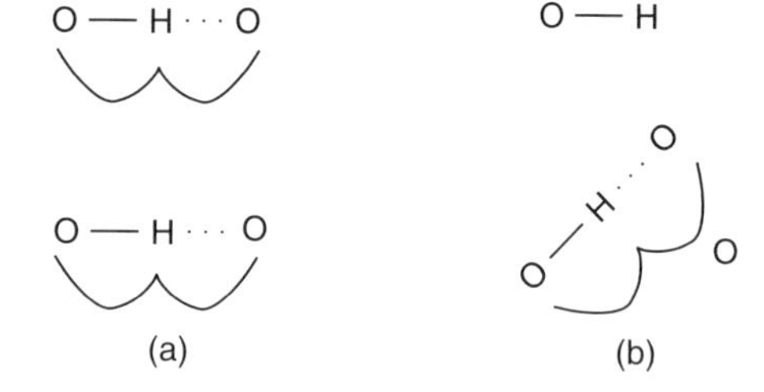

Figure 22. Disruption of two hydrogen bonds connecting two *n*ABA (or *n*AOBA) molecules into the dimer and the formation of one crossed hydrogen bond, which unifies two previously unconnected *n*ABA (or *n*AOBA)—that is, the transition of a hydrogen atom from the intradimer state to the inerdimer state. (From Refs. 57 and 58.)

are quantized site functions that describe the superposition of intradimer vibrations $\rho(\mathrm{O}\cdots\mathrm{H})$ and $\alpha_{\perp}(\mathrm{O}\cdots\mathrm{H})$, and $\hat{B}^{+}(\hat{B})$ is the Bose operator of creation (annihilation) of these vibrations.

By analogy with the operator of current density, considered in the previous sections, we can introduce the operator of proton transfer density between the two different types of hydrogen bonds:

$$j = \frac{1}{\mathcal{V} i\hbar}\left[H_{\text{tun}}, \sum_{l=1}^{2} R_l \hat{a}_l^{+} \hat{a}_l\right] \tag{425}$$

In expression (425) $R_2 - R_1 = g$ where the g is the length of proton jump.

The equation of motion of the statistical operator, which describes the system studied, is

$$i\hbar \frac{\partial \rho}{\partial t} = [H, \rho] \tag{426}$$

If we assume that H_{tun} is a perturbation ($|V| \ll \hbar\omega_{l\mathbf{q}}$, $|E_{1,2}|$, $k_{\text{B}}T$), we derive from Eq. (426), setting in the first approximation $H_{\text{tun}} = 0$, that

$$\rho_0 = \frac{e^{-H_0/k_{\text{B}}T}}{\text{Tr}e^{-H_0/k_{\text{B}}T}} \tag{427}$$

The correction to ρ_0 caused by the interaction Hamiltonian (422) is

$$\rho_{\text{tun}} = -\frac{i}{\hbar}\int_{-\infty}^{t} d\tau e^{-i\frac{t-\tau}{\hbar}H_0}[H_{\text{tun}}, \rho_0]e^{i\frac{t-\tau}{\hbar}H_0} \tag{428}$$

Expressions (425) and (428) make it possible to calculate the proton transfer density by the following formula, which is known also from the previous sections:

$$I = \text{Tr}(\rho_{\text{tun}} j) \tag{429}$$

(here $\mathscr{V}$ is the effective volume of the molecular system in question; see Fig. 22). The overall result is [57] (see also Ref. 36)

$$I = 2^{3/2}\pi^{1/2}\hbar^{-2} n_{\text{ass}} g |V|^2 \sinh\frac{|\bar{E}_1| - |\bar{E}_2|}{k_B T} \exp\left(-\frac{E_a}{k_B T}\right) \times \left[\sum_\alpha |\Delta|^2 \omega_\alpha^2 \operatorname{cosech}(\hbar\omega_l / k_B T)\right]^{-2} \tag{430}$$

where n_{ass} is the concentration of associates and

$$E_a = k_B T \sum_\alpha |\Delta|^2 \tanh\frac{\hbar\omega_\alpha}{4k_B T} - \frac{3}{2}(|\bar{E}_1| - |\bar{E}_2|) \tag{431}$$

is the activation energy. In expressions (430) and (431) we have neglected the dispersion, put $|\Delta|^2 = u_1^2 - u_2^2$, and written the renormalized energies $\bar{E}_1$ and $\bar{E}_2$ of polaron shift for the two types of hydrogen bonds.

Let us estimate I for the mesomorphic 7AOBA crystal. Parameters can be chosen as follows [57,58,60]. The temperature of phase transition "solid crystal 2 $\rightarrow$ solid crystal 1," $T = T_c$, is 320 K (see Fig. 21); the distance g is 0.26 nm; and the energy difference $|\bar{E}_1| - |\bar{E}_2|$ equals $2k_B T_c$. The major contribution to the activation transition in the small polaron model is introduced by low-frequency polarized phonons and hence the inequality $2k_B T > \hbar\omega_l$ should be held. In our case there are two comparatively intensive polarized vibrations of the benzoic ring: $\gamma(\text{CCC}) = 175\ \text{cm}^{-1}$ and $\beta(\text{CCH}) = 290\ \text{cm}^{-1}$. The degree of deformation of the dimer ring (Fig. 20a) is described by the constant $|\Delta|^2$; this parameter characterizes the bandwidths in infrared absorption spectra of the proton subsystem at frequencies $\omega_1 = \gamma(\text{CCC})$ and $\omega_2 = \beta(\text{CCH})$. In addition, the bandwidths have significant dependence on the length of radicals (in particular, the replacement of radical H for $(CH_2)_4CH_3$ leads to the bandwidth broadening of 1.5 in the cyclic dimers of carbonic acids [299]). Since by definition $|\Delta|^2 \gg 1$, we can set $|\Delta|^2 = 40$ (during which the polaron shift $|\Delta|^2\hbar\omega_{1,2}$ is still too small by comparison with the bond energy $|E_{1,2}| \approx 4\,\text{eV}$). The concentration of p-alkyl acids in a dish was $4 \times 10^{21}\ \text{cm}^{-3}$; in the "solid-state 1" phase the concentration of open associates was 40% of this value (see Fig. 21). Therefore, the concentration of open associates, which is entered in expression (430), is $n_{\text{ass}} = 1.6 \times 10^{21}\ \text{cm}^{-3}$.

The square of matrix element

$$|V|^2 = \iint d\mathbf{r}\, d\mathbf{r}' |\psi_1(\mathbf{r})|^2 |\psi_2(\mathbf{r})|^2 \langle\langle\Phi|W|\Phi\rangle\rangle^2 \tag{432}$$

remains the most undetermined value. Symbols $\langle\langle\ldots\rangle\rangle$ mean the thermal averaging by the statistical operator P, where

$$P = \frac{\exp[-\hbar\Omega(\hat{B}^{+}\hat{B} + \frac{1}{2})/k_{\rm B}T]}{\operatorname{Tr}\exp[-\hbar\Omega(\hat{B}^{+}\hat{B} + \frac{1}{2})/k_{\rm B}T]} \tag{433}$$

The averaging (433) reduces the matrix element (432) to the form

$$|V|^2 = V_0^2 \coth^2 \frac{\hbar\Omega}{2k_{\rm B}T_c} \tag{434}$$

Let us set here $\Omega = 2\pi \times 40\ \text{cm}^{-1}$—that is, the mean value of frequency for the superposition of the intradimer vibrations $\rho(\text{O}\cdots\text{H})$ and $\alpha_{\perp}(\text{O}\cdots\text{H})$. Thus the matrix element V_0, which is constructed only on proton wave functions $\psi_{1,2}(\mathbf{r})$, is the fitting parameter. With regard to the numerical values of Ω and T_c, we get from expression (434): $V \simeq 11.6\, V_0$. In as much as the following inequalities should be held, $V_0 \ll \hbar\Omega,\ \hbar\omega_{1,2},\ k_{\rm B}T$, we can put $V_0 = 10^{-23}$ J (i.e., $\simeq 10^{-2}\, k_{\rm B}T_c$).

Using expression (430) for the proton transfer density I, we can readily write the expression for the proton transfer rate

$$K = I\mathscr{A} \tag{435}$$

where $\mathscr{A}$ is the effective area of the cross section of reaction channel characterizing the interaction of transferring proton with an oxygen of the neighboring dimer.

Note that as follows from expression (430) in the range close to the phase transition, the temperature dependence of proton transfer rate (435) is expressed as $T^{1/2}\exp(-\text{const}/T)$. Here, the first factor is responsible for the creation of open associates, and it prevails over the second one for $T > T_c$. Below T_c in the first factor the exponent $1/2$ remains; however, the absolute value of the second factor increases, which accounts for the lack of open associates in the "solid state 2" phase in this temperature range.

If we set $\mathscr{A} \sim 0.1 \times 0.1\ \text{nm}^2$ and insert numerical values of all the parameters mentioned above into expression for I (430), we will obtain from expression (435) the following meaning of the proton transfer rate:

$$K \sim 10^{-2}\ \text{s}^{-1} \tag{436}$$

On the other hand, in the experiments [57,58,60] the samples, when heated to T_c (see Fig. 21), exhibited open associates whose formation was completed in two to three minutes. Thus in fact the calculated value of K qualitatively agrees with the experimental value.

The number of acids merging into polymer associates varied from 2 to 7 and could extend to even higher numbers. Thus such a polymerization is similar to some kind of the clusterization of alkyl- and alkoxybenzoic acids in the solid state. We will return to this problem in Section VIII.

VI. QUANTUM COHERENT PHENOMENA IN STRUCTURES WITH HYDROGEN BONDS

A. Mesoscopic Quantum Coherence and Tunneling in Small Magnetic Grains and Ordered Molecules

Over a period of years there has been considerable interest in the phenomenon of mesoscopic quantum tunneling and coherence revealed in various small-size systems. In particular, the phenomenon has been observed in optically trapped ions [300,301], superconductors in which the order parameter shows the phase difference across Josephson junctions [302], spin-domains in atomic Bose–Einstein condensates [303], and mesoscopic magnetic clusters and grains [304]. Many physical properties of small-size systems being governed by quantum effects are radically distinguished from properties of macroscopic samples. Besides the usual one-particle quantum effects (for instance, tunneling), some low-dimension structures demonstrate a correlated quantum behavior of several particles such as the mesoscopic tunneling of a cooperative magnetic moment observed by the magnetic relaxation measurements in magnetic molecules and small grains [305–308] (note that a mechanism of magnetization reversal was originally proposed by Bean and Livingston [309] much earlier).

Coffey and co-workers [310–312] have conducted detailed studies of the wideband dielectric response of polar molecules and the rotational motion of single-domain ferromagnetic particles in the presence of an external magnetic field with regard to the inertia of the particles. In particular, Coffey and co-workers have developed a generalized approach based on the transformed Langevin equation, which allowed them to calculate the dipole moment of tagged molecules and investigate its motion followed by oscillations of the applied field. They have examined [313,314] the rotational Brownian motion of two- and three-dimensional rotators and have shown how the Langevin equation is transformed into an equation for the dipole moment. The contribution of single-domain particle inertial effects to resonance in ferrofluids has also been analyzed [315]. The study [316] of the nonaxially symmetric asymptotic behavior of a potential having minima at $\theta = (0,\ \pi)$ at the escape from the left to the right of the potential allowed the precise calculation of the prefactor in the

Néel–Brown model; however, the nonlinear field effects have also been very significant. The further search for the accurate description of the macroscopic quantum tunneling of magnetization **M** needs correct asymptotic formulas for the reversal time for magnetocrystalline anisotropy potentials, which have been calculated by Coffey and collaborators as well [315–318].

Similar mesoscopic quantum effects take place also in short hydrogen-bonded chains and in small clusters, which include hydrogen bonds. The phenomenon of large proton polarizability and fast oscillations of the polarization of the chain were studied experimentally by Zundel and co-workers [6,249,288,289]. A theoretical study of macroscopic tunneling of the chain polarization has been conducted in Refs. 319–322.

To gain a better understanding the collective proton dynamics, let us consider an isolated chain schematically shown in Fig. 23. The proton subsystem is found here in the degenerate ground state, and the chain as a whole can take two configurations with different proton polarizations of hydrogen bonds. The mesoscopic quantum tunneling of polarization from one configuration to the other one stems from quasi-one-dimensional proton sublattice dynamics [319]. However, the rotating motion of ionic or orientational defects might relieve the degeneration of polarization in principle as well [161,323–325]. In this instance we meet the over-barrier reconstruction of polarization of the chain,

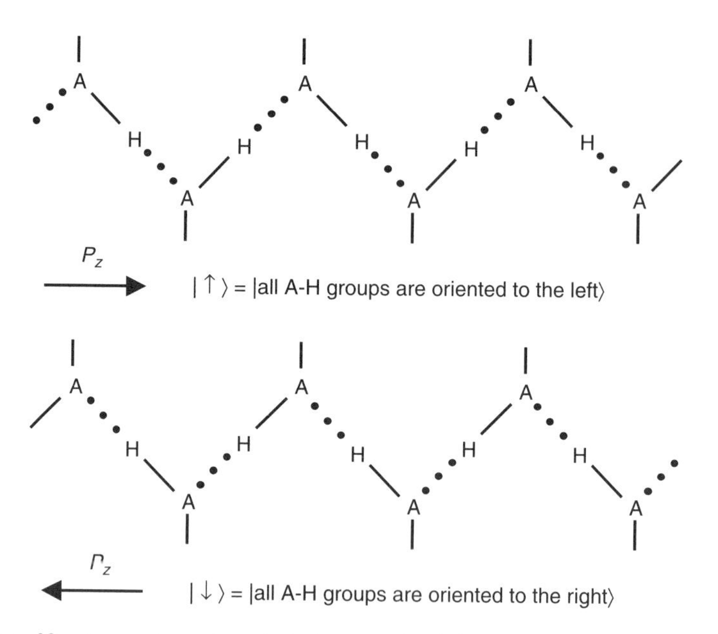

Figure 23. Two configurations of the hydrogen-bonded chain, which are characterized by opposite polarizations. (From Ref. 319.)

and the corresponding solution in terms of a soliton formation is similar to the Bloch wall in ferromagnetics, which is typical for rather macroscopic samples. Traditionally, small clusters are single domains in which the total polarization is fixed and features two equilibrium orientations along and against the principal axis of the chain (Fig. 23). Let us denote these configurations of the chain by symbols $|\uparrow\rangle$ and $|\downarrow\rangle$. Our goal is to demonstrate a possibility of the spontaneous tunnel transformation of a short chain (or a single domain) from configuration $|\uparrow\rangle$ to configuration $|\downarrow\rangle$. Such a behavior of the chain means that protons in the hydrogen bonds coherently move from the left side to the right side; or, in other words, the total chain polarization tunnels between the two opposite directions P_z and $-P_z$, as shown in Fig. 23.

Let us turn to the main properties of the energy spectrum, which should help us to shed light on the chain polarization transition and a role of the Coulomb interaction that realizes the phenomenon. In the simplest case of an isolated hydrogen-bonded chain, the Hamiltonian H of the proton subsystem can be written in the form of two parts: (a) the potential energy H_C corresponding to the Coulomb interaction between protons and (b) the kinetic energy H_{tun} describing the tunneling transfer of a proton along the hydrogen bond:

$$H = H_C + H_{tun} \tag{437}$$

$$H_C = U \sum_l (\hat{a}^+_{R,l}\hat{a}_{R,l}\hat{a}^+_{L,l+1}\hat{a}_{L,l+1}) \tag{438}$$

$$H_{tun} = J \sum_l (\hat{a}^+_{R,l}\hat{a}_{L,l} + \hat{a}^+_{L,l}\hat{a}_{R,l}) \tag{439}$$

Here $\hat{a}^+_{R(L),l}(\hat{a}_{R(L),l})$ is the Fermi operator of creation (annihilation) of a proton in the left (right) well of the two-well potential of the lth hydrogen bond, J is the overlap, or tunnel integral, and U the Coulomb repulsion energy between protons in the neighbor hydrogen bonds. At first let us assume that tunnel integral J is smaller than Coulomb interaction U, so that our Hamiltonian can be represented only by the potential part $H = H_C$ that corresponds to the zero order of the perturbation theory. The energy spectrum has a simple structure (Fig. 24a). The doubly degenerated ground state corresponds to the two chain configurations $|\uparrow\rangle = |10, 10, \ldots, 10\rangle$ and $|\downarrow\rangle = |01, 01, \ldots, 01\rangle$; that is, the protons are oriented to the left or to the right. Excited states of the chain stipulated by unordered proton configurations in the hydrogen bonds (e.g., $|01, 10, \ldots, 10\rangle$) are separated from the ground state by Coulomb gap U. Let the tunneling Hamiltonian H_{tun} be a small perturbation, which thus intensifies the energy degeneration mixing the ground and excited states (Fig. 24b).

We will treat the splitting of the ground state related to the tunnel transition of the chain as a whole, which occurs between the two configurations with opposite proton polarizations. To make these coherent tunnel transitions of

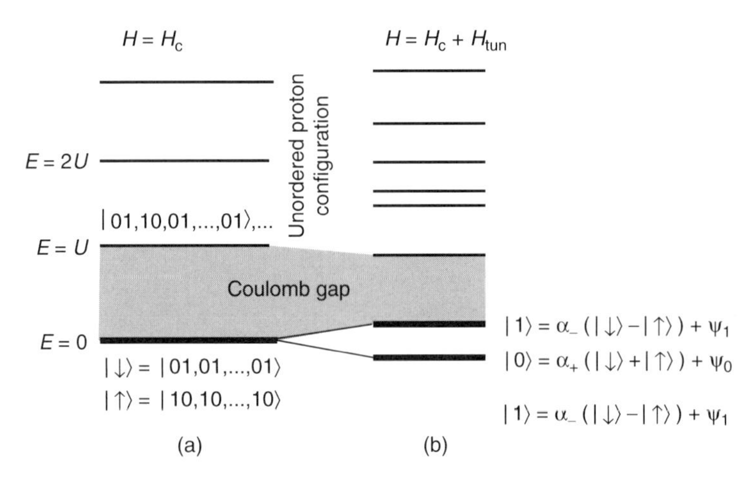

Figure 24. Energy spectrum of the proton subsystem: (a) Energy levels determined only by Coulomb repulsion of protons (H_C) in nearest hydrogen bonds. The doubly degenerated ground level corresponds to the two states of the chain with opposite proton polarizations. (b) The energy splitting caused by a small perturbation H_{tun}. The ground state splitting is associated with the coherent tunnel dynamics of protons. In the case of the strong Coulomb interaction, the Coulomb gap separates the excited state from the ground one, which ensures the coherent tunneling.

protons possible, the energies of the two lowest states must be separated by a gap from excited states. Obviously, when the local tunnel integral J is small in comparison with U, the Coulomb gap preserves such transitions. However, if the kinetic energy of the system has an order of the Coulomb repulsion, the gap between the ground and excited states will be virtually reduced to zero. This in turn will destroy the coherent proton dynamics. Thus an important prerequisite to the correlated tunnel dynamics is the strong Coulomb interaction between protons.

Another important condition for such transitions is low temperature at which thermal fluctuations do not blur the ground state. The ground state will be split when temperature increases, though the value of splitting will still be less than the Coulomb gap U. Thermal fluctuations are also able to generate the possibility of the chain repolarization between configurations $|\uparrow\rangle$ and $|\downarrow\rangle$. Thus, at some temperature the coherent tunnel regime of the cooperative proton oscillations (temperature independent) should change to a classical regime based on the mechanism of over-barrier proton transitions.

The matrix element, or the probability of coherent transition between states, includes the full information about the transition. Let us denote wave functions of the two degenerate ground states (at $J = 0$) as $|\uparrow\rangle = |10, 10, \ldots, 10\rangle$ (the chain's protons are oriented to the left) and $|\downarrow\rangle = |01, 01, \ldots, 01\rangle$ (the chain's protons are oriented to the right). Let $|l\rangle$ be the eigenstate of the Hamiltonian $H = H_C + H_{tun}$, $H|l\rangle = E_l|l\rangle$. The probability corresponding to the transition from $|\uparrow\rangle$ to $|\downarrow\rangle$ is

$$|\langle\uparrow|e^{-\frac{i}{\hbar}Ht}|\downarrow\rangle|^2 = \sum_l |\langle\uparrow|l\rangle\langle l|\downarrow\rangle|^2 + 2\sum_{l>n}\langle\uparrow|l\rangle\langle l|\downarrow\rangle\langle\uparrow|n\rangle\langle n|\downarrow\rangle\cos\left(\frac{E_n - E_l}{\hbar}t\right) \tag{440}$$

As can readily be seen, at low temperature the oscillations between the $|\uparrow\rangle$th and the $|\downarrow\rangle$th states take place with the frequency that corresponds to the energy splitting of the ground state:

$$\Gamma^2 \sim \langle\uparrow|0\rangle\langle 0|\downarrow\rangle\langle\uparrow|1\rangle\langle 1|\downarrow\rangle\cos\left(\frac{E_1 - E_0}{\hbar}t\right) \tag{441}$$

where $|0\rangle$ and $|1\rangle$ are the two lowest levels. The transitions $|\uparrow\rangle \leftrightarrow |\downarrow\rangle$ include both the single-domain repolarization and transitions through unordered proton configurations. The wave functions of the two lowest levels have the following forms:

$$\begin{aligned} |0\rangle &= \alpha_+(|\uparrow\rangle + |\downarrow\rangle) + \Psi_0 \\ |1\rangle &= \alpha_-(|\uparrow\rangle - |\downarrow\rangle) + \Psi_1 \end{aligned} \tag{442}$$

Terms Ψ_0 and Ψ_1 describe unordered proton configurations in the bonds, so that the probability of single-domain chain oscillations becomes $\sim |\alpha_+|^2|\alpha_-|^2 \cos[(E_1 - E_0)t/\hbar]$. When Ψ_0 and Ψ_1 are small in comparison with the α terms, one can treat the chain repolarization as that of a single domain. Contribution of Ψ_0 and Ψ_1 in (442) becomes more appreciable at the increasing of the integral J that leads to the destroying the correlated proton dynamics. For example, in the case of a chain consisting only of two bonds, which is characterized by periodical boundary conditions, we have

$$\begin{aligned} |0\rangle &= \alpha_+(|\uparrow\rangle + |\downarrow\rangle) + \beta_+(|01, 10\rangle + |10, 01\rangle) \\ |1\rangle &= \frac{1}{\sqrt{2}}(|\uparrow\rangle - |\downarrow\rangle) \end{aligned} \tag{443}$$

where

$$2\alpha_+^2 + 2\beta_+^2 = 1, \qquad \beta_+ = \alpha_+\frac{E_0}{2J}, \qquad E_2 = \tfrac{1}{2}[U + J - ((U + J)^2 + 8J^2)]^{1/2} \tag{444}$$

If the local tunnel integral J is small compared with Coulomb interaction U, the contribution on the side of the term proportional to β_+ in expression (443) is negligible (its contribution is of the order of $J/U \ll 1$). Hence in this case we can consider the tunnel repolarization of the chain as a repolarization of a single domain. Increasing of γ involves an increase of the β term, which will suppress the coherent tunneling of protons.

In such a manner the coherent oscillation is restricted by two factors. First, it is the strong Coulomb correlation between particles which suppresses the local tunneling of separate protons ($U \gg J$). Second, the small value of J decreases the proton mobility. However, the increase of the proton mobility due to the enlargement of the J will diminish the coherence in proton oscillations. Note that similar correlated tunneling of particles can occur in the systems with the strong Coulomb interaction—for instance, in small ferroelectric clusters and chain-like structures with charge-ordering (Coulomb crystals). Another example of a molecular system in which similar tunnel oscillations can be observed is a short chain of *trans*-isomeric polyacetylene that features the periodic change of single and double bonds, [326,327] (Fig. 25).

Having estimated the tunneling rate of the aforementioned proton transitions, we shall draw the analogy to ferroelectric systems. For this purpose we should represent the system of hydrogen bonds in the framework of the pseudo-spin formalism (see Section II.D). However, before the detailed consideration of the chain repolarization, let us touch upon the effect of the coherent quantum tunneling in small magnetic grains and the quasi-classical approach to the calculation of the tunneling rate, which is employed below.

Let us briefly state the most important features associated with small magnetic grains [306–309,328–330]. Small ferromagnetic particles with size about 15 nm usually consist of a single magnetic domain. The total magnetization is fixed in one of several possible directions of the so-called easy magnetization, which is determined by the crystalline anisotropy and the shape of a magnetic particle. In the particular case of the easy-axes anisotropy, the total magnetization $\mathbf{M}$ can be specified by two equilibrium states, which differ only by the vector orientation. Transitions of the particle magnetization between these states occur spontaneously. At high temperature T, the moment jumps from one orientation to another over the anisotropy barrier. Below some $T = T_c$, thermal processes are essentially frozen, but oscillating of the total magnetization does not disappear. Quantum mesoscopic tunneling in the systems in question was evidenced by the magnetic relaxation measurements with a sharp

Figure 25. Possible tunnel transitions between the two different configurations of a *trans*-polyene chain.

crossover to the temperature-independent magnetic relaxation for low temperature [306–309,328–330]. The magnetic relaxation time that characterized the oscillations did not depend on temperature and was finishing when T trends to zero. Measurements of the susceptibility conducted on small antiferromagnetic grains with size 7 nm showed (a) the coherent quantum tunneling of the sublattice magnetization through the anisotropy barrier even below $T_c \simeq 0.2$ K [328,329] and (b) the tunneling of the total magnetic moment of the mesoscopic $Mn_{12}O_{12}$ magnetic molecule below 2 K [307,308]. Thus, the mentioned experiments in fact demonstrated the total magnetic moment (in ferromagnetics) or the Néel vector (in antiferromagnetic) tunnels between the two energy minima. The phenomenon of mesoscopic (macroscopic) quantum tunneling (or the decay of a metastable state) and the mesoscopic quantum coherence (or the resonance between degenerate states) arise owing to the coherent behavior of individual magnetic moments, whose quantity is rather macroscopic as it varies from 10^4 to 10^6. Typical frequencies of coherent tunneling fall in the range 10^6 to $10^8\ \mathrm{s}^{-1}$ (see, e.g., Refs. 331 and 332), though in the case of the hydrogen-bonded chain the frequency of coherent tunneling of protons can reach $10^{12}\ \mathrm{s}^{-1}$ [249,288,289].

Such a tunnel switching of the magnetization can be described by the so-called "one-domain" approximation, when the total magnetization vector $\mathbf{M}$ is taken as a main dynamic variable with fixed absolute value M_0. Then the total energy density, or the anisotropy energy E, is obtained from the spin-Hamiltonian H using a spin coherent state $|n\rangle$ chosen along the direction $\mathbf{n}$ [332,333]:

$$\langle n|H|n\rangle = \mathscr{V} \cdot E(\theta, \varphi) \tag{445}$$

where $\mathscr{V}$ is the volume of a magnetic particle (i.e., domain), and θ and φ are polar coordinates of the vector $\mathbf{n}$. Minima of the E correspond to equilibrium directions of the particle magnetization, $\mathbf{n}_1$, $\mathbf{n}_2, \ldots$, which are separated by the appropriate energy barriers. In order to calculate the probability of the magnetization tunneling between the aforementioned equilibrium directions, we have to consider a matrix element of type (440), which can be represented by the spin-coherent-path integral [332,333]

$$\Gamma = \langle n_1|e^{-\frac{i}{\hbar}Ht}|n_2\rangle = C\int\{d\mathbf{n}\}\exp\left\{-\frac{\mathscr{S}_E[\mathbf{n}(\tau)]}{\hbar}\right\} \tag{446}$$

where C is a normalization factor and $\mathscr{S}[\mathbf{n}(\tau)]$ is the imaginary time, or Euclidean action [306,332,333]

$$\mathscr{S}_E[\mathbf{n}(\tau)] = v\int_{-T/2}^{T/2} d\tau\left\{-i\frac{M_0}{\lambda}\frac{d\varphi}{d\tau}(\cos\theta - 1) + E(\theta,\varphi)\right\}, \qquad t = i\tau \tag{447}$$

where $\lambda \equiv e\gamma/2mc$ and γ is the gyromagnetic ratio. Below we restrict our consideration by the quasi-classical approximation, which allows expression (446) to be rewritten as follows:

$$\Gamma \approx Ae^{-\frac{\mathscr{S}_E}{\hbar}} \tag{448}$$

$\mathscr{S}_E$ is the Euclidean action written for a classical trajectory corresponding to the sub-barrier rotation of the particle magnetization **M**. This is a typical instanton trajectory that satisfies the equations of motion

$$\begin{aligned} i\frac{M_0}{\lambda}\sin(\theta)\dot{\theta} &= \frac{\partial E(\theta,\varphi)}{\partial\varphi} \\ i\frac{M_0}{\lambda}\sin(\theta)\dot{\varphi} &= -\frac{\partial E(\theta,\varphi)}{\partial\theta} \end{aligned} \tag{449}$$

Prefactor A (i.e., the Van Vleck determinant) in Eq. (448) relates to fluctuations around the classical path. It should be noted that the utilization of the "single-domain" approximation (445) partly ignores a contribution on the side of unordered spin configurations to wave functions of the ground states; that is, any term similar to Ψ_0 in expression (442), the decoherence factor, is omitted. Thus, the spin–spin interaction results in the coherent behavior of a magnetic one-domain particle.

Being interested in the effect of tunnel repolarization, one can apply the semiclassical approximation, which is the most general, though is more crude at the same time. The essence of the approach is the following. We are not interested in forces, which hold a hydrogen-bonded chain. We are interested only in the possibility of transition of the polarization of the chain from the state when the polarization vector is directed to the right of the state when that is vectorial to the left. The two states are characterized by the same energy. However, in order that the polarization comes from one state to the other, the polarization should overcome some barrier. It can be made by means of either the over-barrier transition or tunnel one. We will not take an interest in a substructure (local barriers) of the aforementioned super barrier. The barrier will be determined from the anisotropy dependence of the polarized energy on the polarization vector, much as in the case of the magnetization tunneling in ferromagnetics. This allows the direct calculation of the probability of repolarization in the quasi-classical approximation, and then parameters that characterize the barrier are phenomenological constants.

We assume that the interaction between protons in the neighbor hydrogen bonds plays a role of the exchange pseudo-spin interaction, which gives rise to the coherent proton tunneling. In such a way, we will not treat the combined probability of individual protons, but will consider the probability of coherent proton tunneling determined by the interference effect. The mesoscopic quantum coherence phenomenon can manifest itself in the measurement of the total dipole moment (i.e., polarization) **P** of a chain.

In the framework of our model, the dipole operator takes the form $P = d\sum_l (\hat{a}^+_{R,l}\hat{a}_{R,l} - \hat{a}^+_{L,l}\hat{a}_{L,l})$, where d is the dipole moment of a hydrogen bond. If we drop thermal fluctuations and dissipation, we can write the correlator, which takes into account successive measurements of the P separated by time interval Δt:

$$\langle P(t)P(t+\Delta t)\rangle \sim P_0^2\Gamma^2\cos\left(\frac{E_1 - E_0}{\hbar}\Delta t\right) \tag{450}$$

where $\langle P^2\rangle = P_0^2$. Here Γ is the matrix element [Γ^2 is written explicitly in expression (441)], and the angular brackets denote the quantum mechanical averaging.

Short hydrogen-bonded chains are characterized by a large proton polarizability, and they play a role of typical proton channels in biomembranes. That is why it is reasonable to assume that coherent proton transitions in the chain can be involved in the mechanism of real proton transport along the chain [327].

The coherent repolarization can be realized by means of the coherent tunnel proton motion along hydrogen bonds [319,320] (i.e., coherent tunnel repolarization of bonds A—H $\cdots$ A $\leftrightarrow$ A $\cdots$ H—A) and/or owing to the proton rotation around heavy atoms [320,321] (coherent tunnel reorientation of ionic groups A—H $\cdots$ A—H $\leftrightarrow$ H—A $\cdots$ H—A). The latter becomes possible when protons are tightly bound with the ionic groups, so that the probability of proton transfer along the hydrogen bond is lower than that of the orient proton motion around a heavy atom.

From the preceding, it may be seen that the probability of coherent tunneling is specified by a nonmonotonic dependence on J. This brings about an interesting "isotopic effect" [322,327]: The probability of coherent tunnel transitions for heavy particles can be higher than that for light particles. An analogous effect occurs in the case of the proton–phonon interaction [320], which, on the one hand, can effectively suppress local proton tunneling J (due to the increasing the effective mass of a particle) and, on the other hand, enlarge the probability of collective tunneling of protons.

Let us now consider these two possibilities and discuss the influence of phonons, which modulate the distance between heavy atoms A $\cdots$ A, on the coherent tunneling rate.

B. Two Possible Mechanisms of Coherent Tunneling of the Repolarization of Hydrogen-Bonded Chain

Let us calculate the tunneling rate of correlated proton dynamics associated with oscillations of the chain's total polarization using the quasi-classical (WKB)

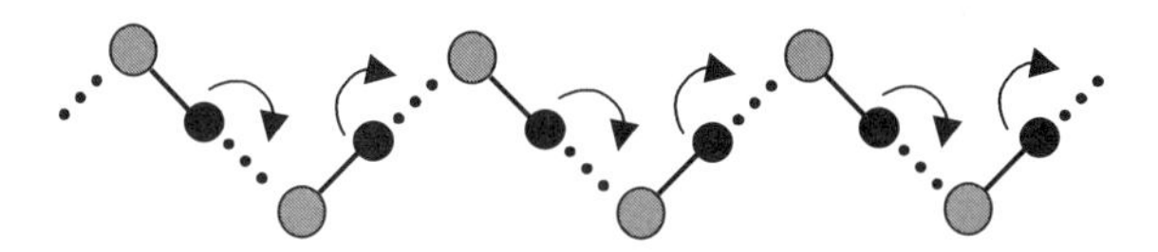

Figure 26. Chain repolarization as the coherent proton tunneling.

approximation [319,320]. We can distinguish the two following options: (i) The chain repolarization is realized owing to the coherent tunneling of protons along hydrogen bonds, and (ii) the chain repolarization occurs due to the rotation of protons around heavy backbone atoms.

In the first case (Fig. 26) a quasi-one-dimensional hydrogen-bonded chain is characterized by the two different configurations of the chain, $|\uparrow\rangle$ and $|\downarrow\rangle$ (Fig. 23), which possess the same energy. Let us assume that the hydrogen bonds are exemplified by the two-well potential, and let the two-level model in which protons are distributed between the ground state and an excited state be achieved. We can then introduce the Fermi operators of creation (annihilation) of a proton in the right and the left well of the *k*th hydrogen bond, $\hat{a}^+_{R,l}(\hat{a}_{R,l})$ and $\hat{a}^+_{L,l}(\hat{a}_{L,l})$, respectively. Since one hydrogen bond is occupied by only one proton—that is, $n_{R,l} + n_{L,l} = \hat{1}$, where $n_{R,l}(n_{L,l})$ is the proton number operator in the right (left) well of the *l*th hydrogen bond—we can pass to the pseudo-spin presentation of the proton subsystem as expressions (70)–(72) prescribe.

In this presentation, the bond polarization can be written as $p^z = 2S^z d$, where d is the dipole moment of the hydrogen bond. Then the Hamiltonian (437) of the proton subsystem is transmitted to [compare with expression (73)]

$$H = -\hbar\Omega \sum_l S^x_l - U \sum_l S^z_{l-1} S^z_l \tag{451}$$

where Ω is the tunneling frequency of a proton in the hydrogen bond [note that $\hbar\Omega = 2J$, where J is the tunnel integral written in expression (439)] and U is the Coulomb energy of a couple of protons located in neighbor hydrogen bonds. If the chain satisfies the periodical boundary conditions, the following equations of motion for the operator $\mathbf{S}_l$ are obtained:

$$\hbar \frac{dS^x_l}{dt} = U(S^y_l S^z_{l+1} + S^z_{l-1} S^y_l) \tag{452}$$

$$\hbar \frac{dS^y_l}{dt} = \hbar\Omega S^z_l - U(S^x_l S^z_{l+1} + S^z_{l-1} S^x_l) \tag{453}$$

$$\hbar \frac{dS^z_l}{dt} = -\hbar\Omega S^y_l \tag{454}$$

We recall that our key interest is the study of the tunneling dynamics of the chain polarization or, equivalently, the total pseudo-spin **S** of the chain. Following Ref. 332, we can estimate, using the quasi-classical approximation, the tunneling rate $\hbar\tau$, or the splitting energy between the states $|\uparrow\rangle$ and $|\downarrow\rangle$:

$$\hbar\tau = p\hbar\omega_p \left(\frac{\mathcal{S}_E}{2\pi\hbar}\right)^{1/2} \exp\left(-\frac{\mathcal{S}_E}{\hbar}\right) \tag{455}$$

where ω_p is the oscillating frequency of the **S** in the well, (i.e., the so-called small-angle pseudo-spin precession), p is the dimensionless prefactor, which can be about 10, and I_E is the Euclidean action for the sub-barrier rotation of the **S**. Using the time-dependent mean field approximation and taking into account our assumption that the chain is a single domain, we can write the classical equation for the **S** in the following form [96]:

$$\hbar\frac{d\mathbf{S}}{dt} = -\mathbf{S} \times \frac{\partial E}{\partial \mathbf{S}} \tag{456}$$

where the anisotropy energy

$$E = -\hbar\Omega S^x - U \cdot (S^z)^2 \tag{457}$$

Vector **S** characterizes the polarization of one-domain chain. In the spherical coordinate system we have $\mathbf{S} = S(\sin\theta\cos\phi, \sin\theta\sin\phi, \cos\theta)$ and equation (456) is reduced to

$$\begin{aligned} \hbar S\frac{d\theta}{dt}\sin\theta &= \frac{\partial E}{\partial\phi} \\ \hbar S\frac{d\phi}{dt}\sin\theta &= -\frac{\partial E}{\partial\theta} \end{aligned} \tag{458}$$

The classical action that corresponds to Eqs. (458) has the form

$$\mathcal{S} = N\int dt\left(\hbar S\frac{d\phi}{dt}(\cos\theta - 1) - E(\theta,\phi)\right) \tag{459}$$

where N is the number of hydrogen bonds. The anisotropy energy

$$E(\theta,\phi) = US^2\sin^2\theta - \hbar\Omega\, S\sin\theta\cos\phi + \frac{\hbar^2\Omega^2}{4U} \tag{460}$$

reaches minima at the following values of the pseudo-spin: $S_{\uparrow} = S(\sin\theta_0,\ 0,\ \cos\theta_0)$ and $S_{\downarrow} = S(\sin\theta_0,\ 0, -\cos\theta_0)$, where

$$\sin\theta_0 = \frac{\hbar\Omega}{2US} \tag{461}$$

They are the minima, which determine the two equilibrium polarizations of the chain. The minima are equal in magnitude but opposite in sign and direction to each other. The energy of the states is chosen to be equal to zero.

In such a way we have reduced our consideration to the problem of one-domain ferromagnetic particle with anisotropy energy described by equations (457) or (460). The problem of the macroscopic and mesoscopic quantum coherence which correspond to the tunnel switching of the magnetic moment between the two equilibrium directions was studied by Garg and Kim [332] in detail. This means that we may apply their results to calculate the tunneling frequency in the case of the hydrogen-bonded chain. Thus the Euclidean action $\mathcal{S}_E$ becomes

$$\mathcal{S}_E = N\int d\tau\left(-i\hbar S\frac{d\phi}{d\tau}(\cos\theta - 1) + E(\theta, \phi)\right) \tag{462}$$

where $\tau = it$ and the extremum of $\mathcal{S}_E$ is reached at the solution of equations

$$i\hbar S\frac{d\theta}{d\tau}\sin\theta = \frac{\partial E}{\partial\phi} \tag{463}$$

$$i\hbar S\frac{d\phi}{d\tau}\sin\theta = -\frac{\partial E}{\partial\theta} \tag{464}$$

Using the energy conservation law written for a classical path, $E = 0$, one can obtain the relationship between $\sin\theta$ and $\cos\phi$:

$$\sin^2\left(\frac{\phi}{2}\right) = -\frac{(\sin\theta - \sin\theta_0)^2}{4\sin\theta\sin\theta_0} \tag{465}$$

The sub-barrier path that conforms to the switching the motion of $\mathbf{S}$ from the state $S_{\uparrow}$ at $\tau = -\infty$ to the state $S_{\downarrow}$ at $\tau = \infty$ is defined by the instanton solution of Eqs. (464) and (465):

$$\cos\theta = -\cos\theta_0\tanh(\omega_p\tau) \tag{466}$$

$$\sin\phi = \frac{i}{2}\frac{\cot^2\theta_0\,\mathrm{sech}^2(\omega_p\tau)}{[1 + \cot^2\theta_0\,\mathrm{sech}^2(\omega_p\tau)]^{1/2}} \tag{467}$$

where $2\omega_p = \Omega \cot \theta_0$ is the small oscillation frequency in the well. Calculating the action for this trajectory, we get

$$\frac{1}{\hbar}\mathcal{S}_E(\cos\theta_0) = 2\,SN\left(-\cos\theta_0 + \frac{1}{2}\ln\left(\frac{1+\cos\theta_0}{1-\cos\theta_0}\right)\right) \tag{468}$$

where N is the number of hydrogen bonds in the chain. The tunneling rate is determined by the WKB exponent dependent only on $\mathcal{S}_E$ and the prefactor calculated in Ref. 332. Using these results, we can represent the splitting energy $\hbar\tau_N$ for such transitions as

$$\hbar\tau_N = 8US\left(\frac{SN}{\pi}\right)^{1/2}\left(\frac{x^5}{1-x^2}\right)^{1/2}\left(\frac{1-x}{1+x}\right)^{x/2}\exp\left\{-\frac{\mathcal{S}_E}{\hbar}\right\}, \qquad x = \cos\theta_0 \tag{469}$$

The behavior of τ_N/Ω as a function of $\hbar\Omega/U$ is demonstrated in Fig. 27. It can easily be seen that the frequency of the coherent tunnel repolarization may come close to the frequency of an individual proton that tunnels in the hydrogen bond. Note that the coherent tunnel repolarization depends nonmonotonically on Ω.

The critical temperature T_c, which correlates with the crossover from the thermal to the quantum repolarization mechanism, can be easy to estimate. The

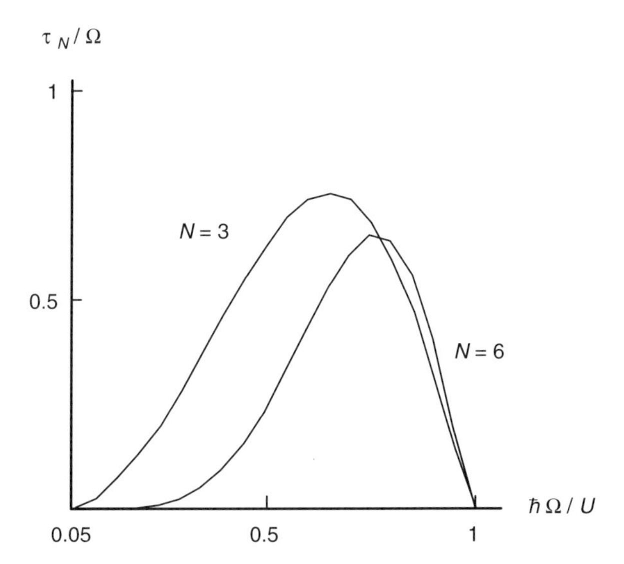

Figure 27. Normalized coherent tunneling rate τ_N/Ω as a function of the parameter $\hbar\Omega/U$. N is the number of hydrogen bonds in the chain.

probability of switching of the chain polarization via the thermal activation is proportional to $\exp(-\Delta E/k_B T)$, where $\Delta E = S^2 N(1 - \sin\theta_0)^2$ is the barrier height. When this expression is compared with that of the tunneling probability, it is apparent that

$$\exp(-\Delta E/k_B T_c) \approx \exp(-\mathcal{S}_E/\hbar) \tag{470}$$

and we immediately gain the critical temperature

$$T_c \approx \frac{\hbar \Delta E}{k_B \mathcal{S}_E} \tag{471}$$

Let us now proceed to a study of the second scenario of cooperative proton transitions, namely, the coherent tunnel motion of protons around heavy ions [320,321]. The pattern is schematically shown in Fig. 28. Such a motion of protons can be realized if protons are tightly bound with the ion groups, so that the probability of proton transfer along the hydrogen bond becomes lower than that of orientational proton motion of the group A—H. To illustrate the feasibility of such transitions, we shall treat simplest models of orientation oscillations of the ionic groups in the hydrogen-bonded chain [135,323–325]. In the two-level approximation, these models can be reduced to the model of an easy-axes ferromagnetic with the transversal external field. In Ref. 325 a model of the orientational kink defect in a quasi-one-dimensional ice crystal was proposed. In the model, the main dynamic variable was the angle between the direction of O—H bond and the principal axis of the chain. The potential energy was determined by the interaction of neighbor water molecules, which were specified by the two-well potentials, and the potential minima corresponded to the equilibrium orientations ($\cdots$O—H$\cdots$ and $\cdots$H—O$\cdots$) of water molecules in the chain. In the pseudo-spin representation, the secondary quantized Hamiltonian of such a system can be reduced to a model of the easy-axes ferromagnetic with the transversal external field [96].

However, the model by Stasyuk et al. [135] described in Section II.A is more attractive because it allows an estimation of the frequency of coherent orientational tunnel transitions taking into account both the proton dynamics of the hydrogen bonds and the reorientational processes of A—H groups. Let us

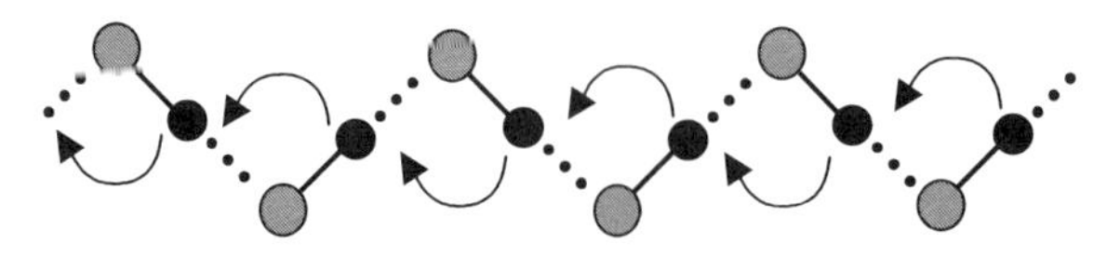

Figure 28. Cain repolarization caused by the motion of protons around backbone heavy atoms.

treat transitions between the chain configurations $|\uparrow\rangle$ and $|\downarrow\rangle$ caused by the coherent tunnel reorientation of groups A–H starting from their Hamiltonian [135] [see expression (120)]. We may choose the Hamiltonian in the form

$$H = H_0 + H_{\rm tun} + H_{\rm rot} + H_{\rm C} \tag{472}$$

$$H_0 = \sum_l [w'(1 - n_{R,l})(1 - n_{L,l+1}) + w n_{R,l} n_{L,l+1} + \varepsilon(1 - n_{R,l}) n_{L,l+1} + \varepsilon n_{R,l}(1 - n_{L,l+1})] \tag{473}$$

Here w, w', and ε are the energies of proton configurations in the minima of the potential near an ionic group; $n_{R,l}(n_{L,l})$ is the proton number operator that characterizes the occupation of the right (left) well of lth hydrogen bond. The Hamiltonian $H_{\rm tun}$ in expression (472) describes the tunnel transition between two proton states in the same hydrogen bond:

$$H_{\rm tun} = -J \sum_l (\hat{a}^+_{L,l} \hat{a}_{R,l} + \hat{a}^+_{R,l} \hat{a}_{L,l}) \tag{474}$$

where J is the tunnel integral. Orientational transitions of ionic groups—that is, (A–H) $\leftrightarrow$ (H–A), Fig. 28—can be described as a pseudo-tunnel effect. So the corresponding Hamiltonian is

$$H_{\rm rot} = -\Upsilon_{\rm rot} \sum_l (\hat{a}^+_{R,l} \hat{a}_{L,l+1} + \hat{a}^+_{L,l+1} \hat{a}_{R,l}) \tag{475}$$

where $\Upsilon_{\rm rot}$ is the pseudo-tunnel integral caused by the orientational transitions of protons. The term $H_{\rm C}$ in expression (472) includes the Coulomb interaction between protons and electron pairs in the same hydrogen bonds (see Section II.A) [135,334]:

$$H_{\rm C} = U_{(D\text{-})} \sum_l n_{R,l} n_{L,l} + U_{(L\text{-})} \sum_l (1 - n_{R,l})(1 - n_{L,l}) \tag{476}$$

Let the chain obey the periodical boundary conditions and let protons be strongly connected with heavy atoms, so that tunnel proton transitions along the hydrogen bond are negligible in comparison with the reorientation motion of A–H groups. This means that only one proton is localized near each heavy atom in the chain; that is, equality $n_{R,l} + n_{L,l+1} = \hat{1}$ holds, which makes it possible to introduce the following pseudo-spin operators:

$$S_l^x = \tfrac{1}{2}(\hat{a}^+_{R,l} \hat{a}_{L,l+1} + \hat{a}^+_{L,l+1} \hat{a}_{R,l}) \tag{477}$$

$$S_l^y = \tfrac{1}{2}(\hat{a}^+_{R,l} \hat{a}_{L,l+1} - \hat{a}^+_{L,l+1} \hat{a}_{R,l}) \tag{478}$$

$$S_l^z = \tfrac{1}{2}(\hat{a}^+_{R,l} \hat{a}_{R,l} - \hat{a}^+_{L,l+1} \hat{a}_{L,l+1}) \tag{479}$$

This allows us to construct the Hamiltonian for the orientational motion of ionic groups in the hydrogen-bonded chain:

$$H_{\text{rot}} = -\hbar\Omega_{\text{rot}} \sum_{l} S_l^x - U_{\text{rot}} \sum_{l} S_{l-1}^z S_l^z \tag{480}$$

where $\hbar\Omega_{\text{rot}} = 2\Upsilon_{\text{rot}}$ and $U_{\text{rot}} = U_{(D\text{-})} + U_{(L\text{-})}$ specifies the energy of Bjerrum's D and L defects.

The Hamiltonian (480) of orientational oscillations of ionic groups in the hydrogen-bonded chain can be related to the model of easy-axis ferromagnetic in transversal external field $2\Upsilon_{\text{rot}}$. The Hamiltonian (480) resembles the Hamiltonian (451) in outward appearance, and this means that we can reduce the problem to the previous one. However, we are interested in the explicit form of parameters Ω_{rot} and U_{rot}. For this purpose we should start from the appropriate classical Hamiltonian that describes the motion of an oriental defect in the hydrogen bonded chain [325]:

$$H = \frac{1}{2} \sum_{l} mr^2 \dot{\vartheta}_l^2 + \chi(\vartheta_{l+1} - \vartheta_l)^2 + U[1 - (\vartheta_l/\vartheta_0)^2]^2 \tag{481}$$

where ϑ_l is the angle of orientation of lth A—H group, as illustrated in Fig. 29; $\pm\vartheta_0$ are the angles that conform to the two different equilibrium orientations of the A—H group in the chain; m is the reduced mass of a proton in the A—H dipole; r is the length of the aforementioned dipole; U denotes the barrier height between the two equilibrium orientations of the dipole; and χ is the constant that characterizes the interaction of neighbor dipoles. Quantization of the Hamiltonian (481) by the scheme described above results in the Hamiltonian form (480).

Thus starting from the simplest pseudo-spin model of proton dynamics in the hydrogen bond, we have studied a possibility of spontaneous tunnel oscillations of the polarization of a short hydrogen-bonded chain. The phenomenon can be affected by two reasons: (a) the coherent motion of protons along the hydrogen bonds and (b) the coherent motion of protons around heavy backbone atoms.

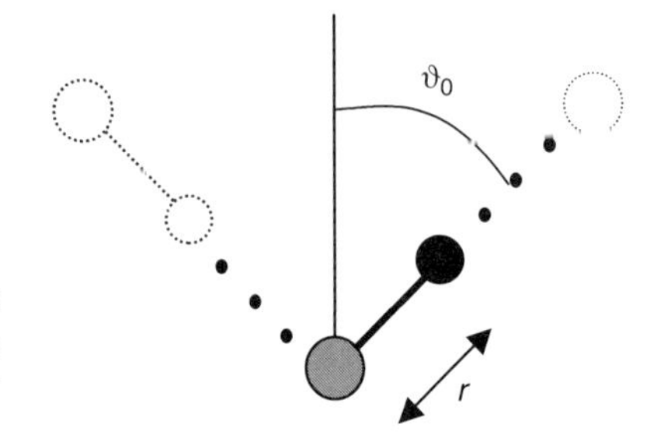

Figure 29. Orientation of the group A—H in the hydrogen-bonded chain. The angle $\pm\vartheta_0$ corresponds to two different equilibrium orientations of the A—H group.

The phenomenon is a typical example of the mesoscopic quantum coherence, which has been observed in small magnetic grains [305,309,328,329,332,335] and mesoscopic magnetic molecules [307,308]. Besides, a large proton polarizability of the hydrogen-bonded chain revealed by Zundel and co-workers [6] is the direct empirical corroboration of the coherent proton motion in the chain. The frequency of the coherent tunnel motion can be close to that of the tunnel motion of a proton in the single hydrogen bond, 5×10^{12} s^{-1}. Experimental estimation of the tunneling rate of a hydrogen-bonded chain, $\tau_{exper} \geq 10^{12}$ s^{-1}, was achieved by Zundel and collaborators [249,288,289]. The maximum value of the tunneling rate calculated in the framework of our model is also about $\tau \sim 10^{12}$ s^{-1}. The value of τ significantly depends on the number N of hydrogen bonds in the chain. For instance, assigning typical numerical values to the parameters $U/\hbar \sim 8 \times 10^3$ cm^{-1}, $\Omega \simeq 50$ to 200 cm^{-1}, and $N = 3$, we obtain the following estimate from expression (469): $\tau \sim 10^5$ to 10^{11} s^{-1}.

C. Can Coherent Tunneling of Heavy Particles Be More Probable than That of Light Particles? The Role of Proton–Phonon Coupling

The tunneling rate (469) as a function of the tunnel integral is specified by a nonmonotone behavior (Fig. 27). Such a dependence of the probability of tunneling of the chain polarization can produce an interesting effect, namely, that the coherent tunneling of heavier particles becomes greater than that of light particles. It seems reasonable that only a part of the total energy—namely, the kinetic energy, which induces the resonance tunnel integral J—depends on the particle's mass ($J \propto \exp(-\sqrt{m}\ldots)$). However, the smaller the mass of a particle, the greater its mobility—that is, greater the value of J. Light particles are more sensitive to fluctuations, which are able to destroy the strong correlation between particles. That is why, in the case of the lightest particles, when $J \sim U$, fluctuations will strongly drop the probability of coherent tunnel transitions. On the other hand, the mobility of heavy particles is rather small, which should result in a low frequency of coherent transitions. Thus, the probability of coherent tunnel dynamics tends virtually to zero for both the lightest and the heaviest particles and, consequently, should have a maximum in an intermediate range, giving rise to a peculiar isotope effect. A similar phenomenon may appear due to the potential coupling of protons with local vibrations.

Since dissipation can bring about the degradation of the tunneling rate, we shall investigate a possible coupling of protons with backbone atoms. In some instances, for example, in the case of a mesoscopic magnetic molecule $Mn_{12}O_{12}$, the coherent spin tunneling occurs with exchange of phonons [336]. In the case of the hydrogen-bonded chain, we can take into account the interaction between protons and acoustic vibrations of hydrogen bonds, which modulate the distance $A \cdots A$ [320]. The modulations alter the anisotropy energy (see, e.g., Ref. 96).

The interaction potential between the protons and the atom vibrations can be written as [323]

$$H_{\text{p-ph}} = \sum_{l} V(x_l)\rho_l, \qquad V(x_l) = \chi(x_l^2 - x_0^2) \tag{482}$$

where χ is the coupling constant, ρ_l is the additional stretching of the lth hydrogen bond, x_l is the proton coordinate with relation to the center of the two-well potential, and $\pm x_0$ are the coordinates of the minima of the wells. In the mean field approximation the Hamiltonian of the spin–phonon system can be written as

$$\begin{aligned} H = & -\hbar\Omega \sum_{l} S_l^x - U \sum_{l} S_{l-1}^z S_l^z + \sum_{q} \hbar\omega_q \hat{b}_q^+ \hat{b}_q \\ & + \sum_{l;q} (2V_{RL} S_l^x + V_{RR}) \tau_{q,l} (\hat{b}_{-q}^+ + \hat{b}_q) \end{aligned} \tag{483}$$

where $\hat{b}_q^+ (\hat{b}_q)$ is the Bose operator of creation (annihilation) of a phonon with the wavenumber q and the frequency ω_q:

$$\begin{aligned} V_{RL} &= \langle \psi_R | V(x) | \psi_L \rangle \\ V_{RR} &= \langle \psi_R | V(x) | \psi_L \rangle \end{aligned} \tag{484}$$

are the matrix elements of the interaction potential $V(x)$ (482); here $|\psi_R\rangle$, $|\psi_L\rangle$ are the wave functions of a proton localized in the right and the left potential well, respectively. The interaction parameter $\tau_{q,l}$ is expressed as

$$\tau_{q,l} = \sqrt{\hbar / 2MN\omega_q} \exp(iglq) \tag{485}$$

where g is the chain constant, M is the atom (or heavy ion) mass, and N is the number of hydrogen bonds in the chain.

The effective anisotropy energy of the chain takes the form [320]

$$E^{\text{eff}} = -\hbar(\Omega + \Omega_0) S^x - U(S^z)^2 - B(S^x)^2 \tag{486}$$

where

$$\hbar\Omega_0 = 2\frac{V_{RR} V_{RL}}{M\omega_0^2}, \qquad B = 2\frac{V_{RL}^2}{M\omega_0^2} \tag{487}$$

The energy (486) can be rewritten in the spherical coordinates

$$E^{\text{eff}}(\theta,\phi) = (U-B)S^2(\sin\theta - \sin\theta_*)^2 + BS^2(1-\cos^2\phi)\sin^2\theta + 2(U-B)S^2(1-\cos\phi)\sin\theta\sin\theta \tag{488}$$

where

$$\sin\theta_* = \frac{\hbar(\Omega+\Omega_0)}{2S(U-B)} = \sin\theta_0 \frac{1+\Omega_0/\Omega}{1-B/U} \tag{489}$$

and $\sin\theta_0$ is defined in expression (461).

The spin–phonon interaction renormalizes the tunneling frequency of a proton, $\Omega \to \Omega + \Omega_0$ (the sign of the Ω_0 can be both positive and negative), which, in turn, changes the interaction between the spins. The effective anisotropy energy (488) has two minima: the first one at $\phi = 0$, $\theta = \theta_*$ and the second one at $\phi = 0$, $\theta = \pi - \theta_*$. These two solutions determine the two possible equilibrium directions of the chain polarization at the same value of the energy. We may set $E^{\text{eff}} = 0$ along the aforementioned directions.

The energy conservation law makes it possible to obtain the following expression for the classical path of the polarization:

$$\cos\phi = \frac{\sqrt{1+s}\sqrt{s\sin^2\theta + \sin^2\theta_*} - \sin\theta_*}{s\sin\theta} \tag{490}$$

where $s = B/(U-B)$. Expression (490) is reduced to expression (465) if one puts $B,\ \Omega_0 \to 0$. Combining (490) with the equation of motion (463), we get the instanton that moves from θ_* to $\pi - \theta_*$:

$$\cos\theta(\tau) = \cos\theta_* \tanh(\omega_*\tau)\frac{\cosh(\omega_*\tau)}{\cosh(\omega_*\tau)+\eta} \tag{491}$$

where

$$\eta = \sqrt{\frac{(1+s)\sin^2\theta_*}{s+\sin^2\theta_*}} \quad \text{and} \quad \omega_* = (\Omega+\Omega_0)\sqrt{1+s}\cot\theta_* \tag{492}$$

The instanton obtained corresponds to the subbarrier path of the tunnel switching of the vector **S** moving between the two equilibrium directions, defined by the anisotropy energy (488). The imaginary-time action associated with the solution obtained takes the form

$$\frac{\mathcal{S}_E^{\text{eff}}}{\hbar} = 2SN\left[-\delta^{-1/2}\arcsin\left(\frac{\delta^{1/2}}{(1+\delta)^{1/2}}\cos\theta_*\right) + \frac{1}{2}\ln\frac{1+\sqrt{1-B/U}\cos\theta_*}{1-\sqrt{1-B/U}\cos\theta_*}\right] \tag{493}$$

where

$$\delta = \frac{1}{\sin^2 \theta_0} \frac{B}{U} \left(1 - \frac{B}{U}\right) \tag{494}$$

The behavior of the normalized action $\mathcal{S}_E^{\text{eff}}/\mathcal{S}_E$ as a function of the ratio Ω_0/Ω at different values of B/U is shown in Fig. 30. The function $\mathcal{S}_E$ determined in expression (468) is the imaginary-time action that has been obtained without an influence of phonons on the tunneling of the polarization. In general case, the values of Ω_0 and B cannot be treated as independent parameters; the smaller the value of Ω_0, the smaller that of B. Assuming that inequalities $\Omega_0/\Omega \ll 1$ and $B/(U \sin^2 \theta_0) \ll 1$ hold, the action $\mathcal{S}_E^{\text{eff}}$ can be rewritten in the form

$$\mathcal{S}_E^{\text{eff}} \approx \mathcal{S}_E - 2\hbar SN \left[\frac{\Omega_0}{\Omega} + \frac{B}{U}\left(1 + \frac{1}{6}\cot^2 \theta_0\right)\right] \tag{495}$$

It is obvious from expression (495) that the chain vibrations effect a decrease of the tunneling rate when $\hbar\Omega_0 \geq -B/3 \sin \theta_0$ (note that $\sin \theta_0 \ll 1$; for typical material parameters we have $\sin \theta_0 \sim 10^{-2}$–$10^{-3}$). In particular, expression

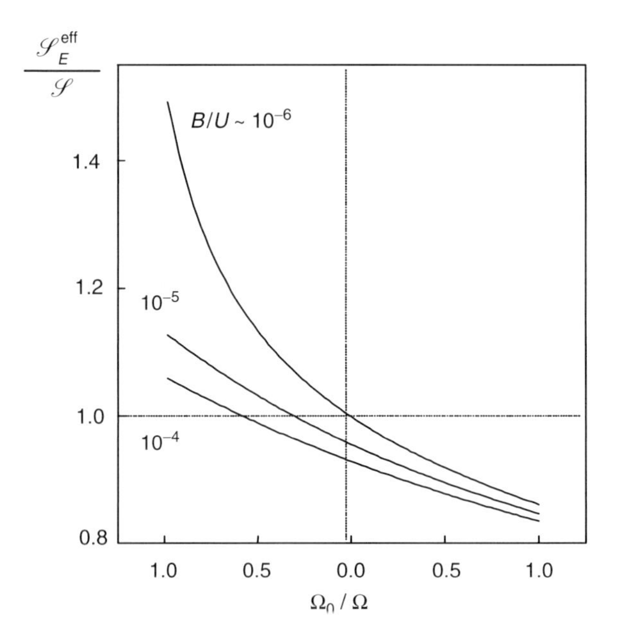

Figure 30. Role of the proton–phonon interaction in the coherent tunnel repolarization of a short chain. $\mathcal{S}_E^{\text{eff}}$ is the Euclidean renormalized action of phonons, and $\mathcal{S}_E$ is their nonrenormalized action. The bond vibrations can effectively decrease the local proton tunneling ($\Omega_0/\Omega < 0$) and at the same time increase the coherent tunneling of protons ($\mathcal{S}_E^{\text{eff}} < \mathcal{S}_E$).

(489) shows that the distance between the energy minima is reduced ($\sin\theta_* \geq \sin\theta_0$) if the spin–phonon interaction is accounted for. As this takes place, the barrier height between the minima is dropped on the value of

$$\Delta E(\Omega_0, B) = S^2 N(U - B)(1 - \sin\theta_*)^2 \leq S^2 NU(1 - \sin\theta_0)^2 = \Delta E$$

While B is always positive, Ω_0 can change in sign. The sign of Ω_0 depends on the type of the wave function $\psi_R(x)(\psi_L(x))$ of a proton localized in the right (left) potential well. The reason for the sign change can be roughly understood if we analyze the proton–phonon potential (482). The behavior of values $(x^2 - x_0^2)$, $\psi_R(x)\psi_L(x)$, and $\psi_R^2(x)$ as functions of x is schematically shown in Fig. 31. It is easily seen that the matrix element $V_{RL} = \langle\psi_R|x^2 - x_0^2|\psi_L\rangle$ is always negative, while $V_{RR} = \langle\psi_R|x^2 - x_0^2|\psi_R\rangle$ changes in sign: $V_{RR} < 0$ when the maximum of $\psi_R^2(x)$ is close to $x = 0$ and $V_{RR} > 0$ when the function $\psi_R^2(x)$ is shifted toward the right to the well minimum x_0. $\Omega_0 \leq 0$ when V_{RR} and V_{RL} are not the same sign. Therefore, the introduction of phonons decreases the tunneling frequency of individual protons; however, the coherent tunneling frequency can increase as it follows from the action (495).

Thus, the proton–phonon coupling being inserted into the initial Hamiltonian is able to suppress or enlarge the coherent tunnel repolarization of the chain. The realization of this or that option depends not only on the spin–phonon interaction, but also on the form of the two-well potential of the hydrogen bond.

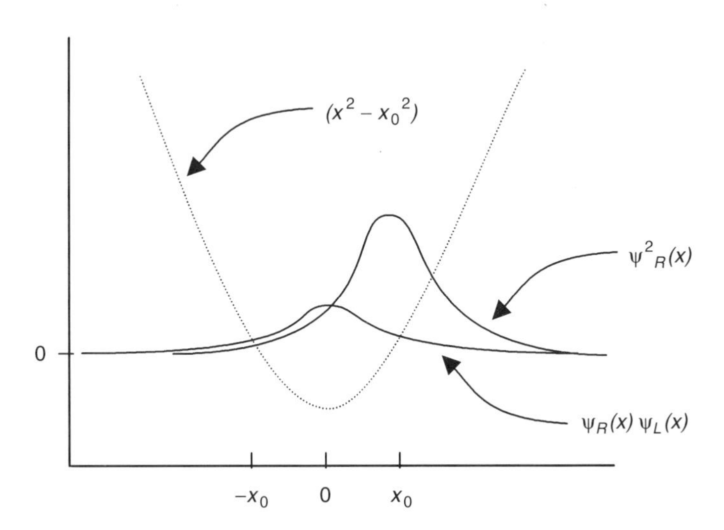

Figure 31. Schematic behavior of the functions $\psi_l(x)$, $\psi_r(x)$, $\psi_R^2(x)$, and $(x^2 - x_0^2)$, where $\psi_L(x)$ and $\psi_R(x)$ are wave functions of a proton localized in the left and right wells of the two-well potential, respectively; $\pm x_0$ are coordinates of the minimums of the right and the left wells.

When protons are tightly bound with the backbone ions, phonons will reduce the coherent tunnel rate. By contrast, when protons are localized near the potential barriers, the coherent tunnel rate will increase.

VII. UNUSUAL PROPERTIES OF AQUEOUS SYSTEMS

A. Organization and Thermodynamic Features of Degassed Aqueous Systems

The remarkable treatise by Eisenberg and Kauzman [139], early computer simulations based on realistic intermolecular potentials like Rahman's and Stillinger's [337,338], and the detailed random network model by Sceats and Rice [339], which took into account important peculiarities of the intermolecular interaction, allowed a significant advance in the study of water, which is believed to be a very nonordinary substance. A detailed quantitative theory of the structure of water rests on the model of a continuous tetrahedrally coordinate network of hydrogen bonds [339–342] (see also Ref. 343). The structural features of the model of a disordered network are specified in the description of the short-range order of liquid water. Changes in the potential energy of OH oscillators under the effect of the hydrogen bonds are interpreted as a result of action of a certain potential field: the potential of the disordered network. It depends only on the water molecules of the instantaneous configuration of the water molecules. In other words, it is assumed that the network of hydrogen bonds is quasi-static within the time interval of approximately 10^{-11} s (so-called *V* structure). The model of the disordered network leaves out of account the existence of large fluctuations brought about by cooperative motion of the water molecules, though such cooperative fluctuations, as was emphasized by Sceats and Rice, [340] must exist.

It has recently been pointed out by Rønne et al. [344] that "the structure and dynamics of liquid water constitute a central theme in contemporary natural science [345–353]." Modern theoretical considerations are aimed at (a) a detailed description of an electronic structure model of hydrogen bonding, applied to water molecules (see, e.g., Ref. 354), (b) models that involve a certain critical temperature where the thermodynamic response functions of water diverge (see, e.g., Ref. 353), and (c) models that presuppose a coexistence between two liquid phases [344]: a low-density liquid phase at the low-pressure side and a high-density liquid phase at the high-pressure side (see also Refs. 355–357).

Meanwhile, experimentalists have measured and analyzed, in particular, the dielectric response of liquid water in the frequency range from 10 up to 1000 cm^{-1} [344,358]. An analysis of the dielectric spectra of water shows the availability of complex permittivity in the microwave region and two absorption bands in the far-infrared region (the maxima are recorded at 200 and 700 cm^{-1}). Generally, two-component models provide a simple way of accounting for many thermodynamic anomalies of liquid water [344]. The major results obtained are the following [344]: The dielectric relaxation is successfully represented by a biexponential model with a fast (<300 fs) and a slow (>2ps) decay time; the slow decay time is consistent with structural relaxation of water; and the temperature dependence of the slow relaxation time allows the modeling from a singularity point at 228 K.

Below we would like to state results [359,360] obtained on samples of bidistillated water (and exceptionally pure water obtained by means of ionic gum), which were partly degassed. These results correlate very well with the recent study on the two-component water model, the existence of a singularity temperature point, and the existence of thermodynamic anomalies of water.

In 1987, Zelepukhin and Zelepukhin [361,362] established that the removal of part of the gases dissolved in water under normal temperature and pressure conditions change biological activity of the water. Degassed water is absorbed appreciably better by the leaves of plants; and when acting on biological objects, such water stimulates their respiration and enzymatic activity. The stimulating effect manifests itself even if only a few percent of the gases contained in the aqueous system are removed from it (under normal conditions, water contains approximately 30 mg/liter of air gases). In order to elucidate the biophysical and physiological mechanisms of action of degassed water, some of its physical and electrophysical properties were studied [359,360].

The starting component was distilled water that reached equilibrium with the gases of the air (usually this takes 3 days). Degassed water was prepared in two modifications: by heating to 90°C and subsequent cooling down to 20°C (a flask was cooling with running water), or by boiling for 30 minutes and subsequent cooling down to 20°C. Different modifications of water contained the following quantities of oxygen at 20°C: the starting (equilibrium) water, 9.05 mg/liter; water degassed at 90°C, 5.2 mg/liter; water degassed by boiling for 30 minutes, 2–3 mg/liter.

1. *Experimental Results*

Since the resonance bandwidth in the proton nuclear magnetic resonance (^{1}H NMR) spectrum is in inverse dependence on the mobility of the molecules [363], this technique can provide reliable information on the degree of structurization of different modifications of water. The width of the ^{1}H NMR spectrum lines for different modifications of water was measured with a high-resolution spectrometer (Bs-467, Tesla). The resolving power of the instrument was 2×10^{-8}, and the sensitivity expressed by the signal-to-noise ratio for 1% C_6H_6-CH_2CH_3 is 100:1 at the working frequency of 60 MHz. The averaged results are presented in Table I. As is seen in Table I, the samples of the water degassed at 90°C have a

TABLE I
Magnetic Resonance Studies

Modification of Water	Bandwidth (Hz)	The Experimental Error
Equilibrium (control)	3.6	0.18
Degassed (at 90°C)	4.3	0.19
Degassed (by boiling for 30 minutes)	3.5	0.18

reliable greater width of the resonance absorption band, approximately by 20%, compared with the equilibrium water. Consequently, the mobility of the H_2O molecules of this modification of water is lower by the same percentage; hence, structurization in this modification is more pronounced. There are some grounds to suppose that water subjected to prolonged boiling features a somewhat smaller structurization than the equilibrium water (the measured error did not allow one to make a more definite statement).

Water degassed at 90°C with the following cooling was called degassed structural water. Water subjected to prolonged boiling and subsequently cooled was called degassed water with disordered structure. We will use these definitions below.

The optical density of the aforementioned water modifications was measured in the ultraviolet region with the help of a spectrometer at 188.6, 189, and 190 nm wavelengths. At these wavelengths the difference in the optical density of the degassed and equilibrium water proved to be maximum. The optical density was measured in absolute units on a control sample (equilibrium water) and on a test samples in succession. The results are presented in Table II. As can easily be seen from Table II, the optical density of degassed structural water decreases with certainty in the ultraviolet region. When degassed water was kept in thermostat in contact with air for 2–3 days, the optical density of the samples gradually approaches the values of equilibrium water.

TABLE II
Optical Density

	Unit of Optical Density		
Wavelength (nm)	Equilibrium Water (Control)	Degassed Water (at 90°C)	Degassed Water with Disordered Structure (Boiled for 30 minutes)
188.6	0.218	0.200	0.186
189.0	0.189	0.174	0.158
190.0	0.130	0.117	0.104

The electrical conductivity of the water samples was measured with the help of a slide wire bridge on D.C. in an electrolytic cell with platinized electrodes. The value of the electrical conductivity of degassed water was approximately equal to that of equilibrium water (the difference is not certain), 2.49×10^{-7} S/cm.

The pH value and the redox potential ($E_{\text{r.p.}}$, in mV) were measured as well and then the values of $E_{\text{r.p.}}$ was converted into the hydrogen index. The pH value in degassed water increases reliably compared with the control sample (Table III). In degassed structured water these changes are insignificant, whereas in degassed water with disordered structure they are significant (10%). When degassed water is kept in closed glass flask for 3 days, its pH remains higher than that of equilibrium water. When degassed water is kept for 3 days in an open glass flask, elevated pH values are preserved in structured water, whereas in water with a disordered structure the pH value practically reaches equilibrium. The growth of the water pH after degassing may be connected with the removal of the CO_2 gas; therefore, the connection between the changes of the water pH and the changes of the structure of the water may only be indirect.

The redox potential $E_{\text{r.p.}}$ of degassed water reliably declines (Table III). The greatest difference is observed in degassed water with a disordered structure. The lowered value of the redox potential is preserved in degassed water also on the third day of its remaining in a closed or open vessel.

Note that the change of the redox potential in degassed water is indicative of a change in its thermodynamic properties. There is a direct relationship between the value of $E_{\text{r.p.}}$ and the change of the free energy of the system studied [364]:

$$\Delta G = -nFE_{\text{r.p.}} \tag{496}$$

where n is the number of the electrons transported in the redox reaction, and F is the Faraday number. Calculations in accordance with formula (496) show that the value of free electrochemical energy of equilibrium water at 19°C is equal to 13.4 kJ/mol, that of degassed structural water is 12.73 kJ/mol, and that of degassed water with a disordered structure is 10.22 kJ/mol.

TABLE III
pH and Redox Potential

	Modification of Water		
Index	Equilibrium	Degassed, Structured	Degassed with Disordered Structure
pH	5.29	5.44	5.82
E, mV	340	322	307

Consequently, the change of the redox potential makes it possible to determine the change of the free energy, the thermodynamic state of water on transition from the equilibrium state to the activated one. These changes for degassed water are essential and reliable. It was the major sensitive test that the Zelepukhins exploited for an indication of biologically active water working in the area of crop production.

2. *Thermodynamics*

The conception of configuration (relaxation) contributions to the thermodynamic properties that are caused by structural changes of liquid water at different temperature and pressure is well known [301,304]. On the other hand, structural changes in water are caused by the change in the potential energy connected with interaction of the molecules—that is, with the change in the energy of hydrogen bonds.

Water subjected to experimental investigations always contains gases of the air. As a rule, the effect produced by this factor on the structure and properties of water is not taken into account. However, gases can be taken into consideration within the scope of thermodynamic solutions. Zelepukhins [359, 361] proposed to consider water that is kept in a thermostat and has reached equilibrium with the gases of the air at atmospheric pressure (usually this takes several days) as the standard state of liquid water. Kittel [365] and Pauling [366] employed such standardization of water for the thermal function and freezing point of water. From chemical thermodynamics it is known (see, e.g., Refs. 367 and 368) that the formation of solutions is accompanied by a reduction of free energy and is a spontaneous process. Therefore, when gases are removed from water, which is in equilibrium with the air, the free energy and thermodynamic activity of the water must increase. Hence, particular calculations of the thermodynamic parameters of degassed water should be performed just from this standpoint. The role of gases dissolved in the water should be regarded as the fundamental condition for the aqueous system to be in equilibrium with the environment.

The change of the free energy and the enthalpy are connected by the Gibbs–Helmholtz equation

$$\Delta G = \Delta H - T\Delta S \tag{497}$$

which is essential for the direction of chemical reactions. The chemical system can change its state spontaneously only if this change is accompanied by the negative quantity ΔG—that is, if reversible reactions do the work. The condition of chemical equilibrium of the system is $\Delta G = 0$; then from Eq. (497) follows the equation

$$\Delta H = T\Delta S \tag{498}$$

which relates the enthalpy factor ΔH and the entropy factor ΔS. According to the conception of Karapetiants [368], the change in the system enthalpy ΔH reflects, in the main, the tendency of molecular interactions to combine particles into aggregates (associates), whereas the change in the entropy ΔS reflects the opposite tendency toward chaotic disposition of the particles—that is, toward their disaggregation (disassociation). Thus, according to Karapetiants, these two tendencies compensate for each other in the state of chemical equilibrium defined by Eq. (498).

On the other hand, the tendency to association displayed by the molecules depends exclusively on the magnitude of their intermolecular interaction; and the measure of energy of intermolecular interaction is, apparently, the heat capacity C of the system, because it is just the heat capacity that characterizes the degree of heating of the substance. Therefore, it would be more appropriate to regard the product $C\Delta T$ as the associating index of the substance (for instance, at $p = \text{const}$, the increment of enthalpy $\Delta H = C_p \Delta T$). Then, since the entropy determines directly the degree of the system disorder, the product $S\Delta T$ can be referred to as disordering index. If we could write an equation that combines terms $C_{p(V)}\Delta T$ and $S\Delta T$, we would be able to examine a peculiar order/disorder of water system based on the weight of each of the said terms.

It should be noted that quite recently Langner and Zundel [369] have proposed an approach to the description of proton transfer equilibria in hydrogen bonds, which in some aspects is similar to that stated herein. They have treated the proton transfer equilibria $\text{AH} \cdots \text{B} \rightleftarrows \text{A}^- \cdots \text{H}^+\text{B}$ as a function of the $\Delta \text{p}K_\text{a}$—that is, the $\text{p}K_\text{a}$ of the base minus the $\text{p}K_\text{a}$ of the acid. They represented ΔH and ΔS as the sum of the intrinsic quantities ΔH_0 and ΔS_0, which describe the behavior of the systems studied in gas phase, and the external quantities ΔH_I and ΔS_I, which depicted the influence of environment. This allowed them to investigate the transition from $\Delta H = 0$ to $\Delta G = 0$ [note that ΔH and ΔG are linked by Eq. (497)] with increasing $\Delta \text{p}K_\text{a}$ and accounted for the reason of this effect. The effect appeared owing to the large negative interaction entropy term ΔS_I arising from the large order around the polar structure, which shifts the equilibria strongly to the left-hand side. In another system a single-minimum potential has been found; the proton potential has, on average, been symmetrical, whereas the proton is still largely on the left-hand side. In this case the large negative interaction entropy ΔS_I due to the order of the environment has also been drawn to explain the result (the proton is found at the acceptor B).

Coming back to our consideration, let us treat the Gibbs potential

$$G = H - TS \tag{499}$$

As it directly follows from expression (499), the total change of G has the form

$$\Delta G = \Delta H - S\Delta T - T\Delta S \tag{500}$$

Replacing ΔH for $C_p\Delta T$, we get instead of Eq. (500)

$$\Delta G = (C_p - S)\Delta T - T\Delta S \tag{501}$$

(it is assumed here that p = const, but we can write down an analogous equation for the process at V = const). The first summand in Eq. (501), that is,

$$\Delta G_{\text{in.in.}} = (C_p - S)\Delta T \tag{502}$$

can formally be called [359] the change of the Gibbs potential associated with intermolecular interaction (or the configuration potential), which should be taken into account at nonisothermic processes. Equation (502) is very convenient for the determining the degree of structurization of aqueous system caused by its heating and/or cooling. Indeed, $\Delta G_{\text{in.in.}}$ is directly connected with entropy S, but the residual entropy that is a part of the total entropy of the system studied is just a property of amorphous compounds.

It is easy to find numerical values of $\Delta G_{\text{in.in.}}$ from Eq. (502) by using the tabular data for C_p [370] and S [339,367,371,372]. The calculation performed in accordance with Eq. (502) has shown that for equilibrium water, $\Delta G_{\text{in.in.}}$ has positive values up to the temperature of 318 K (45°C), above which the sign of the potential change becomes negative. In other words, in the aqueous system, changes take place in the thermodynamic parameters of the intermolecular interaction (the second "melting" point, i.e. the melting point of the associates present in the liquid).

For equilibrium water, $\Delta G_{\text{in.in.}}$ turns to zero at 318 K, and at this temperature the curves of S and C_p intersect (Fig. 32). At the point of equilibrium obtained, the increment of the enthalpy, according to Eq. (498), is equal to

$$\Delta H|_{\text{eq.}} = T_{c0}\Delta S = 3.39\,\text{kJ/mol} \tag{503}$$

The absolute value of the equilibrium enthalpy at 318 K is given by

$$H|_{\text{eq.}} = \Delta H|_{\text{eq.}} \frac{T_{c0}}{\Delta T} - n\Delta H_{\text{m}} = 17.92\,\text{kJ/mol} \tag{504}$$

(the numerical value of ΔH_{m} was borrowed from Ref. 139). The value $H|_{\text{eq.}}$ exactly coincides with the value of the activation energy of self-diffusion for the molecules of water, employed in Ref. 373, and is close to the value of the hydrogen bond in water, equal to 18.84 kJ/mol, cited in Ref. 366.

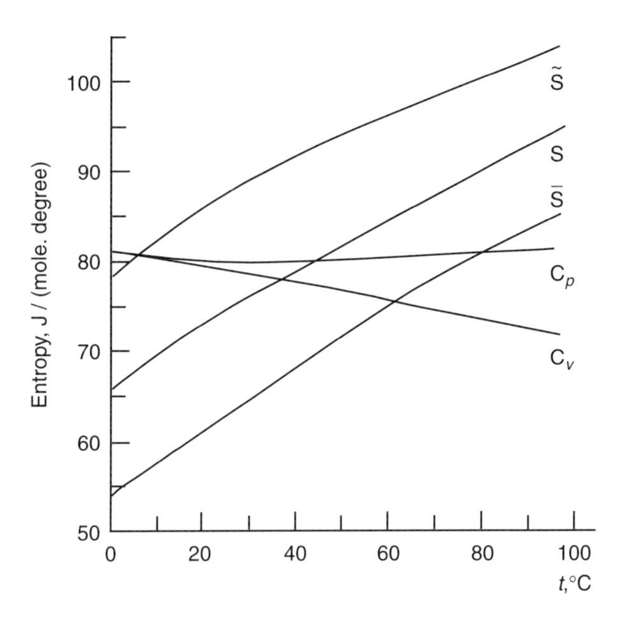

Figure 32. Entropy as a function of temperature for different modifications of water: equilibrium (S), degassed structured ($\bar{S}$), degassed with disordered structure ($\tilde{S}$). The heat capacities of water C_p and C_V as functions of T are plotted as well. (From Ref. 359.)

This approach enables the introduction of the concept of structurization of fluid—in particular, of water—with the help of the parameters C_p and S, whose numerical values for different liquids can be found in the chemical engineers' handbooks. The coefficient $\mathcal{K}$ of fluid structurization can be determined from Eq. (502); that is, putting $\Delta G_{\text{in.in.}} = 0$ we have

$$\mathcal{K} = C_p/S \tag{505}$$

For equilibrium water, $\mathcal{K} > 1$ up to $T_{c0} = 318$ K; at $T > T_{c0}$ the inequality $\mathcal{K} < 1$ takes place. Consequently, at $T < T_{c0}$, equilibrium water as a whole is structured (the enthalpy factor prevails over the entropy factor), whereas at $T > T_{c0}$ it is nonstructured (the entropy factor prevails over the enthalpy factor). For comparison, we shall cite the values of $\mathcal{K}$ for some liquids, calculated from the published reference data for C_p and S at 25°C [374]: 1.14 for heavy water; 0.815 for hydrogen peroxide; 0.19 for ammonia. At this temperature, $\mathcal{K} = 1.073$ for equilibrium water.

Let us temporarily introduce the notion of a certain "structurization potential" and apply it for the consideration of thermodynamic parameters of degassed water. This notion must be based on the main thermodynamic potentials and should include such peculiarities as strong intermolecular interaction, thermal expansion, and so on. In the past, theoretical [375] and experimental [376] investigations of the macrophysics properties of liquid water made it possible to establish considerable strengthening of water (by several orders of magnitudes!) in the case of specific degassing, namely, when large bubbles of gases and air present in the water were dissolved under the effect of applied pressure. A more compact arrangement of the water molecules upon application of pressure is pointed out also in monographs [377,378]. At normal conditions, incorporation of gases into the solvent disturbs the mutual ordering of the solvent molecules and is accompanied by increase of entropy [367]. Thus, on degassing of the aqueous system, its entropy must reduce in the main, and the intermolecular interaction must increase.

In the general case, not all the energy of the system during an isochoric process can be transformed into work ($-\Delta G$). Part of the energy ("bound" energy, proportional to the residual entropy δS) is not used [368]; then we may write

$$\delta S \Delta T = V \Delta p - (-\Delta G_{\text{in.in.}}) \tag{506}$$

Let us denote the entropy of the maximum-degassed aqueous system by $\bar{S}$. Then, insofar as the residual entropy δS is greater than 0, in the first, linear approximation we have $\delta S \simeq S - \bar{S}$. In the case of an isochoric process the work $V\Delta p$ can be represented as

$$V \Delta p = T \Delta S - \Delta U \tag{507}$$

In the linear approximation $\Delta \bar{S} = \Delta S$; in the same approximation at $V = \text{const}$ the change of internal energy can be estimated as $\Delta U = C_V \Delta T$. Substituting expressions (502) and (507) with regard to explicit forms of δS and ΔU into Eq. (506), we derive the expression for the entropy of the maximum-degassed associated aqueous system

$$\bar{S} \simeq 2S - C_p + C_V - \Delta S \frac{T}{\Delta T} \tag{508}$$

From here on we shall call the maximum-associated aqueous system maximum-structured, whereas the "bound" energy $\delta S \Delta T$ can be spoken of as of "structurization potential" $\Delta\alpha$. The calculation of the value $\Delta\alpha$ from formula (506) shows that $\Delta\alpha \simeq R\Delta T$, where R is the gas constant. The calculation of the entropy from formula (508) yields the estimate $\bar{S} \simeq S - R$ (the $\bar{S}$ versus T plot is presented in Fig. 32).

The change of the Gibbs "potential" $\Delta G_{\text{in.in.}}$ in the discussed modification of water can be written in a form analogous to expression (441):

$$\Delta G_{\text{in.in.}} = (C_p - \bar{S})\Delta T \tag{509}$$

Since $\bar{S} < S$ and $\bar{C}_p = C_p$, the structurization coefficient (505) for the given aqueous system at one and the same temperature will be greater than the equilibrium system.

The plots of $\bar{S}$ and C_p as functions of temperature intersect at 85°C (see Fig. 32). Thus, at $T_c = 358$ K we have $\Delta\bar{G}_{\text{in.in.}} = 0$; consequently, in this state the equilibrium value of the enthalpy increment [see (498)] is expressed as

$$\Delta\bar{H}|_{\text{eq.}} = T_c\Delta\bar{S} = 6.41 \text{ kJ/mol} \tag{510}$$

The absolute value of equilibrium enthalpy at 358 K is given by

$$\bar{H}|_{\text{eq.}} = H|_{\text{eq.}} + (\Delta\bar{H}|_{\text{eq.}} - \Delta H|_{\text{eq.}}) \tag{511}$$

where $\Delta H|_{\text{eq.}}$ and $H|_{\text{eq.}}$ are determined in expressions (503) and (504), respectively. Substituting the numerical values into (511), we get

$$\bar{H}|_{\text{eq.}} = 20.93 \text{ kJ/mol} \tag{512}$$

Thus the enthalpy of intermolecular bonding—that is, energy of the hydrogen bond in the equilibrium state in the maximum-structured water—is 3 kJ/mol greater than in the case of equilibrium water.

In the case of degassing carried out by way of long-term boiling, degassed water becomes disordered, hydrogen bonds become deformed, and the structural phase must be nearly absent. Water activated in such a manner has an enhanced dissolvability with respect to different salts (see also, e.g., Ref. 379) and is characterized by higher Gibbs potential. Consequently, we can presume that compared with equilibrium water, the entropy increases: $S \rightarrow \tilde{S} > S$ in degassed water with a disordered structure.

The maximum possible increase of entropy can be arrived at from the energy conservation law:

$$V\Delta p - (-\Delta\tilde{G}_{\text{in.in.}}) = 0 \tag{513}$$

where the change of the Gibbs "potential" of intermolecular interaction for the given modification of water is given by

$$\Delta\tilde{G}_{\text{in.in.}} = (C_p - \tilde{S})\Delta T \tag{514}$$

Let us put $\tilde{S} = S + \delta S$ and substitute this expression into Eq. (514). Then combining Eqs. (514) and (513), and taking Eq. (507) into account, we obtain the expression for estimating the entropy of a maximum-degassed disordered aqueous system:

$$\tilde{S} \simeq \Delta S \frac{T}{\Delta T} + C_p - C_V \tag{515}$$

The calculation shows that $\tilde{S} \simeq S + R$. The structurization coefficient for the given water $\mathscr{K} = 1$ at about 6°C. At this temperature, $T_c = 279$ K, the curves $\tilde{S}$ and C_p in Fig. 32 intersect and

$$\Delta\tilde{H}|_{\text{eq.}} = T_c \Delta\tilde{S} = 0.64\,\text{kJ/mol} \tag{516}$$

correspondingly, $\tilde{H}|_{\text{eq}} = 15.16$ kJ/mol, i.e., the energy of hydrogen bond of water with a disordered structure is 2.76 kJ/mol smaller than in the case of equilibrium water.

To better understand the changes taking place in the intermolecular interactions in the aqueous system, Table IV gives the calculated values of the coefficient $\mathscr{K}$ (the index of a peculiar structurization) for different modifications of water. The value $\mathscr{K}$ for equilibrium water was calculated from formula (505); for degassed structural water it was calculated from the formula $\mathscr{K} = C_p/\bar{S}$; and for degassed water with disordered structure, it was calculated from the formula $\mathscr{K} = C_p/\tilde{S}$.

It is nearly impossible to degas water completely and preserve its structure. In order to evaluate experimentally the degree of structurization of the aqueous system and, consequently, evaluate the parameters of intermolecular interaction

TABLE IV
Coefficient of Water Structurization $\mathscr{K}$

Temperature of Water (°C)	Coefficient $\mathscr{K}$			
	Equilibrium Water	Degassed, Structured Water	Snowmelt Water	Degassed Water with Disordered Structure
0	1.19	1.41	1.29	1.03
10	1.14	1.26	1.22	0.98
20	1.09	1.27	1.16	0.94
30	1.05	1.22	1.12	0.91
40	1.02	1.16	1.08	0.89
50	0.99	1.11	1.04	0.88
60	0.96	1.07	1.00	0.86
80	0.91	1.02	0.95	0.83
100	0.87	0.97	0.91	0.80

as well, the characteristics of the heat of evaporation were employed [359]. It was comparatively easy to investigate artificially prepared snowmelt water (samples contained 6.9 mg/liter oxygen at 20°C). It was found that for this modification of water, $\Delta G_{\text{in.in.}}^{(\text{exper})} = 0$ at 60°C; in this situation, $\Delta \bar{H}^{(\text{exper})}|_{\text{eq}} =$ 4.69 kJ/mol and $\bar{H}^{(\text{exper})}|_{\text{eq}} = 19.26$ kJ/mol. It will readily be seen that the enthalpy in snowmelt water is greater by approximately 2 kJ/mol than that in equilibrium water. The values of the coefficient $\mathscr{K}$ for snowmelt water are listed in Table IV; the equality $\mathscr{K} = 1$ is attained at about 60°C. This is a direct indication that the degree of structurization of snowmelt water is greater than that in equilibrium water.

3. *Organization of Water System*

Recent studies partly mentioned in the introduction to this section adhere chiefly to two liquid phase models, one of which is similar to the V-structure that appeared in the coordinate network of hydrogen bonds in Refs. 139 and 339. Spectral investigation (see, e.g., Refs. 341,342, and 380) allowed one to subdivide tentatively the diversity of hydrogen bonds in a single network into an ensemble of strong, approximately tetrahedrally directed hydrogen bonds and an ensemble of weak, appreciably disordered hydrogen bonds. Quite recently, Kaivarainen [381] has constructed a quantitative theory of liquid state, mesoscopic molecular Bose condensation in water and ice in the form of coherent clusters. Mechanisms of the first- and second-order phase transitions, related to such clusters formation, their assembly, and symmetry change, have been suggested as a consequence of computer calculation (300 parameters have been taken into account). His theory, indeed, unifies dynamics and thermodynamics on microscopic, mesoscopic, and macroscopic scales. Chaplin [382] is an advocate of the icosahedron structure of water; he presents an icosahedral cluster model and has made all explanations for 37 anomalies of water. In the past, a peculiar kind of clusters in the single network of hydrogen bonds of water was also recorded (by X-ray technique [383] and neutron scattering technique [384,385]). Very interesting results were obtained by Gordeev and Khaidarov [385]: They detected density fluctuations, supposedly globules to 3 nm in size comprising up to 10^3 molecules, and noted that the bulk water could be regarded as a polycrystal-ferroelectric with a domain structure in time $\sim$100 ps (see also Luck [386]).

Thus many investigations support an idea about ordered regions in water. Since the water–gaseous solution is characterized by the temperature of structurization T_c discussed above, we can hypothesize that the water network features an order parameter. As a first crude approximation, in Ref. 359 an ordering of OH groups of water molecules was treated in the framework of a simplified model in which water molecules were located in knots of the Ising lattice. The model was based on the Blinc's formalism stated briefly in

Section II.D above. The pseudo-spin operator S_l^z [see formulas (68) and (72)] corresponded to two possible projections of the coordinate of the ith proton onto the z axis. The order parameter obeyed the equation

$$\langle S^z \rangle = \frac{1}{2} \tanh \frac{\langle S^z \rangle (\mathcal{J} + \delta\varepsilon)}{2k_B T} \tag{517}$$

where $2\mathcal{J}$ is the energy of cooperation of the protons and

$$2\delta\varepsilon = (\Delta\bar{H}|_{\text{eq.}} - \Delta\tilde{H}|_{\text{eq.}}) = 5.77\,\text{kJ/mol} \tag{518}$$

is the difference between the energy of hydrogen bond in the maximum-structured water (510) and maximum-nonstructured water (516). Tending $\langle S^z \rangle$ to zero, we obtain from Eq. (517) the sought temperature of structural (phase) transition of the protons from the ordered to the mismatched state:

$$T_c = (2\mathcal{J} + 2\delta\varepsilon)/8k_B \tag{519}$$

The temperature (519) then was successfully compared [359] (with accuracy of about 5%) with the temperature of structurization T_c obtained above for the considered modifications of water.

We will return to the problem of cluster formation in water in Section VIII.

B. Determination of Water Structure by Pulsed Nuclear Magnetic Resonance (NMR) Technique

To verify the results on the existence of the critical temperature T_c in the three different modifications of water, there were measured [360] spin-lattice relaxation rates for protons in water, assuming that the latter contains both tightly and loosely bound molecules.

The spin-lattice relaxation time T_1 as a function of temperature T in liquid water has been studied by many researchers [387–393], and in all the experiments the dependence $T_1(T)$ showed a distinct non-Arrhenius character. Other dynamic parameters also have a non-Arrhenius temperature dependence, and such a behavior can be explained by both discrete and continuous models of the water structure [394]. In the framework of these models the dynamics of separate water molecules is described by hopping and drift mechanisms of the molecule movement and by rotations of water molecules [360]. However, the cooperative effects during the self-diffusion and the dynamics of hydrogen bonds formation have not been practically considered.

The spin-lattice relaxation time was measured with an ISSH-2-13 coherent nuclear quadrupole resonance spectrometer-relaxometer equipped with a Tesla BS 488 electromagnet (magnetic field strength is 1.6×10^5 A/m) to realize the pulsed NMR regime. The probe pulse duration was several tens of microseconds and tat of the front no more than 0.5 μs. Owing to the use of the water sample volume of 0.6 cm^3, we achieved a signal-to-noise ratio on the order of 50. The signal from the sample was recorded and accumulated in a measuring computing unit. The data array involving the values of signal amplitudes and corresponding values of the pulse repetition rate was processed with a computer by the least squares method to determine spin-lattice relaxation time T_1.

The dependence of the reciprocal of the spin-lattice relaxation time T_1^{-1} on temperature T for water is generally approximated [390–394] by the sum of two exponents:

$$T_1^{-1} = A\exp(E_1/k_B T) + B\exp(E_2/k_B T) \tag{520}$$

where E_1 and E_2 are the activation energies of movement (rotational or translational) of water molecules in the low-temperature and high-temperature regions, respectively; A and B are constants or are functions of T. At the same time, the temperature dependence of T_1 can also be considered in the single-exponential approximation in the high-temperature ($T > 320$ K) and low-temperature ($T < 240$ K) regions. In this case the value of the activation energy at low temperature determined from the slope of dependence of $\ln T_1$ on T approaches the activation energy of ice (60 kJ/mol), whereas at high temperature it nears the energy of the hydrogen bond rupture (14.6 kJ/mol) [392].

Aside of the activation energies E_1 and E_2 characterizing the motion of water molecules in loosely and tightly bound structures (dominating in the high-temperature and low-temperature regions, respectively)—that is, the energy (E_1) of hydrogen bond rupture and the energy (E_2) corresponding to the continuous network of hydrogen bonds—we are interested in the contribution of each structure in the region of moderate temperatures. The study of the influence of the past history of water and of its treatment on the energetic characteristics is also of interest, because it is precisely the effect of the past history of the $T_1(T)$ that can explain the scatter of E_1 and E_2 values observed in different papers (see, e.g., Refs. 389 and 392). However, the use of the two-structure model complicates experimental data processing because one has to consider additionally the contribution from each structure; that is, the temperature dependence of coefficients A and B in Eq. (520) must be taken into account.

To process the non-Arrhenius temperature dependence of T_1, we use an approach different from that described by Eq. (520). The method, which is describing, required data in a wide temperature range (from $-30°$ to 180°C) and rather prolonged computer calculations. The temperature dependence $T_1(T)$ can be broken in a small temperature range (from 5° to 70°C) into two intervals divided by a temperature T_c and approximated in each interval by a single exponential (Fig. 33). In this case the value of the effective activation energy E_a

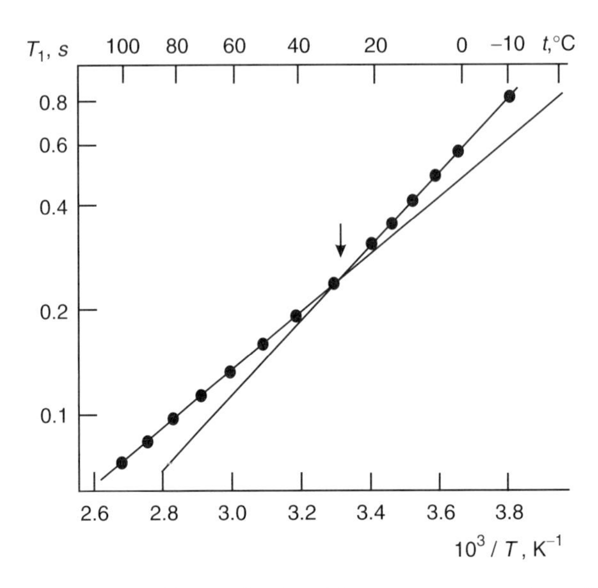

Figure 33. Single-exponential approximation of the temperature dependence of the spin-lattice relaxation rate in two temperature intervals. (From Ref. 360.)

in the region of $T < T_c$ is more than that in the region of $T > T_c$. The point $T = T_c$ plays the role of the critical temperature of water transition from the dominating tightly bound component ($T < T_c$) to the dominating loosely bound component ($T > T_c$).

The large value of E_a suggests that tightly bound water molecules greatly contribute to the spin-lattice relaxation rate in this temperature range. The closer the E_a value to $E_1 \simeq 14.6$ kJ/mol, the closer the water structure involved to the system with a single hydrogen bond per molecule. The greater the E_a, the greater the contribution of the tightly bound molecules that form a structure with an energy value E_a determined in the single-exponential approximation that allows one to estimate the effect of water treatment on the degree of molecule binding.

A glass ampoule 7 mm in diameter was approximately half-filled with water under study. The air was pumped from the ampoule by a vacuum pump, and then the ampoule was sealed and placed into a detector.

Measurements of the spin-lattice relaxation rates for protons in a freshly distilled water (taken from a flask of volume 1 liter) showed that $E_a \sim 19.5$ kJ/mol and $T_c = 284$ K (or $t_c = 11$°C). On standing in a flask at room temperature, the structuring of the water system increased. Even after several days, measurements showed a considerable increase in the values of E_a and T_c, whereas after 40 days these parameters reached $E_a = 21.9$ kJ/mol and $t_c = 52$°C (Fig. 34).

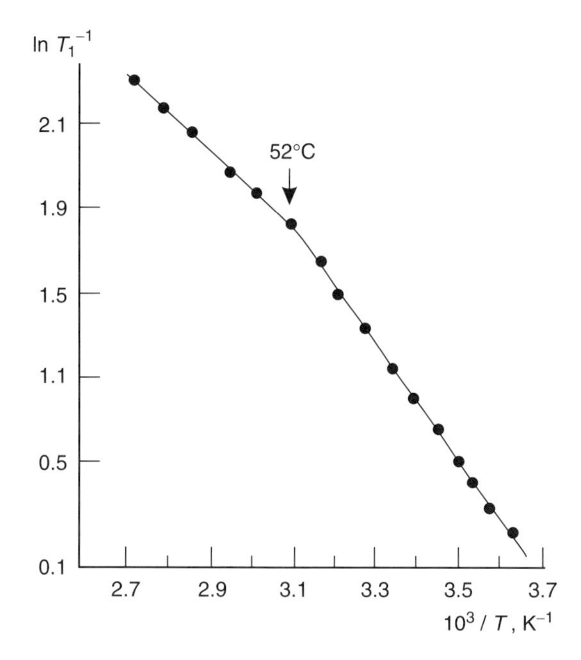

Figure 34. Temperature dependence of the spin-lattice relaxation rate for protons in water settled for 40 days at room temperature and exposed to the surrounding air. (From Ref. 360.)

When water taken from a flask had parameters $E_{a0} = 20.6$ kJ/mol and $t_{c0} = 20°C$, two extra portions of water were taken and the disordering in one of them and the structuring in the other were preformed. The value of the activation energy $\tilde{E}_a$ for the disordered water modification proved to be less than the initial one E_{a0}, namely, $\tilde{E}_a = 19.7$ kJ/mol (disordered by durational boiling), $\tilde{E}_a = 18.9$ (disordering by vacuum pumping), and $\tilde{E}_a = 15.5$ kJ/mol (disordering by passing air). The temperature $\tilde{t}_c$ for these disordered modifications was not determined because of the limitation of our method (it could not have fixed t_c below 10°C).

Water structuring results in an increase of the activation energy ($\bar{E}_a = 22$ kJ/mol) and of the transition temperature ($\bar{t}_c = 28°C$) compared to the initial values of E_{a0} and t_{c0}. Figure 35 shows the results of measurements of the spin-lattice relaxation rates for another series of samples.

In the first crude approximation the transition temperature has been described by Eq. (519). It is obvious that the value $8k_BT$ characterizes the degree of binding of water molecules at $T < T_c$. On the other hand, the energy E_a obtained in this chapter should determine the barrier height for hopping of water molecules during translational diffusion.

Table V presents the values of E_a and $8k_BT$ [see expression (519)] for different water modifications from the first series of samples. If the model [359]

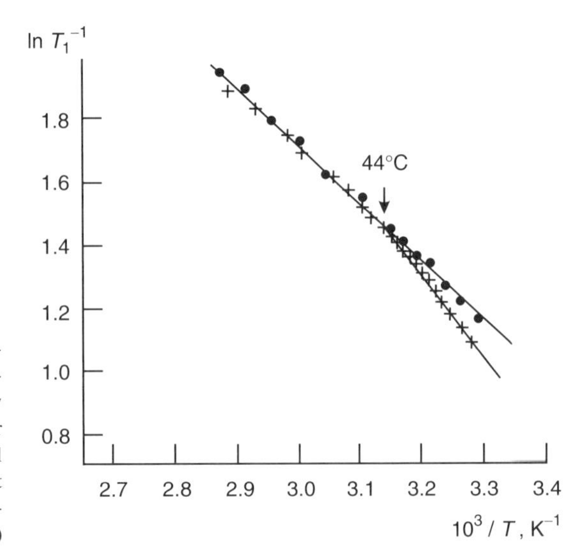

Figure 35. Temperature dependence of the spin-lattice relaxation rate for a freshly distilled water (•) and water allowed to stand in air for several days and thermally treated at 90°C—that is, for degassed structured water (+). (From Ref. 360.)

discussed in the previous subsection is assumed to be valid, inequality $E_a > 8\,k_B T$ can account for the fact that the self-diffusion of water molecule involves the surrounding groups of molecules. In particular, self-diffusion of this type in water was considered in the model of collective molecular motion [395,396]. The possibility of a diffuse random walk of groups of molecules (with a lifetime of about 100 ps) was conjectured in Ref. 385 (see also Ref. 386).

Thus, investigations of the spin-lattice relaxation rates in the water system showed directly that water by its nature is a metastable liquid. The degree of binding of its molecules depends essentially on the past history of the water system involved, and therefore in studies of dynamic parameters one has to take into account the metastable structural state of the water system. The water structure is considerably determined by atmospheric gases. Their removal from water without distortion of its structure results in the ordering of intermolecular interactions (water structuring), whereas bubbling with air, on the contrary,

TABLE V
Numerical Estimates of the Model Parameters

t_c(°C)	E_a kJ/mol	$8\,k_B T$ kJ/mol
52	21.9	21.6
11	19.5	18.8
20	20.6	19.5
28	22	20

weakens intermolecular interactions in water (water disordering). The water structure is also disordered after prolonged boiling. The pulsed nuclear magnetic resonance technique allowed the direct determination of the temperature T_c of peculiar structural transition in water. The activation energy E_a is also an important parameter. Its value may indicate the dominating contribution of one or another self-diffusion mechanism of water molecules in each specific modification of the water system.

C. Water-Dependent Switching in Continuous Metal Films

In moist surrounding air, sandwiched diodes Ag/BN/Si/Al [397] and planar diodes with continuous films of Ag [398] and Au [399] are characterized by a switching current that depends on the frequency and form of alternating voltage. For example, in Ref. 399 the voltage frequency was varied in the range from 0.1 to 100 Hz, the voltage reached 4.3 V, the strength of the applied field $\mathscr{E}$ is $\sim 10^4$ V/cm (possibly, $\mathscr{E}$ in the diode was high as a result of the microstructure of the films, i.e., small metallic inlands) [399], current density I is $\sim 10^6$ A/cm^2, and the curve $I(\mathscr{E})$ had the form $I \propto \mathscr{E}^{\alpha}$, where $0 < \alpha < 1$. The presence of the switching current is associated with the formation of a water layer on the film in the conditions of the moist surrounding space.

Earlier experiments [400,401] (see also Refs. 402 and 403) showed that hydrated colloidal particles are characterized by polarization whose nature is determined by spontaneous orientation of the polar molecules of water on the surface of particles leading to formation of the giant momentum **P**. The energy advantage of formation of **P** is shown in Ref. 404, which is in agreement with a detailed study of the dielectric anomalies in thin water layers conducted in Ref. 405. Spontaneous polarization of the monolayers of polar molecules is theoretically substantiated in Refs. [406,407].

Based on the known properties of layers of water molecules with regions of spontaneous polarization and taking into account the results stated in the section above, we have proposed [408] a detailed microscopic theory of switching current as a polarization surface current in the aforementioned regions (i.e., current of water domains).

In the experiments [397,398] the strength of the field applied to the films was low ($\mathscr{E} \leq 10^4$ V/cm) in comparison with the strength of the intramolecular field ($\mathscr{E} \sim 10^7$ V/cm); therefore, the detected higher switching current cannot be linked with induction of an additional moment (proportional to $\alpha\mathscr{E}$) on water molecules. There is another possibility—to link the switching current with polarization current determined by rotation of the water molecules in the applied field, taking into account the friction of molecules of H_2O on the surface of the metallic film.

Spontaneous polarization of the layer of water molecules on the metallic film cannot only be static. Since in bulk water some kinds of polycrystal-ferroelectric domains with lifetime $\tau_{\rm dom} \sim 100$ ps were observed by neutron scattering technique [385] (and the literature on dynamic clusters is quite extensive), we assumed that two-dimensional domains could exist as well, and due to the strong bonding with the hydrophilic surface the lifetime of the domains is several orders of magnitude higher (cf. Ref. 409). The vector of polarization **P** of these "twinkling" domains on the surface of the film is the direction of the external fields. This holds only in the case in which the field over the period $\tau_{\rm dom}$ is static; that is, the inequality $\Omega \ll \tau_{\rm dim}^{-1}$ must be fulfilled for the frequency Ω of alternating voltage.

Let us consider the cooperated behavior of the dipoles (i.e., O—H $\cdots$ O bonds) in the domain on the film using Ising's model with regard to friction—that is, phonons of the film. In Ising's model the ordering of dipoles is characterized by the mean value of the z component of the pseudospin $\langle S^z \rangle$.

Thus let the electric field that is applied to the film be oriented along the z axis. The polarization current appears as a result of rotation of the dipoles in the direction of the field; that is, the dipoles periodically change the orientation and are now straight and in parallel direction and antiparallel direction in relation to the z axis. These positions of the dipoles in Ising's model are linked with two projections of the zth component of the pseudospin $\langle S^z \rangle$: 1/2 and $-1/2$, respectively. Since the water molecules are situated on the surface of the film, we should include in the consideration the interaction with the film, which in turns introduces the interaction with the film phonons.

The form of the operators of the pseudospin is given in expressions (66)–(68). The pseudospin operators obey rules (69), which represent the switching relations in this situation studied. The total Hamiltonian for the dipoles in the domain in the mean field approximation is retained in the form

$$H = H_0 + H_{\rm tun} + H_{\mathscr{E}}(t) \tag{521}$$

$$H_0 = \sum_{l,l'} \mathscr{I}_{ll'} S_l^z \langle S_{l'}^z \rangle + \sum_{\mathbf{q}} \hbar\omega_{\mathbf{q}} \left(\hat{b}_{\mathbf{q}}^{+} \hat{b}_{\mathbf{q}} + \frac{1}{2} \right) - \sum_{l,l';\mathbf{q}} S_l^z \langle S_{l'}^z \rangle \hbar\omega_{\mathbf{q}} [u_{ll'}(\mathbf{q}) \hat{b}_{\mathbf{q}}^{+} + u_{ll'}(\mathbf{q}) \hat{b}_{\mathbf{q}}] \tag{522}$$

$$H_{\rm tun} = W \sum_l S_l^x \tag{523}$$

$$H_{\mathscr{E}}(t) = -\sum_l S_l^z d_0 \mathscr{E}_0 \exp(-i\Omega t) \tag{524}$$

The first term in expression (522) describes the interaction of dipoles, the second term depicts the phonon energy of the film, and the last term characterizes the interaction of the dipoles with the film and allows for small displacements from equilibrium positions of water atoms caused by interaction with the film's surface atoms or ions. The Hamiltonian (523) specifies the variation of the dipole spontaneous orientation from parallel to antiparallel in relation to the x axis and vice versa; H_{tun} is regarded as a perturbation. The Hamiltonian (524) describes the interaction of dipoles with the external field $\mathscr{E}(t) = \mathscr{E}_0 \exp(-i\Omega t)$, where $\mathscr{E}_0$ and Ω are the amplitude and the frequency of the field. W is the "kinetic" energy required for reorientation of the dipole, and d_0 is the dipole moment of the O—H bond. $\mathscr{J}_{ll'}$ is the interaction energy between the lth and the l'th hydrogen bonds; $u_{ll'}(\mathbf{q}) = u\exp[i\mathbf{q} \cdot (\mathbf{R}_l - \mathbf{R}_{l'})]$; u is the dimensionless value describing the displacement of an atom/ion of the film owing to its hydration; $\mathbf{R}_l$ is the radius vector of the lth dipole; $\hat{b}^+_{\mathbf{q}}\,(\hat{b}_{\mathbf{q}})$ is the Bose operator of creation (annihilation) of a phonon with the wave vector $\mathbf{q}$ and the energy $\hbar\omega_{\mathbf{q}}$. Since the phenomenon of the water-dependent switching of current was detected [399–401] for films of Ag, Al, and Au, and the ions of these atoms are strongly hydrated, it must be assumed that $u > 1$ or even $u \gg 1$, like in the situation of a small polaron.

Using the small polaron model, we can easily diagonalize the Hamiltonian (521) with respect to the phonon variables by canonical transformation [compare with expression (255)]

$$\bar{H} = e^{-S} H e^{S}, \qquad S = \sum_{l,l'} S^z_l [u_{ll'}(\mathbf{q})\hat{b}^+_{\mathbf{q}} - u^*_{ll'}(\mathbf{q})\hat{b}_{\mathbf{q}}] \tag{525}$$

This leads us to the following explicit form of the total Hamiltonian:

$$\bar{H} = \bar{H}_0 + \bar{H}_{\text{tun}} + H_{\mathscr{E}}(t) \tag{526}$$

$$\bar{H}_0 = \sum_{l,l'} 2\bar{\mathscr{J}}_{ll'} S^z_l \langle S^z_{l'} \rangle + \sum_{\mathbf{q}} \hbar\omega_{\mathbf{q}} \left(\hat{b}^+_{\mathbf{q}} \hat{b}_{\mathbf{q}} + \frac{1}{2} \right) \tag{527}$$

$$H_{\text{tun}} = W \sum_l \bar{S}^x_l \tag{528}$$

where

$$2\bar{\mathscr{J}}_{ll'} = 2\mathscr{J}_{ll'} - \sum_{\mathbf{q}} |u_{ll'}(\mathbf{q})|^2 \hbar\omega_{\mathbf{q}} \tag{529}$$

$$\bar{S}^x_l = e^{-S} S^x_l e^{S} \tag{530}$$

Thus, it is evident that the fluctuation formation of domains with polarization **P**, oriented along the applied field $\mathscr{E}$ (z axis), is more advantageous from the viewpoint of energy. Variation of the direction of the field is accompanied by the reorientation of the domains; and since the molecules of water are linked with the lattice of the metallic film, this repolarization process is extended in time to the measured values t. In the Hamiltonian (526) the reorientation of the dipole—that is, the presence of the current of repolarization domains—is caused by operator (528), and the operator $H_{\mathscr{E}}(t)$ (524) specifies the direction of current. In the case $\bar{H}_{\text{tun}}, H_{\mathscr{E}}(t) \to 0$—that is, when the energy W of "spontaneous" reorientation of the dipole, amplitude $\mathscr{E}_0$, and frequency Ω of the field tend to zero—polarization current should not occur.

Current density is calculated by the method described in Sections III.D to III.I:

$$I = \operatorname{Im} \operatorname{Tr}(\rho_{\mathscr{E}} j) \tag{531}$$

where symbol Im takes into account the phase shift of current in comparison with the field $\mathscr{E}(t)$. In expression (531),

$$j = \frac{1}{\mathscr{V} i\hbar}[H_{\text{tun}}, D] \tag{532}$$

is the operator of current density where $\mathscr{V}$ is the typical volume of an elementary cell,

$$D = -d_0 \sum_l S_l^z \tag{533}$$

is the operator of the dipole moment, and

$$\rho_{\mathscr{E}} = -\frac{i}{\hbar} e^{i\Omega t} \int_{-\infty}^{t} d\tau e^{-i(t-\tau)\bar{H}/\hbar} [H_{\mathscr{E}}(t-\tau), \rho_1] e^{i(t-\tau)\bar{H}/\hbar} \tag{534}$$

is the operator of the density matrix. The correction to the density matrix caused by the Hamiltonian $\bar{H}_{\text{tun}}$ (528) is

$$\rho_1 = -\frac{e^{-\bar{H}_0/k_B T}}{\operatorname{Tr} e^{-\bar{H}_0/k_B T}} \int_0^{1/k_B T} d\lambda e^{\lambda \bar{H}_0} H_{\text{tun}} e^{-\lambda \bar{H}_0} \tag{535}$$

Furthermore, in the exponents in Eq. (534) we will ignore the operators $\bar{H}_{\text{tun}}$ and $H_{\mathscr{E}}$, and we assume that $\bar{H} \to \bar{H}_0$. Substituting expressions from (532) to (535) into the expression for the current (531), we obtain

$$\begin{aligned} I = {} & \frac{W^2 d_0^2 \mathscr{E}_0 \sin \Omega t}{\mathscr{V} \hbar^2 \operatorname{Tr} e^{-\bar{H}_0/k_B T}} i \sum_l \int_0^\infty d\tau \int_0^{1/k_B T} d\lambda \operatorname{Tr}\{e^{\bar{H}_0/k_B T} \\ & \times (S_l^z e^{(-i\tau/\hbar+\lambda)\bar{H}_0} \bar{S}_l^x e^{-(-i\tau/\hbar+\lambda)\bar{H}_0} \bar{S}_l^y - e^{(-i\tau/\hbar+\lambda)\bar{H}_0} \bar{S}_l^x e^{-(-i\tau/\hbar+\lambda)\bar{H}_0} S_l^z \bar{S}_l^y)\} \end{aligned} \tag{536}$$

Here the designations

$$\bar{S}_l^x = \frac{1}{2}(e^{-\hat{\beta}}\hat{a}_{+,l}^{+}\hat{a}_{-,l} + \hat{a}_{-,l}^{+}\hat{a}_{-,l}e^{\hat{\beta}})$$
$$\bar{S}_l^x = \frac{1}{2i}(e^{-\hat{\beta}}\hat{a}_{+,l}^{+}\hat{a}_{-,l} + \hat{a}_{-,l}^{+}\hat{a}_{+,l}e^{\hat{\beta}} \tag{537}$$

are introduced [compared with expressions (66)–(68)] where

$$\hat{\beta}_l \equiv \sum_{l;\mathbf{q}} [u_{ll'}(\mathbf{q})\hat{b}_{\mathbf{q}}^{+} - u_{ll'}^{*}(\mathbf{q})\hat{b}_{\mathbf{q}}] \tag{538}$$

After operator transformations and calculation of the integrals, we obtain for the density current [407]

$$I = \frac{W^2 n d_0^2 \mathscr{E}_0 \sin\Omega t}{\hbar^2 \omega \bar{\mathscr{J}}}\left(\frac{2\pi}{|u|^2 \operatorname{cosech}(\hbar\omega/k_{\mathrm{B}}T)}\right)^{1/2} \times \exp\left(-\frac{E_{\mathrm{a}}}{k_{\mathrm{B}}T}\right)\sinh\frac{\langle S^z\rangle\bar{\mathscr{J}}}{2k_{\mathrm{B}}T} \tag{539}$$

where the activation energy (height of the barrier, separated into opposite orientations of the dipole) is given by

$$E_{\mathrm{a}} = k_{\mathrm{B}}T|u|^2 \tanh\frac{\hbar\omega}{k_{\mathrm{B}}T} - \frac{1}{2}\bar{\mathscr{J}}\langle S^z\rangle \tag{540}$$

and $2\bar{\mathscr{J}}$ is the energy of a hydrogen bond in the domain (i.e., $2\bar{\mathscr{J}} = 2\bar{\mathscr{J}}_{l,l-1}$).

In calculating expression (539), we examined only the nearest neighbors and ignored the dispersion; consequently, ω is the characteristic frequency of the phonons on the surface of the metallic film, and n is the concentration of the dipoles—that is, O–H bonds in the domain on the film.

The equation for the order parameter is

$$\langle S^z\rangle = \frac{\mathrm{Tr}\{S^z \exp[-(\bar{H}_0 - N d_0 \mathscr{E}_0 S^z)/k_{\mathrm{B}}T]\}}{\mathrm{Tr}\exp[-(\bar{H}_0 - N d_0 \mathscr{E}_0 S^z)/k_{\mathrm{B}}T} = \frac{1}{2}\tanh\frac{\langle S^z\rangle \mathscr{J}_{\mathrm{coll}} - N d_0 \mathscr{E}_0}{2k_{\mathrm{B}}T} \tag{541}$$

Note that in Eq. (541) the Hamiltonian $H_{\mathscr{E}}(t)$ (524) is regarded as written at the saddle point τ_0, since $\mathrm{Re}\exp(-i\Omega\tau) \to \cos\Omega\tau_0 \simeq 1$ and N is the mean number of the dipoles for mean polarization of $\mathbf{P}$ of the domain.

Evaluating I, we shall substitute into (539) the characteristic values of the parameters. The concentration n is approximately $3 \times 10^{28}\,\mathrm{m}^{-1}$ [410], and $d_0 \simeq 10^{-29}\,\mathrm{C \cdot m}$. Energy W can be found from the time of direct relaxation of water $\nu^{-1} \simeq 10^{-11}$ s [385], and then $W = h\nu \simeq 6 \times 10^{-23}$ J. Like in the problem of a small polaron, we accept $|u|^2 = 10$. In this case, it is realistic to assume that the phonon frequency is $\omega \approx 10^{12}\,\mathrm{s}^{-1}$. The value $2\mathcal{J}$ should correspond to the energy of a hydrogen bond for water, and the presence of spontaneous polarization in the layer should indicate structurization of water. Therefore we put $2\mathcal{J} \simeq 20.93$ kJ/mol (see Section VI.A). Besides we set $T = 300$ K and $\langle S^z \rangle < 1$. Comparing the parameters on the basis of the order of magnitude, we can greatly simplify expression (539) for the required current density

$$I = \frac{\sqrt{2\pi}}{16} \frac{W^2 n d_0^2 \mathscr{E}_0}{|u|\hbar (k_{\mathrm{B}} T)^2} \langle S^z \rangle \sin \Omega t \tag{542}$$

The numerical solution of Eq. (541) for the order parameter $\langle S^z \rangle$ as a function of $\mathscr{E}_0$ at $2\mathcal{J} \simeq 20.93$ kJ/mol, $k_{\mathrm{B}}T = 2.53$ kJ/mol (i.e., $T = 300$ K), $d_0 = 10^{-29}$ C·m, and $N = 100$ is given in Fig. 36. Information regarding the size of the domain can be obtained from the experiments with scattering neutrons in water [387]: They examined stable groupings of water molecules, including up to 10^3 molecules, and estimated their size at 2.5 to 3 nm. Expression (542) describes the density of current in one such domain. In experiment [399] the width of the examined film was equal to 7 mm. Consequently, the maximum number of water domains, which might be distributed along the film

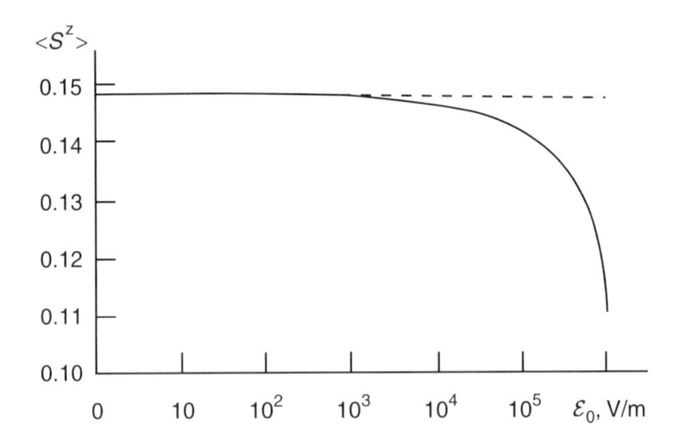

Figure 36. Numerical solution of Eq. (541) for $\langle S^z \rangle$ as a function of applied electric field $\mathscr{E}_0$. (From Ref. 408.)

width, could be evaluated as $\mathcal{N}_{\max} = 7\,\mathrm{nm}/3\,\mathrm{nm} \approx 10^6$. Thus, the maximum total polarization current for the film at $\mathcal{E}_{0\max} = 10^6$ V/m is given by

$$I_{\max} = \mathcal{N}_{\max} I \tag{543}$$

Substituting into expressions (542) and (543) all previously mentioned parameters, we obtain an estimate of the maximum current density: $I_{\max} \approx 10^6$ A/cm^2. This value is in complete agreement with the experimental results by Muller and Pagnia [399]. It is evident that the dependence $I_{\max}(\mathcal{E}_0)$ is specified by the dependence of $\langle S^z \rangle$ on $\mathcal{E}_0$ (Fig. 36), and this behavior of polarization current also corresponds to that detected in experiment [399]. It is evident that the number of domains on the surface of the film is a stochastic quantity (it is possible that it can also depend in a certain manner on the voltage applied to the film). In this case, the results (542) and (543) are in agreement with the conclusions drawn by Muller and Pagnia [399], who reported that this phenomenon on the whole is of the statistical nature.

VIII. CLUSTERING IN MOLECULAR SYSTEMS

A. Clusterization as Deduced from the Most General Statistical Mechanical Approach

Belotsky and Lev [411] proposed a very new approach to the statistical description of a system of interacting particles, which made allowance for spatial nonhomogeneous states of particles in the system studied. However, in the case when the inverse operator of the interaction energy cannot be determined, one should search for the other method, which, nevertheless, will make it possible to take into account a possible nonhomogeneous particle distribution. In Refs. 412–415, systems of interacting particles were treated from the same standpoint [411]; however, the number of variables describing the systems in question was reduced and a new canonical variable, which characterized the nascent nonhomogeneous state (i.e., cluster), automatically arose as a logical consequence of the particles behavior.

Having considered whether nonhomogeneous states (i.e., clusters) may appear spontaneously in systems with hydrogen bonds, we first should describe the methodology developed in Refs. 411–414. We shall start from the construction of the Hamiltonian for a system of two types of interacting particles (note that systems with hydrogen bonds consist of the minimum of two types of atoms, namely, hydrogen and oxygen).

Let atoms/molecules (called particles below) form a 3D lattice and let $n_s = \{0, 1\}$ be filling number of the sth lattice site. The energy for a system of two types of particles (sorts A and B) can be written in the form [415]

$$H = \sum_{\mathbf{r},\mathbf{r}'} \mathcal{U}_{\mathrm{AA}}(\mathbf{r},\mathbf{r}')c_{\mathrm{A}}(\mathbf{r})c_{\mathrm{A}}(\mathbf{r}') + 2\sum_{\mathbf{r},\mathbf{r}'} \mathcal{U}_{\mathrm{AB}}(\mathbf{r},\mathbf{r}')c_{\mathrm{A}}(\mathbf{r})c_{\mathrm{B}}(\mathbf{r}') + \sum_{\mathbf{r},\mathbf{r}'} \mathcal{U}_{\mathrm{BB}}(\mathbf{r},\mathbf{r}')c_{\mathrm{B}}(\mathbf{r})c_{\mathrm{B}}(\mathbf{r}') \quad (544)$$

where $\mathcal{U}_{ij}$ is the potential of the interaction of particles of types i and j ($i, j = \mathrm{A}, \mathrm{B}$), which occupy notes in Ising's lattice described by the radius vectors $\mathbf{r}$ and $\mathbf{r}'$; $c_{\mathrm{A,B}}(\mathbf{r})$ are the random functions ($c = \{0, 1\}$) that satisfy condition

$$c_{\mathrm{A}}(\mathbf{r}) + c_{\mathrm{B}}(\mathbf{r}) = 1 \quad (545)$$

Expression (544) can be written as follows:

$$H = H_0 + \frac{1}{2}\sum_{\mathbf{r},\mathbf{r}'} \mathcal{U}(\mathbf{r},\mathbf{r}')c_{\mathrm{A}}(\mathbf{r})c_{\mathrm{A}}(\mathbf{r}') \quad (546)$$

where

$$H_0 = \frac{1}{2}\sum_{\mathbf{r}} [1 - 2c_{\mathrm{A}}(\mathbf{r})] \sum_{\mathbf{r}'} \mathcal{U}_{\mathrm{BB}}(\mathbf{r},\mathbf{r}') + \sum_{\mathbf{r}} c_{\mathrm{A}}(\mathbf{r}) \sum_{\mathbf{r}'} \mathcal{U}_{\mathrm{AB}}(\mathbf{r},\mathbf{r}') \quad (547)$$

$$\mathcal{U}(\mathbf{r},\mathbf{r}') = \mathcal{U}_{\mathrm{AA}}(\mathbf{r},\mathbf{r}') + \mathcal{U}_{\mathrm{BB}}(\mathbf{r},\mathbf{r}') - 2\mathcal{U}_{\mathrm{AB}}(\mathbf{r},\mathbf{r}') \quad (548)$$

Let us rewrite the Hamiltonian (546) in the form

$$H = H_0 - \frac{1}{2}\sum_{\mathbf{r},\,\mathbf{r}'} V_{\mathbf{r}\,\mathbf{r}'}c(\mathbf{r})c(\mathbf{r}') + \frac{1}{2}\sum_{\mathbf{r},\,\mathbf{r}'} U_{\mathbf{r}\,\mathbf{r}'}c(\mathbf{r})c(\mathbf{r}') \quad (549)$$

where index A is omitted at the function $c(\mathbf{r})$ and the following definitions

$$V_{\mathbf{rr}'} = \mathcal{U}_{\mathrm{AB}}(\mathbf{r},\mathbf{r}') \quad (550)$$

$$U_{\mathbf{rr}'} = \frac{1}{2}[\mathcal{U}_{\mathrm{AA}}(\mathbf{r},\mathbf{r}') + \mathcal{U}_{\mathrm{BB}}(\mathbf{r},\mathbf{r}')] \quad (551)$$

are introduced.

If the potentials $V_{\mathbf{rr}'}$, $U_{\mathbf{rr}'}$ are greater than 0, then, as seen from the form of the Hamiltonian (549), the second term on the right-hand side corresponds to the effective attraction and the third term conforms to the effective repulsion. The Hamiltonian (549) can be represented as the model of ordered particles, which feature a certain nonzero parameter σ:

$$H(n) = \sum_s E_s n_s - \frac{1}{2}\sum_{s,s'} V_{ss'} n_s n_{s'} + \frac{1}{2}\sum_{s,s'} U_{ss'} n_s n_{s'} \tag{552}$$

where E_s is the additive part of the particle energy in the sth state. The main point of our approach is the initial separation of the total atom/molecular potential into two terms: the repulsion and attraction components. So, in the Hamiltonian (552), $V_{ss'}$ and $U_{ss'}$ are respectively the paired energies of attraction and repulsion between particles located in the states s and s'. It should be noted that the signs before the potentials in expression (552) specify proper signs of the attractive and repulsive paired energies, and this means that both functions $V_{ss'}$ and $U_{ss'}$ in expression (552) are positive. The statistical sum of the system

$$Z = \sum_{\{n\}} \exp(-H(n)/k_B T) \tag{553}$$

may be presented in the field form

$$Z = \int_{-\infty}^{\infty} D\phi \int_{-\infty}^{\infty} D\psi \sum_{\{n\}} \exp\left[\sum_s (\psi_s + i\phi_s) n_s - \sum_s \tilde{E}_s n_s - \frac{1}{2}\sum_{s,s'} (\tilde{U}_{ss'}^{-1} \phi_s \phi_{s'} + \tilde{V}_{ss'}^{-1} \psi_s \psi_{s'})\right] \tag{554}$$

due to the following representation known from the theory of Gauss integrals:

$$\exp\left(\frac{\rho^2}{2}\sum_{s,s'} \mathcal{W}_{ss'} n_s n_{s'}\right) = \mathrm{Re}\int_{-\infty}^{\infty} D\vartheta \exp\left[\rho \sum_s n_s \vartheta_s - \frac{1}{2}\sum_{s,s'} \mathcal{W}_{ss'}^{-1} \vartheta_s \vartheta_{s'}\right] \tag{555}$$

where $D\vartheta \equiv \prod_s \sqrt{\det \| \mathcal{W}_{ss'} \|} \sqrt{2\pi} d\vartheta_s$ implies the functional integration with respect to the field ϑ; $\rho^2 = \pm 1$ in relation to the sign of interaction (+1 for attraction and −1 for repulsion). The dimensionless energy parameters $\tilde{V}_{ss'} = V_{ss'}/k_B T$, $\tilde{U}_{ss'} = U_{ss'}/k_B T$, and $\tilde{E}_s/k_B T$ are introduced into expression (554). Furthermore, we will use the known formula

$$\frac{1}{2\pi i}\oint dz z^{N-1-\sum_s n_s} = 1 \tag{556}$$

which makes it possible to settle the quantity of particles in the system, $\sum_s n_s = N$, and, consequently, we can pass to the consideration of the canonical ensemble of N particles. Thus the statistical sum (554) is replaced for

$$Z = \mathrm{Re}\frac{1}{2\pi i}\oint dz \int D\phi \int D\psi \exp\left\{ -\frac{1}{2}\sum_{s,s'} (\tilde{U}_{ss'}^{-1}\phi_s\phi_{s'} + \tilde{V}_{ss'}^{-1}\psi_s\psi_{s'}) \right.$$
$$\left. + (N-1)\ln z \right\} \times \sum_{\{n_s\}=0}^{1} \exp\left\{ \sum_s n_s(\psi_s + i\phi_s - \tilde{E}_s) - \ln z \right\} \tag{557}$$

Summing over n_s, we obtain

$$Z = \mathrm{Re}\frac{1}{2\pi i}\int D\phi \int D\psi \oint dz e^{\mathscr{S}(\phi,\psi,z)} \tag{558}$$

where

$$\mathscr{S} = \sum_s \left\{ -\frac{1}{2}\sum_{s'} (\tilde{U}_{ss'}^{-1}\phi_s\phi_{s'} + \tilde{V}_{ss'}^{-1}\psi_s\psi_{s'}) \right.$$
$$\left. + \eta \ln\left| 1 + \frac{\eta}{z} e^{-\tilde{E}_s} e^{\psi_s} \cos\phi_s \right| \right\} + (N-1)\ln z \tag{559}$$

Here, the symbol η characterizes the kind of statistics: Bose ($\eta = +1$) or Fermi ($\eta = -1$). Let us set $z = \xi + i\zeta$ and consider the action $\mathscr{S}$ on a transit path that passes through the saddle point at a fixed imaginable variable $\mathrm{Im}z = \zeta_0$. In this case, $\mathscr{S}$ may be regarded as the functional that depends on the two field variables ϕ and ψ; and the fugacity ξ equals $e^{-\mu/k_B T}$, where μ is the chemical potential.

In a classical system the mean filling number of the sth energy level obeys the inequality

$$n_s = \frac{1}{\xi} e^{-\tilde{E}_s} = e^{(\mu - E_s)/k_B T} \ll 1 \tag{560}$$

(note the chemical potential $\mu < 0$ and $|\mu|/k_B T \gg 1$). By this means, we can simplify expression (559) expanding the logarithm into a Taylor series with respect to the small second member. As a result, we get the action that describes the ensemble of interacting particles, which are subjected to the Boltzmann statistics:

$$\mathscr{S} \cong \sum_s \left\{ -\frac{1}{2}\sum_{s'} (\tilde{U}_{ss'}^{-1}\phi_s\phi_{s'} + \tilde{V}_{ss'}^{-1}\psi_s\psi_{s'}) + \frac{1}{\xi} e^{-\tilde{E}_s} e^{\psi_s} \cos\phi_s \right\} + (N-1)\ln \xi \tag{561}$$

The extremum of functional (561) is realized at the solutions of the equations $\delta\mathcal{S}/\delta\phi = 0$, $\delta\mathcal{S}/\delta\psi = 0$, and $\delta\mathcal{S}/\delta\xi = 0$, or explicitly

$$\sum_{s'} \tilde{U}_{s'}^{-1}\phi_{s'} = -\frac{2}{\xi} e^{-\tilde{E}_s} e^{\psi_s} \sin\phi_s \tag{562}$$

$$\sum_{s'} \tilde{V}_{s'}^{-1}\psi_{s'} = -\frac{2}{\xi} e^{-\tilde{E}_s} e^{\psi_s} \cos\phi_s \tag{563}$$

$$\frac{1}{\xi}\sum_{s'} e^{-\tilde{E}_{s'}} e^{\psi_{s'}} \cos\phi_{s'} = N - 1 \tag{564}$$

If we introduce the designation

$$\mathcal{N}_s = \frac{1}{\xi} e^{-\tilde{E}_s} e^{\psi_s} \cos\phi_s \tag{565}$$

we will easily see from Eq. (565) that the sum $\sum_s \mathcal{N}_s$ is equal to the number of particles in the system studied. So the combined variable $\mathcal{N}_s$ specifies the quantity of particles in the sth state. This means that one may treat $\mathcal{N}_s$ as the variable of particle number in a cluster. Using this variable we can rewrite the action (561) as a function of only one variable $\mathcal{N}_s$ and the fugacity ξ (see technical details in Ref. 414):

$$\mathcal{S} = -2\sum_{s,s'}\left[\tilde{V}_{ss'}\mathcal{N}_s\mathcal{N}_{s'} + \tilde{U}_{ss'}\mathcal{N}_s\mathcal{N}_{s'}\left(\frac{e^{-2\tilde{E}_s+4\sum_{s'}\tilde{V}_{ss'}\mathcal{N}_{s'}}}{\xi^2\mathcal{N}_s^2} - 1\right)\right] + \sum_s \mathcal{N}_s(1+\ln\xi) \tag{566}$$

If we put the variable $\mathcal{N}_s = \mathcal{N} = \text{const}$ in each of the clusters, we may classify the $\mathcal{N}$ as the number of particles in a cluster [413]. Thus the model deals with particles entirely distributed by clusters.

It is convenient now to pass to the consideration of one cluster and change the discrete approximation to the continual one. The transformation means the passage from the summation over discrete functions in expression (566) to the integration of continual functions by the rule

$$\begin{aligned}\sum_s f_s &= \frac{1}{\mathcal{V}}\int_{\text{cluster}} d\vec{x}\, f(\vec{x}) \\ &= \frac{1}{\mathcal{V}}\int_0^{2\pi} d\phi \int_0^{\pi} d\theta \sin\theta \int_1^{R/g+1} dx\, x^2 f(x) \\ &= \frac{3}{\mathcal{N}}\int_1^{(\mathcal{N}+1)^{1/3}} dx\, x^2 f(x)\end{aligned} \tag{567}$$

where $\mathcal{V} = 4\pi R^3/3$ is the volume of a cluster, R/g is the dimensionless radius of a cluster where g is the mean distance between particles in a cluster, and x is the dimensionless variable ($x = [1, R/g]$); therefore, the number of particles in the cluster is linked with R and g by the relation $(R/g)^3 \cong \mathcal{N}$. We assume that $\mathcal{N} \gg 1$ and set the upper limit in the integral (567) equal to $\mathcal{N}^{1/3}$.

If we introduce the designations

$$a = \frac{3}{\mathcal{N}} \int_1^{\mathcal{N}^{1/3}} dx\, x^2 \tilde{U}, \qquad b = \frac{3}{\mathcal{N}} \int_1^{\mathcal{N}^{1/3}} dx\, x^2 \tilde{V} \tag{568}$$

we will arrive at the action for one cluster written in the simple form

$$\mathcal{S} = 2\mathcal{N}(a - b) - 2\frac{1}{\xi^2} a e^{-2\tilde{E} + 4b\mathcal{N}} + \mathcal{N} \ln \xi \tag{569}$$

The second term on the right-hand side of expression (569) is the smallest one (owing to the inequality $e^{-2\tilde{E}}/\xi^2 \ll 1$) and is omitted hereafter.

Thus we shall start from the action

$$\mathcal{S} = 2\mathcal{N}(a - b) + \mathcal{N} \ln \xi \tag{570}$$

The extremum (minimum) of the action (570) is reached at the meaning of $\mathcal{N}$, which is found from the equation $\delta\mathcal{S}/\delta\mathcal{N} = 0$ and satisfies the inequality $\partial^2\mathcal{S}/\delta\mathcal{N}^2 > 0$. The value $\mathcal{N}_{\text{extr}}$ obtained in such a way will correspond to the number of particles that give a cluster.

B. Cluster Formation in Solid Phase of Alkyl- and Alkoxybenzoic Acids

As we mentioned in Section V, near the temperature of phase transition to the metaphase the solid state of alkyl- and alkoxybenzoic acids is unstable in relation to the formation of hydrogen-bonded open associates. The associates merge in polymer chains, which can include over 10 acid molecules. Let us treat how the statistical model described above can account for such a behavior of the associates.

We shall consider potentials that describe the interaction between dimers ($\mathcal{U}_{\text{AA}}$), between dimers and open associates ($\mathcal{U}_{\text{AB}}$), and between open associates ($\mathcal{U}_{\text{BB}}$) [see expressions (548)–(550)]. It is reasonable to assume that the potentials $\mathcal{U}_{\text{AA}}$ and $\mathcal{U}_{\text{AB}}$ are typical van der Waals; that is, $\mathcal{U}_{\text{AA}} = -C_{\text{AA}} \cdot (g/r)^6$ and $\mathcal{U}_{\text{AB}} = -C_{\text{AB}} \cdot (g/r)^6$, where C_{ij} are constants. Notice that the two peculiarities are important: (i) open associates can be regarded as impurities to the main matrix (dimers), and (ii) open associates feature a larger dipole moment in comparison with dimers. These two circumstances (see Ref. 282) allow us to conclude that open associates interact as dipoles, and hence we can simulate the interaction between them as follows (see also Ref. 177):

$$\mathcal{U}_{\text{BB}} = -\frac{2d^2}{4\pi\,\varepsilon_0\, r^3} \tag{571}$$

Here d is the dipole moment of the open associate formed by alkyl- or alkoxybenzoic acid molecules, and r is the distance between two interacting associates.

Now we can construct potentials of effective attraction (550) and repulsion (551), namely, $V(r) = \mathcal{U}_{\text{AB}}$ and $U(r) = (\mathcal{U}_{\text{AA}} + \mathcal{U}_{\text{BB}})/2$. Substituting these parameters into expressions for a and b (568), we get instead of the action (570)

$$\mathcal{S} = -\frac{d^2}{2\pi\varepsilon_0 g^3 k_{\text{B}} T}\ln\mathcal{N} + \frac{C_{\text{AA}} - 2C_{\text{AB}}}{2k_{\text{B}}T}\left(\frac{1}{\mathcal{N}} - 1\right) + \mathcal{N}\ln\xi \tag{572}$$

where g is the effective distance between molecules.

The equation $\delta\mathcal{S}/\delta\mathcal{N} = 0$ that determines the minimum of the $\mathcal{S}$ as a function of $\mathcal{N}$ is the following:

$$\mathcal{N}^2 - \frac{d^2}{2\pi\varepsilon_0 g^3 k_{\text{B}} T \ln\xi}\mathcal{N} - \frac{C_{\text{AA}} - 2C_{\text{AB}}}{2k_{\text{B}}T\ln\xi} = 0 \tag{573}$$

In the first approximation the solution to Eq. (573) is equal to

$$\mathcal{N} \cong \frac{d^2}{2\pi\varepsilon_0 g^3 k_{\text{B}} T \ln\xi} \tag{574}$$

The number of ABA/OABA molecules, which enter in the cluster, can be easily estimated. Indeed, let the dipole moment of the open associate be equal to 7D, that is, $(7/3) \times 10^{-29}$ C$\cdot$m (note that long molecules linked through hydrogen bonds are characterized by a large proton polarizability [6]); the distance between two associates, g, is approximately 0.26 nm; the temperature of the phase transition, T_c, is 320 K. Substituting these values into expression (574) we obtain

$$\mathcal{N} \simeq \frac{130}{\ln\xi} \tag{575}$$

Let us now evaluate the fugacity ξ, which should satisfy the inequality $\xi \gg 1$ since we employ the classical statistics. For the evaluation, let us use the expression for the chemical potential μ of ideal gas (see, e.g., Ref. 416):

$$\mu = 3k_{\text{B}}T\ln(\langle\lambda\rangle n^{1/3}) \tag{576}$$

where $\langle\lambda\rangle = h/(3mk_BT)^{1/2}$ is the de Broglie thermal wavelength of particles of gas and n is the concentration of the particles. As a particle, we take one open associate; hence the particle mass m is approximately 1500 m_p, where m_p is the proton mass and $n = n_{ass} = 1.4 \times 10^{27}$ m^{-3} (see Section V). These meanings allow the calculation of the chemical potential by expression (576): $\mu \simeq -16\, k_BT$. Consequently, $\ln\xi \equiv \ln[\exp(-\mu/k_BT)] \simeq 16$. Substituting this value into the expression (575), we get the estimate of the number of open associates that are merged into a cluster: $\mathcal{N} \approx 8$. This result is in good agreement with the observed quantity of ABA/AOBA molecules that enter in a polymer chain (see Section V).

C. Clustering of H_2O Molecules in Water

In the case of ABA/AOBA acids, we have neglected the repulsive interaction between acid molecules due to their large size. In the case of the interaction of water molecules, we can keep the repulsion at least at the first stage of consideration. But what kind of potentials should we choose in the case of water? Since the hydrogen bond in many aspects is similar to the ionic interaction, we may assume that the pair potential between two water molecules combines both van der Waals type of interaction and the electrostatic energy. Let us choose the potential as the sum of the Lennard-Jones potential and the ionic crystal potential:

$$W_{H_2O-H_2O} = \varepsilon\left[\left(\frac{1}{r/g}\right)^{12} - \left(\frac{1}{r/g}\right)^{6}\right] + \lambda_{\text{ion rep.}} \exp(-r/r_0) - \alpha\frac{e^2}{4\pi\varepsilon_0\varepsilon\, r} \tag{577}$$

Here ε is the bound energy linked oxygen and hydrogen in the water molecule and g is the distance between O and H in the molecule; $\lambda_{\text{ion rep.}}$ is the constant that describes short-range repulsion and r_0 is a typical radius of this repulsion force; α is Madelung's constant, which, as a rule, falls within the range from unit to two (see, e.g., Kittel [177]); ε is the dielectric constant (which, however, is absent in the case of ionic crystals). Having chosen the pair potential (577), we can now represent the repulsive potential U and the attractive potential V, which enter the Hamiltonian (552), as follows:

$$U = \varepsilon\left(\frac{1}{r/g}\right)^{12} + \lambda_{\text{ion rep.}} \exp(-r/r_0) \tag{578}$$

$$V = \left(\frac{1}{r/g}\right)^{6} + \alpha\frac{e^2}{4\pi\varepsilon_0\varepsilon\, r} \tag{579}$$

Thus the functions a and b determined in expression (568) become

$$a = \frac{1}{\mathcal{N}}\left[-\frac{\varepsilon}{3k_{\rm B}T}\left(\frac{1}{\mathcal{N}^3} - 1\right)\right] \tag{580}$$

$$b = \frac{1}{\mathcal{N}}\left[-\frac{\varepsilon}{k_{\rm B}T}\left(\frac{1}{\mathcal{N}} - 1\right) + \frac{3\alpha e^2}{8\pi\varepsilon_0\varepsilon g k_{\rm B}T}(\mathcal{N}^{2/3} - 1)\right] \tag{581}$$

(the term proportional to $\lambda_{\text{ion rep.}}$ is omitted owing to its extremely small contribution). If we substitute a (580) and b (581) in expression (570) for the action $\mathcal{S}$, we obtain

$$\mathcal{S} = -\frac{2\varepsilon}{3k_{\rm B}T}\left(\frac{1}{\mathcal{N}^3} - 1\right) - \frac{2\varepsilon}{k_{\rm B}T}\left(\frac{1}{\mathcal{N}} - 1\right) + \frac{3\alpha e^2}{8\pi\,\varepsilon_0\,\varepsilon g k_{\rm B}\,T}(\mathcal{N}^{2/3} - 1) + \mathcal{N}\ln\xi \tag{582}$$

Retaining the major terms, expression (582) is reduced to

$$\mathcal{S} \simeq -\frac{3\alpha e^2}{8\pi\varepsilon_0\varepsilon g k_{\rm B}T}(\mathcal{N}^{2/3} - 1) + \mathcal{N}\ln\xi \tag{583}$$

The equation for the minimum of the $\mathcal{S}$ (i.e., $\delta\mathcal{S}/\delta\mathcal{N} = 0$) is

$$-\frac{\alpha e^2}{4\pi\varepsilon_0\varepsilon g k_{\rm B}T}\frac{1}{\mathcal{N}^{1/3}} + \ln\xi = 0 \tag{584}$$

The solution to Eq. (584) is

$$\mathcal{N} = \left(\frac{\alpha e^2}{4\pi\varepsilon_0\varepsilon g k_{\rm B}T\ln\xi}\right)^3 \tag{585}$$

Let us now assign numerical values to the parameters g (lattice constant), α (Madelung's constant), ε (permittivity), and ξ(fugacity) (other parameters are elementary charge e, dielectric constant ε_0, and the temperature T, which can be set at 300 K). For liquid water we can set $g = 0.281$ nm; for the estimation let α be 1.3 and let ε be 4 (though $\varepsilon = 81$ for bulk water). Stobbe and Peschel [405], based on literature data [417–419] and their own studies, have concluded that the dielectric constant ε varies from 5 to 10 for the absorbed water layer with thickness from 1 nm to 12 nm. A cluster of water molecules with the radius on the order of $10g$ just falls under the aforementioned water layer thickness.

However, as was mentioned in Ref. 405, the water structure near a solid surface owing to its orienting power appears to be anisotropic, and therefore the dielectric constant of oriented water becomes a tensor. In our case we do not deal with any orienting surface, and that is why we may expect that ε is less than 5. Using expression (576), we can evaluate the chemical potential: $\mu \simeq -6.75\, k_B T$ (for water at the normal conditions $n \simeq 3 \times 10^{28}\, \text{m}^{-3}$ and $\langle \lambda \rangle \simeq 0.034$ nm) and hence $\ln \xi \simeq 6.75$. Inserting all these numerical values into expression (585), we obtain $\mathcal{N} \approx 800$. This result is in line with the experimental data [383–386] discussed in Section VII. At the same time, in the case of ice, water molecules do not form any clusters: In the crystal state the interaction between water molecules is significantly stronger and hence the fugacity ξ should approach unity. Therefore, Eq. (584) will not have any cluster solution.

Thus, using methods of statistical mechanics we have shown that the homogeneous water network is unstable and spontaneously disintegrates to the nonhomogeneous state (i.e., peculiar clusters), which can be treated as an ordinary state of liquid water. However, the investigation of the dynamics of clusters is beyond of the approach described. The number $\mathcal{N}$ of water molecules that enter a cluster is a function of several parameters, namely, α, ε, and ξ (ξ, in turn, is a function of μ). It seems reasonable to assume that the main variation should undergo the dielectric constant ε and the chemical potential μ. Indeed, μ depends on temperature, volume, and total number of particles of the system studied (moreover, μ includes, though indirectly, the residual entropy). In a water system the total number of particles varies (in particular, owing to the dilution of gases of air). ε can be altered owing to introduced polar molecules and ions. Besides, external field sources (electromagnetic radiation, ultrasound, temperature, and others) can also strongly influence the water system changing conditions of the cluster formation treated above; in particular, these would cause the time relaxation to change over a long period of time.

IX. SUMMARY

In the present work the specific physical effects in structures with hydrogen bonds have received the bulk of our attention. Starting from the microscopic physics viewpoint, we have analyzed (a) the behavior of protons in a single hydrogen bond embedded in a surrounding matrix, (b) transfer and transport properties of protons in hydrogen-containing compounds, (c) proton dynamics caused by the influence of external fields, (d) the rearrangement of hydrogen bonds stipulated by temperature, (e) coherent phenomena that occur in hydrogen-bonded chains, and (f) clustering in systems with hydrogen bonds. Considerable attention has been given to the description of theoretical methods developed and employed in resolving the problems, which had their origins in very detailed experimental research.

Particular emphasis has been placed on two antithetical urgent problems: proton transfer incorporating acoustic phonons, which is intensively studied by Trommsdorf and co-workers [15,16,21,22,78–84], and the proton dynamics that is very decoupled from the backbone lattice, which is investigated by Fillaux and collaborators [1,7–11,14,49,110–114].

Much attention has been given to the developmet and application of the small polaron model to problems associated with proton transport in systems with hydrogen bonds; it is this model that the authors have employed for a long time. In the framework of the polaron model, we have studied various aspects connected with the motion of protons both along a hydrogen-bonded chain and in the bulk compounds. In particular, we have treated two different mechanisms of superionic conductivity for the $M_3H(XO_4)_2$ class of crystals and the $NH_4IO_3 \cdot 2NHIO_3$ crystal. The small polaron model has also found use in the problem of rearrangement of hydrogen bonds and the formation of open associates in alkyl- and alkoxybenzoic acids.

The mechanism of the polariton absorption by strong symmetric hydrogen bonds has successfully been applied for the description of anomalous of the infrared spectra of the α-$KIO_3 \cdot HIO_3$ crystal. A complex investigation has been done at the disclosing of the molecular mechanism of the bacteriorhodopsin functioning.

Coherent phenomena experimentally revealed in hydrogen-bonded chains by Zundel and collaborators [3–6] (i.e., the availability of a large proton polarizability in hydrogen-bonded chains) has allowed us to construct the microscopic mechanism of fast oscillations of the chain polarizability considering the phenomenon as a macroscopic tunneling similar to the tunneling of the magnetization in ferromagnetic particles.

Rigorous experimental results obtained on degassed water modifications and the thermodynamic analysis of the corresponding aqueous systems make it possible to conclude that the impact of dissolved air on the structure of aqueous systems is very significant (for instance, though the gases of the air in standard water account only for 0.003% of its mass, the change of entropy after the degassing of water is quite great and can reach 15% [359]). More research should be done to account for a microscopic mechanism of changes brought about by the gases.

A new approach to the problem of clustering of particles (atoms or molecules) in condensed media has been applied to systems with hydrogen bonds. The aforementioned statistical mechanical approach has allowed us to investigate the spatial nonhomogeneous distribution of interacting particles starting from the initially homogeneous particle system. The major peculiarity of the concept is that it separates the paired potential to two independent components: the attractive potential and the repulsive one, which in turn should feature very different dependence on the distance from particle. We have treated two problems: (a) clusterization of open associates in the solid phase of alkyl- and alkoxybenzoic acids and (b) clusterization of water molecules in liquid water. The method employed has enabled us to calculate the number of molecules that enter the cluster. Besides, the study conducted has revealed an important role of the chemical potential in the problem of clustering in the two mentioned cases.

APPENDIX A: STRETCHING AND BENDING ENERGIES AS FUNCTIONS OF $R_{O \cdots O}$

Novak [70] tabulated the dependence of the O—H stretching excitation energies (ν_{OH}) of many compounds on the distance R_{OH}. The rate of increase of ν_{OH} is greatest in the range $0.24 < R < 0.26$ nm. Ikeda et al. [105] have recently proposed a new model of H—O—H hydrogen bond dynamics, which makes it possible to calculate the hydrogen stretching and bending mode energies as functions of the hydrogen-bond length R. Their result agrees well with the experimental values presented in Ref. 70.

In the model of Ikeda et al. [105] the O—H—O bonds have been considered as fixed at a separation of R, and the hydrogen has interacted with the lattice system through the oxygens. The Hamiltonian was written in the form [106,107]

$$H = H_p + H_q + H'_{pq} \tag{A1}$$

$$H_p = -\frac{\hbar^2}{2m}\frac{d^2}{dx^2} + U_{\text{unperturb}}(X) \tag{A2}$$

$$H_q = -\frac{\hbar^2}{2M}\frac{d^2}{dq^2} + \frac{1}{2}M\omega^2 q^2 \tag{A3}$$

$$H'_{pq} = \gamma q x \tag{A4}$$

Here $X = (x, y, z)$; H_p and H_q are the total Hamiltonians of the hydrogen bond and the lattice, respectively (it is supposed that the O—H—O bond is decoupled from the lattice system); m is the hydrogen mass; q and M are the coordinate in the lattice system and the effective mass of the lattice's atom, respectively. H'_{pq} is an interaction Hamiltonian describing the interaction between the hydrogen and the lattice; x is the coordinate of the hydrogen along the hydrogen bond; γ is a constant.

Adopting the adiabatic approximation, Ikeda et al. [105] derived the wave functions of the hydrogen and the lattice, $\varphi(X; q)$ and $\vartheta(q)$, and the corresponding eigenvalues (i.e., energies E and ε) from the equations

$$\left[-\frac{\hbar^2}{2m}\frac{d^2}{dX^2} + U_{\text{unperturb}}(X) + \gamma q x\right]\varphi(X;q) = E(q)\varphi(X;q) \tag{A5}$$

$$\left[-\frac{\hbar^2}{2M}\frac{d^2}{dq^2} + G(q)\right]\vartheta(q) = \varepsilon(q)\vartheta(q) \tag{A6}$$

Here $G(q)$ is the modified lattice potential that is defined as

$$G(q) = \frac{1}{2}M\omega^2 q^2 + E_0(q) \tag{A7}$$

where $E_0(q)$ is the ground-state energy of hydrogen. The unperturbed hydrogen potential function is given as [106,108]

$$\begin{aligned} H_{\text{unperturb}}(x,y,z) = {} & \frac{f}{2}(\alpha_1^2 + \alpha_2^2) + \frac{g}{2}(\beta_1^2 + \beta_2^2) + V([(x-x_1)^2 + y^2 + z^2]^{1/2}) \\ & + V([(x-x_2)^2 + y^2 + z^2]^{1/2}) \end{aligned} \tag{A8}$$

where x_1 and x_2 are the positions of two oxygen at the equilibrium value $x = \pm R/2$. $\alpha_{1,2}$ and $\beta_{1,2}$ are the bending angles of the hydrogen defined with respect to both oxygens (below $i = 1, 2$):

$$\alpha_i = \sin^{-1}\{y/[(x-x_i)^2 + y^2]^{1/2}\} \tag{A9}$$

$$\beta_i = \sin^{-1}\{z/[(x-x_i)^2 + z^2]^{1/2}\} \tag{A10}$$

$V(r)$ is the Morse potential defined as

$$V(r) = V_0[e^{-2a(r-r_0)} - 2e^{-a(r-r_0)}] \tag{A11}$$

All of the parameters, especially expression (A8), have been determined for KH_2PO_4 shape KD_2PO_4. The shape of $U_{\text{unperturb}}(X)$ in expression (A8) is given as a function of R. Hence $U_{\text{unperturb}}$ depends on $E_0(q;H)$, $E_0(q;D)$ and $\lambda q_c(H)$, $\lambda q_c(D)$. It has been found that λq_c increases with R, and this changes the effective hydrogen potential $[U_{\text{unperturb}}(X) + \lambda q_c x]$.

In order to calculate the R dependence of the stretching excitation energies across a wide range of R ($0.24 < R < 0.29$ nm), they additionally introduced higher-order bending terms, $\zeta(\alpha_1^4 + \beta_1^4 + \alpha_2^4 + \beta_2^4)$, into Eq. (A8). Using calculated values of $\lambda q_c(H)$ and $\lambda q_c(D)$, we have calculated the stretching ν_{OH} and bending δ_{OH} and γ_{OH} energies. The obtained values have been compared with Novak's summary of the experimental data. The calculated values are indeed totally consistent with the observations. Besides, at $R \approx 0.26$ nm a strong Fermi resonance has been predicted as a general future of any hydrogen bond compound with 0.26 nm.

In the case of the R dependence of ν_{OD} for deuterated compounds, a Fermi resonance is predicted as well, but at $R \cong 0.255$ nm. It has been found that the potential $\lambda q_c(H)$ differs from $\lambda q_c(D)$ greatly; and because of that, the H and D isotopomers feature different properties in the ranges of $R < 0.243$ nm and $R > 0.26$ nm. $\lambda q_c(D) > \lambda q_c(H)$, which means that a stronger effective distortion takes place in the D potentials and in O—H—O bonds. These strong distortions depress the ground-state energy of deuterium and therefore depress the reaction rate of D isotopomers even in the range expected from simple mass consideration conducted in Ref. 109.

APPENDIX B: A POSSIBLE MECHANISM OF SONOLUMINESCENCE

Sonoluminesce is the glow seen from bubbles in a liquid under ultrasound. The experiment [123] allowed the investigation of a single bubble. The band of light is broad, $200 \leq \lambda \leq 800$ nm [124]. The light is radiated during each acoustic cycle, which lasted from 40 to 350 ps [125,126]. A number of studies have been devoted to explain the phenomenon (see, e.g., Refs. 127–134). Willison [122] has considered the motion of a proton in the hydrogen bond in cold water quantum mechanically, and the radiation from the tunneling charge has been treated classically.

The one-dimensional quadratic potential $V = \frac{1}{2}kx^2$ has been used for the description of covalent binding. The ground-state wave functions for a simple harmonic oscillator, ψ_L and ψ_R, have been used to describe the proton in the left and right wells. The force constant k has been determined from the stretch-mode vibrational transitions for water occurring at 3700 cm^{-1}. The ground-state energy for the proton is 0.368×10^{-19} J. The tunneling barrier is $\Delta E = 4 \times 10^{-19}$ J.

First-order perturbation theory makes it possible to estimate the transition rate between ψ_L and ψ_R. Initializing the system in ψ_L, all of the wave function amplitude moves to ψ_R in the 18 ns after the step perturbation is applied as asserted in Ref. 122. Then, reduction of the distance between potential wells by 0.025 nm reduces the time to move all of the population to less than 4 ps. The making and breaking of hydrogen bonds that takes place during a phase transition is a "switch" that turns "on" the tunneling current. It is assumed that the intensity of the radiated light will depend on the number of tunneling events that occur during the phase transition.

The wavelength of the emissions then will depend on the time Δt that the proton needs to tunnel between binding sites. This time may be shorter than the transition rate. The time can be estimated from the uncertainty principle: $\Delta t \sim 2.2 \times 10^{-16}$ s.

The current impulse of a proton tunneling event is written as a Gausssian impulse:

$$i(t) = \frac{2e}{\sqrt{2\pi\Delta t}} \exp\left(-\frac{2t^2}{\Delta t^2}\right) \tag{B1}$$

The current impulse (B1) can be expressed as Fourier component amplitudes, $2e\exp[-(\pi\nu\Delta t^2)]$ per second. In such a presentation, ν plays the role of frequency (in hertz) of radiated light.

The spectral radiance for the wavelength λ is written as [122]

$$R(\lambda) = \frac{4\pi\sqrt{\mu}\, ce^2\delta^2}{3\sqrt{\varepsilon}\,\lambda^4 e^{\frac{c^2 h^2}{2\lambda^2 \Delta E^2}}} \tag{B2}$$

where δ is the charge tunneling distance; ΔE is the height of the potential barrier; ε and μ are the permittivity and permeability of a medium (i.e., water in our case), respectively. For the estimated tunneling barrier ($\Delta E = 4.64 \times 10^{-19}$ J) the maximum wavelength is $\lambda_{\text{peak}} = 213$ nm.

All wavelengths are radiated simultaneously by a current impulse. The total number of photons emitted by a single tunneling event is estimated by dividing $R(\lambda)$ by the energy per photon (hc/λ) and integrating over the observed interval of wavelengths. The calculation yields 10^{-9} photons per one tunneling event radiated into a band between 200 and 800 nm.

In the case of sonoluminescence, typically 10^6 photons are radiated for a bubble collapse. This can be satisfied by 10^{15} tunneling events that happen incoherently. The sample of water incorporated by one bubble included approximately 10^{16} molecules of H_2O, which is quite enough for the number of tunneling events needed for the phenomenon of sonoluminescence observed.

APPENDIX C: DIAGONALIZATION OF PHONON VARIABLES

Let us sum over l and q (or k) in the second term of expression (203) with allowance for $\langle n_l \rangle = n = \text{const}$, $u_l(q) = N^{-1/2} u_q \exp(iqR_l)$, and $\upsilon_l(k) = N^{-1/2} \upsilon_k \exp(ikR_l)$. Then let us write Hamiltonian (192) as the sum of two terms:

$$H = \mathscr{H}_0 + \mathscr{H}_{01} \tag{C1}$$

$$\mathscr{H}_{01} = \sum_l E\hat{a}_l^+ \hat{a}_l + \frac{1}{2}\sum_q \hbar(\omega_0(q) + \omega_{\text{ac}}(q)) - \sum_l \hbar\omega_0(q)\hat{a}_l^+ \hat{a}_l [u_l(q)\hat{b}_q^+ + u_l^*(q)\hat{b}_q] + \sum_{l,m} J(R_m)\hat{a}_{l+m}^+ \hat{a}_l \tag{C2}$$

The second term in expression (197), symmetrized in q, can be represented as

$$\mathscr{H}_0 = \sum_{\alpha,\beta;q>0} \left\{ \frac{1}{2}[R_{\alpha\beta}(q)\hat{x}_\alpha^+(q)\hat{x}_\beta^+(-q) + R_{\alpha\beta}^*(q)\hat{x}_\alpha^+(-q)\hat{x}_\beta^+(q)] + [S_{\alpha\beta}(q)\hat{x}_\alpha^+(q)\hat{x}_\beta(q) + S_{\alpha\beta}^*(q)\hat{x}_\alpha^+(-q)\hat{x}_\beta(-q)] + [R_{\alpha\beta}^*(q)\hat{x}_\alpha(q)\hat{x}_\beta(-q) + R_{\alpha\beta}(q)\hat{x}_\alpha(-q)\hat{x}_\beta(q)] \right\} \tag{C3}$$

where $\alpha = 1,2$; $\beta = 1,2$; $\hat{x}_1(q) = \hat{b}_q$, $\hat{x}_2(q) = \hat{B}_q$

$$\begin{aligned} ||R_{\alpha\beta}(q)|| &= \begin{Vmatrix} 0 & h\chi n u_q \upsilon_q \\ h\chi n u_q^* \upsilon_q^* & 0 \end{Vmatrix} \\ ||S_{\alpha\beta}(q)|| &= \begin{Vmatrix} \hbar\omega_0(q) & h\chi n u_q \upsilon_q \\ h\chi n u_q^* \upsilon_q^* & \hbar\omega_{\text{ac}}(q) \end{Vmatrix} \end{aligned} \tag{C4}$$

By means of the Bogolyubov–Tyablikov transformation [172,173]

$$\begin{aligned} \hat{b}_q &= \sum_\iota [\hat{b}_\iota(q) f_{q\iota}^{(1)} + \hat{b}_\iota^+(-q) g_{q\iota}^{(1)*}] \\ \hat{B}_k &= \sum_\iota [\hat{b}_\iota(k) f_{k\iota}^{(2)} + \hat{b}_\iota^+(-k) g_{k\iota}^{(2)*}] \end{aligned} \tag{C5}$$

the Hamiltonian (C3) is diagonalized:

$$\mathscr{H}_0 = -\sum_{\iota=1}^{2}\sum_{\alpha=1}^{2}\sum_q \hbar\Omega_\iota(q)|g_{-q\iota}^{(\alpha)}|^2 + \sum_{\iota=1}^{2} \hbar\Omega_\iota(q)\hat{b}_\iota^+(q)\hat{b}_\iota(q) \tag{C6}$$

The eigenvalues of $\mathscr{H}_0$ are determined from the equations

$$\begin{aligned} \hbar\Omega_\iota(q) f_{q\iota}^{(\alpha)} &= \sum_\beta [S_{\alpha\beta}(q) f_{q\iota}^{(\beta)} + R_{\alpha\beta}(q) g_{-q\iota}^{(\beta)}] \\ -\hbar\Omega_\iota(q) f_{q\iota}^{(\alpha)} &= \sum_\beta [S_{\alpha\beta}(-q) g_{q\iota}^{(\beta)} + R_{\alpha\beta}(-q) f_{-q\iota}^{(\beta)}] \end{aligned} \tag{C7}$$

The functions $f_{q\iota}^{(\alpha)}$ and $g_{q\iota}^{(\alpha)}$, where $\iota = 1,2$, are determined from Eq. (C7). The form of these functions should meet the requirement that passing from the operators $\hat{b}_{1,2}$ back to the operators $\hat{b}_q$, $\hat{B}_q$, the Hamiltonian (C2) in the limit $\chi \to 0$ must be reduced to that without charge carrier interaction with acoustic phonons; that is, the coefficients of the terms $\hat{a}_l^+ \hat{a}_l (\hat{B}_q^+ + \hat{B}_q)$ must be equal to zero. The explicit form of the function $f_{ql}^{(\alpha)}$ and $g_{ql}^{(\alpha)}$ is presented in Ref. 165.

Returning to Hamiltonian (C1) in which the operator $\mathscr{H}_0$ is given by (C5) and the operator $\mathscr{H}_{01}$ is given by (C2), expressed through the new operators $\hat{b}_\iota$ (here $\iota = 1, 2$), we obtain Hamiltonian (205). In expression (205) the energy is counted from

$$\frac{1}{2}\hbar \sum_q \left[\omega_0(q) + \omega_{\text{ac}}(q) - \sum_\iota \hbar\Omega_l(q) \left(1 + \sum_\alpha |g^\alpha_{-q\iota}|^2 \right) \right]$$

APPENDIX D: PROTON BIFURCATION AND THE PHONON MIXTURE

In the ammonium triiodate hydrogen $NH_4IO_3 \cdot 2HIO_3$ crystal, the proton bifurcation for $T > T_c = 120$ K is treated as the origin of a mixture of low-frequency phonon mode ($\omega_1 = 99\,\text{cm}^{-1}$) and high-frequency intracellular mode ($\omega_2 = 756\,\text{cm}^{-1}$, which characterizes the collective vibrations of IO_3 pyramids). Following Ref. 47, let us study a mechanism of the change of intensity of the two aforementioned modes with temperature in the range $T_c < T < T_0$, where $T_c = 120$ K and $T_0 = 213$ K.

The interaction between an incident electromagnetic field and a polar crystal has the form $H_{\text{int}} = \mathbf{P} \cdot \mathscr{E}(\mathbf{r}, t)$, where $\mathscr{E}(\mathbf{r}, t) = \mathscr{E}_0 e^{i\mathbf{qr} - i\omega t}$. If the crystal features two active optical modes, the components of its polarization can be presented as follows [47]:

$$P_s(\mathbf{q}, \omega) = \frac{-1 + \varepsilon_{ss'}(\mathbf{q}, \omega)}{4\pi} \mathscr{E}_{0s'} \tag{D1}$$

where the permittivity of the crystal is equal to

$$\varepsilon_{ss'}(\mathbf{q}, \omega) = \varepsilon_{ss'}^{(0)} + \sum_{j=1}^{2} d_{js}(\mathbf{q}) d_{js'}(\mathbf{q}) \times \left[\frac{2\omega(\omega^2 - \omega_j^2(\mathbf{q}))}{(\omega^2 - \omega_j^2(\mathbf{q}))^2 + 4\omega^2\eta^2} + i\eta \frac{4\omega^2}{(\omega^2 - \omega_j^2(\mathbf{q}))^2 + 4\omega^2\eta^2} \right] \tag{D2}$$

Here $d_{js}(\mathbf{q})$ refers to the components of dipole moments of the crystal cell.

Upon the first phase transition ($T > T_c$), when a proton in a unit cell is bifurcated, the proton deforms the lattice and interacts with crystal vibrations. The interaction leads to the rearrangement of the phonon spectrum. Corresponding transformations of the $\omega_j(\mathbf{q})$ values can be obtained in the framework of the small polaron model. For the crystal studied, the $\omega_{1,2}(\mathbf{q})$ change only by a few cm^{-1} when the temperature passes the point $T = T_c$ on the temperature scale; and further on, for $T > T_c$, they remain constant ($\omega_1 = 99\ \text{cm}^{-1}$ and $\omega_2 = 756\ \text{cm}^{-1}$). However, according to the experimental results, in the ammonium triiodate hydrogen crystal these two modes that provide the proton polaron motion have anomalous temperature behavior, unlike all the other modes of the crystal. Consequently, in this case the proton should influence the intensity of atomic vibrations of the crystal.

The first phase transition leads to very interesting changes in the lattice of the crystal studied. For $T > T_c$, since each proton is bifurcated, it does not have a strictly fixed position but migrates ceaselessly between two hydrogen bonds. Protons in a polar crystal are strongly connected with the most polarizable longitudinal optic mode—that is, the 99-cm^{-1} mode in our case. Then with two possible positions of the proton the neighboring sites move in different potential wells. However, the frequency of proton jumps is very large, on the order of 10^{13}–$10^{14}\ \text{s}^{-1}$. Therefore the sites have not managed to occupy new equilibrium positions. So, the stationary state of the lattice is not fixed, but fluctuates uninterruptedly. A similar statement is true for IO_3 pyramids in the cell, whose collective behavior is described by the 756 cm^{-1} mode. Thus, the proton bifurcation can be considered as a distinctive fluctuation source that induces an additional displacement $\delta u_{\mathbf{l}}$ of the cellular sites. The intensity $\mathcal{I}_1$ of the source is directly connected with the lattice mode, 99 cm^{-1}; that is, the real fluctuation displacement of the **l**th site should be described by the term $\delta u_{\mathbf{l}}(\mathcal{I}_1)$. This term can be represented in the form $\delta u_{\mathbf{l}} = u_{2\mathbf{l}} g_{\mathbf{l}}(\mathcal{I}_1)$, where $g_{\mathbf{l}}(\mathcal{I}_1)$ is a function of the intensity of the phonon field of the 99-cm^{-1} mode and $u_{2\mathbf{l}}$ is the value that describes the displacement of pyramid IO_3 caused by the 756-cm^{-1} mode. Below these two modes are the first mode and the second mode, respectively.

Let us introduce the Hamiltonian function for a model cubic lattice

$$\Delta\mathcal{H} = K + U \tag{D3}$$

Here the kinetic and potential energies are, respectively,

$$K = \frac{m}{2}\sum_{\mathbf{l}} \delta\dot{u}_{\mathbf{l}}^2, \qquad U = \frac{\gamma}{2}\sum_{\mathbf{l}} (\delta u_{\mathbf{l}} - \delta u_{\mathbf{l}-\mathbf{a}})^2 \tag{D4}$$

In expression (D4), m is the mass of the IO_3 pyramid, γ is the effective elasticity constant of the model lattice, and **a** is the lattice vector. The transition to new collective vibrations $A_{\mathbf{q}}$, which characterize a collective fluctuation motion of the IO_3 pyramids, could be made by means of the canonical transformation

$$\delta u_{\mathbf{l}} = \frac{1}{\sqrt{N}}\sum_{\mathbf{q}} A_{\mathbf{q}} \exp(i\mathbf{l}\cdot\mathbf{q}) \tag{D5}$$

where $A_{\mathbf{q}} = A^*_{\mathbf{q}}$ and N is the quantity of the IO_3 pyramids in the lattice. In the new representation the kinetic energies take the form

$$K = \frac{m}{2}\sum_{\mathbf{q}} \dot{A}_{\mathbf{q}}\dot{A}_{-\mathbf{q}}, \qquad U = \frac{m}{2}\sum_{\mathbf{l}} \Delta\omega^2(\mathbf{q})A_{\mathbf{q}}A_{-\mathbf{q}} \tag{D6}$$

where

$$\Delta\omega^2(\mathbf{q}) = 4\frac{\gamma}{m}\sin^2\left(\frac{1}{2}\mathbf{a}\cdot\mathbf{q}\right) \tag{D7}$$

Furthermore, we can write the Lagrange function as $L = K - U$ and find the generalized momentum $P_{\mathbf{q}} = \partial L/\partial \dot{A}_{\mathbf{q}} = m\dot{A}_{-\mathbf{q}}$. The classical energy (D3) as a function of the generalized variables $A_{\mathbf{q}}$ and $P_{\mathbf{q}}$ has the form

$$\Delta\mathscr{H} = \frac{1}{2}\sum_{\mathbf{q}}\left[\frac{1}{m}P_{\mathbf{q}}P_{-\mathbf{q}} + m\Delta\omega^2(\mathbf{q})A_{\mathbf{q}}A_{-\mathbf{q}}\right] \tag{D8}$$

A change of variables from $A_{\mathbf{q}}$ and $P_{\mathbf{q}}$ to operators $\hat{A}_{\mathbf{q}}$ and $\hat{P}_{\mathbf{q}}$ transforms the energy (D8) to the energy operator ΔH.

Usually, the permutation relations

$$[\hat{A}_{\mathbf{q}}, \hat{P}_{\mathbf{q}'}]_- = i\delta_{\mathbf{qq}'}, \quad [\hat{A}_{\mathbf{q}}, \hat{A}_{\mathbf{q}'}]_- = [\hat{P}_{\mathbf{q}}, \hat{P}_{\mathbf{q}'}]_- = 0 \tag{D9}$$

hold. However, in our case the displacement $\delta u_{\mathbf{l}}$ is a compound value: $\delta u_{\mathbf{l}} = u_{2\mathbf{l}}g_{\mathbf{l}}(I_1)$. Therefore the collective variables $A_{\mathbf{q}}$ should have the same structure; this can be presented as follows:

$$A_{\mathbf{q}} = \mathscr{A}_{2\mathbf{q}}e^{\frac{1}{2}\alpha N_{1\mathbf{q}}} \tag{D10}$$

where $\mathscr{A}_{2\mathbf{q}}$ is the generalized combined variable of the motion the motion of IO_3 pyramids, $N_{1\mathbf{q}}$ is the quantity of phonons with the wave vector $\mathbf{q}$ of the first mode, and α is the coefficient that characterizes the degree of influence of this lattice mode on the second (i.e., intracellular) mode. In other words, α is the constant of coupling between the two kinds of phonons. Note that the function of the fluctuation intensity $g_{\mathbf{q}}(\mathscr{I}_1)$ written in the $\mathbf{q}$-representation is obviously in complete agreement with the expression for the intensity $\mathscr{I}$ of a normal phonon mode. The expression for $\mathscr{I}$ is determined by the thermodynamic averaging:

$$\mathscr{I} = \prod_{\mathbf{q}} \exp\langle \hat{b}^+_{\mathbf{q}}\hat{b}_{\mathbf{q}}\rangle$$

(see, e.g., Ref. 88). By this means, the transition from variables $A_{\mathbf{q}}$ and $P_{\mathbf{q}}$ to the corresponding operators may be performed in the following manner:

$$A_{\mathbf{q}} \rightarrow \hat{\mathscr{A}}_{2\mathbf{q}} e^{\frac{1}{2}\alpha \hat{N}_{1\mathbf{q}}} \tag{D11}$$

$$\begin{aligned} P_{\mathbf{q}} &\rightarrow m(\hat{\mathscr{A}}_{2,-\mathbf{q}} e^{\frac{1}{2}\alpha \hat{N}_{1\mathbf{q}}})^{\cdot} \\ &= \hat{\mathscr{P}}_{2\mathbf{q}} e^{\frac{1}{2}\alpha \hat{N}_{1\mathbf{q}}} + \frac{1}{2}\alpha m \hat{\mathscr{A}}_{2,-\mathbf{q}} e^{\frac{1}{2}\alpha \hat{N}_{1\mathbf{q}}} \dot{\hat{N}}_{1\mathbf{q}} \end{aligned} \tag{D12}$$

Here one takes into account that the generalized variable $\mathscr{P}_{2\mathbf{q}}$ corresponds to $m\dot{\mathscr{A}}_{2,-\mathbf{q}}$; $\hat{N}_{1\mathbf{q}}$ is the operator of the phonon quantity for the first mode. Using relations (D9), which should be valid for the operators $\hat{\mathscr{A}}_{2\mathbf{q}}$ and $\hat{\mathscr{A}}_{2\mathbf{q}}$, we obtain the permutation relations for compound operators (D11) and (D12):

$$[\hat{\mathscr{A}}_{\mathbf{q}}, \hat{\mathscr{P}}_{\mathbf{q}'}]_- = i\hbar \delta_{\mathbf{q}\mathbf{q}'} e^{\alpha \hat{N}_{1\mathbf{q}}}, \qquad [\hat{\mathscr{A}}_{\mathbf{q}}, \hat{\mathscr{A}}_{\mathbf{q}'}]_- = [\hat{\mathscr{P}}_{\mathbf{q}}, \hat{\mathscr{P}}_{\mathbf{q}'}]_- = 0 \tag{D13}$$

(here one sets $\dot{\hat{N}}_{1\mathbf{q}} = i\hbar[\hat{N}_{1\mathbf{q}}, \sum_{j;\mathbf{q}} \hat{b}^{+}_{j\mathbf{q}} \hat{b}_{j\mathbf{q}}]_- = 0$). Then one can pass from these operators $\hat{\mathscr{A}}_{\mathbf{q}}$ and $\hat{\mathscr{P}}_{\mathbf{q}}$ to the Bose operators for phonons $\hat{b}^{+}_{j\mathbf{q}}$ and $\hat{b}_{j\mathbf{q}}$ ($j = 1,\ 2$), which satisfy standard permutation relations

$$[\hat{b}^{+}_{j\mathbf{q}}, \hat{b}_{j'\mathbf{q}'}]_- = \delta_{jj'} \delta_{\mathbf{q}\mathbf{q}'}, \qquad [\hat{b}_{j\mathbf{q}}, \hat{b}_{j'\mathbf{q}'}]_- = 0 \tag{D14}$$

The transition can be made via the following rules:

$$\hat{\mathscr{A}}_{\mathbf{q}} = \left(\frac{\hbar}{2m\Delta\omega(\mathbf{q})}\right)^{1/2} (\hat{b}^{+}_{2,-\mathbf{q}} + \hat{b}_{2\mathbf{q}}) e^{\frac{1}{2}\alpha \hat{b}^{+}_{1,\mathbf{q}} \hat{b}_{1\mathbf{q}}} \tag{D15}$$

$$\hat{\mathscr{P}}_{\mathbf{q}} = i\left(\frac{m\hbar\Delta\omega(\mathbf{q})}{2}\right)^{1/2} (\hat{b}^{+}_{2\mathbf{q}} - \hat{b}_{2,-\mathbf{q}}) e^{\frac{1}{2}\alpha \hat{b}^{+}_{1,\mathbf{q}} \hat{b}_{1\mathbf{q}}} \tag{D16}$$

Substitution of the variables $A_{\mathbf{q}}$ and $P_{\mathbf{q}}$ on the right-hand side of the Hamiltonian (D8) the operators $\hat{\mathscr{A}}_{\mathbf{q}}$ and $\hat{\mathscr{P}}_{\mathbf{q}}$ from expressions (D15) and (D16) converts the Hamiltonian function (D3) into the fluctuation Hamiltonian

$$\Delta\mathscr{H} = \sum_{\mathbf{q}} \hbar\Delta\omega(\mathbf{q}) \left(\hat{b}^{+}_{2\mathbf{q}} \hat{b}_{2\mathbf{q}} e^{\frac{1}{2}\alpha \hat{b}^{+}_{1,\mathbf{q}} \hat{b}_{1\mathbf{q}}} + \frac{1}{2} \right) \tag{D17}$$

Since the unperturbed Hamiltonian of the crystal is

$$\mathscr{H}_0 = \sum_{j;\mathbf{q}} \hbar\omega_j(\mathbf{q}) \left(\hat{b}^{+}_{j\mathbf{q}} \hat{b}_{j\mathbf{q}} + \frac{1}{2} \right) \tag{D18}$$

the total Hamiltonian becomes

$$\mathscr{H} = \mathscr{H}_0 + \Delta\mathscr{H} \tag{D19}$$

Now we can consider the electromagnetic field absorption by the lattice mode (the first mode) and the intracellular one (the second mode) when the modes are perturbed by described fluctuations. We suppose that the operator $\Delta\mathscr{H}$ in the Hamiltonian (D19) is a small perturbation; that is, the fluctuation energy is smaller than the energy of regular vibrations.

The absorption of the lattice mode is derived by the formula

$$\langle \mathbf{P}(\mathbf{r},\ t)\rangle = \mathrm{Tr}\{\rho_{\mathrm{int}}\hat{\mathbf{P}}(\mathbf{r},\ t)\} \tag{D20}$$

where the operators of the crystal polarization and the correction to the statistical operator are, respectively,

$$\hat{\mathbf{P}}(\mathbf{r}) = \frac{1}{\mathscr{V}}\sum_{j;\mathbf{q}}\left(\frac{\hbar}{2\gamma_j(\mathbf{q})\omega_j(\mathbf{q})}\right)^{1/2}\frac{\mathbf{q}}{|\mathbf{q}|}e^{i\mathbf{q}\cdot\mathbf{r}}(\hat{b}^{+}_{j\mathbf{q}} + \hat{b}_{j\mathbf{q}}) \tag{D21}$$

$$\rho_{\mathrm{int}} = \rho_0 - \frac{i}{\hbar}\int_{-\infty}^{t}[H_{\mathrm{int}}(\tau), \rho_0]\ d\tau, \qquad \rho_0 = \frac{e^{-H_0/k_{\mathrm{B}}T}}{\mathrm{Tr}e^{-H_0/k_{\mathrm{B}}T}} \tag{D22}$$

where the operator of the interaction between the incident electromagnetic field and the operator of the crystal polarization is given by

$$H_{\mathrm{int}} = \hat{\mathbf{P}}\cdot\mathscr{E}(t) \tag{D23}$$

Having calculated the absorption by the lattice mode, we should substitute the operator ρ_0 for the following:

$$\tilde{\rho}(\mathscr{H}_0 + \Delta\mathscr{H}) \simeq \rho_0(\mathscr{H}_0) - \rho_0(\mathscr{H}_0)\int_0^{1/k_{\mathrm{B}}T} d\lambda e^{\lambda\mathscr{H}_0}\Delta\mathscr{H}e^{-\lambda\mathscr{H}_0} \tag{D24}$$

Then calculations by formula (D20) have shown [47] that the permittivity that represents the lattice mode—that is, the term with $j = 1$ in expression (D2)—should be supplemented with the factor $1 + f_1(\mathbf{q}, T)$, where

$$f_1(\mathbf{q},\ T) = -\frac{\hbar\Delta\omega(\mathbf{q})}{k_{\mathrm{B}}T[\exp(\hbar\omega_2(\mathbf{q})/k_{\mathrm{B}}T) - 1]}\exp\left(\alpha\coth\frac{\hbar\omega_1(\mathbf{q})}{2k_{\mathrm{B}}T}\right) \tag{D25}$$

The Hamiltonian of the interaction between the electromagnetic wave and the fluctuations of the intercellular mode in the crystal has the form analogous to expression (D23):

$$H_{\mathrm{int}}^{(\mathrm{fl})}(t) = -\sum_{\mathbf{q}} \mathbf{d}_{\mathrm{fl}}(\mathbf{q}) \cdot \mathscr{E}_0(\hat{V}_{\mathbf{q}}^{+} - \hat{V}_{\mathbf{q}})e^{i\omega t+\eta t} \tag{D26}$$

Here $\hat{V}_{\mathbf{q}}^{+}(V_{\mathbf{q}})$ is the effective operator for the creation (annihilation) of fluctuations of the intracellular mode:

$$\hat{V}_{\mathbf{q}} = \hat{b}_{2\mathbf{q}}e^{\alpha\hat{b}_{1\mathbf{q}}^{+}\hat{b}_{1\mathbf{q}}} \tag{D27}$$

The operator of the fluctuation dipole moment is

$$\hat{\mathbf{P}}_{\mathrm{fl}}(\mathbf{r}, t) = \sum_{\mathbf{q}} \mathbf{d}_{\mathrm{fl}}(\mathbf{q})e^{i\mathbf{q}\cdot\mathbf{l}}(\hat{V}_{\mathbf{q}}^{+} + \hat{V}_{\mathbf{q}}) \tag{D28}$$

In expressions (D26) and (D28), $\mathbf{d}_{\mathrm{fl}}(\mathbf{q})$ is the effective dipole moment of the fluctuations of the intracellular mode. In this case the polarization is defined as

$$\langle \mathbf{P}_{\mathrm{fl}}(\mathbf{r}, t)\rangle = \mathrm{Tr}\{\rho_{\mathrm{fl}}\hat{\mathbf{P}}_{\mathrm{fl}}(\mathbf{r}, t)\} \tag{D29}$$

where the statistical operator is expressed as

$$\tilde{\rho}_{\mathrm{fl}} = \tilde{\rho}_0 - \frac{i}{\hbar}\int_{-\infty}^{t} d\tau [H_{\mathrm{int}}^{(\mathrm{fl})}(\tau), \tilde{\rho}_0] \tag{D30}$$

In the present case we may neglect secondary effects of an expansion of $\tilde{\rho}_0(\mathscr{H})$ in terms of $\Delta\mathscr{H}$. Hence we assume that in Eq. (D30), $\tilde{\rho}_0 \simeq \rho_0$. This approximation allows the calculation of the polarization (D29). The result shows that the dielectric function of the intracellular mode fluctuation obtained from expression (D29) differs of the second term in expression (D2) for the permittivity only by the factor

$$f_2(\mathbf{q}, T) = \frac{d_{(\mathrm{fl})2j}(\mathbf{q})\, d_{(\mathrm{fl})2j'}(\mathbf{q})}{d_{2j}(\mathbf{q})\, d_{2j'}(\mathbf{q})} \exp\left(\alpha \coth\frac{\hbar\omega_1}{2k_{\mathrm{B}}T}\right) \tag{D31}$$

Thus, the overall result for the imaginary part of the permittivity is

$$\mathrm{Im}\,\varepsilon(\mathbf{q},\ \omega,\ T) = \frac{4\omega^2}{(\omega^2 - \omega_1^2(\mathbf{q}))^2 + 4\omega^2\eta^2}[1 + f_1(\mathbf{q},\ T)] + \frac{4\omega^2}{(\omega^2 - \omega_2^2(\mathbf{q}))^2 + 4\omega^2\eta^2}[1 + f_2(\mathbf{q},\ T)] \tag{D32}$$

The investigation of the absorption [47] has shown that in the temperature interval between two phase transitions ($T_c = 120$ K to $T_0 = 213$ K) the relative intensities of the absorption maximum (for the lattice 99-cm^{-1} mode) and the scattering maximum (for the intracellular 756-cm^{-1} mode) have changed very markedly. The first mode decreased by approximately a factor of two at 213 K. The second one increased by approximately a factor of five at 213 K. Since the imaginary part of the permittivity defines the absorption (scattering) maximum, the expression (D32) has to describe the real temperature behavior of the two aforementioned anomalous modes.

The analysis of the behavior of $\text{Im}\varepsilon_{1(2)}$ conducted in Ref. 47 in fact correlates very well with the experimental data. Consequently, in the $NH_4IO_3 \cdot 2HIO_3$ crystal the phonon–phonon coupling between the active lattice mode and the intracellular one is realized due to the proton bifurcation. The availability of moderate phonon–phonon coupling α ($\alpha = 0.95$ [47]) permits the intracellular mode to "live" on the energy of phonons of the lattice mode. Since the lattice mode is more powerful than the intracellular mode [47], the latter has taken energy out of the former through the phonon–phonon coupling. So, at the second phase transition (i.e., 213 K), the power of the lattice mode can no longer hold proton polarons very strongly and they become more mobile. Thus, the crystal goes into a superionic state in which the band polaron conductivity prevails (see Section III.I).

Acknowledgments

We are thankful to Professors G. Zundel, W. Coffey, V. Gaiduk, F. Fillaux, H.-P. Trommsdorff, W. A. P. Luck, G. Peschel, P. L. Huyskens, A. C. Legon, I. V. Stasyuk, and A. Kaivarainen, as well as Dr. E. Shadchin, who kindly placed at our disposal reprints of their papers quoted in the present work.

References

1. F. Fillaux and J.-P. Perchard, *J. Chim. Phys.* **Hors série**, 91 (1999).
2. E. D. Isaacs, A. Shukla, P. M. Platzman, D. R. Hamann, B. Barbiellini, and C. A. Tulk, *Phys. Rev. Lett.* **82**, 600 (1999).
3. G. Zundel, *Hydration and Intermolecular Interaction—Infrared Investigations of Polyelectrolyte Membranes*, Academic Press, New York, 1969 and Mir, Moscow, 1972.
4. G. Zundel and H. Metzger, *Z. Physik. Chem.* (*Frankfurt*) **58**, 225 (1968).
5. G. Zundel and H. Metzger, *Z. Physik. Chem.* (*Leipzig*) **240**, 50 (1969).
6. G. Zundel, *Adv. Chem. Phys.* **111**, 1 (2000).
7. F. Fallaux, J. Tomkinson, and J. Penfold, *Chem. Phys.* **124**, 425 (1988).
8. F. Fallaux, J. P. Fontaine, M. H. Baron, G. J. Kearly, and J. Tomkinson, *Chem. Phys.* **176**, 249 (1993).
9. F. Fallaux, J. P. Fontaine, M. H. Baron, G. J. Kearly, and J. Tomkinson, *Biophys. Chem.* **53**, 155 (1994).
10. F. Fillaux, N. Leygue, J. Tomkinson, A. Cousson and W. Paulus, *Chem. Phys.* **244**, 387 (1999).
11. F. Fillaux, *J. Mol. Struct.* **511–512**, 35 (1999).

12. T. Horsewill, M. Johnson, and H. P. Trommsdorff, *Europhys. News* **28**, 140 (1997).
13. C. Rambaud, and H. P. Trommsdorff, *Chem. Phys. Lett.* **306**, 124 (1999).
14. F. Fillaux, *Physica D* **113**, 172 (1998).
15. H. P. Trommsdorff, *Adv. Photochem.* **24**, 147 (1998).
16. H. P. Trommsdorff, *Optoélectronique Moléculaire*, Observatoire Français des Techniques Avancées, Arago 13, Masson, Paris, 1993, Chapter VIII, p. 247.
17. C. Rambaud, A. Oppenlander, M. Pierre, H. P. Trommsdorff, and J. C. Vial, *Chem. Phys.* **136**, 335 (1989).
18. J. Bardeen, *Phys. Rev. Lett.* **6**, 57 (1961).
19. C. Herring, *Rev. Mod. Phys.* **34**, 341 (1970).
20. C. P. Flynn and A. M. Stoneham, *Phys. Rev. B* **1**, 3967 (1970).
21. J. L. Skinner and H. P. Trommsdorff, *J. Chem. Phys.* **89**, 897 (1988).
22. R. Silbey and H. P. Trommsdorff, *Chem. Phys. Lett.* **165**, 540 (1990).
23. V. Benderskii, V. I. Goldanskii, and D. E. Makarov, *Phys. Rep.* **233**, 195 (1993).
24. G. N. Robertson and M. C. Lawrence, *Chem. Phys.* **62**, 131 (1981).
25. A. V. Skripov, J. C. Cook, T. J. Udovic, and V. N. Kozhanov, *Phys. Rev. B* **62**, 14099 (2000).
26. R. Baddour-Hadjean, F. Fillaux, F. Floquet, S. Belushkin, I. Natkaniec, L. Desgranges, and D. Grebille, *Chem. Phys.* **197**, 81 (1995).
27. H. J. Morowitz, *Am. J. Physiol.* **235**, R99 (1978).
28. J. Teissié, M. Prats, P. Soucaille, and J. F. Tocanne, *Proc. Natl. Acad. Sci. USA* **82**, 3217 (1985).
29. J. Vanderkooy, J. D. Cuthbert, and H. E. Petch, *Can. J. Phys.* **42**, 1871 (1964).
30. A. D. Reddy, S. G. Sathynarayan, and G. S. Sastry, *Solid State Commun.* **43**, 937 (1982).
31. J. D. Cuthbert and H. E. Petch, *Can. J. Phys.* **41**, 1629 (1963).
32. S. Suzuki and Y Makita, *Acta Crystallogr. B* **34**, 732 (1978).
33. P. M. Tomchuk, N. A. Protsenko, and V. Krasnoholovets, *Biol. Membr.* **1**, 1171 (1984) (in Russian).
34. P. M. Tomchuk, N. A. Protsenko, and V. Krasnoholovets, *Biochem. Biophys. Acta* **807**, 272 (1985).
35. V. Krasnoholovets, N. A. Protsenko, P. M. Tomchuk, and V. S. Guriev, *Int. J. Quant. Chem.* **33**, 327 (1988).
36. V. Krasnoholovets, N. A. Protsenko and P. M. Tomchuk, *Int. J. Quant. Chem.* **33**, 349 (1988).
37. V. Krasnoholovets, V. B. Taranenko, P. M. Tomchuk, and N. A. Protsenko, *J. Mol. Struct.* **355**, 219 (1995).
38. I. V. Stasyuk, N. Pavlenko, and D. Hilczer, *Phase Transitions* **62**, 135 (1997).
39. I. V. Stasyuk, N. Pavlenko, and M. Polomska, *Phase Transitions* **62**, 167 (1997).
40. N. Pavlenko, M. Polomska, and B. Hilczer, *Cond. Matter. Phys.* **1**, 357 (1998).
41. I. V. Stasyuk and V. Pavlenko, *J. Phys. Cond. Matter* **10**, 7079 (1998).
42. V. Stasyuk, V. Pavlenko, and B. Hilczer, *J. Korean Phys. Soc.* **32**, S24 (1998).
43. G. A. Puchkovska and Yu. A. Tarnavski, *J. Mol. Struct.* **267**, 169 (1992).
44. Yu. Tarnavski, G. A. Puchkovska, and J. Baran, *J. Mol. Struct.* **294**, 61 (1993).
45. Yu. Tarnavski, Thesis, Institute of Physics, Kyiv, Ukraine, 1993.
46. V. Krasnoholovets, G. A. Puchkovska, and Tarnavski, *Khim. Fiz.* **12**, 973 (1993) (in Russian); English translation: *Sov. J. Chem. Phys.* **12**, 1434 (1994).

47. V. Krasnoholovets, *J. Phys. Cond. Matter* **8**, 3537 (1996).
48. G. A. Puchkovska and Yu. A. Tarnavski, *J. Mol. Struct.* **403**, 137 (1997).
49. F. Fillaux, *Solid State Ionics* **125**, 69 (1999).
50. W. A. P. Luck and M. Fritzsche, *Z. Phys. Chem.* **191**, 71 (1995).
51. P. L. Huyskens, *J. Mol. Struct.* Special Issue Sandorfy **297**, 141 (1993).
52. P. L. Huyskens, D. P. Huyskens, and G. G. Siegel, *J. Mol. Liquids* **64**, 283 (1995).
53. K. Nelis, L. Van den Berge-Parmentier, and F. Huyskens, *J. Mol. Liquids* **67**, 157 (1995).
54. W. A. P. Luck, *Angew. Chem.* **29**, 92 (1980).
55. W. A. P. Luck, *Opt. Pur. Appl.* 18, 71 (1985).
56. W. A. P. Luck, in *Intermolecular Forces. An Introduction to Modern Methods and Results*, P. L. Huyskens, W. A. P. Luck, and T. Zeegers-Huyskens, eds., Springer-Verlag, Berlin, 1991, p. 317.
57. V. Krasnoholovets, G. A. Puchkovska, and A. A. Yakubov, *Ukr. Fiz. Zh.* **37**, 1508 (1992) (in Russian).
58. V. Krasnoholovets, G. A. Puchkovska, and A. A. Yakubov, *Khim. Fiz.* **11**, 806 (1992) (in Russian).
59. V. Krasnoholovets and B. Lev, *Ukr. Fiz. Zh.* **39**, 296 (1994) (in Ukrainian).
60. V. Krasnoholovets, G. A. Puchkovska, and A. A. Yakubov, *Mol. Cryst. Liq. Cryst.* **265**, 143 (1995).
61. L. M. Babkov, E. Gabrusenoks, V. Krasnoholovets, G. A. Puchkovska, and I. Khakimov, *J. Mol. Struct.* **482–483**, 475 (1998).
62. V. Krasnoholovets, I. Khakimov, G. Puchkovska, and E. Gabrusenoks, *Mol. Cryst. Liq. Cryst.* **348**, 101 (2000).
63. N. D. Gavrilov and O. V. Mukina, *Neorg. Mater.* **33**, 871 (1997) (in Russian).
64. M. A. Kovner and V. A. Chuenkov, *Izv. Akad. Nauk SSSR Ser. Fiz.* **14**, 435 (1950).
65. D. Hadzi, *Hydrogen Bonding*, Papers Symposium, Ljubljana 1957 (1959).
66. Y. Marechal and A. Witkowski, *J. Chem. Phys.* **48**, 3697 (1968).
67. E. R. Lippinkott and R. Schröder, *J. Am. Chem. Soc.* **78**, 5171 (1956).
68. T. Matsubara and E. Matsubara, *Prog. Theor. Phys.* **67**, 1 (1982).
69. S. Tanaka, *Phys. Rev. B.* **42**, 1088 (1990).
70. A. Novak, *Struct. Bond.* (Berlin) **18**, 177 (1977).
71. J. P. Sethna, *Phys. Rev. B* **24**, 698 (1981).
72. J. P. Sethna, *Phys. Rev. B* **25**, 5050 (1982).
73. A. J. Bray and M. A. Moore, *Phys. Rev. Lett.* **49**, 1545 (1982).
74. A. L. Fetter and J. D. Walecka, *Quantum Theory of Many-Particle Systems*, McGraw-Hill, New York, 1971, p. 390.
75. S. Nagaoka, T. Terao, F. Imashiro, A. Saika, N. Hirota, and S. Hayashi, *J. Chem. Phys. Lett.* **79**, 4694 (1983).
76. J. M. Clemens, R. M. Hochstrasser, and H. P. Trommsdorff, *J. Chem. Phys.* **80**, 1744 (1984).
77. G. R. Holtom, R. M. Hochstrasser, and H. P. Trommsdorff, *Chem. Phys. Lett.* **131**, 44 (1986).
78. D. F. Brougham, A. J. Horsewill, and H. P. Trommsdorff, *Chem. Phys.* **243**, 189 (1999).
79. V. A. Benderskii, E. V. Vetoshkin, S. Yu. Grebenshchikov, L. von Laue, and H. P. Trommsdorff, *Chem. Phys.* **219**, 119 (1997).

80. V. A. Benderskii, E. V. Vetoshkin, L. von Laue, and H. P. Trommsdorff, *Chem. Phys.* **219**, 143 (1997).
81. V. A. Benderskii, E. V. Vetoshkin, and H. P. Trommsdorff, *Chem. Phys.* **234**, 153 (1998).
82. V. A. Benderskii, E. V. Vetoshkin, *Chem. Phys.* **234**, 173 (1998).
83. V. A. Benderskii, E. V. Vetoshkin, and H. P. Trommsdorff, *Chem. Phys.* **244**, 299 (1999).
84. V. A. Benderskii, E. V. Vetoshkin, I. S. Irgibaeva, and H. P. Trommsdorff, *Chem. Phys.* **262**, 369 (2000); 262, 393 (2000).
85. A. C. Legon, *Chem. Soc. Rev.* **22**, 153 (1993).
86. H. Haken, *Quantenfieldtheorie des Festkörpers*, B. G. Teubner, Stuttgart, 1973; Russian translation: *Quantum Field Theory of Solids*, Nauka, Moscow, 1980, p. 47.
87. T. Holstein, *Ann. Phys. (NY)* **8**, 325 (1959); **8**, 343 (1959).
88. Yu. A. Firsov, ed., *Polarons*, Nauka, Moscow, 1975 (in Russian).
89. S. F. Fischer, G. L. Hofacker, and M. A. Rather, *J. Chem. Phys.* **52**, 1934 (1970).
90. I. I. Roberts, N. Apsley, and R. W. Munn, *Phys. Rep.* **60**, 59 (1980).
91. A. O. Azizyan and M. I. Klinger, *Doklady Acad. Nauk SSSR* **242**, 1046 (1978) (in Russian); *Theor. Math. Fiz.* **43**, 78 (1980) (in Russian).
92. M. I. Klinger and A. O. Azizyan, *Fiz. Tekhn. Poluprovodn.* **13**, 1873 (1979) (in Russian).
93. D. L. Tonks and R. N. Silver, *Phys. Rev. B* **26**, 6455 (1982).
94. R. Bilnc and D. Hadzi, *Mol. Phys.* **1**, 391 (1958).
95. R. Blinc, *J. Chem. Phys. Solids* **13**, 204 (1960).
96. R. Blinc and B. Žecš, *Soft Mode in Ferroelectrics and Antiferroelectrics*, North-Holland, Amsterdam, American Elsevier, New York, 1974.
97. P. G. de Genn, *Solid State Commun.* **1**, 150 (1963).
98. M. Tokunaga and T. Matsubara, *Prog. Theor. Phys.* **35**, 581 (1966).
99. R. Silbey and H. P. Trommsdorff, *Chem. Phys. Lett.* **165**, 540 (1990).
100. S. S. Rozhkov, E. A. Shadchin, and S. P. Sirenko, *Teoret. Eksperim. Khim.* **35**, 343 (1999) (in Russian).
101. E. D. Isaacs, A. Shukla, and P. M. Platzman, *Phys. Rev. Lett.* **82**, 600 (1999).
102. T. Matsubara and K. Kamiya, *Prog. Theor. Phys.* **58**, 767 (1977).
103. A. P. Petrov and A. B. Khovanski, *Zh. Vyssh. Mat. Mat. Fiz.* **14**, 292 (1974) (in Russian).
104. E. A. Shadchin and F. I. Barabash, *J. Mol. Struct.* **325**, 65 (1993).
105. S. Ikeda, H. Sugimoto, and Y. Yamada, *Phys. Rev. Lett.* **81**, 5449 (1998).
106. Y. Yamada and S. Ikeda, *J. Phys. Soc. Jpn.* **63**, 3691 (1994).
107. Y. Yamada, *J. Phys. Soc. Jpn.* **63**, 3756 (1994).
108. H. Sugimoto and S. Ikeda, *Phys. Rev. Lett.* **67**, 1306 (1998).
109. R. P. Bell, *Tunnel Effect in Chemistry*, Chapman and Hall, London, 1980.
110. F. Fillaux, B. Nicolaï, M. H. Baron, A. Lautié, J. Tomkinson, and G. J. Kearly, *Ber. Bunsenges. Phys. Chem.* **102**, 384 (1998).
111. F. Graf, R. Meyer, T. K. Ha, and R. R. Ernst, *J. Chem. Phys.* **75**, 1914 (1981).
112. B. H. Meier, F. Graf, and R. R. Ernst, *J. Chem. Phys.* **76**, 767 (1982).
113. R. Meyer and R. R. Ernst, *J. Chem. Phys.* **86**, 784 (1987).
114. S. Nagaoka, T. Terao, F. Imashiro, A. Saika, and S. Hayashi, *J. Chem. Phys.* **79**, 4694 (1983).

115. A. J. Pertsin and A. I. Kitaigorodsky, *The Atom–Atom Potential Method*, Springer Series in Chemical Physics, Springer, Berlin, 1987.
116. A. Griffin and H. Jobic, *J. Chem. Phys.* **75**, 5940 (1981).
117. H. Jobic and H. Lauter, *J. Chem Phys.* **88**, 5450 (1988).
118. S. W. Lovesey, *Theory of Neutron Scattered from Condensed Matter*, Vol. I: *Nuclear Scattering*, Clarendon Press, Oxford, 1984.
119. F. Fillaux, *Chem. Phys.* **74**, 405 (1983).
120. C. Cohen-Tannoudji, B. Diu, and F. Laloë, *Méchanique Quantique*, Vol. 1, Herman, Paris, 1977, p. 576.
121. S. Ikeda and F. Fillaux, *Phys. Rev. B* **59**, 4134 (1999).
122. J. R. Willison, *Phys. Rev. Lett.* **81**, 5430 (1998).
123. D. F. Gaitan, L. A. Crum, C. C. Church, and R. A. Roy, *J. Acoust. Soc. Am.* **91**, 3166 (1992).
124. R. Hiller, S. J. Putterman, and B. P. Barber, *Phys. Rev. Lett.* **69**, 1182 (1992).
125. B. Compf, R. Günther, G. Nick, R. Pecha, and W. Eisenmenger, *Phys. Rev. Lett.* **79**, 1405 (1997).
126. R. A. Hiller, S. J. Putterman, and K. R. Weninger, *Phys. Rev. Lett.* **80**, 1090 (1998).
127. L. A. Crum and T. J. Maluta, *Science* **276**, 1348 (1997).
128. J. Schwinger, *Proc. Natl. Acad. Sci. USA* **90**, 2105 (1993).
129. C. Eberlein, *Phys. Rev. Lett.* **76**, 3842 (1996).
130. A. Lambrecht, M. T. Jaekel, and S. Reynaud, *Phys. Rev. Lett.* **78**, 2267 (1997).
131. N. Garsia and A. P. Levanyuk, *JETP Lett.* **64**, 907 (1996); *Phys. Rev. Lett.* **78**, 2268 (1997).
132. L. S. Bernstein and M. R. Zakin, *J. Phys. Chem.* **99**, 14619 (1995).
133. T. Lepoint et al., *J. Acoust. Soc. Am.* **101**, 2012 (1997).
134. A. Prosperetti, *J. Acoust. Soc. Am.* **101**, 2003 (1997).
135. I. V. Stasyuk, O. L. Ivankiv, and N. I. Pavlenko, *J. Phys. Studies* **1**, 418 (1997).
136. M. Hubman, *Z. Physik B* **32**, 127 (1979).
137. R. Hassan and E. Campbell, *J. Chem. Phys.* **97**, 4326 (1992).
138. I. V. Stasyuk and A. L. Ivankiv, *Ukr. Fiz. Zhurn.* **36**, 817 (1991) (in Ukrainian); *Mod. Phys. Lett. B* **6**, 85 (1992).
139. D. Eisenberg and W. Kauzman, *The Structure and Properties of Water*, Gidrometeoizdat, Leningrad, 1975 (Russian translation).
140. M. Kunst and J. M. Warman, *Nature* **288**, 465 (1980).
141. N. Sone, M. Yoshida, H. Hirata, and Y. Kagava, *J. Biol. Chem.* **252**, 2956 (1977).
142. H. Okamoto, N, Sone, H. Hirata, and Y. Kagawa, *J. Biol. Chem.* **252**, 6125 (1977).
143. K. Sigrist-Nelson and A. Azzi, *J. Biol. Chem.* **255**, 10638 (1980).
144. B. Hilczer and A. Pawłowski, *Ferroelectrics* **104**, 383 (1990).
145. A. Pawłowski, Cz. Pawlaczyk, and B. Hilczer, *Solid State Ionics* **44**, 17 (1990).
146. A. Pietraszko, K. Łukaszewicz, and M. A. Augusyniak, *Acta Crystallogr. C* **48**, 2069 (1992); **49**, 430 (1993).
147. T. Fukami, K. Tobaru, K. Kaneda, K. Nakasone, and K. Furukawa, *J. Phys. Soc. Jpn.* **63**, 2829 (1994).
148. W. Salejda and N. A. Dzhavadov, *Phys. Status Solidi B* **158**, 119 (1990); **158**, 475 (1990).

149. N. Pavlenko, ICMP-98-26E Preprint of the Institute for Condensed Matter Physics National Academy of Science of Ukraine, 1998 (see also on http://www.icmp.lviv.ua).

150. I. V. Stasyuk and N. Pavlenko, ICMP-98-30U Preprint of the Institute for Condensed Matter Physics National Academy of Science of Ukraine, 1998 (see also on http://www.icmp.lviv.ua).

151. B. V. Merinov, N. B. Bolotina, A. I. Baranov, and L.A. Shuvalov, *Cristallografiya* **33**, 1387 (1988) (in Russian).

152. B. V. Merinov, A. I. Baranov, and L. A. Shuvalov, *Cristallografiya* **35**, 355 (1990) (in Russian).

153. B. V. Merinov, M. Yu. Antipin, A. I. Baranov, A. M. Tregubchenko, and L.A. Shuvalov and Yu. T. Struchko, *Cristallografiya* **36**, 872 (1992) (in Russian).

154. R. Kubo, *Can. J. Phys.* **34**, 1274 (1956); *J. Phys. Soc. Jpn.* **12**, 570 (1957).

155. A. Pawlowski, Cz. Pawlaczyk, and B. Hilczer, *Solid State Ionics* **44**, 17 (1990).

156. J. Grigas, *Microwave Dielectrics spectroscopy of Ferroelectrics and Related Materials*, Gordon and Beach Publishers, 1996.

157. A. V. Belushkin, C. J. Carlile, and L. A. Shuvalov, *Ferroelectrics* **167**, 83 (1995).

158. J. Nagle, M. Mille, and H. J. Morowitz, *J. Chem. Phys.* **72**, 3959 (1980).

159. J. F. Nagle and H. J. Morowitz, *Proc. Natl. Acad. Sci. USA* **75**, 298 (1978).

160. E.-W. Knapp, K. Schulten, and Z. Schulten, *Chem. Phys.* **46**, 215 (1980).

161. V. Ya. Antonchenko, A. S. Davydov, and A. V. Zolotariuk, *Phys. Status Solidi B* **115**, 631 (1983).

162. S. Yomosa, *J. Phys. Soc. Jpn.* **51**, 3318 (1982); **52**, 1866 (1983).

163. P. M. Tomchuk, V. Krasnoholovets, and N. A. Protsenko, *Ukr. Fiz. Zh.* **28**, 767 (1983) (in Russian).

164. V. Krasnoholovets, P. M. Tomchuk, and N. A. Protsenko, Preprint 9/83, Institute of Physics Akademie Nauk UkrSSR, Kyiv, 1983 (in Russian).

165. V. Krasnoholovets and P. M. Tomchuk, *Phys. Status Solidi B* **123**, 365 (1984).

166. N. D. Sokolov, in *Hydrogen Bond*, Nauka, Moscow, 1975, p. 65 (in Russian).

167. H. Merz and G. Zundel, *Biochim. Biophys. Res. Commun.* **101**, 540 (1981).

168. S. Scheiner, *J. Chem. Soc. Faraday Trans. II* **103**, 315 (1981).

169. E. G. Weidemann and G. Zundel, *Z. Naturforsch.* **25a**, 627 (1970).

170. R. Janoschek, E. G. Weidemann, H. Pfeiffer, and G. Zundel, *J. Am. Chem. Soc.* **94**, 2387 (1972).

171. A. S. Davydov, *The Theory of Solid*, Nauka, Moscow, 1976 (in Russian).

172. S. V. Tyablikov, *Methods of Quantum Theory of Magnetism*, Nauka, Moscow, 1965.

173. K. Hattori, *Phys. Rev. B* **23**, 4246 (1981).

174. H. Eigelhardt and N. Riel, *Phys. Lett.* **14**, 20 (1965).

175. M. A. Lampert and P. Mark, *Current Injection in Solids*, Academic Press, New York, 1970.

176. V. Krasnoholovets and P. M. Tomchuk, *Phys. Status Solidi B* **131**, K177 (1985).

177. C. Kittel, *Introduction to Solid State Physics*, Nauka, Moscow, 1978 (Russian translation).

178. V. Krasnoholovets and P. M. Tomchuk, *Phys. Status Solidi B* **130**, 807 (1985)

179. V. Krasnoholovets and P. M. Tomchuk, *Phys. Status Solidi B* **138**, 727 (1986).

180. V. N. Schmidt, J. E. Drumcheller, and F. L. Howell, *Phys. Rev. B* **12**, 4582 (1971).

181. Z. Schulten and K. Schulten, *Eur. Biophys. J.* **11**, 149 (1985).

182. B. I. Stepanov and V. P. Gribkovskii, *Introduction to the Theory of Luminescence*, (Academic Science Belorus SSR Publishing, Minsk, 1963 (in Russian).

183. V. L. Gurevich, *The Kinetics of Phonon System*, Nauka, Moscow, 1980 (in Russian).
184. J. W. Tucker and V. W. Rampton, *Microwave Ultrasonic in Solid State Physics*, Mir, Moscow, 1975 (Russian translation).
185. A. S. Bakay and N. P. Lazarev, *Fiz. Tverd. Tela* **26**, 2504 (1984).
186. Yu. M. Kagan, in *Defects in Insulating Crystals*, V. M. Tushkevich and K. K. Shvarts, eds., Zinatne Publishing House, Riga; Springer-Verlag, Berlin, 1981, p. 17.
187. P. S. Peercy, *Phys. Rev. B* **12**, 2725 (1975).
188. M. C. Laurence and G. N. Robertson, *J. Phys. C* **13**, L1053 (1980).
189. V. V. Bryksin, *JETP* **100**, 1556 (1991) (in Russian).
190. Yu. Kagan and M. I. Klinger, *JETP* **70**, 255 (1976) (in Russian).
191. V. Krasnoholovets, *Ukr. Fiz. Zhurn.* **38**, 740 (1993) (in Ukrainian).
192. H. Haken, *Quantenfeldtheorie des Festkörpers*, B. G. Teubner, Stuttgart, 1973; Russian translation: Nauka, Moscow, 1980.
193. A. I. Baranov, G. F. Dobrzhanskii, V. V. Ilyukhin, V. S. Ryabkin, N. I. Sorokina, and L. A. Shuvalov, *Kristallografiya* **26**, 1259 (1981) (in Russian).
194. A. I. Baranov, L. A. Muadyan, A. A. Loshmanov, L. E. Fykin, E. E. Rider, G. F. Dobrzhanskii, and V. I. Simonov, *Kristallografiya* **29**, 220 (1984) (in Russian).
195. A. I. Baranov, G. F. Dobrzhanskii, V. V. Ilyukhin, V. I. Kalinin, V. S. Ryabkin, and L. A. Shuvalov, *Kristallografiya* **24**, 280 (1979) (in Russian).
196. G. A. Puchkovska and Yu. A. Tarnavski, *J. Mol. Struct.* **267**, 169 (1992).
197. D. F. Baisa, A. I. Barabash, E. A. Shadchin, A. I. Shanchuk, and V. A. Shishkin, *Kristallografiya* **34**, 1025 (1989) (in Russian).
198. Z. M. Alekseeva, G. A. Puchkovska, and Yu. A. Tarnavski, *Structural Dynamic Processes in Disordered Media*, abstracts of papers read at the conference, Samarkand, 1992, Pt. 1, p. 91 (in Russian).
199. I. P. Aleksandrova, Ph. Colomban, F. Denoyer, et al., *Phys. Status Solidi A* **114**, 531 (1989).
200. Ph. Colomban and J. C. Batlot, *Solid State Ionics* **61**, 55 (1993).
201. A. Baranov, *Izvestia AN SSSR, Ser. Fiz.* **51** 2146 (1987) (in Russian).
202. D. F. Baisa, E. D. Chesnokov, and Shamchuk, *Fiz. Tverd. Tela* **32**, 3295 (1990) (in Russian).
203. M. B. Salamon, ed., *Superionic Conductor Physics*, Zinatne, Riga, 1982 (in Russian).
204. Yu. Gurevich and Yu. I. Kharkats, *Advances in Science and Technology*, Vol. 4, Series in Solid-State Chemistry, VINITI, Moscow, 1987, (in Russian).
205. D. Hadzi, *Chimia* **26**, 7 (1972).
206. S. Bratos, J. Lascombe, and A. Novak, in *Molecular interactions*, Vol. 1, H. Ratajczak and W. J. Orville-Thomas, eds., Wiley, New York, 1980, p. 301.
207. J. Baran, *J. Mol. Struct.* **162**, 211 (1987).
208. M. Szafran and Z. Dega-Szafran, *J. Mol. Struct.* **321**, 57 (1994).
209. S. Bratos and H. Ratajczak, *J. Chem. Phys.* **76**, 77 (1982).
210. K. Unterderweide, B. Engelen, and K. Boldt, *J. Mol. Struct.* **322**, 233 (1994).
211. H. Ratajczak and A. M. Yaremko, *Chem. Phys. Lett.* **243**, 348 (1995).
212. G. Zundel, in *The Hydrogen Bond*, P. Schuster, G. Zundel, and C. Sandorfy, eds., North Holland, Amsterdam, 1972, p. 685.
213. A. Barabash, T. Gavrilko, V. Krasnoholovets, and G. Puchkovska, *J. Mol. Struct.* **436–437**, 301 (1997).

214. A. S. Davydov, *Theory of Molecular Excitons*, Nauka, Moscow, 1968 (in Russian).

215. M. P. Lisitsa and A. M. Yaremko, *Fermi-Resonance*, Naukova Dumka, Kyiv, 1984 (in Russian).

216. D. Ostrovski, M. Ya. Valakh, T. A. Karaseva, Z. Latajka, H. Ratajczak, and M. A. Yaremko, *Opt. Spektrosk.* **78**, 422 (1995) (in Russian).

217. A. Barabash, J. Baran, T. Gavrilko, K. Eshimov, G. Puchkovska, and H. Ratajczak, *J. Mol. Struct.* **404**, 187 (1997).

218. A. M. Petrosian, A. A. Bush, V. V. Chechkin, A. F. Volkov, and Yu. N. Venevtsev, *Ferroelectrics* **21**, 525 (1878).

219. E. N. Treshnikov, A. A. Loshmanov, V. R. Kalinin, V. V. Ilyukhin, I. I. Yamzin, L. E. Fykin, V. Ya. Dudarev, and S. P. Solov'ev, *Koordinat. Khimia* **4**, 1903 (1978); **5**, 263 (1979) (in Russian).

220. D. Oesterhelt and W. Stoeckenius, *Nature* **233**, 149 (1971); *Proc. Natl. Acad. Sci. USA* **70**, 2853 (1973).

221. R. H. Lozier, W. Niederberger, R. A. Bogomolni, S. Hwang, and W. Stoeckenius, *Biochim. Biophys. Acta* **440**, 545 (1976).

222. W. Stoeckenius, R. H. Lozier, and R. A. Bogomolni, *Biochim. Biophys. Acta* **505**, 215 (1979).

223. R. Henderson, J. M. Baldwin, R. A. Ceska, F. Zemlin, E. Beckmann, and K. H. Downin, *J. Mol. Biol.* **213**, 899 (1990).

224. A. N. Dencher, G. Papadopoulos, D. Dresselhaus, and G. Büldt, *Biochim. Biophys. Acta* **1026**, 51 (1990).

225. G. Papadopoulos, A. N. Dencher, D. Oesterhelt, H. J. Plöhn, G. Rapp, and G. Büldt, *J. Mol. Biol.* **214**, 15 (1990).

226. A. N. Dencher, D. Dresselhaus, G. Zaccai, and G. Büldt, *Proc. Natl. Acd. Sci. USA* **86**, 7876 (1989).

227. M. H. Koch, A. N. Dencher, D. Oesterhelt, H. J. Plöhn, G. Rapp, and G. Büldt, *EMBO J.* **10**, 521 (1991).

228. T. Mogi, L. J. Stern, T. Marti, B. H. Chao, and H. G. Khorana, *Proc. Natl. Acad. Sci. USA* **85**, 4148 (1988).

229. J. Soppa, J. Otomo, J. Straub, J. Tittor, S. Meessen, and D. Oesterhelt, *J. Biol. Chem.* **264**, 13049 (1989).

230. R. Rammelsberg, G. Huhn, M. Lübben, and K. Gerwert, *Biochemistry* **37**, 5001 (1998).

231. S. Stoeckenius and R. A. Bogomolni, *Annu. Rev. Biochem.* **51**, 587 (1982).

232. R. R. Birge and T. M. Cooper, *Biophys. J.* **42**, 61 (1983).

233. J. Olejnik, B. Brzezinski, and G. Zundel, *J. Mol. Struct.* **271**, 157 (1992).

234. G. Zundel, *J. Mol. Struct.* **322**, 33 (1994).

235. H. G. Khorana, *J. Biol. Chem.* **263**, 7439 (1988).

236. D. Oesterhelt, *Biochem. Int.* **18**, 673 (1989).

237. S. P. A. Fodor, J. B Ames, R. Gebhard, E. M. M. Van der Berg, W. Stoeckenius, J. Lugtenburg, and R. A. Mathies, *Biochemistry* **27**, 7097 (1988).

238. R. R. Birge, *Biochem. Biophys. Acta* **1016**, 293 (1990).

239. J. F. Nagle and S. Tristran-Nagle, *J. Membr. Biol.* **74**, 1 (1983).

240. A. Lewis, *Proc. Natl. Acad. Sci. USA* **75**, 549 (1978).

241. H. Merz and G. Zundel, *Biochem. Biophys. Res. Commun.* **101**, 540 (1981).

242. P. Tavan, K. Schulten, and D. Oesterhelt, *Biophys. J.* **47**, 415 (1985).

243. D. Oesterhelt and J. Tittor, *Trends Biochem. Sci.* **14**, 57 (1989).
244. M. A. El-Sayed, in *Biophysical Studies of Retinal Proteins*, University of Illinois Press, Champaign, 1987, p. 174.
245. S. P. Balashov, R. Govindjee, and T. G. Ebrey, *Biophys. J.* **60**, 475 (1991).
246. S. W. Lin and R. A. Mathies, *Biophys. J.* **56**, 633 (1989).
247. A. Ikagami, T. Kouyama, K. Kinosita, H. Urabe, and J. Otomo, in *Primary Processes in Photobiology*, Springer-Verlag, New York, 1987, p. 173.
248. B. Brzezinski, J. Olejnik, and G. Zundel, *J. Chem. Soc. Faraday Trans.* **90**, 1095 (1994).
249. G. Zundel and B. Brzezinski, *Proton Transfer in Hydrogen-Bonded Systems*, Plenum, New York, 1992, p. 153.
250. M. P. Heyn, R. J. Cherry, and U. Muller, *J. Mol. Biol.* **117**, 607 (1977).
251. J. Y. Huang and A. Lewis, *Biophys. J.* **55**, 835 (1989).
252. K. J. Rothschild, P. Roepe, P. L. Ahl, T. N. Earnest, R. A. Bogomolni, S. K. Das Gupta, C. M. Mulliken, and J. Herzfeld, *Proc. Natl. Acad. Sci. USA* **83**, 347 (1986).
253. L. Eisenstein, S.-L. Lin, G. Dollinger, K. Odashima, J. Termini, K. Konno, W.-D. Ding, and K. Nakanishi., *J. Am. Chem. Soc.* **109**, 6860 (1987).
254. B. Brzezinski, H. Urjasz, and G. Zundel, *Biochem. Biophys. Res. Commun.* **89**, 819 (1986).
255. H. Richter, L. Brown, J. Needleman, and R. Lany, *Biochem.* **35**, 4054 (1996).
256. J. Hebrle and N. A. Dencher, *FEBS Lett.* **277**, 277 (1990).
257. P. Hildebrandt and M. Stockburger, *Biochemistry* **23**, 5539 (1984).
258. G. Popandopoulos, N. A. Dencher, Zaccai, and G. Buld, *J. Mol. Biol.* **214**, 15 (1990).
259. G. Dollinger, L. Eisenstein, S.-L. Lin, K. Nakanishi, and J. Termini, in *Biophysical Studies of Retinal Proteins*, Univiversity of Illinois Press, Champaign, IL, 1987, p. 120.
260. M. S. Braiman, T. Mogi, T. Marti, L. J. Stern, H. G. Khorana, and K. J. Rotshild, *Biochemistry*, **27**, 8516 (1988).
261. L. Eisenstein, S.-N. Lin, G. Dollinger, J. Termini, K. Odashima, and K. Nakanishi, in *Biophysical Studies of Retinal Proteins*, Univiversity of Illinois Press, Champaign, IL, 1987, p. 149.
262. R. R. Birge, *Computer*, November, 56 (1992).
263. L. Eisenstein, S.-N. Lin, G. Dollinger, K. Odashima, J. Termini, K. Konno, W.-D. Ding, and K. Nakanishi, *J. Am. Chem. Soc.* **109**, 6860 (1987).
264. K. Bagley, G. Dollinger, L. Eisenstein, A. K. Singh, and L. Linanyi, *Proc. Natl. Acad. Sci. USA* **79**, 4972 (1982).
265. N. A. Dencher, *Photochem. Photobiol.* **38**, 753 (1983).
266. J. E. Drahein and J. Y. Cassim, *Biophys. J.* **41**, 331 (1983).
267. J. Czege, *FEBS Lett.* **242**, 89 (1988).
268. L. A. Drachev, A. D. Kaulen, and V. V. Zorina, *FEBS Lett.* **243**, 5 (1989).
269. J. Czege and L. Reinisch, *Photochem. Phobiol.* **54**, 923 (1991).
270. M. Sheves, A. Albeck, N. Friedman, and M. Ottolenghi, *Proc. Natl. Acad. Sci. USA* **83**, 3262 (1986).
271. M. Engelhard, K. Gerwert, B. Hess, W. Kreutz, and F. Siebert, *Biochemistry* **24**, 400 (1985).
272. K. Gerwert, G. Souvignier, and B. Hess, *Proc. Natl. Acad. Sci. USA* **87**, 9774 (1990).
273. K. J. Rothschild, M. S. Brainman, Y.-W. He, Th. Marti, and H. G. Khorana, *Biochemistry* **29** 18985 (1990).

274. T. Koboyashi, H. Ohtani, J. Iwai, A. Ikegami, and H. Uchiki, *FEBS Lett.* 162, 197 (1983).
275. A. N. Dencher, G. Popadopoulos, D. Dresselhaus, and G. Buld, *Biochim. Biophys. Acta* **1026**, 51 (1990).
276. H. J. Koch, N. A. Dencher, D. Oesterchelt, H.-J. Plohn, G. Rapp, and G. Buld, *EMBO J.* **10**, 521 (1991).
277. J. Y. Huang, Z. Chen, and A. Lewis, *J. Phys. Chem.* **93**, 3324 (1989).
278. R. R. Birge and C.-F. Zhang, *J. Chem. Phys.* **92**, 7178 (1990).
279. T. Ya. Paperno, V. V. Pozdnyakov, A. A. Smirnova, and L. M. Elagin, *Physics–Chemistry Methods of Investigation in Organic and Biological Chemistry*, Prosveshchenie, Moscow, 1977 (in Russian).
280. S. P. Balashov and Litvin, *Biofizika* **26**, 557 (1981) (in Russian).
281. A. I. Akhieser and S. V. Peletminsky, Methods of Statistic Physics Nauka, Moscow, 1977 (in Russian).
282. V. M. Agranovich and M. D. Galanin, *Electronic Excitation Energy Transfer in Condensed Media*, Nauka, Moscow, 1978 (in Russian).
283. A. V. Sharkov, Yu. A. Matveev, S. V. Chekalin, and A. V. Tsakulev, *Doklady Acad. Nauk SSSR* **281**, 466 (1985) (in Russian).
284. J. Dobler, W. Zinth, W. Kaiser, and D. Oesterhelt, *Chem. Phys. Lett.* **144**, 215 (1988).
285. R. A. Mathies, C. H. Brito Cruz, W. T. Pollard, and C. V. Shank, *Science* **240**, 777 (1988).
286. S. L. Dexheimer, Q. Wang, L. A. Peteanu, W. T. Pollard, R. A. Mathies, and C. V. Shank, *Chem. Phys. Lett.* **188**, 61 (1992).
287. R. R. Birge, T. M. Cooper, A. F. Lawrence, M. B. Masthay, and C. Vasilakis, *J. Am. Chem. Soc.* **111**, 4063 (1989).
288. F. Bartl, G. Decker-Hebestreit, K. H. Altendorf, and G. Zundel, *Biophys. J.* **68**, 104 (1995).
289. B. Brzezinski, P. Radzievski, J. Olejnik, and Zundel, *J. Mol. Struct.* **323**, 71 (1993).
290. J. Heberle, J. Riesle, G. Thiedemann, D. Oesterhelt, and N. A. Dencher, *Nature (London)* **370**, 379 (1994).
291. G. W. Gray and B. Jones, *J. Chem. Soc. (London)* **12**, 4179 (1953).
292. L. M. Babkov, G. A. Puchkovska, S. P. Makarenko, and T. A. Gavrilko, *IR Spectroscopy of Molecular Crystals with Hydrogen Bonds*, Naukova Dumka, 1989 (in Russian).
293. C. Manohar, V. K. Kelkar, and J. V. Yakhmi, *Mol. Cryst. Liquid Cryst.* **49** (Letters), 99 (1978).
294. L. M. Sverdlov, M. A. Kovner, E. P. Krainov, *Vibrational Spectra of Polyatomic Molecules*, Nauka, Moscow, 1970.
295. J. H. Schachtschneider and R. G. Snyder, *Spectrochim. Acta* **19**, 117 (1963).
296. A. V. Iogansen, in *Hydrogen Bond*, Nauka, Moscow, 1981, p. 112 (in Russian).
297. A. S. N. Rao, C. R. K. Murty, and T. R. Reddy, *Phys. Status Solidi A* **68**, 373 (1981).
298. J. D. Bernal and D. Crowfood, *Trans. Farad. Soc.* **29**, 1032 (1933).
299. P. Excoffon and Y. Mareshal, *Spectrochim. Acta A* **28**, 269 (1972).
300. I. V. Bargatin, B. A. Grishanin, and V. N. Zadkov, *Uspekhi Fiz. Nauk* **171**, 625 (2001) (in Russian).
301. D. L. Haycock, P. M. Alsing, I. H. Deutsch, J. Grondalski, and P. S. Jessen, *Phys. Rev. Lett.* **85**, 3365 (2000).
302. J. Clarke et al., *Science* **239**, 992 (1988).

303. D. M. Stamper-Kurn, H.-J. Miesner, A. P. Chikkatur, S. Inouye, J. Stenger, and W. Ketterle, *Phys. Rev. Lett.* **83**, 661 (1999).

304. W. Wernsdorfer, E. Benoit, D. Mailly, O. Kubo, K. Hasselbach, A. Benoit, D. Mailly, O. Kubo, H. Nakamo, and B. Barbara, *Phys. Rev. Lett.* **79**, 4014 (1997).

305. E. M. Chudnovsky and J. Tejada, *Macroscopic Quantum Tunneling of the Magnetic Moment*, Cambridge University Press, Cambridge, England, 1998.

306. E. M. Chudnovsky, L. Gunther, *Phys. Rev. Lett.* **60**, 661 (1988).

307. R. Sessoli, D. Gatteschi, A. Caneschi, and M. Novak, *Nature* (London) **365**, 141 (1993).

308. J. R. Friedman, M. P. Sorchik, J. Tejada, and R. Zilo, *Phys. Rev. Lett.* **76**, 3830 (1996).

309. C. P. Bean and J. D. Livingston, *J. Appl. Phys.* **30**, 120S (1959).

310. W. T. Coffey, *J. Mol. Liq.* **51**, 77 (1992); *J. Chem. Phys.* **99**, 3014 (1993); **107**, 4960 (1997).

311. W. T. Coffey, V. I. Gaiduk, B. M. Tsetlin, and M. E. Walsh, *Physica A* **282**, 384 (2000).

312. P. C. Fannin and W. P. Coffey, *Phys. Rev. E* **52**, 6129 (1995).

313. Yu. P. Kalmykov, W. T. Coffey, and J. T. Waldron, *J. Chem. Phys.* **105**, 2112 (1996).

314. W. T. Coffey, Y. P. Kalmykov, and E. S. Massawe, *Phys. Rev. E* **48**, 699 (1993).

315. W. T. Coffey and Y. P. Kalmykov, *J. Magn. Magn. Mater.* **164**, 133 (1996).

316. H. Kachkachi, W. T. Coffey, D. S. F. Crothers, A. Ezzir, E. C. Kennedy, M. Noguès, and E. Tronc, *J. Phys. Condens. Matter* **12**, 3077 (2000).

317. W. T. Coffey, D. S. F. Crothers, J. L. Dormann, L. J. Geoghegan, E. C. Kennedy, and W. Wernsdorfer, *J. Phys. Condens. Matter* **10**, 9093 (1998).

318. W. T. Coffey, D. S. F. Crothers, J. L. Dormann, L. J. Geoghegan, E.C. Kennedy, *J. Magn. Magn. Mater.* **173**, L219 (1997).

319. P. M. Tomchuk and V. Krasnoholovets, *J. Mol. Struct.* **416**, 161 (1997).

320. P. M. Tomchuk and S. P. Lukyanets, *J. Mol. Struct.* **513**, 35 (1999).

321. P. M. Tomchuk and S. P. Lukyanets, *Cond. Matter Phys.* **1**, 203 (1998).

322. P. M. Tomchuk and S. P. Lukyanets, *Phys. Status Solidi B* **218**, 291 (2000).

323. A. S. Davydov, *Solitons in Molecular Systems*, Reidel, Dordrecht, 1986.

324. A. I. Sergienko, *Phys. Status Solidi B* **144**, 471 (1987).

325. O. E. Yanovskii and E.S. Kryachko, *Phys. Status Solidi B* **147**, 69 (1988).

326. R. B. Laughlin, *Rev. Mod. Phys.* **71**, 863 (1999).

327. R. D. Fedorovich, D. S. Inosov, O. E. Kiyaev, S. P. Lukyanets, P. M. Tomchuk, A. G. Naumovets, *Proc. SPIE* (2001), in press.

328. D. D. Awschalom, J. F. Smyth, G. Grinstein, D. P. DiVinchenzo, and D. Loss, *Phys. Rev. Lett.* **68**, 3092 (1992).

329. J. Tejada, X. X. Zhang, E. del Barco, J. M. Hernandez, and E. M. Chudnovsky, *Phys. Rev. Lett.* **79**, 1754 (1997).

330. M. J. O'Shea and P. Perera, *J. Appl. Phys.* **76**, 6174 (1994).

331. S. Washburn, R. A. Webb, and S. M. Faris, *Phys. Rev. Lett.* **54**, 2712 (1985).

332. A. Garg and G.-H. Kim, *Phys. Rev. B* **45**, 12921 (1992).

333. H.-B. Braun and D. Loss, *Phys. Rev. B* **53**, 3237 (1996).

334. R. Hassan and E. Campbell, *J. Chem. Phys.* **97**, 4326 (1992).

335. L. Weil, *J. Chem. Phys.* **51**, 715 (1954).

336. P. Politi, A. Rettori, F. Hartmann-Boutrn, and J. Villain, *Phys. Rev. Lett.* **75**, 537 (1995).

337. A. Rahman and F. H. Stillinger, *J. Am. Chem. Soc.* **95**, 7943 (1973).
338. F. H. Stillinger, *Adv. Chem. Phys.* **31**, 1 (1975).
339. M. G. Sceats and S. A. Rice, *J. Chem. Phys.* **72**, 3236 (1980); **72**, 3248 (1980); **72**, 3260 (1980).
340. M. G. Sceats and S. A. Rice, *Water and Aqueous Solutions at Temperature Below 0° C*, Naukova Dumka, Kyiv, 1985, pp. 76, 346 (Russian translation).
341. G. G. Malenkov, *Physical Chemistry. Contemporary Problems*, Ya. M. Kolotyrkin, ed., Khimia, Moscow, 1984, p. 41 (in Russian).
342. Y. I. Naberukhin, *Zh. Strukt. Khim.* **25**, 60 (1984) (in Russian).
343. V. Y. Antonchenko, A. S. Davydov, and V. V. Ilyin, *The Essentials of Physics of Water*, Naukova Dumka, Kyiv, 1991 (in Russian).
344. C. Rønne, P.-O. Åstrand, and S. R. Keiding, *Phys. Rev. Lett.* **82**, 2888 (1999).
345. *Water: A Comprehensive Treatise*, Vols. 1–7, F. Franks, ed., Plenum Press, New York, 1972, p.181.
346. C. A. Angell, *Annual Review of Physical Chemistry*, Vol. 34, Annual Reviews, Palo Alto, 1983, p. 593.
347. C. A. Angell, *Nature (London)* **331**, 206 (1988).
348. P. H. Poole, F. Sciorino, U. Essmann, and H. E. Stanley, *Nature (London)* **360**, 324 (1992).
349. N. Agmon, *J. Phys. Chem.* **100**, 1072 (1996).
350. H. Tanaka, *Nature (London)* **380**, 328 (1996).
351. *Supercooled Liquids*, Vol. 676, J. T. Fourkas, D. Kivelson, U. Mohanty, and K. A. Nelson, eds., American Chemical Society, Washington, D.C., 1997.
352. S. Woutersen, U. Emmerichs, and H. J. Bakker, *Science* **278**, 658 (1997).
353. O. Mishina and H. E. Stanley, *Nature (London)* **392**, 164 (1998); **396**, 329 (1998).
354. J. Ortega, J. P. Lewis, and O. F. Sankey, *Phys. Rev. B* **50**, 10516 (1994).
355. P. H. Poole, F. Sciortino, T. Grande, H. E. Stanley, and C. A. Angell, *Phys. Rev. Lett.* **73**, 1632 (1994).
356. F. Scortino, P. H. Poole, U. Essmann, and H. E. Stanley, *Phys. Rev. E* **55**, 727 (1997).
357. C. T. Moynhian, *Mater. Res. Soc. Symp. Proc.* **455**, 411 (1997).
358. V. I. Gaiduk and V. V. Gaiduk, *Russ. J. Phys. Chem.* **71**, 1637 (1997).
359. V. D. Zelepukhin, I. D. Zelepukhin, and V. Krasnoholovets, *Khimich. Fiz.* **12**, 992 (1993) (in Russian); English translation: *Sov. J. Chem. Phys.* **12**, 1461 (1994).
360. A. V. Kondrachuk, V. Krasnoholovets, A. I. Ovcharenko, and E. D. Chesnokov, *Khimich. Fiz.* **12**, 1006 (1993) (in Russian); English translation: *Sov. J. Chem. Phys.* **12**, 1485 (1994).
361. V. D. Zelepukhin and I. D. Zelepukhin, *A Clue to "Live" Water*, Kaynar, Alma-Ata, 1987 (in Russian).
362. T. H. Maugh II, *Science* **202**, 414 (1978).
363. D. Freifelder, *Physical Biochemistry*, Mir, Moscow, 1980 (Russian translation).
364. L. I. Antropov, *Theoretical Electrochemistry*, Vysshaya Shkola, Moscow, 1980 (in Russian).
365. C. Kittel, *Statistical Thermodynamics*, Nauka, Moscow, 1977 (Russian translation).
366. L. Pauling, *Nature of the Chemical Bond and the Structure of Molecules and Crystals: An Introduction to Modern Structural Chemistry*, 3rd ed., Cornell University Press, New York, 1960.
367. G. A. Krestov, *The Thermodynamics of Ionic Processes in Solutions*, Khimia, Leningrad, 1973 (in Russian).

368. M. K. Karapetiants, *Introduction to the Theory of Chemical Processes*, Vysshaya Shkola, Moscow, 1981 (in Russian).
369. R. Langner and G. Zundel, *Can. J. Chem.* **79**, 1376 (2001).
370. A. A. Aleksandrov and M. S. Trakhtenberg, *The Thermophysical Properties of Water at Atmospheric Pressure*, Izdatel'stvo Standartov, Moscow, 1980 (in Russian).
371. S. L. Rivkin and A. A. Aleksandrov, *The Thermophysical Properties of Water and Water Vapor*, Energia, Moscow, 1977 (in Russian).
372. N. B. Vargaftik, *Handbook on the Thermophysical Properties of Fluids* Nauka, Moscow, 1972 (in Russian).
373. O. Y. Samoilov, *The Structure of Aqueous Solutions of Electrolytes and Hydration of Ions*, Izdatel'stvo AN SSSR, Moscow, 1955 (in Russian).
374. V. A. Rabinovich and Z. Y. Khavin, *Concise Chemical Handbook*, Khimiya, Moscow, 1978 (in Russian).
375. Y. B. Zeldovich, *Zh. Exp. Teor. Fiz.* **12**, 525 (1942) (in Russian).
376. E. N. Harvey, W. D. McElroy, and A. H. Whiteley, *J. Appl. Phys.* **18**, 162 (1947).
377. Y. E. Geguzin, *Bubbles*, Nauka, Moscow, 1985 (in Russian).
378. A. M. Blokh, *The Structure of Water and Geological Processes*, Nedra, Moscow, 1969 (in Russian).
379. N. F. Bondarenko and E. Z. Gak, *Electromagnetic Phenomena in Natural Water*, Gidrometeoizdat, Leningrad, 1984 (in Russian).
380. G. V. Yuknevich, *Zh. Struct. Khim.* **25**, 60 (1984); Preprint No. 86-102P, (Institute Theoretical Physics Ukrainian Academy of Science, Kyiv, 1986 (in Russian).
381. A. Kaivarainen, *Am. Inst. Phys. Conf. Proc. (New York)* **573**, 181 (2001); *arXiv.org e-print archive* physics/0102086.
382. M. Chaplin, http://www.sbu.ac.uk/water.
383. I. Z. Fisher and B. N. Adamovich, *Zh. Strukt. Khim.* **4**, 818 (1963) (in Russian).
384. P. A. Egelstaff, *Adv. Phys.* **11**, 203 (1962).
385. G. P. Gordeev and T. Khaidarov, *Water in Biological Systems and Their Components*, Leningrad State University, Leningrad, 1983, p. 3; (in Russian).
386. W. A. R. Luck, *Angew. Chem. Int. Engl.* **19**, 28 (1980).
387. R. Hausser, *Z. Naturforsch.* **18**, 1143 (1963).
388. D. W. G. Smith and J. G. Powles, *Mol. Phys.* **10**, 451 (1966).
389. J. C. Hindman, Svirmickas, and M. Wood, *J. Chem. Phys.* **59**, 1517 (1973).
390. J. C. Hindman, *J. Chem. Phys.* **60**, 4483 (1974).
391. J. Jonas, T. De Fries, and D. J. Wilbur, *J. Chem. Phys.* **65**, 582 (1976).
392. E. W. Lang and H. D. Ludeman, *J. Chem. Phys.* **67**, 718 (1977).
393. N. A. Melnichenko and V. I. Chizhik, *Zh. Strukt. Khim.* **22**, 76 (1981).
394. V. V. Mank and Lebovka, *NMR Spectroscopy of Water in Heterogeneous Systems*, Naukova Dumka, Kyiv, 1988 (in Russian).
395. V. I. Jaskichev, *Adv. Mol. Relaxation Interact. Processes* **24**, 157 (1982).
396. V. I. Jaskichev, V. P. Sanygin, and Y. P. Kalmukov, *Zh. Fiz. Khim.* **57**, 645 (1983).
397. T. Kimura, K. Yamamoto, T. Shimizu, and S. Yugo, *Thin Solid Films* **70**, 351 (1980).
398. K. Nanaka and M. Iwata, *Thin Solid Films* **81**, L85 (1981); **86**, 279 (1981).
399. R. Muller and H. Pagnia, *Matt. Lett.* **2**, 283 (1984).

400. N. A. Tolsoi, *Dokl. Akad. Nauk SSSR* **110**, 893 (1955) (in Russian).

401. N. A. Tolstoi and A. A. Spartakov, *Kolloid. Zh.* **28**, 580 (1966) (in Russian).

402. E. McCafferty and A. C. Zettlemoyer, *Discuss. Farad. Soc.* **52**, 239 (1971).

403. R. McIntosh, *Dielectric Behavior of Physically Absorbed Gases*, Marcel Dekker, New York, 1966.

404. N. A. Tolstoi, A. A. Spartakov, and A. A. Tolstoi, *Kolloid. Zhurn.* **28**, 881 (1966) (in Russian).

405. H. Stobbe and G. Peschel, *Colloid. Polym. Sci.* **275**, 162 (1997).

406. B. V. Deriagin and Y. V. Shulepov, *Surf. Sci.* **81**, 149 (1979).

407. I. R. Yukhnovsky and Y. V. Shulepov, Preprint ITP No. 83-103E, Institute of Theoretical Physics, Kyiv (1983).

408. V. Krasnoholovets and P. M. Tomchuk, *Ukr. Fiz. Zh.* **36**, 1392 (1991) (in Russian); English translation: *Ukr. J. Phys.* **36**, 1106 (1991).

409. W. Drost-Hansen, *Phys. Chem. Liquids* **7**, 243 (1978).

410. N. I. Lebovka and V. V. Mank, in *The State of Water in Different Physical Chemical Conditions*, Leningrad University, Leningrad, 1986, p. 84 (in Russian).

411. E. D. Belotsky and B. I. Lev, *Theor. Math. Phys.* **60**, 120 (1984) (in Russian).

412. V. Krasnoholovets and B. Lev, *Ukr. Fiz. Zh.* **33**, 296 (1994) (in Ukrainian).

413. B. I. Lev and A. Y. Zhugaevich, *Phys. Rev. E* **57**, 6460 (1998).

414. V. Krasnoholovets and B. Lev, *Cond. Math. Phys.*, submitted (also cond-mat/0210131).

415. A. G. Khachaturian, *The Theory of Phase Transitions and the Structure of Solid Solutions*, Nauka, Moscow, 1974 (in Russian).

416. V. B. Kobyliansky, *Statistical Physics*, Vyscha Schkola, Kyiv, 1972 (in Ukrainian).

417. J. S. Metzik, V. D. Perevertaev, V. A. Liopo, G. T. Timoshchenko, and A. B. Kiselev, *J. Colloid Interface Sci.* **43**, 662 (1973).

418. A. Gupta and M. M. Shamura, *J. Colloid Interface Sci.* **149**, 392 (1992).

419. W. A. P. Luck, *Structure of Water and Aqueous Solutions*, Verlag Chemie, Weinheim, 1974.

AUTHOR INDEX

Numbers in parentheses are reference numbers and indicate that the author's work is referred to although his name is not mentioned in the text. Numbers in *italic* show the pages on which the complete references are listed.

Abramowitz, M., 14(55), 27(55), 55-56(55), 70-71(55), *95*
Ackermann, J., 45(129), *97*
Adamovich, B. N., 501(383), 522(383), *547*
Adamowicz, L.: 45(133), 50(133), *97*; 102(61-63), 110(142), *140, 142*
Adamowsky, J., 65(171), *98*
Agmon, N., 490(349), *546*
Agranovich, V. M., 456(282), 518(282), *544*
Ahl, P. L., 448(252), *543*
Ahlrichs, R., 112(151), *142*
Akheiser, A. I., 452(281), *544*
Albeck, A., 448(270), *543*
Alberts, I. L., 102(15), *138*
Aleksandrov, A. A., 496(370-371), *547*
Aleksandrova, I. P., 429(199), *541*
Alekseeva, Z. M., 428(198), *541*
Alivisatos, A. P., 81(186), *98*
Allen, L., 192(40), 205(40), *266*
Allen, W. D., 107-108(124), 124-125(236), *142, 145*
Alsing, P. M., 469(301), 494(301), *544*
Altendorf, K. H., 456(288), 470(288), 485(288), *544*
Alvarez, M. M., 81(188), *98*
Alvarez-Collado, J. R., 271(31), 296(31), *346*
Ames, J. B., 446(237), *542*
Amestony, P. R., 47(141), *97*
Amos, R. D., 102(19-20), 115(208), 118(225-226), *139, 144*
Anderson, J. B., 103(80), *140*
Anderson, K., 112(154-155), *143*
Andrews, J. S., 115(208), 118(225-226), *144*
Angell, C. A., 490(346-347), *546*
Antipin, M. Yu., 395(153), 400(153), *540*
Antonchenko, V. Ya., 400(161), 470(161), 490(343), *540, 546*
Antropov, L. I., 493(364), *546*
Apsley, N., 368(90), *538*
Aquilanati, V., 270(11-12,14), 272(42), *345–347*
Arfken, G. B., 279(52), 310(52), 313(52), *347*
Armour, E. A. G., 50(147), *97*
Arnold, D. W., 127(237), *145*
Arteca, G. A., 39(113), *96*
Ashoori, R. C., 81(195), *98*
Asmis, K. R., 133(254), *145*
Åstrand, P.-O., 490(344), 491(344), *546*
Auer, A. A., 102(44), 119(44), 121(234), *139, 145*
Augusyniak, M. A., 390(146), *539*
Avery, J., 5(29,35), 6(29), 11(44), *94*
Avron, J. E., 203(45-46), *266*
Awschalom, D. D., 474-475(328), 485(328), *545*
Azizyan, A. O., 368(91-92), *538*
Azzi, A., 389-390(143), *539*

Babkov, L. M., 356(61), 360-361(292), 465(292), *537, 544*
Bacic, Z., 270(15), *346*
Baddour-Hadjean, R., 359(26), *536*
Bagley, K., 448(264), *543*
Baisa, D. F., 428(197), 429-430(202), *541*
Bakay, A. S., 421(185), *541*
Baker, J. D., 33(102), 38(102), 44(102), 52(102), *96*
Bakker, H. J., 490(352), *546*
Balashov, S. P., 446(245), 448(245), 451(280), *543–544*
Baldwin, J. M., 445-446(223), *542*
Bally, T., 105(102,104), *141*
Baloitcha, E., 271(21), 291(21), *346*
Bandrauk, A. D., 215(70), *267*
Banin, U., 81(196), *98*
Barabara, B., 469(304), 494(304), *545*

Barabash, F. I., 374(104), 428(197), 438(213), 444(213,217), *538, 541–542*
Baran, J., 355(44), 438(207), 444(217), *536, 541–542*
Baranov, A. I., 395(151-153), 400(152-153), 425(193), 428(193-195), 429(193,201), 436(193,195), *540–541*
Barber, B. P., 526(124), *539*
Barber, M. N., 3(23), 19(23), *94*
Barbiellini, B., 353(2), *535*
Bardeen, J., 354(18), *536*
Bargatin, I. V., 469(300), *544*
Barnes, L. A., 130(245), *145*
Barnett, R. N., 103(81), *141*
Baron, M. H., 353(8-9), 374(110), 375(8-9,110), 378(8-9), 523(8-9), *535, 538*
Bartl, F., 456(288), 470(288), 485(288), *544*
Bartlett, R. J., 102(3,5,13,16,18-19,21, 27,29,38,40-41), 103(38,40,71-72), 104(89,91), 105(109), 107(3,120), 108(127), 109(38,40,127-128,131-133), 110(71-72,89,135,137,143), 112(152), 113(185,187,190-191,196-197), 116(210), 118(223-224,228), 127(238), 129(243), 130(109), 134(258), 135(238,260,265), *138, 140–146*
Bates, D. R., 43(120), *96*
Batlot, J. C., 429(200), *541*
Bawendi, M. G., 81(187), *98*
Baxter, D. V., 2(6), *93*
Baym, G., 22(83), *95*
Beale, P. D., 21(74), *95*
Bean, C. P., 469(309), *545*
Beckmann, E., 445-446(223), *542*
Beddoni, A., 270(13), *346*
Bednarek, S., 65(171), *98*
Bell, J. L., 295(56), *348*
Bell, R. P., 526(109), *538*
Bellac, M. L., 75(177), *98*
Bellissard, J., 151(4), *265*
Belotsky, E. D., 513(411), *548*
Belushkin, A. V., 400(157), *540*
Belushkin, S., 354(26), *536*
Benderskii, V., 354(23), 359(23), 364(79-84), 365(83), 366(83-84), 523(79-84), 532(82), *536–538*
Benderson, B., 2(1), *93*
Benenti, G., 91(209), *99*
Benguria, R., 44(124), *96*
Benoit, A., 469(304), 494(304), *545*
Benoit, E., 469(304), 494(304), *545*
Bergmann, K., 212(63), 214(69), 215(71-72), 219(72), 225(63), 226(63,69,74), 227(69), 235(81), 246(84), *267*
Berkovits, D., 43(123), *96*
Berman, P. R., 207(59), *266*
Bernal, J. D., 464(298), *544*
Bernstein, L. S., 526(132), *539*
Berry, M. V., 203(44), 204(53-54), 205(53), *266*
Berry, R. S., 3(20), *93*
Beverly, K. C., 81(193), 82(198-199), *98*
Bhatia, A. K., 34(106), *96*
Bhattacharyya, B., 82(204), *98*
Bialynicki-Birula, I., 154(7-8), 156-157(7-8), 161(7), 262(8), *265*
Bialynicki-Birula, Z., 154(7), 156-157(7), 161(7), *265*
Billy, N., 46(138), *97*
Binkley, J. S., 102(12), *138*
Binney, J. J., 2(2), *93*
Birge, R. R., 446(232,238), 447(238), 448(262), 451(238,278), 456(287), 457(238,262), *542–544*
Blekher, P., 166(19), 260(19), *265*
Blinc, R., 368(94-96), 369(96), 479(96), 482(96), 486(96), *538*
Blokh, A. M., 498(378), *547*
Boaretto, E., 43(123), *96*
Bobomolni, R. A., 445(221-222), 446(221-222,231), 448(252), *542–543*
Bohn, J. L., 3(17), *93*
Boldt, K., 438(210), *541*
Bolotina, N. B., 395(151), *540*
Bomble, Y., 136(269), *146*
Bondarenko, N. F., 499(379), *547*
Bone, R. G. A., 115(208), *144*
Borden, W. D., 104(94), 105(104), 127(94), *141*
Bordon, W. D., 105(103), *141*
Bradforth, S. E., 127(237), *145*
Braiman, M. S., 448(260), 449(273), *543*
Brändas, E., 39(111-112), 55(163), *96–97*
Bratos, S., 438(206,209), *541*
Braun, H.-B., 475(333), *545*
Bray, A. J., 361(73), *537*
Bressanini, D., 80(182), *98*
Breuer, H. P., 213(66), *267*
Brito Cruz, C. H., 456(285), *544*
Brooks, B. R., 102(10), *138*
Brougham, D. F., 364(78), 523(78), *537*

Brown, L., 448(255), *543*
Brown, W. B., 50(147), *97*
Brueckner, K. A., 105(105), 107(105), *141*
Bruijn, N. G., 12(46), *94*
Brunet, J.-P., 271(18), 290-291(18), *346*
Bryskin, V. V., 422(189), *541*
Brzezinski, B., 446(233,248-249,254), 448-449(233), 451(233), 456(289), 470(249,289), 475(249,289), 485(249,289), *542–544*
Bulboaca, I., 13(48), *94*
Buld, G., 445-446(224-227), 448(224,258), 451(258,275-276), *542–544*
Bulirsch, R., 27(89), 34(89), 40(89), *96*
Burkhardt, T. W., 84(206), *98*
Burrows, B . L., 65(170), *98*
Bush, A. A., 444(218), *542*
Byrd, E. F. C., 114(201-202), *144*

Cabrera, M., 6(39), *94*
Califano, S., 298(57), *348*
Camblong, H. E., 3(18), 16(18), *93*
Campbell, E., 381(137), 485(334), *539, 545*
Caneschi, A., 469(307), 474-475(307), 485(307), *545*
Cao, Y. W., 81(196), *98*
Cardy, J. L., 3(25), 19-21(25), 63(25), 78(25), *94*
Carignano, M., 3(21), *93*
Carini, J. P.: 2(3,6), 4(3), 22(3), *93*
Carlile, C. J., 400(157), *540*
Carney, G. D., 270(5), *345*
Carrington, T., 271(24-25,28), 272(24-25,46), 291(24-25,28), 292(28,55), *346–347*
Carruthers, P., 156(12), *265*
Carter, S., 270(6), *345*
Cassim, J. Y., 448(266), *543*
Castro, E. A., 39(113), *96*
Cavagnero, B. D., 3(17), *93*
Cavalli, S., 270(11-12,14), 272(42), *345–347*
Cederbaum, L. S.: 3(12), 16(61), 45(12), *93, 95*; 113(163,170-171), 120(230), 127(239), 137(270-272), *143, 145–146*
Certain, P. R., 55(160), *97*
Ceska, R. A., 445-446(223), *542*
Chaim, K.-H., 207(59), *266*
Chakraborty, T., 81(185), *98*
Chakravorty, S. J., 43(117), *96*
Chandre, C., 168(26-27), *266*
Chao, B. H., 446(228), *542*
Chaplin, M., 501(382), *547*
Chapuisat, X., 271(16-20), 280(17), 290(17-18), 291(17-20), 312(65), *346, 348*
Chatterjee, A., 5(33), *94*
Chaudhari, P. R., 82(198), *98*
Chechkin, V. V., 444(218), *542*
Chekalin, S. V., 456(283), *544*
Chelkowski, S., 215(70), *267*
Chen, Z.: 45(132), *97*; 451(277), *544*
Chepilko, N., 312(62), *348*
Cherry, R. J., 447(250), 451(250), *543*
Chesnokov, E. D., 429-430(202), 491(360), *541, 546*
Chihzhik, V. I., 502-503(393), *547*
Chikkatur, A. P., 469(303), *545*
Child, M. S., 270(7), *345*
Chiles, R. A., 105(107), 129(107), *141*
Christiansen, O., 102(42), 107(123), 108(125), 109(134), *139, 142*
Chu, S. I., 151(5-6), 166(5-6,17), *265*
Chuang, Y.-Y., 102(53), *140*
Chudnovsky, E. M., 469(305-306), 474-475(306,329), 485(305,329), *545*
Chuenkov, V. A., 356(64), *537*
Church, C. C., 526(123), *539*
Cioslowski, J., 113(175), *143*
Cizek, J., 104(96), 107(116-117), 108(116), 119(96,229), *141, 145*
Cizek, M., 66(172), *98*
Clarke, J., 469(302), *544*
Clemens, J. M., 362(76), *537*
Coester, F., 107(113-114), *141*
Coffey, W. T., 469(310-316), 470(315-318), *545*
Cohen, E. R., 52(154), *97*
Cohen, M., 65(170), *98*
Cohen, R. D., 133(253), *145*
Cohen-Tannoudji, C.: 154(10), 175(10), *265*; 377(120), *539*
Cole, L. A., 42(116), *96*
Cole, S. J., 110(135), *142*
Collier, C. P., 81(189), 90(207), *98–99*
Colomban, Ph., 429(199-200), *541*
Colwell, S. M., 271(30), 296(30), *346*
Comeau, D. C., 113(191), *144*
Compagno, G., 154(11), *265*
Compf, B., 526(125), *539*
Compton, R. N.: 3(12), 45(12), *93*; 16(61), *95*
Continentino, M. A., 91(210), *99*
Cook, J. C., 359(25), *536*
Cooper, F., 13(51), *94*
Cooper, T. M., 446(232), 456(287), *542, 544*

Corben, H. C., 288(54), *348*
Cotton, F. A., 300(59), *348*
Cousson, A., 353(10), 375(10), 523(10), *535*
Crawford, T. D., 102(8), 105(111), 110(136), 124-125(236), 127(241), 128(242), 129-130(241), *138, 141–142, 145*
Cremer, D., 102(4,13,17), 107(121), *138, 142*
Crothers, D. S. F., 469(316), 470(316-318), *545*
Crowfood, D., 464(298), *544*
Crum, L. A., 526(123,127), *539*
Curtis, A. R., 47(140), *97*
Curtiss, L. A., 103(73-75), *140*
Cuthbert, J. D., 355(29,31), 389(29), 408(29), 415(29), *536*
Czege, J., 448(267,269), *543*

Da Costa, R. C. T., 312(61), *348*
Dalgaard, E., 113(183), *143*
Dallos, M., 102-103(2), 112(2), *138*
Danileiko, M. V., 219(73), 226(73), *267*
Das Gupta, S. K., 448(252), *543*
DasSarma, S., 81(190), *98*
Datta, B., 113(177), *143*
Davidson, E. R.: 43(117), *96*; 104(94), 106(112), 112(112), 117(212), 118(227), 127(94), 135(263), *141, 145*
Davis, J. P., 204(49), 220(49), *266*
Davis, T. A., 47(141), *97*
Davydov, A. S., 400(161), 403(171), 439(171,214), 442(214), 470(161,323), 482(323), 486(323), 490(343), *540, 542, 545–546*
Decker-Hebestreit, G., 456(288), 470(288), 485(288), *544*
De Fries, T., 502-503(391), *547*
Dega-Szafran, Z., 438(208), *536, 541*
DeGenn, P. G., 368(97), *538*
Delande, D., 46(138), *97*
Del Barco, E., 474-475(329), 485(329), *545*
Deleuze, M., 135(261-262), *145*
Delhalle, J., 135(261), *145*
Delon, A., 3(14), *93*
Dencher, A. N., 445-446(224-227), 448(224,256,258,265), 451(258,275-276), 460(290), *542–544*
Denis, J. P., 135(261), *145*
Denoyer, F., 429(199), *541*
Deriagin, B. V., 507(406), *548*
Derrida, B., 21(75,77), 25(77), *95*
De Seze, L., 21(77), 25(77), *95*
Desgranges, L., 354(26), *536*
Deutsch, I. H., 469(301), 494(301), *544*
Dexheimer, S. L., 456(286), *544*
Diercksen, G., 117(211), *144*
Ding, C. F., 45(128), *97*
Ding, W.-D., 448(253,263), *543*
Dion, C., 198(43), 235(43), *266*
Dirac, P. A. M., 157(14), *265*
Diu, B., 377(120), *539*
DiVinchenzo, D. P., 474-475(328), 485(328), *545*
Dobler, J., 456(284), *544*
Dobrzhanskii, G. F., 425(193), 428(193-195), 429(193), 436(193,195), *541*
Dobson, C. M., 105(100), *141*
Doi, M., 78(179), *98*
Dollinger, G., 448(253,259,261,263-264), *543*
Domany, E., 5(28), *94*
Domb, C., 63(166), *97*
Domcke, W., 113(163,171), 120(230), *143, 145*
Dormann, J. L., 470(317-318), *545*
Dorst, L., 322(78), *349*
Downin, K. H., 445-446(223), *542*
Dowrick, N. J., 2(2), *93*
Drachev, L. A., 448(268), *543*
Drahein, J. E., 448(266), *543*
Drake, G. W. F., 34(104), 38(110), 39(114), *96*
Drese, K., 204-205(47,56), 235(47), *266*
Dresselhaus, D., 445-446(224,226), 448(224), 451(275), *542, 544*
Drost-Hansen, W., 508(409), *548*
Drumcheller, J. E., 419(180), *540*
Dudarev, V. Ya., 444-445(219), *542*
Duff, I. S., 47(141), *97*
Dunning, T. H. Jr., 102(47,49,51,56-58), 103(83), *139–141*
Dupont, J.-M., 154(9), 262(9), *265*
Dupont-Roc, J., 154(10), 175(10), *265*
Duxbury, P. M., 21(74), *95*
Dyck, R. S., 52(155), *97*
Dykhne, A. M., 204(48), 220(48), *266*
Dykstra, C. E., 105(107), 129(107), *141*
Dzhavadov, N. A., 390(148), *539*

Earnest, T. N., 448(252), *543*
Eberlein, C., 526(129), *539*
Ebrey, T. G., 446(245), 448(245), *543*
Edwards, S. F., 78(179), *98*
Egelstaff, P. A., 501(384), 522(384), *547*
Eigelhardt, H., 408(174), *540*

Eisenberg, D., 381(139), 490(139), 496(139), 501(196), *539*
Eisenmenger, W., 526(125), *539*
Eisenstein, L., 448(253,259,261,263-264), *543*
Elagin, L. M., 451(279), *544*
Elander, N., 55(163), *97*
Elbert, S. T., 111(149), *142*
Eleonsky, V. M., 13(53), *94*
Eletskii, V. L., 27(91), *96*
Elgart, A., 203(45-46), *266*
Eliasson, L. H., 176(33-34), 260(33-34), *266*
Ellison, G. B., 105(100), *141*
El-Sayed, M. A., 446(244), *543*
Eltschka, C., 16(59), *95*
Emmerichs, U., 490(352), *546*
Emrich, K., 113(182), *143*
Engelen, B., 438(210), *541*
Engelhard, M., 449(271), *543*
Epele, L. N., 3(18), 16(18), *93*
Ernst, R. R., 374(111-113), *538*
Eshimov, K., 444(217), *542*
Esry, B. D., 3(17), *93*
Essmann, U., 490(348,356), *546*
Excoffon, P., 467(299), *544*
Ezzir, A., 469(316), 470(316), *545*

Fabrikant, I. I., 3(17), *93*
Fanchiotti, H., 3(18), 16(18), *93*
Fannin, P. C., 469(312), *545*
Faris, S. M., 475(331), *545*
Farnham, D. L. Jr., 52(155), *97*
Fasse, E. D., 315(73), *348*
Faucher, O., 193(42), 235(42), *266*
Fedorovich, R. D., 474(327), 477(327), *545*
Feller, D., 102(46), *139*
Fernández, F. M., 39(113), *96*
Feschbach, H., 15(56), *95*
Fetter, A. L., 362(74), *537*
Fillaux, F., 352(1), 353(7-11,14), 354(26), 355(1), 374(110), 375(8-10,110,119), 378(7-9,119), 379-380(119), 381(121), 523(1,7-11,14), *535, 538–539*
Firsov, Yu. A., 368(88), 397(88), 410(88), 417(88), 419(88), 425(88), 428(88), 432(88), 432(88), 434(88), *538*
Fischer, S. F., 368(89), 419(89), *538*
Fisher, A. J., 2(2), *93*
Fisher, I. Z., 501(383), 522(383), *547*
Fisher, M. E., 3(22), 19(22), *94*
Fitzgerald, G. B., 102(13,16,19,21), *138–139*
Fleischhauer, M., 214(69), 226-227(69), *267*
Floquet, F., 354(26), *536*
Flügge, S., 13(50), *94*
Flynn, C. P., 354(20), 359(20), 363(20), 368(20), *536*
Fodor, S. P. A., 446(237), *542*
Fontaine, J. P., 353(8-9), 375(8-9), 378(8-9), 523(8-9), *535*
Fourkas, J. T., 490(351), *546*
Fox, D. J., 103(73), *140*
Frank, W. M., 16(58), *95*
Frantz, D. D., 6(37), *94*
Frederick, J. H., 281(53), 312(63), 317(53), *347–348*
Freed, K. F., 113(167), *143*
Freifelder, D., 491(363), *546*
Freund, D. E., 33(102), 38(102), 44(102), 52(102), *96*
Friedman, J. R., 469(308), 474-475(308), 485(308), *545*
Friedman, N., 448(270), *543*
Friedrich, B., 193(41), *266*
Friedrich, H., 16(59), 54(158), *95, 97*
Frisch, M. J., 103(79), *140*
Froelich, P., 55(161), *97*
Frolov, A. M., 50(144), *97*
Fukami, T., 390(147), *539*
Fukutome, H., 104(97), *141*
Fulden, P., 82(201), *98*
Furukawa, K., 390(147), *539*
Fykin, L. E., 428(194), 444-445(219), *541–542*

Gabrusenoks, E., 356(61-62), *537*
Gaiduk, V. I., 469(311), 490(358), *545–546*
Gaiduk, V. V., 490(358), *546*
Gaitain, D. F., 526(123), *539*
Gak, E. Z., 499(379), *547*
Galanin, M. D., 456(282), 518(282), *544*
Galbraith, H. W., 272(44), *347*
Galindo, A., 13(49), *94*
Gantmacher, F. R., 21(72), *95*
García Canal, C. A., 3(18), 16(18), *93*
Garg, A., 475(332), 479-481(332), 485(332), *545*
Garrett, B. C., 102(55), *140*
Garsia, N., 526(131), *539*
Garwan, M. A., 43(121), 45(121), *96*
Gatteschi, D., 469(307), 474-475(307), 485(307), *545*

Gauss, J., 102(7,14,17,22-24,30-37,42-44), 110(139), 117(216,219), 118(219-220, 223-224,228), 119(44), 127(238), 135(260,264), 136(267-268), *138–139, 142, 144–146*
Gavrilko, T., 438(213), 444(213,217), 460-461(292), 465(292), *541–542, 544*
Gavrilov, N. D., 356(63), *537*
Gaw, J. F., 102(19-20), *139*
Gdanitz, R., 112(151), *142*
Gebhard, R., 446(237), *542*
Geertsen, J., 113(173,187), *143*
Geguzin, Y. E., 498(377), *547*
Geoghegan, L. J., 470(317-318), *545*
Gerratt, J., 121(233), *145*
Gerwert, K., 446(230), 448(230), 449(271-272), 460(230), *542–543*
Ghosh, S., 113(184), *143*
Gilbert, M. M., 111(149), *142*
Girvin, S. M., 2(3), 4(3), 22(3), *93*
Glumac, Z., 21(82), *95*
Goddard, J. D., 102(10), *138*
Golab, J. T., 113(172), *143*
Goldanskii, V. I., 354(23), 359(23), *536*
Goldenfeld, N., 32(94), 59(94), *96*
Goldstein, H., 272(51), 288(51), *347*
Goldstone, J., 107(115), *141*
Goodson, D. Z., 6(40), *94*
Gordeev, G. P., 501(385), 502(385), 512(385), 512(385), 522(385), *547*
Goscinski, O.: 5-6(29), 39(111-112), *94, 96*; 113(158), *143*
Govindjee, R., 446(245), 448(245), *543*
Graboske, H. C., 28(92), *96*
Gradshteyn, I. S., 16(60), 55(60), *95*
Graf, F., 374(111-112), *538*
Gray, G. W., 460(261), *544*
Grebenshchikov, S. Yu., 364(79), 523(79), *537*
Grebille, D., 354(26), *536*
Grémaud, B., 46(138), *97*
Gribkovskii, V. P., 419(182), *540*
Griffin, A., 375(116), *539*
Grifoni, M., 235(78), *267*
Grigas, J., 399(156), *540*
Grinstein, G., 474-475(328), 485(328), *545*
Grischowsky, D., 226(75), *267*
Grishanin, B. A., 469(300), *544*
Grondalski, J., 469(301), 494(301), *544*
Grosse, H., 38(107), *96*
Gruber, G. R., 313(68-69), *348*
Gruner, G., 2(6), *93*
Grynberg, G., 154(10), 175(10), *265*
Guérin, S., 154(9), 171(29), 176(38), 193(42), 235(42), 205(58), 211-212(62), 214(67-68), 219-221(58), 226(74), 235(62,67,79,82), 236(83), 246(38,84-85), 248(38), 262(9), *265–267*
Gunther, L., 469(306), 474-475(306), *545*
Günther, R., 526(125), *539*
Gupta, A., 521(418), *548*
Gurevich, V. L., 420(183), *541*
Gurevich, Yu., 437(204), *541*
Guriev, V. S., 355(35), 446(35), 451-452(35), *536*

Ha, T. K., 374(111), *538*
Hadder, J. E., 312(63), *348*
Hadzi, D., 356(65), 368(94), 438(205), *537–538, 541*
Haken, H., 367(86), 422(192), *538, 541*
Hald, K., 137(273), *146*
Halfmann, T., 215(71), 215(72), 219(72), 226(74), *267*
Halkier, A., 102(50), *140*
Halonen, L., 270(7-8), 271(39), 300(58), 314-315(39), 317(39), *345, 347–348*
Hamann, D. R., 353(2), *535*
Hampel, C., 103(84), 105(110), 129(110), *141*
Handy, N. C.: 102(15,19-20), 105(108), 107(122), 115(208), 118(225-226), 129(108), *138–139, 141–142, 144*; 270(6), 271(23,30), 291(23), 296(30), *345–346*
Hänggi, P., 235(78), *267*
Harding, L. B., 133(257), *145*
Harrison, R. J., 102(13,19,21,57), *138–140*
Hartmann-Boutrn, F., 485(336), *545*
Harvey, E. N., 498(376), *547*
Harwood, D. J., 28(92), *96*
Hassan, R., 381(137), 485(334), *539, 545*
Hasselbach, K., 469(304), 494(304), *545*
Hatano, N., 67(173), *98*
Hatfield, B., 22(85), *95*
Hättig, C., 137(273), *146*
Hattori, K., 405(173), 411(173), 528(173), *540*
Hausser, R., 502(387), 512(387), *547*
Havel, T., 315(70-71), 336(71), *348*
Hawrylak, P., 81(184), *98*
Hayashi, S., 362(75), 374(114), *537–538*
Haycock, D. L., 469(301), 494(301), *544*
He, Y.-W., 449(273), *543*

He, Z., 107(121), *142*
Head-Gordon, M., 103(70,73), 105(108), 107(119), 110(70), 113(192-193), 114(201-204), 129(108,201), 131(250), 132(250-251), 133(252), *140–142, 144–145*
Heath, J. R., 82(199), 90(207), *98–99*
Heberle, J., 448(256), 460(290), *543–544*
Heckert, M., 117(216), *144*
Hehre, W. J., 115(207), *144*
Helgaker, T. U., 102(25,50,60), 112(156), 113(189), *139–140, 143*
Heller, J., 312(64), *348*
Hempel, C., 117(217), *144*
Henderson, R., 445-446(223), *542*
Henderson, W., 2(6), *93*
Henkel, M., 27(90), *96*
Henrichs, S. E., 81(189), *98*
Herman, M. F., 113(167), *143*
Hernandez, J. M., 474-475(329), 485(329), *545*
Herring, C., 354(19), *536*
Herschbach, D. R.: 5(29,35-36), 6(29,37,40), *94*; 193(41), 235(41), *266*
Herzfeld, J., 448(252), *543*
Hess, B., 449(271-272), *543*
Hestenes, D., 271(32-34), 272-273(50), 277(50), 305(32,34), 306(50), 307(34), 315(32-34,72-73), 318(32,50,75-77), 321(32), 330(50), 335(50), 336(32,34,50), 343(32,34), *347–349*
Heun, O., 117(216), *144*
Heyn, M. P., 447(250), 451(250), *543*
Hilczer, B., 355(40,42), 390(40,42,144-145), 399(155), *536, 539–540*
Hilczer, D., 355(38), 390-391(38), 394(38), *536*
Hildebrandt, P., 448(257), *543*
Hill, R. N., 33(102), 38(102,108), 44(102), 50(108), 52(102,108), *96*
Hiller, R., 526(124,126), *539*
Hindman, J. C., 502(389-390), 503(390), *547*
Hirao, K., 112(153), *143*
Hirata, H., 389(141-142), *539*
Hirata, S., 104(91), 131(250-251), 132(251), *141, 145*
Hirota, N., 362(75), *537*
Hirsch, J. E., 82(202), *98*
Ho, T. S., 166(17), *265*
Ho, Y. K., 50(143), *97*
Hochstrasser, R. M., 362(76-77), *537*
Hofacker, G. L., 368(89), 419(89), *538*
Hoffman-Ostenhof, M., 16(64), 38(64), 52(64), *95*
Hoffman-Ostenhof, T., 16(64), 38(64), 52(64), *95*
Hogreve, H., 43(119), 45(129), 46(136), *96–97*
Hohenberg, P., 110(145), *142*
Holstein, T., 368(87), *538*
Holthaus, M., 204-205(47,55), 208(60), 210-211(60), 213(66), 235(47,77), *266–267*
Holthausen, M. C., 110(147), *142*
Holtom, G. R., 364(77), *537*
Horácek, H., 66(172), *98*
Horsewill, A. J., 364(78), 523(78), *537*
Horsewill, T., 353(12), *536*
Hotop, H., 43(122), *96*
Hounkonnou, M. N., 271(21), 291(21), *346*
Howell, F. L., 419(180), *540*
Howland, J., 151(3), *265*
Hoy, A. R., 270(9), *345*
Hrusak, J., 130(247), *145*
Hu, X., 67(173), *98*
Huang, J. Y., 447(251), 451(277), *543–544*
Hubac, I., 117(218), *144*
Hubbard, J., 82(200), *98*
Hubman, M., 381(136), *539*
Huhn, G., 446(230), 448(230), 460(230), *542*
Hunziker, W., 44(124), *96*
Hwang, J.-T., 204(50), *266*
Hwang, S., 445(221), 446(221), *542*
Hylleraas, E. A., 33(96), 34(105), *96*

Ikagami, A., 446(247), *543*
Ikeda, S., 381(121), 524(105-106), 525(106,108), *538–539*
Ikegami, A., 450(274), *544*
Ilyin, V. V., 490(343), *546*
Ilyukhin, V. V., 425(193), 428(193,195), 429(193), 436(193,195), 444-445(219), *541–542*
Imashiro, F., 362(75), 374(114), *537–538*
Inosov, D. S., 474(327), 477(327), *545*
Inouye, S., 469(303), *545*
Iogansen, A. V., 463(296), *544*
Irgibaeva, I. S., 364(84), 366(84), 523(84), *538*
Isaacs, E. D., 353(2), 371-372(101), *535, 538*
Ivankiv, O. L., 381(135,138), 382(135), 383-384(138), 387-389(135), 482-483(135), *539*
Ivanov, I. A., 33-34(103), *96*

Iwai, J., 450(274), *544*
Iwata, M., 507(398), *547*
Iwata, S., 130(247), *145*

Jacak, L., 81(184), *98*
Jaekel, M. T., 526(130), *539*
Janoschek, R., 402(170), *540*
Janssen, C. J., 117(214), *144*
Jaskichev, V. I., 506(395-396), *547*
Jauslin, H. R., 154(9), 166(18-20,26-27), 171(29), 176(37-38), 193(42), 205(58), 211-212(62), 214(67-68), 219-221(58), 235(42,62,67,82), 236(83), 246(38,85), 248(38), 260(19), 262(9), *265–267*
Jayatilaka, D., 115(208), *144*
Jensen, H. J. A., 113(189), *143*
Jessen, P. S., 469(301), 494(301), *544*
Jeziorski, B., 102(61), *140*
Jobic, H., 375(116-117), *539*
Johnson, M., 353(12), *536*
Jonas, J., 502-503(391), *547*
Jones, B., 460(261), *544*
Jongeward, G., 78(180), *98*
Jørgensen, P., 102(25-26,50,60), 107(123), 108(125), 109(134), 112(156), 113(159-160,164,188-189), 130(248), 133(256), 137(273), *139–140, 142–143, 145–146*
Joye, A., 204(51-52,55-56), 220(51-52), *266*
Just, B., 208(60), 210-211(60), *266*

Kachkachi, H., 469(316), 470(316), *545*
Kadanoff, L. P., 22(83-84), *95*
Kagan, C. R., 81(187,191), *98*
Kagan, Yu. M., 421(186), 422(190), *541*
Kagawa, Y., 389(141-142), *539*
Kais, S., 2(7), 3(8-11,21), 4(10), 5(7,30-32,3), 6(7), 7(30-32,42-43), 13(52,54) 16(62), 17(11,66), 23(10), 25(8-11,25,87-88), 27(10,88), 28(88), 29(93), 30(93), 31(88,92), 32(93), 34(8,87), 35(87), 38(87), 39-40(9,87), 41(87), 42(93), 43(62,118), 45(62), 46(11,118), 45(127), 47(11), 49(11), 51(11,66,87), 53(156), 55(156), 63(11,32,156), 73(11), 75(156,174), 83(205), *93–98*
Kaiser, W., 456(284), *544*
Kaivarainen, A., 501(381), *547*
Kalinin, V. I., 428(195), 436(195), 444-445(219), *541–542*
Kallay, M., 104(92), *141*
Kalmukov, Y. P., 506(396), *547*
Kalmykov, Yu. P., 469(312-315), 470(315), *545*
Kamiya, K., 372(102), *538*
Kaneda, K., 390(147), *539*
Kaplan, L., 312(64), *348*
Karapetiants, M. K., 494-495(368), 498(368), *547*
Karaseva, T. A., 442(216), *542*
Kastner, M. A., 81(194), *98*
Kato, T., 33(97), 38(109), 64(97), *96*
Katori, M., 67(173), *98*
Katriel, J., 5(28), *94*
Katz, D., 81(196), *98*
Kaulen, A. D., 448(268), *543*
Kauppi, E., 272(43), *347*
Kauzman, W., 381(139), 490(139), 496(139), 501(196), *539*
Kearly, G. J., 353(8-9), 374(110), 375(8-9,110), 378(8-9), 523(8-9), *535, 538*
Kedziora, G. S., 102-103(2), 112(2), *138*
Keiding, S. R., 490(344), 491(344), *546*
Kelkar, V. K., 460(293), *544*
Keller, A., 172(32), 193(32), 235(32), *266*
Kendall, R. A., 102(57), *140*
Kendrick, B. K., 272(49), *347*
Kennedy, E. C., 469(316), 470(316-318), *545*
Kern, C. W., 270(5), *345*
Ketterle, W., 469(303), *545*
Khachaturian, A. G., 513-514(415), *548*
Khahre, A., 13(51), *94*
Khaidarov, T., 501(385), 502(385), 512(385), 512(385), 522(385), *547*
Khakimov, I., 356(61-62), *537*
Kharkats, Yu. I., 437(204), *541*
Khavin, Z. Y., 497(374), *547*
Khorana, H. G., 446(228,235,260), 449(273), *542–543*
Khoury, J. T., 81(188), *98*
Khovanski, A. B., 374(103), *538*
Kim, G.-H., 475(332), 479-481(332), 485(332), *545*
Kimura, T.: 312(60), *348*; 507(397), *547*
Kinghorn, D. B., 45(133), 50(133), *97*
Kinosita, K., 446(247), *543*
Kiselev, A. B., 521(417), *548*
Kitaigorodsky, A. I., 375(115), *539*
Kittel, C., 410(177), 494(365), 519-520(177), *540, 546*
Kivelson, D., 490(351), *546*

Kiyaev, O. E., 474(327), 477(327), *545*
Klaus, M., 12(47), 15(47), 27-28(47), 31(47), 59(47), *94*
Klein, W., 21(78), *95*
Kleinert, H., 75(175), *98*
Klinger, M. I., 368(91-92), 422(190), *538, 541*
Klopper, W., 102(50,66-68), 103(69), *140*
Knapp, E.-W., 400(160), 415(160), *540*
Knight, R. E., 33(98), *96*
Knowles, P. J., 107(122), 117(217), 118(225-226), *142, 144*
Kobayashi, R., 130(248), *145*
Koboyashi, T., 450(274), *544*
Kobushkin, A., 312(62), *348*
Kobyliansky, V. B., 519(416), *548*
Koch, H., 102(50), 107(123), 108(125), 109(134), 113(188-189), 120(230), 130(248), *140, 142–143, 145*
Koch, H. J., 451(276), *544*
Koch, M. H., 445-446(227), *542*
Koch, W., 110(147), *142*
Kohn, W., 110(145-146), *142*
Kondrachuk, A. V., 491(360), *546*
Konno, K., 448(253,263), *543*
Köppel, H., 120(230), *145*
Korolev, V. G., 13(53), *94*
Korolkov, M., 208(61), 235(61), *267*
Korsc, H. J., 55(162), *97*
Kouyama, T., 446(247), *543*
Kovner, M. A., 356(64), 461(294), *537, 544*
Kowalski, K., 114(205-206), *144*
Kozhanov, V. N., 359(25), *536*
Kozlowski, P. M., 102(61-63), *140*
Krainov, E. P., 461(294), *544*
Kraka, E., 128(242), *145*
Kramer, B., 92(211), *99*
Krasnoholovets, V., 355(33-37,46-47), 356(57-62), 359(57-60), 400(163-165), 406-408(165), 409(33-34,165,176), 411(178), 414-415(178), 416(178-179), 419(165), 422(191), 429-430(46), 431(46,57-58,191), 432(163), 434(164), 437(47), 438(213), 444(213), 446(35-37), 449-450(37), 451-452(35), 458(34,37), 459(37), 461(57-58,60), 462(57-58), 463(57), 464(57-58,60), 467(36,57-58,60), 468(57-58,60), 470(319), 477(319), 491(359-360), 496-497(359), 501(359), 502(359-360), 504(360), 505(359-360), 507(408), 512(408), 513(412,414), 517(414), 523(359), 529(47,165), 533(47), 535(47), *536–537, 540–541, 545–546, 548*
Kravchhenko, Y. P., 3(15), *93*
Krestov, G. A., 494(367), 496(367), 498(367), *546*
Kreutz, W., 449(271), *543*
Krikorian, R., 176(35-36), 260(35-36), *266*
Krishnan, R., 102(11-12), *138*
Krivec, R., 50(145), *97*
Kryachko, E. S., 470(325), 482(325), 484(325), *545*
Krylov, A. I., 113(198-199), 114(200-202), *144*
Kubo, O., 469(304), 494(304), *545*
Kubo, R., 398(154), *540*
Kucharski, S. A., 103(71-72), 104(89), 107(120), 110(71-72,89,137,143), 114(206), *140–142, 144*
Kümmel, H., 107(114), *141*
Kunst, M., 383(140), 389(140), *539*
Kuntz, H., 204(51), 220(51), *266*
Kutzelnigg, W., 102(67-68), 103(69), *140*

Laidig, W. D., 102(10,13,19,21), *138–139*
Laloë, F., 377(120), *539*
Lambrecht, A., 526(130), *539*
Lampert, M. A., 408(175), *540*
Lanczos, C., 47(142), *97*
Land, D. J., 16(58), *95*
Landau, L. D., 213(64), *267*
Lang, E. W., 502-503(392), *547*
Langner, R., 495(369), *547*
Lany, R., 448(255), *543*
Lascombe, J., 438(206), *541*
Lasota, A., 78(178), *98*
Lassaut, M., 13(48), *94*
Latajka, Z., 442(216), *542*
Lauderdale, W. J., 118(223-224), *144*
Laughlin, C., 65(170), *98*
Laughlin, R. B., 474(326), *545*
Laurence, M. C., 422(188), *541*
Laurenzi, B. J., 45(130), 46(134), *97*
Lauter, H., 375(117), *539*
Lautié, A., 374-375(110), *538*
Lavorel, B., 193(42), 235(42), *266*
Lawrence, A. F., 456(487), *544*
Lawrence, M. C., 354(24), *536*
Lazarev, N. P., 421(185), *541*
Lebovka, S., 502-503(394), 512(410), *547–548*

Lebowitz, J. L.: 63(166), *97*; 166(18-19), 260(19), *265*
Lee, H. L., 2(6), *93*
Lee, M. S., 133(252), *145*
Lee, S. B., 21(81), *95*
Lee, T. D., 19(68-69), *95*
Lee, T. J., 102(6,19,28), 105(111), 109(129), 110(138), 132(251), *138–139, 141–142, 145*
Legon, A. C., 366(85), *538*
Leininger, M. l., 107-108(124), *142*
Lengsfield, B. H., 102(45), *139*
Lepoint, T., 526(133), *539*
Lester, W. A. Jr., 103(81-82), *140–141*
Lev, B., 356(59), 359(59), 513(412-414), 517(413-414), *537, 548*
Levanyuk, A. P., 526(131), *539*
Levine, R. D., 3(16), 81(183), 82(197), 83(205), *93, 98*
Lévy-Leblond, J. M., 156(13), *265*
Lewis, A., 446(240), 447(251), 451(277), *542–544*
Lewis, J. P., 490(354), *546*
Leygue, N., 353(10), 375(10), 523(10), *535*
Li, H., 318(77), *349*
Li, X., 117(215), *144*
Liao, P. F., 226(75), *267*
Lieb, E. H., 16(63), 42(115), 45(126), *95–96*
Lieberman, M. A., 3(15), *93*
Light, J. C., 270(15), *346*
Lin, C. D., 50(146), *97*
Lin, S.-L., 448(253,259,261,263), *543*
Lin, S. W., 446(246), 447(246), *543*
Linanyi, L., 448(264), *543*
Linderberg, J., 113(157), 135(157), *143*
Lindh, R., 130(245), *145*
Lineberger, W. C., 43(122), *96*
Liopo, V. A., 521(417), *548*
Lipowski, A., 67(173), *98*
Lippinkott, E. R., 358(67), *537*
Lipscomb, W. N., 121(235), *145*
Lischka, H., 102-103(2), 112(2), *138*
Lisitsa, M. P., 439(215), *542*
Litherland, A. E., 43(121), 45(121), *96*
Littlejohn, R. G., 270(13-14), 272(42,47), 276(47), 289-290(47), 303(13), *346–347*
Litvin, S., 451(280), *544*
Livingston, J. D., 469(309), *545*
Loeser, J. G., 6(38-39), *94*
Lombard, A., 270(14), *346*
Lombard, R. J., 13(48), *94*
Loshmanov, A. A., 428(194), 444-445(219), *541–542*
Loss, D., 474(328), 475(328,333), 485(328), *545*
Louck, J., 272(44), *347*
Lounesto, P., 271(35), *347*
Lovas, R. G., 3(19), *93*
Lovesey, S. W., 375(118), *539*
Löwdin, P. O., 104(93), 115(93), *141*
Loy, M. M. T., 226(75), *267*
Lozier, R. H., 445(221-222), 446(221-222), *542*
Lübben, M., 446(230), 448(230), 460(230), *542*
Luck, W. A. R., 501(386), 506(386), 521(419), 522(386), *547–548*
Ludeman, H. D., 502-503(392), *547*
Lugtenburg, J., 446(237), *542*
Lukaszewicz, K., 390(146), *539*
Lukka, T., 271(29), 272(43), 295-296(29), *346–347*
Lukyanets, S. P., 470(320-322), 474(327), 477(320-322,327), 482(320-321), 486(320), *545*

Macek, J. H., 3(17), *93*
Mackey, M. C., 78(178), *98*
MacKinnon, A., 92(211-212), *99*
Mailly, D., 469(304), 494(304), *545*
Maitra, N. T., 312(64), *348*
Makarenko, S. P., 460-461(292), 465(292), *544*
Makarov, D. E., 354(23), 359(23), *536*
Makita, Y., 355(32), *536*
Makri, N., 75(176), *98*
Malenkov, g. G., 490(341), 501(341), *546*
Mall, M. F., 45(130), *97*
Malmqvist, P.-A., 112(154-155), *143*
Maluta, T. J., 526(127), *539*
Mandelzweig, V. B., 50(145), *97*
Mank, V. V., 502-503(394), 512(410), *547–548*
Manohar, C., 460(293), *544*
Manz, J., 208(61), 235(61), *267*
Mardis, K. L., 272(45), *347*
Marechal, Y., 356(66), 357-359(66), 438(66), *537*
Mareshal, Y., 467(299), *544*
Mark, P., 408(175), *540*
Markovich, G., 81(189), *98*
Marti, T., 446(228,260), 449(273), *542–543*
Martin, A., 46(137), 50(149), *97*

Martin, J. M. L., 102(48), 103(76), 110(141), *139–140, 142*
Martin, R. L., 135(263), *145*
Massawe, E. S., 469(314), *545*
Masthay, M. B., 456(487), *544*
Mathies, R. A., 446(237,246), 447(246), 456(285), *542–544*
Matsubara, E., 358(68), *537*
Matsubara, T., 358(68), 368-369(98), 372(102), *537–538*
Matveev, Yu. A., 456(283), *544*
Maugh, T. H., 491(362), *546*
Mayer, M., 127(239), *145*
McCafferty, E., 507(402), *548*
McElroy, W. D., 498(376), *547*
McIntosh, R., 507(403), *548*
McKoy, V., 113(166), *143*
McLean, A. D., 133(257), *145*
McWeeny, R., 117(211), *144*
Mead, C. A., 272(49), *347*
Meesen, S., 446(229), *542*
Meier, B. H., 374(112), *538*
Melnichenko, N. A., 502-503(393), *547*
Merinov, B. V., 395(151-153), 400(152-153), *540*
Merz, H., 400(167), 446(240), *540, 542*
Merzbacher, E., 23(86), 26(86), *95*
Metzger, H., 353(4-5), 523(4-5), *535*
Metzik, J. S., 521(417), *548*
Meyer, H.-D.: 64(168), *97*; 137(272), *146*
Meyer, R., 374(111,113), *538*
Mezei, J. Z., 3(19), *93*
Miani, A., 300(58), *348*
Midtdal, J., 33(100), 38-39(100), *96*
Mielke, S. L., 102(55), *140*
Miesner, H.-J., 469(303), *545*
Milde, F., 92(213), *99*
Mileti, G., 204(52), 220(52), *266*
Mille, M., 400(158), 415(159), *540*
Millo, O., 81(196), *98*
Mills, A. P., 50(148), *97*
Mills, I. M.: 121(233), *145*; 270(9), *345*
Minami, K., 67(173), *98*
Mishina, O., 490(353), *546*
Mitchell, K. A., 272(42), 312(66), *347–348*
Mitroy, J., 3(19), *93*
Mladenovic, M., 271(26-270), 291-292(26-27), *346*
Mogi, T., 446(228,260), *542–543*
Mohanty, U., 490(351), *546*
Moiseyev, N., 53(156), 54(157), 55(156-157, 160-162), 56-57(157), 63(156), 75(156), *97*
Møller, C., 107(118), *142*
Monkhorst, H. J., 102(61), 113(180,183), *140, 143*
Montgomery, J. A., 103(77-79), *140*
Monti, F., 154(9), 262(9), *265*
Moore, M. A., 361(73), *537*
Morgan, J. D. III, 33(102), 38(102), 44(102), 52(102), *96*
Moritz, M. J., 16(59), *95*
Morowitz, H. J., 355(27), 400(158-159), 401(159), 415(158), *536, 540*
Morse, P. M., 15(56), *95*
Mott, N. F., 90(208), *99*
Moynihan, C. T., 490(357), *546*
Muadyan, L. A., 428(194), *541*
Mukherjee, D., 113(177,181,184), *143*
Mukherjee, P. K., 113(181), *143*
Mukhopadhyay, D., 113(177), *143*
Mukina, O. V., 356(63), *537*
Muller, R., 507(399), 512-513, *547*
Müller, T., 102-103(2), 112(2), *138*
Muller, U., 447(250), 451(250), *543*
Mulliken, C. M., 448(252), *543*
Munn, R. W., 368(90), *538*
Murray, C., 118(227), *145*
Murray, C. B., 81(187,191), *98*
Murty, C. R. K., 464(297), *544*
Musial, M., 104(89), 110(89), *141*

Naberukhin, Y. I., 490(342), 501(342), *546*
Nadeau, M.-J., 43(121), 45(121), *96*
Nagaoka, S., 362(75), 374(114), *537–538*
Nagle, J. F., 400(158-159), 401(159), 415(158), 446(239), *540, 542*
Najfeld, I., 315(70-71), 336(71), *348*
Nakamo, H., 469(304), 494(304), *545*
Nakanishi, H., 21(79-81), *95*
Nakanishi, K., 448(253,259,261,263), *543*
Nakasone, K., 390(147), *539*
Nakatsuji, H., 113(194-195), *144*
Nanaka, K., 507(398), *547*
Narevicius, E., 55(162), *97*
Natkniec, I., 354(26), *536*
Nauenberg, M., 32(95), *96*
Naumovets, A. G., 474(327), 477(327), *545*
Nauts, A., 271(16-18,20), 280(17), 290(17-18), 291(17-18,20), 312(65), *346, 348*
Needleman, J., 448(255), *543*

Neirotti, J. P., 3(8-9), 13(54), 25(8-9,54,87-88), 27-28(88), 31(88), 34(8,87), 35(87), 38(87), 39-40(9,87), 41(87), 43(118), 46(118), 51(87), *93–96*
Nelson, K., 490(351), *546*
Neogrady, P., 117(218), *144*
Nesbet, R. K., 105(106), 115(209), *141, 144*
Neuhauser, D., 55(162), *97*
Neumark, D. M., 127(237), 133(254), *145*
Newman, M. E. J., 2(2), *93*
Newton, R. G., 54(159), *97*
Nick, G., 526(125), *539*
Nicolai, B., 374-375(110), *538*
Niederberger, W., 445(221), 446(221), *542*
Nielsen, H. H., 270(2), *345*
Nienhuis, B., 32(95), *96*
Nieto, M. M., 156(12), *265*
Nightingale, M. P., 20(70), 21(73), 24(70), *95*
Nigra, P., 3(21), *93*
Nirmal, M., 81(191), *98*
Noga, J., 102(38), 103(38,69,71), 104(98), 109(38,132), 110(71,135), *139–142*
Noguès, M., 469(316), 470(316), *545*
Nomnomura, Y., 67(173), *98*
Nooijen, M., 104(91), 113(196-197), 118(222), 135(265-266), *141, 144, 146*
Novak, A., 360(70), 374(70), 438(206), 469(307), 474-475(307), 485(307), 524(70), *537, 541, 545*
Nygard, J., 3(14), *93*

Ochterski, J. W., 103(77-79), *140*
Odashima, K., 448(253,261,263), *543*
Oddershede, J., 113(164,173), *143*
Oesterhelt, D., 445(220,225,227), 446(225,227,229,236,242-243), 447(220), 451(276), 456(284), 460(290), *542–544*
Ogawa, N., 312(62), *348*
Öhrn, N. Y., 113(157,168-169), 135(157), *143*
Ohtani, H., 450(274), *544*
Ohtani, T., 312(60), *348*
Okamoto, H., 389(142), *539*
Olejnik, J., 446(233,248), 448-449(233), 451(233), 456(289), 470(289), 475(289), 485(289), *542–544*
Oliphant, N., 110(142), *142*
Olsen, J., 102(60), 104(90), 107(123), 108(125), 112(156), *140–143*
Olson, J., 102(50), *140*
Oppenlander, A., 354(17), *536*
Ortega, J., 490(354), *546*
Ortiz, J. V., 113(168-169,174-175,178-179), 137(174,178-179), *143*
O'Shea, M. J., 474-475(330), *545*
Ostrovski, D., 442(216), *542*
Otomo, J., 446(229,247), *542–543*
Ottolenghi, M., 448(270), *543*
Ovcharenko, A. I., 491(360), *546*

Pagnia, H., 507(399), 512-513, *547*
Paldus, J., 104(96), 113(186), 117(215), 119(96,229), *141, 143–145*
Paperno, T. Ya., 451(279), *544*
Paramonov, G. K., 208(61), 235(61), *267*
Parlett, B. N., 51(153), *97*
Parr, R. G., 110(144), 129(144), *142*
Parsons, A. F., 105(101), *141*
Pascual, P., 13(49), *94*
Pasinski, A., 3(14), *93*
Passante, R., 154(11), *265*
Pauling, L., 494(366), *546*
Paulus, W., 353(10), 375(10), 523(10), *535*
Pavlenko, N., 355(38-42), 381-382(135), 387-389(135), 390(38-42), 391(38), 394(38), 395(149-150), 397-399(149-150), 400(150), 437(149-150), 482-483(135), *536, 539–540*
Pawlaczyk, Cz., 390(145), 399(155), *539–540*
Pawlowski, A., 390(145), 399(155), *539–540*
Pearson, R. G., 127(240), *145*
Pecha, R., 526(125), *539*
Pechukas, P., 204(49-50), 220(49), *266*
Peercy, P. S., 422(187), *541*
Pekeris, C. L., 47(139), 51(151), *97*
Peletminsky, S. V., 452(281), *544*
Penfold, J., 353(7), 378(7), 523(7), *535*
Perchard, J.-P., 352(1), 355(1), 523(1), *535*
Perdew, J. P., 42(116), *96*
Perera, P., 474-475(330), *545*
Perevalov, V. I., 171(31), *266*
Perevertaev, V. D., 521(417), *548*
Perez-Conde, J., 82-83(203), *98*
Persico, F., 154(11), *265*
Persson, B. J., 102(64-65), *140*
Pertsin, A. J., 375(115), *539*
Peschel, G., 507(405), 521(405), *548*
Pesonen, J., 271(36-41), 296(36), 300(58), 314-315(39), 317(39), *347–348*
Petch, H. E., 355(29,31), 389(29), 408(29), 415(29), *536*

Peteanu, L. A., 456(286), *544*
Peterson, K. A., 102(51,55,59), 103(83), 105(110), 129(110), *140–141*
Petersson, G. A., 103(77-79), *140*
Petrosian, A. M., 444(218), *542*
Petrov, A. P., 374(103), *538*
Pfeuty, P., 82-83(203), *98*
Pfister, C.-Ed., 204(51-52,55), 220(51-52), *266*
Picharad, J.-L., 91(209), *99*
Pickup, B. T.: 113(158), *143*; 135(261-262), *145*
Piecuch, P., 114(205-206), *144*
Pierre, M., 354(17), *536*
Pietraszko, A., 390(146), *539*
Pittner, L., 38(107), *96*
Pitzer, R. M., 102-103(2), 112(2), 121(235), *138, 145*
Platzman, P. M., 353(2), 371-372(101), *535, 538*
Plesset, M. S., 107(118), *142*
Plöhn, H. J., 445-446(225,227), 451(276), *542, 544*
Politi, P., 485(336), *545*
Pollard, W. T., 456(285-286), *544*
Polomska, M., 355(39-40), 390(39-40), *536*
Poole, C., 272(51), 288(51), *347*
Poole, P. H., 490(348), 490(354-355), *546*
Popadopoulos, G., 445-446(224-225), 448(224,258), 451(258,275), *542–544*
Pople, J. A., 102(11-12), 103(70,73-75), 105(108), 107(119), 115(207,209), 118(225), 129(108), *138, 140–142, 144*
Popov, V. S., 5(35), 27(91), *94, 96*
Pöschl, G., 80(181), *98*
Powles, J. G., 502(388), *547*
Pozdnyakov, V. V., 451(279), *544*
Prats, M., 355(28), *536*
Primas, H., 171(30), *266*
Privman, V., 3(24), 19(24), 25-26(24), 73(24), *94*
Prosperetti, A., 526(134), *539*
Protsenko, N. A., 355(33-37), 400(163-164), 409(33-34), 432(163), 434(164), 446(35-37), 449-450(37), 451-452(35), 458(34,37), 459(37), 467(36), *536, 540*
Puchkovska, G. A., 355(43-44,46,48), 356(57-58,60-62), 359(57-58,60), 428(196,198), 429(46,196), 430(46), 431(46,57-58), 438(213), 444(213,217), 460(292), 461(57-58,60,292), 462(57-58), 463(57), 464(57-58,60), 465(292), 467-468(57-58,60), *536–537, 541–542, 544*
Pulay, P., 102(9), 103(87), *138, 141*
Purvis, G. D., 108-109(127), *142*
Putterman, S. J., 526(124,126), *539*

Rabinovich, V. A., 497(374), *547*
Radom, L., 115(207), *144*
Radzievski, P., 456(289), 470(289), 475(289), 485(289), *544*
Raggio, G.: 59(165), *97*; 161(16), *265*
Raghavachari, K., 103(70,73-75), 105(108), 107(119), 110(70), 129(108), *140–142*
Rahman, A., 490(337), *546*
Rambaud, C., 353(13,17), *536*
Rammelsberg, R., 446(230), 448(230), 460(230), *542*
Rampton, V. W., 421(184), *541*
Rao, A. S. N., 464(297), *544*
Rapp, G., 445-446(225,227), 451(276), *542, 544*
Rassolov, V., 103(75), *140*
Ratajczak, H., 438(209,211), 442(211,216), 444(217), *541–542*
Rather, M. A., 368(89), 419(89), *538*
Rau, A. R. P., 3(17), *93*
Rebane, T. K., 45(131), 46(135), *97*
Reddy, A. D., 355(30), 389(30), 408(30), *536*
Reddy, T. R., 464(297), *544*
Redfern, P. C., 103(75), *140*
Reid, J. K., 47(140), *97*
Reinhardt, W. P., 33(101), *96*
Reinisch, L., 448(269), *543*
Reinsch, M., 270(13), 272(47), 276(47), 289-290(47), 303(13), *346–347*
Reisler, H., 104(99), *141*
Remacle, F., 3(16), 81(183), 82(197), 90(207), *93, 98–99*
Replogel, E. S., 107(119), *142*
Rettori, A., 485(336), *545*
Reynaud, S., 526(130), *539*
Reynolds, P. J., 21(78-79), 80(182), *95, 98*
Rice, J. E., 102(19-20,28), 109(129), *139, 142*
Rice, J. R., 51(152), *97*
Rice, S. A., 490(339-340), 496(339), 501(339), *546*
Richard, J., 46(137), *97*
Richard, J.-M., 16(65), *95*
Richter, H., 448(255), *543*
Rickes, T., 212(63), 215(72), 219(72), 225-226(63), *267*
Rico, R. J., 113(192-193), *144*
Rider, E. E., 428(194), *541*

Riel, N., 408(174), *540*
Riesle, J., 460(290), *544*
Rittby, M.: 55(163), *97*; 113(187), *143*
Rivkin, S. L., 496(371), *547*
Rizzo, A., 133(255), *145*
Roberts, I. I., 368(90), *538*
Robertson, G. N., 354(24), 422(188), *536, 541*
Rockwood, A., 318(77), *349*
Roepe, P., 448(252), *543*
Rogers, F. J., 28(92), *96*
Romanenko, V. I., 219(73), 226(73), 235(80), 238(80), *267*
Romer, R. A., 92(213), *99*
Rønne, C., 490(344), 491(344), *546*
Roos, B. O., 111(148), 112(154-155), *142–143*
Ross, B., 111(150), *142*
Rost, J. M., 51(150), 53(150), *97*
Rost, R., 3(14), *93*
Rothschild, K. J., 448(252,260), 449(273), *543*
Roy, R. A., 526(123), *539*
Rozhkov, S. S., 371-372(100), 374(100), *538*
Ruedenberg, K., 111(149), *142*
Ryabkin, V. S., 425(193), 428(193,195), 429(193), 436(193,195), *541*
Ryzhihk, I. M., 16(60), 55(60), *95*

Sachdev, S., 2(4), 22(4), 34(4), *93*
Sadeghpour, H. R., 3(17), *93*
Sadlej, A., 112(154), *143*
Saebø, S., 103(187), *141*
Saeh, J. c., 121(231), 124(231), 127(231), 129(231), 136(231), *145*
Safko, J., 272(51), 288(51), *347*
Saika, A., 362(75), 374(114), *537–538*
Salamon, M. B., 437(203), *541*
Salejda, W., 390(148), *539*
Saleur, H., 21(75), *95*
Salter, E. A., 102(16), 129(243), *139, 145*
Sambe, H., 151(2), *265*
Samoilov, O. Y., 496(372), *547*
Sampaio, J. F., 81(193), 82(199), *98*
Sample, J. L., 81(193), 82(198), *98*
Sankey, O. F., 490(354), *546*
Sanygin, V. P., 506(396), *547*
Sastry, G. S., 355(30), 389(30), 408(30), *536*
Sathynarayan, S. G., 355(30), 389(30), 408(30), *536*
Sauerwein, R. A., 75(174), *98*
Saxe, P., 102(10), *138*
Sceats, M. G., 490(339-340), 496(339), 501(339), *546*
Schachtschneider, J. H., 461(295), *544*
Schaefer, H. F., 102(8,10,28,39), 103(39), 105(111), 107-108(124), 109(39,129), 117(214), 124-125(236), 129(244), 130(246), *138–139, 141–142, 144–145*
Schatz, G. C., 81(192), *98*
Scheiner, A. C., 102(28), 109(129), *139, 142*
Scheiner, S., 401(168), *540*
Scheller, M. K.: 3(12), 45(12), *93*; 16(61), *95*
Scherer, w., 168(20-25), 171(20-25), *266*
Scherr, C. W., 33(98), *96*
Schirmir, J., 113(170-171), 137(270-271), *143, 146*
Schlegel, H. B., 102(11-12), 118(221), *138, 144*
Schleyer, P. v. R., 115(207), *144*
Schmidt, M. W., 111(149), *142*
Schmidt, V. N., 419(180), *540*
Schreiber, M., 92(211,213), *99*
Schröder, R., 358(67), *537*
Schulten, K., 400(160), 415(160), 419(181), 446(242), *540, 542*
Schulten, Z., 400(160), 415(160), 419(181), *540*
Schütz, M., 103(85-86,88), *141*
Schwartz, C., 102(54), *140*
Schwegler, E., 102(63), *140*
Schwinberg, P. B., 52(155), *97*
Schwinger, J., 526(128), *539*
Sciortino, F., 490(348), 490(354-355), *546*
Scuseria, G. E., 102(6,28,39), 103(39), 109(39,129), 110(138), 113(173), 129(244), *138–139, 142–143, 145*
Sekino, H., 113(185), 129(243), *143, 145*
Sergeev, A. V., 16(62), 43(62), 45(62), *95*
Sergienko, A. I., 470(324), 482(324), *545*
Serra, P., 2(7), 3(8-10), 4(10,26), 5(7,30-32), 6(7), 7(30-32), 13(54), 21(76), 23(10), 25(8-9,54,87-88), 27(10,88), 28(88), 29(92), 30(93), 31(88,93), 31-32(93), 34(8,87), 35(87), 38(87), 39-40(9,87), 41(87), 42(93), 43(118), 46(118), 51(87), 53(156), 55(156), 63(10,32,156), 66(26), 73(10), 75(156), *93–97*
Sessoli, R., 469(307), 474-475(307), 485(307), *545*
Seth, M., 102-103(2), 112(2), *138*
Sethna, J. P., 360(71-72), 363(71-72), *537*
Shadchin, E. A., 371-372(100), 374(100,104), 428(197), *538, 541*

Shahar, D., 2(3), 4(3), 22(3), *93*
Sham, L. S., 110(146), *142*
Shamura, M. M., 521(418), *548*
Shanchuk, A. I., 428(197), 429-430(202), *541*
Shank, C. V., 456(285-286), *544*
Sharkov, A. V., 456(283), *544*
Shavitt, I., 102(1-2), 103(2), 106(1), 112(2), *138*
Shepard, R., 102-103(2), 111(148), 112(2), *138, 142*
Shepelyansky, D. L., 91(209), *99*
Sherrill, C. D., 107-108(124), 114(200-202), 129(201), 133(252-253), *142, 144–145*
Sheves, M., 448(270), *543*
Shewell, J. R., 313(67), *348*
Shi, Q., 3(11), 7(41), 17(11,66), 45(11,127), 46-47(11), 49(11), 51(11,66), *93–96*
Shirley, J. H., 151(1), 154(1), 161(1), *265*
Shishkin, V. A., 428(197), *541*
Shore, B. W., 184(39), 192(39), 205(39), 212(63), 214(69), 215(71-72), 219(72), 225(63), 226(63,69,74), 227(69), 235(76), 246(84), *266–267*
Shukla, A., 353(2), 371-372(101), *535, 538*
Shulepov, Y. V., 507(406-407), 411(407), *548*
Shuvalov, L. A., 395(151-153), 400(152-153,157), 425(193), 428(193,195), 429(193), 436(193,195), *540–541*
Sibert, E. L. (III), 272(45), *347*
Siebert, F., 449(271), *543*
Siegbahn, P. E. M., 111(150), *142*
Siegert, A. F. J., 64-66(167), 69(167), *97*
Sigrist-Nelson, K., 389-390(143), *539*
Sil, S., 82(204), *98*
Silbey, R., 354(22), 371(99), *536, 538*
Silver, D. M., 109(131), *142*
Silver, R. N., 368(93), *538*
Simandiras, E. D., 102(19-20), *139*
Simon, B., 11(45), 12(47), 15(47), 16(64), 23(45), 27-28(47), 31(47), 38(64), 52(64), 59(47,164), *94–95, 97*
Simonov, V. I., 428(194), *541*
Simons, J., 102(26), 113(159,161-162,165), *139, 143*
Sinanoglu, O., 109(130), *142*
Singh, A. K., 448(264), *543*
Sirenko, S. P., 371-372(100), 374(100), *538*
Skinner, J. L., 354(21), 359-360(21), 363(21), 374(21), 523(21), *536*
Skobczyk, G., 271(32), 305(32), 315(32), 318(32), 321(32), 336(32), 343(32), *347*
Skripov, A. V., 359(25), *536*
Smirnova, A. A., 451(279), *544*
Smith, D. W. G., 502(388), *547*
Smyth, J. F., 474-475(328), 485(328), *545*
Snyder, R. G., 461(295), *544*
Sobcyzk, G., 318(74), *348*
Sokolov, N. D., 400(166), *540*
Solov'ev, S. P., 444-445(219), *542*
Somasundram, K., 107(122), *142*
Sondi, S. L., 2(3), 4(3), 22(3), *93*
Sone, N., 389(141-142), *539*
Soppa, J., 446(229), *542*
Sorchik, M. P., 469(308), 474-475(308), 485(308), *545*
Sørensen, G. O., 270(10), *345*
Sorokina, N. I., 425(193), 428-429(193), 436(193), *541*
Soucaille, P., 355(28), *536*
Souvignier, G., 449(272), *543*
Spartakov, A. A., 507(401,404), *548*
Spirko, V., 114(206), *144*
Sprandel, L. L., 270(5), *345*
Spruch, L., 45(132), *97*
Stafford, C. A., 81(190), *98*
Stamper-Kurn, D. M., 469(303), *545*
Stanley, H. E.: 7(41), 21(78,80), *94–95*; 490(348,353,355-356), *546*
Stanton, J. F., 102(30-36,42-43), 105(109), 108(126), 110(136,139-140), 113(190), 116(210), 117(213), 118(223-224,228), 121(231-232), 124(126,231,236), 125(236), 127(231,238,241), 128(126,242), 129(126,231,241), 130(109,241,249), 134(259), 135(238,260,264), 136(126,231,267-269), *139, 141–146*
Stasyuk, I. V., 355(38-39,41-42), 381(135,138), 382(135), 383-384(138), 387-389(135), 390(38-39,41-42), 391(38), 394(38), 395(150), 397-400(150), 437(150), 482-483(135), *536, 539–540*
Stegun, I. A., 14(55), 27(55), 55-56(55), 70-71(55), *95*
Stehle, P., 288(54), *348*
Stenger, J., 469(303), *545*
Stepanov, B. I., 419(182), *540*
Stern, L. J., 446(228,260), *542–543*
Steuerwald, S., 215(72), 219(72), *267*
Stevens, R. M., 121(235), *145*

Stilck, J. F., 21(76), *95*
Stillinger, D. K., 5(27), *94*
Stillinger, F. H.: 5(27), 16(57), 33(99), *94–96*; 490(337-338), *546*
Stobbe, H., 507(405), 521(405), *548*
Stockburger, M., 448(257), *543*
Stoeckenius, W., 445(220-222), 446(221-222,231,237), 447(220), *542*
Stoer, J., 27(89), 34(89), 40(89), *96*
Stoneham, A. M., 354(20), 359(20), 363(20), 368(20), *536*
Straub, J., 446(229), *542*
Strey, G., 270(9), *345*
Struchko, Yu. T., 395(153), *540*
Sugarno, R., 312(60), *348*
Sugimoto, H., 524(105), 525(108), *538*
Sukhhatme, U., 13(51), *94*
Sung, R., 207(59), *266*
Sung, S. M., 5(36), *94*
Suominen, K.-A., 226(76), *267*
Surjan, R. J., 104(92), *141*
Sutcliffe, B. T., 271(22), 272(48), 291(22), *346–347*
Suzuki, M., 67(173), *98*
Suzuki, S., 355(32), *536*
Sverdlov, L. M., 461(294), *544*
Svirmickas, M., 502(389), *547*
Szafran, B., 65(171), *98*
Szafran, M., 438(208), *536, 541*
Szalay, P. G., 102(2,35), 103(2), 112(2,152), 117(216,219), 118(219-220), *138–139, 142, 144*
Szalewicz, K., 102(61), *140*

Takahashi, M., 113(186), *143*
Tan, A. L., 6(39), *94*
Tanaka, H., 490(350), *546*
Tanaka, S., 360(69), 372(69), *537*
Taranenko, V. B., 355(37), 446(37), 449-450(37), 458-459(37), *536*
Tarantelli, F., 137(272), *146*
Tarnavski, Yu. A., 355(43-46), 428(196,198), 429(46,196), 431(46), *536, 541*
Tavan, P., 446(242), *542*
Taylor, B. N., 52(154), *97*
Taylor, H. S., 64(169), *98*
Taylor, P. R., 102(64-65), 111(150), *140, 142*
Taylor, T. R., 133(254), *145*
Teissié, J., 355(28), *536*
Tejada, J., 469(305,308), 474-475(308,329), 485(305,308,329), *545*
Teller, E., 80(181), *98*
Temkin, A., 34(106), *96*
Terao, T., 362(75), 374(114), *537–538*
Termini, J., 448(253,259,261,263), *543*
Thiedemann, G., 460(290), *544*
Thirring, W., 17(67), *95*
Thomas, L. D., 64(169), *98*
Thompson, C. J., 20(71), *95*
Thouless, D. J., 104(95), *141*
Timoshchenko, G. T., 521(417), *548*
Tittor, J., 446(229,243), *542–543*
Tobaru, K., 390(147), *539*
Tocane, J. F., 355(28), *536*
Tokunaga, M., 368-369(98), *538*
Tolstoi, A. A., 507(404), *548*
Tolstoi, N. A., 507(400-401,404), *548*
Tomchuk, P. M., 355(33-37), 400(163-165), 406-408(165), 409(33-34,165,176), 411(178), 414-415(178), 416(178-179), 419(165), 429(165), 432(163), 434(164), 446(35-37), 449-450(37), 451-452(35), 458(34,37), 459(37), 467(36), 470(319-322), 474(327), 477(319-322,327), 482(320-321), 486(320), 507(408), 512(408), 529(165), *536, 540, 545, 548*
Tomkinson, J., 353(7-10), 374(110), 375(8-10,110), 378(7-9), 523(7-10), *535, 538*
Tonks, D. L., 368(93), *538*
Trakhtenberg, M. S., 496(370), *547*
Tregubchenko, A. M., 395(153), 400(153), *540*
Treshnikov, E. N., 444-445(219), *542*
Tristran-Nagle, S., 446(239), *542*
Trommsdorff, H. P., 353(12-15), 354(15-17,21-22), 359(21-22), 360(21), 363(21), 364(76-81,83-84), 365(83), 366(83-84), 371(99), 374(21), 523(14-16,21-22,78-81,83-84), *536–538*
Tronc, E., 469(316), 470(316), *545*
Trucks, G. W., 102(16,18,29), 103(70,74), 105(108), 110(70), 129(108), *139, 141*
Truhlar, D. G.: 102(52-53), *140*; 272(49), *347*
Tsakulev, A. V., 456(283), *544*
Tsetlin, B. M., 469(311), *545*
Tsipis, C. A., 5(35), *94*
Tuck, A. F., 105(100), *141*
Tucker, J. W., 421(184), *541*

Tulk, C. A., 353(2), *535*
Tyablikov, S. V., 403(172), 528(172), *540*
Tyuterev, V. G., 171(31), *266*

Uchiki, H., 450(274), *544*
Udovic, T. J., 359(25), *536*
Unanyan, R., 176(38), 235(81-82), 236(83), 246(38,84-85), 248(38), *266–267*
Unterderderweide, K., 438(210), *541*
Urabe, H., 446(247), *543*
Urban, M., 109(132), 110(135), 117(218), *142, 144*
Urjasz, H., 448(254), *543*
Uski, V., 92(213), *99*
Uzelac, K., 21(82), *95*

Vaida, V., 105(100), *141*
Vainberg, V. M., 27(91), *96*
Valakh, M. Ya., 442(216), *542*
Valiron, P., 104(98), *141*
Van, C. L., 154(8), 156-157(8), 161(8), 262(8), *265*
Van der Berg, E. M. M., 446(237), *542*
Vanderkooy, J., 355(29), 389(29), 408(29), 415(29), *536*
van Leuven, J. M. J., 84(206), *98*
Van Vleck, J. H., 168(28), *266*
Van Voorhis, T., 114(203-204), *144*
Vardi, A., 215(71), *267*
Varga, K.: 3(19), *93*; 50(145), *97*
Vargaftik, N. B., 496(372), *547*
Vasilakis, C., 456(487), *544*
Venevtsev, Yu. N., 444(218), *542*
Vetoshin, E. V., 354(23), 359(23), 364(79-84), 365(83), 366(83-84), 523(79-84), 532(82), *537–538*
Vial, J. C., 354(17), *536*
Villain, J., 485(336), *545*
Vitanov, N. V., 212(63), 214(69), 215(72), 219(72), 225(63), 235(76), 226(63,69), 227(69), 235(81), *267*
Volkov, A. F., 444(218), *542*
von Laue, L., 364(79-80), 523(79-80), *537–538*
Von Neumann, J., 157(15), *265*
Von Niessen, W., 113(170-171), 137(271), *143, 146*
Voss, D., 2(5), *93*

Waintal, X., 91(209), *99*
Waldron, J. T., 469(313), *545*
Walecka, J. D., 362(74), *537*
Walsh, M. E., 469(311), *545*
Walter, O.: 64(168), *97*; 137(270), *146*
Wang, J. X., 82(205), *98*
Wang, L., 3(13), *93*
Wang, L. S., 45(128), *97*
Wang, Q., 456(286), *544*
Wang, X., 3(13), *93*
Wang, X. B., 45(128), *97*
Wang, X. G., 271(28), 291-292(28), *346*
Wang, X.-G., 292(55), *348*
Warman, J. M., 383(140), 389(140), *539*
Washburn, S., 475(331), *545*
Watkins, E., 55(161), *97*
Watson, J. K. G., 270(3-4), *345*
Watts, J. D., 102(18,40), 103(40,71), 109(40), 110(71), 109(133), 118(223-224), *139–140, 142, 144*
Weaver, A., 127(237), *145*
Webb, R. A., 475(331), *545*
Wei, H., 271(24-25), 272(24-25,46), 291(24-25), *346–347*
Weikert, H.-G., 137(272), *146*
Weil, L., 485(335), *545*
Weinhold, F., 55(160), *97*
Weisshaar, J. C., 133(257), *145*
Weninger, K. R., 526(126), *539*
Wenzel, K., 102(61), *140*
Werner, H.-J., 103(84-85), 105(110), 111(149), 117(217), 129(110), *141–142, 144*
Wernsdorfer, W., 469(304,317), 494(304), *545*
Whetten, R. L., 81(188), *98*
Whiteley, A. H., 498(376), *547*
Wilbur, D. J., 502-503(391), *547*
Willison, J. R., 381(122), 526-527(122), *539*
Wilson, A. K., 102(49-50), *140*
Wilson, E. B., 270-272(1), 291-292(1), 295(1), 316(1), 345(1), *345*
Wintgen, DS., 51(150), 53(150), *97*
Witkowski, A., 356(66), 357-359(66), 438(66), *537*
Witten, E., 5(34), *94*
Wolinski, K., 112(154), *143*
Wols, A., 81(184), *98*
Wolynes, P. G., 78(180), *98*
Wood, M., 502(389), *547*
Woon, D. E., 102(51,58), *140*
Woutersen, S., 490(352), *546*
Wu, T. T., 46(137), *97*

Xantheas, S. S., 102(47), *139*
Xie, Y., 130(246), *145*

Yabjshita, S., 102-103(2), 112(2), *138*
Yakhmi, J. V., 460(293), *544*
Yakubov, A. A., 356(57-58,60), 359(57-58,60), 461(57-58,60), 462(57-58), 463(57), 464(57-58,60), 467-468(57-58,60), *537*
Yamada, Y., 524(105-107), 525(106), *538*
Yamaguchi, Y., 102(10), 130(246), *138, 145*
Yamamoto, K., 507(397), *547*
Yamzin, I. I., 444-445(219), *542*
Yan, L., 207(59), *266*
Yan, Z., 34(104), 39(114), *96*
Yang, C. N., 19(68-69), *95*
Yang, W., 110(144), 129(144), *142*
Yanovskii, O. E., 470(325), 482(325), 484(325), *545*
Yaremko, A. M., 438(211), 439(215), 442(211,216), *541–542*
Yarkony, D. R., 102(45), *139*
Yatsenko, L., 176(38), 193(42), 212(63), 214(67-68), 215(71-72), 219(72-73), 225(63), 226(63,73-74), 235(67,80,82), 238(80), 246(38), 248(38), *266–267*
Yeager, D. L., 113(167,172,176), 133(255-256), *143, 145*
Yeomans, J., 21(74), *95*
Yomosa, S., 400(162), *540*
Yoshida, M., 389(141), *539*
Yugo, S., 507(397), *547*
Yukhnovsky, I. R., 507(407), 411(407), *548*
Yuknevich, G. V., 501(380), *547*

Zabolitzky, J. G., 102(61), *140*
Zaccai, G., 445-446(226), 448(258), *542*
Zadkov, V. N., 469(300), *544*
Zakin, M. R., 526(132), *539*
Zecs, B., 368-369(96), 479(96), 482(96), 486(96), *538*
Zeldovich, Y. B., 498(375), *547*
Zelepukhin, I. D., 491(359,361), 496-497(359), 501-502(359), 505(359), 523(359), *546*
Zelepukhin, V. D., 491(359,361), 496-497(359), 501-502(359), 505(359), 523(359), *546*
Zemlin, F., 445-446(223), *542*
Zener, C., 213(65), *267*
Zettlemoyer, A. C., 507(402), *548*
Zhang, C.-F., 451(278), *544*
Zhang, X. X., 474-475(329), 485(329), *545*
Zhang, Z., 102-103(2), 112(2), *138*
Zhao, X.-L., 43(121), 45(121), *96*
Zhislin, G. M., 44(124), *96*
Zhugaevich, A. Y., 513(413), 517(417), *548*
Ziegler, R., 318(76), *349*
Zilo, R., 469(308), 474-475(308), 485(308), *545*
Zinth, W., 456(284), *544*
Zivi, S., 161(16), *265*
Zolotariuk, A. V., 400(161), 470(161), *540*
Zorina, V. V., 448(268), *543*
Zundel, G., 353(3-6), 400(167), 402(169-170), 419(6), 438(212), 446(6,233-234,241,248-249), 448(6,233,241,254), 449(233), 451(233), 456(288-289), 457(6), 470(6,249,288-289), 475(249), 485(6,249,288-289), 495(369), 519(6), 523(3-6), *535, 540–544, 547*

SUBJECT INDEX

Acoustic phonons, hydrogen bonds
polaronic chain conductivity, 404–409
proton transfer, 360–366, 420–423
Acoustic vibrations, hydrogen bonds, anharmonicity influence, 409–410
Adiabatic approximation
dressed eigenenergy surface topology, stimulated Raman adiabatic passage (STIRAP), 226–235
hydrogen bonds, polaronic chain conductivity, 403–409
O–H–O hydrogen bond, stretching and bending energies, 524–526
open-shell molecules, symmetry breaking, molecular orbital response, 120–127
Adiabatic elimination, dressed Floquet Hamiltonian, dynamical resonances, 183–184
Adiabatic Floquet theory. *See also* Floquet Hamiltonian
laser pulses, quantum dynamics, 198–214
diabatic *vs.* adiabatic dynamics, eigenenergy and avoided crossings, 212–214
dressed Schrödinger equation, chirped pulses, 199–201
dressed states adiabatic/diabatic evolution, 201–212
Hamiltonian equations, 263–265
Adiabatic passage
dressed eigenenergy surface topology
chirped pulse/ SCRAP analysis, 215–217
optimization, 219–222
robustness properties, 217–219
Floquet Hamiltonian, coherent photon exchange, 164–166
Alkyl-/alkloxybenzoic acids
cluster formation, 518–520
mesomorphic transformations and proton subsystem dynamics, 460–469
molecular associates, 460–464
open associates formation mechanism, 464–469
A matrix eigenvalues, open-shell molecules, symmetry breaking, molecular orbital response, 124–127
Amino acid residues, hydrogen bonds, bacteriorhodopsin microscopic physics, 447–451
Ammonium triiodate crystal model, hydrogen bonds, superionic phase transitions, 428–437
Amplitude of probability, finite-size scaling (FSS), near-threshold states, 13–16
Anderson MIT, finite-size scaling (FSS), quantum dots, 89–92
Angular momentum, vibration-rotation Hamiltonians, Lagrangian formalism, 289–291
Anharmonicity influence, hydrogen bonds
polaron/polariton transport properties, 409–410
proton transfer, 420–423
Anisotropic properties, hydrogen bonds
coherent tunneling, repolarization mechanisms, 479–485
proton ordering, 389–394
proton-phonon coupling, 486–490
Annihilation operator
finite-size scaling (FSS), quantum dots, 85–92
Floquet Hamiltonian
coherent states, 261–263
quantized cavity dressed states, 155–158
hydrogen bonds
bacteriorhodopsin retinal evolution, 452–457
mesoscopic coherence and tunneling, 471–477
orientational-tunneling model, one-dimensional molecular systems, 383–389

Annihilation operator (*Continued*)
polaronic chain conductivity, 404–409
proton ordering, 369–371
proton-phonon coupling, 486–490
proton transfer, 417–423
open-shell molecules, quantum mechanics, size-consistent methods, 108–113
symmetric O–H–O hydrogen bond, polariton effect, 439–445
Antiphase vibrations, hydrogen bonds, superionic phase transitions, 396–400
Aqueous systems. *See also* Water molecules
hydrogen bonds
continuous metal films, water-dependent switching, 507–513
degassed systems organization and thermodynamics, 490–502
experimental results, 491–494
thermodynamics, 494–501
pulsed nuclear magnetic resonance, 502–507
Arrhenius coordinates, hydrogen bonds, superionic phase transitions, ammonium triiodate crystal model, 429–437
Asymmetry parameter, finite-size scaling (FSS), near-threshold states, large-dimensional limits, 7–8
Asymptotic behavior
finite-size scaling (FSS)
crossover phenomena, quantum systems, 59–63
near-threshold states, short-wave potentials, 14–16
quantum criticality, path integral techniques, 79–81
quantum phase transitions and stability, two-electron atoms, 38–39
spatial finite-size scaling (SFSS), 67–76
Atomic beam deflection, two-level topological quantization, bichromatic selectivity, 237–245
dressed eigenenergy construction, 242–243
effective Hamiltonian, 237–239
eigenenergy surface topology, 239–242
photon exchange dynamics, 244–245
Atomic systems, finite-size scaling (FSS)
classical statistical mechanics, 19–21
multicritical points, 63–65
near-threshold states, 5–18
few-body potentials, 16–18
finite dimension space, critical phenomena, 8–18
large-dimensional limit phase transitions, 5–8
one-particle central potentials, 10–16
path integral quantum criticality, 75–81
lattice system mapping, 75–77
Pöschl-Teller potential, 77–81
quantum dots, 81–92
quantum mechanics, 21–32
equations, 23–27
extrapolation and basis set expansions, 27–29
Schrödinger equation data collapse, 29–32
statistical mechanics and classical analogies, 21–23
quantum system crossover phenomena, 59–63
quantum system resonances, 53–58
research background, 2–4
spatial scaling (SFSS), 65–75
stability, quantum phase transitions and, 33–53
critical nuclear charges, 42–45
diatomic molecules, 45–50
three-body Coulomb systems, phase diagrams, 50–53
three-electron atoms, 39–42
two-electron atoms, 33–39
Averaging procedures, Floquet Hamiltonian, Kolmogorov-Arnold-Moser (KAM) transformations, 170–171
Avoided crossing
adiabatic Floquet evolution, 201–202
diabatic/adiabatic dynamics, 212–214
dressed eigenenergy surface topology, robustness, 218–219

Bacteriorhodopsin, hydrogen bonds
proton ejection, 457–460
proton path active site, 445–451
retinal light-excitation, 451–457
Basis set expansions
finite-size scaling (FSS)
quantum mechanics, 27–29
quantum phase transitions and stability
diatomic molecules, 49–50
two-electron atoms, 34–39
quantum system resonances, 57–58

open-shell molecules, quantum mechanics, research background, 102–105
vibration-rotation Hamiltonians, three-dimensional geometric algebra, 323–326
Bending vibrations
hydrogen bonds, quantum mechanics, 371–374
O–H–O hydrogen bond, 524–526
Bernal's hypothesis, alkyl-/alkloxybenzoic acids, mesomorphic transformations, 464
Berry geometrical phase, adiabatic Floquet theory, 202–206
Bessel function
finite-size scaling (FSS), near-threshold states, short-wave potentials, 14–16
Floquet Hamiltonian, coherent states, 262–263
spatial finite-size scaling (SFSS), 70–76
Bichromatic pulses, state-selectivity, 235–255
three-level systems, 245–255
two-level system, atomic beam deflection topological quantization, 237–245
Bilinear phonon-phonon interaction, hydrogen bonds, anharmonicity influence, 409–410
Biradicals, open-shell molecules, quantum mechanics, research background, 105
Bivectors, vibration-rotation Hamiltonians
geometric algebra, 320–321
law of sines, 326–327
rotations, 328–333
three-dimensional geometric algebra, 325–326
Bjerrum defect, hydrogen bonds
coherent tunneling, repolarization mechanisms, 484–485
orientational-tunneling model, one-dimensional molecular systems, 381–389
polaronic chain conductivity, 400–409
Blinc's formalism, aqueous systems models, 501–502
Block-diagonal partitioning, dressed Floquet Hamiltonian, dynamical resonances, 181–183
Block renormalization group (BRG) technique, finite-size scaling (FSS), quantum dots, 82–92
Body-fixed frame, vibration-rotation Hamiltonians
geometric rotations, 332–333
Lagrangian formalism, 274–291
Body-frame component, vibration-rotation Hamiltonians, measuring vectors, 293–309
Bogolyubov-Tyablikov canonical transformation, hydrogen bonds, polaronic chain conductivity, 403–409
Boltzmann distribution
hydrogen bonds, superionic phase transitions, ammonium triiodate crystal model, 435–437
molecular clusters, statistical mechanics, 516–518
Bond-breaking problem, open-shell molecules, 114–119
Bond vibrations, hydrogen bonds, 356–360
Bond-z frame, vibration-rotation Hamiltonians, constrained quantization, 315–317
Born-Oppenheimer approximation
finite-size scaling (FSS)
quantum phase transitions and stability, diatomic molecules, 45–60
three-body Coulomb systems, phase diagrams, 52–53
Floquet Hamiltonian, dynamical resonances
diatomic molecule rotational excitation, 193–194
effective Hamiltonian for vibrations, 196
high-frequency perturbation, 174
open-shell molecules, symmetry breaking, molecular orbital response, 120–127
vibration-rotation Hamiltonians, 272
Bose operators
alkyl-/alkloxybenzoic acids, open associates formation, 466–469
hydrogen bonds
bacteriorhodopsin retinal evolution, 452–457
Coulomb correlations, 411–416
polaronic chain conductivity, 403–409
proton bifurcation and phonon mixture, 532–535
proton ordering, 369–371
proton-phonon coupling, 486–490
proton polaron, 367–368
superionic phase transitions, 397–400
molecular clusters, statistical mechanics, 516–518
symmetric O–H–O hydrogen bond, polariton effect, 439–445

Boson coupling, hydrogen bonds, tunneling transition, coupled protons, 377–381
Boundary conditions, spatial finite-size scaling (SFSS), 70–76
Brillouin zones, Floquet Hamiltonian, eigenvector/eigenvalue structure, 257–260
Broad resonances, finite-size scaling (FSS), quantum systems, 53–58
Brownian path
finite-size scaling (FSS), quantum criticality, path integral techniques, 78–81
hydrogen bonds, mesoscopic coherence and tunneling, 469–477
Brueckner coupled-cluster (B-CC) method, open-shell molecules, pseudo-Jahn-Teller (PJT) effect, 129–134
Brueckner orbitals, open-shell molecules, quantum mechanics, Slater determinant, 105–113
Bulirsch-Stoer algorithm, finite-size scaling (FSS)
quantum mechanics, 28–29
quantum phase transitions and stability
three-electron atoms, 40–42
two-electron atoms, 34–39

Canonical quantization, vibration-rotation Hamiltonians, 309–313
constrained quantization, 312–313
unconstrained quantization, 309–312
Canonical transformation, hydrogen bonds
polaronic chain conductivity, 403–409
superionic phase transitions, 397–400
Carbon-carbon bonds, hydrogen bonds, bacteriorhodopsin retinal excitation, 451–457
Carboxylic acids, hydrogen bonds, bond vibrations, 358–360
Cartesian coordinates, vibration-rotation Hamiltonians
Lagrangian formalism, 273–291
limits of, 271
scalar chain rules, 291–292
CASPT2 technique, open-shell molecules, quantum mechanics, 112–113
CASSCF. *See* Complete active space self-consistent field (CASSCF)
CC. *See* Coupled-cluster (CC) approximation
CCSD. *See* Coupled-cluster (CC) singles and doubles (CCSD) technique
CCSD(T) approximation
open-shell molecules, quantum mechanics
research background, 103–105
size-consistent methods, 110–113
spin contamination, 114–119
open-shell molecules, symmetry breaking, molecular orbital response, 121–127
CCSDT approximation, open-shell molecules, quantum mechanics
size-consistent methods, 109–113
spin contamination, 116–119
CCSDTQP approximation, open-shell molecules, quantum mechanics
research background, 103–105
size-consistent methods, 110–113
Centrosymmetric molecules, open-shell molecules, symmetry breaking, molecular orbital response, 126–127
Chain polarization, hydrogen bonds
bacteriorhodopsin retinal excitation, 451–457
coherent tunneling of repolarization, 477–485
mesoscopic coherence and tunneling, 470–477
polaronic systems, 400–409
proton ejection, 457–460
proton-phonon coupling, 487–490
proton transfer influence, 416–423
Chain rules, vibration-rotation Hamiltonians
geometric algebra, vector derivatives, 302–309
geometric calculus, directional derivatives, 337–339
scalar chain rules, 291–292
Charge carrier migration, hydrogen bonds, polaronic chain conductivity, 401–409
Chirped laser pulses
adiabatic Floquet theory, dressed Schrödinger equation, 199–203
dressed eigenergy surface topology, adiabatic passage, 215–217
Christoffel symbols, vibration-rotation Hamiltonians, constrained quantization, 313
CISD technique, open-shell molecules
pseudo-Jahn-Teller (PJT) effect, 132–134
quantum mechanics, 106–113
Classical mechanics

finite-size scaling (FSS), atomic and molecular systems, 19–21
quantum mechanics, 21–23
vibration-rotation Hamiltonians, 272–309
Lagrangian source, 272–291
measuring vectors, 292–309
scalar chain rules, 291–292
Closed-shell molecules
reference functions, 135–137
spin contamination, 114–119
Coherent anomaly, spatial finite-size scaling (SFSS), 67–76
Coherent states
dressed eigenenergy surface topology
half-scrap superposition, 222–226
stimulated Raman adiabatic passage (STIRAP), 234–235
Floquet Hamiltonian
adiabatic passage, photon exchange in, 164–166
operators, 261–263
photon exchange, 162–163
semiclassical formalism, 159–160
expectation values, 161–162
hydrogen bonds
alkyl- and alkoxybenzoic acids, mesomorphic transformations and proton subsystem dynamics, 460–469
molecular associates, 460–464
open associates formation mechanism, 464–469
aqueous systems
continuous metal films, water-dependent switching, 507–513
degassed systems organization and thermodynamics, 490–502
experimental results, 491–494
thermodynamics, 494–501
pulsed nuclear magnetic resonance, 502–507
bacteriorhodopsin
proton ejection, 457–460
proton path active site, 445–451
retinal light-excitation, 451–457
molecular clustering, 513–522
alkyl- and alkoxybenzoic acid solid phase cluster formation, 518–520
H_2O waters in water, 520–522
statistical mechanical approach, 513–518
phonon variable diagonalization, 527–529
polaron/polariton transport properties, 381–445
anharmonicity influence, 409–410
Coulomb correlations and electric field local heterogeneities, 411–416
external influences, proton conductivity, 416–422
orientational-tunneling model, one-dimensional molecular system, 381–389
polaronic chain conductivities, 400–409
proton ordering model, 389–394
resonance integral vibration fluctuations, 422–428
superionic phase conductivity, ammonium triiiodate crystal, 428–437
superionic phase transition, proton conductivity, 394–400
symmetric O· · ·H· · ·O bonds, 437–445
proton bifurcation and phonon mixture, 529–535
quantum coherent phenomena
chain repolarization, tunneling mechanisms, 477–485
mesoscopic quantum coherence and tunneling, 469–477
proton-phonon coupling, 485–490
quantum mechanics
acoustic photon proton transfer, 360–366
bending vibrations, 371–374
bond vibrations, 356–360
proton ordering, 368–371
proton polaron, 366–368
tunneling transition, coupled protons, 374–381
sonoluminescence mechanism, 526–527
stretching and bending energies, 524–526
structural physical effects, 352–356
Complete active space self-consistent field (CASSCF), open-shell molecules, 111–113
spin contamination, 114–119
Complex scaling techniques, finite-size scaling (FSS), quantum system resonances, 56–58
Compton scattering, hydrogen bonds, research applications, 353–356

Configuration interaction (CI), open-shell molecules, pseudo-Jahn-Teller (PJT) effect, 132–134
Constrained quantization, vibration-rotation Hamiltonians, 312–313
Contact transformations. *See* KAM transformations
Continuous metal films, water-dependent switching, 507–513
Continuous phase transition, finite-size scaling (FSS), three-electron atoms, 39–42
Contravariant measuring tensors, vibration-rotation Hamiltonians, 288–291, 293–309
  covariant measuring vectors and, 342–343
Control process, laser pulses, research background, 149–150
Cooperative proton transitions, hydrogen bonds, coherent tunneling, repolarization mechanisms, 482–485
Coriolis elements, vibration-rotation Hamiltonians
  Lagrangian formalism, 290–291
  unconstrained quantization, 313
Coulombic limit, finite-size scaling (FSS), quantum mechanics, 27–29
Coulomb interactions
  finite-size scaling (FSS)
    near-threshold states, large-dimensional limits, 6–8
    quantum phase transitions and stability, 33–53
      critical nuclear charges, 42–45
      diatomic molecules, 45–50
      three-body Coulomb systems, phase diagrams, 50–53
      three-electron atoms, 39–42
      two-electron atoms, 33–39
  hydrogen bonds
    coherent tunneling, repolarization mechanisms, 478–485
    mesoscopic coherence and tunneling, 471–477
    orientational-tunneling model, one-dimensional molecular systems, 383–389
    polaron/polariton transport properties, 411–416
Counterion neutralization, hydrogen bonds, bacteriorhodopsin microscopic physics, 449–451
Coupled-cluster (CC) approximation, open-shell molecules
  pseudo-Jahn-Teller (PJT) effect, 128–134
  reference functions, 134–137
  research background, 102–105
  Slater determinant, 105–113
  spin contamination, 114–119
Coupled-cluster (CC) singles and doubles (CCSD) technique, open-shell molecules, quantum mechanics
  size-consistent methods, 108–113
  spin contamination, 116–119
  reference functions, 136–137
  research background, 102–105
Coupled perturbed Hartree-Fock (CPHF) theory, open-shell molecules, symmetry breaking, molecular orbital response, 121–127
Coupled protons, hydrogen bonds, tunneling mechanisms, 374–381
Coupling constants, hydrogen bonds
  polariton effect, symmetric O–H–O hydrogen bond, 442–445
  superionic phase transitions, ammonium triiodate crystal model, 436–437
Covariant measuring vectors, vibration-rotation Hamiltonians
  contravariant measuring vectors and, 342–343
  Lagrangian formalism, 279–291
Creation operators
  Floquet Hamiltonian, quantized cavity dressed states, 155–158
  open-shell molecules, quantum mechanics, size-consistent methods, 108–113
Critical behavior, finite-size scaling (FSS)
  crossover phenomena, quantum systems, 59–63
  near-threshold states, 8–18
  quantum phase transitions and stability
    diatomic molecules, 45–50
    nuclear charges, *N*-electron atoms, 42–45
  research background, 3–4

"Crossed" hydrogen bonds, alkyl-/alkloxybenzoic acids, open associates formation, 465–469
Crossover phenomena, finite-size scaling (FSS), quantum systems, 59–63
Cross products, vibration-rotation Hamiltonians
  shape coordinates, 344–345
  three-dimensional geometric algebra, 324–326
Current density calculation
  hydrogen bonds
    polaronic chain conductivity, 404–409
    polaron vibration fluctuations, 425–428
    proton ejection, 459–460
    proton transfer, 418–423
    superionic phase transitions, ammonium triiodate crystal model, 432–437
  water-dependent switching, continuous metal films, 510–513
Curvilinear internal coordinates, vibration-rotation Hamiltonians, 270–272
Cutoff radius, spatial finite-size scaling (SFSS), 66–76
Cyclic dimerization, alkyl-/alkloxybenzoic acids, mesomorphic transformations, 461–464

Data collapse, finite-size scaling (FSS)
  quantum phase transitions and stability, three-electron atoms, 40–42
  Schrödinger equation, quantum mechanics, 29–32
Debye frequency, hydrogen bonds
  acoustic phonon proton transfer, 362–366
  anharmonicity influence, 410
Deformation potential approximation, hydrogen bonds, acoustic phonon proton transfer, 362–366
Degassed aqueous systems, hydrogen bonds, organization and thermodynamics, 490–502
  experimental results, 491–494
  thermodynamics, 494–501
Degeneracy
  adiabatic Floquet evolution
    lifting and creation, 201–202
    resonant laser lifting, 206–212
  dressed eigenenergy surface topology, half-scrap coherent state superposition, 223–226
Density functional theory (DFT), open-shell molecules, quantum mechanics
  pseudo-Jahn-Teller (PJT) effect, 132–134
  size-consistent methods, 110–113
Detuning strategies, dressed eigenergy surface topology, stimulated Raman adiabatic passage (STIRAP), 227–235
Diabatic dynamics, adiabatic Floquet evolution, eigenenergy and avoided crossings, 212–214
Diagonalization transformation
  eigenvectors and, 260–261
  hydrogen bonds
    phonon variables, 527–529
    polariton effect, symmetric O–H–O hydrogen bond, 442–445
    polaronic chain conductivity, 403–409
Diatomic molecules
  finite-size scaling (FSS), quantum phase transitions and stability, 45–50
  rotational excitation, Floquet Hamiltonian dynamical resonances, 193–198
    Born-Oppenheimer approximation, 193–194
    ground electronic surface, Raman processes, 194–198
Differential coupling equations, hydrogen bonds, bacteriorhodopsin retinal evolution, 454–457
Differentials, vibration-rotation Hamiltonians, geometric calculus, directional derivatives, 338–339
Differential scanning calorimetry, hydrogen bonds, superionic phase transitions, ammonium triiodate crystal model, 428–437
Dihedral angles, vibration-rotation Hamiltonians
  geometric rotations, 334
  Lagrangian formalism, 274–291
Dilatation analyticity, finite-size scaling (FSS), quantum phase transitions and stability, two-electron atoms, 33–39
Dimensional matrix, vibration-rotation Hamiltonians, Lagrangian formalism, 289–291
Dimensionless average potential, hydrogen bonds, bending vibrations, 373–374
Dimensionless symmetry, hydrogen bonds, acoustic phonon proton transfer, 364–366

Direct chirping, dressed eigenergy surface topology, adiabatic passage, 215–217
Directional derivatives, vibration-rotation Hamiltonians
classical mechanics, infinitesimal displacement, 297–298
geometric calculus, 336–339
shape coordinates, 344–345
Displaced harmonic oscillator, hydrogen bonds, proton polaron, 367–368
Displacement coordinates, vibration-rotation Hamiltonians, geometric algebra, 300–309
DMC calculations, open-shell molecules, quantum mechanics, 113
Dot products, vibration-rotation Hamiltonians
geometric algebra, 321–323
Lagrangian formalism, 283–284
Double-well potential, hydrogen bonds
acoustic phonon proton transfer, 360–366
orientational-tunneling model, one-dimensional molecular systems, 382–389
research overview, 354–356
tunneling transition, coupled protons, 374–381
Dressed eigenenergies, two-level topological quantization, atomic beam deflection, 242–243
Dressed Floquet states, adiabatic/diabatic evolution, 201–212
adiabatic evolution, 202–204
laser field resonance, degeneracy lifting, 206–212
nonresonant deviations, 204–206
Dressed Hamiltonians
eigenvector/eigenvalue structure, 257–260
Floquet Hamiltonian partitioning, 179–190
adiabatic elimination, 184–185
enlarged space partitioning, 187–190
formulation, 179–183
high-frequency partitioning, 185–187
Dressed Schrödinger equation
adiabatic Floquet theory
adiabatic evolution, 202–204
adiabatic theorem, 263–265
chirped laser pulses, 199–202
three-level bichromatic pulse selectivity, dynamics, 252–255
Dressed surface topology, eigenenergy crossings, 214–235
adiabatic passage optimization, 219–222
adiabatic passage robustness, 217–219
chirped pulse and SCRAP adiabatic passages, 215–217
coherent resonance superposition half-scrap, 222–226
stimulated Raman adiabatic passage (STIRAP) and variations, 226–235
Dykhne-Davis-Pechukas formula
adiabatic Floquet evolution, 204–206
dressed eigenenergy surface topology, adiabatic passage optimization, 219–222
Dynamical electronic polarizability tensor, diatomic molecule rotational excitation, ground electronic surface, Hilbert subspace, 195
Dynamical phase, adiabatic Floquet theory, 202–204
Dynamical resonances, Floquet Hamiltonian, 167–198
diatomic molecule rotational excitation, 193–198
dressed Hamiltonians, partitioning techniques, 179–190
high-frequency perturbation theory, 171–174
nonperturbative resonant transformation, 174–178
perturbation theory - KAM techniques, 167–171
two-photon RWA, 190–192

Easy magnetization, hydrogen bonds, mesoscopic coherence and tunneling, 474–477
Eckart potential
finite-size scaling (FSS), near-threshold states, 13–16
vibration-rotation Hamiltonians, research background, 272
Effective Hamiltonian
diatomic molecule rotational excitation, 193–198
Born-Oppenheimer approximation, 193–194
ground electronic surface, Raman processes, 194–198
dressed Hamiltonian partitioning, 180–183
enlarged space partitioning, 187–190

hydrogen bonds, bond vibrations, 359–360
three-level bichromatic pulse selectivity, 246–247
two-level topological quantization, atomic beam deflection, 237–239
two-photon quasi-resonant atomic processes, 190–192
Eigenenergy crossings
adiabatic Floquet evolution, diabatic/adiabatic dynamics, 212–214
dressed surface topology, 214–235
adiabatic passage optimization, 219–222
adiabatic passage robustness, 217–219
chirped pulse and SCRAP adiabatic passages, 215–217
coherent resonance superposition half-scrap, 222–226
stimulated Raman adiabatic passage (STIRAP) and variations, 226–235
three-level bichromatic pulse selectivity, 247–249
two-level topological quantization, atomic beam deflection, 239–242
dressed eigenenergies, 242–243
Eigenvalues
Floquet Hamiltonian
dressed Hamiltonian, 257–260
Kolmogorov-Arnold-Moser (KAM) transformations, 169–170
semiclassical formalism, 153–154
hydrogen bonds
orientational-tunneling model, one-dimensional molecular systems, 385–389
tunneling transition, coupled protons, 375–381
Eigenvectors
diagonalization transformation, 260–261
Floquet Hamiltonian
dressed Hamiltonian, 257–260
semiclassical formalism, 153–154
hydrogen bonds
orientational-tunneling model, one-dimensional molecular systems, 385–389
tunneling transition, coupled protons, 375–381
Electrical conductivity, degassed aqueous systems, hydrogen bonds, 493–502
Electric field local heterogeneities, hydrogen bonds, polaron/polariton transport properties, 411–416
Electronic excitation, hydrogen bonds, bacteriorhodopsin retinal evolution, 451–457
Electron propagator theory (EPT), open-shell molecules, quantum mechanics, 113
reference functions, 136–137
Electron small polaron theory, hydrogen bonds, proton transfer influence, 417–423
Enthalpy mechanisms
alkyl-/alkloxybenzoic acids, mesomorphic transformations, 461–464
degassed aqueous systems, thermodynamics, 496–502
Entropy mechanisms, degassed aqueous systems, thermodynamics, 498–502
EOM-CC technique
open-shell molecules, pseudo-Jahn-Teller (PJT) effect, 128–134
open-shell molecules, quantum mechanics, size-consistent methods, 109–113
EOMEA-CC method, closed-shell molecules, 135–137
EOMIP-CC techniques, closed-shell molecules, 135–137
Essential states, Floquet Hamiltonian, dynamical resonances, dressed Hamiltonian partitioning, 180–183
Euclidean action, hydrogen bonds
coherent tunneling, repolarization mechanisms, 479–485
mesoscopic coherence and tunneling, 475–477
proton-phonon coupling, 487–490
Euler angles, vibration-rotation Hamiltonians
classical mechanics, infinitesimal displacement, 297–298
constrained quantization, 314–317
geometric algebra, 303–309
Lagrangian formalism, 274–291
rotations, 329–333
unconstrained quantization, 310–312
Exciton-vibration interaction, hydrogen bonds, bacteriorhodopsin retinal evolution, 452–457
Expansion rates, vibration-rotation Hamiltonians
classical mechanics, infinitesimal displacement, 296–298
geometric algebra, 323

Exponential *ansatz*, coupled-cluster (CC) singles and doubles (CCSD) technique, 108–113
Extrapolation techniques, finite-size scaling (FSS), quantum mechanics, 27–29

Faraday number, degassed aqueous systems, 493–502
Far-infrared absorption spectrum, hydrogen bonds, superionic phase transitions, ammonium triiodate crystal model, 428–437
FCI. *See* Full configuration interaction (FCI) energy
Fermi-Dirac statistics, hydrogen bonds, proton ordering, 391–394
Fermion coupling, hydrogen bonds, tunneling transition, coupled protons, 377–381
Fermi operator
  hydrogen bonds
    bacteriorhodopsin retinal evolution, 452–457
    coherent tunneling, repolarization mechanisms, 478–485
    Coulomb correlations, 411–416
    mesoscopic coherence and tunneling, 471–477
    polaronic chain conductivity, 403–409
    proton ordering, 369–371
    superionic phase transitions, ammonium triiodate crystal model, 433–437
  molecular clusters, statistical mechanics, 516–518
Ferroelectric crystals, hydrogen bonds
  mesoscopic coherence and tunneling, 471–477
  proton ordering, 368–371
Few-body potentials, finite-size scaling (FSS)
  multicritical points, 63–65
  near-threshold states, 16–18
  Schrödinger equation data collapse, 29–32
Feynman's path integral, finite-size scaling (FSS), 4
  quantum mechanics, 22–23
Field induced resonances. *See* Dynamical resonances
Finite-dimensional space, finite-size scaling (FSS), critical phenomena, near-threshold states, 8–18
Finite-size scaling (FSS), atomic and molecular systems
  classical statistical mechanics, 19–21
  crossover phenomena, quantum systems, 59–63
  multicritical points, 63–65
  near-threshold states, 5–18
    few-body potentials, 16–18
    finite dimension space, critical phenomena, 8–18
    large-dimensional limit phase transitions, 5–8
    one-particle central potentials, 10–16
  path integral quantum criticality, 75–81
    lattice system mapping, 75–77
    Pöschl-Teller potential, 77–81
  quantum dots, 81–92
  quantum mechanics, 21–32
    equations, 23–27
    extrapolation and basis set expansions, 27–29
    Schrödinger equation data collapse, 29–32
    statistical mechanics and classical analogies, 21–23
  quantum phase transitions and stability, 33–53
    critical nuclear charges, 42–45
    diatomic molecules, 45–50
    three-body Coulomb systems, phase diagrams, 50–53
    three-electron atoms, 39–42
    two-electron atoms, 33–39
  quantum system resonances, 53–58
  research background, 2–4
  spatial scaling (SFSS), 65–75
First-order method (FOM)
  finite-size scaling (FSS)
    quantum mechanics, 24–27
    quantum phase transitions and stability, three-electron atoms, 39–42
    quantum system resonances, 53–58
  hydrogen bonds, sonoluminescence, 526–527
First-order phase transitions, finite-size scaling (FSS) equations, quantum mechanics, 23–27
Floquet Hamiltonian. *See also* Adiabatic Floquet theory
  bichromatic selectivity, 235–255
    three-level systems, 245–255

two-level system, atomic beam deflection
topological quantization, 237–245
eigenvector/eigenvalue structure, 257–260
laser pulses, quantum dynamics, 150–166
coherent states, 261–263
dressed Hamiltonian eigenvectors and eigenvalues, 257–260
eigenvectors and diagonalization transformations, 260–261
Hamiltonian-dynamical resonances, 167–198
diatomic molecule rotational excitation, 193–198
dressed Hamiltonians, partitioning techniques, 179–190
high-frequency perturbation theory, 171–174
nonperturbative resonant transformation, 174–178
perturbation theory - KAM techniques, 167–171
two-photon RWA, 190–192
interaction representation and coherent states, 158–162
multiple lasers, 166
photon emission and absorption, 162–166
semiclassical model and evolution of, 255–257
Fock operator
Floquet Hamiltonian
coherent states, 262–263
quantized cavity dressed states, 155–158
semiclassical coherent states, 159–160
open-shell molecules, quantum mechanics, spin contamination, 115–119
Force constants, open-shell molecules, symmetry breaking, molecular orbital response, 123–127
FORS-SCF. *See* Fully optimized reaction space self-consistent field (FORS-SCF)
Fourier component
Floquet Hamiltonian, eigenvectors and diagonalization transformations, 154
hydrogen bonds
polariton effect, symmetric O–H–O hydrogen bond, 443–445
proton ordering, 393–394
Fourier transform-infrared (FT-IR) spectroscopy, hydrogen bonds, bacteriorhodopsin microscopic physics, 446–451
Franck-Condon factor, hydrogen bonds, bond vibrations, 359–360
Free molecules, vibration-rotation Hamiltonians, classical mechanics, 273–291
Fugacity properties, alkyl-/alkloxybenzoic acids, cluster formation, 519–520
Full configuration interaction (FCI) energy, open-shell molecules, quantum mechanics
pseudo-Jahn-Teller (PJT) effect, 133–134
Slater determinant, 105–113
spin contamination, 114–119
Fully optimized reaction space self-consistent field (FORS-SCF), open-shell molecules, quantum chemistry, 111–113
Functional path variation, vibration-rotation Hamiltonians, geometric calculus, directional derivatives, 338–339

Gaussian basis sets, finite-size scaling (FSS), quantum phase transitions and stability, diatomic molecules, 49–50
Gaussian profile, hydrogen bonds, tunneling transition, coupled protons, 379–381
Gauss integrals, molecular clusters, statistical mechanics, 515–518
Geometric algebra, vibration-rotation Hamiltonians, 317–342
classical mechanics, 298–309
geometrical relations, 326–335
angles, 332–334
projections, 327–328
rotations, 328–332
spherical trigonometry, 334–335
geometric calculus, 335–342
directional derivative, 336–339
vectorial differentiation, 339–342
Lagrangian formalism, 278–291
research background, 271–272
sums and products, 318–323
expansion rules, 323
magnitude, 323
multiplication, 321–322
vector additions, 318
vector multiplication, 319–321
three-dimensional basis representation, 323–326

Geometric calculus, vibration-rotation Hamiltonians, 335–342
  directional derivative, 336–339
  vectorial differentiation, 339–342
Gibbs-Helmholtz equation, degassed aqueous systems, thermodynamics, 494–502
Gibbs potential, degassed aqueous systems, thermodynamics, 495–502
Global adiabatic passage, dressed eigenenergy surface topology, 215
  robustness, 218–219
Grade involution, vibration-rotation Hamiltonians, geometric algebra, 322
Green's function, hydrogen bonds
  orientational-tunneling model, one-dimensional molecular systems, 386–389
  polariton effect, symmetric O–H–O hydrogen bond, 443–445
Green's function QMC, open-shell molecules, quantum mechanics, 112–113
  reference functions, 137
Grid methods, finite-size scaling (FSS), multicritical points, 64–65
Grotthus mechanism, hydrogen bonds, orientational-tunneling model, one-dimensional molecular systems, 381–389
Ground electronic surface, diatomic molecule rotational excitation, effective Hamiltonian, 194–195
Ground-state energy, finite-size scaling (FSS), quantum phase transitions and stability, two-electron atoms, 34–39

Half-SCRAP coherent superposition, dressed eigenenergy surface topology, 225–226
Hamiltonian equations
  finite-size scaling (FSS)
    multicritical points, 63–65
    near-threshold states, one-particle central potentials, 10–16
    path integral lattice mapping, 76–77
    quantum dots, 83–92
    quantum mechanics, 23–27
      extrapolation and basis-set expansions, 27–29
    quantum phase transitions and stability
      diatomic molecules, 45–50
      two-electron atoms, 33–39
      quantum system resonances, 54–58
    three-body Coulomb systems, phase diagrams, 51–53
Floquet Hamiltonian
  bichromatic selectivity, 235–255
    three-level systems, 245–255
    two-level system, atomic beam deflection topological quantization, 237–245
  eigenvector/eigenvalue structure, 257–260
  laser pulses, quantum dynamics, 150–166
    coherent states, 261–263
    dressed Hamiltonian eigenvectors and eigenvalues, 257–260
    eigenvectors and diagonalization transformations, 260–261
    Hamiltonian-dynamical resonances, 167–198
      diatomic molecule rotational excitation, 193–198
      dressed Hamiltonians, partitioning techniques, 179–190
      high-frequency perturbation theory, 171–174
      nonperturbative resonant transformation, 174–178
      perturbation theory - KAM techniques, 167–171
      two-photon RWA, 190–192
    interaction representation and coherent states, 158–162
    multiple lasers, 166
    photon emission and absorption, 162–166
  semiclassical model and evolution of, 255–257
hydrogen bonds
  acoustic phonon proton transfer, 361–366
  bond vibrations, 356–360
  orientational-tunneling model, one-dimensional molecular systems, 384–389
spatial finite-size scaling (SFSS), 66–76
vibration-rotation Hamiltonians
  canonical quantization, 309–313
    constrained quantization, 312–313
    unconstrained quantization, 309–312
  classical mechanics, 272–309
    Lagrangian source, 272–291
    measuring vectors, 292–309

scalar chain rules, 291–292
co- and contravariant measuring vectors, 342–343
geometric algebra principles, 317–342
geometrical relations, 326–335
angles, 332–334
projections, 327–328
rotations, 328–332
spherical trigonometry, 334–335
geometric calculus, 335–342
directional derivative, 336–339
vectorial differentiation, 339–342
sums and products, 318–323
expansion rules, 323
magnitude, 323
multiplication, 321–322
vector additions, 318
vector multiplication, 319–321
three-dimensional basis representation, 323–326
research background, 270–272
shape coordinate properties, 343–345
volume elements, 313–317
Harmonic approximation, hydrogen bonds, bending vibrations, 373–374
Harmonic-oscillator function, hydrogen bonds
bond vibrations, 357–360
tunneling transition, coupled protons, 378–381
Hartree-Fock (HF) approximation, finite-size scaling (FSS), near-threshold states, large-dimensional limits, 6–8
Hartree-Fock self-consistent field (HFSCF) approximation
open-shell molecules
pseudo-Jahn-Teller (PJT) effect, 127–134
spin contamination, 113–119
symmetry breaking, molecular orbital response, 121–127
open-shell molecules, quantum mechanics
research background, 102–105
Slater determinant, 105–113
Heisenberg's uncertainty principle
finite-size scaling (FSS), 2–4
Floquet Hamiltonian, dynamical resonances, KAM transformation averaging, 171
Hellmann-Feynman theorem
finite-size scaling (FSS)
near-threshold states
one-particle central potentials, 11–16
scattering problems, 13–16
three-body Coulomb systems, 18
quantum mechanics, 26–27
spatial finite-size scaling (SFSS), 68–76
Hermite polynomials
finite-size scaling (FSS), quantum system resonances, 54–58
hydrogen bonds, tunneling transition, coupled protons, 378–381
Hessian matrix, finite-size scaling (FSS), near-threshold states, large-dimensional limits, 6–8
HF-SCF, open-shell molecules, spin contamination, 113–119
High-frequency partitioning
diatomic molecule rotational excitation, 197–198
Floquet Hamiltonian, 185–187
High-frequency perturbation theory, Floquet Hamiltonian, dynamical resonances, 171–174
High-resolution cryoelectron microscopy, bacteriorhodopsin, hydrogen bonds, proton path, 445–451
Hilbert space
diatomic molecule rotational excitation
effective Hamiltonian for vibrations, 196
ground electronic surface, Hilbert subspace, 194–195
Floquet Hamiltonian
dressed Hamiltonian partitioning, 180–183
adiabatic elimination, 183–185
eigenvector/eigenvalue structure, 257–260
enlarged space partitioning, 187–190
quantized cavity dressed states, 154–158
semiclassical formalism, 151–154
semiclassical model, 159
theoregical background, 150–151
semiclassical model and Floquet evolution, 256–257
two-photon quasi-resonant atomic processes, 191–192
Hopping activation energy, hydrogen bonds
Coulomb correlations, 412–416
proton ejection, 457–460
proton transfer, 418–423
superionic phase transitions, ammonium triiodate crystal model, 432–437

Internal coordinates, vibration-rotation Hamiltonians, geometric algebra, vector derivatives, 301–309
Iogansen method, alkyl-/alkloxybenzoic acids, mesomorphic transformations, 461–464
Ising's model
  molecular clusters, statistical mechanics, 514–518
  water-dependent switching, continuous metal films, 508–513
Iterative perturbation algorithm, Floquet Hamiltonian, dynamical resonances, KAM technique, 167–169

Jacobian matrices, finite-size scaling (FSS), near-threshold states, one-particle central potentials, 11–16
Jacobi vectors, vibration-rotation Hamiltonians
  scalar chain rules, 292
  unconstrained quantization, 312–313
Jahn-Teller effect. *See also* Pseudo-Jahn-Teller (PJT) effect

Kinetic energy operator
  hydrogen bonds, superionic phase transitions, ammonium triiodate crystal model, 434–437
  vibration-rotation Hamiltonians, 270–272
    constrained quantization, 312–313
    unconstrained quantization, 309–312
Kohn-Sham self-consistent-field (KS/KS-SCF) equations, open-shell molecules, quantum mechanics, size-consistent methods, 110–113
Kolmogorov-Arnold-Moser (KAM) transformations, Floquet Hamiltonian, dynamical resonances, 167–171
  averaging interpretation, 171
  construction process, 169–170
  dressed Hamiltonian partitioning, 179–190
  high-frequency perturbation, 171–174
  iterative perturbation algorithm, unitary transformation, 167–169
  nonperturbative resonant transformation, 176–178
  partitioning techniques, 179–183
Kronecker delta
  open-shell molecules, symmetry breaking, molecular orbital response, 122–127
  vibration-rotation Hamiltonians, Lagrangian formalism, 288–291
Kubo linear response theory, hydrogen bonds, superionic phase transitions, 397–400

Lagrangian equation
  hydrogen bonds, proton bifurcation and phonon mixture, 531–535
  vibration-rotation Hamiltonians
    classical mechanics, 272–291
    limits of, 271
    unconstrained quantization, 310–312
Laguerre polynomial, finite-size scaling (FSS)
  quantum mechanics, 27–29
  quantum phase transitions and stability, diatomic molecules, 47–50
  Schrödinger equation data collapse, 31–32
  three-body Coulomb systems, phase diagrams, 51–53
Lanczos procedure, finite-size scaling (FSS), three-body Coulomb systems, phase diagrams, 51–53
Landau phenomenological theory, hydrogen bonds, proton ordering, 390–394
Landau-Zener formula
  adiabatic Floquet evolution
    diabatic/adiabatic dynamics, eigenenergy and avoided crossings, 213–214
    Dykhne-Davis-Pechukas formula, 204–206
  dressed eigenenergy surface topology, adiabatic passage robustness, 217–219
Langevin equation, hydrogen bonds, mesoscopic coherence and tunneling, 469–477
Laplacian operator
  hydrogen bonds, bacteriorhodopsin retinal evolution, 455–457
  vibration-rotation Hamiltonians
    constrained quantization, 312–313
    spherical trigonometry, 335
    unconstrained quantization, 310–312
Large-dimensional limits, finite-size scaling (FSS), near-threshold states, 5–8
Laser pulses, quantum dynamics
  adiabatic Floquet theory, 198–214
    dressed Schrödinger equation, chirped pulses, 199–201
    dressed states adiabatic/diabatic evolution, 201–212
    Hamiltonian equations, 263–265

Laser pulses, quantum dynamics (*Continued*)
bichromatic pulses, state-selectivity, 235–255
three-level systems, 245–255
two-level system, atomic beam deflection topological quantization, 237–245
dressed eigenenergy surface topology, 214–235
adiabatic passage optimization, 219–222
adiabatic passage robustness, 217–219
chirped pulse and SCRAP adiabatic passages, 215–217
coherent resonance superposition half-scrap, 222–226
stimulated Raman adiabatic passage (STIRAP) and variations, 226–235
Floquet Hamiltonian, 150–166
coherent states, 261–263
dressed Hamiltonian eigenvectors and eigenvalues, 257–260
eigenvectors and diagonalization transformations, 260–261
Hamiltonian-dynamical resonances, 167–198
diatomic molecule rotational excitation, 193–198
dressed Hamiltonians, partitioning techniques, 179–190
high-frequency perturbation theory, 171–174
nonperturbative resonant transformation, 174–178
perturbation theory - KAM techniques, 167–171
two-photon RWA, 190–192
interaction representation and coherent states, 158–162
multiple lasers, 166
photon emission and absorption, 162–166
quantized cavity dressed states, 154–158
semiclassical formalism, 151–154
semiclassical Hamiltonian and evolution, 255–257
research background, 149–150
Lattice systems
finite-size scaling (FSS)
path integral mapping, 75–77
quantum dots, 83–92
hydrogen bonds
proton bifurcation and phonon mixture, 533–535
superionic phase transitions, ammonium triiodate crystal model, 436–437
Law of sines, vibration-rotation Hamiltonians, three-dimensional geometric algebra, 326–327
Lennard-Jones potential, water molecule clustering, 520–522
Level lines, dressed eigenenergy surface topology, adiabatic passage optimization, 219–222
Light-excited retinal transfer, hydrogen bonds, bacteriorhodopsin microscopic physics, 451–457
Linear oscillators, hydrogen bonds, tunneling transition, coupled protons, 376–381
Linear variation, finite-size scaling (FSS), quantum mechanics, 24–27
Lippinkott-Schröder potential, hydrogen bonds, bond vibrations, 358–360
Liquid phase models, aqueous systems, 501–502
Local diabatic evolution, dressed eigenenergy surface topology, 215
Lone pair systems, hydrogen bonds, orientational-tunneling model, one-dimensional molecular systems, 383–389
Lorentzian profiles, hydrogen bonds, polariton effect, symmetric O–H–O hydrogen bond, 444–445

Madelung's constant, water molecule clustering, 521–522
Magnitudes, vibration-rotation Hamiltonians, geometric algebra, 323
Many-body perturbation theory (MBPT)
open-shell molecules
pseudo-Jahn-Teller (PJT) effect, 127–134
symmetry breaking, molecular orbital response, 125–127
open-shell molecules, quantum mechanics
research background, 102–105
size-consistent methods, 107–113
spin contamination, 116–119
Matrix diagonalization, open-shell molecules, quantum mechanics, 112–113
MCSCF. *See* Multiconfigurational SCF (MCSCF)
Measuring vectors, vibration-rotation Hamiltonians

classical mechanics
geometrical algebra approach, 298–309
infinitesimal approach, 295–298
co- and contravariant measuring vectors, 342–343
geometric algebra, 304–309
Mesomorphic transformations, alkyl-/alkloxybenzoic acids, 460–469
molecular associates, 460–464
open associates formation mechanism, 464–469
Mesoscopic quantum coherence, hydrogen bonds, small magnetic grains and ordered molecules, 469–477
Metric tensor elements, vibration-rotation Hamiltonians, Lagrangian formalism, 286–287
Microscopic physics, bacteriorhodopsin, hydrogen bonds
proton ejection, 457–460
proton path active site, 445–451
retinal light-excitation, 451–457
Mobile velocity, vibration-rotation Hamiltonians, geometric algebra, 305–309
Molecular associates, alkyl-/alkloxybenzoic acids, mesomorphic transformations, 460–464
Molecular clustering, hydrogen bonds, 513–522
alkyl- and alkoxybenzoic acid solid phase cluster formation, 518–520
H_2O waters in water, 520–522
statistical mechanical approach, 513–518
Molecular motion, vibration-rotation Hamiltonians, classical mechanics, 273–291
Molecular orbitals, open-shell molecules
MCSCF calculations, 112–113
symmetry breaking, 119–127
Molecular structures
finite-size scaling (FSS)
classical statistical mechanics, 19–21
dianions, quantum phase transitions and stability, 45
multicritical points, 63–65
near-threshold states, 5–18
few-body potentials, 16–18
finite dimension space, critical phenomena, 8–18
large-dimensional limit phase transitions, 5–8
one-particle central potentials, 10–16
path integral quantum criticality, 75–81
lattice system mapping, 75–77
Pöschl-Teller potential, 77–81
quantum dots, 81–92
quantum mechanics, 21–32
equations, 23–27
extrapolation and basis set expansions, 27–29
Schrödinger equation data collapse, 29–32
statistical mechanics and classical analogies, 21–23
quantum system crossover phenomena, 59–63
quantum system resonances, 53–58
research background, 2–4
spatial scaling (SFSS), 65–75
stability, quantum phase transitions and, 33–53
critical nuclear charges, 42–45
diatomic molecules, 45–50
three-body Coulomb systems, phase diagrams, 50–53
three-electron atoms, 39–42
two-electron atoms, 33–39
hydrogen bonds
alkyl- and alkoxybenzoic acids, mesomorphic transformations and proton subsystem dynamics, 460–469
molecular associates, 460–464
open associates formation mechanism, 464–469
aqueous systems
continuous metal films, water-dependent switching, 507–513
degassed systems organization and thermodynamics, 490–502
experimental results, 491–494
thermodynamics, 494–501
pulsed nuclear magnetic resonance, 502–507
bacteriorhodopsin
proton ejection, 457–460
proton path active site, 445–451
retinal light-excitation, 451–457
molecular clustering, 513–522

Molecular structures (*Continued*)
alkyl- and alkoxybenzoic acid solid phase cluster formation, 518–520
H_2O waters in water, 520–522
statistical mechanical approach, 513–518
phonon variable diagonalization, 527–529
polaron/polariton transport properties, 381–445
anharmonicity influence, 409–410
Coulomb correlations and electric field local heterogeneities, 411–416
external influences, proton conductivity, 416–422
orientational-tunneling model, one-dimensional molecular system, 381–389
polaronic chain conductivities, 400–409
proton ordering model, 389–394
resonance integral vibration fluctuations, 422–428
superionic phase conductivity, ammonium triiiodate crystal, 428–437
superionic phase transition, proton conductivity, 394–400
symmetric O···H···O bonds, 437–445
proton bifurcation and phonon mixture, 529–535
quantum coherent phenomena
chain repolarization, tunneling mechanisms, 477–485
mesoscopic quantum coherence and tunneling, 469–477
proton-phonon coupling, 485–490
quantum mechanics
acoustic photon proton transfer, 360–366
bending vibrations, 371–374
bond vibrations, 356–360
proton ordering, 368–371
proton polaron, 366–368
tunneling transition, coupled protons, 374–381
sonoluminescence mechanism, 526–527
stretching and bending energies, 524–526
structural physical effects, 352–356
open-shell molecules, quantum mechanics
molecular orbitals and "symmetry breaking," 119–127
pseudo-Jahn-Teller effect, 127–134
reference functions, 134–137
research background, 101–105
spin contamination, 113–119
techniques, summary, 105–113
Møller-Plesset perturbation theory (MPPT), open-shell molecules, quantum mechanics
research background, 102–105
size-consistent methods, 107–113
Monte Carlo techniques, open-shell molecules, quantum mechanics, 112–113
Morse potential, hydrogen bonds, bending vibrations, 371–374
Mott metal-insulator transition (MIT), finite-size scaling (FSS), quantum dots, 82–92
MR-ACPF calculations, open-shell molecules, 112–113
MR-AQCC calculations, open-shell molecules, 112–113
MRCI calculations, open-shell molecules, 112–113
spin contamination, 119
Multiconfigurational (MC) techniques
open-shell molecules, quantum chemistry, Slater determinants, 111–113
open-shell molecules, quantum mechanics, pseudo-Jahn-Teller (PJT) effect, 133–134
open-shell molecules, spin contamination, 119
Multiconfigurational SCF (MCSCF)
open-shell molecules, 111–113
quantum mechanics, pseudo-Jahn-Teller (PJT) effect, 133–134
Multicritical points, finite-size scaling (FSS), quantum systems, 63–65
Multiphonon resonances, Floquet Hamiltonian, nonperturbative treatment, 175–178
Multiple laser dynamics, Floquet Hamiltonian, 166
Multiplication generalization, vibration-rotation Hamiltonians, geometric algebra, 321–322
Multivector, vibration-rotation Hamiltonians
geometric calculus, directional derivatives, 341–342
geometric rotations, 330–333
Mute resonances, dressed eigenergy surface topology, 216–217

Narrow resonances, finite-size scaling (FSS), quantum systems, 53–58
Near-instabilities, open-shell molecules, symmetry breaking, molecular orbital response, 124–127
Near-threshold states, finite-size scaling (FSS), atomic and molecular systems, 5–18
  few-body potentials, 16–18
  finite dimension space, critical phenomena, 8–18
  large-dimensional limit phase transitions, 5–8
  one-particle central potentials, 10–16
Néel-Brown model, hydrogen bonds, mesoscopic coherence and tunneling, 469–477
N-electron atoms, finite-size scaling (FSS), quantum phase transitions and stability, 42–45
Neutronographic analysis, hydrogen bonds, superionic phase transitions, ammonium triiodate crystal model, 428–437
Neville-Richardson analysis, finite-size scaling (FSS), quantum phase transitions and stability, two-electron atoms, 33–39
Newton-Raphson method, finite-size scaling (FSS), multicritical points, 64–65
Nonadiabatic transitions
  adiabatic Floquet theory, 203–204
  dressed eigenergy surface topology, stimulated Raman adiabatic passage (STIRAP), 232–234
Non-Hermitian eigenvalues, finite-size scaling (FSS), multicritical points, 64–65
Noninteracting electron model, finite-size scaling (FSS), near-threshold states, large-dimensional limits, 5–8
Nonlinear molecules, vibration-rotation Hamiltonians, Lagrangian formalism, 274–291
Nonlinear resonances. *See* Dynamical resonances
Nonperturbative resonant transformation, Floquet Hamiltonian, 174–178
Nonresonant deviations, adiabatic Floquet theory, 204–206

O–H–O hydrogen bond
  bending vibrations, 374
  polariton effect in symmetric crystals, 437–445
  stretching and bending energies, 524–526
One-dimensional molecular systems, hydrogen bonds, orientational-tunneling model, 381–389
"One-domain" approximation, hydrogen bonds, mesoscopic coherence and tunneling, 475–477
One-particle central potentials, finite-size scaling (FSS), near-threshold states, 10–16
Open associates, alkyl-/alkloxybenzoic acids, hydrogen bond rearrangements, 464–469
Open-shell molecules, quantum mechanics
  molecular orbitals and "symmetry breaking," 119–127
  pseudo-Jahn-Teller effect, 127–134
  reference functions, 134–137
  research background, 101–105
  spin contamination, 113–119
  techniques, summary, 105–113
Operator transformations, finite-size scaling (FSS), quantum dots, 85–92
Optical density, degassed aqueous systems, hydrogen bonds, 492–502
Optimized-orbital coupled-cluster (OO-CC) method, open-shell molecules, pseudo-Jahn-Teller (PJT) effect, 129–134
Ordered molecules, hydrogen bonds, mesoscopic coherence and tunneling, 469–477
Orientational defect, hydrogen bonds, orientational-tunneling model, one-dimensional molecular systems, 381–389
Orientational-tunneling model, hydrogen bonds, one-dimensional molecular systems, 381–389
Orthonormal basis sets, finite-size scaling (FSS), quantum mechanics, 27–29
Orthonormal molecule-fixed axis system, vibration-rotation Hamiltonians, Lagrangian formalism, 274–291

Particle weight, hydrogen bonds, proton-phonon coupling, 485–490
Partition function, finite-size scaling (FSS)
  path integral lattice mapping, 75–77
  quantum mechanics, 21–23

Partitioning techniques
dressed Hamiltonians, 179–190
adiabatic elimination, 184–185
enlarged space partitioning, 187–190
formulation, 179–183
high-frequency partitioning, 185–187
Floquet Hamiltonian, 187–189
Path integral techniques
finite-size scaling (FSS)
quantum criticality, 75–81
lattice system mapping, 75–77
Pöschl-Teller potential, 77–81
quantum mechanics, 22–23
research background, 4
hydrogen bonds, acoustic phonon proton transfer, 360–366
Pauli exclusion principle, hydrogen bonds, tunneling transition, coupled protons, 376–381
Perron-Frobenius theorem, finite-size scaling (FSS), atomic and molecular systems, 21
Perturbation theory
adiabatic Floquet evolution, 204–206
Floquet Hamiltonian, dynamical resonances
high-frequency perturbation theory, 171–174
nonresonant unitary transformation, KAM technique, 167–171
hydrogen bonds
bending vibrations, 372–374
sonoluminescence, 526–527
open-shell molecules, pseudo-Jahn-Teller (PJT) effect, 127–134
open-shell molecules, quantum mechanics
Slater determinant, 105–113
variants of, 112–113
Perturbative instanton approach, hydrogen bonds
acoustic phonon proton transfer, 364–366
proton-phonon coupling, 487–490
Phase diagrams, finite-size scaling (FSS)
crossover phenomena, quantum systems, 59–63
three-body Coulomb systems, 50–53
Phase transitions
finite-size scaling (FSS)
near-threshold states, large-dimensional limits, 5–8
quantum criticality, path integral techniques, 78–81
quantum phase transitions and stability,
atomic and molecular systems, 33–53
critical nuclear charges, 42–45
diatomic molecules, 45–50
three-body Coulomb systems, phase diagrams, 50–53
three-electron atoms, 39–42
two-electron atoms, 33–39
research background, 2–4
hydrogen bonds
proton bifurcation and phonon mixture, 530–535
superionic phase transitions
ammonium triiodate crystal model, 428–437
proton conductivity, 394–400
proton ordering, 390–394
Phenomenological renormalization (PR) equation, finite-size scaling (FSS)
atomic and molecular systems, 20–21
quantum dots, 82–92
quantum mechanics, 24–27
quantum phase transitions and stability, two-electron atoms, 35–39
pH levels, degassed aqueous systems, hydrogen bonds, 493–502
Phonon subsystems, hydrogen bonds
diagonalization transformation, 527–529
polaron vibration fluctuations, 423–428
proton bifurcation and, 529–535
Photocycle kinetics, hydrogen bonds, bacteriorhodopsin microscopic physics, 447–451
Photon emission and absorption
Floquet Hamiltonian, 162–166
photon exchange, 162–163
relative number invariance, 164
two-level topological quantization, atomic beam deflection, 244–245
Photon field initial states, Floquet Hamiltonian, semiclassical formalism, 160–161
Photon number operator, Floquet Hamiltonian, quantized cavity dressed states, 155–158
π-electron excitation, hydrogen bonds, bacteriorhodopsin retinal evolution, 451–457
Planck function, hydrogen bonds, bacteriorhodopsin retinal evolution, 456–457

Proton transfer, hydrogen bonds (*Continued*)
orientational-tunneling model, one-dimensional molecular system, 381–389
polaronic chain conductivities, 400–409
proton ordering model, 389–394
resonance integral vibration fluctuations, 422–428
superionic phase conductivity, ammonium triiiodate crystal, 428–437
superionic phase transition, 394–400
symmetric O···H···O bonds, 437–445
proton bifurcation and phonon mixture, 529–535
quantum coherent phenomena
chain repolarization, tunneling mechanisms, 477–485
mesoscopic quantum coherence and tunneling, 469–477
proton-phonon coupling, 485–490
quantum mechanics
acoustic photon proton transfer, 360–366
bending vibrations, 371–374
bond vibrations, 356–360
proton ordering, 368–371
proton polaron, 366–368
tunneling transition, coupled protons, 374–381
sonoluminescence mechanism, 526–527
stretching and bending energies, 524–526
structural physical effects, 352–356
Pseudo-Jahn-Teller (PJT) effect, open-shell molecules
quantum mechanics, 127–134
research background, 135–137
Pseudoospin operators, hydrogen bonds, proton ordering, 369–371
Pseudo-spin operators
hydrogen bonds
mesoscopic coherence and tunneling, 476–477
orientational-tunneling model, one-dimensional molecular systems, 381–389
water-dependent switching, continuous metal films, 508–513
Pseudosystems, finite-size scaling (FSS), path integral lattice mapping, 76–77
Pseudo-tunneling effect, hydrogen bonds
coherent tunneling, repolarization mechanisms, 483–485
orientational-tunneling model, one-dimensional molecular systems, 383–389
Pulsed nuclear magnetic resonance, aqueous system structural analysis, 502–507
Pump-Stokes sequence
bichromatic pulse selectivity, three-level systems, 247–249
dressed eigenenergy surface topology, half-SCRAP coherent state superposition, 225–226
dressed eigenergy surface topology, stimulated Raman adiabatic passage (STIRAP), 232–234
Pythagorean theorem, vibration-rotation Hamiltonians, geometric algebra, 319–321

Qicun Shi research, finite-size scaling (FSS), quantum phase transitions and stability, diatomic molecules, 50
Quantized cavity dressed states, Floquet Hamiltonian, laser pulses, quantum dynamics, 154–158
Quantum chemistry, open-shell molecules, quantum mechanics, 105–113
Quantum coherence, hydrogen bonds
chain repolarization, tunneling mechanisms, 477–485
mesoscopic quantum coherence and tunneling, 469–477
proton-phonon coupling, 485–490
Quantum criticality, finite-size scaling (FSS), path integral techniques, 75–81
lattice system mapping, 75–77
Pöschl-Teller potential, 77–81
Quantum dots, finite-size scaling (FSS), 81–92
Quantum mechanics
finite-size scaling (FSS)
atomic and molecular systems, 21–32
equations, 23–27
extrapolation and basis set expansions, 27–29
Schrödinger equation data collapse, 29–32
statistical mechanics and classical analogies, 21–23
research background, 2–4
hydrogen bonds
acoustic photon proton transfer, 360–366

bending vibrations, 371–374
bond vibrations, 356–360
proton ordering, 368–371
proton polaron, 366–368
tunneling transition, coupled protons, 374–381
open-shell molecules
molecular orbitals and "symmetry breaking," 119–127
pseudo-Jahn-Teller effect, 127–134
reference functions, 134–137
research background, 101–105
spin contamination, 113–119
techniques, summary, 105–113
Quantum Monte Carlo (QMC) method, open-shell molecules, quantum mechanics, 112–113
pseudo-Jahn-Teller (PJT) effect, 133–134
Quantum phase transitions and stability, finite-size scaling (FSS), atomic and molecular systems, 33–53
critical nuclear charges, 42–45
diatomic molecules, 45–50
three-body Coulomb systems, phase diagrams, 50–53
three-electron atoms, 39–42
two-electron atoms, 33–39
Quasi-energy operator, Floquet Hamiltonian
photon exchange, 163
semiclassical formalism, 152–154
Quasi-resonant processes
adiabatic Floquet evolution
degeneracy lifting, 206–212
dynamical degeneracy lifting, 211–212
dressed eigenergy surface topology, stimulated Raman adiabatic passage (STIRAP), 228–235
effective Hamiltonian, two-photon RWA, 190–192
Quasi-restricted Hartree-Fock (QRHF), open-shell molecules, quantum mechanics, reference functions, 134–137
Quasi-velocity, vibration-rotation Hamiltonians, Lagrangian formalism, 277–291

Rabi frequency
adiabatic Floquet evolution, nonresonant deviations, 205–206
bichromatic selectivity, 235–255
dressed eigenergy surface topology
adiabatic passage, chirped pulse/SCRAP analysis, 215–217
stimulated Raman adiabatic passage (STIRAP), 227–235
three-level bichromatic pulse selectivity, dynamics, 250–255
eigenenergy surface topology, 247–249
two-level topological quantization, atomic beam deflection
effective Hamiltonian, 237–239
eigenenergy surface topology, 239–242
two-photon quasi-resonant atomic processes, 192
Radau vectors, vibration-rotation Hamiltonians, scalar chain rules, 292
Radicals, open-shell molecules, pseudo-Jahn-Teller (PJT) effect, 130–134
Raman spectroscopy
diatomic molecule rotational excitation, single laser ground electronic state, 194–198
hydrogen bonds
bacteriorhodopsin microscopic physics, proton path, 446–451
superionic phase transitions, ammonium triiodate crystal model, 429–437
Random-phase approximation (RPA), open-shell molecules, symmetry breaking, molecular orbital response, 123–127
Rate constant, hydrogen bonds, acoustic phonon proton transfer, 363–366
Rayleigh-Ritz expansion, finite-size scaling (FSS)
quantum mechanics, 23–27
research background, 4
Rayleigh-Schrödinger power series, Kolmogorov-Arnold-Moser (KAM) transformations, Floquet Hamiltonian, dynamical resonances, 169
Reciprocality condition, vibration-rotation Hamiltonians, measuring vectors, 294–309
Redox potential, degassed aqueous systems, hydrogen bonds, 493–502
Reference functions, open-shell molecules, quantum mechanics, 134–137
Reflections, vibration-rotation Hamiltonians, 328

Relative number invariance, Floquet Hamiltonian, photon emission and absorption, 164
Relaxation contribution, open-shell molecules, symmetry breaking, molecular orbital response, 120–127
Renormalization group (RG) methods
finite-size scaling (FSS), quantum dots, 82–92
Floquet Hamiltonian, dynamical resonances, nonperturbative resonant transformation, 176–178
Repolarization mechanisms, hydrogen bonds, coherent tunneling, 477–485
Resonance integral, hydrogen bonds, polaron vibration fluctuations, 423–428
Resonances
adiabatic Floquet evolution, degeneracy lifting, 206–212
dressed eigenenergy surface topology, coherent state superposition, 222–226
finite-size scaling (FSS), quantum systems, 53–58
Restricted open-shell Hartree-Fock (ROHF)
open-shell molecules, symmetry breaking, molecular orbital response, 121–127
spin contamination, 117–119
Retinal light excitation, hydrogen bonds, bacteriorhodopsin microscopic physics, 451–457
Riemannian differential equation, vibration-rotation Hamiltonians
constrained quantization, 312–313
unconstrained quantization, 310–312
Robustness, dressed eigenenergy surface topology, adiabatic passage, 217–219
Rotating wave transformation (RWT)
dressed eigenergy surface topology, adiabatic passage, chirped pulse/SCRAP analysis, 215–217
Floquet Hamiltonian, dynamical resonances, 167
nonperturbative resonant transformation, 175–178
three-level bichromatic pulse selectivity, 246–247
two-level topological quantization, atomic beam deflection
dressed eigenenergies, 242–243
effective Hamiltonian, 239
two-photon quasi-resonant atomic processes, 190–192
Rotational degrees of freedom, vibration-rotation Hamiltonians
classical mechanics, infinitesimal displacement, 296–298
geometric algebra, 303–309
Rotational excitation, effective Hamiltonian, diatomic molecules, 193–198
Born-Oppenheimer approximation, 193–194
ground electronic surface, Raman processes, 194–198
Rotations, vibration-rotation Hamiltonians, geometrical algebra, 328–333
Rotor parameters, vibration-rotation Hamiltonians, Lagrangian formalism, 278–283
Rydberg states, finite-size scaling (FSS), quantum phase transitions and stability, *N*-electron atoms, 44–45

Scalar chain rules, vibration-rotation Hamiltonians, 291–292
Scaling laws, finite-size scaling (FSS), near-threshold states, 13–16
Schiff base protonation, hydrogen bonds, bacteriorhodopsin microscopic physics, 447–451
Schrödinger equation. *See also* Dressed Schrödinger equation
adiabatic Floquet theory
degeneracy lifting, 207–212
dressed equation, chirped laser pulses, 199–201
bichromatic selectivity, 237–255
dressed eigenenergy surface topology, robustness, 219
finite-size scaling (FSS)
near-threshold states, one-particle central potentials, 10–16
quantum mechanics, 23–27
data collapse, 29–32
Floquet Hamiltonian
multiple laser dynamics, 166
semiclassical formalism, 151–154
interaction representation, 158–159
theoretical background, 150–151
hydrogen bonds
acoustic phonon proton transfer, 365–366
proton polaron, 367–368

laser pulses, quantum dynamics, control process research, 149–150
open-shell molecules, quantum mechanics
pseudo-Jahn-Teller (PJT) effect, 133–134
size-consistent methods, 108–113
semiclassical model and Floquet evolution, 256–257
spatial finite-size scaling (SFSS), 65–76
three-level bichromatic pulse selectivity, dynamics, 249–255
vibration-rotation Hamiltonians, unconstrained quantization, 311–312
Second-order perturbation
hydrogen bonds, acoustic phonon proton transfer, 362–366
open-shell molecules, pseudo-Jahn-Teller (PJT) effect, 127–134
Semiclassical model
Floquet evolution and, 255–257
Floquet Hamiltonian
coherent states, expectation values, 161–162
interaction representation and coherent states, 158–162
laser pulses, quantum dynamics, 151–154
hydrogen bonds, mesoscopic coherence and tunneling, 476–477
three-level bichromatic pulse selectivity, 245–255
dynamics, 249–255
Shape coordinates, vibration-rotation Hamiltonians
geometric algebra, 299–309
Lagrangian formalism, 273–291
properties, 343–345
Shape-invariant potentials, finite-size scaling (FSS), near-threshold states, 13–16
Shape-type resonances, finite-size scaling (FSS), quantum systems, 53–58
Short-range potentials, finite-size scaling (FSS), near-threshold states, 12–16
Siegert method, spatial finite-size scaling (SFSS), 69–76
Size-consistent methods, open-shell molecules, quantum mechanics, 107–113
Slater determinants, open-shell molecules, quantum chemistry, 105–113
spin contamination, 113–119
Small magnetic grains, hydrogen bonds, mesoscopic coherence and tunneling, 469–477
Small polaron models
hydrogen bonds
polaron vibration fluctuations, 423–428
superionic phase transitions, ammonium triiodate crystal model, 432–437
water-dependent switching, continuous metal films, 509–513
Sonoluminescence, hydrogen bonds, 526–527
Space-charge-limited density, hydrogen bonds, Coulomb correlations, 415–416
Spatial finite-size scaling (SFSS), atomic and molecular systems, 65–75
Spherical trigonometry, vibration-rotation Hamiltonians, geometric algebra, 334–335
Spin contamination
open-shell molecules, 113–119
open-shell molecules, quantum mechanics, research background, 104–105
Spin-lattice relaxation, aqueous system structural analysis, 502–507
Spin multiplicity, open-shell molecules, quantum mechanics, 137
Spin-phonon interaction, hydrogen bonds, proton-phonon coupling, 485–490
Spin-related selection rule, hydrogen bonds, tunneling transition, coupled protons, 378–381
Square-integrable functions, finite-size scaling (FSS), quantum system resonances, 56–58
Stark chirped rapid adiabatic passage (SCRAP)
dressed eigenenergy surface topology, half-SCRAP coherent state superposition, 225–226
dressed eigenergy surface topology, 215–217
Stark shift, two-photon quasi-resonant atomic processes, 192
State-selectivity, bichromatic pulses, 235–255
three-level systems, 245–255
two-level system, atomic beam deflection topological quantization, 237–245
Static electronic polarizability, diatomic molecule rotational excitation, ground electronic surface, Hilbert subspace, 195
Statistical mechanics
finite-size scaling (FSS), atomic and molecular systems, 21–23
molecular clusters, hydrogen bonds, 513–518

Stimulated Raman adiabatic passage (STIRAP)
adiabatic Floquet theory, evolutionary mechanisms, 205–206
dressed eigenenergy surface topology
coherent superposition of states, 234–235
related processes, 226–235
research background, 214–215
unique state transfer, 228–234
three-level bichromatic pulse selectivity,
dynamics, 249–255
effective Hamiltonian, 245–247
eigenenergy surface topology, 247–249
Stokes-pump sequence
bichromatic pulse selectivity, three-level systems, 247–255
dressed eigenenergy surface topology, half-SCRAP coherent state superposition, 225–226
dressed eigenergy surface topology, stimulated Raman adiabatic passage (STIRAP), 234–235
Stretching coordinates
hydrogen bonds, bond vibrations, 356–360
O–H–O hydrogen bond, 524–526
polariton effect, symmetric O–H–O hydrogen bond, 437–445
Structurization potential, degassed aqueous systems, thermodynamics, 497–502
Superadiabatic formula, adiabatic Floquet evolution, 204–206
Superconvergent iterative algorithm, Kolmogorov-Arnold-Moser (KAM) transformations, Floquet Hamiltonian, dynamical resonances, 168–169
Superionic phase transitions, hydrogen bonds
ammonium triiodate crystal model, 428–437
proton conductivity, 394–400
proton ordering, 390–394
Superposition of states
adiabatic Floquet evolution, degeneracy lifting, 210–212
dressed eigenenergy surface topology
half-SCRAP coherent superposition, 222–226
stimulated Raman adiabatic passage (STIRAP), 234–235
Switching currents, water-dependent switching, continuous metal films, 507–513
Symmetry breaking
finite-size scaling (FSS), near-threshold states, large-dimensional limits, 6–8
open-shell molecules
molecular orbital response, 119–127
research background, 134–137
spin contamination, 118–119
"Symmetry breaking," open-shell molecules, quantum mechanics, research background, 104–105

Tamm-Dancoff approximation (2*hp*-TDA), open-shell molecules, reference functions, 135–137
Tauberian theorem, finite-size scaling (FSS), near-threshold states, one-particle central potentials, 12–16
Taylor expansion
finite-size scaling (FSS), Schrödinger equation data collapse, 30–32
molecular clusters, statistical mechanics, 516–518
vibration-rotation Hamiltonians
classical mechanics, infinitesimal displacement, 297–298
geometric calculus, directional derivatives, 339
shape coordinates, 344–345
Temperature dependence
aqueous system structural analysis, spin-lattice relaxation, 502–507
hydrogen bonds
acoustic phonon proton transfer, 364–366
mesoscopic coherence and tunneling, 472–477
proton transfer, 420–423
superionic phase transitions, 399–400
Tetra-atomic molecule, vibration-rotation Hamiltonians, constrained quantization, 315–316
Thermal bath, hydrogen bonds
acoustic phonon proton transfer, 364–366
proton polaron, 366–368
Thermodynamic limit
degassed aqueous systems, 493–502
finite-size scaling (FSS), atomic and molecular systems, 19–21
hydrogen bonds, degassed aqueous systems, 490–501

Thomas-Fermi statistical model, finite-size scaling (FSS), near-threshold states, large-dimensional limits, 5–8
Three-body Coulomb systems, finite-size scaling (FSS)
near-threshold states, 16–18
phase diagram, 50–53
Three-dimensional geometric algebra, vibration-rotation Hamiltonians, basis set representations, 323–326
Three-electron atoms, finite-size scaling (FSS), quantum phase transitions and stability, 39–42
Three-level systems, bichromatic pulse selectivity, 245–255
dynamics, 249–255
effective Hamiltonian, 246–247
eigenenergy surface topology, 247–249
Torsion-vibration Hamiltonians, hydrogen bonds, acoustic phonon proton transfer, 366
Transfer matrix, finite-size scaling (FSS)
atomic and molecular systems, 20–21
path integral lattice mapping, 76–77
quantum criticality, path integral techniques, 78–81
Transition amplitudes, finite-size scaling (FSS), quantum mechanics, 22–23
Transport properties, hydrogen bonds, polaron/polariton systems, 381–445
anharmonicity influence, 409–410
Coulomb correlations and electric field local heterogeneities, 411–416
external influences, proton conductivity, 416–422
orientational-tunneling model, one-dimensional molecular system, 381–389
polaronic chain conductivities, 400–409
proton ordering model, 389–394
resonance integral vibration fluctuations, 422–428
superionic phase conductivity, ammonium triiiodate crystal, 428–437
superionic phase transition, proton conductivity, 394–400
symmetric O–H–O bonds, 437–445
Triatomic molecule, vibration-rotation Hamiltonians
constrained quantization, 315–316
geometric algebra, 308–309
Lagrangian formalism, 285–291
unconstrained quantization, 312–313
Trig pulse, adiabatic Floquet theory, dressed Schrödinger equation, 200–202
Triple box component, vibration-rotation Hamiltonians, measuring vectors, 294–309
Triplet instability, open-shell molecules, symmetry breaking, molecular orbital response, 123–127
Truncated basis, finite-size scaling (FSS), quantum dots, 85–92
Truncated wave function, finite-size scaling (FSS), quantum mechanics, 25–27
Tunneling mechanisms, hydrogen bonds
acoustic phonon proton transfer, 361–366
coupled protons, 374–381
orientational-tunneling model, one-dimensional molecular system, 381–389
polaron vibration fluctuations, 424–428
proton-phonon coupling, 485–490
proton transfer, 421–423
research overview, 353–356
small magnetic grains and ordered molecules, 469–477
sonoluminescence, 526–527
superionic phase transitions, 397–400
Two-dimensional lattice models, finite-size scaling (FSS), atomic and molecular systems, 20–21
Two-dimensional phase space, vibration-rotation Hamiltonians, 287–291
Two-electron atoms, finite-size scaling (FSS), quantum phase transitions and stability, 33–39
Two-level topological quantization, bichromatic selectivity, 237–245
dressed eigenenergy construction, 242–243
effective Hamiltonian, 237–239
eigenenergy surface topology, 239–242
photon exchange dynamics, 244–245
Two-photon quasi-resonant atomic processes, effective Hamiltonian, 190–192

Ultrasound, hydrogen bonds, proton transfer, 421–423
Unconstrained quantization, vibration-rotation Hamiltonians, 309–312

Unique state transfer, dressed eigenenergy surface topology, stimulated Raman adiabatic passage (STIRAP), 228–234
Unitary transformation
adiabatic Floquet evolution, degeneracy lifting, 208–212
Floquet Hamiltonian, dynamical resonances
dressed Hamiltonian partitioning, 180–183
high-frequency partitioning, 185–187
KAM technique, 167–171
hydrogen bonds, Coulomb correlations, 413–416
Unnatural parity states, finite-size scaling (FSS), quantum phase transitions and stability, two-electron atoms, 38–39
Unrestricted Hartree-Fock (UHF) approximation
open-shell molecules, quantum mechanics, spin contamination, 115–119
open-shell molecules, symmetry breaking, molecular orbital response, 121–127

van der Waals interactions, alkyl-/alkloxybenzoic acids, cluster formation, 518–520
Vector additions, vibration-rotation Hamiltonians, geometric algebra, 318
Vector decomposition, vibration-rotation Hamiltonians, projections, 327–328
Vector derivative, vibration-rotation Hamiltonians, constrained quantization, 315–317
Vector differentiation, vibration-rotation Hamiltonians, geometric calculus, directional derivatives, 340–342
Vectorial differentiation, vibration-rotation Hamiltonians, geometric calculus, directional derivatives, 339–342
Vector multiplication, vibration-rotation Hamiltonians, geometric algebra, 319–321
Vibrational derivatives, vibration-rotation Hamiltonians, Lagrangian formalism, 280–291
Vibration-rotation Hamiltonians
canonical quantization, 309–313
constrained quantization, 312–313
unconstrained quantization, 309–312
classical mechanics, 272–309
Lagrangian source, 272–291
measuring vectors, 292–309
scalar chain rules, 291–292
co- and contravariant measuring vectors, 342–343
geometric algebra principles, 317–342
geometrical relations, 326–335
angles, 332–334
projections, 327–328
rotations, 328–332
spherical trigonometry, 334–335
geometric calculus, 335–342
directional derivative, 336–339
vectorial differentiation, 339–342
sums and products, 318–323
expansion rules, 323
magnitude, 323
multiplication, 321–322
vector additions, 318
vector multiplication, 319–321
three-dimensional basis representation, 323–326
research background, 270–272
shape coordinate properties, 343–345
volume elements, 313–317
Vibrations
diatomic molecule rotational excitation, effective Hamiltonian, 196
hydrogen bonds, polaron fluctuations, 423–428
Volume elements, vibration-rotation Hamiltonians, constrained quantization, 313–317

Wannier orbitals, finite-size scaling (FSS), quantum dots, 83–92
Water molecules
clustering mechanaisms, 520–522
hydrogen bonds
bacteriorhodopsin microscopic physics, 448–451
continuous metal films, water-dependent switching, 507–513
degassed systems organization and thermodynamics, 490–502
experimental results, 491–494
thermodynamics, 494–501
Waugh-Fedin relation, hydrogen bonds, superionic phase transitions, ammonium triiodate crystal model, 429–437
Wigner crystals, finite-size scaling (FSS), 2

Yukawa potential, finite-size scaling (FSS)
quantum mechanics, 27–29
Schrödinger equation data collapse, 31–32

Zero-field resonances
adiabatic Floquet evolution, 201–202
Floquet Hamiltonian, nonperturbative treatment, 174–178
Zeroth-order wave functions, open-shell molecules, quantum mechanics, Slater determinants, 111–113
Z parameter, finite-size scaling (FSS), 2–4